Systems Engineering and Analysis of Electro-Optical and Infrared Systems

Systems Engineering and Analysis of Electro-Optical and Infrared Systems

William Wolfgang Arrasmith

CRC Press
Taylor & Francis Group
Boca Raton London New York

CRC Press is an imprint of the
Taylor & Francis Group, an **informa** business

MATLAB® is a trademark of The MathWorks, Inc. and is used with permission. The MathWorks does not warrant the accuracy of the text or exercises in this book. This book's use or discussion of MATLAB® software or related products does not constitute endorsement or sponsorship by The MathWorks of a particular pedagogical approach or particular use of the MATLAB® software.

CRC Press
Taylor & Francis Group
6000 Broken Sound Parkway NW, Suite 300
Boca Raton, FL 33487-2742

First issued in paperback 2017

© 2015 by Taylor & Francis Group, LLC
CRC Press is an imprint of Taylor & Francis Group, an Informa business

No claim to original U.S. Government works

ISBN-13: 978-1-4665-7992-7 (hbk)
ISBN-13: 978-1-138-89345-0 (pbk)

Visit the Taylor & Francis Web site at
http://www.taylorandfrancis.com

and the CRC Press Web site at
http://www.crcpress.com

Contents

Section II Application of Systems Engineering Tools, Methods, and Techniques to Optical Systems

Preface

So why write this book? There are already a variety of excellent texts that discuss the field of systems engineering in a general fashion. Similarly, there are many excellent books about imaging, detection, and optical systems from a technical point of view. In the former case, these general systems engineering–oriented books do an excellent job of describing fundamental systems engineering principles, tools, techniques, and methodologies, but often they stop short on giving specific discipline-centric, technical examples and implementations in an integrated fashion. There are many excellent examples of general systems engineering texts from various publishers. An excellent example is Benjamin S. Blanchard and Wolter J. Fabrycky's book titled *Systems Engineering and Analysis*.

On the other end of the spectrum (pardon the pun), there are many excellent technical books on optical detection and imaging but the connections to the systems engineering discipline are tenuous and brief. An excellent example of this kind of book is the one by R. D. Hudson, *Infrared Systems Engineering*, wherein abides a wealth of technical information with regard to infrared optical systems, but the systems engineering content is relegated to only a few pages at the beginning of the book. This is somewhat understandable since the development and description of systems engineering processes has a different flow than the development of technical information in an optical or imaging book. For this book, we wanted to achieve three goals:

1. We wanted to write an optical systems engineering book that uses fundamental systems engineering principles throughout the book to drive the optical/imaging technical content.

2. The optical/imaging technical content has to be sufficient to allow the practicing engineer to understand and analyze optical systems and to be able to apply the optical technical content to modern optical systems.

3. We wanted an integrated case study to serve as a running example of how the systems engineering content is applied to optical systems and also to provide examples of how to apply the optical systems technical content. This also provided the opportunity to incorporate professional nuances and applications-related discussions that normally are out of place in a purely technical discussion.

In addition to these three goals, we also wanted to provide coverage of enterprise architecture methods, systems of systems, and family of systems and the role systems engineering takes in these paradigms. We also wanted to introduce the reader to model-based systems engineering (MBSE) methods and tools and provide some examples of how these tools are used throughout the systems engineering life cycle. Finally, we wanted to present some coverage of modern-day electro-optical and infrared systems.

Why the title? We chose the title *Systems Engineering and Analysis of Electro-Optical and Infrared Systems* to show first of all that this is a systems engineering–oriented text. By emphasizing "analysis", we state that this book is intended to have enough theory, analytical content, and technical depth, to allow the practicing engineer to analyze optical systems from both a systems and a technical perspective. We use the words electro-optical and infrared (EO/IR) systems because this characterization is well known in the U.S. Department of Defense (DoD) and the intelligence, surveillance, and reconnaissance communities. We could have used the combination of "visible and infrared systems" instead of "electro-optical and infrared systems" in the title since we predominantly focus on optical systems in the visible and infrared part of the electromagnetic spectrum. However, according to *Merriam-Webster's Dictionary*, "electro-optical" means "relating to or being an electronic device for emitting, modulating, transmitting, or sensing light." This definition has more of a systems interpretation and for this reason as well as for name recognition, we chose to use "electro-optical and infrared systems" in the title. Also, the last word in the title is "systems" to emphasize that we are providing a systems-level technical focus in this book. We should mention at this point that unless otherwise stated, we use the term "optical" to describe applicable parts of the ultraviolet (UV), visible, and infrared (IR) parts of the electromagnetic spectrum. If we need to distinguish between the UV, visible, IR, and other parts of the electromagnetic spectrum, we will make that clear within the text.

Our organizational approach is to start each chapter with a focused discussion that sequentially describes the systems engineering process. These discussions are meant to be general and directly applicable to optical systems. Our intent here is not to exhaustively cover general systems engineering topics but rather to present the key systems engineering methodologies, techniques, and tools needed to drive the optical analysis and design processes that are presented in this text. We therefore do not intend nor find it necessary to replicate large portions of the generic systems engineering texts that are currently prevalent in the literature. We do intend to capture the inherent systems engineering methodologies contained in the leading current texts and journals and apply them to optical systems. We will selectively use appropriate processes and tools to outline the fundamental systems engineering practices in modern times and use these to drive the focus of our optical systems engineering and analysis.

Just as with the systems engineering content of this book, we do not try to exhaustively replicate technical optics books that are widely available on the market. This would be redundant and unnecessary and require too much space. Instead, we focus on optical systems analysis, metrics, and design considerations that are needed by the optical systems engineer. The level of detail in our "optical building blocks" content would be more in line with a modernization and topical expansion of the classic text from Richard D. Hudson, Jr. titled *Infrared Systems Engineering*. Although light in emphasizing modern fundamental systems engineering principles—an issue that is remedied in this book—the Hudson text has the right level of detail for the optical systems engineer. We will provide a similar level of detail but also include an expanded set of topics, modernized material, and an expansion from just infrared to visible and infrared systems.

In each chapter, we link the systems engineering content with the optical systems analysis and design content. We call this content optical building blocks, and we sequentially develop the necessary insights throughout this book. The linking of the driving systems engineering principles to the optical content was challenging since the logical flow of systems engineering content development is different than that of optical systems analysis. For instance, in the early systems engineering conceptual phase, you need practically all of the optical building blocks material to fully conduct an optical systems analysis. We solved this linking issue by abstractly linking the topics covered in the systems engineering content with the optical systems analysis content in each chapter. For instance, in Chapter 1, where we *introduce* enterprise architecture, systems of systems, family of systems, acquisition process, and systems engineering process, we correspondingly *introduce* the optical systems engineering, fundamental units, and radiometric quantities in Section 1.2.

Each subsequent chapter has a logical connection between the systems engineering content and the optical systems analysis content, and by the time we get to Chapter 12, the essential fundamental aspects of optical systems analysis have been developed and subsequently more specialized topics are addressed.

We end each chapter with an integrated case study that serves as a running example of how the systems engineering content in that chapter is used to drive the optical systems analysis and design processes. The integrated case study uses the current chapter material as well as previous chapter materials to apply both the systems engineering principles and the optical systems analysis principles and methods. We use a fictitious company that bids on and consequently wins a contract with homeland security to provide day/night optical systems capability on an unmanned aerial vehicle (UAV) for border patrol applications. This company's piece of the development is part of a larger enterprise activity and so enterprise architectural issues, systems-of-systems issues, systems engineering issues, and optical systems analysis and design issues can be, and are, addressed in the running integrated case study. In essence, the integrated case study ties together the systems engineering content with the optical systems analysis and design material of the book. It also serves to illustrate how to apply the material and provides analytical examples.

We now describe the contents and layout of this book. Our book can be grouped into two major sections. Section I includes Chapters 1 through 4 and covers where systems engineering fits within the larger context of an enterprise, systems of systems, and family of systems. This block also describes key aspects of the systems engineering process and, in each chapter, links these processes to corresponding optical building blocks and case studies.

Chapter 1 is an introductory chapter and introduces the concepts of systems, enterprises, systems of systems, and family of systems. In Section 1.2, we introduce fundamental optical parameters, definitions, and radiometric quantities. The integrated case study introduces and sets up the running case study that will provide applications examples throughout the book.

Chapter 2 is focused on understanding enterprises through their enterprise architecture models and methodologies. The systems engineering process is seen to be complementary to enterprise architecture methods. We present the purpose of enterprise architectures, a short historical background, types of architectures, the roles they play in multientity development environments, and a short description of some common enterprise architectures. We also discuss some enterprise modeling tools and present a specific description of a useful enterprise architectural framework—the Department of Defense Architectural Framework (DoDAF 2.0). Just as the enterprise architecture is the model that unites the various entities through its architectural framework, linear and nonlinear systems models unite and integrate various aspects of the optical system. In Section 2.2, we present linear and nonlinear optical systems models and the integrated case study provides applications of our results.

Chapter 3 focuses on systems of systems and family of systems and shows how the system engineering process is used over the systems development lifecycle (SDLC). In Section 3.2, we show that the optical system itself can be considered as a systems of systems and we define the elements that make up this systems of systems. The integrated case study, as it does in other appropriate chapters of this book, will once again integrate the systems engineering content with the optical building blocks content and serve as a section where we apply our results.

Chapter 4 addresses and present an overview of the concept of MBSE. We provide a short overview of some MBSE tools and showcase MagicDraw as a representative example of an

MBSE tool. One use of MBSE is to link requirements from the enterprise level down to the system component level. Often, as in the case of MathWork's Simulink®, the requirements model can be directly linked to hardware and software elements in a rapid-prototyping configuration. In Section 4.2, we develop the basic optical principles, tools, and techniques needed to develop optical models. The understanding of the basic optical principles and methods allows sophisticated optical technical models to be developed that integrate with the MBSE tools and become part of the MBSE paradigm. In the integrated case study, we use the analytical hierarchy process (AHP) to select an analytical tool and demonstrate the importance of understanding the basic optical system.

Section II covers an in-depth look at the systems engineering process itself. Chapters 5 through 16 each sequentially present critical steps in the systems engineering process. As in Section I, we take a similar approach for each chapter and link the systems engineering content to the optical building blocks subject matter. The integrated case study is again used to provide a means to show examples and applications of the chapter material. The integrated case study builds on the material from the previous case studies and has at its disposal the previous optical systems analytical material.

Chapter 5 describes the problem definition phase and addresses stakeholder identification, stakeholder needs analysis, the concept of operations (CONOPS) and its connection to the enterprise architectural framework, stakeholder requirements, project scope, and goals and objectives. We introduce the quality functional deployment and use MagicDraw as a representative requirements management tool. Just as the problem definition step is the starting point for the systems engineering methodology, understanding optical sources is the starting point for understanding, modeling, and analyzing optical systems. In Section 5.2, we develop the analytical framework for understanding optical sources in the visible and infrared parts of the electromagnetic spectrum.

In Chapter 6, we address feasibility studies, trade studies, and alternative analysis. We present the analytical hierarchy method as a choice means for analyzing alternatives. We also show the connections between feasibility and risk and feasibility and requirements. In order to establish technical feasibility, or show infeasibility, of a particular optical systems concept, the propagation of radiation from the source to a distant detection plane needs to be understood. In Section 6.2, we discuss optical radiation and its propagation.

Chapter 7 deals with systems requirements, and we discuss the requirements generation process and its connection with optical systems. We address the optical systems functional, nonfunctional, and inverse requirements and also discuss how to write good optical systems requirements. We make the case for the need of requirements documents and present the value of MBSE in requirements change management. Just as requirements are central to the success of a project, product, system, or service, filtering and modulation devices are often the essential ingredients in the success of an optical system. Section 7.2 describes various filtering and modulation mechanisms and systems requirements, and filtering and modulation methods are applied in the integrated case study.

In Chapter 8, we evaluate the maintenance and support aspects of our optical system. We develop early reliability, maintainability, and availability models and consider preplanned product improvement (PPPI), logistics support, and the eventual optical system retirement and disposal. The optical system will likely have maintenance and support requirements for its critical systems, components, assemblies, and parts. One of the primary system components is the optical detection system. Section 8.2 provides an introduction to optical detection mechanisms and detection system metrics. Here, we will present analytical information necessary to understand the detection system at a fundamental level, and discuss important maintenance and support considerations along with other critical optical systems sensor components.

In Chapter 9, we establish the technical performance measures and metrics (TPM&Ms) for an optical system. We discuss measures of performance (MOPs) and key performance parameters (KPPs) and discuss their connection to the optical systems requirements. The performance of an optical system can be quickly gauged by understanding its TPMs. An understanding of the performance characteristics of the optical sensor as well as system and sensor noise is essential in determining how the optical sensor will interact with the rest of the optical system. Therefore, Section 9.2 addresses system noise, sensor noise, and important sensor characteristics. The integrated case study in this section applies the analytical material in this section along with previous results to determine the optical systems sensor TPMs, MOPs, KPPs, and performance characteristics.

In Chapter 10, we conduct a functional analysis of our optical system. We present the functional analysis process and show how it links to a set of driving requirements. We conduct a functional decomposition of the requirements and use functional block diagrams (FBDs) and functional flow diagrams (FFDs) to document the functional decomposition. We also make the connection to MBSE tools such as Simulink. The functional analysis process can be used to capture and illustrate the functional aspects of the required optical system. Environmental conditioning functions may include the consideration of functions to control temperature, humidity, dust, radiation levels, light levels, ambient, and transmitted noise levels, and pressure. In Section 10.2, we look at environmental conditioning considerations for optical systems.

Chapter 11 deals with requirements allocation for our optical system. We discuss the requirements allocation process and its connection to the functional analysis step. We also present some common TPMs that are useful in describing high-level attributes in many optical systems and discuss allocating appropriate TPMs such as weight, space, power, reliability metrics, and noise specifications.

In Chapter 12, we look at systems design. We present the three-stage systems engineering design process (conceptual design, preliminary design, and detailed design and development) and discuss the relationship between the design process and key aspects of the previous chapters. We emphasize the analysis, synthesis, and optimization steps of the design process. We will use the KPPs, MOPs, and TPMs to develop an analytical method for evaluating our optical systems performance in Section 12.2.

Chapter 13 deals with construction, manufacturing, and production. In this chapter, we look at useful systems engineering methods that can be applied in production, manufacturing, and construction processes. We look at statistical analysis, quality engineering techniques, and process control methods and their applications to optical systems engineering. We also show how optical systems themselves are fundamental as calibration and measurement tools in the production, manufacturing, and construction phase of the SDLC. After the optical system is produced, it needs to be tested and characterized.

Chapter 14 presents systems test and evaluation. We cover the systems qualification process and present essential aspects of optical systems testing. It's a small world! Accurate testing and characterization of optical systems and components often requires highly sophisticated testing methods and tools. In Section 14.2, we describe optical systems testing and characterization methods. We also provide insight to understanding and controlling the optical systems test environmental conditions. We then cover the optical test bench and describe optical calibration methods and standards and also contemporary optical testing tools.

Chapter 15 explores the optical systems utilization phase (i.e., operational use and support, and system retirement, and disposal phases). We discuss systems modification through the implementation of the engineering change proposal. We also describe the PPPI process and block upgrades. We also provide an introduction to reliability, maintainability, and availability as they pertain to the utilization phase, the need for logistics support, and an emphasis on providing expected *services*. Optical systems are often built in a modular and software-adaptable fashion, which permits effectively modifying the performance aspects of the optical system while it is still operational in the field. For example, when considering the optical system's image/signal processing capabilities, by using open-architecture and modular design principles, and standardization during the optical system's design and development, upgrades and enhancements are implementable in a more efficient and cost-effective manner. In Section 15.2, we present some important signal processing considerations and discuss some signal processing methods that are useful when working with optical systems. The integrated case study applies the systems engineering methods in the utilization phase and signal processing methods to the Fantastic Imaging Technologies (FIT) UAV optical system.

In Chapter 16, we discuss system retirement and disposal. In this final chapter of this section, we look at issues that arise at the end of an optical system's life. We discuss disposal versus salvage costs, hazardous materials and special handling, and security considerations. We also talk of the need for collecting lessons learned and updating a historical database with critical systems information for use in lifecycle costing, decision and risk management, reliability analysis, and systems modeling and optimization activities. Supporting an optical system over its life cycle is challenging enough when the optical system is earthbound with direct access to the optical system components by the maintenance and service teams. What about when the optical system is not readily available such as in hard-to-access remote environments like space? Section 16.2 describes optical systems in space with an emphasis on systems support and retirement issues.

With regard to the organization of the chapters themselves, they are usually broken down into three parts consisting of a front section that defines systems engineering principles and methods and applies them to optical systems in general. This general optical systems section is then linked to an Optical Systems Building Block section that provides technical focus and presents the technical details of optical systems engineering. The Optical Systems Building Blocks are intended to be stand-alone sections as much as possible, but they should be sequentially read for best effect. In other words, the Optical Systems Building Blocks in each chapter may draw upon information from previous chapters. The experienced reader may go directly to the chapter of interest, but the learner or trainee should go through the chapters sequentially. Each chapter ends with an integrated case study wherein the principles, tools, and techniques of the chapter are applied.

In addition to this material, we also supply some useful appendices at the end of each chapter (if needed) and also at the end of the book that include important mathematical relationships, useful data, tables, and information needed to understand the technical content of this book.

This book is written for a wide variety of readers. Our intended audience is foremost the practicing systems engineer working on optical systems or systems of systems as part of a larger enterprise. The book is also meant for the general engineer that is working on or with optical systems. Optical scientists as well as specialty engineers will find this book of interest to gain insights into optical systems from a systems engineering perspective. Program and project managers that are responsible for optical systems will find useful metrics, analytical tools, and connections between managerial and technical drivers, using systems development methods throughout the SDLC. This book should also be useful to technicians that are working on optical systems. Additionally, this text will also help graduate students in systems engineering; optical systems; science, technology, engineering, and mathematics (STEM) curriculum; or other multidisciplinary systems curricula that involve optical systems. This book should also be of interest to international professional colleagues.

Additional material is available from the CRC website: http://www.crcpress.com/product/isbn/9781466579927.

Acknowledgments

This has been a long journey and there are many who deserve credit and thanks for helping get this project to completion. Foremost, I thank God for giving me peace and strength throughout this project and helping me throughout my life. Nothing would have been accomplished if not for His grace and support. I also thank my family, in particular Lena, who has supported me throughout the project and has spent countless hours on her own while I stared at a computer screen. I also thank my children Christina (and her husband Carlos) and Kari and my grandkids Christian and Tristan, who are wonderful beings and bring Lena and me great joy. I also thank my parents Bill Black (MD) and Hannelore Black, my brother Gordon, and sister Katrin for their love, encouragement, and support over the years and for being great people. Also, thanks to my extended family out there, all my relatives, and great friends (in particular John, Julie, Eric, Jackie, Dan, Denise, Paul, Nancy, Barry, Vee, Nitaya, Yolli, and Bill). Also thanks to Pastor Mark and everyone at Calvary Chapel.

Thanks to Dr. Muzaffar Shaikh who gave me the time to work on this project and who is an amazing and kind individual. Thanks to Barry, Danny, Aldo, Adrian, and especially Arlene, who lights up the office, is a source of endless humor, and is a gentle spirit (touched with a bit of sass). Finally, I must thank the many students over the years that have directly helped me on this project in one form or another. From helping with graphics, to generating figures and equations, to conducting focused and directed research, to doing related projects, analysis, and reviews, and to helping with the material itself, many thanks for your hard work and many contributions. I consider this our project. Last but not least, I thank Taylor & Francis Group for giving me the opportunity to develop this material.

MATLAB® is a registered trademark of The MathWorks, Inc. For product information, please contact:

The MathWorks, Inc.
3 Apple Hill Drive
Natick, MA 01760-2098 USA
Tel: 508-647-7000
Fax: 508-647-7001
E-mail: info@mathworks.com
Web: www.mathworks.com

Author

 **William W. Arrasmith** received his PhD in engineering physics from the Air Force Institute of Technology (AFIT) in Dayton, Ohio, in 1995. He earned an MS in electrical engineering from the University of New Mexico in Albuquerque, New Mexico, in 1991. He received his BS in electrical engineering from Virginia Tech in Blacksburg, Virginia, in 1983. Currently, Dr. Arrasmith is a professor of engineering systems at the Florida Institute of Technology (FIT) in Melbourne, Florida. Prior to FIT, he served in the United States Air Force for over 20 years, culminating with a rank of Lt. Colonel.

During his time in the Air Force, Dr. Arrasmith held several positions including chief, Advanced Science and Technology Division, Applied Technology Directorate at the Air Force Technical Applications Center; assistant professor, Weapons and Systems Engineering Department, United States Naval Academy; program manager, Physics and Electronics Directorate, Air Force Office of Scientific Research; director, Flood Beam Experiment, Air Force Research Laboratory (Kirtland Air Force Base); and project engineer, Teal Ruby Systems Program Office, Space Division.

Dr. Arrasmith is a member of Phi Kappa Phi, Tau Beta Pi, and the American Society of Engineering Education (ASEE) and has two national patents pending. He received the President's Award for Service at FIT in 2013 and the Walter Nunn Excellence in Teaching Award in the College of Engineering at FIT in 2010.

Section I

The Modern Optical Systems Engineering Landscape: Systems Engineering in Relation to the Enterprise, System-of-Systems, and Family-of-Systems

1

Introduction to Systems Engineering

Scientists investigate that which already is;
Engineers create that which has never been.

—Albert Einstein

1.1 Systems Engineering in the Modern Age

The introduction and application of systems engineering methods, principles, and techniques into the workplace, and their continued support over the years, have made a significant impact on systems development efforts. These efforts can be seen in the government, industrial, commercial, and academic sectors in the United States, as well as its allies, and in various countries throughout the world. In the United States, systems engineering principles and methods are especially prevalent in the Department of Defense (DOD) and its suppliers, partners, and defense contractors. Within the DOD and many of its allies, systems engineering principles are an integral part of their systems development efforts, acquisitions, and support strategies. Outside of the defense sector, many commercial and industrial entities are implementing fundamental systems engineering principles and methods and are making them part of their core business practices. Notable examples are the automotive, transportation, and space industries. Government, corporate, private, and educational sectors have applied systems engineering methods with great success to large, complex programs and small projects alike. This chapter introduces fundamental systems engineering concepts that directly relate and apply to optical systems development activities. This book focuses on optical systems as the subject matter for applying systems engineering principles, methods, and techniques. We will connect systems engineering concepts and methodologies with optical detection methods that are predominantly, but not singularly, optical imaging applications and show how the systems engineering principles and methods drive the optical systems development, support, and eventual retirement and disposal. By optical, unless otherwise stated, we mean systems that operate in the ultraviolet (UV), visible, and infrared (IR) parts of the electromagnetic spectrum. We first start this section with a short history of systems engineering and provide fundamental definitions. Afterward, we follow with a discussion about the modern-day systems engineer and the need for diversity. Afterward, we introduce the overarching concept of an enterprise, its associated enterprise architecture (EA), and its relationship to systems engineering. The EA discussion is followed with an introduction of the systems-of-systems (SOS) and family-of-systems (FOS) concepts. The next section discusses the formal acquisition process in the United States known as the Joint Capabilities Integration and Development System (JCIDS). We end the first section of this chapter with a look at some fundamental systems engineering principles and methods that are applied throughout the systems development life cycle (SDLC).

The second section of this chapter focuses on optical systems engineering building blocks. These are the fundamental optical systems analytical methods and techniques needed to understand optical systems engineering. The majority of the chapters in this book have an Optical Systems Engineering Building Blocks section; these sections are interrelated and sequential. In this section, we provide introductory information such as units and fundamental concepts that will be needed for later chapters. We introduce the terminology and concepts necessary to lay the foundation for a wide array of optical systems and optical systems engineering applications. The last section of this chapter introduces our optical systems engineering case study that will be used throughout this book. Just as the optical systems engineering building blocks, the case study is intended to be sequential and evolves from chapter to chapter. The intent is to integrate the systems engineering concepts, principles, and techniques, with the optical systems engineering analytical methods in a running, applications-oriented case study. In the next section, we provide a short history of the field of systems engineering along with some essential definitions.

1.1.1 Short History of Systems Engineering

In a loose sense, systems engineering can be traced back to ancient days when systems were first engineered. Ancient architectures qualify as systems and so the construction of the pyramids, the Great Wall of China, the Roman aqueducts, and other ancient architectures, and these are all examples of engineered systems engineering. However, from a historical perspective, the bulk of the development of modern-day thinking on systems engineering principles, methods, tools, and techniques occurred relatively recently. Systems engineering involves the organized, optimized, and requirements-driven development of systems with an emphasis on the entire system, over its entire life cycle, and with a focus on integration.

According to the International Council on Systems Engineering (INCOSE), the systems engineering term itself had its origins at Bell Telephone Laboratories in the early 1940s (Schlanger 1956; Hall 1962; Fagan 1978), and some of the essential systems engineering concepts within Bell Labs trace back to the early 1900s. The DOD entered into the picture in the late 1940s and applied systems engineering to missile defense systems (Goode and Machol 1957). During World War II, the U.S. Army Air Force formed the RAND Corporation. World War II saw the advent of major systems engineering applications, but it was not until 1950 that the systems engineering discipline, as we know it today, was first taught. This initial attempt at teaching systems engineering began at the Massachusetts Institute of Technology (MIT) by Mr. Gilman who was the director of systems engineering at Bell Labs (Hall 1962). By 1951, Fitts presented the idea of allocating system functions down to the physical elements (Fitts 1951). RAND developed the fundamentally important area of systems analysis in 1956 (Goode and Machol 1957). In 1962, Hall included five phases in his description of systems engineering (Hall 1962). These phases were as follows:

1. Conducting system studies along with appropriate planning activities.
2. Exploratory planning: Well-known systems engineering functions were accomplished.
3. Development planning: A repetition of phase two occurs but at a more detailed level.
4. Development studies.
5. Current engineering: Current engineering takes place during the operational phase of the system when the system gets refined.

Since the 1960s, there has been a robust development in the systems engineering discipline. Some more recent notable events include the establishment of the Defense Acquisition University (DAU) in 1971, the creation of the Air Force Center of Excellence for systems engineering that resides at the Air Force Institute of Technology (AFIT), and the establishment of the International Council on Systems Engineering in 1990. Some notable publications include MIL-STD-499 (DOD 1969) in 1969, Andrew Sage's systems engineering text in 1992 (Sage 1992), IEEE Std 1220-1998 (which looked at the application and management of systems engineering) (IEEE 1998), ANSI/EIA-632-1999 (which focused on systems engineering processes) (ANSI 1999), the systems engineering handbook by INCOSE in 2007 (DAU 2001), and Blanchard and Fabrycky's books on *Systems Engineering and Analysis* (first edition was in 1981 and fifth edition in 2010) (Blanchard and Fabrycky 2010). Other recent significant events include the establishment and advancement of EA methods, capability maturity model integration (CMMI) methods, life-cycle costing (LCC) methods, and model-based systems engineering (MBSE) methods.

CMMI is a process-centric methodology that is useful in establishing a set of organizational assessment benchmarks and best practices. The CMMI methodology started as a project that involved a productive collaboration between commercial and governmental entities and the Software Engineering Institute (SEI) from Carnegie Mellon. Organizations such as the National Defense Industrial Association and the Office of the Secretary of Defense (OSD) were strong supporters.

According to Eisenberger and G. Lorden, the concept of systems LCC became popular in the 1960s when the DOD began to recognize problems associated with awarding contracts based on price alone. Studies of weapon systems and other procurements revealed that acquisition costs were typically smaller than costs of ownership such as the cost of labor and materials required to operate and maintain the system.

In MBSE, the idea is to use integrated models at all levels of the development process and wherever possible to improve integration across disciplines and the integration of diverse systems engineering activities across the systems life cycle itself. There are various excellent MBSE tools available such as No Magic's MagicDraw™ that, among other things, has an EA plug-in that provides a model-based EA framework. It can also integrate with their own (or other) standard requirements management tools such as IBM's Dynamic Object-Oriented Requirements System (DOORS™) and other MBSE tools. Many of the MBSE tools come in an integrated suite that allows modeling of systems engineering methods and techniques such as EA methods, requirements engineering and requirements management methods, business process modeling, and decision and risk analysis modeling. EA methods and MBSE are discussed further in Chapter 4. The adoption of "agile" methods, lean engineering, "Six Sigma" methods, and integrated life-cycle product development tools, along with decision and risk analysis methods, techniques, and tools, contributed significantly to the modern understanding and evolution of the systems engineering discipline. These methods will be discussed throughout the text. We next present some of the recent definitive events that evolved the systems engineering discipline.

1.1.2 Some Definitions of Systems Engineering

When we think of the word *system*, what comes to mind? According to Merriam-Webster's Collegiate Dictionary, a system is "A regularly interacting or interdependent group of items forming a unified whole" (Merriam-Webster 2003). Taking this definition, we notice that to qualify as being a system, there must be interacting items and they have to depend on each other in such a way that they are in some fashion unified. This definition is quite broad and fits many items from ordinary to extraordinary. For instance, a coffee cup can be thought to be a system. It has a handle (item 1) that serves as an interface between the hand and the coffee, a flattened surface (item 2) that serves as the interface between the cup and a surface, and a liquid containing element (item 3) that serves as the receptor for our delightful beverage. On the other extreme, the galaxies themselves can each be thought of as a system that contains billions of stars, dust, planets, black holes, quasars, plasmas, electromagnetic radiation, and more. These elements within the galaxy influence each other through physical interactions such as gravitational and electromagnetic effects and unify to produce the galaxies themselves. Other systems are satellite systems, medical systems, military systems, quality control systems, inspection systems, manufacturing system, and optical systems, to name a few.

An observation at this point may be that a system is whatever we define it to be, as long as it has interdependent elements that are unified in some fashion. This means that a U.S. guided missile cruiser could be thought of as a system in its role as a maneuverable U.S. Navy surface ship that has a specific role with well-defined functions in the fleet. Equally valid, the cruiser's Aegis combat system itself could also be considered a system, collecting real-time telemetry of targets and providing that information to the ship's command structure. Systems engineering involves the principles, methodologies, and techniques involved in engineering a system as we define it. When reading systems engineering texts, the definition for systems engineering varies from one text to another. However, the common elements in the definition of systems engineering are that systems engineering

1. Deals with the entire system
2. Applies over its entire life cycle, from conceptual design to retirement/disposal
3. Focuses on integration

We can consider these points when we think of engineering a system.

The system must be defined in its environment, so its boundaries and relational context to other systems or entities can be established. The word *"entities"* implies membership in an encompassing EA. Our system may interact with other entities, and if so, EA methods become useful. Once we define what we mean by system, we can further subdivide the system into smaller functional pieces called *subsystems*. This functional decomposition is hierarchical and depends on the system itself and often on who is subdividing the system. The subsystems can then be further divided into *components/assemblies/pieces/parts*. We will use the term *components* to refer to this level of system decomposition.

An example of an optical system as we define it is shown in Figure 1.1, which presents a satellite system as a collection of subsystems. Notice that this satellite has optical components as subsystems (primary mirror, secondary mirror, solar array, sun sensor, instrument module). Furthermore, other parts of the satellite may be critical components that directly affect the performance of the optical system (baffle, light shield, power, signal conditioning, radiation shielding, environmental conditioning, and data communications).

Therefore, if it suits us, we can define or classify this satellite to be a space-based optical system. On the other hand, this satellite could be part of a larger system wherein the satellite's optical data play a key role. In this sense, the satellite is an entity in a larger enterprise, and the satellite may be considered to be part of an SOS. In essence, we define the system, its constituent parts, and its surrounding environment.

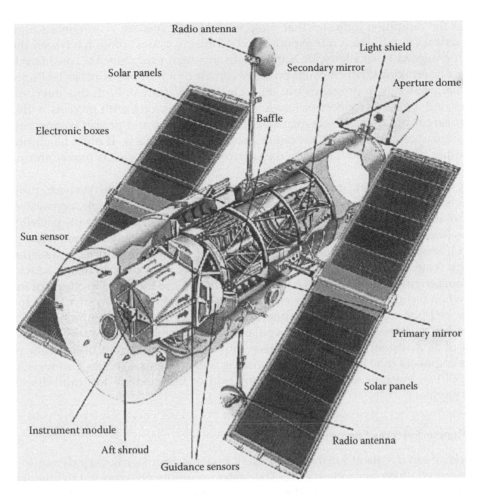

FIGURE 1.1
Hubble Space Telescope architecture. (Reprinted from NASA, Hubble Space Telescope diagram, 2014, NASA Quest website: http://quest.arc.nasa.gov/hst/photos-i.html, accessed November 3, 2014; http://quest.arc.nasa.gov/hst/images/HSTdigram.gif. With permission.)

In addition to our understanding and definition of *systems*, we want to distinguish between the words *program* and *project* and also between *service*, *product*, and *process*. *Merriam-Webster's Collegiate Dictionary* defines a *program* as "a plan or system under which action may be taken toward a goal" (Merriam-Webster 2003). In the DOD, the term usually means a large, complex development or acquisition undertaking, and program management then relates to the management activities associated with realizing the program. For example, a high-budget, complex satellite activity in the DOD to develop a new imaging capability would be typically classified as a development or acquisition program. Interestingly, a definition for *project* according to *Merriam-Webster's Collegiate Dictionary* is "a planned undertaking: as

a. A definitely formulated piece of research

b. A large usually government-supported undertaking

c. A task or problem engaged in usually by a group of students to supplement and apply classroom studies"

Item *b* in the definition indicates that projects and programs are synonymous and some people state as much. From our experience, however, we most often have seen the label of *project* assigned to smaller-scale activities. In this text, programs are considered large, complex undertakings, which could possibly consist of a number of projects. Projects are similar undertakings to programs but on a much smaller scale. With this interpretation, you may find a junior engineer or low-level manager working with projects within confides of a larger program. Project manager, project engineer, and project engineering then have a similar connotation to their larger program counterparts. It must be emphasized that in other disciplines and sectors, the interpretation of the words *project* and *program* may be different.

The systems engineering discipline does not only concern itself with systems, programs, or projects but also with services, products, and processes. For example, one system may be organized in accordance with a traditional DOD acquisition program paradigm but may also have service or process aspects to it. Another system may be completely service oriented like a help desk support service or technical writing support service system.

In large systems, very often there are components of each present in the activity, and the systems engineer will be involved in all aspects! We use the word product (or material goods) when we are considering systems and activities with commercial aspects to it. For example, cell phones, iPads, and commercial off-the-shelf (COTS) computers are considered products, whereas an optical surveillance system or weapon system is more aptly thought of as a system that may have been implemented through a program or project. Systems engineers may find themselves working with systems, services, products, or processes (SSPPs). We now look at the unique world of the modern-day, multidisciplinary systems engineer.

1.1.3 Diverse Systems Engineer

The working environment for the current-day systems engineer is very dynamic, multidisciplinary, challenging, and exciting. The systems engineer is expected to interact proficiently with SSPPs, various stakeholders, administrators, managers, support staff, technical experts, and technical points of contact in different fields, officials, and support contractors (whether internal or external to the organization). The modern-day systems engineer needs to be capable in highly technical environments as well as proficient in managerial environments. Sometimes, the expectations for a systems engineer are to have lots of engineering expertise (big "E") and a little managerial proficiency (little "m") such as in a typical junior systems engineer. Often, junior-level systems engineers are assigned to understand and work on lower levels of the functional system components and so require more engineering technical proficiency than managerial proficiency (big "E" and little "m"). In contrast, senior systems engineers usually support upper management and so focus more on engineering management activities (big "M" and little "e"). This does not mean that senior systems engineers have little engineering experience, just that their focus often is more on engineering management activities than on lower-level engineering details. Also, some organizations also have a technical track that has promotion opportunities to senior levels. However, promotion opportunities to the highest levels of an organization in many cases involve a deliberate transition from big E type of activities to big M type of activities. To effectively focus on "big M" activities, senior systems engineers in larger organizations usually have junior systems engineers that provide lower-level technical support to upper management. Senior systems engineers are then free to focus on higher-level technical issues and can effectively provide systems engineering managerial support to the program manager and senior administration of an organization or enterprise.

The type of support expected depends on the needs of the organization, the resources available to the organization, and the capabilities of the particular systems engineer.

For example, large organizations with big pools of systems engineers are able to use senior and junior engineers to support large, complex programs and systems. In smaller organizations, however, there may not be a large pool of systems engineers or even a small pool of systems engineers for that matter, and so the system engineer may be forced to provide both technical and managerial supports (big "M" and big "E"). This would be impractical for large organizations dealing with large, complex projects. Smaller companies and organizations tend to use the big "M" and big "E" approach due to resource constraints. The systems engineer must be able to fulfill multiple roles including both the technical and managerial.

In the case of the optical systems engineer, the expected technical proficiency has to do with optical SSPPs. For proficiency with optical systems, the optical systems engineer needs strong mathematical skills such as familiarity with linear and nonlinear system theory, integral and differential calculus, probability theory, statistics, stochastic processes, complex math, optimization methods, functional analysis, frequency analysis, Fourier transform theory, Laplace transforms, Z-transforms, and wavelet transforms as basic skills. There is great importance given to math since it is the unifying language between different technical disciplines. The optical systems engineer will have to work with optical designers, mechanical engineers, optical testers, software engineers, hardware experts, optical scientists, optical technicians, and a host of other subject-matter experts. It is also beneficial to have experience with technical topics such as modern optics, lasers, electromagnetics, electrical circuits, computer architectures, atmospheric turbulence physics with regard to its interaction with light, statistical optics (SO), Fourier optics, nonlinear optics, detectors, material properties, signal/image processing, basic mechanics, system noise analysis, and optical systems modeling methods. We will provide the required basic introductory level technical understanding in Section 1.2 throughout this text.

With regard to systems management, familiarity is expected with requirements management and engineering, modeling and simulation methods, engineering design principles, feasibility analysis, trade studies, functional analysis, requirements allocation, configuration management methods, interface control procedures, technical performance measures and metrics, decision and risk analysis, cost, schedule, performance, risk assessment/management, qualification planning and execution, and system optimization methods. The optical systems engineer must also be proficient in, and aware of, the following representative concepts, processes, and issues:

• Reliability	• Maintainability	• Availability
• Human factors	• Quality	• Logistics
• Engineering ethics	• Value engineering	• LCC
• Concurrent engineering	• Standards	• Benchmarking
• Supportability	• Intellectual property	• Producibility
• Safety	• Security	• Disposability
• Contract management/law	• Flexibility	• Environmental interactions
• Performance	• Liability	• Suitability

The optical systems engineer must ensure that these considerations are properly addressed in the system's requirements development process and throughout the system's development life cycle. In the next section, we provide a brief overview of the EA concept and its connection to systems engineering.

1.1.4 Introduction to the Enterprise and Its Architectural Description

This section provides a brief introduction to the EA concept and establishes a connection with the systems engineering discipline. A more detailed look at EA methods is provided in Chapter 2. EA methods and their interaction with the systems engineering discipline are becoming critically necessary with the ever-increasing complexity of modern-day technical systems. Although it might be argued that an *enterprise* fits our general definition of a system—and so EA methods should be part of the systems engineering discipline—the reality is that EA methods have evolved on their own. EA methods play a fundamental and complementary role in defining, understanding, developing, and supporting systems that are part of a larger enterprise activity.

According to The Open Group (TOGAF 2012), an *enterprise* is "any collection of organizations that has a common set of goals and/or a single bottom line." One important observation is that we are dealing with multiple organizations. These organizations likely have different motivations, agendas (open and hidden), resources, commitments, and capabilities. They are also likely to have some independence and autonomy but are linked by the common goal(s). The EA methods define the enterprise strategic goals, establish the framework of cooperation between various organizations within the enterprise, and provide for a governance structure between various *entities* making up the enterprise. An example would be various agencies, components, and organizations making up the DOD. These DOD organizations have separate budgets and a degree of autonomy but cooperate to achieve the strategic goals set by the DOD and U.S. leadership. Note that the DOD itself may be part of a larger enterprise consisting of the U.S. Government and its allies concerned with common defense issues. Each organization in the enterprise is an entity in the DOD enterprise, and their cooperative interactions are structured by the enterprise's architectural framework—in this case, the Department of Defense Architectural Framework (DODAF). Each of these organizations has a key role to play in the overall enterprise. Some organizations provide services, others provide systems, others ensure that processes are followed, still others produce products, and some produce mixtures of SSPPs.

It is the task of the enterprise architectural framework to provide the structure or framework that guides the interactions of the constituent entities. The EA framework does not build the EA architecture itself; it just provides the interaction protocols and tools for jointly building the architecture in an integrated and unified manner. One way this is accomplished is through a collection of predefined views and architectural framework artifacts. For example, many architectural frameworks have viewpoints such as operational views, system views, and service views that are integrated with each other. These EA framework products provide the necessary information required by enterprise stakeholders. The EA typically provides the highest level of integration available to the systems engineer. The exception is when the enterprise is itself part of a larger enterprise and so consequently has a higher-level structure, an enterprise of enterprises. An example would be the U.S. Government structure interacting on the world stage. There would possibly be several different enterprises, each with their own architectural framework, which would be expected to interact efficiently with each other. The enterprise is typically the highest level that the systems engineer has exposure to. In this text, we will assume that the enterprise is the highest level of organizational structure that exists for our applications and needs. If the enterprise were part of a larger enterprise, then the EA methods outlined in Chapter 2 would still apply, and so there is no loss in generality by adopting our "one-enterprise" approach. The EA framework provides the structure for defining the interaction environment of the constituent SSPPs that make up the enterprise.

As an example, one of the key EA artifacts is the concept of operations (CONOPS) for the enterprise. The EA CONOPS shows the critical systems, actors, processes, and the interaction environment itself. Actors are "A person, organization, or system that has a role that initiates or interacts with activities" (TOGAF 2014). We now look at the concepts of SOS and FOS in context to the EA and also the systems engineering discipline.

1.1.5 Introduction to SOS and FOS

In the previous section, we pointed out that the EA typically contains the highest level of integration that is available to the systems engineer. At the EA level, all the entities in the enterprise are defined as well as the expected interactions between these entities whether we are considering SSPPs or mixtures of these elements. The enterprise then can consist of many separate systems that work together in a unified way. The enterprise can, and often does, contain SOS. We think of the SOS concept as a larger system that is itself made up of a collection of systems as shown in Figure 1.2.

The larger system then has an overall purpose that can only be achieved by the interaction of the systems that make it up. An example would be a U.S. Navy ship that contains many systems (radar systems, weapon systems, communications systems, damage control systems, defense systems, and so forth). All of these systems cooperate harmoniously together to provide the necessary capabilities to the ship commander and crew to fulfill the intended missions assigned by higher command. In this sense, the ship is an SOS, and the constituent systems—HF communications, UHF communications, X-band communications, damage control systems, radar systems, weapon systems, power systems, control

FIGURE 1.2
The SOS concept. (From Wikimedia Commons, Global information grid operational view-1, http://commons.wikimedia.org/wiki/File:Gig_ov1.jpg, accessed November 8, 2014.)

systems, optical systems, and as such—are individual systems that make up the SOS. Note that the U.S. Navy ship can be thought of as part of a larger enterprise consisting of early warning satellites, the command structure, other response elements, intelligence services, and support elements. The enterprise itself can be thought of as having SOS as part of its integral components. However, in addition to the SOS component, the enterprise also has services, products, and processes to make up its whole. As an example, the DOD enterprise consists of its collection of individual systems, collaborative SOS, support systems (e.g., weapon systems, communications systems, navigation systems, early warning systems, force protection systems, logistics systems, maintenance systems), products (e.g., doctrine, tactics, capabilities, military and humanitarian operations, peacekeeping, protection, airmen, soldiers, and sailors), and enabling processes and services (e.g., command and control, intelligence, surveillance, reconnaissance, finance, medical, legal, supply, training, administration, engineering). This list is not intended to be all-inclusive but just to serve as an illustration how the SOS concept fits into the overall EA paradigm. In other words, we define an SOS to suit our needs and the situation at hand. The EA then defines the interrelationships of the systems and SOS to each other and in relationship to the enterprise itself. "*Family of Systems* (FOS) is defined as a set of systems that provide similar capabilities through different approaches to achieve similar or complementary effects" (ODUSD(A&T) SSE 2008). The difference between the FOS and SOS definitions is a matter of scope. The FOS provides similar capabilities, whereas SOS provides an increased capability from its constituent systems. In other respects, it would be hard to tell one from the other.

An example of an FOS would be a collection of assets that hunt submarines. There are different assets that all provide the same or similar capability to detect submarines, surface vessels, other submarines, air assets, and specialized sensors. In this scenario, these assets are part of an FOS. In contrast, the U.S. Navy ship we mentioned earlier, working with satellite systems, command-and-control systems, intelligence systems, ground systems, air systems, and integrated communications systems, may be part of an SOS that provides a new, improved, sea-based rapid response capability. The SOS in the latter case is distinct from the FOS since it provides a new capability that is not achievable by the individual systems themselves.

In the following section, the U.S. joint capability integration, development system, and the process of acquiring systems in the DOD are introduced. Most of the jobs relating to systems engineering in the United States are related to the defense industry. A thorough understanding of the DOD acquisition process is relevant and important for those working on or with U.S. defense systems. The next section will provide a brief introduction to the acquisition process used in the DOD.

1.1.6 U.S. JCIDS and the DOD Acquisition System

This section provides a basic overview of the acquisition process to better understand the interrelationship between EA methods, the formal defense capability acquisition process, and the systems engineering discipline. Some good resources for further information are the Defense Acquisition Guidebook (DAU 2012) and material from the DAU (DAU 2012). Figure 1.3 shows the relationship between the U.S. Defense Acquisition System (DAS); the planning, programming, budgeting, and execution (PPBE) process; and the JCIDS. From the figure, you can see that there are three distinct entities that have their own oversight structure. For the PPBE process, the oversight comes from the Deputy Secretary of Defense. For the JCIDS process, the oversight comes from the Vice Chairman of the Joint Chiefs of Staff and the Joint Requirements Oversight Council (VCJCS/JROC). The JCIDS process is

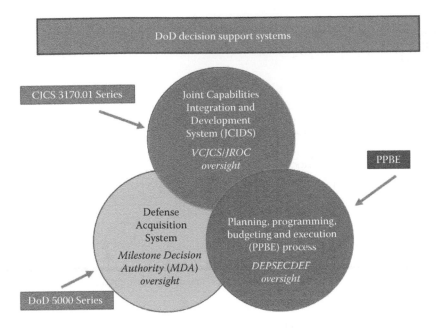

FIGURE 1.3
Relationship of the JCIDS with the defense acquisition system and PPBE process. (From Defense Acquisition Portal, https://dap.dau.mil/aphome/Pages/Default.aspx, accessed November 8, 2014.)

to a large degree a precursor to the traditional DAS in that it ensures that acquired systems are capability focused and have the required compatibilities with the systems of other services, partners, and allies. There are checks and balances to ensure that the DAS is only engaged after other more cost-effective capability solutions such as doctrine, organization, training, materiel, leadership, personnel, and finance (DOTMLPF) are exhausted.

The JCIDS is governed by a series of instructions from the Chairman of the Joint Chiefs of Staff—CJCS 3170.01 Series instructions. The DAS has oversight from the Milestone Decision Authority (MDA) and is governed by the DOD 5000 Series of directives and instructions. The Federal Acquisition Regulations (FARs) are also prominent as a legal, directive, policy, and process defining entity. While the systems engineer may interact with all of these entities, they mostly are engaged in the DAS itself. Systems engineering principles, methods, and techniques are very much a part of defense systems development and acquisition activities.

Figure 1.3 shows the relationship between the JCIDS and DAS. The PPBS is integrated over the entire systems acquisition process and throughout the systems development, support, and retirement phases, which allow optimization of technical performance, development, and support schedule, along with the associated SSPP activities, over all life-cycle phases.

The acquisition process starts with high-level strategic guidance shown at the top left of Figure 1.4. This guidance is itself developed from higher-level strategic guidance such as national policies and is subject to the FARs. A series of joint—multiservice activities that include appropriate representation from air force, army, navy, marine corps, and coast guard stakeholders—concept evaluations, capability assessments, and gap analysis are conducted that lead to an Initial Capabilities Document (ICD). If the MDA is convinced that a material solution is called for in the ICD, then the MDA and appropriate stakeholders make a Materiel Development Decision (MDD). The MDD initiates an analysis

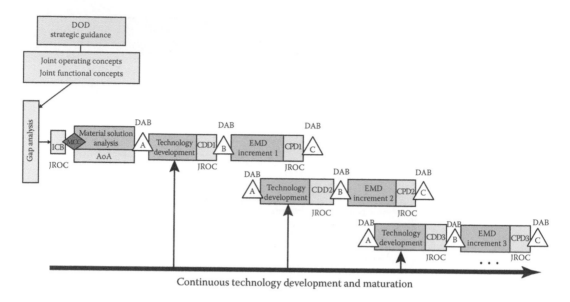

FIGURE 1.4

The requirements and acquisition process flow. (Reprinted from ACQWeb, 2014, http://www.acq.osd.mil/asda/docs/dod_instruction_operation_of_the_defense_acquisition_system.pdf. With permission.)

of potential materiel solutions—materiel solution analysis and an analysis of alternatives (AoA). In this part of the acquisition process, promising technologies from various sources are investigated to include foreign developments, government laboratories academia, centers of excellence, and industry. Once a materiel solution approach is determined, then the decision is documented in the Acquisition Decision Memorandum (ADM). The ADM and the results of the AoA generate a proposed solution from the lead DOD component that carries out the AoA, and a life-cycle funding profile, generated from the PPBE, marks the end of the materiel solution analysis phase and the start of the technology development phase.

The triangles in Figure 1.4 are formal milestone events that are consistent with the systems engineering development cycle (e.g., conceptual design phase, preliminary design phase, and detailed design phase). Depending on program cost levels, the oversight for these milestone events is the Defense Acquisition Board, which is chaired by the MDA. There are different approval levels depending on the overall research, development, test, and evaluation (RDT&E) costs of the program, and the approval levels are based on Acquisition Categories (ACATs) that range from I to IV (with two subcategories for ACAT I programs). As an example, ACAT II programs are those programs that do not qualify as an ACAT I system but are still considered major systems. Major systems are those systems in which the RDT&E costs in fiscal year (FY) 2000 in dollars equal or exceed $140 million. At this stage, it is not as important to delve into the different ACATs, as it is to know that the MDA depends on the RDT&E cost structure of the acquisition program. Where appropriate, the JROC reviews the JCIDS products—Capability Development Document (CDD) and the Capability Production Document (CPD). For major systems development activities, after the JROC reviews the CDD and the MDA approves milestone B—Preliminary Design Review (PDR)—the acquisition proceeds to the engineering, manufacturing, and development (EMD) phase. The next phase in the acquisition process is the CPD review by the JROC and milestone C—Critical Design Review (CDR), approval by the MDA. The overlapping bars in Figure 1.4 illustrate evolutionary development. It is noteworthy to point out that the DOD 5000.02 directive calls out that the DOD EA "shall underpin all information architecture development." Accordingly, any information-related element of an acquisition, development, or activity that must comply with the DOD 5000.02 directive must be subject to the DODAF (current version as of this writing 2.02).

From a business perspective, Figure 1.5 shows the contract process where businesses typically interact and interface with the defense acquisition process. Three phases make up the contracting process: (1) acquisition planning, (2) source selection, and (3) contract administration. Note that these perspectives are from the government side. In the acquisition planning phase, the need for the system is established using the processes discussed in Figure 1.4. The results of these processes are shown in the top block of Figure 1.5 labeled requirement determination. The requirements for the proposed solution that are established are the requirement for a DAS solution. Subsequently, a request for proposal (RFP) is initiated by the government and is posted on government websites such as http://www.fedbizopps.gov, http://www.grants.gov, and http://www.sbir.gov, for larger efforts, academic, and small business opportunities, respectively. Once the RFP or broad agency announcement has been released, the business and academic community can choose to respond.

The business and/or academic community will use their internal processes to decide whether or not a particular solicitation is worth pursuing. An institution may go through the following representative phases as part of their business development process: opportunity creation, identification, qualification, pursuit, proposal development, proposal review, and proposal submission. Many things are considered in the opportunity creation phase such as evaluating the market need, establishing the business case for a proposed venture, and projecting opportunities for business growth. The identification phase looks at validating the creation opportunities and finding available opportunities that are in line with the strategic vision and core competencies of the business. The qualification phase

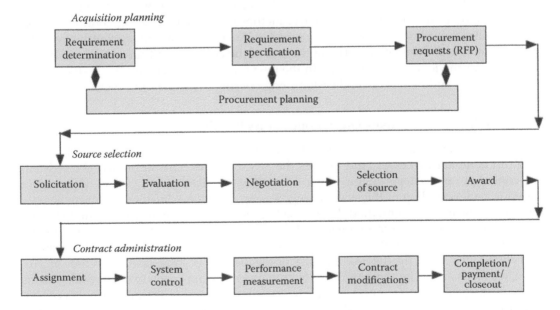

FIGURE 1.5
The contracting process. (Reprinted from Defense Acquisition University Press, *Systems Engineering Fundamentals*, January 2001, Fort Belvoir, VA, http://www.dau.mil/pubs/pdf/SEFGuide%2001–01.pdf, accessed April 24, 2014. With permission.)

evaluates the capabilities of the business entity to ensure that there are sufficient technical depth, breadth, and resources to accomplish the work. Collaborations and partnerships are often established at this point to form the "right team" for the effort. Pursuit involves establishing contact with the RFP point of contact (POC) to obtain clarification and available guidance if possible. Once the proposal development phase is completed, there should be a proposal review effort. The proposal review effort consists of independent groups (red team and blue team) that critically review the proposal. The red team tries to challenge the proposal, and the blue team defends it. The purpose of this process is to improve the proposal prior to submitting it to the government/stakeholder for official review. After suggested improvements from the red/blue team reviews are integrated into the proposal, the proposal is then submitted to the contracting office where it becomes a solicitation as shown in the middle block of Figure 1.5.

The government evaluates the submitted solicitation. The government's evaluation process has life-cycle cost estimation as well as competency metrics. The government will use systems engineers and field experts to evaluate the solicitations. Based on the reviews, the government may enter into negotiations with business entities. After the negotiation step, the source selection process determines which business entity or entities are chosen for the work outlined in the RFP. Awarding the contract follows this step. There are several different types of contract vehicles that the government uses such as firm fixed price (FFP), cost plus fixed fee (CPFF), and cost plus incentive fee (CPIF). Which of the contract vehicles is chosen depends on the nature of the task and the amount of risk the government is willing to take on the effort. Once the contract award is made, this ends the source selection phase of the contracting process.

The last row of Figure 1.5 shows the contracting process steps after the contract award has been made. In this contract administration phase, contract oversight agencies such as the Defense Contract Administration Service and the Defense Contract Audit Agency ensure that proper administration and compliance are followed by the business entity. This is done by establishing system control methods and by measuring contract performance. Contract modifications are also conducted if necessary as well as contract completion, payment, and closeout activities at the end of the contract. In this section, we briefly learned about the JCIDS and the DAS. In the next section, we focus on the systems engineering process and look at general systems engineering principles and methods.

1.1.7 Introduction to Systems Engineering across the Life Cycle

This section provides an introduction to the systems engineering methodology, illustrates some fundamental processes, and defines the SDLC phases. Our approach is consistent with contemporary systems engineering professional practices and consistent with current academic understanding of the systems engineering discipline. Further, we draw heavily from the applications and methodologies used in the U.S. defense industry. Our approach is general and adaptable to commercial and other applications as detailed in the following. This defense-centric application orientation is adopted since the majority of U.S. systems engineering jobs are in the defense industry or related to it in some fashion. Where appropriate, we focus and tailor the general systems engineering approach to the optical systems engineering application area and discipline. In order to provide the appropriate context, certain illustrations will remain at the general level, whereas others will be adapted to fit optical systems engineering scenarios throughout the optical systems engineering life cycle. This approach is used throughout the text and is designed to provide introductory level information. More detail can be found in Chapters 3 through 16 of this book.

The systems engineering discipline is fundamentally important in developing large, complex systems, FOS, SOS, and related elements of an EA. Systems engineering methods and processes can also be adapted and tailored for other organizations and activities to include small businesses, commercial, industrial, governmental, and academic applications (Arrasmith 2007). The systems engineering discipline consists of principles, methodologies, processes, and techniques for optimally bringing systems into existence, supporting them throughout their life cycle, and then providing for efficient, safe, environmentally friendly system disposal or retirement. In other words, systems engineering brings complex systems into being and supports them throughout their life cycle and sees to their disposal in an integrated, optimal fashion.

The evolutionary development of complex SSPPs typically follows an RDT&E-oriented, phased structure such as the traditional systems engineering life-cycle phases: (a) conceptual design phase, (b) preliminary design phase, (c) detailed design phase, (d) manufacturing/production/construction phase, (e) operations and support phase, and (f) system retirement/disposal phase. Notice that we are using high-level categories, and these categories will be detailed later in Chapter 3. There are a variety of useful process models that help the systems engineer understand a system and that break the system into a manageable set of developmental steps. These models are intended to furnish us with a general view of the systems engineering processes that lead to the successful development of a system. We will briefly discuss some of the more often used process models starting with the generic systems engineering across the life-cycle waterfall diagram and adapt it for optical systems engineering.

This waterfall process model outlines the essential and sequential steps in the systems engineering process. Notice that the steps from optical systems requirements to optical systems design integration are sequential steps but are repeated at each of the three SE design phases (conceptual design, preliminary design, and detailed design). This iteration is shown in the figure by the dashed lines to the left of the individual blocks. We will discuss this more in Chapter 3. As we can see in Figure 1.6, the first step in this waterfall model is to define the problem.

This single block has a number of activities, processes, and events that are associated with it, but for brevity and clarity's sake, we focus only on the high-level detail in this introduction. The next step involves establishing an architectural framework for our system to guide how our system interacts and interfaces with other systems in our enterprise. This step should happen early in the systems engineering development life cycle to ensure that enterprise-related technical issues, systems issues, and governance and strategic drivers are considered in the systems requirements and any derived potential solutions. At this stage, the feasibility studies and trade studies are focused on the stakeholder requirements and are initially assessed at a high and broad level. For example, each stakeholder requirement is assessed not only for technical feasibility but also for organizational, motivational, political, operational, financial, and environmental feasibility. We must also determine whether or not the potential feasible solution to the requirement is sustainable within the enterprise over the systems life cycle.

The system-level requirements for the candidate optical system are shown in the optical systems requirements block. General systems engineering texts would label this as "Systems Requirements," but since this book is geared toward engineering optical systems, we add the "optical" term as a distinction. Notice that the steps in the waterfall model that start with optical systems requirements and end with optical systems design integration are boxed. These represent core systems engineering processes and are repeated at every design phase. For instance, if we start with system-level

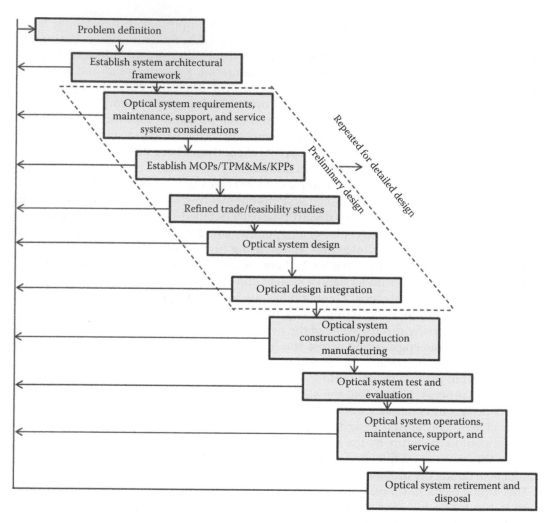

FIGURE 1.6
The optical systems engineering development across the life-cycle waterfall model. (Adapted from Blanchard, B.S. and Fabrycky, W.J., *Systems Engineering and Analysis*, 5th edn., Pearson, London, U.K., 2010.)

requirements, we consider the system-level maintenance, support, and service implications associated with those requirements.

We also determine appropriated measures of performance (MOPs), technical performance measures and metrics (TPM&Ms), and key performance parameters (KPPs) for our system. These are necessary to effectively test and qualify our system and to effectively communicate the systems performance capabilities to system stakeholders. Afterward, we conduct a functional analysis that is based on the system-level requirements. This functional analysis holistically evaluates the system-level requirements and defines the subsystem functions needed to satisfy these requirements. The system requirements are then allocated into functional partitions (e.g., subsystems) determined by the functional analysis and requirements allocation processes.

The engineering team then designs and integrates the designs based on the functional decomposition. The connection between requirements and design can be broken into two distinguishable steps: (1) generate requirements and (2) design based on the generated requirements. As an example, our optical systems requirements are used to drive the design effort once they have been baselined and put into configuration control. The design process for our optical system starts with a functional decomposition of the optical systems requirements. In the functional analysis process, subsystems are defined and analyzed that functionally partition our technical response to the optical systems requirements. For example, we might choose to partition an optical imaging system into power, control, communications, environmental conditioning, signal processing, image processing, pointing and tracking, and optical subsystems.

The optical systems may be further partitioned into daytime optical subsystems and nighttime optical subsystems, and each of these could be further subdivided into optical telescope, image formation optics, optical detector, detector housing, and detector cooling subsystems. Of course, this decomposition is not unique, and other forms of decomposition are equally valid. The important thing is that the functional decomposition is complete in the sense that all the necessary functions needed to satisfy the optical systems requirements have been defined. The optical system's functional partitions are then grouped together in the requirements allocation process, and the design process continues with further functional analysis, trade studies, modeling, and simulation as needed.

When analyzing the subsystems of our candidate optical imaging system, it is necessary to derive and generate new requirements at the subsystem level. These newly derived and generated requirements along with the allocated system-level requirements form the basis for the subsystem-level requirements and associated requirements documents. This process repeats once the subsystem requirements (B specifications) are baselined. Once all of the requirements generation/design phases are complete, the optical system goes to construction, manufacturing, or production. Subsequently, the complete optical system is tested, accepted, and supported throughout the optical systems life cycle. At the end of the optical systems life cycle, it is retired or disposed of in a cost-effective, environmentally friendly, safe, legal, and ethically sound manner.

Figure 1.7 shows another useful systems engineering process model—the "V" process model. This model is effective in showing sequential, hierarchical, and complementary life-cycle systems engineering activities that may be separated in time. We can see the traditional systems engineering design phases in the gray rectangular box. The conceptual design phase starts with problem definition and ends with the preparation of the systems requirements document—the A specification—for our optical system. The systems requirements document is formally approved at the system segment review, and this event starts the preliminary design phase of the SDLC. The preliminary design phase uses the approved A specification to drive the preliminary design. The result of this phase is the production of preliminary design documents and the preparation of subsystem-level requirements documents—B specifications—at the PDR. Upon approval of the B specifications, the last design phase—the detailed design and development phase—is started. In this last design phase, a detailed design, associated specifications, drawings, and detailed requirements based on the A and B specifications are developed. During this stage, prototypes are built and tested. This phase ends with the approval of all detailed design documentation at the CDR. The end of this phase signals the start of the production, manufacturing, and/or construction phase. This phase is shown at the bottom of the V-diagram.

After production, the systems engineering development process continues with the subsequent testing of the "as-built" system as shown on the right side of the V-diagram. Notice that there is a corresponding formal test block for each level of requirements (e.g., system level, subsystem level, and component level). Establishing a good set of TPM&Ms, MOPs, and KPPs during the requirements generation process is critical in effectively testing the

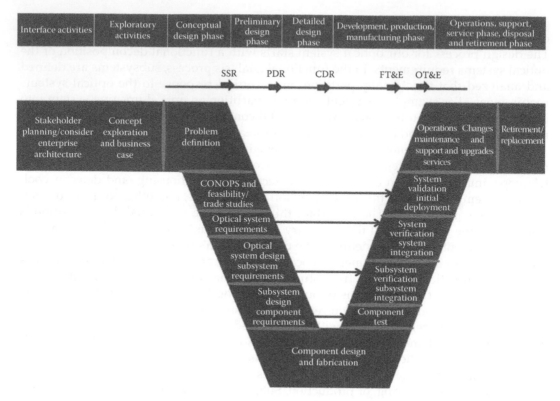

FIGURE 1.7
The V-process model illustrating the connection between system-level, subsystem-level, and component-level requirements and test throughout the traditional systems engineering life-cycle phases.

developed system. The system-level formal test is often called functional test and evaluation (FT&E), and the formal test that signals the end of the systems development process is called operational test and evaluation (OT&E). OT&E is often conducted in the field under operational (or simulated) conditions. Once the system passes the OT&E, the development activity is complete, and the system transitions to the operational, maintenance, support, and service phase. In this phase, the systems engineer is involved in evaluating the systems performance, providing preplanned product improvement (PPPI) upgrades for the system, and implementing changes according to a change and configuration control methodology. An example of this would be generating and responding to engineering change proposals (ECPs). The final phase of the SDLC is the system retirement and disposal phase, which should go smoothly if it was properly considered early on in the SDLC.

1.1.8 Transition to Optical Systems Building Blocks

We provided an introduction to systems engineering and modern-day systems and EA methodologies and processes. In the next section, we will begin to build the optical systems building blocks that will be necessary to understand and successfully analyze optical systems. Most chapters of this book will have an optical building blocks section that relates to the chapter content in some way but also is largely independent. The intent of the optical building blocks sections in each chapter is to provide the technical details needed to understand and analyze optical systems. They are intentionally designed as related, but stand-alone, sections that are sequentially progressive. By this, we mean that subsequent optical building block sections in later chapters may depend on principles, methods, and understanding developed in previous sections.

We have attempted to make each section stand-alone as much as possible, but due to the complexity of optical systems in general, some progressive material development and interdependence of sections are unavoidable. Nevertheless, the technical content in these optical building block sections is essential for the practicing systems engineer, working professional, or student that is working with optical systems. Just as we provided an introduction to Systems Engineering in the Modern Age section, we now provide introductory material needed to understand and analyze optical systems in the following section.

1.2 Optical Systems Building Blocks: Introduction, Systems of Units, Optical Systems Methodologies, and Terminology

In this section, we start with a short introduction and background of important discoveries associated with the technical aspects of optical systems engineering. We then provide an overview of our adopted optical systems engineering systems of units. Subsequently, we introduce optical systems methodologies, which are useful in understanding, describing, modeling, and analyzing optical systems. We also want to provide different perspectives in modeling optical systems and discuss when particular approaches and methods apply. For instance, we want to introduce terms that are essential to the reader's understanding of different types of propagation models. We can model and analyze the propagation of visible and IR light using several different approaches. The chosen approach depends on the physical situation. In the appropriate chapter, we discuss the means for deciding which approach to take and also key results from each approach.

A full treatment of just one of these areas could easily fill this text, and so we focus on results that can be directly applied by the practicing engineer and optical systems manager. Some examples of different approaches to modeling and analyzing the propagation of optical waves in the visible and IR part of the electromagnetic spectrum are as follows: vector electromagnetic theory, scalar electromagnetic theory, SO, geometrical optics, Fourier optics, or radiation transport analysis. Each approach has a conceptual framework that needs to be understood by the practicing optical systems engineer or optical systems manager, and we attempt to provide introductory level details in this section. We end this section with a description of essential terminology and concepts that will be needed in later chapters. We will use International System of Units (SI) throughout the text. We first start with a little background information on key developments in the optics field.

1.2.1 Introduction to Optical Systems

For millennia, the only known portion of the electromagnetic spectrum was that of visible light. The IR region of the electromagnetic spectrum was discovered when early scientists and scholars began to investigate what was beyond the red end of the rainbow of colors that were identified with light. In 1665, Sir Isaac Newton determined the basic principles of light and color and the use of the prism. He also discovered that without any correction, *diffraction-limited* performance could not be achieved by an optical system that is imaging at visible wavelengths with an aperture greater than a few tens of centimeters (Newton 1952).

Diffraction-limited performance, in the context of the spatial resolution of a given imaging system, refers to the best possible spatial resolution attainable using classical electromagnetic theory. In essence, it is the resolution of a perfect imaging system—no optical systems aberrations or noise, no atmospheric effects—and so represents a benchmark in analyzing the performance of an optical system. During Newton's time, it was also known that when viewing a point source—an object that is smaller than the spatial resolution of a given optical system—from a distance, the spreading of the light is broader than if the point source were closer. Newton looked at starlight using a telescope and discovered that the spreading of the light—the observed point spread function—was broader than when telescope was used to look at a similar point source in the laboratory. This was an early indication of the effects of atmospheric turbulence on optical imaging systems. Unfortunately, scientific equipment, photographic, and camera technology did not evolve sufficiently to do anything about this until the 1900s.

During the 1800s, Sir Friedrich Wilhelm Herschel discovered a region of the spectrum beyond the region of visible light that emitted heat. Later on, this part of the electromagnetic spectrum was dubbed "invisible rays" or "dark heat" and is today known as the IR part of the electromagnetic spectrum. Herschel set out to investigate the colors of the spectrum that were accountable for heating objects. Hershel knew that sunlight produced the colors red, orange, yellow, green, blue, indigo, and violet (ROYGBIV) with the use of a prism, and so he ventured to measure the temperature associated with each of these colors. By means of a prism and thermometers, he measured the temperatures of the different colors to observe where the increase in temperature occurred as he moved the thermometer throughout the prism. He discovered that the temperature reached its highest levels when it was past the red light of the prism in a region where there was no color. The radiation causing this heating was not visible; as a result, he concluded that these invisible rays had some kind of "dark heat." In 1860, Gustav Kirchhoff found that good absorbers are also good radiators, which lead to the idealized blackbody model. In 1879, Jozef Stefan discovered that radiated power from a blackbody source is related to its temperature. In 1884, Ludwig Boltzmann derived the same result from theoretical thermodynamic calculations leading to the Stefan–Boltzmann law. As will be seen later, this law is fundamentally important in estimating the expected spatial power density (W/m^2) emitted from an ideal (blackbody) source. In 1893, Wilhelm Wien derived Wien's law using thermodynamic principles showing that the shape of the thermal radiation that is emitted from a blackbody, as a function of wavelength, is essentially the same for all temperatures. He also showed that the peak of this thermal distribution moves toward lower wavelengths as temperature increases. Lord Rayleigh (aka John William Strutt) and Sir James Jeans developed the Rayleigh–Jeans law that attempted to analytically describe thermal radiation from a blackbody as a function of wavelength. Their law was shown to work for higher wavelengths but diverged from physical observations in the lower-wavelength regime. In 1900, Max Planck empirically found the correct blackbody radiation law that works for all wavelengths. Also in 1900, Heinrich Rubens and Ferdinand Kurlbaum experimentally verified Planck's work that leads to the establishment of quantum theory. In 1941, Andrey Nikolaevich Kolmogorov published a series of mathematical papers on atmospheric turbulence (Kolmogorov 1941). In 1961, Valer'ian Il'ich Tatarski developed a model for wavefront perturbations that were based on Kolmogorov's work. In 1965, Dave Fried developed a parameter r_0 using statistical methods that presented a metric related to the quality of "seeing" through the atmosphere. His parameter is called the Fried parameter and represents the diameter of a patch of atmosphere over which the atmosphere can be thought of as being somewhat the same. Effectively, for telescope diameters that are larger than r_0,

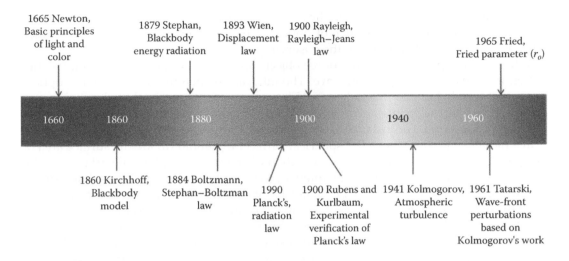

FIGURE 1.8
Timeline of important optical systems developments.

adaptive optics and/or atmospheric turbulence compensation methods are required to obtain the best resolution out of the imaging system. Figure 1.8 presents a timeline of key technical contributions to the optical systems field relating to remote detection systems that are affected by the Earth's atmosphere.

After the 1900s, we only show a few contributions that relate to remote optical detection systems that are affected by atmospheric turbulence. We intend this to be an introductory level timeline that is relevant to remote optical detection systems such as ground-to-ground, ground-to-air, air-to-air, and air-to-ground imaging systems found on a variety of optical sensing platforms. A more complete list of key contributions is included with each of the subsequent chapters. Also, in the later part of this timeline, we focus on key developments that impact optical remote detection systems and also key results in understanding and modeling the effects of atmospheric turbulence.

Since 1965, a tremendous amount of activity and research has been done in adaptive optics (real-time, hardware-dominant solutions), atmospheric turbulence compensation methods (traditionally non-real-time, software-dominant solutions), and hybrid methods (a mix of the two approaches) to correct for the effects of atmospheric turbulence in optical images. For many well-designed optical imaging systems, the effects of atmospheric turbulence are the limiting factor in the optical systems spatial resolution.

1.2.2 Overview of Adopted Systems of Units

In this section, we provide an overview of our adopted system of units. It is important to focus on terminology early on since many terms can be confusing when speaking to someone in another technical field. Some terms have different meanings depending on which technical discipline uses the term. Much confusion can be avoided by using optical systems–related terms correctly. For example, the word intensity is used in a variety of ways in literature and in practice. The "intensity of the target" is often used in practice and in literature when the target is fully resolved by the optical system. In other words, the target system has observable dimensions of length, width, and possibly depth. In this case, the term intensity is not appropriate. The actual units for intensity are Watts/steradian, and so the term *intensity* technically only applies to point sources. As mentioned earlier, for our purposes, a point source is a source of optical radiation whose dimensions are smaller than the spatial resolution limits of the observing optical system.

In the case of a point source, the source object can be thought of as a radiating "point" that emits an expanding spherical wave. The intensity is then that portion of the optical power that is radiated into a steradian (sr). The steradian is determined by dividing a part (or whole part) of the spherical surface area of the expanding optical wave by its separation distance squared. For instance, if we wanted to know how many steradians our point source radiates into over the entire spherical surface, we would divide the surface area of the sphere ($4\pi r^2$) by the square of the separation distance of the spherical surface and the point source (r^2) and obtain 4π sr. The intensity is the optical power in Watts that radiates into this sphere. Similarly, if we were interested in finding the intensity in a hemisphere about the point source, we would want to know how much optical power is being radiated into the 2π sr of the hemisphere. In essence, the word intensity applies to point sources. As previously stated, this term is often misused in conversations and even in published works in the sense that the word intensity is used to also describe optical power associated with *extended objects*—objects that have larger spatial dimensions than the resolution limits of an observing optical system. This abuse is so frequent that many optical scientists and engineers have learned to turn a blind eye to it. A more appropriate term in the extended object case is radiant emittance, which has units of W/m^2. To avoid these issues, it is helpful to use an established system of units and define terms early on.

There are different systems of units for describing properties and characteristics of observable physical phenomena like electromagnetic, seismic, acoustic, and infrasound waves. Which system is used depends on the convention of the technical discipline and also the problem at hand. For convenience, and in consideration of our international audience, we will use SI units throughout this book. With regard to optical systems, there are two SI systems of units that are particularly useful when it comes to the measurement of electromagnetic radiation—photometric units and radiometric units. Photometry typically involves measuring light at individual wavelengths and includes the response of the human eye as a filtering function within its formulation. Radiometry usually deals with measuring radiation in terms of total power (e.g., integrated over all wavelengths). The radiometric units also have single-wavelength elements, but these are distinguished by adding qualifiers such as *spectro* or *spectral* in front of the radiometric term. For instance, a *radiometer* measures the total optical power across a spectrum of wavelengths, whereas a spectroradiometer measures the optical power within a narrow band of wavelengths (or frequencies) surrounding a particular center wavelength. The spectroradiometer can be tuned to different center wavelengths to provide narrowband optical power readings at differing center wavelengths.

We use the SI-based photometric units when the quantities of interest contain the response of the human visual system, which attenuates or filters light differently at separate wavelengths. Table 1.1 shows the elements for the photometric measurement system.

This set of units is used quite often in specification sheets for optical imaging systems, camera systems, and some detectors. Column 2 shows the corresponding SI symbols. Authors often define and use different symbols in their work, and so the practitioner needs to check the units and dimensions of symbols to understand their proper usage. For example, if an author uses the words *luminous intensity* and the symbol I for a source of optical radiation, but the units end up being Lumens/m^2, then we know that what the author meant was luminous exitance and not luminous intensity. Unit analysis is critical in correctly interpreting the technical context.

TABLE 1.1

Photometric Measurement Elements

Photometric Name	Symbol	SI Units	Dimensions
Luminous flux (power)	Φ_v	Lumen	J
Luminous energy	Q_v	Lumen second	T J
Luminous intensity or candlepower	I_v	Candela (lm/sr)	J
Luminous emittance	M_v	Lux (lm/m²)	$L^{-2} \cdot J$
Luminous incidence or illuminance	E_v	Lux (lm/m²)	$L^{-2} \cdot J$
Luminance, luminous sterance, or photometric brightness	L_v	Candela per square meter (cd/m²)	$L^{-2} \cdot J$
Luminous exposure	H_v	Lux second (lx·s)	$L^{-2} \cdot T \cdot J$
Luminous energy density	ω_v	Lumen second per meter³ (lm·s/m³)	$L^{-3} \cdot T \cdot J$
Luminous efficacy	η	Lumen per Watt (lm/W)	$M^{-1} \cdot L^{-2} \cdot T^3 \cdot J$
Luminous efficiency	V		1

When conducting requirements analysis, one step should be to consider the best consistent system (or systems) of units to use in the application and then rewrite the requirements in that set of units. The dimension column entries in Table 1.1 are useful when comparing table elements in different systems of units. When it comes to radiometric analysis, the system of units in Table 1.2 is quite common.

We will adopt this system of units throughout the majority of the text since these units are well understood and are applicable to a variety of optical systems analysis applications. We will use SI-based photometric units where necessary, but because of their familiarity, we will predominantly use the SI-based radiometric units. Table 1.3 shows the symbol and units of some important physical quantities. The fundamental unit of energy in the SI-based radiometric system of units is the *joule*, which can be thought of as the amount

TABLE 1.2

Radiometric Measurement Elements

Radiometric Name	Symbol	SI Units	Dimensions
Radiant energy	Q_e	Joule	$M \cdot L^2 \cdot T^{-2}$
Radiant flux	Φ_e	Watt	$M \cdot L^2 \cdot T^{-3}$
Spectral power	$\Phi_{e\lambda}$	Watt per meter	$M \cdot L \cdot T^{-3}$
Radiant intensity	I_e	Watt per steradian	$M \cdot L^2 \cdot T^{-3}$
Spectral intensity	$I_e\lambda$	Watt per steradian per meter	$M \cdot L \cdot T^{-3}$
Radiance	L_e	Watt per steradian per square meter	$M \cdot T^{-3}$
Spectral radiance	$L_{e\lambda}$ or L_{ev}	Watt per steradian per cubic meter or Watt per steradian per square meter per hertz	$M \cdot L^{-1} \cdot T^{-3}$ or $M \cdot T^{-2}$
Irradiance	E_e	Watt per square meter	$M \cdot T^{-3}$
Spectral irradiance	$E_{e\lambda}$ or E_{ev}	Watt per cubic meter or Watt per square meter per hertz	$M \cdot L^{-1} \cdot T^{-3}$ or $M \cdot T^{-2}$
Radiant exitance/Radiant emittance	M_e	Watt per square meter	$M \cdot T^{-3}$
Spectral radiant exitance/spectral radiant emittance	$M_{e\lambda}$ or M_{ev}	Watt per cubic meter or Watt per square meter per hertz	$M \cdot L^{-1} \cdot T^{-3}$ or $M \cdot T^{-2}$
Radiosity	J_e or $J_{e\lambda}$	Watt per square meter	$M \cdot T^{-3}$
Radiant exposure	H_e	Joule per square meter	$M \cdot T^{-2}$
Radiant energy density	ω_e	Joule per cubic meter	$M \cdot L^{-1} \cdot T^{-2}$

TABLE 1.3

Symbol and Units of Physical Quantities

Quantity	Symbol	SI Unit
Length	L	1 meter (m)
Mass	m	1 kilogram (kg)
Time	s	1 second (s)
Force	F	1 Newton (N)
Frequency	ν	1 hertz (Hz)
Energy	U	1 Joule (J)
Power	P	1 Watt (W)
Current	I	1 Ampere (A)
Potential	V, ϕ	1 volt (V)

of work that is done by applying 1 N of force over a distance of 1 m—a Newton-meter. Recall that force is given by a mass times its acceleration and 1 N of force is equal to a 1 kg mass experiencing and acceleration of 1 m/s^2. The units of 1 J of energy are then kg m^2/s^2. In terms of electrical current, this equates to 1 A of current running through a 1 Ω resistor for 1 s. *Radiant flux* is radiated optical power and is given in *Watts*. A Watt is given as the time rate of change of energy and so has units of J/s. The spatial power density for an electromagnetic wave has units of W/m^2 and has the term *radiant exitance* or *radiant emittance* when applied to a radiometric source with finite extent.

This term can be interpreted as the amount of optical power emitted from a 1 m^2 area from the source. If the radiant exitance is constant across the emitting surface, it is said to have a *uniform* spatial power density across the surface. Otherwise, it has a *nonuniform* spatial power density distribution or a *spatially varying power density* across the emitting surface. Notice that the units for *irradiance* are the same as those for *radiant emittance*. That is because irradiance refers to the spatial power density received in a distant observation plane, whereas radiant emittance refers to the spatial power density emitted from the source.

In our system of units, radiant intensity refers to the optical power radiated into a solid angle of 1 sr. The term radiant intensity, therefore, applies only to radiating sources that can be properly characterized as point sources and not to radiometric quantities that are in a distant observation plane. For instance, it does not make sense to say the radiant intensity received on an optical detector in a remotely located detection plane was 1 W/m^2 since (a) the units of intensity are W/sr and (b) radiant intensity is an emitted property and not a received property. The term *radiance* is given to characterize a source that has both point source and extended source attributes. For example, the source can have a known finite extent but still be unresolved by an imaging system and so appear like a point source. Radiance is the spatial power density per unit solid angle and so has units of $\text{W/(m}^2 \text{ sr)}$. Note that if the radiance is integrated with respect to the source's surface area, then we arrive at the source's intensity. Likewise, if we integrate the source over the solid angle, we arrive at the source's radiant emittance. If we add the term spectral in front of the radiometric term, this means that we are referring to the radiometric quantity over a narrow spectral band of frequencies. For instance, if we were interested in measuring optical power radiated from a source over a narrow spectral range of frequencies, we would use the term spectral radiant flux. Other non-SI systems of units exist and include luminance units such as the footlambert, millilambert, and stilb and illuminance units such as the footcandle and phot. As previously stated, we adopt SI-based units throughout this book. In the next section, we briefly introduce some optical systems methodologies that are useful in analyzing optical systems. These will be further detailed in subsequent chapters.

1.2.3 Optical Systems Methodologies

There are a variety of very useful analytical methodologies that help us understand optical systems and their performance. In this section, we provide an overview of these methods and discuss in a general sense when it is appropriate to use them. However, before we present these methods, we want to provide some fundamental physical concepts and, from a systems perspective, describe some important characteristics of the electromagnetic spectrum. Many types of radiation are familiar in everyday life, including sunlight, heat, radio waves, x-rays, and microwaves. These types of radiation appear to be very different from one another, but they are all considered to be part of the *electromagnetic spectrum*. In using the term *optical*, we mean electromagnetic radiation that is radiating in the UV, *visible*, or IR part of the electromagnetic spectrum.

Since we will most often be working with visible and/or IR systems, our usage of *optical* typically refers to these parts of the electromagnetic spectrum either individually or together. The correct meaning will be made clear by the context of the discussion. The type of electromagnetic radiation depends on the wavelength or equivalently the frequency of the radiating *electromagnetic wave*. The following equation provides a fundamental physical relationship between the electromagnetic wave's wavelength and its corresponding frequency:

$$\lambda \upsilon = c. \tag{1.1}$$

The frequency is the number of electromagnetic wave cycles per unit of time(s) measured at a point in hertz (Hz). The term c on the right side of the equation is the speed of light in a vacuum (299,792,458 m/s). Equation 1.1 applies at every frequency. In our adopted system of units, the unit of length that applies to the wavelength is meters (m). Because of the extremely short length of an optical wavelength, it is convenient to further characterize and define our unit of length. The optical wavelength is often specified in terms of microns, aka micrometers (μm), or nanometers (nm), for optical radiation, with occasional reference to units of millimicrons (mμ), inverse centimeters, or Angstroms (Å). The following relationship helps in converting between these units, the unit of an inverse centimeter is called a wave number, and it is measured, as the name implies, in terms of an inverse of a centimeter:

$$1\,m = 39.37\,in. = 10^2\,cm = 10^3\,mm = 10^6\,\mu m = 10^9\,m\mu m = 10^9\,nm = 10^{10}\,\overset{\circ}{A}. \tag{1.2}$$

For example, a 1 cm length is equivalent to 1 wave number (e.g., inverse centimeter). Similarly, a 10 cm length is equivalent to 0.1 wave numbers. The wave number unit is obtained by first converting the length unit into centimeters and then taking the reciprocal. A 1 μm wavelength is equal to 10^{-4} cm or equivalently 10,000 wave numbers. Wave number units are often used in fields that perform spectral analysis on materials such as in the field of spectroscopy. Spectroscopy is the study of the interaction of matter with radiated energy (Skoog et al. 2006). As the units for the frequency in Equation 1.1 are in cycles per second, the inverse centimeter unit can be thought of as one cycle per centimeter—a spatial frequency. In some cases, where lambda is part of a narrow band of frequencies or wavelengths, lambda is thought of as the center wavelength of the band. The following equation shows the relationship between a change in frequency and the corresponding change in wavelength:

$$\left|\Delta v\right| = \frac{c}{\lambda^2}\left|\Delta\lambda\right|. \tag{1.3}$$

Solving Equation 1.1 for the frequency, introducing a perturbation on both sides of the equation, and taking the absolute value of the result obtain this expression. Equation 1.3 can then be used to relate a change in frequency to a change in wavelength.

The electromagnetic spectrum can be characterized either by wavelength, frequency, or the wave number of a radiating electromagnetic wave. Figure 1.9 shows useful properties of radiation throughout various parts of the electromagnetic spectrum. The top part of the figure shows the parts of the spectrum for which the Earth's atmosphere is relatively transparent. These transparent regions are good for observing distant objects either within or outside of the atmosphere. The transparent regions are shown with a Y at the top of Figure 1.9, and nontransparent regions are shown with an N. The gray areas are partially transparent. Notice that the UV part of the electromagnetic spectrum is not transparent and so that is a typical reason why we do not build long-range communication and/or detection systems in the UV part of the electromagnetic spectrum. Underneath the top bar, the relative size of the wavelength is shown. On the left side of the figure, there are very long wavelengths, and on the right side, the wavelengths are very short. The wavelength, and frequency component of the electromagnetic wave, is determined by Equation 1.1.

Underneath the qualitative picture of the wavelength, the corresponding region of the electromagnetic spectrum and the numeric value of the wavelength are shown along with a picture of something familiar that represents the size of the wavelength. For instance,

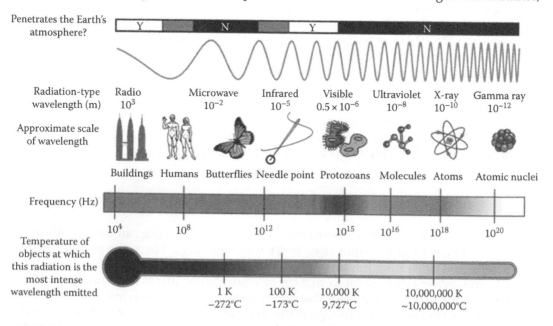

FIGURE 1.9

Characteristics of the electromagnetic spectrum. (From Wikimedia Commons, EM spectrum properties, http://en.wikipedia.org/wiki/File:EM_Spectrum_Properties_edit.svg.)

at the far left of the figure, we see a transparent region of the atmosphere that contains radio waves. A representative wavelength in this part of the electromagnetic spectrum can range from the size of buildings to over 1000 m long. Wavelengths in the microwave part of the electromagnetic spectrum are strongly absorbed and scattered by the Earth's atmosphere, and the wavelengths are approximately the size of a butterfly. In the visible region, the size of the wavelength is on the order of protozoans, and at the far right of the chart, for gamma rays, the wavelength is on the order of the radius of atomic nuclei. Underneath the approximate scale of wavelength, we see the corresponding frequency of the electromagnetic wave according to Equation 1.1. The different types of electromagnetic radiation share common characteristics and obey similar laws of refraction, reflection, diffraction, and polarization.

The different types of radiation also propagate at the same speed, often referred to as the "speed of light." The significant difference is in their frequency or wavelength. Finally, underneath the frequency bar, we see a temperature bar given in kelvins. This bar shows the equivalent temperature of an object that radiates most intensely at the peak of the wavelength shown. The temperature bar at the bottom of Figure 1.9 is important because energy can also be quantified in terms of heat. This is particularly useful in the IR part of the electromagnetic spectrum where the electromagnetic wave is not visible to the human eye. Thermal radiation is a type of electromagnetic radiation emitted by objects because of the object's temperature and material properties. This type of radiation is readily observable in the IR part of the electromagnetic spectrum because the atmosphere is relatively translucent in the IR. Consequently, depending on the temperature of the object, the emitted light due to thermal radiation may dominate over reflected and scattered light in the IR part of the electromagnetic spectrum. This is especially true at night when the reflected component of light is typically minimalized due to the absence of the sun.

Figure 1.10 shows a further subdivision of the IR part of the electromagnetic spectrum and its relationship to other parts of the electromagnetic spectrum. We can see that the IR radiation appears on the electromagnetic spectrum from near infrared (NIR) at 0.7 μm through the extreme IR at 100 μm. This is bounded between visible radiation on the shorter-wavelength side and microwave radiation on the longer-wavelength side. Specific wavelength limits, which further subdivide the IR portion of the spectrum, can be seen in Table 1.4. The NIR, middle-IR (MIR), and far-IR (FIR) ranges are those that are specified in the ISO Standard 20473:2007.

The short-wave IR (SWIR), mid-wave IR (MWIR), long-wave IR (LWIR), and others in Table 1.4 can vary based on convention. We adopt these definitions based on the ISO standard mentioned earlier and a NASA Jet Propulsion Laboratory report but caution the reader that different ranges may be seen in other literature.

In Table 1.4, the last column indicates whether or not the Earth's atmosphere is transparent at those wavelengths. The thermal imaging bands are bands associated with common

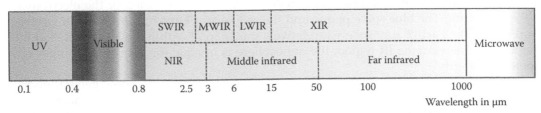

FIGURE 1.10
(See color insert.) Expanded view of the visible and IR parts of the electromagnetic spectrum.

TABLE 1.4

IR Radiation Band Subdivisions

IR Segment	Abbreviation	Range (µm)	Relative Atmospheric Transparency
Near infrared	NIR	0.7–3.0	Transparent
Short-wave infrared	SWIR	1.1–2.5	Transparent
Middle infrared or	MIR	3.0–50	Transparent
mid-wave infrared	MWIR	2.5–6	
First thermal imaging band		3.0–5.5	
Blue spike plume		4.1–4.3	
Red spike plume		4.3–4.6	
Far infrared or	FIR or	50–1,000	Transparent
Long-wave infrared	LWIR	7.0–15.0	
Second thermal band		8–14	
Very long-wave infrared	VLWIR	>15	Opaque
Extreme infrared	XIR	15–100	
Near millimeter		100–1,000	
Millimeter		1,000–10,000	

IR imaging system where relatively good detection of IR radiation is possible. The first thermal imaging band corresponds with MWIR imaging systems. Below this band in the SWIR part of the electromagnetic spectrum, IR imaging systems are typically dominated by reflected light. In the first thermal imaging band, both reflected light and emitted light are important. Therefore, in the MWIR, imaging systems can take advantage of low light levels if they are present, but these imaging systems can also work in complete darkness by detecting a target's emitted radiation. The latter of course depends on the potential target having the right temperature and emission properties to be detected by the MWIR imaging system. The spectral radiant emittance over all wavelengths increases, as the temperature of the system (object) increases. Stefan and Boltzmann found the relationship between an object's radiant emittance and its temperature. Wien found that the peak of the radiant emittance for a given object moves to shorter wavelengths, as its temperature increases. This explains why we can see the hot exhaust of a jet engine with our naked eyes in the visible band of wavelengths, but we cannot see the radiation emitted by the "relatively cool" temperatures of people. For example, a person's temperature and physical properties are such that the peak radiant emittance for a human occurs at about 9.4 µm. Consequently, an LWIR imaging system would be better matched to detect this peak radiant emittance than an MWIR imaging system. We will discuss the Stefan–Boltzmann law and Wien's law more in Chapter 5. The second thermal band is in the long-wave part of the electromagnetic spectrum. Here, imaging systems are designed to detect emitted radiation since emitted light dominates over reflected light in this region of the electromagnetic spectrum. The blue spike plume and the red spike plume have to do with gaseous emissions. The Earth's atmosphere is relatively transparent within the NIR, MIR, and FIR regions; in the extreme IR region, the atmosphere is opaque for paths exceeding a few meters in length.

In order to grasp and properly analyze optical systems concepts, some general understanding of the behavior and properties of the electromagnetic waves is necessary—especially in the optical part of the electromagnetic spectrum. Different mathematical models and physical concepts have been developed throughout the years that help in analyzing and modeling optical waves. We introduce these methods and provide a short overview along with some key physical concepts and considerations. We focus on the key concepts, results, assumptions, and limitations from a systems engineering perspective. Further detail can be found in some of the discipline-centric texts in the reference section.

One physical concept that is important when working with optical systems is the *duality of light*. Sometimes, it is convenient to think of light as a propagating electromagnetic wave. At other times, it is more convenient to think of light as particles. An interesting fact is that light has both particle and wave properties. For our purposes, we work with what is most convenient for a given application. For example, Equation 1.1 shows that the frequency of an electromagnetic wave can be given by the speed of light c divided by the wavelength of the electromagnetic wave. In this case, we are thinking of an electromagnetic *wave*. When the wavelength is in the optical part of the electromagnetic spectrum (UV, visible, and IR), it is referred to as an *optical wave*. We also know that the energy of a *photon* is given by

$$E_v = hv, \tag{1.4}$$

where h is Planck's constant, $h = 6.626 \times 10^{-24}$ m^2 kg/s, and is a fundamental physical parameter.

The photon can be thought of as a quantum of optical energy that behaves like a particle. In some optical systems analysis applications, it is more convenient to think in terms of particles than waves. Some examples would be in understanding low-light detection instruments such as *photon detectors* and *photomultiplier tubes*. Photon detectors can detect very low levels of light—down to a few photons—and photomultiplier tubes can effectively amplify photons that are converted into electrons. In this text, we will adopt either the wave or the photon approach where necessary to conveniently analyze optical system properties.

There are several useful analytical methodologies when it comes to understanding, analyzing, and modeling electromagnetic phenomena in the optical part of the spectrum. Which method applies depends on the application at hand. We will provide a brief introduction to these methods and some guidelines for when to apply them, but we will reserve more detailed comments and mathematical developments for later chapters as necessary. The most rigorous analysis in the subject of optical waves and photons can be accomplished through application of classical electrodynamics (Jackson 1975). Many of the significant results that will be used in our analyses and modeling activities can be derived from fundamental principles using this methodology. Electrodynamics works well for understanding the interaction of light with materials and for relatively well-defined and simple objects such as planes, spheres, crinkled spheres, hemispheres, slabs, cylinders, and combinations of basic shapes. An electrodynamics approach becomes analytically intractable when dealing with very rough, macroscopic objects, and in this regime, computer simulations and other analytical methods become more useful. For manageable boundary conditions, electrodynamics treats electrostatics, magnetostatics, time-varying electromagnetic fields, radiating systems, wave propagation, scattering, diffraction, the interaction of the electromagnetic wave with particles and matter, plasma physics, and more. Electrodynamics applies to homogeneous and nonhomogeneous media, can be applied on extremely small scales—*quantum electrodynamics*—and can also deal with physical problems that involve interactions occurring near the speed of light—*relativistic electrodynamics*. For our applications, we are typically most interested in the radiating, propagating, scattering, diffracting, absorbing, reflecting, transmitting, and material interactive aspects of the electromagnetic field. Additionally, electrodynamics is also useful in describing the electromagnetic wave in a waveguide or in a resonating cavity like in a laser. The fundamental physical equations that describe the behavior of the electromagnetic field are *Maxwell's equations*, and they have both a differential and integral formalism. Additionally, the Helmholtz wave equation can be used to describe electromagnetic wave behavior. Electrodynamics provides a means for understanding the fundamental physical properties of the electromagnetic field and is useful in explaining, deriving, and understanding key characteristics of optical waves.

Fourier optics methods provide a powerful set of analytical tools in the spatial frequency domain that relate the electromagnetic field spectrum, intensity spectrum, or radiant emittance spectrum to corresponding elements in some distant plane such as the imaging system's image plane. Fourier optics methods are a great simplification over the fundamental Maxwell's equations and the Helmholtz equation. The Fourier optics relationships are obtained by simplifying the Rayleigh–Sommerfeld diffraction integrals using a pair of distance constraints. These constraints ensure that the *Fraunhofer diffraction* conditions are established. Fourier optics methods can be derived using classical electrodynamics for the case that the dimensions of the source of the electromagnetic wave *d* are small compared to *z*, the separation distance of the source from the observation plane. More specifically, we must look at the following conditions in order to fully apply Fraunhofer diffraction and consequently Fourier optics methods:

$$\frac{kd^2}{2z} \ll 1 \tag{1.5}$$

and

$$d \ll z, \tag{1.6}$$

where
  k is given by $2\pi/\lambda$
  d is the radial size of the diffracting system
  z is the separation of the diffracting system from the observation plane

As an example, if we were looking at light that is reflected from a 1 m circular plate satellite surface with an optical system that is located 1×10^6 m away, we would have 0.785, which is not much smaller than 1, and so Fraunhofer conditions may not apply. However, in many practical situations, Fraunhofer diffraction assumptions and Fourier optics methods apply for the less stringent case $d/z \ll 1$ (Jackson 1975). As an example for the second criterion, a 15 m diameter satellite at 1×10^6 m separation distance easily satisfies this condition but fails the conditions of Equation 1.5. Nevertheless, Fourier optics methods are routinely applied using the less stringent criterion of $d/z \ll 1$ with excellent results. We advise a somewhat lenient interpretation of Equation 1.5 as a discriminating factor for the applicability of Fourier optics but advise caution in applying Fourier optics methods if this equation is grossly violated. We will consider the second condition—$d/z \ll 1$—as the primary condition and Equation 1.5 to play an important supporting role. The result is useful collections of 2D Fourier transform relationships between an emitting and/or reflecting source and relatable electromagnetic elements in a distant observation/image plane. If Equation 1.5 does not hold, then Fresnel diffraction conditions apply and a somewhat more complicated set of relationships exist. For now, we will proceed with Fourier optics methods where appropriate and address Fresnel propagation where needed.

SO is another powerful analytical tool that is very useful in analyzing and understanding optical systems. With SO methods, we are no longer limited to simple scattering models that are required to make classical electrodynamics calculations tractable. We can use statistical techniques to analyze the properties and characteristics of the optical wave. These methods can be combined with Fourier optics methods to provide a powerful and rigorous mathematical framework for understanding optical wave propagation and the overall detection process. For low light levels, Poisson's statistics are useful in describing photon-counting processes. However, at high levels of light, Gaussian statistics are more commonly used. We will provide more insight into this powerful approach in Chapter 6 where we discuss the propagation of radiation in more detail.

There are a variety of other methods that must be mentioned. Modern optics methods allow us to understand and model optical components such as lenses, filters, optical mounts, coatings, optical wave propagation through these components, and polarization effects to mention a few. These methods are important in designing optical components. *Physical optics methods* include diffraction effects in their analytical approach, whereas *geometrical optics* use simplified ray-tracing methods to approximate optical propagation phenomena. Nonlinear optical (NLO) methods provide an approach for understanding the behavior of optical waves when traditional assumptions of linear, shift-invariant systems do not apply. We will discuss the linear, shift-invariant assumption and NLO models in Chapter 2. We will use results from these methods as needed to help us understand, characterize, and model optical waves and optical systems. As optical systems engineers and optical systems managers, we will focus on fundamental results and depend on our optical engineer for detailed analysis. More detailed information may also be found in the references.

We want to make one last comment about our chosen system of units with regard to electromagnetic applications. There are several systems of units that are useful in working with electromagnetic physical phenomena. These are electrostatic units (esu), electromagnetic units (emu), Gaussian units, Heaviside–Lorentz units, and the rationalized MKSA units (SI). The most common two of these units are the Gaussian units and the rationalized MKSA units. The Gaussian units are more practical when dealing with microscopic issues, and the rationalized MKSA units—our adopted SI units—are more practical for macroscopic problems such as in large-scale engineering applications. We now present some definitions of important terms that will be used throughout the text.

1.2.4 Fundamental Optical Systems Concepts and Terminology

This section contains definitions that will become important in understanding concepts throughout the book. We would like the reader to be able to skip to any subsequent section of the book and be able to understand the content. Of course, it is better to sequentially go through the material since each chapter builds on the preceding one. But we wanted to enable those readers that want to jump straight to a particular topic. For instance, if someone wanted to learn about optical detectors, after reading and understanding this section, they should be able to jump straight to Chapters 8 through 10 and understand the majority of material found there. The reader would not know about any of the preceding material (e.g., sources, optical radiation, basic optical systems), but the material in the detector sections should be understandable. We hope to make each chapter as stand-alone as possible. In order to facilitate this, we need to define some common terminology. In addition to the terminology definition, we provide some additional descriptive detail such as equations and/or insight where appropriate.

Absorptance (α)—The ratio of radiant flux absorbed by a body to radiant flux incident on the body:

$$Absorptance = \frac{\varnothing \; absorbed}{\varnothing \; incident}. \tag{1.7}$$

Acousto-optics—The interaction of light with sound. Acousto-optical devices manipulate light with sound waves.

Active imaging—The imaging system uses a coherent imaging model under man-made illumination conditions such as laser illumination.

Atmospheric turbulence-limited spatial resolution—For most well-designed optical imaging systems that have a large entrance pupil aperture compared to the Fried parameter, r_0, and are in the Earth's atmosphere, atmospheric turbulence effects limit the spatial resolution of the optical imaging system. For this case, the atmospheric turbulence-limited spatial resolution of the imaging system is given by

$$\Delta x_{atm} = 1.22 \frac{\lambda}{r_0} z, \tag{1.8}$$

where
   1.22 results from a circular entrance pupil aperture
   λ is the center wavelength of the optical wave
   z is the separation distance of the observation point from the imaging system's entrance pupil

Adaptive optics—A mostly hardware-based system for real-time correction of atmospheric turbulence effects.

Atmospheric turbulence compensation—Software-dominant methods that correct for atmospheric turbulence in collected data (usually images). These methods are typically much slower than adaptive optics methods and are performed in a post-processing environment.

Brightness—Spatial power density that is independent of location (e.g., can be either at the source or detector). The units for brightness are W/m^2.

Coherent wave—The autocorrelation of the electromagnetic wave is nonzero at more than one point.

Diffraction-limited spatial resolution—This is the theoretically best possible spatial resolution from an imaging system. It is given by

$$\Delta x_d = 1.22 \frac{\lambda}{D} z, \tag{1.9}$$

where
   1.22 comes from using a circular aperture
   λ is the center wavelength of the light in meters
   D is the diameter of the optical system's entrance pupil (typically the primary mirror of a telescope) in meters
   z is the separation distance between the observed object and the imaging system's entrance pupil in meters

The left side of the equation represents a size scale for which smaller spatial details cannot be observed with the given imaging system. It can be considered as the spatial resolution of a theoretically perfect imaging system that has no optical or atmospheric aberrations and no system noise effects.

Electro-optics—These are mostly optical devices that are influenced by electronics. An example is an optical switch.

Emissivity (ϵ)— It is the ability of a surface to emit radiation. It is equal to the radiated energy from a given source (across all wavelengths) divided by the radiated energy of a blackbody source (at the same temperature).

Extended source—If an object that is viewed by an imaging system has spatial dimensions that are larger than the spatial resolution of the imaging system in question, then, ideally, the imaging system can resolve spatial features of the object. In this case, the source object can be thought of as an extended source.

Fourier optics—A methodology that uses the spatial frequency domain to analyze and characterize optical systems.

Fried parameter (*pronounced freed*)—The Fried parameter, r_o, was developed by Dave Fried and represents the diameter of a patch of atmosphere over which the atmosphere can be considered spatially coherent. Another name for the Fried parameter is the atmospheric coherence length. For imaging systems that have an entrance pupil diameter equal to or smaller than the Fried parameter, adaptive optics/ atmospheric turbulence compensation (AO/ATC) methods have little effect. For imaging systems that have a larger diameter than r_o, AO/ATC methods have some observable benefit.

Incoherent wave—The autocorrelation of the wave is nonzero at only one point.

Photonics—A discipline that combines optics with electronics. It involves using electronic devices to control light in either its wave or particle form (photons). The discipline includes optics, optoelectronics, electro-optics, acousto-optics, optical information processing, lasers and quantum electronics, optical electronics and communications, lightwave devices, fiber-optics communications, or lightwave systems (Saleh and Teich 1991).

Optoelectronics—Devices that are mostly electronic but also have optical aspects to them. Some examples include television displays and photodiodes.

Statistical optics—A methodology wherein the interaction of light with matter is considered a fundamentally statistical phenomenon (Goodman 1985). This methodology provides a powerful set of statistically based analytical tools for understanding light propagation, the interaction of light with macroscopic objects, and spatial and temporal coherence effects of the optical wave.

Lightwave optics—A field wherein optical methods are merged with optical signal processing and optical communications devices.

Nonlinear optics—The interaction of light with nonlinear media. Some examples include phase conjugation, two-wave mixing, four-wave mixing, self-focusing, third harmonic generation, and optical amplification.

Passive imaging—The imaging system uses an incoherent imaging model under naturally illuminated conditions (e.g., sunlight, starlight, emitted light in the IR part of the electromagnetic spectrum).

Photometry—The measurement of light's brightness as perceived by the human eye. The sensitivity of human eyes varies with the wavelength of the incident energy, so in photometry, the brightness is measured in terms of radiant power at each wavelength.

Photon flux (Φ_p)—The number of photons emitted per second, measured in *photons/s*.

Point source—If the spatial resolution of a given imaging system is larger than the physical dimensions of a source object viewed by an imaging system, then the imaging system cannot resolve any spatial details of the object. In this case, we can consider the object to have properties of a point source with respect to the imaging system even though the object may have some physical dimensions.

Radiant energy (U, E, Q_e)—This is the energy that is carried by an electromagnetic wave. The units of energy are given in *joules* (J). One joule is equivalent to 1 N of force being applied over 1 m. It is also equivalent to 1 W of power over 1 s.

Radiant flux or radiant power (P, Φ_e)—This is a measure of the change in energy as a function of time rate and has units of *Watts* (W):

$$1\,W = 1\,J/s. \tag{1.10}$$

Radiant emittance or radiant exitance (W, M_e)—Radiant flux emitted from a source per unit area. The units of radiant emittance are W/m^2 or W/cm^2. This quantity is called radiant incidence if measured at the imaging system.

Radiant intensity (J, I_e)—Radiant flux emitted per unit solid angle, measured in W/sr.

Radiance (N, L_e)—Radiant flux emitted per unit solid angle per unit area of source, measured in $W/(m^2 \cdot sr)$.

Radiant photon emittance (Q_p)—The number of photons emitted per second per unit area, measured in $photons/(s \cdot m^2)$.

Radiometry—Measurement of electromagnetic radiation. Measurements are made over a range of wavelengths or equivalently frequencies.

Reflectance (ρ)—The ratio of radiant flux reflected by a body to radiant flux incident on the body,

$$Reflectance = \frac{\varnothing\ reflected}{\varnothing\ incident}. \tag{1.11}$$

Steradian—An angular measurement obtained by dividing the surface area of a portion of a sphere by its radius squared. For instance, if the surface area under consideration is the entire sphere, then the associated angular measurement is 4π sr.

Spectral radiant flux (P_λ)—The radiant flux per unit wavelength interval, measured in $W/\mu m$.

Spectral radiant emittance (W_λ)—Radiant emittance per unit wavelength interval, measured in $W/(m^2 \cdot \mu m)$.

Spectral radiant intensity (J_λ)—Radiant intensity per unit wavelength interval, measured in $W/(sr \cdot \mu m)$.

Spectral radiance (N_λ)—Radiance per unit wavelength interval, measured in $W/(m^2\ sr \cdot \mu m)$.

Spectral irradiance (H_λ)—Irradiance per unit wavelength interval, measured in $W/m^2 \cdot \mu m$.

Total power law—The sum of the absorptance, reflectance, and transmittance of a body is unity. Or, in other words, the radiant flux φ incident on a surface is equal to the sum of the radiant flux absorbed, radiant flux reflected, and radiant flux transmitted:

$$\alpha + \rho + \tau = 1 \tag{1.12}$$

or

$$\varnothing\ absorbed + \varnothing\ reflected + \varnothing\ transmitted = \varnothing\ incident. \tag{1.13}$$

Transmittance (τ)—The amount of radiant flux that is transmitted divided by the total radiant flux that is incident on the body,

$$Transmittance = \frac{\varnothing\ transmitted}{\varnothing\ incident}. \tag{1.14}$$

Some optical systems engineering building blocks have been introduced in this chapter that present a conventional system of units and that introduce standard optical systems terminology. This chapter provides some basic concepts and methodologies that should help those readers that find it necessary to jump ahead to later chapters. The best approach is to sequentially read the material in this text. That way, the proper foundation and contextual elements can be put in place. The intent of this section was not to provide an all-inclusive set of definitions but just the key definitions needed for general understanding. The next section provides an integrated case study that applies some of the concepts learned in this chapter.

1.3 Integrated Case Study: Introduction to Our Optical Systems Engineering Case Study

In this section, a case study is included that relates to the topics discussed in the previous two sections. This section demonstrates how the systems engineering approach is applied in the optics field. The intention behind this case study is to help the reader understand how systems engineering actually works in the professional environment and illustrate some of its advantages. Also, we want to use a single integrated case study to show how different aspects of the systems engineering discipline are applied to the same scenario. We adopt a fictitious, mid-size, expanding company called Fantastic Imaging Technologies (FIT) to demonstrate both the higher-level systems engineering concepts and principles and their connection to systems engineering–oriented optical technical analysis, modeling, feasibility, and trade studies. Table 1.5 provides a list of the fictional main characters and their roles in this case study.

FIT is a mid-size imaging company located at Melbourne, FL, in the United States. FIT develops imaging systems that operate in the visual and IR regions of the electromagnetic spectrum. The company has distinct techniques for ground-to-ground, ground-to-air,

TABLE 1.5

Cast of Main Characters for Integrated Case Study

Actor	Role	Affiliation
Dr. Bill Smith	CEO	FIT
Dr. Tom Phelps	CTO (optics expert)	FIT
Mr. Garry Blair	CIO (EA)	FIT
Mr. Karl Ben	Senior systems engineer	FIT
Jennifer O.	Systems engineer	FIT
Ron S.	Systems engineer (new employee)	FIT
Ginny R.	UAV program manager	FIT
Arlene G.	Business development	FIT
Marie C.	Requirements management/CM	FIT
Carlos R.	Quality manager	FIT
Lena A.	Logistics analyst	FIT
Phil K.	Software engineering	FIT
George B.	Hardware specialist	FIT
Amanda R.	Optical engineer (optics lead)	FIT
Christina R.	Optical technician	FIT
Kari A.	Test manager	FIT
Malcolm P.	Production manager	FIT
Rodney B.	Field service	FIT
Julian F.	Product service	FIT
Steven E.	Technical support and IT	FIT
Andy N.	Maintenance/support	FIT
Warren M.	Agency liaison	FIT
Doris	Executive administrative assistant	FIT
Wilford Erasmus	Chief of operations and acquisitions U.S. Customs and Border Protection Primary stakeholder	DHS
Dr. Glen H.	Technology specialist	DHS
Jean H.	Operations management	DHS
Kyle N.	User	DHS
Ben G.	Depot maintenance	DHS
Simon Sandeman	Chief, acquisitions and procurements	Central Intelligence Agency
Rebecca H.	Special customer	Unnamed organization

and air-to-ground imaging scenarios. FIT also has a dedicated research and development team concentrating on improving the performance of long-distance imaging systems with atmospheric turbulence compensation and adaptive optics methods.

Now, let me introduce you to some of FIT's leadership. The chief executive officer (CEO) of the company is Dr. Bill Smith. He is a retired U.S. Air Force officer with over 25 years of experience with optical detection methods, systems engineering, and the defense acquisition process. The chief technical officer (CTO) of the company is Dr. Tom Phelps who is an expert in optics with an emphasis on optical imaging systems. The chief information officer (CIO) of the company is Mr. Garry Blair. Dr. Bill Smith is a strong proponent of the systems engineering discipline and is insistent about always following systems engineering methods in all the projects within the company. From the beginning, FIT has applied state-of-the-art systems engineering principles, methods, tools, and techniques in all internal and external programs and projects. The focus on systems engineering is never limited to the products, projects, programs, and systems of the company but also includes FIT's processes and services. Systems thinking and the systems engineering methodology is embedded in the company's culture and core values.

FIT takes a holistic approach to product development and support. From the start, FIT has been actively engaged in an optical system's development over its entire life cycle (RDT&E, production/manufacturing, operational support, maintenance, service, and system retirement/disposal). MBSE methods are practiced for requirements management and the design phases of the system. The company uses both DOORS and MagicDraw as its customer-specific requirements management tools and uses Zemax™ in its optical design suite. The company also has some internally developed and patented algorithms in a MATLAB®-based rapid prototyping environment that do high-speed atmospheric turbulence compensation. These algorithms have been ported to various high-speed hardware architectures such as the field-programmable gated arrays (FPGAs) and have been developed using modular design methods and state-of-the-art security and service protocols.

As the company started growing and started handling multiple projects, the normal information flow between the top-level management team and the technical staff in the company became noticeably less efficient and strained. The normal budgeting and accounting processes also became more difficult to manage. This resulted in an increase in budgeting and scheduling problems within the company. To handle this situation, the CIO proposed adopting an EA framework for the management, integration, and documentation of the company's strategic vision, core business processes, technical processes, information flows, systems, operations, and standards. The CEO and board of directors approved this EA approach, and a baseline EA was established along with a projected development path and review cycle. The EA made it possible to clearly identify the goals and objectives of the company. It helped to define the organizational structure, the information flow, the decision flow, the interactive structure of the people, and the products and processes that are critical to the company's operations and growth. With the implementation of an EA, it became easier to follow up on each of the projects in the company, and also the decision-making process was much simplified and more effective.

FIT recently received an RFP from the U.S. Department of Homeland Security (DHS). The project is to develop an optical subsystem for an unmanned aerial vehicle (UAV) in order to identify human bipedal traffic crossing the United States–Mexico border. This system was required to have a standoff ground range of at least 5 miles for daytime and nighttime conditions. The system also needs to have the capability to do near-real-time image processing and send images to a remotely located mission control center (MCC) within 5 s of the image capture. The optical system must also be capable of bidirectional communication and control for local as well as over long distance. Here is a scene from the discussion that happened between the CEO, Dr. Bill Smith; CIO, Mr. Garry Blair; and CTO, Dr. Tom Phelps, on the project.

Bill: "Hello everybody, I appreciate that you both could make it for the meeting even though this is such short notice. Before deciding about whether or not to go forward with the proposal effort for the DHS RFP, we need to take a look at the current obligations and resources of the company. We need to discuss where our resources are currently allocated. Also, do we have the proper resources and infrastructure to take on this project? Do we have the staff in house with the needed skill sets? How many employees will we need to work on this project?

Also what will be our strengths and weaknesses for winning the proposal? This is a huge project for our company, so we should be aware of the associated risks."

Tom: "Yes I agree with you Bill. Before jumping into it, we have to do some homework and make sure we are ready and capable of taking up this huge new task. When I saw this opportunity, I got really excited feeling that this is the break that would propel our company to the next level. Our company has been in the imaging industry for a while now focusing largely on commercial projects. These projects were fitting for the company as we were focusing on optimizing our atmospheric turbulence compensation methods. Now we have a leadership position in atmospheric turbulence compensation methods, and some new doors are beginning to open for us. So if we are all in agreement that we have enough resources to support this effort, it is time for us to take the next step in becoming the leaders in our industry. This project will be a major step in that direction and, if successful, will open up a new market base.

For this project, we have the right technology for the imaging system. This project presents an opportunity for us to take our system to the next level and integrate onto a UAV. I understand that we are involved in other projects, but this represents a growth opportunity we can't ignore. The proposal team will put together a very competitive package. The proposal activity itself should only take 2–3 months of effort, so by the time we start the project, some of our other projects will be over, and we will have people to work on the DHS project. In parallel, we can also put some feelers out for potential hires should we win the bid. There's a good university in the local area; we could make some starting inquiries there."

Garry: "I agree with you Tom. Now that we have our enterprise architectural framework set up, and all the goals, objectives, policies, processes, systems, operations, and standards of the company are being laid out clearly, it is easier to see what we have and make the choice. This project clearly matches with the transition we planned for the company while setting up the EA. The company has aligned itself now to take on a project for a government organization. In 1996, the U.S. government passed the Cohen–Clinger Act, which makes it mandatory for all the federal agencies to follow EA methods for their activities. This requirement trickles down to contractors that support these government agencies such as us. Since we already have implemented an enterprise architectural framework, it will be an advantage on our side when submitting our solicitation to the DHS. From a technology standpoint, FIT will also have an advantage which none of our competitors have—we have the best high-speed, atmospheric turbulence compensation technique in the industry! This project calls for long-distance imaging, so the effect of the atmosphere on the image is unavoidable for sufficiently large-aperture optical systems. Our systems will have longer standoff ranges or higher spatial resolution at equivalent standoff ranges than our competitors. We have the technology edge! We should have a better proposal than our competition; now we will just have to keep the life-cycle costs down."

Bill: "Since we all agree that we should go for the DHS project, let's start working on the proposal as soon as possible. This is a new step for us, and so we may need some extra time to ramp up. Tom, you are in charge of the proposal. You and Garry need to make sure that we have all the necessary architectural framework artifacts in the proposal. The proposal will have to be completed within 2 months from today. Should you need any additional details contact Mr. Wilford Erasmus. He is our DHS POC for this project. For proposal discussions, I want you to have two teams: red and blue. The blue team will discuss the pros, and the red team will discuss the cons. I would like to see a weekly status report. We will meet every 2 weeks to discuss our progress. Let's do this! Good luck!!"

After the meeting, Tom set up a team to work on the proposal effort for the DHS project under Mr. Karl Ben, a senior systems engineer in the company. They started the proposal effort by first identifying the stakeholders. Here is another discussion that happened 2 weeks later between Bill and Tom during the first review of the proposal effort.

Tom: "We started the effort of identifying the stakeholders for this project. The team identified many potential stakeholders apart from DHS—like CIA, FBI, other foreign customers—but we decided to focus solely on the DHS requirements for this proposal effort."

Bill: "That's a good decision. Let us focus only on the DHS requirements right now, and we can think about secondary stakeholders after winning the proposal. If we win, we can use our modular design principles and our rapid prototyping methodology to efficiently pursue opportunities with the secondary stakeholders. Have you come up with an idea of how we are going to address the primary stakeholder requirements in the RFP?"

Tom (pointing to the presentation slide containing the CONOPS): "Yes, we have come up with a CONOPS (Tom points to Figure 1.11 on the screen). Our electro-optical and infrared imaging system within the UAV will be searching for

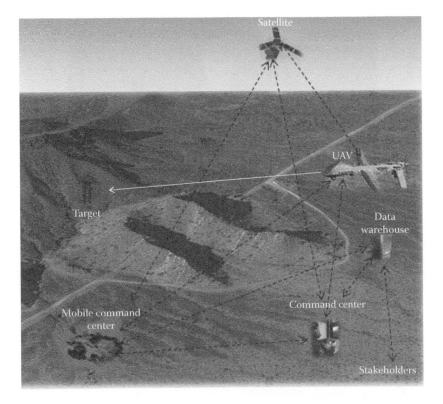

FIGURE 1.11
CONOPS of UAV border patrol mission.

human traffic by using the UAV's conventional scanning mechanism for the imaging solution. Our images will be processed for atmospheric turbulence compensation onboard the UAV and also to make them compatible with the UAV's standard communications protocol. The processed images will then be sent to the mission control center via the communication mechanism available on the UAV. The integrated imaging system will include two different subsystems: a telescopic camera system that works in the visible region of the electromagnetic spectrum for daytime imaging and a telescopic IR camera for nighttime imaging."

Bill: "Can you please explain how each of the systems mentioned above are supposed to work?"

Tom: "OK. Let me start with our nighttime optical imaging solution. All materials radiate and have a radiation profile that peaks at a particular wavelength in the electromagnetic spectrum. Since the target of our optical imaging system is human, humans will be the source of radiation. (Tom shows Figure 1.12 of a full human body.)

The human body emits radiation in the IR part of the electromagnetic spectrum. So if we know the peak wavelength at which human bodies emit IR radiation, we can use an optical detector that is optimized for detecting humans around that wavelength. There will be risk factors to consider, but assuming that there isn't a lot of absorption, noise from the optical system, and clutter from other objects at, or close to, this wavelength, then we should be able to

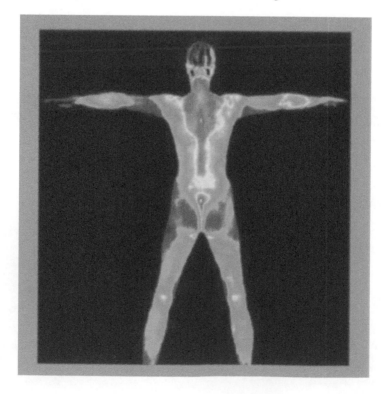

FIGURE 1.12
(See color insert.) IR image of human body. (Reprinted from NASA, 2014, http://commons.wikimedia.org/w/index.php?title=File:Atmospheric_window_EN.svg&page=1. With permission.)

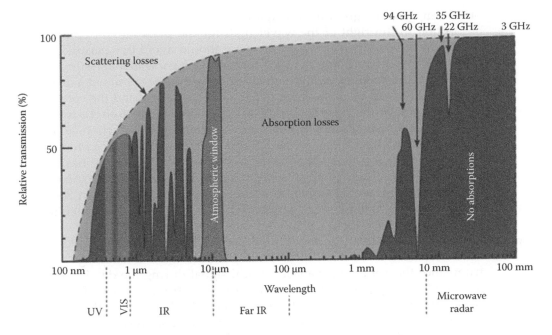

FIGURE 1.13

Atmospheric effects on transmittance. (Reprinted from Cepheiden, Absorption spectrum during atmospheric transition of electromagnetic radiation. An atmospheric transmission 'window' can be seen between 8–14 μm (700–1250 cm⁻¹), 2009, http://commons.wikimedia.org/w/index.php?title=File:Atmospheric_window_EN.svg&page=1. With permission.)

design our optical system to take full advantage of the peak radiation emitted from the human target of interest. One important thing here is to make sure that the optical detector captures signals only from the desired spectral region in the electromagnetic spectrum. There is a trade-off between the bandwidth of the system, the detector noise, the received irradiance on the detector, and background noise. This should be done as part of our design process where we construct a radiometric model, analyze the imaging scenario, determine the optimal detector characteristics, and then design appropriate optical or signal filters for our CONOPS-driven scenarios.

Another thing to be taken into consideration is the effect of the atmosphere in the region in which the human body emits radiation. If we use a narrowband detector in the wrong part of the electromagnetic spectrum, we may lose a good portion of the available signal from our target of interest due to atmospheric absorption."

(Tom then jumped to the following slide that shows how the atmosphere and different compounds in the atmosphere affect transmitted light, as shown in Figure 1.13. He then continued his explanation.)

Tom: "We know that the human body has its peak radiation at a wavelength of about 9.5 μm. Here, we see that there is a loss in the optical power being transmitted at, and around, this wavelength—a little over 20% from this chart. We will have to get a more detailed chart of the atmospheric transmission in the vicinity of 9.5 μm, but this gives us a ballpark result for now. There is also notable absorption to the right of the wavelength for the peak radiant emittance of our prospective human emitter, and so we should avoid this part of the electromagnetic spectrum. Now, a major concern we have is to make sure that our optical detector receives enough signals from the source to put out a detectable response. For that, the radiant emittance from the source and the corresponding irradiance at the detector is calculated. If the signal strength is low but within detectable limits, appropriate amplification, noise reduction, and signal processing methods need to be used. Since the radiant emittance produced by humans (and warm-blooded animals) is typically far above the background in nighttime desert environments, we can go with a wide instantaneous field of view (IFOV) imaging system as a queuing sensor. We can then zoom our imaging system to get higher-spatial-resolution images once the queuing sensor tells us where to look."

Tom: "A similar radiometric argument can be made for our daytime solution. Here, the detector has to be matched to the visible part of the electromagnetic spectrum since the dominant available optical signal during the day is from scattered and reflected visible light from the target and from the scene. The queuing sensor idea may not work since there is so much background clutter during the day. There may be some potential with the queuing approach if we can adapt and incorporate some of the latest automatic target recognition algorithms, but we may need to go to a scanning approach to get acceptable spatial resolutions as our primary method. We are still evaluating this scenario. Also, we can use our real-time atmospheric turbulence compensation methods to enhance the spatial resolution of our optical system or provide a further standoff range than our competitors. I think DHS will like that."

Bill: "How much spatial resolution do you need for the DHS CONOPS?"

Tom: "A rule of thumb is that we need to have at least 10 picture elements (pixels) across our target of interest to properly distinguish whether we are looking at a human or something else. Remember that this is actual resolution across the target of interest (TOI), and so we will have to conduct an analysis to see for what CONOPS-driven scenarios our atmospheric turbulence compensation methods are required to maximize the resolution of our optical system."

Tom: "All I have talked about so far are the technical basics for this project. In the next part of the effort, we will have to develop an optical systems model and begin to analyze our optical scenario in more detail. We also have to do some preliminary calculations to see what effects that we can expect on our imaging system spatial resolution due to the atmosphere. As you know, for many well-designed optical systems, atmospheric effects are often the principal limiting factor in the optical systems spatial resolution. We also have to model the imaging scenarios that we expect from our CONOPS and start to get a handle on look angles, expected slew rates for the pointing and tracking system, and the basic optics needed for our design models.

I was talking to Garry earlier today, and he was mentioning that we should consider some actions relative to our EA. He said it wasn't anything that couldn't wait until the next meeting, but he seemed pretty excited."

Bill: "This looks like a great start; I am quite happy with the work so far. I will see you and Garry at the next review."

1.A Appendix: Acronyms

AFIT	Air Force Institute of Technology
CDR	Critical Design Review
CEO	Chief executive officer
CIO	Chief information officer
CMMI	Capability Maturity Model Integration
CONOPS	Concept of operations
COTS	Commercial off-the-shelf
CPFF	Cost plus fixed fee
CPIF	Cost plus incentive fee
CTO	Chief technical officer
DAS	Defense Acquisition System
DAU	Defense Acquisition University
DHS	Department of Homeland Security
DOD	Department of Defense
DODAF	Department of Defense Architectural Framework
DOORS	Dynamic Object Oriented Requirements System
DOTMLPF	Doctrine, organization, training, materiel, leadership, personnel, and finance
EA	Enterprise architecture
ECPs	Engineering change proposals
emu	Electromagnetic units
esu	Electrostatic units
FAR	Federal Acquisition Regulations
FFP	Firm fixed price
FIR	Far infrared
FIT	Fantastic Imaging Technologies
FoS	Family-of-systems
FPGAs	Field-programmable gated arrays
FT&E	Functional test and evaluation
IFOV	Instantaneous field of view
INCOSE	International Council on Systems Engineering
JCIDS	Joint Capabilities Integration and Development System
JROC	Joint Requirements Oversight Council
KPPs	Key performance parameters
LCC	Life-cycle costing
LWIR	Long-wave infrared
MBSE	Model-based systems engineering
MCC	Mission control center
MDA	Milestone Decision Authority
MIR	Middle infrared
MIT	Massachusetts Institute of Technology
MOPs	Measures of performance
MWIR	Mid-wave infrared
NIR	Near infrared
NLO	Nonlinear optical
OSD	Office of the Secretary of Defense

OT&E	Operational test and evaluation
PDR	Preliminary Design Review
POC	Point of contact
PPBE	Planning, programming, budgeting, and execution
PPPI	Preplanned product improvement
RDT&E	Research, development, test, and evaluation
RFP	Request for proposal
SDLC	Systems development life cycle
SEI	Software Engineering Institute
SI	International System of Units
SI	Rationalized MKSA units
SO	Statistical optics
SOS	Systems of systems
SSPPs	Systems, services, products, or processes
SSR	System Segment Review
SWIR	Short-wave infrared
TPM&Ms	Technical performance measures and metrics
UAV	Unmanned aerial vehicle
VCJCS	Vice chairman of the joint chiefs of staff

References

ACQWeb. 2014. Requirements and Acquisition Process Flow. Acquisition Web Portal: http://www.acq.osd.mil/asda/docs/dod_instruction_operation_of_the_defense_acquisition_system.pdf (accessed November 10, 2014).

ANSI/EIA-632-1999 (R2003). 2003. Processes for Engineering a System. American National Standards Institute and Electronic Industries Alliance. Washington. http://webstore.ansi.org/ (accessed November 10, 2014).

Arrasmith, W.W. 2007. A systems engineering entrepreneurship approach to complex, multi-disciplinary university projects. Paper presented at *National Conference of the American Society of Engineering Education*, Honolulu, HI, June 23–27, 2007.

Blanchard, B.S. 2008. *System Engineering Management*, 4th edn. New York: John Wiley & Sons.

Blanchard, B.S. and W.J. Fabrycky. 2010. *Systems Engineering and Analysis*, 5th edn. London, U.K.: Pearson.

Cepheiden. 2009. Atmospheric Window. Wikimedia Commons website: http://commons.wikimedia.org/w/index.php?title=File:Atmospheric_window_EN.svg&page=1 (accessed November 10, 2014).

DAU. 2012. *Online Defense Acquisition Guidebook*. Defense Acquisition University website. https://dag.dau.mil/Pages/Default.aspx (accessed April 24, 2014).

DAU. 2014. Integration of the DOD Decision Support System. Defense Acquisition University website: https://acc.dau.mil/CommunityBrowser.aspx?id=488288&lang=en-US (accessed November 10, 2014).

Defense Acquisition Portal. 2014. https://dap.dau.mil/aphome/Pages/Default.aspx (accessed November 8, 2014).

Defense Acquisition University. 2001. *Systems Engineering Fundamentals*. http://www.dau.mil/pubs/pdf/SEFGuide%2001–01.pdf (accessed April 24, 2014).

Defense Acquisition University Press. 2001. *Systems Engineering Fundamentals*. Fort Belvoir, VA: Defense Acquisition University Press.

Department of Defense. 1969. MIL-STD-499 Military standard system engineering management. http://www.everyspec.com/MIL-STD/MIL-STD-0300–0499/MIL-STD-499_10376/ (accessed April 4, 2014).

Fagan, M.D. 1978. *A History of Engineering and Science in the Bell System: National Service in War and Peace.* Murray Hill, NJ: Bell Laboratories.

Fitts, P.M. 1951. *Human Engineering for an Effective Air Navigation and Traffic Control System.* Washington, DC: National Research Council. http://www.dtic.mil/cgi-bin/GetTRDoc?AD=ADB815893 (accessed April 4, 2014).

GIG. 2014. Global Information Grid OV-1. Wikimedia Commons website: http://commons.wikimedia.org/wiki/File:Gig_ov1.jpg (accessed November 8, 2014).

Goode, H.H. and R.E. Machol. 1957. *Systems Engineering.* New York: McGraw-Hill.

Goodman, J.W. 1985. *Statistical Optics.* New York: Wiley-Interscience.

Grants.Gov. 2014. http://www.grants.gov (accessed November 8, 2014).

Hall, A.D. 1962. *A Methodology for Systems Engineering.* Princeton, NJ: Van Nostrand.

IEEE. 1998. Standard for application and management of the systems engineering process— Description. IEEE Std 1220-1998. https://standards.ieee.org/findstds/standard/1220-2005.html (accessed April 24, 2014).

Jackson, J.D. 1975. *Classical Electrodynamics.* New York: John Wiley & Sons.

Kolmogorov, A.N. 1941a. Dissipation of energy in the locally isotropic turbulence. *Proceedings of the USSR Academy of Sciences,* 32: 16–18.

Kolmogorov, A.N. 1941b. The local structure of turbulence in incompressible viscous fluid for very large Reynold's numbers. *Proceedings of the USSR Academy of Sciences,* 30: 301–305.

Merriam-Webster's Collegiate Dictionary, 11th edn. 2003. Springfield, MA: Merriam-Webster.

NASA. 2007. EM Spectrum Properties. Wikimedia Commons website: http://en.wikipedia.org/wiki/File:EM_Spectrum_Properties_edit.svg (accessed November 10, 2014).

NASA Thermal. 2014. A Diagram of the EM Spectrum. Wikipedia website: http://commons.wikimedia.org/wiki/File:Fullbody_03.jpg (accessed November 10, 2014).

NASA. 2014. Hubble space telescope diagram. NASA Quest website: http://quest.arc.nasa.gov/hst/photos-i.html (accessed November 9, 2014).

Newton, I. 1952. Optics. In *Great Books of the Western World,* R.M. Hutchins (ed.). Chicago, IL: Encyclopedia of Britannica, 377–550.

Office of the Deputy Under Secretary of Defense for Acquisition and Technology. 2008. *Systems Engineering Guide for Systems of Systems,* Version 1.0. Washington, DC: ODUSD(A&T)SSE.

Sage, A.P. 1992. *Systems Engineering.* New York: Wiley.

Saleh, B.E.A. and M.C. Teich. 1991. *Fundamentals of Photonics.* New York: Wiley-Interscience.

Schlager, J. 1956. Systems engineering: Key to modern development. *IRE Transactions,* EM-3: 64–66.

SEF. 2014. Systems Engineering Fundamentals Guide. Defense Acquisition University weblink: http://www.dau.mil/publications/publicationsDocs/SEFGuide%2001-01.pdf (accessed November 10, 2014).

Skoog, D.A., F.J. Holler, and S.R. Crouch. 2006. Principles of Instrumental Analysis. 6th edn. Wadsworth Australia: Thomson Brooks/Cole.

TOGAF. 2012. The TOGAF welcome screen. http://pubs.opengroup.org/architecture/togaf8-doc/arch/ (accessed April 24, 2014).

TOGAF. 2014. Definitions. The Open Group Definitions web page: http://pubs.opengroup.org/architecture/togaf9-doc/arch/chap03.html (accessed November 10, 2014).

U.S. General Services Administration. 2014. Federal Business Opportunities. http://www.fedbizopps.gov (accessed November 8, 2014).

U.S. Government. 2014. SBIR/STTR. http://www.sbir.gov (accessed November 8, 2014).

2

Enterprise Architecture Fundamentals

> Some regard private enterprise as if it were a predatory tiger to be shot. Others look upon it as a cow that they can milk. Only handfuls see it for what it really is—the strong horse that pulls the whole cart.
>
> —**Winston Churchill**

Enterprise architecture (EA) methods are becoming an essential element in forging successful partnerships and collaborations, between multiorganizational entities working on complex programs, projects, and efforts. These methods are used to apply architecture principles and practices needed to guide large and/or expanding organizations to evolve from simple to a complex organizational structure through business, information, process, and technology changes to execute core enterprise strategies.

The Institute of Electrical and Electronics Engineers (IEEE) defines architecture as "architecture is the fundamental organization of a system embodied in its components, their relationships to each other, and to the environment, and the principles guiding its design and evolution." The main aim of EA is to capture/establish the core business, organizational and technical integration, standards, and governing aspects of an enterprise and focus on the planning, design, and evolution toward a future state of the enterprise. The EA methodology also provides a strategic and business-driven framework for making systems and infrastructure development decisions, provides an enterprise modeling capability that integrates with model-based systems engineering (MBSE) tools, and captures the vision and direction of an enterprise. The main three perspectives of EA are the regulation-oriented perspective, design-oriented perspective, and pattern-oriented perspectives.

The "blueprint analogy" is often used in describing and understanding EA. Imagine building a room without blueprints. This can be compared to developing business resources and systems without EA (Bernard 2012). Systems have become more complex over the past century presenting organizations with new opportunities and challenges. Stakeholders use systems and architectural concepts, principles, and procedures to understand and manage the composition and evolution of the enterprise. Stakeholders can use architectural descriptions to improve communication and cooperation regarding business strategies to achieve desired goals. Architectural frameworks and description languages are used to organize conventions and common practices of architecture in various communities and application domains. EA acts as the "collaboration force between the aspects of business planning, business operations, automation and enabling technological infrastructure of the business" (de Vries 2010, pp. 17–29).

EA outlines organizational mission, goals, and objectives and translates business vision and strategy by improving and communicating key requirements, processes, and models to enable change throughout the enterprise. This includes the information on how the architecture should adapt in response to the change in the mission.

EA provides a structure for management (Bernard 2012) and aims to define clear relationships from business and/or strategic goals and objectives to performance improvements for the enterprise. To be able to achieve the desired performance improvements, the EA must evolve and be progressively applied to the entire enterprise. The EA and its principles and methods must be supported by top management and must be progressively implemented throughout the entire enterprise. EA methods help facilitate global thinking in technology and systems selection and provide control mechanisms, standardization, and governance mechanisms to the enterprise. The EA should also project and capture the impact of performance improvements. Metrics should be established that incorporate stakeholder feedback, capture the value added, and improve the quality of EA products, services, and processes (Schekkerman 2008).

2.1 High-Level Integrated Model

In the previous analogy of the blueprints for a house, we described the importance of knowing the "big picture" (the blueprints) before embarking on a venture (building the house). To extend this analogy a little further, if building a house with blueprints represents building a product, process, system, or service with systems engineering (SE) principles, then EA methods would be equivalent to urban planning activities. The house (system) that is developed by the blueprints (SE methods) is part of a greater enterprise (the city/ state/country subject to the urban planning) that has its own governance structure, processes, systems, and standards (the EA). The EA captures the governance, policies, strategic direction, technical systems, standards, and processes (the urban plan and associated actors and processes) and integrates the activities of the entities making up the enterprise (individual architect/builders). In essence the EA provides a high-level integrated model for its constituent members (entities). These principles and methods work for massively large organizations such as the Department of Defense (DOD), federal agencies, and even international cooperative efforts. They also apply to smaller organizations that integrate into larger organizations or that plan on evolving into large organizations.

2.1.1 Purpose of Enterprise Architecture

Initially, EA tries to improve the overall efficiency of an organization or business, by considering and implementing structural improvements within an organization, the centralization and/or integration of business processes, and the data flow and ensuring the justification of information technology (IT) investments. The EA approach can be used at multiple levels of an organization by using strategic drivers to align business practices and processes, standardize and integrate ITs and systems, streamline decision making, and manage business transformation activities (GAO 2010). EAs are built on an architecture framework. There are many EA frameworks to choose from (e.g., The Open Group Architecture Framework [TOGAF]). TOGAF describes a method for developing an "architectural vision" also referred to as a business with an objective. If a vision is articulated, a road map in the form of transitional architectures is constructed to provide guidance for changing from the present to the future state. Similar methods are described in other EA frameworks. Each of the enterprise architectural frameworks exhibits uniqueness. TOGAF, for instance, has strong defining methods for architectural development, while others have a strong definition of EA artifacts and taxonomy (Zachman 1987, pp. 276–292).

Enterprises will face changes and challenges, which may include mergers, acquisitions, innovations, new technologies, new business models, deregulation, and increased global competition. Advances in e-commerce, globalization of networks, virtual enterprises, and resource availability put a further strain on corporations to manage informational technology and data flow. As technology evolves at an ever-increasing rate, the business environment becomes increasingly dynamic. In order to improve the chance of business survival in today's competitive global environment, enterprises need to evolve and become more agile to allow them to adapt to change quickly and effectively.

2.1.2 Historical Background

In his article entitled *A Framework for Information Systems Architecture* released in 1987, J.A. Zachman (Zachman 1987, pp. 276–292) describes the challenges and the vision of EA as a way to manage the complexity of disturbed systems as they evolve and the addition of needed capabilities into these entities. He envisioned a holistic approach to systems architecture that valued each important aspect that would allow businesses to adapt to change. The costs and the success of a business that depends on its information system require a disciplined approach to managing those systems. This broad view method for architecture systems was first called an "information systems architectural framework" and later coined "enterprise architecture framework" (Zachman 1987, pp. 276–292).

In 1994, the U.S. DOD used Zachman's work in creating an approach known as the Technical Architecture Framework for Information Management (TAFIM) (DOD 1994) that promised to align business needs with technical projects to ensure the achievement of strategic goals and objectives. In 1996, the U.S. Congress issued a bill known as the Clinger and Cohen Act of 1996 or the Information Technology Management Reform Act (ITMRA) (Clinger and Cohen Act 1996). This act established chief information officers (CIOs) for government agencies. The federal agencies were required to focus on the results and the streamlining of IT procurement processes and placed emphasis on the selection processes and management of IT projects to improve business value. Simplified acquisition procedures were allowed if project costs were under $5 million.

The CIO council began to work on the Federal Enterprise Architecture Framework (FEAF) in 1998 and was able to release Version 1.1 by September 2009 (Hite 2004). FEAF discussed segment architectures as a starting point for smaller parts of an organization. Transitioning from the CIO, the Office of Management and Budget (OMB) became responsible for the FEAF in 2002. OMB changed the name of FEAF to Federal Enterprise Architecture (FEA), and its function will be discussed further later. Though the Federal Government placed significant emphasis on EA, the progress was rather slow and high-profile failures were documented. The General Accountability Office (GAO) reported that in 8 years after the implementation of the Clinger and Cohen Act, only 20 out of 96 agencies had started establishing the groundwork necessary for managing architectures. Furthermore, the GAO determined that a number of failures occurred within agencies' application of EA methods, including the Federal Bureau of Investigation (FBI) (GAO 2005a), DOD (Hite 2006), and the National Aeronautics Space Administration (NASA) (GAO 2005b).

TAFIM was revoked 4 years after its introduction, and The Open Group assumed the responsibility of documenting a new standard known as "The Open Group Architecture Framework (TOGAF)" released in 2003. Around this time, the DOD also created their own architectural framework named the "Department of Defense Architecture Framework (DODAF)." These two standards continued to expand and were improved for years along with the FEA. Figure 2.1 shows the evolution of the EA methodology over the years.

2.1.3 Types of Architectures

Many varieties of EA frameworks are in use today such as Gartner, TOGAF, the FEAF, and the Zachman EA Framework. Each of these frameworks has advantages and

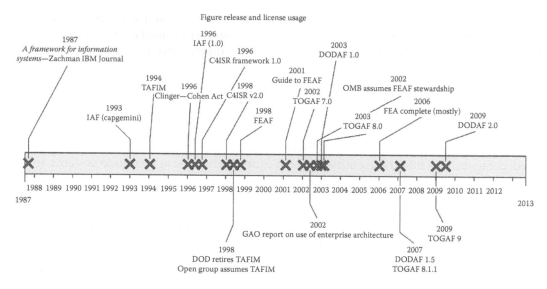

FIGURE 2.1
EA timeline. (Courtesy of Chris Lannon.)

disadvantages that need to be considered in certain scenarios. Each type is designed to address the problems of all businesses including the relationship of business strategy to IT. We start with providing some of the key features for each of the major architecture frameworks.

The first EA type is the Gartner approach created by the Gartner Group. The Gartner Group is known for its IT research as a consulting organization (deVries 2010, pp. 17–29). Their approach to EA is considered more of a practice than a methodology or process. This EA practice is highly collaborative and promotes communications between business owners, IT specialists, and technology engineers.

The second EA framework is TOGAF. This EA framework integrates four different architectures into its approach. These architectures are

- **Business architecture**—Describes the processes able to meet business goals
- **Application architecture**—Deals with the how applications interact
- **Data architecture**—Concerned with the organization of data
- **Technical architecture**—Concerned with the infrastructure of hardware and software in the enterprise (Hornford 2011)

The third type of architecture is the FEA. This architecture unites a number of federal government agencies under one common architecture. FEA has a well-defined architectural process and taxonomy that is usable as a possible methodology or as an evaluation tool of results achieved by executing the process. It has five reference models to measure performance: business, components, service, technical, and data. Furthermore, it gives perspectives in viewing the EA, processes creating EA, taxonomy, and approaches for measuring success (Whitehouse 2007). Finally, the Zachman Framework mostly contains taxonomy on how EA is structured and organized. Zachman uses a catalog approach by organizing items into classifications that allow stakeholders to make better decisions (O'Rourke et al. 2003).

2.1.3.1 Relative Comparisons

In this section, we make a subjective comparison between the TOGAF, FEA, Zachman, and the Gartner EA frameworks. The intent is to highlight the relative strengths and weaknesses of these major EA frameworks. The categories and rankings come from Sessions (Sessions 2007) from a posting in the Microsoft Developer Network (MSDN) library (Sessions 2007). This comparison is a useful example, and the results are shown practically verbatim as follows. Some of these criteria may not be directly relevant for your organization or enterprise, but these evaluations provide a place to start when considering an architectural framework for your own organization. The comparative analysis is as follows:

Each EA framework methodology is ranked according to the following criteria:

1: Does a very poor job in this area
2: Does an inadequate job in this area
3: Does an acceptable job in this area
4: Does a very good job

1. **Taxonomy completeness.** Refers to the efficiency of using the indicated EA framework methodology to classify the various architectural artifacts. This is almost the entire focus of Zachman. None of the other methodologies focuses as much on this area.

   Ratings:
Zachman:	**4**	TOGAF:	**2**	FEA:	**2**	Gartner:	**1**

2. **Process completeness.** Refers to the efficiency of a methodology's guidance through a systematic process in creating EA. This is almost the entire focus of TOGAF, with its Architecture Development Method (ADM).

   Ratings:
Zachman:	**1**	TOGAF:	**4**	FEA:	**2**	Gartner:	**3**

3. **Reference model guidance.** Refers to the usefulness of the methodology in building a relevant set of reference models. This is almost the entire focus of FEA. TOGAF also provides support; however, the TOGAF reference model guidance is not as complete as the FEA reference model guidance.

   Ratings:
Zachman:	**1**	TOGAF:	**3**	FEA:	**4**	Gartner:	**1**

4. **Practice guidance.** Refers to how much the methodology facilitates assimilating the mindset of an EA approach into the organization and develops a culture in which the EA is valued and used. This is a primary focus of Gartner's architectural practice.

   Ratings:
Zachman:	**1**	TOGAF:	**2**	FEA:	**2**	Gartner:	**4**

5. **Maturity model.** Refers to how much guidance the methodology gives in assessing the effectiveness and maturity of different organizations within the enterprise in using EA.

   Ratings:
Zachman:	**1**	TOGAF:	**1**	FEA:	**3**	Gartner:	**2**

6. **Business focus.** Refers to the methodology's focus on using technology to drive business value specifically defined as either reduced expenses and/or increased income. Ratings:
   Zachman: **1**   TOGAF: **2**   FEA: **1**   Gartner: **4**
7. **Governance guidance.** Refers to the methodology's understanding and creating an effective governance model that guides stakeholder interactions. Ratings:
   Zachman: **1**   TOGAF: **2**   FEA: **3**   Gartner: **3**
8. **Partitioning guidance.** Refers to how well the methodology will guide effective autonomous partitions of the enterprise, which is an important approach to managing complexity. Ratings:
   Zachman: **1**   TOGAF: **2**   FEA: **4**   Gartner: **3**
9. **Prescriptive catalog.** Refers to how well the methodology guides in setting up a catalog of architectural assets that can be reused in future activities. Ratings:
   Zachman: **1**   TOGAF: **2**   FEA: **4**   Gartner: **2**
10. **Vendor neutrality.** Refers to how likely you are to get locked in to a specific consulting organization by adopting this methodology. A high rating here indicates low vendor lock in. Ratings:
    Zachman: **2**   TOGAF: **4**   FEA: **3**   Gartner: **1**
11. **Information availability.** Refers to the amount and quality of free or inexpensive information about this methodology. Ratings:
    Zachman: **2**   TOGAF: **4**   FEA: **2**   Gartner: **1**
12. **Time value.** Refers to the length of time you will likely be using this methodology before you start using it to build solutions that deliver high business value. Ratings:
    Zachman: **1**   TOGAF: **3**   FEA: **1**   Gartner: **4**

Sessions (2007)

As can be seen by these ratings, the various EA frameworks have different strengths and weaknesses. Which EA framework should be used depends on the relative importance of the evaluation categories to the organization/enterprise. The choice also often depends what EA framework is used by the stakeholders. For instance, the DOD uses the DODAF.

2.1.4 Architecture Type Roles in Multiple-Entity Development Process

The goal of each architecture type is to determine a framework that allows the flow of information, provide understanding how change affects the complete system or enterprise, and help manage that change with minimal impact or consequences. While each architecture type takes a different approach, they all have the same goals. The role of these architectures is to effectively integrate systems from multiple entities across the entire enterprise and provide a holistic, optimized approach in entity interactions. An EA aids in effectively incorporating and integrating new capabilities into the enterprise. In turn, new capabilities affect the overall enterprise with respect to data flow, IT infrastructure, and system/company resources.

2.1.5 Architectural Framework

Architectural frameworks define the organization or structure of an organization (Minoli 2008). These frameworks should be used as guiding principles for an EA that defines the procedures for creating, analyzing, and applying architecture descriptions within an organization utilizing business processes, technology, and IT infrastructure to determine the ways the new capabilities affect the organization. Architectural frameworks typically contain three components (Minoli 2008):

1. **Views**—Used for information flow and to allow stakeholders to visualize relationships between subsystems or other systems within the organization.
2. **Methods**—Supplies the necessary steps a user needs to obtain a specific goal. Multiple methods can be present within a framework and can consist of processes, tools, or rules.
3. **Approaches**—Defines how to accomplish a desired result.

Many open-source and proprietary frameworks are available for general and specific uses. Each architectural framework should be considered for applicability and fit with respect to the enterprise. The industry applies these frameworks to build EAs that facilitate evolving the business vision, mission, goals, and new desired capabilities to meet stakeholder needs. Architectural frameworks provide the components, processes, and procedures necessary to support the increase in complexity of system of systems architectures (Minoli 2008) and define the interrelationships of these systems, allowing the definition of business goals and a more holistic view of the enterprise. This provides a mechanism for understanding the capabilities of the enterprise and responding to new opportunities to meet business needs. Several EA framework types are described in the following sections.

2.1.5.1 Enterprise Architecture Cube

The Enterprise Architecture Cube (EA3) is a procedure to document the enterprise with different levels of detail. There are two opposing views that formulate this approach: the first view, called "AS IS," shows the EA as currently defined (Bernard 2012), and the second view, called "TO BE," seeks to identify the future state of the enterprise. The EA3 framework is built around a cube structure that defines the architecture management and transition plans between the "AS IS" and the "TO BE" views (Bernard 2012). When applying the EA framework (EA3), there are six core elements that are necessary in the completion of the framework (Bernard 2012):

1. *Enterprise architecture documentation framework*—Identifies the scope and relationships between various documentation levels
2. *Implementation methodology*—Details the current (AS IS) and the future (TO BE) views of the architectures and also includes specific steps to maintain the architecture
3. *Processes for architecture governance*—Manage planning, decision making, and process oversight for changes to the architecture
4. *Documentation artifacts*—Details the types and methods of documentation used in each subsystem, which includes strategic analysis, security, plans, and controls
5. *Architecture repository*—Examples are databases, websites, or other software that stores the documentation necessary for the framework
6. *Proven methods*—The best practices that are implemented for the entire architecture

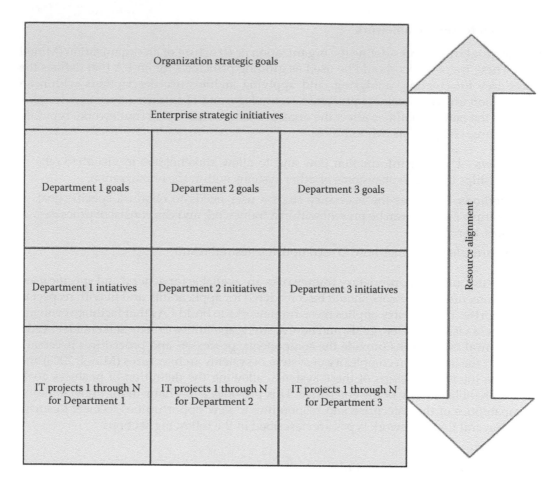

FIGURE 2.2
EA3 framework as it relates to business field. (Courtesy of Chris Lannon.)

Each of these core elements defines the EA3 and together allows the EA framework to be used to identify new capabilities that move the enterprise from the current to the future state. Figure 2.2 shows the EA3 framework as it relates to business objectives.

2.1.5.2 DODAF

The DODAF is intended to ensure that the architecture descriptions are relatable between and among each organization's operational, systems, and technical architectural views and are comparable across organizational boundaries. The framework provides both the rules and guidance for developing and presenting architecture descriptions. The framework provides direction on how to describe architectures; it does not provide guidance in how to construct or implement a specific architecture.

The DODAF Version 2.0 is the encompassing, comprehensive framework and conceptual model that enables the development of architectures to facilitate the ability of DOD managers at all levels to make key decisions more effectively through organized information sharing across the department, joint capability areas (JCAs), mission, component, and program boundaries (DoDAF 2014). The DOD created this framework in response to the

Clinger and Cohen Act of 1996 to regulate the CIO responsibilities with that of developing EAs for the DOD (DoDAF 2014). Conformance with DODAF processes allows for the reuse of information and defining the enterprise as a whole, in providing common artifacts across the enterprise for common understanding.

DODAF's latest version (2.0) switched focus from individual products to the process of data flows through the enterprise's systems. This change in focus was meant to facilitate the understanding of data flows and guide the collection and organizing of data. The enterprise is defined in the different viewpoints provided by the architectural framework. The DODAF viewpoint allows users to understand the architectural description that is consistent with a particular purpose, mission, or objective (DoDAF 2014). Each viewpoint provides a different purpose with different content, structure, and level of detail. These varying viewpoints allow the architectural descriptions to address the specific needs and purposes of the stakeholders.

DODAF supports six core processes that are designed to provide uniform processes in critical defense-related decision-making areas (DoDAF 2014). The six core processes are as follows:

1. Joint capability integration and development system (JCIDS)
2. Defense acquisition system (DAS)
3. Systems engineering (SE)
4. Planning, programming, budgeting, and execution (PPBE) system
5. Operations planning
6. Capabilities portfolio management

JCIDS focuses on the joint capabilities needed to meet DOD mission requirements (DoDAF 2014). This core process identifies the necessary capabilities that must be prioritized along with the mission scope, description, training, and personnel to meet established capability gaps. Some EA artifacts include the Capabilities-Based Assessment (CBA), Initial Capabilities Document (ICD), Capability Development Document (CDD), and the Capabilities Production Document (CPD).

The DAS manages the investment in technologies, program, and product support for the National Security Strategy (DoDAF 2014). Processes and artifacts needed for DAS support the framework for translating mission needs and technology opportunities into stable, affordable, and well-managed acquisition programs. Acquisition programs are advanced through the enterprise based on completing milestones that are significant to program phases.

DODAF artifacts for DAS include road maps and integrated plans that are utilized to provide capability assessments, define investment plans, and guide systems development for resource allocation and input to the defense planning and budgeting reviews (DoDAF 2014).

Many DOD programs are required to use a robust SE approach per policy directives like the Sambur memo dated January 6, 2003 (Sambur 2003). The SE approach aligns with the DAS processes to provide an optimized approach to systems development. A systems engineering plan (SEP) is created to describe the SE technical approach including activities, resources, measures (and metrics), and applicable performance incentives. SE processes are used throughout the systems development life cycle.

DODAF provides guidance, direction, and support for program planning, budgeting, and execution by capturing and identifying military capabilities and resource allocations and providing a structure for the program planning process that allows agencies to develop strategies for prioritization and affordability (DoDAF 2014). The DODAF also supports portfolio management for IT infrastructure investments by establishing the

DoDAF V2.0 / DoDAF V1.5	Operational Viewpoint	Systems Viewpoint	Services Viewpoint	All Viewpoint	Standards Viewpoint	Data & Information Viewpoint
AV-1				AV-1		
AV-2				AV-2		
OV-1	OV-1					
OV-2	OV-2					
OV-3	OV-3					
OV-4	OV-4					
OV-5	OV-5a, OV-5b					
OV-6a	OV-6a					
OV-6b	OV-6b					
OV-6c	OV-6c					
OV-7						DIV-2
SV-1		SV-1	SvcV-1			
SV-2		SV-2	SvcV-2			
SV-3		SV-3	SvcV-3a, SvcV-3b			
SV-4a		SV-4				
SV-4b			SvcV-4			
SV-5a		SV-5a				
SV-5b		SV-5b				
SV-5c			SvcV-5			
SV-6		SV-6	SvcV-6			
SV-7		SV-7	SvcV-7			
SV-8		SV-8	SvcV-8			
SV-9		SV-9	SvcV-9			
SV-10a		SV-10a	SvcV-10a			
SV-10b		SV-10b	SvcV-10b			
SV-10c		SV-10c	SvcV-10c			
SV-11						DIV-3
TV-1					StdV-1	
TV-2					StdV-2	

FIGURE 2.3

DoDAF-V2.0-Viewpoints. (From DoD, DoD Architecture Framework Version 2.0. http://en.wikipedia.org/wiki/File:DoDAF-V2.0-Viewpoints.jpg#mediaviewer/File:DoDAF-V2.0-Viewpoints.jpg, 2014.)

framework for capturing and identifying risks and performance measures to meet agency mission and goals. The collective features of an EA establish a knowledge-based approach that allows stakeholders to obtain critical information at the right times to make informed decisions (DoDAF 2014). EA processes continue to evolve with emphasis on a holistic view of the enterprise rather than individual program efforts.

At the core of DODAF 2.0 are its viewpoints that provide a mechanism for stakeholders to view the EA from a particular perspective. The different viewpoints present data at varying levels of detail and from different perspectives. Some viewpoints offer summary information regarding the enterprise as a whole, such as a high-level concept. Other viewpoints offer a deeper dive into a particular subsystem at a more technical level. Yet others indicate the way various subsystems, or individual systems, are interconnected, providing data and control flow diagrams. The different viewpoints are depicted in Figure 2.3, which also shows the relationship of DODAF version 1.5 to the new version 2.0.

2.1.5.3 The Open Group Architectural Framework

TOGAF is used by organizations to improve business efficiency (Opengroup 2014). It focuses on the business' vision and current capabilities to decide the best approach for designing, planning, and implementing information throughout an organization. Practitioners of this approach are open to processes that allow the efficient use of resources (Hornford 2011). As stated in Section 2.1.2, TOGAF had its roots in the DOD as the TAFIM, which was turned over to The Open Group and transformed into an Architectural Development Framework in 2003 (Hornford 2011). TOGAF is typically used

TABLE 2.1

Architecture Types Supported by TOGAF

Architecture Type	Description
Business architecture	Strategy of the business, governance, processes, and organization
Data architecture	Organization's data assets and data management resources
Application architecture	Organization's applications in use, their interfaces, and relationship to core business processes
Technology architecture	SW and HW capabilities required to support the business, data, and application services; IT infrastructure, networks, communications, etc.

Source: Adapted from Barroero, T. et al., *Business Capabilities Centric Enterprise Architecture*, Springer, Berlin, Germany, 2010.

for developing multiple interrelated architectures in conjunction with other frameworks. A major focus is on its ADM that aids in developing EA.

TOGAF helps define architectures by providing a description of a system, with details on its components including interrelationships and guidelines for governance, along with the EA evolution over time. TOGAF offers guidelines for the development of four related types of architectures. The four types of architectures are shown in Table 2.1 (Hornford 2011).

As explained earlier, one of TOGAF's essential features is the AMD. The AMD is segregated into several sequential phases. Each phase allows the enterprise architect to understand what is needed within that phase. The phases are defined as follows (Hornford 2011):

1. *Preliminary Phase*—Defines the architecture principles. It is also used to identify specific frameworks that can/should be used.
2. *Phase A*—Defines the scope, vision, stakeholders, and mission for creating the architecture.
3. *Phase B (Business Architecture)*—Describes the product and service strategies and processes based on the scope and vision from Phase A.
4. *Phase C (Information Systems Architecture)*—Develops the data/application systems services.
5. *Phase D (Technology Architecture)*—Forms the basis of implementation.
6. *Phase E (The Opportunities and Solutions)*—Allows the architect to evaluate the implementation to determine and identify strategic parameters for future capabilities.
7. *Phase F (Migration Planning)*—Assesses the parameters from Phase E and defines cost, benefits, and migration strategies.
8. *Phase G (Implementation Governance)*—Makes recommendations for implementation and provides for change management and deployment once it is completed.
9. *Phase H (Architecture Change Management)*—Continues to monitor new capabilities or changes to technology and business environments. Figure 2.4 shows the complete flow of these processes.

TOGAF also defines a Content Metamodel (displayed in Figure 2.5) that illustrates the existing architecture building blocks and details the interrelationship (Hornford 2011) between vision and drivers to visualize business capabilities and vice versa. Together with the Architectural Development Process and the Content Metamodel, enterprises can determine how new capabilities affect the overall enterprise and drive business needs (Hornford 2011). Figure 2.5 displays a Content Metamodel diagram.

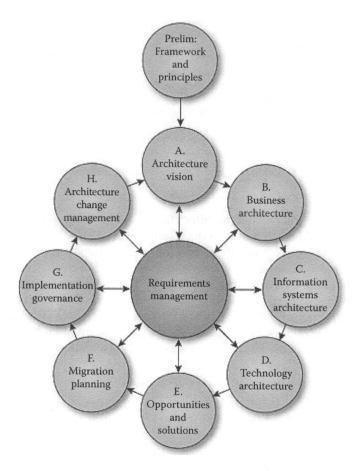

FIGURE 2.4
TOGAF architectural development process. (From Wikipedia, TOGAF Architectural development process is located, on April 13, 2014, on the website http://en.wikipedia.org/wiki/File:TOGAF_ADM.jpg and is considered public domain, 2014.)

2.1.5.4 Federal Enterprise Architectural Framework

The OMB of the U.S. Federal Government created the FEAF, which defines a common approach in coordinating business drivers and goals to formulate a clear vision and strategic direction across the enterprise. This includes reducing waste and duplication methods along with assessing technological impacts and fostering advancements. A lot of money is being spent to support and develop IT infrastructures throughout the federal agencies. FEAF was developed to ensure that programs are taking a more holistic approach in developing various architectures, with emphasis on interrelationships and cost reductions. There are four primary outcomes consistent with using the FEAF:

1. *Service delivery*—Concerned with responding to agency missions and allowing the architecture to meet those needs (Whitehouse 2007).

2. *Functional integration*—Functional integration means interoperability between programs, systems, and services, which requires metacontext and standards to be successful. EA can provide both metacontext across all functional domains

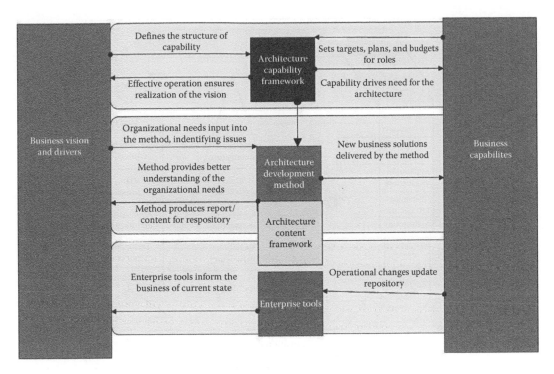

FIGURE 2.5
TOGAF content metamodel. (Courtesy of Chris Lannon.)

(strategic, business, and technology), as well as related standards for the full life cycle of activities in each domain. Program, systems, and services interoperability is foundational for Federal Government organizations to be able to successfully partner in new shared service models that may involve outside providers and new roles for participation (e.g., consumer, developer, or provider). The EA should provide context and be the source of standards for all levels of interoperability (Whitehouse 2007).

3. *Resource optimization*—Defining the best way to use the available resources including asset management (e.g., hardware inventory and software licenses) and configuration management (maintaining and monitoring a documented baseline of users, processes, hardware, and software) that are important elements of resource optimization that an EA also enables (Whitehouse 2007).

4. *Authoritative reference*—Just as the blueprints of a building are the authoritative reference for how the structure will function, the organization's EA provides an integrated, consistent view of strategic goals, mission and support services, data, and enabling technologies across the entire organization, including programs, services, and systems. This allows the EA to document ownership, goals and objectives, and performance metrics (Whitehouse 2007).

Based on the outcomes stated previously, FEAF outlines principles and standards to develop the business, technology, and data using the following models:

- *Performance Reference Model*—Standards to measure the performance and success of IT investments and the impact to agency outcomes (GAO 2001).

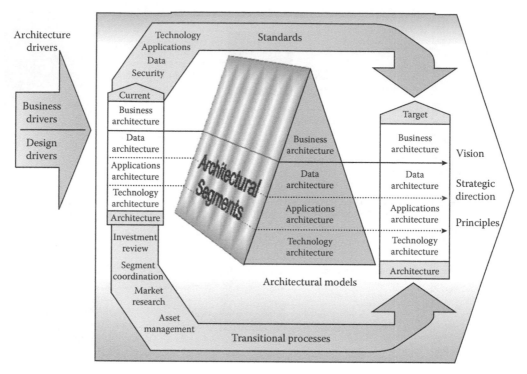

FIGURE 2.6
Structure of the FEAF Components. (From Chief Information Officer Council - A Practical Guide to Federal Enterprise Architecture. Licensed under Public domain via Wikimedia Commons - http://commons. wikimedia.org/wiki/File:Structure_of_the_FEAF_Components.jpg#mediaviewer/File:Structure_of_the_ FEAF_Components.jpg, 2014.)

- *Business Reference Model*—Functional view of the business and its services with a goal to promote collaboration across agencies (GAO 2001).
- *Service Component Reference Model*—Used to identify and classify services that can aid in new capabilities that can be used for reuse in support of new services (GAO 2001).
- *Data Reference Model*—Provides for information sharing and collaboration across agencies by providing uniformity to describing data, facilitation of data discovery, and data sharing (GAO 2001).
- *Technical Reference Model*—A technical model that supports the Service Component Reference Model to enable its capabilities. It aims to reuse technology assets and reclassify existing architectural elements to meet agency mission needs (GAO 2001).

FEAF is similar to other framework architectures; however, it uses architecture or business drivers as input to promote new capabilities, objectives, visions, and principles. It aims to reuse and reduce cost across agencies. The FEAF process is shown in Figure 2.6.

2.1.5.5 *Ministry of Defense Architectural Framework*

The Ministry of Defense Architectural Framework (MODAF) is an internationally recognized architecture framework developed by Ministry of Defense (MOD) in the United

TABLE 2.2

MODAF View Descriptions

View Name	Description
Strategic view	Shows desired business outcome and steps to achieve it
Operational view	Processes and information needed to achieve capability
Service-oriented view	Supports the process of operational view and describes the services by providers
System views	Physical implementation of operational and service-oriented view
Acquisition view	Timelines and milestones for projects to deliver the capability
Technical view	Describes the technical requirements to be applied
All views	Overview description

Source: Adapted from MoD, MOD Architecture Framework, https://www.gov.uk/mod-architecture-framework, 2012.

Kingdom to promote defense planning and change management activities (MoD 2012) and to provide structure and definition to equipment capability. MODAF provides rules and templates that provide visualizations of the business area from different views. These views were defined in seven categories as shown in Table 2.2 (MoD 2012).

These views interact within the MODAF model detailing the interrelationships between the what's, why's, and how's determined by the stakeholder's in trying to accomplish specific tasks (MoD 2012).

2.1.6 Modeling Tools

This section provides a brief overview of architecture development and modeling tools. Architectural tools support the development, presentation, storage, and modeling of the enterprise views of the selected framework. Tools generally provide a graphical view of the system to establish the current (AS IS) model and the future (TO BE) models including new capabilities that affect the organization as a whole.

The uses of automated tools are becoming more extensive due, in part, to the sheer complexity of systems in design and the easy, accessible repository of inherent artifacts and EA building blocks. Some of these tools come in the form of architectural suites, providing modeling tools and repository management, along with multifunctional EA tools. When selecting a tool, organizations should consider the following (van den Berg and van Steenbergen 2006):

- Functionality
- Tool architecture
  - Central or distributed repository
  - Version control
  - Accessibility (web, PC, Mac, Unix)
- Cost of ownership

There are several commercially available and open-source EA tools available on the commercial market. Some representative examples include Rhapsody, IBM's Systems Architect, Essential Project, Modelio, and MagicDraw. Figure 2.7 is an example screen shot from IBM Rhapsody using DODAF. The main aim of choosing a comprehensive EA modeling tool must be discussed and agreed by the enterprise stakeholders. Once the tool selection is made, training will typically be required to get the full utility from the tool.

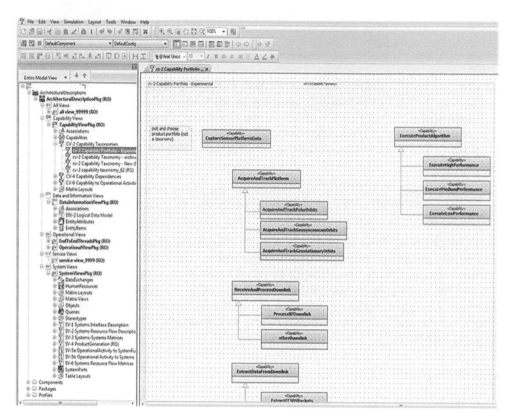

FIGURE 2.7
IBM Rhapsody using DoDAF. (Courtesy of Chris Lannon.)

2.2 Optical Systems Model

In this section, we provide models and an analytical framework for understanding technical aspects of optical systems. In a sense, these frameworks and models have a similar role as the EA has in assimilating, integrating, and defining the relationships between enterprise entities and their artifacts. The analytical frameworks and models describe the technical relationship between key system elements and their observables. By observables, we mean observable quantities of interest such as photometric or radiometric elements of interest, for example, the irradiance on the detector element. The models describe and capture all the essential enterprise or system components and their necessary interactions to properly address the analytical problem at hand. In this chapter, we develop a variety of useful mathematical models and analytical methods that help in our understanding of optical systems. These models are useful for analyzing and simulating the physical phenomena associated with our optical system. Each model has its strengths and weaknesses as discussed later.

Radiometric and photometric models are useful in determining whether or not there is sufficient signal on the detector, as compared with the noise that affects the detector, to successfully conduct an observation. By noise, we mean any unwanted signal from the source, medium between the source and optical system, and optical system itself. Often, these are separated into background and clutter effects, scintillation effects, scattering, atmospheric turbulence effects, aberrations of the receiver optics, detector noise effects, detector readout electronics effects, shock, jitter, environmental effects (pressure, temperature, humidity, radiation, dust), and statistical uncertainty from the random nature of the light itself. We will look at radiometric and photometric models in Chapter 6 and address each of these effects in more detail as needed.

Although good at determining various signal-to-noise characteristics of our optical system, the radiometric and photometric models are not so good at determining spatial effects and possibly temporal correlations that may become significant in some imaging applications. If we are looking at the interaction of light with very simple or idealized objects such as a plane, sphere, or a simple geometric boundary, then a model using the full power of electromagnetic wave theory is workable and most accurate. A simple geometric model assumption is sometimes acceptable for preliminary or rough order of magnitude types of analyses. Electromagnetic theory and derivative methods can also be applied to exceedingly small physical phenomena such as particles; molecules; atoms; subatomic particles such as bosons, fermions, and hadrons; and nanostructures using quantum electrodynamics. Relativistic quantum electrodynamics methods are used when quantum electrodynamics methods involve physical interactions that approach the speed of light. Although modern electromagnetic theory and methods most accurately represent the physics of the interaction of the electromagnetic wave and matter, the computational aspects of applying electromagnetic theory to objects that have complex spatial structures and details become too burdensome, and other models that are based on electromagnetic methods are more useful. For example, most man-made objects are "rough" on the scale of an optical wavelength, and a detailed electromagnetic treatment of optical scattering from a large man-made object like a satellite, or a scene in the instantaneous field of view of an imaging system, is mathematically intractable unless some simplifying assumptions can be made. One such simplifying approach results if the spatial dimensions of the object viewed by the imaging system are small compared to the separation distance of the object from the imaging system. We will discuss simplification of spatial dimension in Section 2.2.3. Another modeling approach that is based on statistical analysis works well if the object is very large compared to the size of the optical wavelength of the imaging/observing system and also if the object is "rough" compared to the optical wavelength. Since most man-made objects in typical imaging scenarios fall into this category (apart from mirrors or reflective surfaces that are treated separately), the statistical optics models are very useful and will be discussed in Section 2.2.4.

This chapter is organized as follows: we will first start with an overview of basic electromagnetic propagation theory to show the basis and general assumptions for the linear (and nonlinear) systems models. We then present some fundamental concepts of linear and nonlinear systems along with some useful properties such as superposition and shift invariance. We then develop a linear systems model and list some useful properties. We extend the model to two spatial dimensions (2D) for visible and infrared imaging applications and three dimensions, two spatial dimensions and one wavelength (or frequency) dimension, for spectral imaging applications such as ultraspectral, hyperspectral, and multispectral imaging systems. These will be discussed in more detail later. We then provide some useful relationships between the spatial domain and spatial frequency domain for both coherent and incoherent imaging systems. We discuss the energy spectrum, power spectrum, generalized pupil function (GPF), impulse response, point spread function (PSF), optical transfer function (OTF), modulation transfer function (MTF), coherent transfer function (CTF), and some useful associated properties. We finish our presentation of coherent and incoherent imaging models by discussing some sampling effects and illustrating how to properly represent a continuous signal or image on a computer. We then move on to a statistical optics analytical and modeling approach and start with some important terminology, concepts, and results from statistics and probability theory. As previously stated, this methodology is very useful in modeling very large optical systems with respect to the size of the optical wavelength. As before, our approach will be focused on the optical systems engineer and optical systems manager, and so we will focus on the key points and results but will provide references to more detailed supporting information. A good understanding of the terminology and concepts presented in Chapter 1 is necessary to get the most out of this chapter and the rest of this text.

2.2.1 Fundamentals of Electromagnetic Propagation Theory

For all of the models that we will present, the characteristics of the electromagnetic field, its derivative components, its propagation effects, and interactions with matter are central considerations in their formulation. The differential form of Maxwell's equations describes how electromagnetic waves interact with matter (or propagate through free space) at a given point in space and time. The differential form of Maxwell's equations are given by

$$\nabla \times H = J + \frac{\partial D}{\partial t}, \quad \nabla \cdot D = \rho,$$

$$\nabla \times E = -\frac{\partial B}{\partial t}, \quad \nabla \cdot B = 0,$$

(2.1)

where
  E represents the electric field in V/m
  B is the magnetic induction in teslas
  J is the current density in A/m^2
  D is the displacement in C/m^2
  H is the magnetic field in (ampere turns)/m
  ρ is the charge density in C/m^3
  ∇ is the gradient operator given by

$$\nabla = \frac{\partial}{\partial x}\hat{x} + \frac{\partial}{\partial y}\hat{y} + \frac{\partial}{\partial z}\hat{z}.$$

(2.2)

Here, the "hatted" (^) coordinates are unit vectors in the x, y, and z directions, respectively.

Notice that these are a set of first-order, coupled, electromagnetic equations. The top right of Equation 2.1 is Coulomb's law, the top left is Ampere's law, the bottom right shows that there are no free magnetic poles, and the bottom left is Faraday's law. Although these equations can be solved directly, it is mathematically preferable to decouple these equations and deal with separate scalar and vector quantities. This can be done by defining a scalar potential Φ and a vector potential A and relating these terms to the electric field and magnetic induction by

$$E = \nabla\Phi - \frac{\partial A}{\partial t}, \quad B = \nabla \times A.$$

(2.3)

By making an additional substitution called a gauge transformation for mathematical convenience, we can substitute these results into Maxwell's equations and solve for two decoupled inhomogeneous wave equations:

$$\nabla^2 \Phi - \mu_o \epsilon_o \frac{\partial^2 \Phi}{\partial t^2} = -\frac{\rho}{\epsilon_o},$$

$$\nabla^2 A - \mu_o \epsilon_o \frac{\partial^2 A}{\partial t^2} = -\mu_o J.$$

(2.4)

In these equations, the ∇^2 symbol is the Laplacian operator given by

$$\nabla^2 = \frac{\partial^2}{\partial x^2} + \frac{\partial^2}{\partial y^2} + \frac{\partial^2}{\partial z^2}.$$

(2.5)

This set of wave equations, in Equation 2.3, is mathematically equivalent to Maxwell's equations but have the advantage of being decoupled from each other. Notice that these wave equations have the same structural form as

$$\nabla^2 \Psi_{(\bar{x},t)} - \mu_o \epsilon_o \frac{\partial^2 \Psi_{(\bar{x},t)}}{\partial t^2} = -\rho f\left(\bar{x},t\right),$$

(2.6)

where the term on the right is a driving term that has to do with the presence of sources such as charge densities or current densities. We can solve this equation by removing the explicit time dependence in expressing the solution in terms of its Fourier transform pair

$$\Psi\left(\bar{x},t\right) = \int_{-\infty}^{+\infty} \psi\left(\bar{x},\omega\right) e^{-j\omega t} d\omega,$$

$$F\left(\bar{x},t\right) = \int_{-\infty}^{+\infty} f\left(\bar{x},\omega\right) e^{-j\omega t} d\omega,$$

and the corresponding inverse

$$\psi\left(\bar{x},\omega\right) = \int_{-\infty}^{+\infty} \Psi\left(\bar{x},t\right) e^{j\omega t} dt,$$

$$f\left(\bar{x},\omega\right) = \int_{-\infty}^{+\infty} F\left(\bar{x},t\right) e^{j\omega t} dt.$$

(2.7)

Substituting Equation 2.7 into Equation 2.6, we end up with the inhomogeneous Helmholtz wave equation:

$$\left(\nabla^2 + k^2\right) \Psi\left(\bar{x},\omega\right) = -f\left(\bar{x},\omega\right) \text{in } \mathbb{R}^n.$$

(2.8)

In this equation, k is a term related to the velocity of the wave in the medium in which it is propagating. If we now consider the propagation of the electromagnetic field through free space, then we have no current density J or charge density ρ, and the form of the wave equation in Equation 2.6 becomes

$$\nabla^2 \Psi_{(\bar{x},t)} - \mu_o \epsilon_o \frac{\partial^2 \Psi_{(\bar{x},t)}}{\partial t^2} = 0. \tag{2.9}$$

Since the driving terms shown explicitly on the right side of Equation 2.4 are zero in this case, the corresponding homogeneous Helmholtz equation is given by

$$\left(\nabla^2 + k^2\right) \Psi\left(\bar{x}, \omega\right) = 0, \tag{2.10}$$

where k is given by

$$k = \frac{\omega}{c} = \frac{2\pi}{c} \nu = \frac{2\pi}{\lambda}, \tag{2.11}$$

where
    λ is the electromagnetic wavelength in meters
    ν is the corresponding frequency in Hz

The Helmholtz equation has several solutions that propagate. The simplest of these solutions for the scalar, homogeneous Helmholtz equation is the plane wave given by

$$\underline{u}(P,t) = \Re\left\{U(P)e^{j\phi(P)} e^{j2\pi\nu t}\right\},$$

$$\underline{u}(P,t) = U(P)\cos\left(2\pi\nu t + \phi(P)\right). \tag{2.12}$$

In these equations, the symbol u represents either the electric field or the magnetic induction since both these quantities satisfy the wave equation. The parameter P indicates position, $U(P)$ is the amplitude of the wave at position P, $\phi(P)$ is the phase at position P, and $\Re\{\cdot\}$ indicates taking the real part of the term inside the brackets. The phasor notation in Equation 2.12 is quite useful and will be adopted throughout the book. A useful consequence of using the phasor notation is that the phasors can usually be manipulated through a wide range of analytical calculations and the determination of the real physical result, obtained by taking the real part of the solution, can be deferred until the last step of the calculation. In using u, we will make it clear by the context of our discussion whether we are referring to the electric field or the magnetic induction. The plane wave solution to the homogeneous Helmholtz equation is itself quite useful since expanding waves that are far enough away from their source point are well approximated by plane waves. A physical example of this would be starlight outside of our solar system. Stars are so far away that they are well approximated as point sources. By the time the light from the star reaches us, the electric field that is carrying the information from the star is well approximated by a plane wave.

   The solutions to Maxwell's equations and the Helmholtz wave equation accurately describe the behavior of the electromagnetic field at a particular point in space and time; however, we need to get expressions that relate the interactions of the electromagnetic field with extended objects. We also need to understand how these fields propagate to, and interact with, some remote observational system. An example of this scenario is an imaging system that is observing a resolvable object, such as an earthbound telescope viewing a lioness on the plains of the Serengeti. For this case, an integral representation based on the Helmholtz wave equation is most practical. We will briefly describe this approach and provide some references for a more detailed explanation. Our aim is to arrive at the main results and show the connection to our modeling methodology. In transforming the Helmholtz wave equation into an integral expression, Green's theorem is very useful (Goodman 2005):

$$\iiint_V \left(\underline{G}(P)\nabla^2\underline{U}(P) - \left(\underline{U}(P)\nabla^2\underline{G}(P)\right)\right) dV = \iint_S \underline{G}(P)\frac{\partial \underline{U}(P)}{\partial n} - \underline{U}(P)\frac{\partial \underline{G}(P)}{\partial r} dS, \quad (2.13)$$

where the function $G(P)$ is the well-known Green's function at spatial position P. Notice that we have dropped the time dependency in Equation 2.12 since we are interested in spatial effects and since this is a separable function in its phasor form. We will bring in this dependency later when necessary. In Green's theorem, $G(P)$ and $U(P)$ can be any two complex-valued functions. The integration is over a volume V that is contained within a surface S. The partial derivative on the right side of Equation 2.13 is in the outward normal direction of the surface S. For our purposes, a free-space Green's function is given by

$$G(P_0) = \frac{\exp(jkr_{os})}{r_{os}}, \quad (2.14)$$

where r_{os} is the distance from a source point s to an observational point P_0. This Green's function describes a spherically expanding wave that originates at point s. Note that an assumption here is that the observational point is not directly on the source point since this term would tend toward infinity. In remote observation applications, the term r_{os} is never zero. The choice of this spherically expanding Green's function is motivated by the work of the Dutch physicist Christiaan Huygens who noted that the electromagnetic field at some distant observational point from an extended object can be determined by superimposing spherically expanding wavelets from each source point in the observational plane (Born and Wolf 1970). By substituting the Green's function in Equation 2.14 into Green's theorem along with the spatial component of our phasor in Equation 2.12, evaluating the partial derivatives for Green's function, and simplifying, we get the integral theorem of Helmholtz and Kirchhoff, an intermediate result:

$$U(P_0) = \iint_S \left\{ \frac{\partial \underline{U}(P)}{\partial n}\left[\frac{\exp(jkr_{os})}{r_{os}}\right] - \underline{U}(P)\frac{\partial}{\partial n}\left[\frac{\exp(jkr_{os})}{r_{os}}\right]\right\} ds$$

$$= \iint_S \left\{ \frac{\partial \underline{U}(P)}{\partial n}G - \underline{U}(P)\frac{\partial G}{\partial n}\right\} ds. \quad (2.15)$$

This equation was used by Kirchhoff to describe the observed electromagnetic field in a distant observational plane due to diffraction effects from an aperture. The only nonvanishing source term in Equation 2.15 was in the actual aperture itself. Consequently, the limits of integration in Equation 2.15 are confined to just the area of the aperture. Equation 2.15 was used to describe the diffraction effects of a plane screen, and the Fresnel and Kirchhoff diffraction formula was developed. Although surprisingly accurate experimentally, this

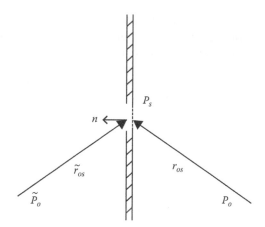

FIGURE 2.8
The geometry for the Rayleigh–Summerfield formulation.

formulation had some theoretical problems in satisfying the Kirchhoff boundary conditions near the aperture. To solve this problem, Rayleigh and Sommerfeld solved Equation 2.15 using a different Green's function:

$$G_{rs}(P_o) = \frac{\exp(jkr_{os})}{r_{os}} - \frac{\exp(jk\tilde{r}_{os})}{\tilde{r}_{os}}, \tag{2.16}$$

where the addition of the second term was found by Sommerfeld to satisfy the appropriate boundary conditions. Figure 2.8 shows the geometry for the Rayleigh–Sommerfeld formulation of diffraction from a plane screen.

The point source on the left is a mirror image of the point source on the right and is 180° out of phase. This choice of Green's function, in addition to satisfying the Kirchhoff boundary conditions, permits expressing the electromagnetic field in a distant observation plane in terms of integration over the clear aperture of the diffracting screen (*Sa*). The point in the clear aperture of the screen is P_s, and the point at the observer is given by P_o. Substituting Equation 2.16 into Equation 2.13 and once again evaluating the partial derivative on the new Green's function in Equation 2.16, and simplifying the result, lead to the Rayleigh–Sommerfeld diffraction integral:

$$u(P_o) = \frac{1}{j\lambda} \iint\limits_{Sa} \underline{u}(P_o) \left[\frac{\exp(jkr_{os})}{r_{os}}\right] \cos(\boldsymbol{n}, \boldsymbol{r}_{os}) dSa. \tag{2.17}$$

This equation is very general and useful in describing the electromagnetic field in a distant plane, due to a source with finite extent, represented by the field within the aperture of the plane screen, to a remote observational point. Let us make some variable substitutions to make the notation of the result in Equation 2.17 more convenient with the notation of our linear systems model presented later. Let the point P_s in the clear aperture of the plane screen be represented by the 2D position vector $\vec{x}_o$ where the subscript "*o*" represents the object (instead of the observer as was previously defined earlier). Let the subscript "1" represent the observational plane so that $\vec{x}_1$ represents a 2D position vector in the plane of the observer. Let the lowercase *u* represent the complex electromagnetic field, and let θ be the angle between the normal surface unit vector $\mathbf{n}$ and $\mathbf{r}_{os}$. The Rayleigh and Sommerfeld diffraction formula then has the following form (Roggemann and Welsh 1996):

$$u_1\left(\vec{x_1}\right) = \frac{1}{j\lambda} \int d\vec{x_o}\, u_o\left(\vec{x_o}\right) \frac{\exp\left(jkr_{o1}\right)}{|\vec{r}_{o1}|} \cos\theta. \tag{2.18}$$

We adopt the notation from Roggemann and Welsh since it is intuitive and practical for a number of imaging concepts and applications. Notice that Equation 2.18 can be written in the following form:

$$u_1\left(\vec{x_1}\right) = \int d\vec{x_o} u_o\left(\vec{x_o}\right) h_d\left(\vec{x_1}, \vec{x_o}\right), \tag{2.19}$$

where

$$h_d\left(\vec{x_1}, \vec{x_o}\right) = \frac{1}{j\lambda} \frac{\exp\left(jkr_{o1}\right)}{|\vec{r}_{o1}|} \cos\theta. \tag{2.20}$$

Equation 2.20 is the *impulse response* of the imaging system, and Equation 2.19 is a superposition integral. The corresponding geometry to Equations 2.19 and 2.20 is shown in Figure 2.9. Notice that the clear aperture on the left side of the diagram contains the object plane complex field distribution.

This complex field distribution can be considered as an extended object source term and is applicable as a model for various physical interpretations. For instance, just as in the derivation earlier, we can assume that we have a source on the left side of a very large screen with an opening at the center. In this case, we know that the complex electromagnetic field distribution in the clear aperture is due to the illuminating source. In addition, we also know that the value of the field is zero outside the clear aperture along the infinite plane. Alternatively, we can think of the complex electromagnetic field in the aperture as consisting of monochromatic emitted or reflected light from an extended object. In both

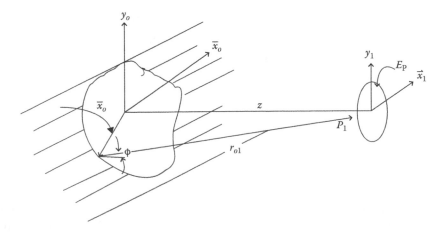

FIGURE 2.9
Superposition integral geometry.

these alternate cases, the source electromagnetic field is zero beyond the spatial extent of the object, and the complex field itself is fully represented across the spatial regions of the object (represented by the finite extent of the aperture). Stated in another way, once we have a spatial description of the field within the source aperture (the object plane), it is irrelevant what generated the field within the aperture. By knowing the field within the aperture, we have the ability to describe the complex electromagnetic field in a separate, distant observation plane. From here on, we use the subscript "o" to denote items of interest in the object plane and the subscript "1" for any items of interest in the observation plane. The position vector $\vec{r}_{o1}$ connects an arbitrary point P_o in the observation plane to an arbitrary point P_1 in the observation plane, and its magnitude is given by

$$r_{o1} = \left[z^2 + \left(x_o - x_1\right)^2 + \left(y_o - y_1\right)^2 \right]^{1/2}. \tag{2.21}$$

We now want to look at some special cases of the Rayleigh–Sommerfeld diffraction formula considering Equation 2.21. It turns out that it is difficult to work with Equation 2.20 because of the square root term in Equation 2.21. However, if the distance z is much larger than the maximum finite extent of the source term $\left|\vec{x}_o\right|_{max}$ and $\lambda \ll \left|\vec{x}_o\right|_{max}$, then the paraxial approximation can be made, and only the lower-order terms in the binomial expansion of Equation 2.21 need to be kept. This condition is satisfied if (Bahaa and Teich 2007)

$$F = \frac{\left|x_o\right|_{max}^2}{\lambda z} \geq 1. \tag{2.22}$$

In this equation, F is the Fresnel number. If this condition is met, then we can approximate the Rayleigh–Sommerfeld diffraction formula with the Fresnel diffraction integral given by

$$u_1\left(\vec{x}_1\right) = \frac{\exp\left(jkz\right)}{j\lambda} \frac{\exp\left[jk\left|\vec{x}_1\right|^2\right]}{2z} \iint\limits_{S_A} dx_o u_o\left(\vec{x}_o\right) \exp\left[j\frac{2\pi}{\lambda z}\left|\vec{x}_o\right|^2\right] \exp\left[j\frac{2\pi}{\lambda z}\left(\vec{x}_o \cdot \vec{x}_1\right)\right] \tag{2.23}$$

Notice that Equation 2.23 can be written in the compact form

$$u_1\left(\vec{x}_1\right) = \iint\limits_{S_A} dx_o u_o\left(\vec{x}_o\right) h_d\left(\vec{x}_1 - \vec{x}_o\right), \tag{2.24}$$

where

$$h_d\left(\vec{x}_1 - \vec{x}_o\right) = \frac{\exp\left(jkz\right)}{j\lambda z} \exp\left\{j\frac{k}{2z}\left[\left|\vec{x}_1 - \vec{x}_o\right|^2\right]\right\}. \tag{2.25}$$

The impulse response shown in this equation is the free-space impulse response (Roggemann and Welsh 1996; Goodman 2005). We have adopted Goodman's assumptions that the distance z is much larger than the angular extent of the object and the distance off of the optical axis and so the cosine term in Equation 2.20 is approximately one and the r_{o1}

term in Equation 2.21 is approximately equal to z. Equation 2.24 is a 2D spatial convolution integral, and Equation 2.22 can be interpreted as a 2D spatial Fourier transform as follows:

$$u_1\left(\overrightarrow{x_1}\right) = \frac{\exp(jkz)}{j\lambda z} \frac{\exp jk\left|\overrightarrow{x_1}\right|^2}{2z} \Im\left\{u_o\left(\overrightarrow{x_o}\right)\exp\left[j\frac{k}{2z}\left|\overrightarrow{x_o}\right|^2\right]\right\}_{\overrightarrow{f}=\frac{\overrightarrow{x_1}}{\lambda z}}, \tag{2.26}$$

where $\Im$ represents the 2D Fourier transform given by the relation

$$G\left(\overrightarrow{f}\right) = \Im\left[g\left(\overrightarrow{x}\right)\right] = \int\int_{-\infty}^{+\infty} d\overrightarrow{x}\, g\left(\overrightarrow{x}\right)\exp\left(-j2\pi\overrightarrow{f}\cdot\overrightarrow{x}\right), \tag{2.27}$$

and the associated 2D inverse Fourier transform

$$g\left(\overrightarrow{x}\right) = \Im^{-1}\left[G\left(\overrightarrow{f}\right)\right] = \int\int_{-\infty}^{+\infty} d\overrightarrow{f}\, G\left(\overrightarrow{f}\right)\exp\left(j2\pi\overrightarrow{f}\cdot\overrightarrow{x}\right). \tag{2.28}$$

The parameter $\overrightarrow{f}$ is a 2D spatial frequency variable that has units of cycles per meter. We use the convention here and throughout the book that lowercase letters represent parameters in the spatial domain and uppercase letters represent parameters in the frequency domain. Equations 2.23 through 2.28 apply if Equation 2.22 is satisfied, and if $z \gg \lambda$, and if the source aperture maximum spatial extent $\left|\overrightarrow{x_o}\right|_{max} \gg \lambda$. Notice that z cannot be too large; otherwise, the Fresnel condition in Equation 2.22 no longer holds. In such a case, we need to look at our second case of interest, which is the Fraunhofer diffraction formulation of the Rayleigh–Sommerfeld diffraction integral.

In the case of Fraunhofer diffraction, we have an additional assumption on the exponential term inside the Fourier transform operator in Equation 2.26. We need to ensure that this exponential term is practically constant in the object plane aperture. In this case, the argument of the exponential term inside the Fourier transform integral tends to zero and the exponential term itself tends to one. This condition is satisfied when the Fresnel number is much less than one:

$$F = \frac{\left|x_o\right|_{max}^2}{\lambda z} \ll 1. \tag{2.29}$$

Notice that this condition is satisfied if the product of λ and z is much larger than the square of the maximum spatial extent of the source aperture in the object plane. If Equation 2.29 holds, we have the Fraunhofer diffraction integral

$$u_1\left(\overrightarrow{x_1}\right) = \frac{\exp(jkz)}{j\lambda z} \frac{\exp jk\left|\overrightarrow{x_1}\right|^2}{2z} \int\int_{SA} d\overrightarrow{x_o}\, u_o\left(\overrightarrow{x_o}\right)\exp\left[j\frac{2\pi}{\lambda z}\overrightarrow{x_o}\cdot\overrightarrow{x_1}\right] \tag{2.30}$$

that has the compact notation

$$u_1\left(\overrightarrow{x_1}\right) = \frac{\exp(jkz)}{j\lambda z} \frac{\exp\left(jk\left|\overrightarrow{x_1}\right|^2\right)}{2z} F\left\{u_o\left(\overrightarrow{x_o}\right)\right\}_{\overrightarrow{f}=\overrightarrow{x_1}/\lambda z}. \tag{2.31}$$

At this point, we should mention something about extending this monochromatic result to multiple wavelengths, or over a spectrum of wavelengths. For multiple wavelengths, or over a broad range of wavelengths, the results are equally valid at each separate wavelength, and the superposition principle still holds. The resulting field in the observation plane would be the superposition of the separate fields at each wavelength or, by extension, the superposition of the fields across the spectrum of wavelengths.

Equations 2.22, 2.25, 2.26, 2.29, and 2.31 are the main results of this section and will be the basis for our linear systems model in the next section. It is worth repeating that these results originate from Maxwell's equations and require only the Huygens superposition principle to establish linearity. Additionally, in order to be valid, some constraints on aperture size with respect to the wavelength and aperture size with respect to the object plane to observation plane distance z must be met. These constraints are typically easily met for most remote imaging applications. In the next section, we present a linear systems imaging model that is based on these results.

2.2.2 Linear and Nonlinear Systems Models

In the previous section, we presented a short development of how the electromagnetic field in a distant observational plane can be related to the electromagnetic field distribution in the object plane. We showed that if the Fresnel conditions are met, the complex electromagnetic field in the observer plane could be obtained by a 2D spatial convolution of the complex electromagnetic field in the object plane with a shift invariant impulse response. We showed also that for both Fresnel and Fraunhofer imaging conditions, the electromagnetic field in the observer plane is determinable by 2D Fourier transforming the spatially modulated object electromagnetic field, in the case of Fresnel diffraction, or by the 2D Fourier transform of the object plane electromagnetic field itself, as in the case of Fraunhofer diffraction. In essence, the 2D Fourier transform plays prominently in the determination of the observed field for both the Fresnel and Fraunhofer diffraction cases.

In order to determine our linear systems model, we need one additional piece of information. We know that since the object plane field and observation plane field are related through a Fourier transform, then in order to recover the image, we need to determine the inverse Fourier transform (or something somewhat equivalent) of the observed field. This can be done using a thin lens. To see how this is accomplished, we first need to describe the effect that a thin lens has on an optical field. The effect of a thin lens on an incident electromagnetic field can be thought of as multiplication by a transparency function

$$u_t\left(\vec{x}\right) = u_i\left(\vec{x}\right)t_l\left(\vec{x}\right), \tag{2.32}$$

where the transparency function itself is given by

$$t_l\left(\vec{x}\right) = \exp\left(-j\frac{k}{2f}\left|\vec{x}\right|^2\right), \tag{2.33}$$

$$u_i\left(\vec{x}\right) = \int\limits_{-\infty}^{+\infty} d\vec{x}\, u_o\left(\vec{x}_o\right)h\left(\vec{x}-\vec{x}_o\right)d\vec{x}_o. \tag{2.33a}$$

The subscript "i" in Equation 2.32 refers to the electromagnetic field that is *incident* on the thin lens. The vector $\vec{x}$ is a position vector in the plane of the lens, and f is the focal length of the thin lens. The subscript "*t*" represents the transmitted electromagnetic wave just after it has passed through the thin lens. If we now apply this concept to the electromagnetic field in the observer plane given by Equation 2.31 by placing a thin lens in the observer plane, assuming that the electromagnetic field in the observer plane is now the source, and propagating a distance equal to the focal length of the thin lens, it is seen that the resulting image will be a scaled version of the original object in the object plane. Working out the details is a fun exercise for the reader. The effect of the lens is to effectively take the 2D Fourier transform of the field in the observer plane. From now on, we will refer to the previously discussed observer plane as the entrance pupil plane of our imaging system. As a point of reference, for a well-designed telescopic imaging system, the primary mirror of the telescope is placed in the entrance pupil of the imaging system. Working through the details just discussed, we find that the electromagnetic field in the focal plane of the thin lens is, apart from a complex scale factor and diffraction effects, a scaled version of the electromagnetic field in the object plane. In effect, the thin lens cancels out the quadratic phase factor in the entrance pupil and takes the 2D Fourier transform of the result. We, therefore, have a scaled version of the Fourier transform, of the Fourier transform of the object field, in the image plane of the optical system. The effect of the second Fourier transform, as opposed to an inverse Fourier transform, is nothing more than inverting the image of the object in the focal plane. We can generalize the thin lens approach to a collection of arbitrary optical elements by using the systems model shown in Figure 2.10.

The left coordinate system in Figure 2.10 is the plane where the object resides. The first aperture to the right of the object is the entrance pupil of the imaging system. The optics of the optical imaging system are treated as a "black box" and can consist of lenses, relay optics, optical filters, polarizers, beam splitters, and a variety of other beam-shaping and beam-forming optics. We will discuss how to characterize and mathematically describe the optics within the "black box" in Chapter 7. The next point of interest in Figure 2.10 is the exit pupil of the imaging system. The distance from the exit pupil to the image plane is the effective focal length f_{eff}. Notice that we have dropped the subscripts distinguishing the object plane from the entrance pupil plane and the image plane. The reason for this is that, in an aberration-free environment, the image is a scaled version of the object, and so the relative image scaling issues that are addressed by the optical systems transverse and lateral magnification are typically addressed separately from the image formation process. For convenience, we then use the same coordinate system in all three planes, object, entrance pupil, and image plane, and separately address magnification. We will make it clear by the context what plane is being discussed. We address magnification issues in Chapter 4.

We now have sufficient information to present our linear, shift invariant systems model. The linearity in our model results from the superposition of the Huygens wavelets and

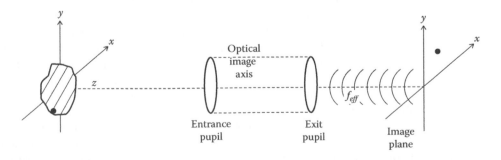

FIGURE 2.10
Optical systems model.

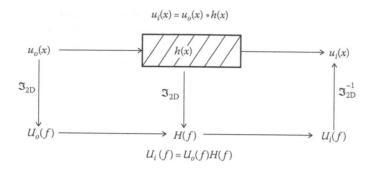

FIGURE 2.11
Linear systems model.

the shift invariance results from the geometrical simplifications resulting from the Fresnel or Fraunhofer approximations. If the shift invariance assumption does not hold, then we would have to use Equation 2.19 as the governing relationship between the object plane electromagnetic field and the observation plane electromagnetic field with Equation 2.20 as the impulse response. Luckily, the shift invariance assumption is valid for a large number of practical imaging applications, and so we can benefit from some mathematical simplifications in both the spatial domain and in the spatial frequency domain. In this section, we develop two types of linear, shift invariant models that have to do with the spatial coherence properties of the electromagnetic field. If the electromagnetic field has a high degree of spatial coherence, as is the case for a laser, we will adopt a model that is linear and shift invariant at the field level. On the other hand, if the electromagnetic field is highly incoherent, as is the case for emitted or reflected electromagnetic waves from optically rough surfaces, then we adopt a model that is linear and shift invariant at the optical power level (e.g., magnitude squared of the electromagnetic field). We first start with the assumption that the electromagnetic field is highly spatially coherent. In this case, the linear, shift invariant imaging model that we will use is illustrated in Figure 2.11.

The relationship between the electromagnetic field in the object plane $u_o(\vec{x})$ and that in the image plane $u_i(\vec{x})$ is given by a 2D spatial convolution of the object electromagnetic field with the impulse response of the optical imaging system $h(\vec{x})$. The asterisk represents convolution, $\Im_{2D}$ represents the 2D Fourier transform operator, $\Im_{2D}^1$ is the 2D inverse Fourier transform, and $H(\vec{f})$ is the CTF. Notice that these relationships hold true at every spatial point $\vec{x}$ and every spatial frequency $\vec{f}$. A physical example wherein this model applies would be a laser-illuminated circular mirror wherein the reflected field is detected at some far distant point (also in space) that satisfies the Fresnel or Fraunhofer criterion previously specified. In this case, the reflected field would be highly coherent and bounded by the finite extent of the mirror. Let us say that the detector is within the Fresnel region of the reflected electromagnetic wave, and no lens is used (the detector is placed directly in the entrance pupil instead of the image plane); then, the impulse response would be given by Equation 2.25, and the electromagnetic field in the entrance pupil can be obtained by either using Equation 2.24 directly or solving the problem in frequency space and then taking the 2D inverse Fourier transform of the result. This latter approach is shown at the bottom of Figure 2.11. The computational burden is typically much less cumbersome by taking the Fourier transform and solving in frequency space than by solving the problem in the spatial domain. However, with the advent of parallel processing hardware architectures, especially 3D hardware architectures and clever image processing algorithms, this may not always be universally true (Arrasmith 2010). Extending our simple example, let us say that the illuminating laser light is constant and the reference phase is zero everywhere. Then, the spatial component of the reflected field is modeled by

$$u_o(\vec{x}) = A\,\mathrm{circ}\left(\frac{|\vec{x}|}{\rho}\right). \tag{2.34}$$

The circ(ax) function is defined as 1 where the quantity ax is less than 1 and zero elsewhere. In effect, we have a constant value A within the clear aperture of a circular mirror with radius ρ. Equation 2.25 gives the corresponding free-space impulse response. This impulse response has a Fourier transform given by (Goodman 2005; Roggemann and Welsh 1996)

$$H_d(\vec{f}) = \begin{cases} \exp(jkz)\exp\left[-j\pi\lambda z|\vec{f}|^2\right], & |\vec{f}| < \dfrac{1}{\lambda} \\ 0, & \text{elsewhere} \end{cases}. \tag{2.35}$$

The "d" subscript denotes that this is the CTF for free-space diffraction. The Fourier transform of Equation 2.34 is seen from Fourier transform tables to be

$$U_o(\vec{f}) = A\frac{\rho^2}{(\lambda z)^2}\frac{J_1\left(2\pi\rho|\vec{f}|/\lambda z\right)}{\left(\rho|\vec{f}|/\lambda z\right)}, \tag{2.36}$$

where the symbol J_1 means the Bessel function of the first kind and of order 1. The field spectrum in the entrance pupil plane is obtained by pointwise multiplying Equation 2.36 by Equation 2.35 everywhere within the clear aperture of the detector in the entrance pupil and then inverse Fourier transforming the result. This procedure is much simpler computationally than calculating the 2D spatial convolution using Equations 2.24, 2.25, and 2.34.

We now want to consider the case for imaging a coherent electromagnetic field. In this case, the detector is placed in the image plane and not in the entrance pupil plane. By making use of the lens function in Equation 2.33 and ensuring that we have paraxial imaging conditions, we can determine the impulse response of the coherent imaging system as (Goodman 2005)

$$h(\vec{x}) = \int\!\!\!\int_{-\infty}^{+\infty} d\vec{f}\,W\left(\vec{f}\lambda f_{eff}\right)\exp\left[-2\pi\vec{f}\cdot\vec{x}\right] \tag{2.37}$$

or

$$h(\vec{x}) = \mathcal{F}\left\{W\left(\vec{f}\lambda f_{eff}\right)\right\}. \tag{2.38}$$

The function $W(\vec{f}\lambda f_{eff})$ is the pupil function and, for the coherent imaging case, takes on the value of 1 inside the clear aperture of the imaging system's entrance pupil and zero elsewhere. The impulse response is then the 2D Fourier transform of the pupil function. The parameter f_{eff} is the effective focal length and is given by the distance from the imaging system's exit pupil to the focal plane. Equation 2.38 is very useful. If we know the shape of the entrance pupil aperture (e.g., the primary mirror of an imaging system's telescope), then we can determine the pupil function by scaling the spatial coordinates $\vec{x}$ in the entrance pupil plane by

$$\vec{x} = \vec{f}\lambda f_{eff}. \tag{2.39}$$

We can then determine the impulse response for the imaging system by simply taking the 2D Fourier transform of the result. As an example, many telescopic imaging systems have a circular primary mirror. As before, we can describe the entrance pupil using the circ (ax) function and then scale the entrance pupil coordinates appropriately according to Equation 2.39:

$$H\left(\vec{f}\right) = \text{circ}\left(\frac{\left|\vec{f}\right|\lambda f_{eff}}{\rho_{ep}}\right). \tag{2.40}$$

This scaled entrance pupil function is the CTF. The quantity ρ_{ep} is the radius of the aperture in the entrance pupil. Since the CTF has radial symmetry across the circular pupil, we show the spatial frequency coordinate as a magnitude. The impulse response is found by taking the 2D inverse Fourier transform of Equation 2.40:

$$h\left(\vec{x}\right) = \frac{\rho_{ep}^2}{\left(\lambda f_{eff}\right)^2} J_1 \frac{\left(\left(2\pi\rho_{ep}^2\left|\vec{x}\right|\right)\big/\left(\lambda f_{eff}\right)\right)}{\left(\left(\rho_{ep}^2\left|\vec{x}\right|\right)\big/\left(\lambda f_{eff}\right)\right)}. \tag{2.41}$$

This equation is, as before, a Bessel function of the first kind with order 1. Notice that if we had a rectangular entrance pupil aperture instead of a circular aperture, then the CTF would be

$$H\left(f_x, f_y\right) = \text{rect}\left(\frac{f_x\lambda f_{eff}}{D_x}\right)\text{rect}\left(\frac{f_x\lambda f_{eff}}{D_y}\right), \tag{2.42}$$

where
  f_x and f_y are spatial frequency components in the x and y directions, respectively
  D_x and D_y are the respective diameters of the rectangular entrance pupil aperture in the
    x and y directions

The function rect (x/b) is defined as 1 for $-b/2 < x < b/2$ and 0 elsewhere. As mentioned earlier, the impulse response is found by taking the 2D inverse Fourier transform and is given by

$$h\left(x, y\right) = \frac{D_x D_y}{\left(\lambda f_{eff}\right)^2} \text{sinc}\left(\frac{x D_x}{\lambda f_{eff}}\right)\text{sinc}\left(\frac{y D_y}{\lambda f_{eff}}\right). \tag{2.43}$$

The sinc (ax) function is given by sin (ax)/ax, and the $x = 0$ point is defined as 1. We will now provide a quick example to show the utility of our modeling system shown in Figure 2.10. Let us assume that we have a highly polished, mirrorlike, circular surface that is illuminated by a laser beam. As before, we can model the source term (e.g., the mirror) with a "circ" function with amplitude "*A*" as in Equation 2.34. The Fourier transform of our electromagnetic source field is the electromagnetic field spectrum and is given by Equation 2.36. Equation 2.40 gives the CTF for an imaging system with a circular entrance pupil with radius ρ_{ep}. The product of the electromagnetic field spectrum and the CTF gives the electromagnetic field spectrum in the image plane. The 2D inverse Fourier transform gives the electromagnetic field in the image plane, and the magnitude squared of the image plane electromagnetic field gives the irradiance in the image plane.

We will now take a look at the opposite of the coherent imaging case, incoherent imaging. For incoherent imaging systems, we assume that the object is spatially incoherent that means that neighboring points further apart from each other than an optical wavelength do not constructively interfere with each other. The effect of this assumption is that a 2D spatial autocorrelation of the object would be nonzero only for an offset of zero, a direct overlap of the object with itself. This of course is strictly speaking not true, but for most man-made objects, this assumption works very well. The reason for this validity is that most man-made objects, which are not highly polished, tend to be rough on the scale of an optical wavelength, and so neighboring points on the object's surface do not constructively interfere with each other. We will make good use of this fact in determining the mathematical model that governs the relationship between the object's brightness (watts/m²) and the irradiance (watts/m²) in the image plane. To determine this relationship, we must recall that brightness in the object plane is proportional to the magnitude squared of the electromagnetic field in the object plane:

$$i_o\left(\vec{x}\right) = u_o\left(\vec{x}\right)u_o^*\left(\vec{x}\right). \tag{2.44}$$

We can relate the object brightness to the image plane irradiance by making use of Equation 2.33a and substituting into Equation 2.44. Taking the expected value of the result, we get the following relationship (Roggemann and Welsh 1996):

$$\left\langle i(\vec{x}) \right\rangle = \int\limits_{-\infty}^{\infty} d\vec{x}'_o \int\limits_{-\infty}^{\infty} d\vec{x}''_o \left\langle u_o\left(\vec{x}'\right)u_o^*\left(\vec{x}''\right)\right\rangle h_i\left(\vec{x}-\vec{x}'_o\right)h_i^*\left(\vec{x}-\vec{x}''_o\right). \tag{2.45}$$

The expected value produces the average irradiance in the image plane. At this stage, we make use of the incoherent imaging assumption. As previously stated, for incoherent imaging conditions, the only nonzero contributing factor to the expectation is when the fields overlap. Mathematically, we can write this as (Goodman 2005)

$$\left\langle u_0\left(\vec{x}'_0\right)u_0^*\left(\vec{x}''\right)\right\rangle = k\left\langle o\left(\vec{x}'_0\right)\right\rangle \delta\left(\vec{x}'_0 - \vec{x}''_0\right). \tag{2.46}$$

The factor *k* is a proportionality constant that provides equivalence between both sides of this equation. The symbol δ is the Dirac delta function. For this equation, we have annotated the integration variables to show clearly in what plane the integration takes place. Substituting Equation 2.46 into Equation 2.45 and carrying out the integrals, we find that

$$i(\vec{x}) = \int\limits_{-\infty}^{+\infty} d\vec{x}_0 o(\vec{x}_0)\left|h_i\left(\vec{x}-\vec{x}_0\right)\right|^2 \tag{2.47}$$

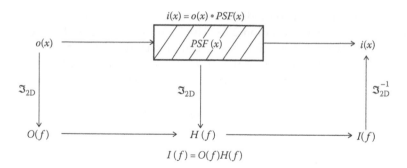

FIGURE 2.12
Incoherent imaging model.

or simply written as

$$i(\vec{x}) = o(\vec{x}) * \left| h_i(\vec{x} - \vec{x}_0) \right|^2. \tag{2.48}$$

We have dropped the expectation values on the left side of the equation for convenience and with the understanding that, unless otherwise stated, we are referring to the average quantities, in this case, the average image irradiance on the left side of Equation 2.47. Likewise, the object brightness within the integral is also an averaged quantity. It is interesting to note that this equation has the same form, 2D spatial convolution, as what we had in the coherent imaging case. The difference is that, for the coherent imaging case, the 2D spatial convolution was at the field level, whereas for the incoherent imaging case, the 2D convolution is at the magnitude squared of the electromagnetic field. At this point, we can present in Figure 2.12 our model for the incoherent imaging case.

Just as before, there is a strong connection with the 2D Fourier transform and its inverse. At the top left of Figure 2.12, we have the object brightness denoted by a lowercase "o." The average irradiance in the image plane is obtained by convolving the PSF with the object brightness. Similar to before, a conventionally, more computationally efficient approach to determining the average image plane irradiance is to first take the 2D Fourier transform of the object brightness to obtain the object brightness spectrum. Then, multiply the object brightness spectrum by the OTF to obtain the image spectrum. As before in the coherent imaging case, we take the inverse Fourier transform of the image spectrum to obtain the average image plane irradiance. Modeling the propagation physics of the object brightness in the spatial frequency domain is most often computationally more efficient; however, the advent and further development of high-speed, parallel processing devices, such as field programmable gated arrays, 3D neural network architectures (Irvine Sensors Corporation n.d.), cellular neural network chips (Arrasmith 2010), and advancements in high-speed, parallel processing, image processing, and atmospheric turbulence compensation methods, offer promising potential in up-and-coming high-speed imaging solutions in both the spatial and frequency domains.

The incoherent imaging model, as depicted in Figure 2.12, applies to a wide variety of *passive* imaging scenarios and applications. By passive imaging, we mean imaging that occurs under natural conditions such as ambient light or natural emissions when operating in the infrared part of the electromagnetic spectrum. In contrast, an *active* imaging system is one in which man-made sources such as lasers and/or radars are used to illuminate the target of interest (TOI). If a high, spatially coherent source is used to illuminate an *optically flat*, smooth on the scale of the optical wavelength, TOI, then the coherent imaging model shown in Figure 2.12 should be used to model the basic propagation and detection physics of the imaging system. An example of this scenario would be a spatially coherent laser that illuminates a mirrorlike surface and is detected in some distant imaging plane under Fresnel or Fraunhofer conditions. It should be noted that in the event a source that has a high degree of spatial coherence illuminates an object that is rough on the scale of an optical wavelength, the reflected light will be spatially incoherent, regardless of the source, and a mixed model should be employed (a coherent propagation model from source to target and an incoherent imaging model for the reflected brightness from the target to the imaging plane).

For the incoherent imaging scenario, since the surface granularity of the TOI is often rough on the scale of the optical wavelength, or the illuminating source itself is spatially incoherent, the incoherent imaging model works exceedingly well in describing the essential characteristics of passive imaging systems or even actively illuminated objects that are rough on the scale of the illuminating wavelength. We will now describe some essential aspects of the incoherent imaging system that are needed in modeling and understanding the performance of the optical imaging system.

Just as in the coherent imaging case, the entrance pupil plays a dominant role in modeling imaging system and atmospheric turbulence effects. The imaging model depicted in Figure 2.12 still applies for the incoherent imaging case. Recall that for many well-designed optical, telescopic imaging systems, the primary mirror of the telescope is placed in the entrance pupil of the imaging system. Due to its importance and its connection to other essential elements of the optical system, we will develop a separate model for the entrance pupil plane of the imaging system. Figure 2.13 provides an illustration of our assumed circular entrance pupil. We use a circular entrance pupil in this illustration since most telescopes have circularly shaped primary mirrors. Although the real and imaginary parts of

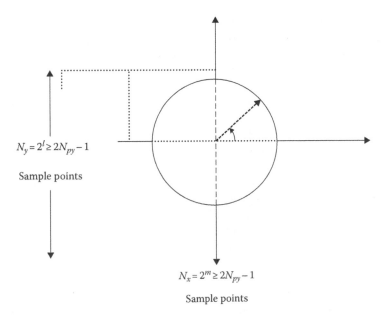

FIGURE 2.13
Sampled circular entrance pupil model.

the electromagnetic wave are continuous within the telescope entrance pupil, to properly model the imaging scenario on a computer, the electromagnetic field within the imaging system's entrance pupil must be correctly spatially (and temporally) sampled. We focus on the spatial sampling requirements for the moment and bring the temporal aspects in the statistical optics section later. The discrete points in the entrance pupil aperture are sample points, in the $\hat{x}$ and $\hat{y}$ directions, which are chosen to obey Nyquist sampling conditions. We will discuss Nyquist sampling conditions further in this chapter.

We are keeping with the convention in dropping the subscripts on the position variables and identifying the proper plane, in this case the entrance pupil, within the context of the dialog supporting the figure. The sampling shown in Figure 2.13 is uniform in both spatial directions. This is not a limiting restriction; however, it is convenient, and so we will continue with this entrance pupil uniform spatial sampling model unless otherwise needed. We have also found it to be convenient to use an odd number of points across the entrance pupil clear aperture so that both the physical center point of the aperture and the aperture edge points along the main coordinate axes coincide with a sample point. In our model, we embed this sampled aperture in a matrix of zeros that has at least $(2N_{px} - 1)$ points in the $\hat{x}$ direction and $(2N_{py} - 1)$ points in the $\hat{y}$ direction. Additionally, we increase the number of sample points in each linear direction to the smallest power of 2 that is larger than the preceding sampling requirements to aid in the efficiency of doing the 2D Fourier transform. In other words,

$$N_y = 2^l, \quad l = \min\{l = 1, 2, 3 \ldots\} \ni 2^l \geq 2N_{py} - 1 \tag{2.49}$$

and

$$N_x = 2^m, \quad m = \min\{m = 1, 2, 3 \ldots\} \ni 2^m \geq 2N_{px} - 1 \tag{2.50}$$

where
  N_{px} and N_{py} are the number of sample points along the diameter in the clear aperture of the entrance pupil in the $\hat{x}$ and $\hat{y}$ directions, respectively
  l and m are integers that are chosen to be as small as possible and that satisfy the right side of Equations 2.49 and 2.50, respectively

N_x and N_y are then the respective total number of entrance pupil sample points in a given linear direction in our model. For example, say that we had 127 sample points across the diameter in the $\hat{x}$ direction. The center point corresponding to $x = 0$ is the 64th point. There are an equal number of points to the left and right of the center point, 63 points. The 1st and 127th point would rest on the edge of the clear aperture. Since we are using uniform sampling, the $\hat{y}$ direction would be identical to the $\hat{x}$ direction for our circular aperture. The number of sample points along the diameter within the clear aperture, in the $\hat{x}$ and $\hat{y}$ directions, respectively, is $N_{px} = N_{py} = 127$. The minimum number of required points in our model is $(2N_{px} - 1) = 253$ points in the $\hat{x}$ direction and $(2N_{py} - 1) = 253$ points in the $\hat{y}$ direction. The smallest l and m that satisfy Equations 2.49 and 2.50 are l and $m = 8$, leading to $N_x = N_y = 256$. To model our entrance pupil aperture, we would have a 256 × 256 array that is 0 everywhere except within a circular aperture that has 127 points along the $\hat{x}$ and $\hat{y}$ diameter. The circular aperture is centered within the array with the center point located on the (129, 129) sample point measured from the top left of the array. Note that if we assign a length to the diameter of the circular entrance pupil diameter, then we can determine sample spacing Δx and Δy. For example, if we had a telescope with a 20 cm diameter primary mirror, then Δx and Δy would be $20/(129 - 1) = 0.15625$ cm. Within the clear aperture of the entrance pupil, we define a GPF as

$$W(\vec{x}) = W_p(\vec{x}) e^{j\Phi(\vec{x})}, \tag{2.51}$$

where

$$W_p(\vec{x}) = \begin{cases} 1 & r(\vec{x}) \leq D/2 \\ 0 & \text{e.w.} \end{cases}, \tag{2.52}$$

where Equation 2.52 is an aperture function that describes the physical aperture of the entrance pupil, in this case 1 inside our circular aperture of diameter D meters. It should be mentioned that the physical aperture can take on values other than 1 for physical situations that have strong amplitude effects, such as in scintillation, but for a broad range of imaging conditions, the constant amplitude assumption is acceptable. If it becomes necessary to include amplitude effects in our imaging model, we would just replace the "1" on the right side of Equation 2.52 with a spatially varying amplitude. The complex exponential in Equation 2.51 is used to describe entrance pupil phase aberrations to include phase aberrations due to the atmosphere. The GPF is shown as a continuous function of spatial position vector $\vec{x}$ but computationally is evaluated at the discrete sample points shown in Figure 2.13. As will be discussed later in the section discussing spatial sampling, if proper sampling conditions are established, it can be shown that an exact copy of the continuous function can be reproduced from the sampled data. For now, we will show the continuous variable $\vec{x}$ but understand that when implemented on the computer, sampled points of our GPF, as well as other functions, will be used to represent their continuous counterparts. We now want to develop some fundamentally important optical SE functions to help characterize and understand our optical system's performance. As will be seen, our approach has many similarities and parallels with the coherent imaging case mentioned earlier.

If we make the variable substitution from Equation 2.39 into Equations 2.50 and 2.51 and take the 2D Fourier transform of our GPF just as in Equation 2.38, we obtain the impulse response for our optical system. By then taking the magnitude squared of the resulting impulse response, we get the PSF for our incoherent imaging system:

$$s(\vec{x}) = \left| h(\vec{x}) \right|^2 = \left| \mathcal{F}\left\{ W\left(\vec{f}\lambda f_{\text{eff}} \right) \right\} \right|^2. \tag{2.53}$$

The PSF describes how the diffraction, atmospheric effects, and system effects "spread" the light across the detector surface in the image plane and thereby decrease the spatial resolution of the imaging system. Even if we had a perfect optical system with no aberrations and we were imaging in a vacuum (no atmospheric aberrations), the diffractive effects of the entrance pupil aperture would still impact our imaging system to some degree. Superresolution methods would need to be employed to potentially mitigate the spatial resolution loss due to diffraction. For most optical systems in the earth's atmosphere, optical systems aberrations and atmospheric turbulence are the dominant effects in the imaging systems loss of spatial resolution. For well-designed optical systems, aberrations due to the optical system itself are minimized, and atmospheric turbulence effects dominate the loss in spatial resolution of the imaging system.

If we take the 2D Fourier transform of the PSF and normalize the result so that the maximum value of the resulting function is 1, for example, by dividing the result of the 2D Fourier transform by its maximum, we obtain the OTF

$$H(\vec{f}) = \frac{\mathcal{F}\{s(\vec{\vec{x}})\}}{\mathcal{F}\{s(\vec{x})\}\big|_{\vec{f}=0}}.$$ (2.54)

The maximum value of the 2D Fourier transform of the PSF is always the zero spatial frequency component of the resulting 2D Fourier transformed function. Another way that the OTF can be determined is by making use of two convenient properties of the Fourier transform. The first property is that the Fourier transform of the product of two functions is the same thing as the spatial convolution of the Fourier transform of each function. The second property is that the complex conjugation of a function has the effect of time/space reversal in the Fourier-transformed space. Combining these two properties and noting that the PSF is the product of the optical systems impulse response with the conjugate of the impulse response, we see that the OTF can also be obtained by an autocorrelation of the GPF:

$$H(\vec{f}) = \frac{\mathcal{F}\{h(\vec{x})h^*(\vec{x})\}}{\mathcal{F}\{h(\vec{x})h^*(\vec{x})\}\big|_{\vec{f}=0}} = \frac{W(\vec{f}\lambda f_{eff}) \otimes W(\vec{f}\lambda f_{eff})}{W(0) \otimes W(0)},$$ (2.55)

where the symbol $\otimes$ means 2D autocorrelation. The method of determining the OTF leading up to Equation 2.54 is typically preferred over Equation 2.55 since the former approach tends to be computationally faster than Equation 2.55 for most conventional applications. From Figure 2.5, and using Equation 2.54 or 2.55, we can get the image spectrum by

$$I(\vec{f}) = O(\vec{f})H(\vec{f}).$$ (2.56)

The image is then obtained by taking the 2D inverse Fourier transform of the image spectrum:

$$i(\vec{x}) = \mathcal{F}^{-1}\{I(\vec{f})\}.$$ (2.57)

We now provide a simple example to illustrate this approach. Let us say that we have a telescopic imaging system that has a diameter of 1 m and an effective focal length f_{eff} of 1 m. Say we want to model the imaging systems performance when it is looking at a binary star system, one with twice the magnitude of the other star. Because of the geometries involved, our earthbound telescope cannot resolve the individual stars themselves, so we can represent them in our model as delta functions. We normalize one star to have a maximum brightness of 1 W/m^2 and the other star to a relative brightness of 0.5 W/m^2. Figure 2.14 shows a MATLAB® model of our TOI.

Experimentally, the PSF can be obtained by looking at a single star. It can be seen from Equation 2.47 that if the object is a point source, we recover the PSF. This PSF would

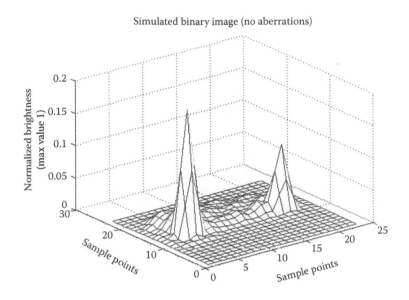

FIGURE 2.14
Object model of a visible binary star system such as α-Centauri A and α-Centauri B.

incorporate the effects of the atmosphere, the entrance pupil aperture, and any optical system effects. If the imaging system were to observe a distant point source object, the effect in the image plane would be to spread the irradiance due to the point source, hence the name PSF. For now, let us assume that we want to determine the effects of the entrance pupil aperture by itself without any atmospheric turbulence or optical system effects. A physical example would be looking at a binary star system with a telescope outside of the earth's atmosphere. Let us also assume, as we have been so far, that we can ignore temporal effects; we will model temporal effects later. Under these assumptions, our modeling results would give us an upper bound on the expected conventional spatial resolution performance of our optical system. As an example, if we use uniform entrance pupil sampling of 127 points across the $\hat{x}$ and $\hat{y}$ directions, then, as previously discussed, we have $N_x = N_y = 256$ samples in our entrance pupil model. Figure 2.15 shows the model for our entrance pupil plane aperture. Notice that by setting the phase term to zero everywhere within the aperture, we effectively eliminate optical system phase effects and atmospheric turbulence effects from our model. By temporarily ignoring system and signal noises, we can isolate the effect of the entrance pupil aperture in our model. The GPF only has real values when the phase is identically zero everywhere within the clear aperture of the entrance pupil. The approximation of a constant phase throughout the entrance pupil only applies to a well-designed optical system that is imaging through a vacuum, like space. In the presence of atmospheric turbulence, the constant entrance pupil plane phase assumption is not valid and the GPF is a complex quantity. System noise and atmospheric turbulence effects can readily be included in our model, but, for now, we want to isolate the effects of the entrance pupil aperture on our imaging system. Notice in Figure 2.15 that our GPF values are nonzero only within the clear aperture of the entrance pupil itself. The values outside of the entrance pupil clear aperture are set to zero since only the irradiance within the clear aperture of the entrance pupil contributes to the formation of the image. Notice also that the maximum value of GPF is one since we have assumed no amplitude effects are present in our imaging system.

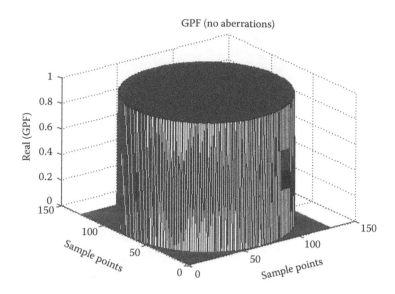

FIGURE 2.15
Model of GPF for no system and atmospheric aberrations.

Taking the 2D Fourier transform according to Equation 2.38, we obtain the impulse response of the optical system. Taking the pointwise magnitude squared of the impulse response gives us the PSF. In the spatial domain, the PSF is an indicator of our imaging system's loss of resolution. For no aberrations present in the imaging system and an infinitely large entrance pupil aperture, the PSF would approach a delta function and, according to Equation 2.47, would exactly replicate the object brightness. In the absence of aberrations in our imaging system, the presence of an entrance pupil with finite spatial extent would effectively spread the received irradiance in the image plane. Therefore, apart from superresolution methods, even with a perfect imaging system with no aberrations or system noises, we cannot exactly reproduce the object brightness because of the presence of the entrance pupil with finite extent. Said another way, the entrance pupil acts as a low-pass filter that effectively removes higher spatial frequency content from our image. Taking the 2D Fourier transform of the PSF and then normalizing by its maximum value yield the OTF.

We now have all the elements necessary to model the performance of our optical imaging system that is viewing a representative, visible binary star system like α-Centauri A and α-Centauri B. The combination of α-Centauri A and α-Centauri B—α-Centauri AB—appears as a single star to the naked eye but are easily resolved into a binary star system with a small telescope. Together, they are the third brightest objects among stars, when excluding our sun. We model each of the binaries as scaled Kronecker delta functions to represent point source like objects with differing brightness and a variable separation between 2 and 22 as. The brightness for α-Centauri A is about 3.038 times that of α-Centauri B, so we can normalize the brightness of α-Centauri A to 1 and α-Centauri B to approximately 1/3. We assume that the telescope is aligned to the center of the binary star system and that the image of the stars lies on the x-axis. We assume that the binaries are separated from each other by 2 as (equivalently 9.69 μrad). The sample point spacing in the object's spatial domain is 0.05 as, and the spatial sampling in the entrance pupil plane is 0.0794 m. We use 500 nm as the center wavelength of our narrowband,

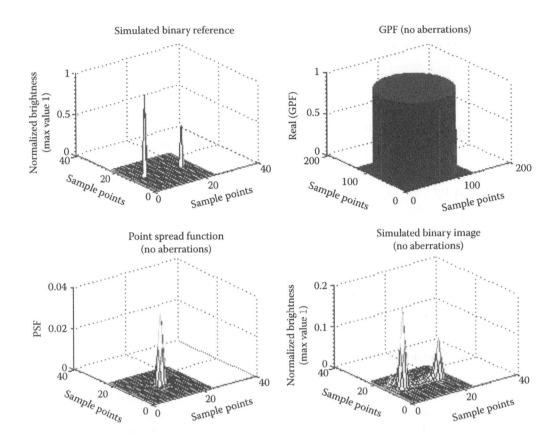

FIGURE 2.16
Simulated binary star system.

filtered, imaging system. Figure 2.16 shows the results for our diffraction-limited optical imaging systems model with a 1 m circular entrance pupil. The modeling was done in MATLAB R2012a.

The top-left figure is a plot of our binary star system wherein the maximum brightness has been normalized to one. The brighter point source object (star) on the left represents α-Centauri A, and the dimmer object on the right represents α-Centauri B. The top-right figure is that of our generalized pupil plane assuming no system noises or atmospheric aberrations. The bottom-left figure is that of our PSF obtained by implementing Equations 2.51 through 2.53. The image obtained by our diffraction-limited imaging system is shown on the bottom right. This simulated image was obtained by first determining the OTF in accordance with Equation 2.54. We then solve for the image by implementing Equations 2.56 and 2.57. We can see that the image is a blurred version of the original object even though we have assumed a noise-free imaging system with no aberrations. If we multiply the image irradiance that falls on the detector by the detector area, we end up with the optical power on the detector. The 2D Fourier transform of the optical "power" gives the *power spectrum* of the image. Likewise, the 2D Fourier transform of the product of the object brightness with the object area produces the power spectrum of the object. In determining the *energy spectrum* of an object or image, the optical power must first be converted into energy units by integrating the optical power over an appropriate time scale. If the total energy spectrum is desired, the power spectrum must be integrated with respect to time from minus to plus infinity. This can only be done if the integrated energy over time is finite. Mathematically, we can define the energy spectral density as

$$\varepsilon\left(\vec{f}\right) = \left|u\left(\vec{f}\right)\right|^2 \tag{2.58}$$

that is only applicable if the temporal integral of the magnitude of the electromagnetic field is finite when integrated from minus infinity to plus infinity (Goodman 1985). If the integrated energy is not finite, then it may be possible to integrate the optical power over a shorter period T and obtain a finite result. If this can be done, we can determine the power spectral density by

$$P\left(\vec{f}\right) = \frac{\left|u\left(\vec{f}\right)\right|^2}{T}. \tag{2.59}$$

Another useful metric is the MTF. This is the magnitude of the OTF and describes the magnitude effects of the OTF on the object spectrum at every applicable spatial frequency. Figure 2.17 shows the MTF for the imaging system viewing our binary star system.

2.2.3 Sampling Considerations

To properly represent a spatially continuous object on a computer, the object is often spatially sampled with the assumption of continuity between sample points. This idea is nothing new to most people. Anyone who has watched television (TV) "sees" a continuous picture but knows that the picture is made up of lots of small picture elements (pixels) that each put out a sampled brightness. Move inches away from the TV display, and often, individual pixels can be seen with the naked eye. However, move a short distance away from the screen, and the picture looks smooth and continuous (for a properly working, good TV). In the previous example of the binary star system, we provided a

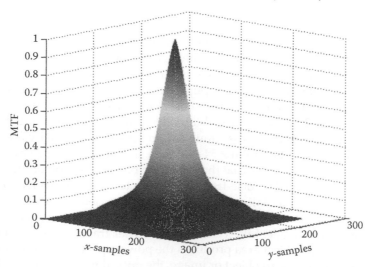

FIGURE 2.17
MTF of imaging system viewing the binary star system.

sampling scheme to facilitate our illustration. But how is the sample spacing determined? To properly use the linear systems methods presented in the previous section, it is necessary to understand spatial and temporal sampling methods. In this section, we provide a short discussion on basic sampling considerations. We will provide both the theoretical sampling requirements dictated by mathematics and also some practical sampling guidelines dictated by experience.

A good point to start in a discussion about spatial sampling is to recall that the relationship between a distant source object and the entrance pupil plane electromagnetic field is often expressible as a 2D Fourier transform relationship. Equation 1.7 gives the diffraction-limited spatial resolution of the optical imaging system for a circular entrance pupil. Conventional telescopes cannot see spatial details smaller than the diffraction-limited resolution. Additionally, it takes a well-designed optical system to provide spatial resolutions that approach the diffraction limit. Consequently, for spatially resolvable objects, the spatial sampling in the object plane, obtained by projecting the dimensions of the imaging system's pixels onto the object, should not exceed the diffraction-limited resolution of the telescope. A similar argument can be made when imagining viewing the imaging system's entrance pupil from the perspective of the object plane. A question might be as follows: what sample spacing in the entrance pupil plane is needed to resolve spatial details due to an object that is contained within a circle of diameter D_o or a rectangle with dimensions D_{ox} and D_{oy} in meters? Since the relationship between the two planes is still a Fourier transform relationship, the required sample spacing in the entrance pupil must be smaller than

$$\Delta X_p = 1.22 \frac{\lambda}{D_o} z \tag{2.60}$$

for the circular case and

$$\Delta X_{p_x} = \frac{\lambda}{D_{ox}} z, \quad \Delta X_{p_y} = \frac{\lambda}{D_{oy}} z \tag{2.61}$$

for the rectangular case. A practical rule of thumb is to have at least two samples in the entrance pupil per required sample spacing given by Equations 2.60 and 2.61. The value D_o is the diameter in meters of a circle that circumscribes the object. The values D_{ox} and D_{oy} are the diameters of a rectangle that just contains the object. The intent is to provide a simple object such as a square, rectangle, or circle that just contains the object. In this manner, the sampling requirements in either the object plane or the entrance pupil plane can be set up, so ensure that available spatial information is captured by the optical imaging system. Because of the nature of the 2D Fourier transform that relates the electromagnetic field from the object plane to the entrance pupil plane, it turns out that the largest spatial dimension of the object to be imaged determines the sampling requirements in the entrance pupil plane (see Equations 2.60 and 2.61) and the largest spatial dimension of the entrance pupil determines the smallest sampling requirements in the object space (see Equation 1.7).

A more formal treatment of sampling involves understanding the Whittaker and Shannon sampling theorem (Goodman 2005). It can be shown that for a band-limited, continuous 2D function, $g(x, y)$, the sampled function

$$g_s(x,y) = \text{comb}\left(\frac{x}{X}\right)\text{comb}\left(\frac{y}{Y}\right)g(x,y) \tag{2.62}$$

where the "comb" functions are given by

$$\text{comb}\left(\frac{X}{T}\right) = \sum_{n=-\infty}^{\infty} \delta\left(X - \frac{n}{T}\right). \tag{2.63}$$

The Kronecker delta function on the right side of this equation has the value 1 at sample separations of T units. This effectively looks like an infinitely long "comb" that has "teeth" with a height of 1 and teeth separation of T. For a 1D signal with finite extent, the comb function in Equation 2.63 is multiplied by a "rect" function:

$$\text{rect}\left(\frac{X}{L}\right) = \begin{cases} 1, & |X| \le \dfrac{L}{2} \\ 0, & \text{e.w.} \end{cases} \tag{2.64}$$

In two dimensions, each "comb" function can be multiplied by its own "rect" function to define the maximum extent of the object. In this way, different sampling in the x and y directions can be accomplished if desired. For uniform sampling, the parameters X and Y in Equation 2.62 are set equal to each other. Figure 2.18 shows the case of a 2D, continuous function, $g(x, y)$, being sampled in the x and y directions with sample spacing of X and Y, respectively.

As can be seen, lower spatial frequencies of the object spectrum are affected less than higher spatial frequencies. Since the OTF pointwise multiplies the object spectrum, the effect of the OTF is to attenuate the higher spatial frequency information in the object spectrum. This higher spatial frequency information contains the fine details in the image. The MTF shows this attenuation and so is a good metric for image quality.

Before leaving this section, we should say a few words about nonlinear optical systems models. The field of nonlinear optics (NLO) is quite rich, interesting, and beyond the scope of this book to treat. Topics such as optical phase conjugation, three-wave and four-wave mixing, second harmonic generation, and optical parametric amplification are examples of nonlinear optical methods that produce intriguing results (Boyd 2008). In NLO, the superposition principle no longer holds, and our linear systems models just mentioned break down. Often, nonlinear optical processes occur in materials where the material properties

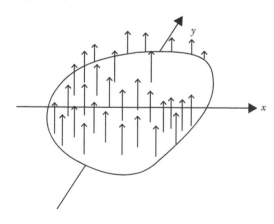

FIGURE 2.18
Two-dimensional, continuous functions.

vary as function of the strength of the electromagnetic field. Typically, very high electric field strengths are required to generate the NLO effect. We consider NLO an advanced and specialized topic that is well treated in the following references (Shen 1998; Anderson 2014). For most practical optical imaging and optical detection scenarios, our linear systems model is sufficiently accurate. However, if high-power lasers interact with nonlinear material such as a barium borate (BBO) crystal, then NLO analytical methods need to be employed to accurately model the optical phenomenon. In this book, we employ NLO methods where needed and space allows and refer the reader to more detailed literature where needed.

The resulting sampled function, $g_s(x, y)$, is given by Equation 2.62. It can be shown that for Nyquist sampling conditions given by

$$X \leq \frac{1}{2B_x} \tag{2.65}$$

and

$$Y \leq \frac{1}{2B_y}, \tag{2.66}$$

the original continuous function can be reproduced *exactly* from the sampled data (Oppenheim and Schafer 1975; Goodman 2005). For the 1D case, a simple reconstruction equation is given by (Oppenheim and Schafer 1975)

$$g(t) = \sum_{k=-\infty}^{\infty} g(kT) \frac{\sin[(\pi/T)(t-kT)]}{(\pi/T)(t-kT)}, \tag{2.67}$$

and for the 2D case, the interpolation formula is given by (Goodman 2005)

$$g(x,y) = \sum_{n=-\infty}^{\infty} \sum_{m=-\infty}^{\infty} g\left(\frac{n}{2B_x}, \frac{m}{2B_y}\right) \text{sinc}\left[2B_x\left(x - \frac{n}{2B_x}\right)\right] \text{sinc}\left[2B_y\left(y - \frac{m}{2B_y}\right)\right], \tag{2.68}$$

and the parameters B_x and B_y are the bandwidths in frequency space. The "sinc" function is given by

$$\text{sinc}(ax) = \frac{\sin(a\pi x)}{a\pi x}, \tag{2.69}$$

where Equation 2.69 can be used in concert with the shift theorem to evaluate Equation 2.68.

The critical assumptions in obtaining the interpolation formulas in Equations 2.67 and 2.68 are that the object can be Fourier transformed and that the object is band limited. For the first assumption, the conditions that lead up to Equations 2.58 and 2.59 need to be satisfied. In essence, in order to be Fourier transformable, the source must be able to be absolutely integrated

$$\int_{-\infty}^{+\infty} \int_{-\infty}^{+\infty} |g(x,y)| dx, dy < \infty, \tag{2.70}$$

or the source must be able to be integrated over a finite period

$$\lim_{T_y \to \infty} \frac{1}{T_y} \int_{-T_y/2}^{T_y/2} \lim_{T_x \to \infty} \frac{1}{T_x} \int_{-T_x/2}^{T_x/2} u^2(x,y) dx dy < \infty. \tag{2.71}$$

In practical cases for real objects with finite extent, this is typically the case. For the temporal component, equivalent to the discussion of the energy spectrum and/or power spectrum earlier, the conditions in order to be Fourier transformable are

$$\int_{-\infty}^{\infty} |g(t)| dt < \infty \tag{2.72}$$

and

$$\lim_{T \to \infty} \frac{1}{T} \int_{-T/2}^{T/2} g^2(t) dt < \infty. \tag{2.73}$$

The second assumption is the basis for the definition of the *bandwidth*, which for our optical system is just the maximum spatial frequency in a given linear direction. The *Nyquist rate* is twice the bandwidth and is the spectral extent of our band-limited signal as shown in Figure 2.19.

For most practical imaging scenarios, the object has finite extent and finite power/energy, and so Fourier transform methods in modeling the optical system are very practical and computationally efficient. To aid in the Fourier-based analysis of optical systems, some useful properties and transform relationships have been developed over time. For convenience, we list the more notable properties and transforms in Table 2.3.

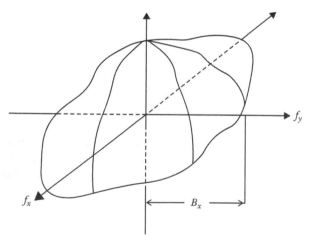

FIGURE 2.19
Band-limited function.

TABLE 2.3

Principles/Theorems

Linearity (1D and 2D)	$\mathcal{F}\left(c_1 u_1 \mp c_2 u_2\right) = c_1 U_2 \mp c_2 U_2$
Shift (1D)	$\mathcal{F}\{u_1(x-a)\} = e^{j2\pi fa} G(f)$
Shift (2D)	$\mathcal{F}\{u_1((x-a, y-b))\} = e^{j2\pi(f_x a + f_y b)} G(f_x, f_y)$
Scaling (1D)	$\mathcal{F}\{u_1(ax)\} = \dfrac{1}{|a|} G\left(\dfrac{f_x}{a}, \dfrac{f_y}{a}\right)$
Scaling (2D)	$\mathcal{F}\{u_1(ax, by)\} = \dfrac{1}{|ab|} G\left(\dfrac{f_x}{a}, \dfrac{f_y}{b}\right)$
Parseval's theorem (1D)	$\displaystyle\int_{-\infty}^{\infty} |u_1(x)|^2\, dx = \int_{-\infty}^{\infty} |U_1(f)|^2\, df$
Parseval's theorem (2D)	$\displaystyle\iint_{-\infty}^{\infty} |u_1(x, y)|^2\, dx dy = \iint_{-\infty}^{\infty} |U_1(f_x, f_y)|^2\, df_x df_y$
Autocorrelation theorem (1D)	$\mathcal{F}\left\{\displaystyle\int_{-\infty}^{\infty} h(\lambda_1 - x) h^*(\lambda_1) d\lambda_1\right\} = |H(f)|^2$
Autocorrelation theorem (1D)	$\mathcal{F}\displaystyle\iint_{-\infty}^{\infty} h(\lambda_1, \lambda_2) h^*(\lambda_1 - x)(\lambda_2 - y) d\lambda_1 \lambda_2 = |H(f_x, f_y)|^2$
Convolution theorem (1D)	$\mathcal{F}\left\{\displaystyle\int_{-\infty}^{\infty} u(\lambda_1) h(x - \lambda_1) d\lambda_1\right\} = U(f) H(f)$
Convolution theorem (2D)	$\mathcal{F}\displaystyle\iint_{-\infty}^{\infty} u(\lambda_1, \lambda_2) h(x - \lambda_2)(y - \lambda_1) d\lambda_1 \lambda_2 = U(f_x, f_y) H(f_x, f_y)$

By using the shift theorem and scaling property of the Fourier transform in combination with the basic functions earlier, spatially complex functions can be modeled and analyzed using a combination of these simple relationships. As an easy example, our binary star system was modeled using two scaled and shifted Kronecker delta functions. A more complex object could be constructed by using individually scaled and shifted Kronecker delta functions that were generated to satisfy the Shannon and Whitaker sampling theorem discussed earlier.

So far, we have presented some linear systems models that deal with fully coherent imaging systems and fully incoherent imaging systems. In addition, we have discussed some concepts related to spatial coherence so far. What models do we use if our imaging model lies somewhere in between these two extremes? How do we model coherence properties in time as well as space? The field of statistical optics provides some relevant and useful methods to answer these questions.

2.2.4 Statistical Optics Models

The field of statistical optics provides a powerful framework for describing, characterizing, and analyzing optical systems. Furthermore, everything we have learned so far still applies and extends to the statistical optics framework. A complete coverage of the field of statistical optics is beyond the scope of this book; however, we will provide some selective models and principles that will add to our present understanding. We will also bring in additional material as needed in subsequent sections. In this section, we predominantly focus on statistical optics methods for describing spatial and temporal coherence. In addition to the previous limiting cases of fully coherent and fully incoherent light, we present a means for treating and modeling partially coherent light. We also present some useful tools that address the coherence properties of light such as the mutual coherence function and the van Cittert and Zernike theorem. We also present some expressions for the limits on the expected signal on a photoelectric detector. A good text for additional details on statistical optics methods can be found in the references (Goodman 1985). We first start with a few mathematical preliminaries. A basic understanding of probability and statistics is assumed.

For high signal levels, a Gaussian density function is often used in modeling, analyzing, and understanding signals of interest. The 1D and 2D probability density functions for a Gaussian signal are given by

$$P_u(u) = \frac{1}{\sqrt{2\pi}\sigma} \exp\left\{-\frac{(u-\bar{u})^2}{2\sigma^2}\right\}$$ (2.74)

and

$$P_{uv}(u,v) = \frac{\exp\left[\frac{-(u^2+v^2-2\rho uv)}{2(1-\rho^2)\sigma^2}\right]}{2\pi\sigma^2\sqrt{1-\rho^2}}.$$ (2.75)

The parameters U and V are random variables that represent random signals of interest. The parameters $\bar{u}$ and σ represent the signal mean and standard deviation, respectively. The parameter ρ in Equation 2.75 is given by

$$\rho \triangleq \frac{\overline{uv}}{\sigma^2},$$ (2.76)

and $\overline{uv}$ is the joint moment of the jointly distributed Gaussian random variables U and V, and the parameter ρ ranges from 0 to 1. In Equation 2.75, we have made a common assumption that the random variables have zero mean and equal variance. If this is not the case, a more general expression for the jointly distributed Gaussian probability density function is given by

$$P_U(\underline{u}) = \frac{1}{2\pi^{n/2}|\underline{C}|^{1/2}} \exp\left\{-\frac{1}{2}(\underline{u}-\bar{\underline{u}})^t |\underline{C}|^{-1} (\underline{u}-\bar{\underline{u}})\right\}.$$ (2.77)

Here
$\underline{C}$ is an $n \times n$ covariance matrix
$\underline{u}$ is a column vector of length n that consists of the random values u
The superscripts "t" and "-1" stand for transpose and inverse, respectively

The covariance matrix consists of "i" rows and "k" columns of covariance values given by

$$\sigma_{ik}^2 = E\left[\left(u_i - \overline{u_i}\right)\left(u_k - \overline{u_k}\right)\right], \tag{2.78}$$

where the "E" is the expectation operator given by

$$\sigma_{ik}^2 = \int\int_{-\infty}^{+\infty} \left(u_i - \overline{u_i}\right)\left(u_k - \overline{u_k}\right) P_{u_i u_k}\left(u_i, u_k\right) du_i du_k. \tag{2.79}$$

In general, expanding the random variables U_1 and U_2 in terms of two parameters t_1 and t_2, such as time t_1 or time t_2, we can write the joint expectation as

$$\overline{u_1 u_2} = E\left[u_1 u_2\right] = \int\int_{\infty}^{+\infty} u_1 u_2 \rho_u\left(u_1, u_2; t_1, t_2\right) du_1 du_2 \tag{2.80}$$

and, in general,

$$\overline{u_1^n u_2^m \ldots u_L^p} = E\left[u_1^n u_2^m \ldots u_L^p\right] = \int\int_{-\infty}^{+\infty} \ldots \int u_1^n u_2^m \ldots u_L^p \rho_{u_1 u_2 \ldots u_L}$$

$$\left(u_1, u_2, \ldots, u_L; t_1, t_2, \ldots, t_L\right) du_1 du_2 \ldots du_L. \tag{2.81}$$

Equation 2.81 is useful in calculating higher-order moments of random variables U_1 and U_2. At this point, it is useful to discuss stationary random processes. A random process is strictly *stationary* if the Lth order probability density function is independent of the time origin (Goodman 1985). It is considered wide-sense stationary (WSS) if the expected value of the random process is independent of time and the correlation of the random process at two different points in time is only a function of the difference in time. In other words,

1. $E[u(t)]$ is independent of time
2. $E[u(t_1)u(t_2)]$ is a function of $\tau = t_2 - t_1$

In Figure 2.20, we see a visual representation of these ideas. The horizontal axis is the temporal axis, and each curve is a particular realization of the random process U. A random process is said to be stationary in increments if the random process, $Y(t) = U(t) - U(t - \tau)$, is stationary for all values of τ. Also, a stationary random process is ergodic if the temporal average is equivalent to the statistical average. This last property is useful if we want to infer statistical attributes of a random process from temporal data. The stationarity of the random process is important since, for nonstationary random processes, we may have different statistical results based on different starting times. Fortunately, there are a variety of optical systems applications where one or more of our stationary assumptions are valid. However, consideration must be given to each application to examine whether or not an assumption of stationarity is warranted.

Notice that every stationary random process is also WSS, but the same cannot be said for the other way around, and every WSS process is not stationary. Additional details and a discussion of other forms of stationarity can be found in the book *Random Signals,*

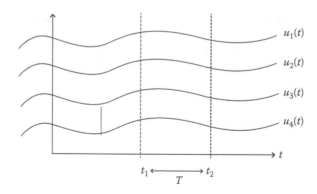

FIGURE 2.20
Random process.

Detection, Estimation, and Data Analysis by K.S. Shanmugan and A.M. Breipohl (Shanmugan and Breipohl 1988). A major reason for mentioning these distinctions is that there are a host of useful analytical properties for random process that satisfy one or more of our stationary and ergodic assumptions. A summary of some of these useful properties is presented in Table 2.4. In this table, $R_{xx}(t)$ is the autocorrelation of the random signal $X(t)$, $R_{xy}(t)$ is the cross-correlation of the random processes $X(t)$ and $Y(t)$, $S_{xx}(f)$ is the power spectral density of the random process $X(t)$, and $S_{xy}(f)$ is the cross-spectral density function (Goodman 1985; Shanmugan and Breipohl 1988).

In Table 2.4, the temporal relationship between the autocorrelation function and its power spectral density is known as the Wiener–Khinchin theorem. Simply put, the power spectral density is the temporal Fourier transform of the autocorrelation of a WSS random process. For example, by taking the temporal Fourier transform of an electromagnetic field that is at least WSS, we end up with its power spectral density. The functions shown in Table 2.4 are quite prevalent and useful in describing, characterizing, and analyzing important optical system properties such as the coherence properties of the light. Both spatial and temporal coherence properties of light are of interest, and these can be described by the relationships shown in Table 2.4. Temporal coherence can be thought of as shifting the electromagnetic wave in the observation plane with respect to itself in time, but not in space, and looking at the correlation properties between the shifted and not shifted versions of the electromagnetic field. Temporal shifts that are within the coherence time of the light are expected to produce sinusoidal variations in the observed intensity pattern at a point in the observation plane. As the temporal shift τ becomes progressively larger, the variations in the observed intensity at the observation point become progressively smaller approaching a constant value. Further increasing τ beyond this point has no effect, and the observed intensity at the observation point remains constant. The temporal separation τ of the electromagnetic field where the observed intensity becomes constant is the temporal boundary where the electromagnetic field becomes temporally incoherent. Experimentally, temporal coherence properties are measured by using a Michelson interferometer as shown in Figure 2.21. The source of optical radiation is shown on the bottom of the figure as S_1. The light propagates to the collimating lens L_1 and then to a partially reflecting beam splitter shown as the tilted rectangle in the center of Figure 2.21. The beam splitter consists of a layer of glass that is shown as the clear area in the figure and a partially reflecting surface that is shown by the cross-hatching.

TABLE 2.4

Useful Attributes of Random Processes

Name	Property	Real/Complex	Stationary
Autocorrelation function	$R_{xx}(\tau) = E[X(t)\,X(t+\tau)]$	Real	WSS
	$R_{xx}(0) = E[X^2(t)]$	Real	WSS
	$R_{xx}(\tau) = R_{xx}(-\tau)$	Real	WSS
	$R_{xx}(0) \ge |R_{xx}(\tau)|$	Real	WSS
	$\lim_{\tau \to \infty} R_{xx}(\tau) = \mu_x^2$	Real	WSS
Cross-correlation function	$R_{xy}(J) = E[X(t)\,X(t+\tau)]$	Real	WSS
	$\sqrt{R_{xx}(0)R_{yy}(0)} \ge |R_{xy}(\tau)|$	Real	WSS
	$|R_{xy}(\tau)| \le \dfrac{1}{2}[R_{xx}(0) + R_{yy}(0)]$	Real	WSS
	$R_{xy}(\tau) = 0,\ \text{orthogonal processes}$	Real	WSS
	$R_{xy}(\tau) = \mu_x \mu_y,\ \text{independant processes}$	Real	WSS
Power spectral density	$s_{xx}(f) = \mathcal{F}\{R_{xx}(\tau)\} = \int_{-\infty}^{\infty} R_{xx}(\tau)e^{-j\pi f t}d\tau$	Real	WSS
	$R_{xx}(\tau) = \mathcal{F}^{-1}\{s_{xx}(f)\} = \int_{-\infty}^{\infty} s_{xx}(f)e^{-j\pi f t}df$	Real	WSS
	$R_{xy}(\tau)$ is even function	Real	WSS
	$s_{xx}(f)$ is real and non negative	Real	WSS
	$s_{xy}(f)$ is even function	Real	WSS
Cross-spectral density	$s_{xy}(f) = \mathcal{F}\{R_{xy}(\tau)\} = \int_{-\infty}^{\infty} R_{xy}(\tau)e^{-j\pi f t}d\tau$	Real	WSS
	$R_{xy}(\tau) = \mathcal{F}^{-1}\{S_{xx}(f)\} = \int_{-\infty}^{\infty} S_{xy}(f)e^{-j\pi f t}df$	Real	WSS
	$\left[(\mathrm{Re}\{S\})\right]_{xy}(f)\}$ is an even function	Complex	WSS
	$\left[(\mathrm{Im}\{S\})\right]_{xy}(f)\}$ is an odd function	Complex	WSS
	$S_{xy}(f) = 0,\ \text{orthogonal processes}$	Complex	WSS
	$S_{xy}(f) = \mu_x \mu_y,\ \text{independent processes}$	Complex	WSS
	$\Gamma_{11}(\tau) = E[u(P_1,t+\tau)u^*(P_1,t)]$	Complex	WSS
	$\Gamma_{11}(0) = I(P_1)$	Real	WSS
	$J_{12} = E[u(P_1,t)u^*(P_2,t)]$	Complex	Stationary
	$\Gamma_{12}(\tau) = E[u(P_1,t+\tau)u^*(P_2,t)]$	Complex	Stationary
	$\mu_{12} = X_{12}(0) = \dfrac{J_{12}}{[J_{11}J_{22}]^{1/2}}$		
	$X_{12}(\tau) = \dfrac{\Gamma_{12}}{[\Gamma_{11}(0)\Gamma_{22}(0)]^{1/2}}$	Complex	Stationary

Notice that the back end of the beam splitter has the partially reflecting optical coating and the front end is an optically transparent material like glass. Half the light is reflected from the back surface of the beam splitter to mirror M_1. The beam splitter transmits the other half of the light toward mirror M_2. The compensator C adjusts the optical path length so that the light in both directions (M_1 and M_2) experiences the same amount of propagation through the glass. The reflected light from M_1 and M_2 is directed toward the collecting lens L_2 that focuses the light from both directions onto the point detector D. The compensator C is designed so that the optical path lengths in both directions are initially

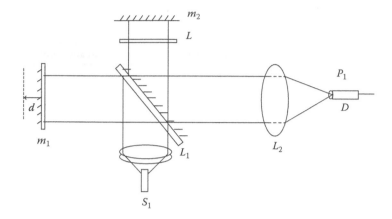

FIGURE 2.21
Source of optical radiation.

exactly the same. As M_1 is moved, the relative optical path length changes as a function of τ. It can be seen that the temporal shift τ is given by

$$\tau = \frac{2d}{c}, \tag{2.82}$$

where
   d is the relative separation of the electromagnetic fields in meters
   c is the speed of light

Dividing d by the speed of light converts the spatial separation into a temporal separation. The expression for the self-coherence function in Table 2.5 provides the observed intensity at observation point P_1, and the expression for the complex degree of coherence can be evaluated with $P_2 = P_1$ for a normalized result with a maximum value of 1 and a minimum value of 0. A technical performance measure that is associated with the temporal coherence of an optical field is the classical visibility given by

$$V = \frac{I_{max} - I_{min}}{I_{max} + I_{min}}, \tag{2.83}$$

where I_{max} and I_{min} are the maximum and minimum intensity values at the observation point P_1, respectively. Note that the complex degree of coherence and the classical visibility will both be one for full temporally coherent fields and zero for full temporally incoherent fields. Intermediate values result from partial temporally coherent fields. A plot of the self-coherence function with respect to τ from a Michelson interferometer is called an interferogram. According to Mandel, the coherence time τ_c can be given by (Goodman 1985)

$$\tau_c \triangleq \int_{-\infty}^{+\infty} |\gamma(\tau)|^2 \, d\tau, \tag{2.84}$$

where the integration is over all possible temporal separations τ. The interpretation of the coherence time is that for time scales equal to or less than the coherence time, the

TABLE 2.5

Useful Temporal and Spatial Coherence Properties

Name	$\Gamma_{11}(\tau)$ Self-coherence function	$\gamma_{11}(\tau)$ Complex degree of self-coherence	$\Gamma_{12}(\tau)$ Mutual coherence function	$\gamma_{12}(\tau)$ Complex degree of coherence	J_{12} Mutual intensity	μ_{12} Complex coherence factor
Source Plane						
Narrow band	$\left\langle u(P_1,t+\tau)u^*(P_1,t)\right\rangle$	$\dfrac{\Gamma_{11}(\tau)}{\Gamma_{11}(0)}$	$\left\langle u(P_1,t+\tau)u^*(P_2,t)\right\rangle$	$\dfrac{\Gamma_{12}(\tau)}{\left[\Gamma_{11}(0)\Gamma_{22}(0)\right]^{1/2}}$	$\left\langle u(P_1,t)u^*(P_2,t)\right\rangle$	$\gamma_{12}(0)$
Coherent case					$A(P_1)A^*(P_2)$	$e^{\left\{j\left[\phi(P_1)-\phi(P_2)\right]\right\}}$
Incoherent case			$\left|\Gamma_{12}(\tau)\right|=0$ $P_1\neq P_2,\tau\neq 0$		$\tilde{\kappa}I(P_1)\delta(\Delta x,\Delta y)$	$\delta(\Delta x,\Delta y)$
Quasimonochromatic case	$I_1 e^{-j2\pi\nu\tau}$	$e^{-j2\pi\nu\tau}$	$J_{12}e^{-j2\pi\nu\tau}$	$\mu_{12}e^{-j2\pi\nu\tau}$	J_{12}	μ_{12}
Image Plane						
Narrow band	$\left\langle u(Q_1,t+\tau)u^*(Q_1,t)\right\rangle$	$\dfrac{\Gamma_{11}(\tau)}{\Gamma_{11}(0)}$	$\left\langle u(Q_1,t+\tau)u^*(Q_2,t)\right\rangle$	$\dfrac{\Gamma_{12}(\tau)}{\left[\Gamma_{11}(0)\Gamma_{22}(0)\right]^{1/2}}$	$\left\langle u(Q_1,t)u^*(Q_2,t)\right\rangle$	$\gamma_{12}(0)$
Incoherent source (I_s) case					$\dfrac{ke^{-j\psi}}{(\bar{\lambda}z)^2}\mathfrak{I}_2(I_s)\Big|_{\substack{f_x=\Delta x/\bar{\lambda}z \\ f_y=\Delta y/\bar{\lambda}z}}$	$\dfrac{e^{-j\psi}}{I_s}\mathfrak{I}_2(I_s)\Big|_{\substack{f_x=\Delta x/\bar{\lambda}z \\ f_y=\Delta y/\bar{\lambda}z}}$
Partially coherent source (I_s) case					$\dfrac{k(\bar{x},\bar{y})e^{-j\psi}}{(\bar{\lambda}z)^2}\mathfrak{I}_2(I_s)\Big|_{\substack{f_x=\Delta x/\bar{\lambda}z \\ f_y=\Delta y/\bar{\lambda}z}}$	$\dfrac{e^{-j\psi}}{I_s}\mathfrak{I}_2(I_s)\Big|_{\substack{f_x=\Delta x/\bar{\lambda}z \\ f_y=\Delta y/\bar{\lambda}z}}$
					$k(\bar{x},\bar{y})=\mathfrak{I}_2\left(\mu(\Delta x_s,\Delta y_s)\right)\Big|_{\substack{f_x=\bar{x}/\lambda z \\ f_y=\bar{y}/\lambda z}}$	

fields are expected to interfere constructively and fringes will be observable in an interferogram. For time scales larger than the coherence time, the interferogram will produce a constant value.

For spatial coherence, an electromagnetic field (reference field) is spatially shifted with respect to itself (shifted field), and the temporal component of the electromagnetic field is kept identical for both the original and space-shifted versions of the electromagnetic field. The spatial correlation properties of the reference field and shifted field can be characterized by the mutual intensity or the complex coherence factor shown in Table 2.4. Experimentally, spatial coherence can be observed in a Young's double-slit experiment.

A source on the left of Figure 2.22 is used to illuminate an opaque screen with two small holes cut into it at P_1 and P_2. The optical path from P_1 to the observing point Q_1 is r_1 and from P_2 to Q_1 is r_2. If the observation point Q_1 is moved up and down the observation plane, fringes will be observed. As the observation point moves further away from the projected center of points P_1 and P_2, the fringes become smaller until they disappear. The point where the fringes disappear is where the fields become spatially incoherent. The classical visibility for the spatial coherence case is given by

$$V = \frac{2\sqrt{I_1 I_2}}{I_1 + I_2} \gamma_{12}(0),\tag{2.85}$$

where I_1 and I_2 are the intensities observed at Q_1 due to P_1 and P_2, respectively. The complex degree of coherence in Equation 2.85 is that due to the source field at the diffracting plane and is evaluated at points P_1 and P_2. The intensity at the observation point Q_1

$$I(Q_1) = I_1(Q_1) + I_2(Q_1) + K_1 K_2^* \Gamma_{12}\left(\frac{r_2 - r_1}{c}\right) + K_1^* K_2 \Gamma_{21}\left(\frac{r_1 - r_2}{c}\right),\tag{2.86}$$

where K_1 and K_2 are complex scaling factors arising from geometrical considerations and the intensities observed at Q_1 due to the fields at P_1 and P_2 and are given by

$$I_1(Q_1) = K_1^2 \Gamma_{11}(0), \quad I_2(Q_1) = K_2^2 \Gamma_{22}(0).\tag{2.87}$$

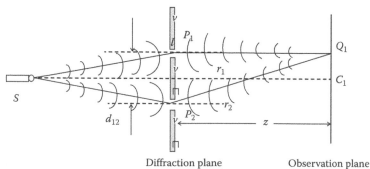

Diffraction plane   Observation plane

FIGURE 2.22
Shows the experimental details of Young's double-slit experiment.

The assumption required for Equations 2.86 and 2.87 to hold are that the pinholes at P_1 and P_2 are small enough so that the electromagnetic field does not vary across each opening and the illuminating light from the source S is narrow band. Additionally, we assume that the paraxial conditions hold as was done in the development of the Fresnel and Fraunhofer electromagnetic fields earlier in this chapter. The fringe spacing in the observation plane is given by

$$L = \frac{\bar{\lambda}z}{d_{12}}, \tag{2.88}$$

where
  d_{12} is the distance between the two pinholes
  z is the separation of the observation plane from the diffracting plane
  $\bar{\lambda}$ is the center wavelength of the narrowband light

Note that the bottom entries of Table 2.5 apply to both spatial and temporal coherence properties of the light. The number of fringes that are observable is given by

$$N = 2\frac{\bar{v}}{\Delta v}, \tag{2.89}$$

where
  Δv is the half-power bandwidth
  $\bar{v}$ is the center frequency of the narrowband light

Analogous to the coherence time τ_c, the coherence length L_c can be thought of as the distance in the observation plane over which the electromagnetic fields are coherent and, therefore, produce fringes. The width of the packet that contains the fringes in the observation plane is given by

$$\Delta l = \frac{2zc}{\Delta v d_{12}}. \tag{2.90}$$

Some remarkable simplifications occur in the mutual intensity and the complex coherence factor when *quasimonochromatic* conditions prevail. For quasimonochromatic light, the half-power bandwidth Δv must be much smaller than the center frequency of light $\bar{v}$, and the difference in the optical path length from the source to the observation point through pinholes P_1 and P_2 must be much smaller than the coherence length of the observed light. An example of this kind of light may be a low-grade laser. For quasimonochromatic conditions, the mutual intensity becomes

$$\Gamma_{12}(\tau) \simeq J_{12}e^{-j2\pi\bar{v}\tau}, \tag{2.91}$$

and the complex coherence factor is given by

$$\Upsilon_{12}(\tau) \simeq \mu_{12} e^{-j2\pi\bar{\nu}\tau}. \tag{2.92}$$

Under paraxial conditions, the observed intensity at the observation point $\vec{Q}_1$ is given by

$$I(\vec{Q}_1) = I_1 + I_2 + 2K_1 K_2 J_{12} \cos\left[\frac{2\pi}{\lambda z}(\vec{d}_{12} \cdot \vec{Q}_1) + \phi_{12}\right]. \tag{2.93}$$

Here, we have generalized our expression slightly so that $\vec{Q}_1$ is a 2D position vector in the observation plane and $\vec{d}_{12}$ is a vector in the diffraction screen plane that corresponds to the difference in positions of the pinholes P_1 and P_2. The phase term ϕ_{12} is associated with the mutual intensity term.

We now have a basic understanding of spatial and temporal coherence. Note that the mutual coherence factor and the complex degree of coherence address both spatial and temporal coherence simultaneously. We want to introduce another useful technical performance metric related to both spatial and temporal coherence of an electromagnetic field. This technical performance metric is very useful in characterizing the effects of the atmosphere on optical systems. The technical performance measure is the structure function given by (Roggemann and Welsh 1996)

$$D(\Delta x, \Delta y) = \overline{\left(\chi(\vec{x}) - \chi(\vec{x} - \Delta\vec{x})\right)^2} + \overline{\left(\psi(\vec{x}) - \psi(\vec{x} - \Delta\vec{x})\right)^2} \tag{2.94}$$

$$= D_\chi(\Delta\vec{x}) + D_\psi(\Delta\vec{x})$$

where $D_\chi(\Delta\vec{x})$ is the log-amplitude structure function, and $D_\psi(\Delta\vec{x})$ is the phase structure function associated with the pupil plane field, and as before, the bar in the expression infers statistical expectation. The mutual intensity and the complex coherence factor deal with spatial coherence, and the self-coherence factor provides information about the temporal coherence of a given electromagnetic field. We also notice that for the complex degree of coherence and the complex coherence factors, fully coherent fields produce a value of 1, fully incoherent fields produce a value of 0, and partially coherent fields produce an intermediate value.

We now want to look at two limiting cases of the mutual coherence function. The first is a fully incoherent source term, and the second case is a fully coherent source term. For the first case, most sources apart from a laser fall into this category. Additionally, even if a source is used with a high degree of spatial coherence, if it is transmitted through, or reflected from, an object that is rough on the scale of an optical wavelength, then the transmitted or reflected light becomes spatially incoherent. A majority of imaging scenarios fall under these conditions, and so it is essential to understand the nature of the mutual coherence function under an incoherent source approximation. Figure 2.23 provides the geometry for visualization. We again assume paraxial imaging conditions, such as those that were used to derive the Fresnel and Fraunhofer diffraction formulas. The source plane is shown at the left of Figure 2.23 with the clear area showing the projection of the source onto the 2D plane. In the case of a transmitting system, the open area on the left side of the figure is the clear aperture of the transmitted electromagnetic field. The distance from the source plane to the observation plane is given by z. A point in the source plane is denoted by P_1, and two points in the observation plane are labeled Q_1 and Q_2. The distance from P_1 to Q_1 is given by

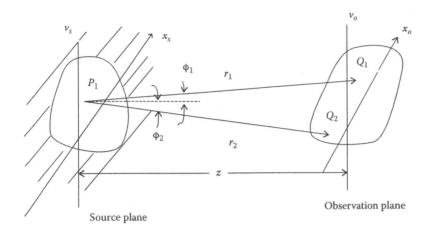

FIGURE 2.23
Transmitting system case.

r_1, and the distance from P_1 to Q_2 is given by r_2. The crosshatched area on the left of Figure 2.23 is the region where the source fields are zero.

We want to take a look at the propagation of the mutual intensity. We will assume that quasimonochromatic imaging conditions hold. In general, the propagation of the mutual intensity is given by (Goodman 1985)

$$J(Q_1,Q_2) = \iint_\Sigma \iint_\Sigma J(P_1,P_2)\exp\left[-j\frac{2\pi}{\lambda}(r_2-r_1)\right]\frac{\xi(\theta_1)}{\overline{\lambda}r_1}\frac{\xi(\theta_2)}{\overline{\lambda}r_2}ds_1ds_2,\tag{2.95}$$

where the integration is over the clear aperture shown on the left side of Figure 2.23, and the angular terms are associated with the imaging geometry. By assuming that paraxial conditions hold just as in the development of our Fresnel and Fraunhofer diffraction formulas, these angular terms approach one, and we get what is known as the van Cittert–Zernike theorem:

$$J(x_1,y_1;x_2,y_2) = \frac{K\left(e^{-j\psi}\right)}{\left(\overline{\lambda}z\right)^2}\int\int_{-\infty}^{+\infty} I(x_s,y_s)\exp\left[j\frac{2\pi}{\overline{\lambda}z}\left((x_2-x_1)x_s+(y_2-y_1)y_s\right)\right]dx_sdy_s.\tag{2.96}$$

In Equation 2.96, the phase term ψ in front of the integrals is given by

$$\psi = \frac{\pi}{\overline{\lambda}z}\left[\left(x_2^2+y_2^2\right)-\left(x_1^2+y_1^2\right)\right],\tag{2.97}$$

and K is a volume normalization term given by

$$K = \frac{\left(\overline{\lambda}\right)^2}{\pi}.\tag{2.98}$$

The van Cittert and Zernike theorem is a remarkable result in that we can determine the mutual intensity in the observation plane by taking the 2D Fourier transform of the object brightness. Normalizing the mutual intensity, we can eliminate a majority of the scale factors in the mutual intensity and arrive at the complex coherence factor:

$$\mu(x_1,y_1;x_2,y_2) = \frac{e^{-j\psi}\int\int_{-\infty}^{+\infty} I(x_s,y_s)\exp\left[j(2\pi/\bar{\lambda}z)(\Delta x x_s + \Delta y y_s)dx_s dy_s\right]}{\int\int_{-\infty}^{+\infty} I(x_s,y_s)dx_s dy_s}, \qquad (2.99)$$

where $\Delta x = x_2 - x_1$ and $\Delta y = y_2 - y_1$. Equations 2.98 and 2.99 are applicable to most sources in the visible and infrared parts of the electromagnetic spectrum except for direct illumination of a highly spatially coherent source such as a laser. For the case of a source with a high degree of spatial coherence, the analytic electromagnetic field can be represented as

$$u(P,t) = A(P,t)e^{-j2\pi\bar{v}t}, \qquad (2.100)$$

where the amplitude term is a complex quantity. This analytic form of the electromagnetic field is often useful in analyzing the electromagnetic field, its propagation, and its interaction with the optical system. It can be shown that the mutual intensity for a coherent source field is given by (Goodman 1985)

$$J_{12} = A(P_1)A^*(P_2), \qquad (2.101)$$

where the complex amplitude terms are evaluated at two spatial points on the coherent source, P_1 and P_2. The associated complex coherence factor is given by

$$\mu_{12} = \exp\left(j\left[\varphi(P_1)-\varphi(P_2)\right]\right), \qquad (2.102)$$

where the phases are obtained by taking the argument of the complex amplitude of Equation 2.100. In the case of Young's experiment, the fringes that would be observed in the observation plane are given by

$$I(Q_1) = I_1(Q_1) + I_2(Q_1) + 2\sqrt{I_1(Q_1)I_2(Q_2)}\cos\left(\frac{2\pi(r_2-r_1)}{\bar{\lambda}}+\varphi(P_2)-\varphi(P_1)\right), \qquad (2.103)$$

where Figure 2.24 is again used for reference of Young's double-slit experiment. To generalize the geometry for the mutual coherence function and its derivate products, we use the following four-point figure.

The angles θ_1 and θ_2 are measured from the normal to the surface at the source point and the line connecting the source point to the corresponding exploratory point in the observation plane. To propagate the mutual coherence function or, when appropriate, the mutual intensity function, from the source plane to the observation plane, several special cases are of interest: (a) narrowband light, (b) narrowband light with quasimonochromatic assumption, (c) coherent light, (d) incoherent light, and (e) partially coherent light. We summarize the main propagation results in Table 2.5 (Goodman 1985).

We see that for both the spatially incoherent and partially coherent image plane cases, the primary propagation integrals obey the van Cittert–Zernike theorem and are 2D spatial

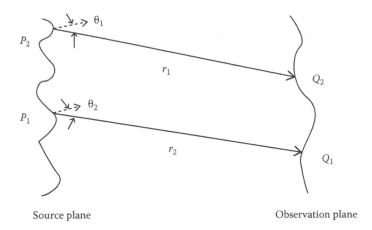

Source plane   Observation plane

FIGURE 2.24
Young's double-slit experiment.

Fourier transforms. Of course, the "k" factor out front of the partially coherent case complicates things somewhat. The condition for which the partially coherent case applies is

$$z > \frac{2D_{max}D_c}{\bar{\lambda}}, \qquad (2.104)$$

where
 D_{max} is the maximum physical extent of the source object
 z is the separation of the source plane from the observation plane
 D_c is the maximum coherence length in the source plane
 $\bar{\lambda}$ is the mean wavelength

Since many sources are well modeled by assuming an incoherent source, we focus some attention on this case. Notice from the equation for the mutual intensity in Table 2.5, if we are interested in describing the intensity in the observation plane for an object that can be modeled as an incoherent source (which is often the case), the observation points Q_1 and Q_2 coincide, and the exponential phase term with Ψ reduces to 1. The relationship between the source intensity and the observed intensity is just a 2D spatial Fourier transform that is readily solvable with either analytical or numerical methods.

We want to end this section of statistical optics models with an overview of some useful probability density functions that can be used to model useful statistical quantities such as the mean, covariance, and even higher moments. An interesting fact, according to the *central limit theorem*, is that for any set of N independent random variables with arbitrary probability density function(s) and finite variance, the summation of a sufficiently large number of these zero mean, unit variance random variables produces a probability density function that approaches a normal distribution with zero mean and unit variance (Craig et al. 1995). For instance, a source that produces a large number of independent photons, high-light-level sources can be modeled as having a Gaussian probability density function such as presented in Equations 2.74 through 2.77. For low light levels, analysis and experimental results indicate that photon arrivals obey Poisson statistics wherein the mean and the variance have the same value. Circularly complex Gaussian statistics have been used to model certain aspects of lasers, while low levels of photons arriving on a detector have been shown to obey a doubly stochastic Poisson process (Andrews and Phillips 1998). We will treat these special cases in more detail later in the book.

A useful tool in determining moments of a known probability density function is the *characteristic* function. The characteristic function is defined as the expected value of the complex exponential of the weighted sums of the random variables:

$$M_u^{(n)}(w_1, w_2, ..., w_n) \triangleq E\left[\exp\left\{j(w_1 u_1 + w_2 u_2 + \cdots + w_n u_n)\right\}\right], \tag{2.105}$$

where n is the highest number of random variables. This expectation takes the form of an n dimensional Fourier transform

$$M_u^{(1)}(w_1) = \int_{-\infty}^{+\infty} \exp\left\{j(w, u)\right\} P_{u_1}(u_1) du_1 \tag{2.106}$$

or

$$M_{u_1 u_2}^{(2)}(w_1, w_2) = \int_{-\infty}^{+\infty}\!\!\int \exp\left\{j(w_1 u_1 + w_2 u_2)\right\} P_{u_1 u_2}(u_1, u_2) du_1 du_2, \tag{2.107}$$

respectively.

Conversely, the corresponding probability density function can be determined from the characteristic function by inverse Fourier transform. If the moments exist, the characteristic function can be used to generate higher-order moments (Goodman 1985). For example,

$$\overline{u_1^n u_2^m} = \frac{1}{j^{(n+m)}} \frac{\partial^{n+m}}{\partial w_{u_1}^n \partial w_{u_2}^m} M_{u_1 u_2}\left(w_{u_1}, w_{u_2}\right)\Big|_{w_{u_1}=0, w_{u_2}=0}, \tag{2.108}$$

where, as before, the bar means statistical expectation. This ability to generate higher-order moments from the characteristic function is very useful, especially in the case of Gaussian random variables, since all the higher-order Gaussian moments can be written as linear combinations of the joint moments. For the case of Gaussian and Poisson probability density functions, the first-order characteristic functions are given by

$$M_{u_1}(w_1) = \exp\left(-\frac{\sigma^2 w_1^2}{2} + j w_1 \overline{u_1}\right) \quad \text{Gaussian} \tag{2.109}$$

and

$$M_{u_1}(w_1) = \sum_{k=0}^{\infty} \frac{\overline{k}^k}{k!} e^{(-\overline{k} + j w_1 k)} \quad \text{Poisson.} \tag{2.110}$$

The nth-order Gaussian characteristic function is given by

$$M_{u_1 \ldots u_n}(\underline{w}) = \exp\left\{j\overline{\underline{u}}^t \underline{w} - \frac{1}{2} \underline{w}^t \underline{C} \underline{w}, \tag{2.111}\right.$$

where the same notation is used as in the case of the multiple-dimensional Gaussian probability density function in Equation 2.77. We end with the 1D probability density function for the Poisson case given by (Goodman 1985)

$$P_{u_1}(u_1) = \sum_{k=0}^{\infty} \frac{\left(\bar{k}\right)^n e^{-\bar{k}}}{k!} \sigma(u_1 - k) \tag{2.112}$$

that has mean and variance given by

$$E[u_1] = \bar{u}_1 \tag{2.113}$$

and

$$\sigma_{u_1}^2 = \bar{u}_1, \tag{2.114}$$

respectively.

This section has provided some fundamental results from the field of statistical optics. These results will be used throughout this book and are complementary to and consistent with previous results. We started this chapter section with Maxwell's equations and showed that the electromagnetic field satisfies either the inhomogeneous or homogenous Helmholtz wave equation. We then showed that the electromagnetic field under paraxial conditions satisfies the Fresnel or Fraunhofer diffraction integrals and that the relationship between the object plane and observation plane is relatable to a 2D Fourier transform. We provided a linear systems model and defined some important optical linear systems terms such as the GPF, impulse response, PSF, and OTF and provided some simple examples. We presented the energy spectrum and power spectrum and discussed sampling requirements in both the object plane and the entrance pupil plane of an imaging system. Lastly, we provided a short overview of some key statistical optics results. We provided probability density functions for two frequently occurring random processes encountered in statistical optics, Gaussian and Poisson. We showed that for high light levels, the Gaussian distribution is a good model to use and for low light levels, Poisson statistics are applicable. We talked about ergodic, stationary, WSS, and stationary in increments processes and their relations to statistical analysis. We discussed temporal and spatial coherence and showed how these properties can be measured using a Michelson interferometer (temporal coherence) and Young's double-slit experiment (spatial coherence). We provided associated analytical expressions for the expected intensity in the observation plane. For Young's double-slit experiment, we presented expressions for the intensity in the observation plane for a fully incoherent source and a fully coherent source. The van Cittert and Zernike theorem was then presented that showed that for a fully incoherent source, the observed mutual intensity in the observation plane is the scaled, 2D spatial Fourier transform of the object brightness. We presented some technical performance metrics such as the correlation coefficient, classical visibility, structure function, coherence time, and coherence length. We also showed the importance of correlation functions to optical systems analysis. We presented correlation functions like the mutual coherence function, mutual intensity, self-coherence factor, complex coherence factor, and the complex degree of coherence and showed how to relate these correlation functions in the source plane to the observation plane. We presented results for these correlation functions under (1) narrowband light; (2) narrowband, quasimonochromatic light; (3) broadband light; (4) incoherent light; and (5) partially coherent light assumptions. We provided expressions for the mean, variance, and a means to generate higher-order moments using the characteristic functions. This section is not intended to provide a comprehensive view of these topics but, instead, to present important and useful optical analysis concepts and methods. We will build upon these basics in later chapters. For a more detailed and comprehensive treatment, please see the provided references.

In the next section, we continue our case study from Chapter 1 and illustrate the tools, techniques, and methods we learned in this chapter. To set the stage, our imaging company, Fantastic Imaging Technologies, has just been informed that they won the Homeland Security contract to build the imaging systems for detecting illegal bipedal border crossings into the United States using an unmanned aerial vehicle (UAV). We continue with our integrated case study.

2.3 Integrated Case Study: Introduction to Enterprise Architecture

This case study illustrates the process of identifying the EA for FIT and how it is used to understand the effects that a new business has on the rest of the company. It also illustrates some key optical system technical considerations. The company reported that with the implementation of EA, decisions became more clear, simplified, and more effective. With FIT being awarded the Homeland Security's UAV optical system effort, Dr. Bill Smith called an executive committee meeting to congratulate the team and discuss plans for the future of the business. This scenario has the following characters: Dr. Bill Smith, FIT chief executive officer (CEO); Dr. Tom Phelps, FIT chief technical officer (CTO) and optics expert; Mr. Garry Blair, CIO and chief enterprise architect; Karl Ben, FIT lead systems engineer; and Ron, S., FIT systems engineer (new hire).

2.3.1 Defining the Enterprise Architecture for FIT

Bill: "Good Morning everyone, I hope everyone is as excited as I am. Let me open the meeting with my congratulations to the team for a job well done! Winning this program will position the company for many future endeavors in this industry. As you know, we have been mainly focused on providing optical imaging for the commercial sector, and this win is a great opportunity for us. One of the ideas in our proposal that stood out was our enterprise architecture model. During the proposal process, we were able to identify new capabilities that would be needed to support this new program and how it would affect our business. We were able to persuade the Department of Homeland Security that we would be able to complete the program on time and within budget. Now we need to deliver! With that in mind, I called this meeting to discuss what the next steps are, and I want to discuss how we can successfully execute this program. I want to open the floor for discussion."

Garry: "I would also like to offer my congratulations to the team on a job well done. During the proposal process, we were able to use our enterprise architecture to map how changes in the proposal would affect the overall company business. This included current projects, staffing resources, infrastructure, information flow, and processes already in place. Winning this effort is largely due to how well we all worked together. Our enterprise architecture is still in its infancy, some systems still need to be implemented, and since being awarded the program, we have noticed some deficiencies that need to be addressed with respect to resources, infrastructure, and tools needed to be successful."

Tom: "A good introduction Garry to the items for discussion. I have put together some slides on the new program and what it means for our company. As we all know, our

TABLE 2.6

FIT Goals and Objectives

Goals	Objectives
Develop and integrate an electro-optic subsystem for a UAV.	Design and manufacture an optical subsystem that integrates with other subsystems becoming a system of systems.
Train U.S. military personnel on use and maintenance of system.	Provide support for optical subsystem to allow for users to execute missions successfully.
Promote sales of optical subsystems to other government agencies.	Increase sales and revenue for the company by 20% over the next 2 years.
Expand company assets with respect to personnel resources, infrastructure, and intellectual property.	Become the industry leader in optical imaging with an emphasis on atmospheric turbulence compensation and adaptive optics methods.

mission is to provide the best optical imaging products for our stakeholders. We also strive to provide our clients with solutions designed to execute their missions effectively by providing reliable, efficient, and rapid solutions to their needs. This first slide shows our goals and objectives. We have also decided to do a feasibility analysis on various aspects of this project, so that we can be confident on the staffing resources, information flow, and various processes already in place" (see Table 2.6).

Bill: "Mr. Phelps, at this stage of the project, why do you want the feasibility study?"

Tom: "That is a good question. Once the requirements for an effort are understood, it is important to conduct a complete feasibility study to determine whether or not the intended effort is doable. The feasibility study should address all the requirements from a holistic perspective."

Bill: "That is a good idea. Can you explain to us what you mean by holistic? Do you mean evaluating the feasibility of the project from more than just the technical perspective?"

Tom: "Yes, exactly! Although other categories can be considered, our feasibility study will target the following six items:

1. Technical feasibility
2. Economic/financial feasibility
3. Organizational feasibility
4. Operational feasibility
5. Motivational feasibility
6. Managerial/political feasibility"

Bill: "Hmmm, doing all those feasibility studies sounds costly."

Tom: "True, this takes some effort. However, it is more costly to miss some critical aspects of the project and have it bite us later on in the development cycle. Also, for now, we have been focusing on the technical and organizational feasibility areas."

Bill: "Ok, so how are things going on the technical front?"

Tom: "We have run all of our models for the typical scenarios and parameters identified in the DHS proposal. We have done a complete radiometric analysis using humans as the target. We spent the majority of the time looking at nighttime operational scenarios and evaluated both a low-light-level and thermal-based infrared imaging system. At this point, a full-blown computational electromagnetic analysis is not warranted. For the daytime visible imaging system solution, we ran our atmospheric turbulence models to see what spatial resolution our imaging system would provide. For the size of the imaging apertures involved in the daytime system, the atmospheric turbulence compensation systems may not strictly be required. However, we may get a further standoff range, which will increase our sensor footprint on the ground. In this way, we may be able to cover more area than a system without the atmospheric turbulence compensating system."

Bill: "Good! What kind of models are we setting up?"

Tom: "Well, we have a radiometric model in SI units that provides the signature levels for expected border crossers. We also have the atmospheric turbulence models that we can turn on or off that provide insight to the spatial resolution of our optical system. We are using statistical optics models to describe the spatial correlation properties of the detected image. We also have generated some basic optical models for the imaging sensor on the UAV host, using their gimbal system. We can model both Fresnel and Fraunhofer propagation. For now, we are assuming that we are integrating into their gimbal system, but we can build our own collection and relay optics if necessary. We also have a geometrical optics model ready for some quick calculations and ray tracing. We also have developed a linear systems model with a circular entrance pupil for incoherent imaging scenarios. We have computer models based on the UAV collecting aperture, the Fresnel zone free-space impulse response, point spread function, and optical transfer function. These models allow us to respond to variations in their CONOPS and allow us to quickly evaluate changes in their requirements."

Bill: "What if we need to use some active imaging methods?"

Tom: "If that becomes necessary, we can quickly generate a coherent imaging model for our optical imaging system on the UAV platform. At this stage, our incoherent imaging models supported by Fourier optics and statistical optics methods are sufficient."

Bill: "Wait, you said you were using a free-space impulse response. That applies to a vacuum and doesn't account for atmospheric effects."

Tom: "That's true. We are using the free-space impulse response to determine an upper limit on the performance of our imaging system. Basically, we want to isolate the entrance pupil effects to provide diffraction-limited, spatial resolution performance estimates. If we build an excellent imaging system, like we always do, we should be able to approach this level of performance if we implemented our atmospheric turbulence compensation system. By having this estimate, we can show how much of an improvement their current optical system would have if they added the atmospheric turbulence compensation subsystem."

Bill: "That makes sense. Mr. Blair, you were mentioning something about deficiencies regarding our enterprise architecture."

Garry: "Yes. Nothing unexpected. However, in order to achieve our business goals and objectives, we needed to find a way to have the ability to unify and integrate business processes across the enterprise and link with external partners. Our documentation needs to be readily available. We need to reduce delivery time and development costs by maximizing reuse. Finally, a common vision was needed for both the business and IT departments. Now that we have won the contract, we need to look at how this win affects the company at an enterprise level and not just at the program level. The new program will put a strain on personnel resources, IT infrastructure, sales and inventory, and so forth."

Bill: "Garry, Do you have specific data that reflect how the new program will affect the enterprise?"

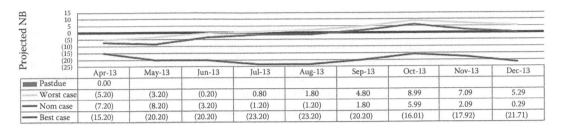

	Apr-13	May-13	Jun-13	Jul-13	Aug-13	Sep-13	Oct-13	Nov-13	Dec-13
Pastdue	0.00								
Worst case	(5.20)	(3.20)	(0.20)	0.80	1.80	4.80	8.99	7.09	5.29
Nom case	(7.20)	(8.20)	(3.20)	(1.20)	(1.20)	1.80	5.99	2.09	0.29
Best case	(15.20)	(20.20)	(20.20)	(23.20)	(23.20)	(20.20)	(16.01)	(17.92)	(21.71)

FIGURE 2.25
Staffing forecast versus new business. (Courtesy of Chris Lannon.)

Garry: "When we started the proposal for Homeland Security, we were concerned with possible technical staffing shortages; some of our current programs are running a little behind schedule. Utilizing the reports from the HR system that tracks staffing resources, we show a staffing shortage of 25 heads over the next few months. Figure 2.25 shows the deficiency."

Bill: "Garry, is this staffing issue going to impact this program?"

Garry: "No. Fortunately, we have the HR system integrated with our manpower tools that allowed us to be aware of the situation and act accordingly. We have already begun hiring to meet the needs, and I have to say, there are some really great candidates."

Bill: "But it will also take time to train the new staff for the project. Have we accounted for the training time?"

Garry: "Yes. As part of the organizational feasibility study, analysis was done on our projected staffing needs. The manpower tool accounts for training and also the learning curve. Our staff did an amazing job of forecasting our resource needs based on our existing and expected workload. The results of the organizational feasibility analysis showed that we could handle the needs of this project."

Tom: "With the success of the HR system integrating with our manpower tools, which allows us to strategically plan for new business, we began researching other areas that would benefit the company and allow us to maintain perspective on our business goals and objectives. Enterprise architecture techniques can be used to help our company in areas such as sales, our inventory system, our cost accounting system, and our training systems, to mention a few. We have begun talking with the various program managers, as stakeholders in the cost accounting system, regarding their needs and the capabilities of the tools needed to support how we do business. In addition, we began looking at the training needs, not only internally but also for the stakeholder. When we looked at the Homeland Security project during the proposal phase, we noticed there were some required integrated logistic capabilities that were needed to win. Part of our response involved stakeholder training with our new optical system in the operational and maintenance phases of the systems development life cycle. We also are getting some interest from other government agencies with respect to our capabilities and how we may benefit them. In our design, we are considering these other government agencies as secondary stakeholders. With that in mind, we should consider training for both internal and external stakeholders. This will be a big undertaking since many of the external stakeholders are using localized tools that range from spreadsheets, to homegrown databases, to sophisticated enterprise architecture suites to store and track data."

Bill: "Due to the success of the HR system and our expanding business base, I want to create a special project to create basic enterprise architecture artifacts for the promising secondary stakeholders. You can start with modifying the OV-1 for their particular applications. We can have these ready as part of the next EA roll-out phase. Garry, you are in charge. After the meeting, I am going to need you, Garry, to set up a meeting with Mr. Karl Ben as the lead systems engineer on the new Department of Homeland Security (DHS) program."

(They started evaluating the current systems and processes implemented in the company and how new capabilities would affect it. The meeting discussion started with the following.)

Karl: "We started evaluating the new configuration items the DHS program was going to need along with new secondary stakeholders, and we determined we may have to implement some new capabilities into our enterprise architecture. I will start with an overview of the program so we can all be on the same page with respect to the concept of operations (CONOPS) and operational view of the system as shown in Figure 2.26."

Karl: "I will explain some of these systems to provide a more detailed picture. The capabilities I wanted to discuss today are the ground control station (GCS), mobile control station (MCS), and the optical imaging system, as it pertains to maintenance and support. As you know, we have a requirement to provide support including training and maintenance of the system we are delivering. In the past, we usually created training modules and did on-site or off-site training for the operators of the system and maintenance personnel. This requires several man-hours, personnel, and travel expenses. We have also had interest in the system from other stakeholders including the FBI, CIA, and some other foreign stakeholders. We are in the process of generating new enterprise architecture artifacts for our Homeland Security stakeholder and will adapt these for our most promising secondary stakeholders. Since many of our prospective customers fall under the DOD, we need to ensure that our EA is compatible with what they expect. Are you familiar with what the DOD expects with regard to EAs?"

Garry: "Yes. The DOD has some specific goals and expectations when it comes to enterprise architecture. The EA should

- Eliminate duplication, incompatibility, and redundancy of systems and business processes
- Manage knowledge as a corporate asset using standard shared information as a driver
- Facilitate effective business decisions
- Capture information once and reuse across the enterprise
- Provide security and protection of sensitive information
- Affect changes through increasing efficiency and economy of scale
- Create consistent EA products with sufficient detail for implementation

We need to make sure our EA accomplishes these items."

Karl: "Is there some guidance for how to accomplish this?"

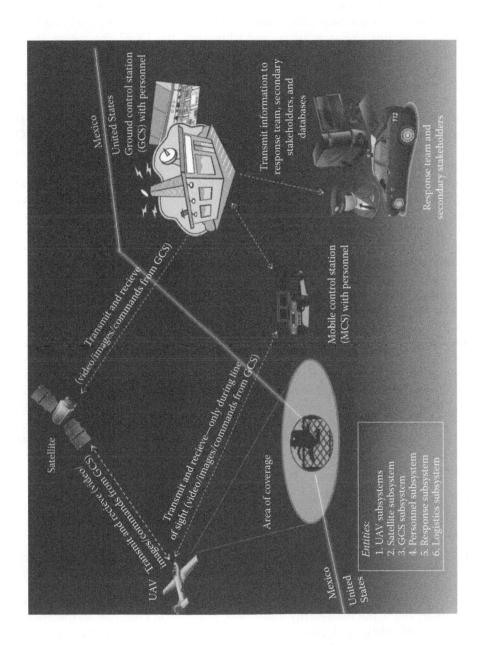

FIGURE 2.26
CONOPS of UAV border patrol mission. (Courtesy of Chris Lannon.)

Garry: "Yes. The DOD has an architectural framework known as the DODAF that facilitates implementing an EA. The DODAF provides the guidance and rules for developing, representing, and understanding architectures based on a common denominator across DOD, joint, and multinational boundaries. It provides insight for external stakeholders into how the DOD develops architectures. The DODAF is intended to ensure that architecture descriptions can be compared and related across programs, mission areas, and ultimately the DOD enterprise, thereby establishing the foundation for analysis that supports decision-making processes throughout the DOD. The DODAF is organized into several 'views.' These views consist of operational views that show the major entities, or nodes, upon which the system is based, system views that describe the systems that support data communication between or within the nodes, and technical standards views that show the standards and conventions upon which the whole system is based. One of the feedback points from DHS as to why we were awarded the contract was that they were looking for a company that had sound EA principles."

Karl: "We have been focused on hiring technical staff, and I believe we will be well positioned for the design and development of the new optical system per the new numbers that came out today from the HR system. With the new talent coming in though, we also should plan for training internally as well."

Karl: "Actually, that is what I wanted to discuss. We are considering new training software to implement across our enterprise. We have been working with your enterprise architecture team in determining the best system for our needs."

Ron: "I am not sure I get the connection of this training software to our EA. Can you explain that please?"

Karl: "Sure, Ron. I will start with an overview of FIT's enterprise architecture process and how we have implemented our enterprise architecture. Remember, in simple terms, an enterprise is an organizational structure of people, processes, and resources acting together to achieve a set of objectives that are expressed as a mission or a set of goals. This organizational structure collects, produces, and uses information to support performance of the business processes as they develop products. The organizational structure uses business processes to achieve objectives. Portions of the business processes are performed by different parts of the organizational structure. Enterprise architecture is a representation of

- The business architecture and the technical architecture of an enterprise
- How the business and technical architectures align to achieve the objectives of the enterprise
- How the technical and business architectures will transform to more effectively achieve current or future mission objectives

While this description of an enterprise uses terms that are business oriented, an enterprise is not necessarily a business, but instead could be an instance of a commercial entity, civil government entity, military entity, an intelligence community entity, or any other stakeholder enterprise representation. The enterprise business architecture is a model that depicts the enterprise business processes, the organizational structure, the information needs, and the products and their interrelationships. The enterprise technical architecture is a model that depicts the enterprise technology and how it supports the business architecture. This

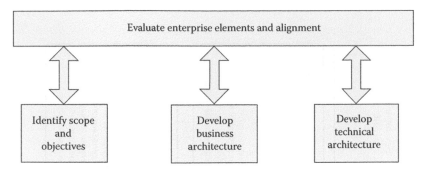

FIGURE 2.27
FIT top-level EA process. (Courtesy of Chris Lannon.)

technology includes applications and support software and computing and infrastructure hardware. Since performing the enterprise architecting tasks is based on heuristics and experience, the process does not describe "how to" do the tasks. We placed emphasis on the set of products needed to adequately define our EA. The slide I am showing now depicts FIT's top-level enterprise architecture process.

As depicted in Figure 2.27, the general concept of operation for the enterprise architecting process is as follows:

- Identify scope and objectives: Develop an understanding of the vision, mission needs, and key requirements of the enterprise architecture; validate FIT's understanding with stakeholders and determine the scope of the EA effort. The result of this activity is the definition of the scope of EA activities, ranging from validating our understanding of an existing stakeholder EA to generating an EA for a stakeholder.

- Develop business architecture: Validate existing or create new process descriptions, information flows, and enterprise security and show how these operate to achieve enterprise objectives.

- Develop technical architecture: Validate existing or define the technology (software and hardware) used by the enterprise and show how this technology supports the business architecture.

- Evaluate enterprise elements and alignments: Evaluate alignment of the business and technical architecture to enterprise objectives and define improvement projects. The improvement projects can be for current objectives or achievement of future objectives.

At each step of the enterprise architecture process for FIT, we have identified activities and products that will be created. A high-level table is shown in Figure 2.28 on the next slide. With this process, we have developed a proposal for the support, training, and maintenance of the DHS optical system. At the high level, we identified the scope and objectives, stakeholders, and governance. Using those, the enterprise architecture team identified improvement opportunities both internally and for stakeholder use. I have put together a proposal for the support, training, and maintenance subsystem that meets the enterprise objectives and goals. The EA team was able to identify the enterprise owner for

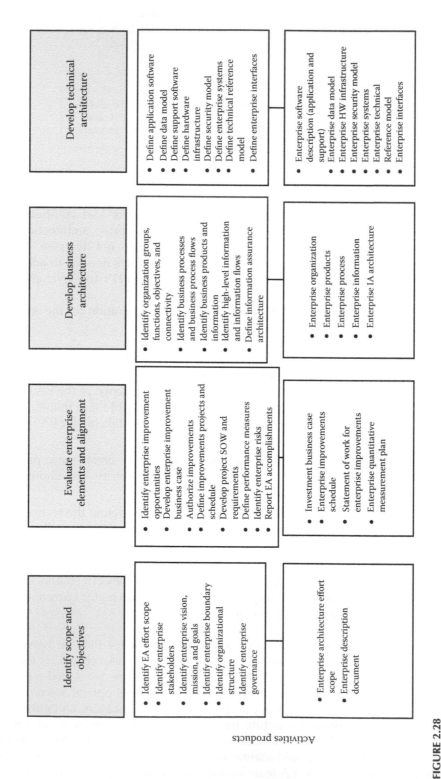

FIGURE 2.28
FIT EA process details. (Courtesy of Chris Lannon.)

this subsystem allowing us to identify the needs early and help meet our own business needs as well. The mission for this particular subsystem is to support, train, and maintain the optical system. This includes training of the operators, maintenance personnel, our own manufacturer staff, technical staff, and field support.

We conducted technical feasibility analysis to evaluate and understand the benefits of the present technical resources of the organization and their applicability to the present and future needs of the proposed system. We conducted the evaluation of the hardware and software packages to see how they will meet our proposed and future needs. We have done some research on training software packages that allows for training documentation to be created and tracked online. The training can run on mobile devices, computers, and printed to hardcopy. Users can ask questions that are sent to an expert, and the answer is then stored with the question online for easy searching later. The training package also allows for curriculums to be created so the user is able to expand upon their knowledge in an easy to follow manner. This software also contains "snippets" or modules of information that allows the user to easily find answers to questions. Simply send an email to the training system account; the system searches the database and provides training that is associated to your desktop or mobile device. The training packages are easily created with the software installed on an expert's computer that allows them to be video recorded or capture input from the keyboard, microphone, or other input device. The training software is similar to YouTube."

Garry: "Sounds too good to be true. What are our current capabilities and what needs do we have for the future?"

Karl: "The training we currently have is in PowerPoint slides and manuals. It is not easy to search or select parts of training that are needed on the fly. For our commercial stakeholders, we travel to the site to help with installation, training, and maintenance. For future government projects, we will be dealing with multiple agencies that will view the training differently. For example, DHS will be primarily using it for border patrol, looking for people crossing the border illegally. They are primarily concerned with immigration issues such as, did someone cross the border illegally? Where the DEA may want to use it for drug trafficking or human trafficking. They may want to capture images of the person that can be stored and searched for identity in another law enforcement database. The local police may have a different CONOPS in mind, for instance, finding a criminal, or searching for a lost person in a search and rescue type of mission. Again, recognition of the person but search through a different database, determine what if any medical attention may be necessary, and answer queries along those lines. These are all high-level descriptions, but you begin to understand that the training scenarios can quickly spiral out of control."

Garry: "Does it integrate with any of the systems we have now?"

Karl: "Yes, the HR system we now have is based on service-oriented architecture (SOA), which is a design methodology for software that allows for collections of separate software modules or services that provide higher functionality when integrated. The SOA technology is also quickly becoming a standard or sought-after enhancement in the DOD area because of the modularization of discrete software components. Sort of like a plug and play of systems, not fully realized yet, but the goal of SOA."

Garry: "What is the return on investment (ROI) for this technology?"

Karl: "The cost of the system is over three million dollars. However, this software will not only be used internally. As we design the system for DHS, all drawing packages, ICDs, and any other documents associated with the design, implementation, and production of the optical system can be uploaded. This allows for the developed information to be incorporated into stakeholder training packages as well. The stakeholder can login to the system and be able to view and/or download training material that is current to the system. They will also have the ability to ask an expert similar to the way our own employees are trained."

Garry: "So this could reduce travel costs and field staffing needs?"

Karl: "Yes, we estimate an ROI in 1.5 years and believe it fits with our company objective of providing high-quality support for our optical systems to our stakeholder base."

Garry: "Ok Karl, I think you have a good start for a package that we can present to Dr. Smith. Let's spend the next couple of weeks with the enterprise architecture team and run through the defined enterprise architecture process. We should identify each of the activities and processes as it pertains to the support and maintenance subsystem and the overall enterprise. These should include the stakeholders, goals and objectives, description, how it aligns with the business and program objectives, and improvement opportunities. We should understand our current capabilities and how the new system will expand upon them to meet and/or exceed our goals."

(Two weeks later, the team had set up a meeting with the CEO [Dr. Smith], CTO [Dr. Phelps], and the CIO [Mr. Blair]. This meeting was to obtain authorization to implement the support and maintenance training module that was identified in the previous section of the case study. The meeting began with a brief overview of the enterprise architecture process as defined in the previous meeting with Mr. Ben and Mr. Blair. We start now with the overview of where the training module fits into the system for the organization.)

Karl: "As seen in the previous overview section of our enterprise architecture for the company, the programs we support, and the infrastructure capabilities we currently have established, we are lacking in some areas that need to be addressed if we are to make the current and future programs successful and allow us to meet our company objectives. The U.S. Government and the agencies we are supplying our system to have placed a high value on the integrated logistics and support components of the optical system. This was seen during the proposal and in conversations with the stakeholders of the contract through our usage of the House of Quality, shown in Figure 2.29. Some of the highest ratings were for ease of use, maintainability, and serviceability. With our current system of documentation control, training design, maintenance, and support, we will have a difficult time responding to the various agencies that are now engaged in discussions with our company for additional enhancements and new programs. We will need to add additional capabilities to our training system to support future company growth and improve the overall training, support, and maintenance regimen.

I am proposing that we incorporate a new training module that allows training to be created and viewed in such a manner that will reduce our on-site travel expenses by 35% and technical staff requirements by 20% and increase communications efficiency throughout the company by 25%."

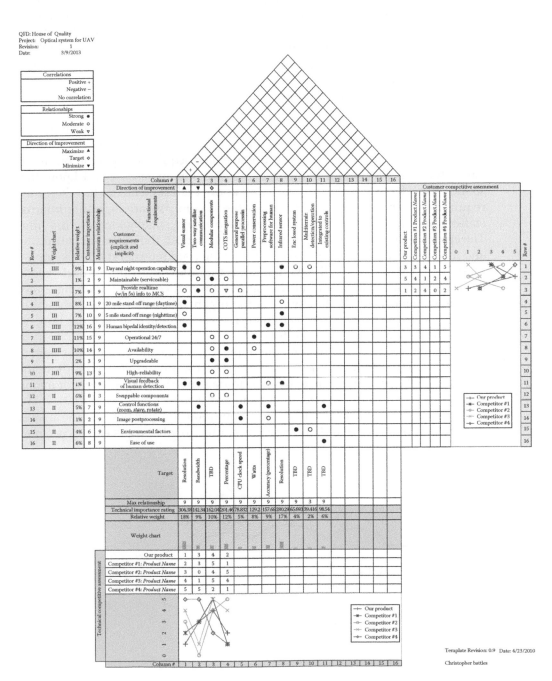

FIGURE 2.29
House of Quality for UAV optical system. (Courtesy of Chris Lannon.)

Bill: "Those are some large returns; how have you been able to identify them?"

Karl: "We identified several areas of training that could be modularized and centralized that we currently do on-site. This frees up our technical staff so that they can focus more on the design activities and other core duties. We also identified several redundant processes and documentation. We were even able to eliminate some repetitive steps and standardize the information flow between functional departments."

Tom: "Karl, you have discussed internal training as well as training for the optical system operators. Who are we targeting as the stakeholders of this system?"

Karl: "The stakeholders of the new systems include DHS, border patrol; this includes the operators and maintenance personnel at the site, technical members of other agencies, business development, procurement and acquisitions people, our internal technical staff, program management, and human resources. I know that sounds like a lot of stakeholders; as we all know, each stakeholder has their own requirements and views into and about the system. We are targeting internal staffing and the DHS, border patrol as our primary stakeholders at this time. The enterprise boundary will be to create and establish the initial set of training and maintenance procedures that can be used to train internal staff. Developing these modules in steps will have the impact of allowing training packages to be developed modularly and at the same time reduce cost of training as we hire new employees. The following is the organizational structure, shown in Figure 2.30, of the new system. The enterprise description also includes how we plan to manage change to this system.

As you can see from the picture, the training module allows for the system to be connected via the cloud infrastructure. The training system will allow the operators of the optical system to have up-to-date information on operator manuals, specific instructions on operations, and access to technical support 24/7. The system is fully searchable. If a user enters a question, the question and the answer become part of the system. This feature is similar to what users experience now

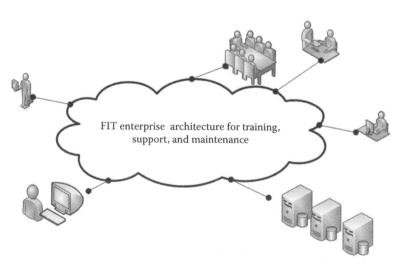

FIGURE 2.30
Organizational view of training support module. (Courtesy of Chris Lannon.)

on discussion forums and behaves the same way. For the maintenance operator, maintenance manuals are also up to date with the latest information. Videos are encoded with keyword searches to identify parts, how-to illustrations, and specifications for various parts. The system keeps track of various training a user has completed and assigns the user to a category, novice, intermediate, or expert based on the training completion and the user's competency on assessments."

Bill: "Is it possible to do this internally for training on systems design processes, coding standards, HR policies, and such? Does it integrate with our current Human Resource system?"

Karl: "The system is actually another module that has been developed by ACME. This is the same vendor that we bought the HR system from and still support. They are developing service-oriented applications that allow for discrete software packages to interact with each other."

Tom: "So what type of reports can we expect to see from this system, and how can we measure the effectiveness of the training?"

Karl: "Each training package is scored by the user. These scores are then tallied by number of users and given an effectiveness rating. If a rating falls below a certain score, the training package is flagged for review and sent back to the originator or a dedicated member of our technical staff for update. The system will also provide metrics on how many questions were asked that could have been alleviated by completing a training module. These metrics are then calculated to formulate decisions on suggested training to the user. There are also more standard metric reports of how many times a user has logged in to the system, how many times a particular training item has been accessed, number of users in the system, and so forth. We could use these metrics to determine types of training users need the most, whether it is video, documentation, or discussion type threads."

Bill: "Ok, I think we all agree that this new system provides us some needed capabilities. These seem to be in line with our requirement to deliver support, training, and maintenance components of the optical system we are developing for the DHS. It integrates with the current HR system and fits in nicely with the overall enterprise concept. What is the cost of the new system and what is the return on investment?"

Garry: "The cost of the new software system is three million dollars; however, because we have provided the ACME company with good references and feedback on the current human resources tool, they have offered us a 10% discount. We have also had the opportunity to set up temporary licenses internally and begin using the system in a small pilot study. The study team had some of our technical employees, Human Resources members, program management, and some business development folks. We also recently had one of the Department of Homeland Security personnel in house for a systems requirements meeting and showed it to them. They were impressed with the layout and ease of use. They also were intrigued by the SOA technology as they have other systems using that functionality. The return on investment we estimate to be 1.5 years. This includes acquisition costs, support costs, labor costs for training internal employees, cost of generating training packages, travel expenses, and securing the data on our data encrypted servers."

Bill: "ACME seems like a reputable company and has provided a system that has allowed us to help win programs by providing sound metrics and reporting capabilities providing an overview of the enterprise and how new capabilities or changes will affect the company's strategic planning. I think we will go ahead and buy the new capabilities and begin implementing it into our DHS optical system program as part of our solution to the support, training, and maintenance requirements. In addition to helping Homeland Security, this will allow us to be a step ahead of our competition and win new business. Karl please go forward with this acquisition."

2.4 Conclusion

Fantastic Imaging Technologies has standardized a process for evolving their EA. FIT's EA process is anchored to a 25-year legacy of SE discipline in requirements analysis and management using industry standard tools. All architecture and system-level elements are contained in a common repository and linked to requirements, thus ensuring long-term maintainability of the complete technical baseline. The system is supported by a skilled support staff and training for internal and external stakeholders. FIT used a tailored EA framework based on DODAF 2.0. FIT systems engineers follow a standard process and workflow to capture, analyze, and model requirements and efficiently interact with other enterprise actors.

Investment Decision Making Is Simplified: FIT's EA uses an EA process to simplify the evaluation of proposed solutions for program or IT infrastructure that provides a common language linking program solutions and IT initiatives with the company's strategic goals, objectives, and core business functions. FIT's holistic approach provides the executive-level staff with a strategic, enterprise view of the complete portfolio.

System Implementation Is Accelerated: FIT's EA practice applies existing processes to accelerate systems design and development and define core business practices, common data elements, applications, and standard system platforms. The processes are used to verify system needs that cross between program areas and internal areas and facilitate communication between program areas and technical staff to define custom requirements.

IT System Diversity Is Reduced: FIT's EA practice uses processes to streamline FIT's IT environment that define core business processes, common data elements, and standard applications and platforms. Standardization allows FIT to reduce cost associated with IT products, for systems maintenance and operation costs, and simplifies training requirements for both internal employees and stakeholder training. IT system diversity is reduced due to the standard-based approach, utilizing SOA technology and techniques.

EA allows entities to respond to the rapid changes in a complex, dynamic, and evolving development environment. Economic and social impacts can be analyzed, and responses to these impacts can be formulated to allow for new capabilities. A study was conducted in 2005 by the National Association of Chief Information Officers (NASCIO), which concluded that since 1999 states have made significant progress in adopting EA processes (NASCIO 2005). While most of these are in the area of technology, other areas of interest are beginning to adopt EA methodologies. These include performance management, process architecture, and business architecture. Many vendors are creating and refining modeling tools to support the usage of EA frameworks and processes.

2.A Appendix: Fourier Transform Pairs

$$u(x) = \int_{-\infty}^{\infty} u(f)e^{j2\pi xf} df \qquad U(f) = \int_{-\infty}^{\infty} u(x)e^{-j2\pi fx} dx$$

$$u(x,y) = \int\int_{-\infty}^{\infty} u(f_x, f_y)e^{j2\pi(f_x x + f_y y)} df_x df_y \qquad U(f_x, f_y) = \int\int_{-\infty}^{\infty} u(x,y)e^{-j2\pi(f_x x + f_y y)} dx dy$$

$\delta(x)$	1
1	$\delta(f)$
$\text{Tri}(x)$	$\text{sinc}^2(f)$
$e^{-\lvert f \rvert}$	$\dfrac{2}{1+(2\pi f)^2}$
$\dfrac{2}{1+(2\pi x)^2}$	$e^{-\lvert f \rvert}$
$\cos(\pi x)$	$\dfrac{1}{2}\delta\left(f-\dfrac{1}{2}\right) + \dfrac{1}{2}\delta\left(f+\dfrac{1}{2}\right)$
$\sin(\pi x)$	$\dfrac{1}{2}\delta\left(f-\dfrac{1}{2}\right) - \dfrac{1}{2}\delta\left(f+\dfrac{1}{2}\right)$
$e^{-\pi x^2}$	$e^{-\pi f^2}$
$J_0(2\pi x)$	$\dfrac{\text{rect}(f/2)}{\pi(1-f^2)^{1/2}}$
$\dfrac{J_1(2\pi x)}{2x}$	$\left(1-f^2\right)^{1/2}\text{rect}(f/2)$
$\text{rect}(x)\,\text{rect}(y)$	$\text{sinc}\, fx\, \text{sinc}\, fy$
$\text{circ}(r)$	$\dfrac{J(2\pi\rho)}{\rho}$
$\delta(r-a)$	$2\pi a J_0(2\pi a\rho)$
$\text{Tri}(x)\text{Tri}(y)$	$\text{sinc}^2(f_x)\,\text{sinc}^2(f_y)$
$e^{j\pi r^2}$	$je^{-j\pi(fx^2+fy^2)}$
$\dfrac{1}{r}$	$\dfrac{1}{\rho}$
$\left[\dfrac{J_1(2\pi r)}{r}\right]^2$	$2\left[\cos^{-1}\left(\dfrac{\rho}{2}\right) - \dfrac{\rho}{2}\sqrt{1-\left(\dfrac{\rho}{2}\right)^2}\,\right]\text{rect}\left(\dfrac{\rho-1}{2}\right)$

References

Anderson, N. and T. Erdogan. Multimodal nonlinear optical (NLO) imaging, 2014. http://www.semrock.com/multimodal-nonlinear-optical-nlo-imaging.aspx (accessed November 14, 2014).

Andrews, L. and R. Phillips. 1998. *Laser Beam Propagation through Random Media*. Bellingham, Washington, SPIE Press.

Arrasmith, W.W. 2010. Novel wavelength diversity technique for high speed atmospheric turbulence compensation. *SPIE's Defense and Security Symposium*, Orlando, FL, May 5–9.

Bahaa, E.A.S. and M.C. Teich. 2007. *Fundamentals of Photonics*, 2nd edn. New York: Wiley.

Barroero, T., G. Motta, and G. Pignatelli. 2010. *Business Capabilities Centric Enterprise Architecture*. Berlin, Germany: Springer.

Bernard, S.A. 2012. *An Introduction to Enterprise Architecture EA3*, 3rd edn. Bloomington, IN: AuthorHouse.

Born, M. and E. Wolf. 1970. *Principles of Optics: Electromagnetic Theory of Propagation, Interference and Diffraction of Light*, 7th edn. Cambridge, U.K.: Cambridge University Press.

Boyd, R.W. 2008. *Nonlinear Optics*, 3rd edn. San Diego, CA: Academic Press.

Clinger and Cohen Act. 1996. (PL 107-347) (See THOMAS [Library of Congress]).

Craig, A.T., J. Hogg, and R.V. Hogg. 1995. *Introduction to Mathematical Statistics*, 5th edn. Upper Saddle River, NJ: Prentice Hall.

deVries, M. 2010. A framework for understanding and comparing enterprise architecture models. *Management Dynamics: Contemporary Research Journal of the Southern Africa Institute for Management Scientist*, 19(2): 17–29.

DOD. 1994. *Technical Architecture Framework for Information Management (TAFIM)*, vols. 1–8, version 2.0. Reston, VA: DISA Center for Architecture.

DoDAF v2.02. 2014. DoD Architecture Framework Version 2.02. DOD Deputy Chief Information Officer. U.S. Department of Defense. http://dodcio.defense.gov/TodayinCIO/DoDArchitectureFramework.aspx (accessed November 14, 2014).

GAO. 2001. *A Practical Guide to Federal Enterprise Architecture*, Version 1.0. CIO Council. http://www.gao.gov/products/P00201 (Publicly Released: February 9, 2001).

GAO. 2005a. FBI is taking steps to develop an enterprise architecture, but much remains to be accomplished. Information Technology. http://www.gao.gov/products/GAO-05-363 (Publicly Released: September 9, 2005).

GAO. 2005b. Some Progress Made toward Implementing GAO Recommendations Related to NASA's Integrated Financial Management Program. Business Modernization. http://www.gao.gov/products/GAO-05-799R (Publicly Released: October 27, 2005).

GAO. 2010. Federal enterprise architecture program EA assessment framework, Version 2.0. Organizational Transformation. http://www.gao.gov/products/GAO-10-846G (Publicly Released: August 5, 2005).

Goodman, J.W. 1985. *Statistical Optics*. New York: Wiley.

Goodman, J.W. 2005. *Introduction to Fourier Optics*, 3rd edn. Greenwood Village, CO: Roberts & Company.

Hite, R.C. 2004. The Federal Enterprise Architecture and Agencies' Enterprise Architectures Are Still Maturing. Information Technology: [GAO Testimony]. http://www.gao.gov/products/GAO-04-798T (Publicly Released: May 19, 2004).

Hite, R.C. 2006. Progress Continues, but Challenges Remain on Department's Management of Information Technology. Homeland Security: [GAO Testimony]. http://www.gao.gov/products/GAO-06-598T (Publicly Released: March 29, 2006).

Hornford, D. 2011. *TOGAF*, Version 9.1. Zaltbommel, the Netherlands: Van Haren Publishing.

Irvine Sensors Corporation. n.d. Development & Sales: Miniaturized infrared and electro-optical devices. http://www.irvine-sensors.com (accessed April 26, 2014).

Lannon, C. 2014. Figures with release for use in chapter 2. See also release file for verification.

Minoli, D. 2008. *Enterprise Architecture from A-Z: Frameworks, Business Process Modeling, SOA, and Infrastructure Technology*. Auerbach Publications (CRC), Boca Raton, FL.

MoD. 2012. MOD Architecture framework. https://www.gov.uk/mod-architecture-framework (accessed November 14, 2014).

NASCIO. 2005. NASCIO Enterprise Architecture Assessment. NASCIO Online. http://www.nascio.org/publications/documents/NASCIO-eaAssessment.pdf (accessed November 14, 2014).

Opengroup. 2014. The Open Group Architecture Framework (TOGAF). TOGAF Online. http://www.opengroup.org (accessed November 14, 2014).

Oppenheim, A. and R. Schafer. 1975. *Digital Signal Processing*. Englewood Cliffs, NJ: Prentice Hall.

O'Rourke, C., N. Fishman, and W. Selkow. 2003. *Enterprise Architecture Using the Zachman Framework*. Boston, MA: Course Technology.

Roggemann, M.C. and B. Welsh. 1996. *Imaging Through Turbulence*. Boca Raton, FL: CRC Press.

Sambur, M.R. 2003. Memorandum: Incentivizing contractors for better systems engineering. January 6, 2003.

Schekkerman, J. 2008. *How to Manage the Enterprise Architecture Practice*. Trafford Publishing, Bloomington, IN.

Sessions, R. 2007. A comparison of the top four enterprise architecture methodologies. Microsoft Developer Network. http://msdn.microsoft.com/en-us/library/bb466232.aspx (accessed November 14, 2014).

Shen, Y.R. 1998. *The Principles of Nonlinear Optics*. Chichester, U.K.: John Wiley & Sons.

Shanmugan, K.S. and A.M. Breipohl. 1988. *Random Signals Detection, Estimation, and Data Analysis*. New York: John Wiley & Sons.

van den Berg, M. and M. van Steenbergen. 2006. *Building an Enterprise Architecture Practice: Tools, Tips, Best Practices, Ready-to-Use Insights*. Dordrecht, The Netherlands, Springer.

Whitehouse. 2007. FEA Consolidated Reference Model Document, version 2.3. http://www.whitehouse.gov/sites/default/files/omb/assets/fea_docs/FEA_CRM_v23_Final_Oct_2007_Revised.pdf (accessed November 14, 2014).

Zachman, J.A. 1987. A framework for information systems architecture. *IBM Systems Journal*, 26(3): 276–292.

3

Systems of Systems, Family of Systems, and Systems Engineering

What is the use of a fine house if you haven't got a tolerable planet to put it on?

—**Henry David Thoreau**

3.1 Overview

Systems are everywhere! From small, familiar objects like a coffee mug to complex objects like a satellite, we are continuously surrounded by systems. Some systems occur naturally, like planet earth, while others are man-made, like an automobile. When classifying each of these items as a system, it is important to remember that a "system" is defined as a combination of things that work together to achieve a common goal. With this definition, it becomes clear that many things can be classified as a system with the proper reference applied.

In early-recorded human history, systems principles were recognized but not fully understood, as evidenced by the use of mysticism to interpret some natural interactions such as seasons, phases of the moon, sunrises, and herd migrations. As human thought developed and advanced, a broader interpretation of the natural systems started to develop, seen by the creation of theories of orbital mechanics, relativity, and quantum mechanics. Eventually, man-made systems interacted with natural systems in significant and observable ways (National Aeronautics and Space Administration Goddard institute for Space Studies 2012). The impacts of some man-made systems on these natural systems became destructive, giving start to the depletion of nonrenewable and limited natural resources. On many occasions, studies into system interactions were initiated to avert further resource exhaustion and damage. For example, the North American Whitetail Deer, Elk, and Bison were once almost hunted into extinction until cattle ranching, as a renewable food source, gained popularity (Evans 1996). Another example is how the clothing industry relied heavily on North American beaver, mink, and cougar, causing them to be hunted and trapped almost into extinction until renewable resources such as sheep ranching, silk worm production, and cotton crops gained popularity (Saundry and EH.net. 2012). However, one of the most recognized and significant impacts of man-made systems on natural systems occurred with the advent of the industrial revolution at the turn of the century. Massive factories' increased smog, soot, and pollution were impacting water and air quality, spurring groups and governments into action to limit industry pollution on the population and the environment. These actions started the modern-day environmental movements and man-made climate change research (Robbins 2007; Lorentz 1972). With better human understanding of the effects that systems have on other systems, insights into the nature and behavior of systems, systems of systems (SoS), and family of systems (FoS) began to emerge.

The earth can be used to examine more closely the definition of a system. The earth is a natural system that contains many elements within it: water, land, gravity, atmosphere, plants, people, and animals. Each of the aforementioned elements brings with it certain characteristics, and all of these elements support each other to achieve harmony within the system, maintaining the ultimate goal of existence. Perturbing one or more of these elements will cause the system to experience a shift that may or may not impact other elements of the system or endanger the system itself. Nevertheless, the degradation or removal of an interacting element within the system would cause turmoil within the system until a new balance is again obtained. An example of this would be a shift in the atmosphere causing a shift in sustainable plant life within a specific region. Eventually, a group of elements affected by it, for example, the plants within that region, would balance with the atmospheric change, whether by adapting or by a different species becoming more prominent.

The earlier example of the automobile as a man-made system also contains many elements within itself: body, tires, engine, seats, steering, fuel, battery, and computer. Although each element will bring certain properties to bear within the system, all of the elements must work together to achieve the ultimate goal of transporting something somewhere. Without the engine, the system will fail; without the tires, the system will fail; and without the fuel, the system will fail. The engine needs fuel, the computer needs energy, the wheels need steering, and the body needs seats; therefore, it takes multiple elements working together to achieve the system goal.

Blanchard said, "Systems are as pervasive as the universe in which they exist. They are as grand as the universe itself or as infinitesimal as the atom. Systems appeared first in natural forms, but with the advent of human beings, a variety of human-made systems have come into existence. In recent decades, humans have begun to understand the underlying structure and characteristics of natural and human-made systems in a scientific way" (Blanchard and Fabrycky 2011). Systems are made up of individual elements. For example, our galaxy as a system has individual elements. These elements are the sun, planets, and moons. The atom, as a system as well, has individual elements: the electrons, nucleus, protons, and neutrons. Each individual element of a system brings its own contributions to the system as a whole, and the interactions within each of these elements are what determine how a system works to achieve a goal. In the end, the system is as we define it. It can be larger or small and it can, itself, be made up of numerous systems—an SoS. As systems engineers, we need to define the system in context to its relation to other systems and its environment.

The remaining sections will discuss in detail the differences between an SoS and an FoS. Also, it will look into the specific example of an optical subsystem on an unmanned aerial vehicle (UAV) and how different components can be treated as either an SoS or FoS. Furthermore, a case study will be presented, integrating the systems engineering (SE) and the UAV technical aspect of this chapter.

3.1.1 Systems of Systems

A system is a group of elements that work together to accomplish a common goal; similarly, an SoS is a group of systems where all systems work interdependently to accomplish a higher-level goal. One formal definition of an SoS is "a collection of task-oriented or dedicated systems that pool their resources and capabilities together to create a new, more complex system, which offers more functionality and performance than simply the sum of the constituent systems" (Wikipedia 2014). Another definition for SoS, as defined in the 2008 Department of Defense (DOD) Defense Acquisition Guidebook (DAG), is "a set or arrangement of systems that results when independent and useful systems are integrated into a larger system that delivers unique capabilities" (ODUSD 2008). Applying these definitions, a quick example could be found in an aircraft navigation system. Sensor suites that take advantage of an existing basic navigation system can be added to the aircraft while also incorporating additional stand-alone sensor systems in such a manner that they complement each other's strengths/weaknesses. All of these systems would then collectively be recognized as one SoS providing its host with robust navigation capabilities. Each system within the SoS has interdependency on the other systems within the system. That is, the inputs and outputs of a system in an SoS have a cause-and-effect relationship on the system's operation as a whole.

Using the previous example of the earth's natural system, the element considered as people could be further defined as a subsystem with elemental composition and functions allocated at the lowest levels. Many distinct societies come together to form this element, and furthermore, these societies are comprised of smaller, independent operating groups of people that work together. The people of the earth are varied by many characteristics: geographic regions, ethnic backgrounds, socioeconomic backgrounds, religious backgrounds, and political backgrounds. However, they all reside on and rely upon the earth system; they have a symbiotic relationship with the higher-level system— the earth. As mentioned before, the impacts that people have on the earth system can directly impact all of the other systems within it, notably the depletion of resources, impacts on other living systems (species), impact on environmental systems, and the climate system.

Looking back, the previous automobile example could be considered an SoS as well. The automobile is composed of smaller systems, each made up of individual parts, altogether functioning as a system. As illustrated in Figure 3.1, these smaller systems include the body system, wheel system (tires), engine system, seating system, steering system, fuel system, power system, and computer system. Each of these smaller systems, with adaptation, could conceivably provide a capability on their own and are themselves made up of smaller components, thereby satisfying our concept of a system. The interdependency of each of these smaller systems within the automobile system is vital to the system as a whole. They operate through working together, thereby inputs and outputs of one system can affect the inputs and outputs of another. For example, if the power steering subsystem suffers a leak in the power steering liquid tank, the car system's wheel (steering system) will become very stiff, thus impeding proper functionality of the wheel system. Both failures will have negative effects on the car's ability to function as a whole.

FIGURE 3.1
Automobile subsystems. (Courtesy of Richard Qualis. With permission.)

Without the ability to steer, tires will not be able to turn; thus, no direction may be given accurately to the car when in motion.

Using the DOD's definition for software SoS, they describe the four primary types of SoS in use today as virtual, collaborative, acknowledged, and directed (ODUSD 2008). The virtual SoS does not use central management, nor does it operate with an agreed-upon purpose for the overall SoS. Collaborative SoS have constituent systems that interact only voluntarily as needed to satisfy a purpose. An acknowledged SoS has common objectives and resource management, but the systems are maintained independently, and they collaborate together to achieve the system's purpose. The last type of system is the directed SoS; it is fully integrated, and its engineered system is geared toward being managed over the long-term operation of the system. However, with the constant desire (and effort) to link systems with information sharing purposes, new approaches are being considered to allow this linkage, without the need of common goals among systems (ODUSD 2008).

Just as a system is a group of elements working in conjunction to accomplish a goal, an SoS is a group of constituent systems working in conjunction to accomplish a larger goal than the individual systems could accomplish on their own. A defining characteristic of the SoS's architecture is the interdependency that exists among the inputs and outputs of the constituent system elements to accomplish the overall SoS goals. The U.S. military recognizes that combining some systems to form larger systems can expand their capabilities, but managing all of those constituent systems to achieve the desired SoS goal would require developing new capabilities.

3.1.2 Family of Systems

Since a system is a group of elements that work together to accomplish a goal, an FoS can be viewed as a group of systems that work independently to accomplish the same goal. The manner in which the subsystems are arranged will determine the output of the overall system (Office of the Under Secretary of Defense for Acquisition, Technology and Logistics). The Office of the Deputy Under Secretary of Defense (ODUSD 2008, 4) stated that the FoS can also be defined as a "set of systems that provide similar capabilities through different approaches to achieve similar or complementary effects." Each system within the FoS operates independently from one another, though they all work to accomplish the overall system goal.

Revisiting the automobile example for the FoS concept, it could be considered to be part of an FoS by looking at the overall system (the car) as part of a larger system, that is, short-range personal transport system. This system has the ability to provide a means for passengers to commute to work and school. As part of its family, there are various systems, including cars, buses, motorcycles, and even bicycles. Comparing the automobile system to the other family members, they are very distinct in capabilities and do not operate with synergy, yet at the same time, all of these systems come together to form the overall personal transport system. These individual systems also have the same goal—provide a personal transport capability for an individual. The system is not paralyzed if one system or vehicle breaks down thanks to each system's independence, and another vehicle may step in to provide the short-range transport capability for the family in need.

Just as a system is a group of elements working in conjunction to accomplish a goal, an FoS is a group of systems that can provide similar or complimentary capabilities through different means. The FoS is different from the SoS concept since the former does not provide expanded functionality or capabilities larger than the sum of its parts.

3.1.3 Systems Engineering Processes across the Systems Development Life-Cycle Phases

The systems engineering (SE) approach takes the long view of how the system under study will be designed and operated over that system's life expectancy. The SE life cycle is commonly divided into phases from the conceptual design to disposal of the system. As depicted in Figure 3.2, the systems development life-cycle (SDLC) phases are the conceptual design phase, the preliminary design phase, the detailed design and development phase, the production phase, the system operations and support phase, and the system retirement and disposal phase. Each of these phases has their own set of rules, best practices, tools, techniques, and methodologies, which we will discuss further ahead.

3.1.3.1 Overall Systems Engineering Development Process

The conceptual design phase of Figure 3.2 is designed to give the system shape by defining the system-level requirements. Block one begins with problem definition and needs identification, both of which are critical processes of this phase. Here, the problem that the system will solve is thoroughly discussed with the stakeholders and the development team. The essential motive is to clarify and take into account the needs and wants of the stakeholders as much as possible and develop the system concept. The needs, and agreed wants, are expanded into a set of stakeholder requirements that are subsequently evolved into systems requirements by the end of the phase. The systems requirements are captured in a system-level requirements document. This requirement document contains both functional and nonfunctional operational requirements. The system operational requirements and the maintenance and support concepts tell us how the system will perform and what is required to keep the system operational through the operations and support phase of the SDLC. Trade studies will be completed to begin architecting the system. Feasibility studies are conducted, typically on every requirement one-on-one to see if they can be accomplished or not. It is important that we have a solid system specification (A spec) before advancing to the next system life-cycle phase.

Block two of Figure 3.2, preliminary design, includes a high-level functional analysis of the system and a subsequent requirements allocation where system-level requirements are allocated to the subsystem level and preliminary designs and midlevel analysis are accomplished. Refined feasibility and trade studies are performed at this stage. A great effort is placed on allocating the system-level requirements to the subsystem level and developing functional block diagrams (FBDs)/functional flow block diagrams (FFBDs) to describe the functional decomposition of the system into its constituent subsystems. The end of the preliminary design phase is where the allocated requirements are used to generate the development specifications (B specs) and these are placed under configuration control (baselined) at the preliminary design review (PDR). There can be many B specs for a given system. These often match the functional decomposition of the system itself and usually are separated into hardware and software development specifications. Further, preliminary specifications such as the C, D, and E specifications may be developed concurrently during this phase as well. These specifications will be discussed in more detail in later chapters.

The detail design block of Figure 3.2 is somewhat similar to the previous one; in this case, higher-level products (e.g., requirements, designs, analyses) are taken to a more detailed and final level. Test plans, test procedures, and test cases are created covering every requirement. The test team eventually executes these test plans/procedures/cases after the system has been produced. In some cases, prototypes are manufactured during

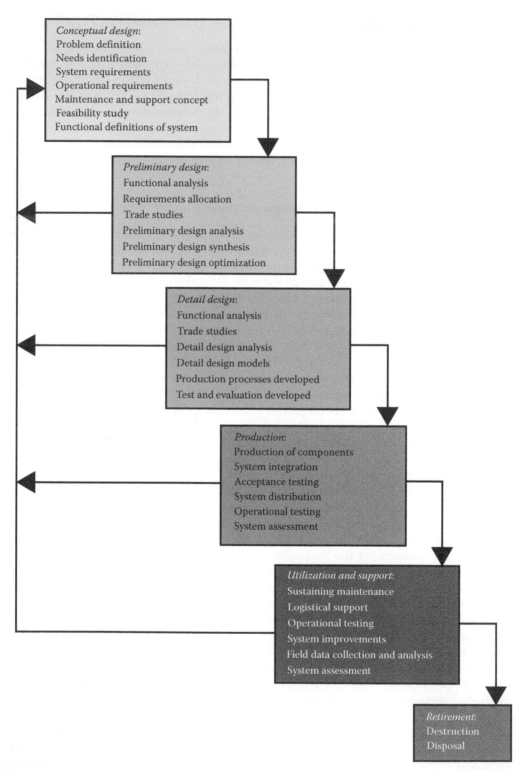

FIGURE 3.2
The SE development process. (Courtesy of Jeffe Baudek. With permission.)

this phase, to further define the system. The product specification (C specification), the process specification (D specification), and the material specification (E specification) are all completed during this phase before heading into the next phase. The milestone event during this phase is the critical design review (CDR) where the product specifications (C specs) are baselined along with the D and E specifications if applicable.

The production phase of Figure 3.2 is where the system is produced, constructed, or manufactured. Changes to the design specifications are documented in a postproduction baseline revision called the "as-built" specifications. The system is then tested to ensure that all of the requirements at each developmental level are satisfied. The test plans, test procedures, and test cases developed during the design phases are executed starting with the most detailed level requirements—the C specifications. The system components are tested against the C specification requirements. Verification testing is used to ensure that each requirement has been individually satisfied. Once the component requirements have been verified, the components are integrated into the system and integration testing occurs. At this stage, the integrated components are tested against the B specification level requirements. Once the integration testing is complete and the integrated components pass the integration tests, then the system-level tests begin. The system-level tests are tested against the system-level requirements contained in the A specification. Once the system passes the system-level tests, operational test and evaluation (OT&E) take place. During OT&E, the stakeholder evaluates if the system performed as envisioned. Validation testing occurs during OT&E wherein the system is tested against real-world/operational conditions. As an analogy, checking a racecar engine's performance requirements on a stand in a laboratory can satisfy verification testing, but validation testing requires checking requirements or groups of requirements with the engine in the car, roaring around the test track! At a successful completion of OT&E, the stakeholder typically takes charge of the system and the developmental effort is considered at an end.

The utilization and support phase of Figure 3.2 block five consists of the system's operation in the stakeholder's environment. This phase is when the maintenance and support plans that were developed in the system's design phases are implemented. Field data are collected and analyzed in the areas of maintenance, system operations, and logistical support for assessment purposes and for use in systems reliability, maintainability, and availability (SRMA) and life-cycle support models. This phase continues throughout the life of the system until the system is ready to be retired from service.

The retirement phase of Figure 3.2 block six consists of the removal from service and satisfactory disposal of the system. The system may have critical components disassembled and destroyed, whereas other components may be salvaged. The space shuttle, for instance, may find a sanitized version of itself on display in a museum somewhere. Component-level disassembly may be required to meet federal, state, or local environmental recycling/disposal requirements or security requirements. Disposal is a crucial design criterion that must not be ignored since eventually all systems face retirement and disposal. Proper planning that includes logistical considerations must be developed early on in the SDLC. These plans must detail how to handle disposal, including the transportation of hazardous materials during disassembly and the safeguarding, handling, and disposal of sensitive/classified information.

3.1.3.2 Systems Engineering Waterfall Process

The spiral model represented in Figure 3.3 is generally intended for large complex programs where it is beneficial to have incremental releases and refinements for each successive loop around the circle (Boehm and Hanson 2000). In the spiral model, a system

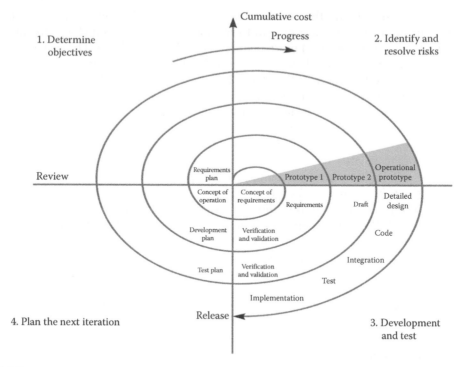

FIGURE 3.3
Spiral model of SE process. (Reprinted from Gansler, J.S. et al., Using spiral development to reduce acquisition cycle times, UMD-AM-08-128, Center for Public Policy and Private Enterprise, College Park, MD, 2008, p. 12, http://www.dtic.mil/dtic/tr/fulltext/u2/a494266.pdf, accessed April 21, 2014. With permission.)

development effort is broken into a set of progressive, spiral activities that have clearly defined exit criteria at the end of each spiral.

For instance, in Figure 3.3, some precursors to the first spiral starting with requirements plan include conceptual formulation activities such as stakeholder requirements generation and the formation of the concept of operations (CONOPS) for the system. Upon successful review of these products, the first spiral starts with the generation of the requirements plan. The requirements plan captures the essential elements of the CONOPS and stakeholder needs and wants. The requirements plan also determines the objectives that need to be satisfied in the requirements generation process. In the top right quadrant of Figure 3.3, system/service/product/or project risks are identified and resolved, and high-level prototypes may be generated as part of risk reduction activities. The main products of this quadrant of this part of the spiral development are a well-defined requirements plan and a high-level risk analysis with acceptably mitigated risks. The next quadrant of this spiral involves the generation, verification, and validation of the systems requirements. This quadrant of the spiral development method is about development and test. Once the requirements are verified, validated, and accepted, planning for the next spiral evolution is accomplished in the lower left quadrant of Figure 3.3. In this case, the development plan is generated. The end of a particular spiral occurs at a review that evaluates the results of the spiral activities. A set of exit criteria are established that must be satisfied in order to proceed to the next spiral. In this case, the results are acceptable versions of the requirements plan, risk analysis, and plan for risk mitigation, any high-level prototypes, the requirements themselves, and the test plan. Assuming that the review is positive,

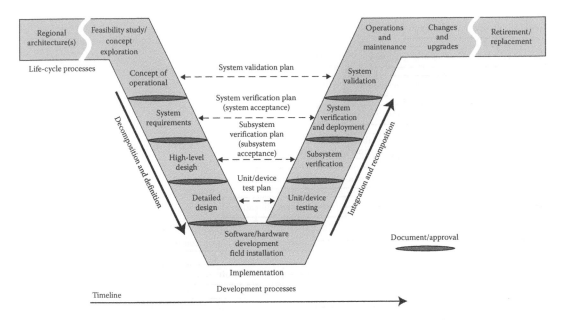

FIGURE 3.4
V Diagram of SE life-cycle process. (Reprinted from Rausch, R. et al., Economic history of the North American fur trade, 1670 to 1870, in: *Encyclopedia of Earth*, C.J. Cleveland, eds., Environmental Information Coalition, National Council for Science and the Environment, Washington, DC, 2007, p. 5. With permission.)

progression to the next spiral is authorized. The spirals continue through successive steps of determining objectives, evaluating and reducing risk, evolving prototypes, developing and testing, and then planning for the next spiral. This model is very effective in projects that involve incremental funding. Funding decisions can often be linked to demonstrated progress at the review points of the spirals.

Figure 3.4 depicts another common SE tool called the V diagram (U.S. Department of Transportation, Federal Highway Administration 2007). The left side of the V shows the systems design process and the evolution from the systems requirements downward to the detailed design. At the bottom of the V diagram, the system is produced. The right side of the V illustrates the testing process starting with verification testing at the unit level (Rausch et al. 2007).

Notice that the unit is tested against the detail design level requirements. Next, the units are integrated into subsystems and integration testing occurs. These subsystems are tested against the high-level design (e.g., B specification level). After integration testing, system-level testing occurs. For system test, the system is tested against the operational requirements given in the A specification. System validation occurs during OT&E, and the system is subsequently fielded.

Figure 3.5 shows the so-called waterfall model. The spiral model, the V diagram, and the waterfall model are all standard models for depicting the SE design process over a product life cycle (Harris Corporation 2012). Each model has broad applications; however, the waterfall model is the methodology that will be used for depicting the iterative design process here. Because the waterfall model has a clear, practical, and sequential process flow, this model will also serve as the organizational model for the rest of this text.

The waterfall model of Figure 3.5 displays the major steps followed in the SDLC. The SE process begins with the problem statement and its definition. The problem definition

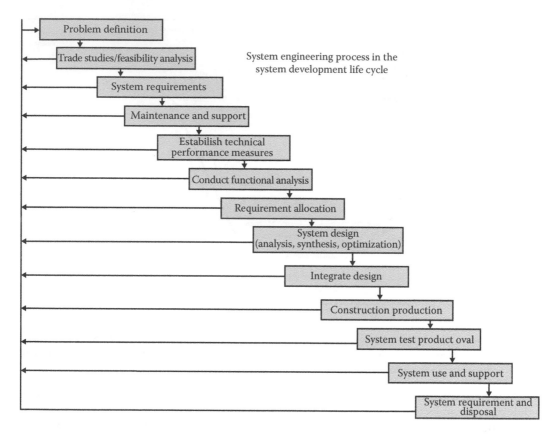

System engineering process in the
system development life cycle

FIGURE 3.5
The waterfall SE process in the system life cycle.

is the most important step in the design process because it is at this point in the process where the development team needs to listen carefully to the stakeholder to clearly identify the problem that the system is intended to solve. This step is accomplished by identifying stakeholder needs and wants through interviews and conversations with the stakeholders and the engineering team. Multiple sessions are usually required in meetings between engineering and stakeholders with each meeting being carefully dissected by the engineering team for further problem definition refinement. A CONOPS is developed to understand the role the system plays in its environment and its relationship to other entities within a given enterprise.

Stakeholder requirements are defined from the stakeholder needs and CONOPS. The systems requirements at this level are very broad, usually with high-level technical performance specifications that focus on "what" the system must do and not "how" to do it. This level of requirement is usually written in the stakeholder's language with engineering intent to refine the requirements at a later date (e.g., feels smooth, lasts long time, runs fast, and works well). In the problem definition phase, a set of goals and objectives are typically developed. The goals are measurable "big deal" type of events and projections that have business or technical focus. An example of a business goal is, "We will capture 10 percent of the local high-resolution imaging market, within 3 years of the delivery of our first sensor." Another might be, "We will achieve an average of 10 percent sales growth annually over the next three years." An example of a technical goal could be, "Our optical imaging systems will provide the highest spatial resolution imaging systems available on the market." Any ambiguities in the high-level stakeholder requirements must be removed in subsequent requirements levels (e.g., system-level requirements and below) by including measurable metrics in the lower-level requirements. For instance, from the previous example, "lasts a long time" would be further refined at the system specification level using reliability-centered maintenance metrics such as the mean time between failure, mean time between repair, and operational availability to name a few.

As an entity in an enterprise, the enterprise architecture that is to be used needs to be established and understood. For instance, if the system is a DOD acquisition, then the DOD Architectural Framework (DODAF) is mandated. The system architecture needs to be developed so that it is in compliance with the guidance in the higher-level enterprise architecture framework.

The system specifications (A specifications) capture the operational requirements for a system. If the system is an optical system, the optical systems requirements capture the functional and nonfunctional operational requirements for the optical system. These requirements are written in operator language and capture all the operational details needed so that the development team can design, build, test, deliver, support, and retire the optical system over its life cycle.

The maintenance and support expectations for the system are discussed to determine what infrastructure, logistics, repair, factory sparing models, and maintainability models the stakeholder needs for their system. This is also the moment when the stakeholder has an opportunity to communicate the operational flow and tempo expected from the system. This helps the engineers determine the reliability the system must have in order to meet these expectations.

Measures of performance (MOPs) and technical performance measures and metrics (TPM&Ms) begin to be established, as are key performance parameters (KPPs). The engineering team evaluates the systems requirements, and they generate associated measures and metrics to indicate systems performance. These measures and metrics are especially important in testing since they are typically used in the test plans, test procedures, and test cases to demonstrate whether or not the system has successfully satisfied the requirements. These measures and metrics are also prioritized with the more important measures, metrics, and associated parameters being separately tracked as MOPs, TPM&Ms, and KPPs. These will be further discussed in a later chapter of this text. In order to communicate the MOPs, TPM&Ms, and KPPs to the stakeholders and to the development team, the quality function deployment (QFD) method is often effective.

The QFD method, otherwise known as the house of quality, can be used to evaluate the requirements based on stakeholder priority. The generic form of the QFD is shown in Figure 3.6. The QFD can also be used to show the results of benchmarking studies wherein performance aspects of the system can be compared to "best in class" systems available on the market.

In the QFD approach, the left side of Figure 3.6 shows the "voice of the customer." Typically, you would list stakeholder requirements or high-level MOPs, KPPs, or TPMs to illustrate what is important to the stakeholder. The customer rankings are accomplished by the stakeholder to show the relative importance of each of the items in question. The top of the QFD shows the technical response to the voice of the customer entries. In this case, product characteristics would contain columns of attributes that would show how product characteristics would address the row entries in the voice of the customer section (e.g., the stakeholder requirements). The right side of Figure 3.6 addresses the competitive evaluation such as benchmarking results. The relationship matrix contains scores that

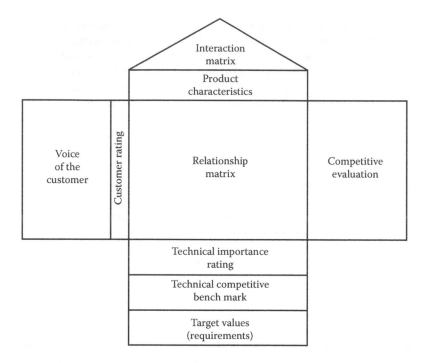

FIGURE 3.6
Sample QFD aka house of quality. (Reprinted from Sherif, J.S. and Tran, T.-L., An overview of the quality development (QFD) technique, Software Assurance Group, Jet Propulsion Laboratory, Pasadena, CA, 1994, p. 9. http://trs-new.jpl.nasa.gov/dspace/bitstream/2014/33621/1/94-1598.pdf, accessed March 30, 2014. With permission.)

relate the voice of the customer to the product characteristics. For example, if the optical system voice of the customer entry were "ease of use," then the product characteristic response could be a "GUI menu for operator." The relationship matrix would score the overall importance of the product characteristic response relative to the weighted "ease of use" row entry. The weighting comes from the stakeholder relative preference rating. At the bottom of the house of quality are the technical importance ratings, which are the column sums. A technical competitive analysis along with requirements-based target values for each technical response (product characteristic column entry) can also be provided. In this fashion, the QFD provides quite a lot of useful information that can be quickly communicated to the stakeholders or the development team.

Once the MOPs, TPM&Ms, and KPPs have been established, trade studies and/or feasibility studies are performed. The system engineer should focus on at least three fundamental feasibility studies: cost, technical, and schedule. Feasibility can be considered the complement of risk. The more feasible something is, the less risky it is and vice versa. Cost feasibility is an assessment of the capability to complete the requirement, task, or activity given the allocated funds. Technical feasibility assesses the state of technology and determines if the requirement, task, or activity can be accomplished with the available technologies. Schedule feasibility is an assessment if the requirement, task, or activity can be completed within the allowed timeframe. If a requirement, task, or activity is found to be unfeasible or have low feasibility, a risk assessment should be accomplished to assess and document the risk. Risk management methods should be employed to mitigate the risk if possible. Trade studies occur when there is a trade-off decision to be made. For example, choosing the type of sensor used in an imaging application, determining the type of data

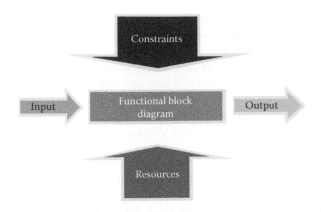

FIGURE 3.7
FFBD for each element. (Courtesy of Jeffe Baudek. With permission.)

exfiltration mechanism, or choosing the best imaging system design can be accomplished by a trade study.

As part of the optical systems design block, a functional analysis is conducted. The functional analysis identifies the necessary functions of each subsystem within the system. FBDs and FFBDs are used to describe the necessary subsystems that are generated from analyzing the system-level requirements. The FBDs are used to define and expand the next lower design level, in this case the subsystem level. FFBDs are used to illustrate processes or activities wherein sequencing and order are important. The functional analysis includes the FFBD where the functional blocks are linked showing how the inputs and outputs of each system element flow within the system design (USAF 1969). For FBDs and FFBDs, functions should have a high-level evaluation of the necessary inputs, the function's constraints, the resources required for the function, and the function's output as illustrated in Figure 3.7.

Each block in a given FBD should be evaluated to ensure that interfaces between functional blocks are properly considered, resources are available to implement the function, all constraints are considered, and there are no missing requirements for the function. As an example, a function for a spaceborne imaging system would require special testing facilities such as shock and vibration testing, thermal cycle testing, radiation testing, and vacuum testing. After the functional analysis is completed, the requirements allocation process commences wherein the functional blocks generated during the functional analysis decomposition are partitioned and requirements are allocated, apportioned, or derived into the lower-level functional blocks and captured in the lower-level requirements document (B specifications).

Further along, in the system design and analysis, the individual system components are developed through an iterative process of model/design, analysis, trade-off, and test. This design process seeks to design and evaluate all alternatives to the design with respect to all of the "ilities": reliability, maintainability, usability, supportability, transportability, flexibility, availability, survivability, testability, producibility, disposability, affordability, and so forth. The result of the system design is a set of components or elements that are ready to be integrated into a system.

In the optical design integration block, the individual subsystem designs are integrated into the overall system design. The integrated design is reviewed and expressed in a lower set of requirements documents, the B specifications. A formal review, the PDR, evaluates the preliminary design and associated requirements documents, and, after a successful review, this marks the end of the preliminary design phase. The "boxed" process shown in Figure 3.5 repeats for the next, more detailed, level of the design, the detailed design phase. This phase starts with the B specification requirements and proceeds in the same fashion as in the preliminary design phase. Namely, the B specifications are used to drive the functional analysis for the lower-level design. After the functional decomposition, requirements allocation is used to derive, allocate, or apportion B spec level requirements to the most detailed level—the component level. The allocated requirements are captured in the product specifications (C specs), process specifications (D specs), and material specifications (E specs), and all are baselined and reviewed at the CDR. Once the design documents are all completed, the optical system is built in the optical system production, construction, and manufacturing block. The optical system then undergoes verification testing for the components and integrated components. Testing starts small with the individual components to meet the lower-level detailed design requirements (C specs). Once component testing is completed and all the components passed the tests, then the components are integrated into higher assemblies (subsystems) and tested against the preliminary design requirements (B specs). The optical systems test block completes with the completed optical system being tested against the A specification requirements and the optical system passing validation testing in the field at OT&E.

In the optical systems operations, maintenance, support, and service block of Figure 3.5, optical systems performance is linked to the reliability, maintenance, and logistical models developed during the conceptual, preliminary, and detailed design phases. During this block, the maintenance infrastructure is implemented and field testing data are compiled and analyzed against the expected models to ensure that the system design is behaving as expected (Rausch et al. 2007). If the maintenance and support design is not meeting requirements, then changes may have to be made to the system design or the maintenance and support models to ensure that systems requirements are being satisfied. This too is an iterative process that consists of evaluating field data to ensure that systems requirements are being satisfied.

Finally, the system retirement and disposal process can often be overlooked within the system design, but it is extremely important. Today's environmental concerns and regulations force good stewardship of the materials used in manufacturing. The technicians, material, and logistic support required to safely disassemble, transport, and dispose of the system components can be a significant cost driver that must not be overlooked in the overall system design.

The SDLC process model depicted in Figure 3.5 is a highly iterative and multidisciplinary process that forces the system engineer to contemplate and plan for many contingencies. The SDLC process starts at the conceptual 30,000 ft level of what problem the system is intended to solve and finishes with a fully functional system that may or may not suffer eventual decommissioning and disposal.

3.2 Optical Systems Building Block: Optical System of Systems Model

In the following section, an SoS model will be described; this encompasses the overall technical components that interact to form an optical system. Similarly to how the previous section was able to fashion the SoS model, and through its use depict the interactions of the constituent members found within the enterprise architecture model, the intent of

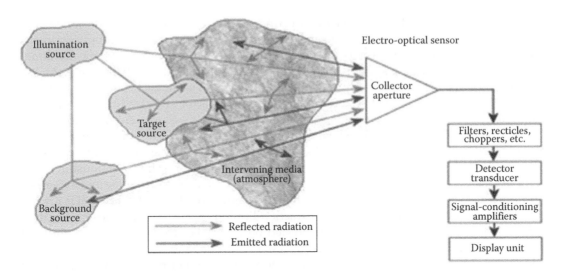

FIGURE 3.8
(See color insert.) General-purpose physical SoS model for analyzing an optical system.

the technical SoS model will be to unravel the interactions found within a specific optical system. Shown within Figure 3.8 is a representation of a general-purpose imaging system, depicting its technical oriented SoS model.

This model may be refined and modified or replaced with other models as necessary to be used with other optical applications such as the following: active imaging (e.g., laser-based imaging), space-based imaging, medical imaging (e.g., magnetic resonance imaging), airborne imaging, optical tomography, spectroscopy, and underwater imaging. An example of a modification to the model required to represent space-based imaging would be to change the intervening medium to a vacuum, instead of atmosphere. Similar modifications become necessary to properly model underwater imaging, with the intervening medium being again changed, this time to water, with the collecting aperture located either in or outside of the water. For an airborne imaging system that is looking at an object in the water, the intervening media would be two layered. The target would be immersed in the water and the immediate intervening media surrounding the target would be water. However, between the surface of the water and the airborne collecting aperture, the intervening media would be air. For a wide variety of optical detection scenarios, the object is at a remote location from the optical detection system and the intervening media is air. As will be seen in subsequent chapters—especially the chapter discussing the effects of the atmosphere on optical systems—the properties of the intervening media can dominantly affect the performance of the optical system. For instance, in well-designed optical imaging systems that operate in the visible part of the electromagnetic spectrum, the atmosphere is often the dominant effect that contributes to the loss of spatial resolution of the optical imaging system (Roggemann and Welsh 1996).

We now describe the essential elements of our optical SoS. Why SoS? We use this terminology to illustrate the multidisciplinary nature of the optical detection problem. For instance, there are a variety of excellent imaging system manufacturers that develop complete imaging systems but do not overly concern themselves with the effects of the atmosphere on the resulting optical image that is captured by their respective imaging system. For example, nearly all camera systems sold to the average consumer do nothing to remove the effects of higher-order atmospheric turbulence from the captured image even though atmospheric turbulence is often the dominant factor affecting spatial resolution in well-designed consumer cameras with sufficiently large apertures. Some cameras come with built-in image stabilization software that corrects low-order atmospheric turbulence—specifically atmospheric tilt. These imaging systems can be considered as "systems" in their own right. We have atmospheric turbulence compensation/adaptive optics systems; beam shaping, beam forming, and relaying optical systems; environmental control and conditioning systems; data communication systems; image processing and analysis systems; signal conditioning systems; command and control systems; power systems; data exfiltration systems; mission planning systems; support systems; and service systems. Also, the sources of optical radiation and the intervening media can often be viewed as systems. The interrelationships of all of these systems are quite complex and require multidisciplinary knowledge and expertise. Additionally, when all of these systems act together, they provide a capability that is greater than the sum of its parts (e.g., a nearly diffraction-limited image that provides higher than conventional spatial resolution to the interested observer). For these reasons, we think of the physical scenario in terms of interacting systems—or an SoS.

In Figure 3.8, the illumination source at the top left can either be a passive source such as the sun, moon, starlight, or infrared (IR) heat emitted from a target, or it can be an active source like a laser, flood light, thermal lamp, or other man-made source. Although the illumination source is shown at the top-left corner, this does not mean that there are geometrical restrictions on the relative position of the illumination source to the target or detector. Rather, the illumination source block of the figure illustrates the possible need for a source of illumination. The green lines in Figure 3.8 show reflected radiation, and the red lines show emitted radiation. The background radiation is radiation that is detected by our optical detection system that is generally from sources other than the one we are interested in. For instance, in an IR imaging system, other heat-producing sources in the imaging system's field of view such as hot pipes, walls of a heated structure, motors, or other living beings produce IR radiation that provides scene-dependent background radiation. As shown in Figure 3.9, the dog acting as a target produces IR radiation; if placed in a scene with another target, the optical detection system would pick up both signals. The intervening media will often scatter, absorb, and refract the radiation from the source as well as the background radiation and will sometimes emit new radiation at different wavelengths. The resulting radiation—both wanted and unwanted radiation—is captured by the collecting aperture of the optical system.

For a well-designed imaging system, the entrance pupil to the imaging system is the collecting aperture itself. Next in the system, subsequent to its entrance through the pupil, the optical beam travels through an application-specific set of optical elements and components. In an imaging application, these elements and components work to shape the optical beam, essentially relaying the entrance pupil brightness to the optical system's image plane and forming the image on the imaging system's detector elements.

To properly shape the spectral response of the radiation prior to it arriving at the detector, filtering may be applied to the incident radiation, manipulating its characteristics throughout its spectral regions and shaping it to include only the desired sections while attenuating the unwanted regions, by applying filter elements such as narrowband, wideband, band-pass, band-reject, low-pass, high-pass, notch, and adaptive filters. Optical hardware such as turning mirrors, corner cubes, beam splitters, lenses, and prisms are used to apply 2D Fourier transformations of the radiation in the entrance pupil and map the result upon an image plane of the small detector found within the optical system. Neutral density filters can be used to control the amount of radiation that falls on the detector to ensure that the detector does not become saturated. This is similar to putting on a pair of sunshades

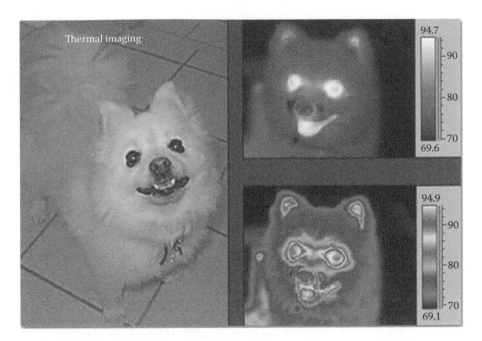

FIGURE 3.9
(See color insert.) IR images of a dog. (Reprinted from NASA. With permission.)

when first walking outdoors to reduce the brightness on the eyes. Items such as reticles and choppers are used to help identify targets of interest (TOIs) and will be discussed in a later chapter. The detector in Figure 3.8 is shown as an electro-optical detector, but the model is not restricted to electro-optical detectors alone since other transducers are equally valid. Electronic transducer hardware is used to detect the change in a specific detectable physical quantity (e.g., pressure, brightness, temperature, moisture, and electromagnetic waves), by converting the energy from the physical change into a much-easier-to-quantify electrical signal. Hence, photoelectric, chemical, mechanical, and thermal optical detectors all fit within our optical SoS paradigm. The output of the detector undergoes various application-specific signal processing prior to appearing on an image/data display. The following tasks would be completed during signal processing as needed: signal conditioning, encryption or decryption, filtering, imagery, and/or signal analysis. Roles and specifics for each of these model elements will be explored deeper within future chapters.

Next, let us further evaluate the radiation in our optical system, in terms of the radiation's source. To better illustrate this, a space-based optical system is depicted within Figure 3.10 as an example of an optical system, along with the various corresponding sources of radiation that would arrive at its entrance pupil plane.

We are assuming that the sun is the illuminating source. The TOIs are the small objects in the water (fish). The signal of interest (SOI) is indicated by the line that represents the reflected sunlight off of our TOI that is captured by the field of view of our optical system. Part of the energy of our illuminating source is transmitted into the water and is subsequently reflected off of our TOI in the direction somewhat toward our optical system. At the water/air interface, the reflected light from our TOI is refracted in such a fashion that the propagation direction of the electromagnetic field associated with it is within the instantaneous field of view (IFOV) of the space-based optical system. Notice that we are focusing on the radiation that is captured by our optical system and are not showing reflected or emitted

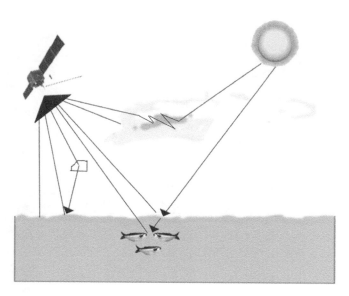

FIGURE 3.10
Sources of radiation. (From California Department of Fish and Wildlife, Marine sportfish identification: Other fishes, http://www.dfg.ca.gov/marine/mspcont7.asp#tomcod, accessed March 30, 2014; http://www.dfg.ca.gov/marine/images/mspc036.jpg.)

radiation that is outside our optical system's IFOV. All other radiation sources are considered background radiation, such as the segment that shows the reflected light off the surface of the water that is reflected into the IFOV of our optical system. Another line segment shows sunlight that is scattered by the earth's atmosphere into the IFOV of our optical system. As will be seen later, the atmosphere itself can generate radiation through absorption and emission processes. Energy at one wavelength can be absorbed and reradiated at another wavelength. A line segment shows this generated radiation that is emitted in the direction of our optical system. Clouds can often be an issue with optical imaging systems since they generate radiation that can sometimes interfere with the optical systems ability to detect the desired TOI. As will be seen later, optical modulation techniques and clutter reduction algorithms are used to often mitigate the effects of unwanted radiation on an optical system. Another segment in Figure 3.10 represents the emitted radiation from clouds intercepted by our optical system. There is also a segment that shows the energy from the clouds that reflect off the surface of the water into the IFOV of our optical system. Finally, there is a segment that shows the radiation that is emitted from the water itself and that is intercepted by the space-based optical system. As can be seen in Figure 3.10, there are a variety of unwanted signals that clutter the IFOV of our space-based optical system example. Additionally, as we learned in the statistical optics section of the preceding chapter, the desired signal can itself have a random component due to the random arrival rates of the photons themselves. These effects cannot be disregarded and instead must be taken into account and incorporated accordingly into the optical system's model. Furthermore, the models depicted within Figures 3.8 and 3.10 could be used to represent other scenarios; with some easily made adaptations, they can be made to apply as general use. A possible adaptation would be to consider a ground-to-ground remote imaging example, depicted by the following perimeter surveillance OV-1 (Figure 3.11).

In the depicted perimeter surveillance mode OV-1, there is a telescope with a digital camera connected to it, as well as a laptop with software installed to handle atmospheric turbulence compensation algorithms.

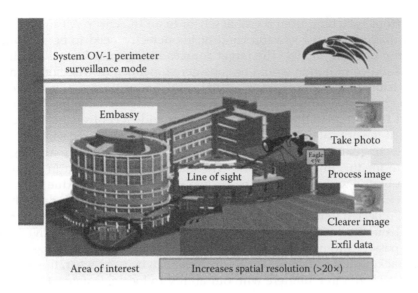

FIGURE 3.11
Perimeter surveillance mode OV-1 with atmospheric turbulence compensating imaging system and data exfiltration.

An example application could be illustrated using a U.S. Embassy located within foreign borders; the telescope would then be focused via local or remote means so that clarity is maximized for a particular *area of interest* (AOI). The AOI would be defined such that any relevant data produced from the telescope are able to satisfy its intended mission, for example, in the case of the U.S. Embassy example, the AOI would be suspicious people outside of the embassy. To provide the desired coverage of this AOI, the ground-to-ground optical imaging system would capture photos of the AOI using the telescope. However, for sufficient aperture sizes, the direct resulting image would be much too blurry, due to atmospheric turbulence causing persons within the image to have insufficient spatial detail for subsequent characterization and/or identification purposes, thus not providing usable information for the U.S. Embassy's intended surveillance. Atmospheric turbulence compensating software resident on the laptop computer removes the atmospheric turbulence and produces an image closer to the best possible image—the diffraction-limited image. The amount of improvement, measured in terms of spatial resolution, increases proportionally with the aperture size of the telescope used with the imaging system. For example, if a telescope was built with its primary mirror size being 8.5 in. (e.g., as viewed through an 8.5 in. telescope), the resulting resolution of the image after atmospheric turbulence compensation is applied would be improved by at least 20 times in comparison to the original uncompensated image.

When this ground-to-ground imaging scenario in terms of Figure 3.8 occurs, we see that the illumination source during the day is the sun. At night, the illumination source could be lights that are strategically located around the embassy. If there are no such lights, then low-light visible imaging technologies, active imaging methods, or IR methods could be used. It should be noted that switching to an IR imaging technology increases the wavelength of the observed light and consequently reduces the spatial resolution of the imaging system. Careful analysis needs to be accomplished to determine whether or not the gains in resolution made by implementing an adaptive optics or atmospheric turbulence compensation methodology offset the loss of resolution due to the increased wavelength from switching from a visible sensor to an IR sensor. All significant sources of radiation that emit in the spectral band of the detector need to be accounted for in developing the imaging model. The target sources in Figure 3.8 are the individuals in the AOI. The source radiation would be considered incoherent whether it was reflected light in the visible part of the electromagnetic spectrum or IR radiation in the IR part of the electromagnetic spectrum. Recall that even if an active source (such as an eye-safe laser) is used to illuminate the AOI—in order to provide a controllable amount of measurable light at the optical detector—the reflected light is considered spatially incoherent since the reflecting surface is optically rough. Therefore, an incoherent imaging model is appropriate for this imaging scenario. The background radiation in Figure 3.8 is all the radiation that is within the imaging system's IFOV that is not due to our intended objects—for example, the suspicious people in front of the embassy. In daylight, our SOIs would be the reflected light from our TOIs. At night, the SOIs would be either the emitted IR radiation from the TOIs or the reflected light from the TOIs due to active illumination or low-light passive illumination. All other radiation captured by the imaging system is considered background radiation. The term clutter can be used when the components of the background radiation interfere with our ability to locate, isolate, classify, characterize, or identify our TOIs. For instance, identifying a person wearing forest camouflage in foliage with an imaging system operating in the visible spectrum, the foliage is considered clutter since it visibly interferes with the ability of separating the human TOI from its surroundings. Similarly, in the case of IR imaging, heat sources that radiate at the same temperature as a human and are in the same IFOV could be considered clutter. Clutter is especially important when attempting to automatically recognize the TOI as in automatic target recognition (ATR) algorithms. Many of these algorithms have clutter rejection components that help them eliminate false targets and help, as the name implies, to reject the clutter in the scene. In Figure 3.8, the intervening media is air. As will be seen in a later section, the atmosphere scatters, absorbs, emits at different wavelengths, and refracts an electromagnetic wave. The scattering and absorption effects, in addition to the ordinary diffraction of the electromagnetic field itself, lead to the attenuation of our SOI. The collecting aperture in Figure 3.8 is the entrance pupil—primary mirror—of our telescope. The filters, reticles, choppers, and associated optics would be used to separate our TOIs from the clutter and place the resulting image on the detector in the optical system's image plane. The output of the detector is typically a low-level current or low voltage that has to be amplified, possibly denoised, and further conditioned to interact with processing and/or display hardware. In our example case, the detector output must also be converted to digital form to be used by the atmospheric turbulence compensating algorithms that reside on the laptop computer. The signal amplifiers from the detector output must amplify the signal so that it is in the range of the analog-to-digital (A/D) converters. The image is then processed to remove the atmospheric turbulence and provide a high-resolution atmospheric turbulence–compensated image. The image may then be encrypted and sent to a separate facility (e.g., exfiltrated) where it is decrypted and evaluated for possible further action.

Our perimeter surveillance OV-1 can also be evaluated from the perspective of an adapted Figure 3.10. In our case, we do not have a TOI in the water and neither is our detector in space. Even so, the different segments can still be evaluated with some adaptations as follows. During the day, the sun is still the primary passive illumination source. Unless there is a pool of water in the AOI of the imaging system and the sun is positioned in such a way that the specular reflection off the water is in the IFOV of the optical imaging system, the direct reflection off the water into the camera as indicated by the ray from the sun, to the water, and up through the path to the imaging system does not apply in the perimeter surveillance scenario. Instead, the equivalent path in our perimeter surveillance application involves the reflection of sunlight off of the ground and all the structures that are subsequently captured by the imaging systems IFOV. Since this application has relatively short range and the angle of the optical system is toward the ground and not the sky, refracted direct illumination from the sun, direct emission from the atmosphere and clouds, and reflected radiation from the ground and from the clouds, as in all corresponding line segments in Figure 3.11, are not significant factors. The last segment we discussed shows direct emissions from the water, which in our case correspond to the ground and surrounding structures that are captured by the IFOV of the optical imaging system. At night, depending on the thermal emission properties of the structures and emitting objects in the imaging systems IFOV, this may or may not be a significant factor. During the day, however, the solar reflection off the background structures and TOIs alike typically dominate over emitted radiation for our typical perimeter surveillance scenario. The segment that reflects light from our TOI that is captured by the imaging system is the radiation that is of most interest. The reflected sunlight/moonlight/starlight or thermal emissions off of the background structures are also of consequence and must be considered in developing the imaging systems model. Further, we are interested in the refractive properties of the atmosphere that affect the TOI reflected electromagnetic fields and that are received by the optical system. At this point, an example may be helpful.

If our optical system in the perimeter surveillance scenario had an 8.5 in. telescope, the entrance pupil would measure 8.5 in. in diameter. The best possible resolution that an optical imaging system attached to this telescope could provide would be the classical diffraction limit

$$\Delta x_d = 1.22 \frac{\lambda}{D} z \tag{3.1}$$

or

$$\Delta x_d = \frac{\lambda}{D} z, \tag{3.2}$$

where
  Factor 1.22 in Equation 3.1 is due to a circular aperture
  λ is the center wavelength of the light in meters received by the optical system
  D is the diameter of the entrance pupil telescope in meters (in this case 8.5 in. or 0.2159 m)
  z is the separation distance of the entrance pupil in meters from our TOIs

Equation 3.2 holds if the entrance pupil of the telescope was square instead of round. For the case of narrowband optical illumination centered at 500 nm and a separation of 1.5 km using a circular primary mirror on the telescope, we would arrive at a diffraction-limited resolution of about 4.24 mm. In the case of a square primary mirror with an 8.5 in. edge length, the diffraction-limited resolution of a well-designed optical system would be about 3.47 mm. These spatial resolution expressions can be considered as an upper limit on the performance of a well-designed optical system in vacuum. Unfortunately, in the case of our example optical system, we are imaging through the atmosphere, and without atmospheric turbulence compensation, the atmosphere is often the limiting factor on an optical system's spatial resolution performance. To quantify the effects of the atmosphere on our imaging system, Dave Fried developed the atmospheric coherence length, or Fried parameter (r_0), that provides a metric as to the seeing quality of the atmosphere (Quirrenbach 1999, 76–78). The Fried parameter can be considered the diameter for which atmospheric turbulence becomes a consideration in the optical systems performance. Entrance pupil diameters that exceed the Fried parameter have their spatial resolution affected by the atmosphere, whereas entrance pupil diameters that are smaller than the Fried parameter have their spatial resolution relatively unaffected by the atmosphere. A set of equivalent expressions for the spatial resolution of the optical imaging system that use the Fried parameter are given by

$$\Delta x_a = 1.22 \frac{\lambda}{r_0} z \tag{3.3}$$

or

$$\Delta x_a = \frac{\lambda}{r_0} z. \tag{3.4}$$

In these equations, the diameter of the entrance pupil of the telescope is replaced with the Fried parameter. Typical values of the Fried parameter for ground-to-ground imaging systems are on the order of 1–4 cm. In our scenario, choosing a Fried parameter of 4 cm for good viewing conditions, the actual spatial resolution for our optical imaging system in the presence of atmospheric turbulence is 0.022875 and 0.01875 m for the square telescope primary mirror. Spatial features that are smaller in scale than the calculated spatial resolutions cannot be resolved by our imaging system. Using the example for the circular telescope aperture discussed previously, we see that facial features that are smaller than 2.2875 cm practically cannot be distinguished by our optical system. The face would be blurry and features such as the eyes and other notable features that are smaller than 2.2875 cm would not be observable. On the other hand, if the atmospheric turbulence was removed from the image, then the spatial resolution for the circular entrance pupil would be on the order of 4.24 mm and much more of the face would become resolvable. If we look at the case for bad viewing, as in a Fried parameter of 1 cm, the resolution increase is even more pronounced. The actual resolution of the optical imaging system is of 0.0915 m for the circular telescope entrance pupil aperture and 0.075 m for the square entrance pupil aperture. The upper limit for the spatial resolution improvement achievable by removing the effects of atmospheric turbulence from the image is given by dividing the expression for the spatial resolution with atmosphere present, Equation 3.3 or 3.4, and by the diffraction-limited resolution of the optical imaging system, Equation 3.1 or 3.2. The result for telescopes with either circular entrance pupil or square entrance pupil is given by

$$\Gamma = \frac{\Delta x_a}{\Delta x_d} = \frac{D}{r_0}. \tag{3.5}$$

For our 8.5 in. telescope, the spatial resolution improvement for good seeing is over a factor of 5, whereas for bad seeing, it is upward of 22. Note that in the presence of atmospheric turbulence, without atmospheric turbulence compensation, telescope diameters larger than the Fried parameter do not produce higher spatial resolution. With atmospheric turbulence compensation methods applied, the upper bound on spatial resolution scales with the telescope entrance pupil diameter.

A crucial limiting factor that must be heeded during optical imaging system design is the spatial resolution attained by the optical imaging system, which must be larger than the detector's pixel size when projected onto the object plane. For instance, in our perimeter surveillance example with a circular aperture, assuming square pixels and in the case of uncompensated atmospheric turbulence with bad viewing (Fried parameter is 1 cm), the projected pixels must be less than 0.0915 m by 0.0915 m. In the case of compensated atmospheric turbulence, the projected pixel size must be less than 0.00424 m by 0.00424 m. If this is not the case, the spatial resolution performance of the optical imaging system will be limited by the pixel size instead of attaining the near diffraction-limited imaging performance of the well-designed optical system (Motorola White Paper 2012). Figure 3.12 is useful in understanding the projected pixel size requirement for our perimeter surveillance scenario. This figure is a general model of an optical system. Shown on the left is the IFOV, which consists of all the pixels of the optical detector, projected onto the object plane. The half angles of the IFOV are given by $m_l * \alpha_l$ and $m_t * \alpha_t$ corresponding to the longitudinal and transverse directions. The parameters m_l and m_t are the longitudinal and transverse magnifications, respectively, and are determined by the specific optical system. We will discuss this more in Chapter 4. For now, these are treated as selectable parameters over a certain range (say 1–25). In typical cases, the IFOV is determined with the assumption that the lateral and transverse magnifications are set to one and then adjusted with respect to the specific magnification, within the acceptable range of possibilities that is required for the specific imaging scenario.

For instance, if the imaging scenario requires the highest resolution possible, then the magnifications are both set to one and the IFOV becomes as small as possible. If the imaging scenario requires wide-area surveillance, then the magnification is set to the highest value (say 25) and the IFOV becomes as wide as possible. The entrance pupil of the telescope is shown to the left of the box labeled "Optical System." The entrance pupil plane is of course defined by the optical system itself, but it is shown in Figure 3.12 as a point of reference. The "Optical System" box treats the optical system as a black box, representing a general optical system. We will use the techniques in our "Basic Optics" chapter section later in the book to describe the specifics of the optical system. For now, we treat it as a "black box" and are only concerned with the entrance pupil and the exit pupil. From the

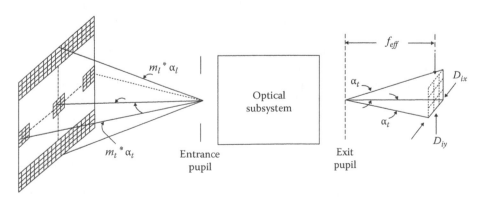

FIGURE 3.12
General model for an optical system.

exit pupil, we have the effective focal length given by f_{eff}, which gives the distance from the exit pupil to the image plane. The detector element is placed in the image plane. The backplane transverse and longitudinal angles are given by, respectively

$$\alpha_t = \tan^{-1} \frac{D_{ix}}{2f_{eff}} \qquad (3.6)$$

and

$$\alpha_l = \tan^{-1} \frac{D_{iy}}{2f_{eff}}, \qquad (3.7)$$

where
   D_{ix} and D_{iy} are the transverse and longitudinal dimensions of the detector element itself, respectively
   f_{eff} is the effective focal length

For a fill factor of 100%, the entire face of the detector area is filled with pixels. This is not always the case since some detectors have their pixel readout electronics connected to the edge of the pixels. Other manufacturers place the readout electronics at the bottom of the pixel and consequently are able to reach very high fill factor percentages. The actual dimensions of the pixels and associated fill factor are determined by the detector element manufacturer, and the optical system is often designed to work with the given detector element. In terms of the detector-limited spatial resolution, the pixel density is an important parameter. If a given detector has n by m pixels in the transverse and longitudinal directions, respectively, in the case of detector-limited spatial resolution, the pixels end up limiting the performance of the optical system. Since this a consequence of inadequate optical system designing, it can be countered with the application of good SE coupled with effective optical systems design practices. An example of this kind of design error is if the projected pixel size on the target plane is larger than the diffraction-limited resolution (in the atmospheric turbulence–compensated imaging case) or larger than the Fried parameter–based resolution obtained when imaging in the presence of atmospheric turbulence. In both cases, the spatial resolution is determined by the pixels and not by other aspects of the optical system or atmosphere itself. In proper optical systems design, the optical system is designed and applied in a fashion that the pixel sampling in the object plane is smaller than the intended spatial resolution given by the optical system. Figure 3.13 provides a 2D view of the projected pixel in the object plane.

In order to calculate spatial sampling in the object plane, both the transverse and longitudinal directions' projected pixel size needs to be determined. In Figure 3.12, we look at one dimension of Figure 3.13 with an expanded view of the pixels in the image plane (right side of the image) and also the plane of the TOI (left side of the image). Let us assume that in this case, the pixels are located along the longitudinal axis of our optical system. If the detector element manufacturer had 4096 pixels in the longitudinal direction and 4096 in the transverse direction with a 100% fill factor, then we would have square pixels. If the detector element has an active area of 1 cm by 1 cm, then the 4096 by 4096 pixels would be uniformly spread across this area. In the longitudinal direction, we would have 4096 pixels as illustrated at the right of Figure 3.13. The figure is exaggerated to show the

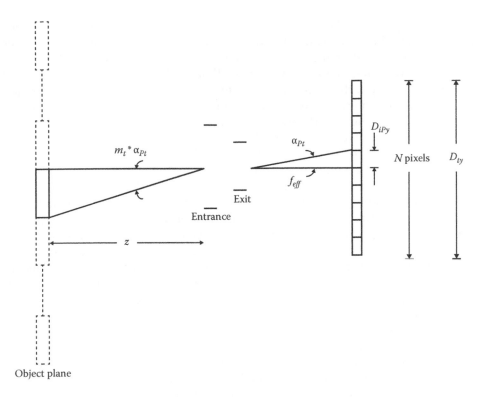

FIGURE 3.13
Projection of pixel size into object space in one linear direction (e.g., horizontal).

relevant geometry. In our example, the linear extent of the detector element in the longitudinal direction is 1 cm. We would have all the pixels adjacent to each other across the 1 cm active area of the detector. Each pixel would have a length of 1/4096 cm per pixels or about 2.44 μm per pixel in the longitudinal direction. The angle α_{pl} can be determined from

$$\alpha_{pl} = \tan^{-1} \frac{D_{iPy}}{2f_{eff}},\qquad(3.8)$$

where
  D_{iPy} is the length of the pixel in the longitudinal direction
  f_{eff} is the effective focal length

Once we know the angle α_{pl}, we can determine the size of the projected pixel in the object plane by noticing that the angle subtended by the projected pixel is just the angle α_{pl} multiplied by the optical systems longitudinal magnification, m_l. We can determine the length of the projected pixel in the object plane from

$$D_{opy} = z\tan(m_l * \alpha_{pl}),\qquad(3.9)$$

where D_{opy} is the length of the projected pixel in the object plane, z is the separation of the object plane from the entrance pupil, and the other parameters are as before. This analysis has to be repeated for the transverse direction. In our example, we have the same number of pixels in the transverse direction as in the longitudinal direction, and so the analysis will be identical to what we have just done. In general, we would replace the subscripts *l* and *y* with *t* and *x*, respectively, in Equations 3.8 and 3.9 and Figure 3.13 and repeat the preceding analysis. We can see that the further away the object plane is from the entrance pupil, the larger the projected pixel will be. If we want to project the smallest pixel size possible, then we use a longitudinal magnification of one. We can also decrease α_{pl} by increasing the effective focal length and/or, if possible, choosing a detector that has a smaller pixel size. Typically, once chosen, the effective focal length and the pixel size are fixed for a generic, simple optical system. As an example, if we used the pixel length of 2.44 μm with a focal length of 0.5 m, the angle α_{pl} would be 2.44 μrad. Using a longitudinal magnification of 1 in order to get the smallest resolution possible, we would get a projected pixel size of about 3.7 mm for objects that are 1500 m away. We see that this result is less than the previous 4.24 mm upper limit needed for the turbulence–compensated imaging example previously discussed. In this case, the projected pixel size is sufficient to properly sample the TOIs in the object plane. It should also be noted that if we did not have an optical system that was capable of compensating for atmospheric turbulence, then the resolution achievable for our optical system with a circular aperture would remain at 91.5 mm, and the additional sampling capability of our optical system would be of no benefit. In this later case, we would have an atmospheric turbulence–limited spatial resolution, as is often the case.

In summary, we presented a general model for an optical system that, similar to the SoS concept in high-level systems modeling methods, provides a comprehensive model for a system's oriented decomposition of a general optical system. We showed how this model could be applied to a generic optical system to evaluate different types of radiation that may be captured by the optical system. We also defined and distinguished between SOIs, background radiation, TOIs, and clutter. We provided an example using optical imaging perimeter surveillance CONOPS and showed a corresponding OV-1. We described how Equation 3.5 could be used to determine whether or not the optical system would benefit from atmospheric turbulence compensation methods. If the result of evaluating Equation 3.5 is larger than one, then atmospheric turbulence methods would provide some benefit in spatial resolution, whereas if it is less than one, then the optical system's spatial resolution would not benefit from atmospheric turbulence compensation. If atmospheric turbulence compensation is required, then Equations 3.1 and 3.2 provide the best possible resolution attainable by a well-designed optical system. The optical system must be designed so that the projected pixel size in the object plane is slightly smaller than that given by Equation 3.1 or 3.2. No additional benefit will be gained by having projected pixels that are smaller that the diffraction-limited spatial resolution size of the optical system. Doing so would just cause additional processing with no spatial resolution benefit. If atmospheric turbulence compensation methods are not used, then the projected pixel size in the object plane must be slightly less than the size dictated by Equations 3.3 and 3.4. To resolve features in the object plane, the projected pixel size and the spatial resolution of the optical system should be a factor of 10 or less than the size of the feature that needs to be resolved. For instance, to resolve a human object sufficiently to tell that it is human and not some other animal or object, a guideline is that 10 resolution cells should be placed across the human object. For a 2 m tall human object, this means that the spatial resolution required to tell that the object is human should be on the order of 20 cm or less. On the other hand, if we wanted to resolve the color of someone's eyeball, we would need resolution cells on the order of 10 across the approximately 1 cm colored portion of the eye (e.g., a required spatial resolution and object plane projected pixel size of 1 mm). The optical system must be designed to address the required resolution for the particular application, as well as the spatial sampling requirements, and the effects of atmospheric turbulence.

In the next section, we provide an example of the SE and optical systems principles methods and techniques learned in this chapter. We see FIT analyzing the particular design parameters required for the UAV example developed in the preceding section.

3.3 Integrated Case Study: Implementing a System of Systems Optical Model

In this section, we continue with our integrated case study and show how the SE principles learned earlier are used in a simulated, realistic technical setting. In this chapter, we see FIT coming to grips with deciding and using the SoS and FoS ideas and applying SE practices to evolve their concepts and technical approach for their primary stakeholder, the Department of Homeland Security (DHS), U.S. Customs and Border Patrol. They are in the midst of an optical propagation analysis of the optical detection scenario. This effort triggered off a series of interactive meetings.

During one of these interactive meeting with the stakeholders, the team identified the problem statement and focused on obtaining and understanding the problem definition. This definition identified what problems DHS was seeking to solve with the system. Along with detailed information gathered at these meetings and from other discussions, the team was able to produce a stakeholder requirements document, which was further analyzed to understand the operational, maintenance, and functional details of the intended optical system.

With a solid understanding of the problem, the team proceeded to perform the needs identification process for the primary stakeholder. The primary stakeholder's need was identified to be a requirement for a general-purpose optical imaging system that was intended to detect illegal bipedal border crossers from an unmanned aerial platform along the U.S.–Mexican border. The intended UAV optical system was in itself part of a larger SoS that included direct line-of-site communications, over the horizon communications, a Mission Command Center (MCC), Mobile Control Center (MobCC), distributed information, and deployable border patrol and other assets.

The team worked aggressively to conduct additional interviews with DHS, which yielded some essential information about the UAV platform so that the interfaces between the optical system and the UAV and the UAV capabilities (e.g., pointing and steering, communications, power, size, and weight restrictions) could be identified. The stakeholder requirements have been identified, and preliminary models are in progress and in various states of completion. These models include a radiometric model, a statistical optics model, a Fourier optics model, a linear systems model, and a geometrical optics model. One immediate use of the geometrical optics model was to determine the projected pixel sizes in the object plane for representative imaging sensors. These representative pixel sizes were used in spatial resolution analysis.

The FIT engineering team has also worked on the initial feasibility study of the stakeholder requirements to make sure that the intended optical system meets the system design criteria and expectations envisioned by the DHS. An FBD of the optical system has been generated, FFBDs are in process, and a secondary examination of the optical system has yielded a good look at what the conceptual design should be for the imaging system.

In one of the internal meetings Jennifer O., systems engineer (Jen), and Mr. Tom Phelps, optics expert (Tom), met to discuss a trade study on the potential detectors to be used for the primary sensors of the UAV optical system.
(Monday morning, Jen walks into Tom's office for a technical meeting.)

Jen: "Morning Tom."

Tom: "Good morning Jen."

Jen: "Thanks for meeting me on such a short notice. Karl asked me to put together a trade study for the detectors, and I figured I could use your knowledge and expertise."

Tom: "Sure, no problem. What you got so far?"

Jen: "So far I've got a list of sensors for object detection. During my studies, I have learned that every imaging system must include a detector of some form. Most detectors exhibit some degree of nonlinearity in their response to incident radiation. Some detectors, such as photographic film, are intrinsically very nonlinear, while others, such as silicon photodiodes, are quite linear over several orders of magnitude if operated properly. All detectors, however, eventually saturate at high radiation levels or display other nonlinearities which may be either global or local (Barrett and Myers 2004, xix)."

Tom: "That's right. Let's go over each list so we can eliminate the obvious ones and then start the trade study from there."

Jen: "Perfect, you read my mind. On the passive list we have…."

Tom: "Sorry to interrupt you, but passive sensors? Although most of these options cost less, they also provide fewer features. We are also at the mercy of the objects themselves, in the case of thermal radiation, providing enough optical power to get a detectable signal. In the visible region, we are dependent on passive sources like the sun to provide enough illumination on the target to be able to see it. We will have to see if we have enough optical power from our Targets of Interest (TOIs) for both day and nighttime operating scenarios. If not, the added complexity of an active source may be justified. Hmm, an interesting one from this category is the stereo imaging camera due to its ability to work well in different types of terrain and its ability to estimate distances given some reference information."

Jen: "Ok. How about optical flow?"

Tom: "Technology is too new to tell. We need something proven and robust for this stakeholder. Leave it in there to see what results we obtain."

Jen: "Alright. As for the active sensors, we have lasers, which have amazing precision and accuracy but add complexity to the system, and may provide size, weight, and power issues (SWaP). Radar is less susceptible to atmospheric effects but is more expensive and doesn't provide comparable spatial resolution. Spectral systems are too costly and complex and may be overkill for this application."

Tom: "I agree."

Jen: "Thanks. At this point we want to evaluate out options based on size and weight, cost, range and accuracy. I've taken the liberty of assigning weights to all. Based on our radiometric analysis that is in progress, we can temporarily defer the discussion on passive sensors for a later date, and address the more complex active sources now. That way, we will be prepared for the worst case scenario of not enough optical power from passive sources at our optical receiver to register a detection."

Tom: "Ok. What does it come down to?"

Jen: "It comes down to Optical Flow."

Tom: "It can't be."

Jen: "Yup, gives us the highest score for size, cost, range and accuracy."

Tom: "I'd say we add a reliability column to this trade study and see how long these sensors have been in the market and what's their sustainability has been like."

Jen: "Makes sense. I'll work on adding that into the trade study, and schedule another meeting with you. We should be prepared to discuss the results of the radiometric study and answer questions in case we have a meeting this week. Karl suspects Bill Erasmus from DHS will contact us soon."

Tom: "Hmm, we need some time to finish the radiometric study. Hopefully it won't come up until we have a chance to complete it. We can't even decide if we can use passive sensors or if we need an active sensing solution until we complete this study. If I find any relevant information I will send it to you."

Jen: "Thanks! Talk to you soon."

(Wednesday morning)

Bill: **(Thinking to self),** "I don't completely understand our technical approach and how it relates to the larger DHS enterprise… Oh no! What if investors ask me questions? I know, I'll get my experts together and have them explain our proposed system in terms of how it addresses our Stakeholder's requirements."

Bill: "Can you please set up a meeting for today at 15:30 with our Systems Engineers, CTO and CIO. Tell them to bring specifics regarding the technical approach used in our UAV contract. Oh, and have them bring their results of the stakeholder requirements analysis."

(Meeting time, Dr. Smith shows up at 15:45)

Attendees to the meeting include Dr. Bill Smith, CEO (Bill); Mr. Karl Ben, Senior Systems Engineer (Karl); Jennifer O., Systems Engineer (Jen); Mr. Tom Phelps, Optics Expert (Tom); and Mr. Gary Blair, Engineering Architecture (Gary).

Bill: "My apologies for being late. I just got off the phone with our sponsor. Thank you everyone for attending. As you may already know, our Stakeholder is the Department of Homeland Security (DHS)/Border Patrol Division (BP). They have identified their need for a system to monitor the US–Mexican border (U.S. Customs and Border Protection 2012). I would like this meeting's focus to be on the overall approach to meet the early-established stakeholder requirements for our UAV contract. First, Karl could you please tell us about the highest level system breakdown?"

Karl: "Absolutely Sir! As you know, the major systems that concern this project are the sensor systems and communications systems, and the other major systems that also affect the UAV operation are the flight system and power system. The UAV, with our optical system in it, is part of a larger system of systems that includes our optical systems, other UAV sensors on multiple UAVs, communications satellites, a Mission Command Center, a Mobile Control Center, deployable assets like helicopters, cars, airplanes, boats, All Terrain Vehicles (ATVs), and assets from other organizations. These major systems work together and are very interdependent, and come together to provide a much larger output than just the sum of each one's individual contribution, thus they are a system of systems. Working together they enable the UAV to operate and fulfill its mission."

Bill: "Karl, before we get into the details of the stakeholder requirements, can you tell everyone how much of the border the optical system is monitoring? What is the typical climate and terrain in the monitored area? What is the typical population expected to be in the monitored area? Why is this border being monitored?"

Karl: "Sure, no problem. We will monitor at 10 'hot spots' along the border that are each 10 km by 10 km with the border in the middle. These are all away from population centers with variations in terrain indicated by the different landscapes."

Bill: "Will the hot spots being monitored have a physical representation of the border, such as a river or a fence? Are the hot spot locations expected to change over time or are they fixed locations?"

Karl: "The hot spots will likely change over time. There are fence-lines, rivers, or natural barriers that separate the countries. We should be prepared to deal with all terrain types along the border (away from population centers)."

Bill: "Tom, I would like to get some background information on the intended optical system. What is the make-up of the optical system, that is, what does and does not lie within our development responsibility for this optical system?"

Tom: "The optical system will consist of the telescope, which is used to change the instantaneous field of view (IFOV), relay optics to image the radiation in the entrance pupil (the telescope irradiance (Watts/m^2) on the primary mirror) onto the detector focal plane. The detectors themselves cover the visible and IR spectral regions. If we want to passively detect a human's thermal radiation in the IR, the peak wavelength should be roughly 9.5 micrometers. If we need an active system (say a laser), we will need to choose/build the laser and provide the power stabilization, power, and pointing and tracking circuitry and optics. Alternatively, we may see if we can leverage existing assets that may be on board the UAV. We may possibly need cooling systems for the detectors to reduce noise. We also need filter wheels to apply notch filters, high-pass filters, low-pass filters, perhaps adaptive filters, and optical modulators (for background suppression), vibration and shock isolators. We also need an Environmental Conditioning unit with temperature regulators, moisture and humidity control, dust mitigation, pressurization, and light shields. We will need to develop anti-reflection coatings for the optical surfaces. Also, the atmospheric turbulence compensation (ATC) system or adaptive optics (AO) system, if we need it, requires light captured at least two wavelengths simultaneously – such as an RGB digital camera. The ATC system also requires a general-purpose parallel processing device (GPPP) and these can be miniaturized. If we go with the AO system, we don't need the GPPP but instead need a wave-front sensor, deformable mirror, or microelectromechanical mirror system (MEMS) along with a controller. We also need a controller for the telescope, a controller for the sensors, an on-board signal-processing unit, and UAV communications unit. We also need an on-board image processing capability that has, at a minimum, a means to select the size of the image frames to transmit back to the MCC or MoCC. We also need to design the interface circuitry for our data with the UAV and hook into the UAV power grid. We assume the gimbal pod is supplied as Government Furnished Equipment (GFE) and we need to interface with that. Raytheon manufactures their Gimbal Pod and that should handle our pointing and tracking needs. Finally, we need to encrypt our data and make sure our data and images are formatted to integrate in the DHS/DOD enterprise architecture."

Tom: "For this project an important aspect is the telescope. 'Telescopes may be broadly classified as refracting or reflecting, according to whether lenses or mirrors are used to produce the image. There are, in addition, catadioptric systems that combine refracting and reflecting surfaces. Telescopes may also be distinguished by the erectness or inversion of the final image, and by either a visual or photographic means of observation (Pedrotti and Pedrotti 1987, 135).'"

Bill: "Great overview of the system. Does the optical system require an on board storage of the raw optical data?"

Gary: "This is a good idea for high value data. We will include that in our technical approach."

Bill: "Tom, is a separate transmission system required for the optics system or can it piggyback on the UAV's communication system? If it can piggyback, what is the bandwidth available for the optics communication system?"

Tom: "I'm an optics guy. I don't know the communications piece. Karl, any ideas or maybe we need to research to see what is doable."

Karl: "We can interface with the existing communications package on the UAV. We will have to build the interface circuitry for our sensor. Also, the UAV has both C-band for line of site and X-band for long-haul communications so we will have to interface with both these systems. We also will need to adjust our data stream on the fly to meet the bandwidth requirements of the UAV. We will also need to share the bandwidth with other systems and so we will need to work closely with DHS's UAV communications engineers."

Bill: "After the video and images are captured, possibly stored, and transmitted 'real time,' will **there** be a data reduction/evaluation process?"

Jen: "No, there are no plans for that. Raw data and 'interesting' mission critical data will be passed on to interested users (e.g. DIA, CIA, FBI) in both raw data form and real-time processed form (either video or still images) if we use the ATM/AO system."

Bill: "Do we have enough detectable optical power or do we need some image intensifiers?"

Tom: [winces], "We are currently doing some research on that, Bill. A little history; image amplification by means of an electronic device was first indicated by Holst *et al.* in 1934. At that time, vacuum-tube technology did not permit realization of the required devices. We had to wait until the late 1940's when, simulated by Chamberlain through his paper on fluoroscope and fluoroscopy in 1942, image intensification was achieved by Coltman, Tol and Oosterkamp, and Teves with the image intensifier as we know it in its present form (Biberman and Nudelman 1999, 150)."

Bill: "All right. Thanks for the image intensifier history, but we need to first understand the radiometry to see whether we even need them. Let's expedite our modeling efforts to see how things look. Do we have any insights on the specific hardware and software that we want to use?"

Karl: "That is a design question and we aren't there yet. This could be an internally developed system or parts could be commercial-off-the-shelf (COTS). We will have to do a trade study to see which is the best way to proceed."

Bill: "Is the system expected to deliver an image and video independently or can we assume that the system will send either an image or a video at any given time?"

Jen: "Predominantly video. We can select images from the video stream to display. Alternatively, we can go into a mode where it only sends an image at selectable time increments. The adaptive optics system will process video data on streaming data at video rates (30 Hz), or on individually selected images, or not at all."

Bill: "Can we assume that the UAV will predominantly communicate to the command center by means of satellite communications?"

Gary: "Yes. However, it will also communicate with a mobile command center, or both simultaneously (one has control—either mobile or command center depending on mission requirements), the other gets an image stream."

Bill: "Thanks Gary. Now I have a high level understand of the technical approach. What methodology are we using to manage the process?"

Karl: "We are using our standard systems engineering processes for developing the system. We are also building the enterprise architecture models so that we can interact with the other entities involved in the DHS Border Patrol CONOPS. In this sense, our optical sensor system is just one of many systems that operate together to provide the DHS the ability to achieve its border patrol mission. In this way, we are part of DHS Border Patrol's system of systems."

Bill: "Are there any constraints on the operating environment?"

Karl: "The stakeholder would like our system to be very adaptable to a variety of different environments. This may be very important for our marketing strategy for secondary stakeholders like downstream customers"

Bill: "Did we receive sufficient high-level technical information from the stakeholder to understand the operational scenario?"

Karl: "Yes, we have sufficient information to refine our feasibility studies and start developing the system-level requirements. Of course, we still need to do some analysis like the radiometric study, but we are ready to proceed. We will schedule some regular technical interchange meetings with the stakeholder to make sure we stay on track."

Bill: "I see there is a need for encryption, what kind of encryption are we talking about?

Karl: "That is correct. We think that 128-bit encryption would suffice, but we will need to confirm this with the client."

Bill: "Let's do this ASAP. We need to tie up some of these loose details before we proceed."

Karl: "Absolutely Bill. I will make a note to contact the client today."

Bill: "How do we plan on measuring our performance?"

Karl: "Well, we have sample values for key metrics provided by the stakeholder's technical team. Also, the stakeholder did provide some expected results in the form of Measures of Performance (MOPs), Key Performance Parameters (KPPs), and Technical Performance Measures and Metrics (TPM&Ms). We need to ensure that our system meets the baselined requirements and these performance metrics."

Bill: "I see here that there is a requirement that our system may be used in other scenarios and that it should be adaptable. Do we have some insight into these other CONOPS?"

Karl: "We were given some information on their Perimeter Surveillance OV-1 and we were able to see one of their systems in operation, and have a good idea what the requirements mean. Our optical sensor system will be designed to easily be implementable as a stand-alone ground to ground imaging system. We will have to take this into account during our design process and use modular design principles."

Tom: "Also, we were provided with Automatic Target Recognition algorithms by our primary stakeholder. With these we can simulate the various alternatives and directly estimate our optical system performance."

Bill: "Team, what about the maintenance, upgrades, pre-planned product improvement, service, availability, reliability, and things along that line. We need to understand those requirements. Karl, can you head the effort to identify these for our next meeting, please?"

Karl: "Yes Bill."

Bill: "For our next meeting, let's also have a list of all the risks, and the corresponding planned mitigation strategies, for the medium to high-risk items. I also want to see the preliminary modeling results, especially the radiometric models. That will tell us if we need an active imaging approach or if we can go with the lower cost passive imaging solution. Based on the radiometric modeling results and spatial resolution analysis, we need to identify alternatives to our approach and conduct a trade study."

Bill: "Thank you all for all the feedback at such short notice. Are there any concerns? Is there anything that we need to address? If not, I will have my secretary schedule a follow up meeting next week same time."

3.4 Summary

Through this chapter, we have provided an introduction to SoS and FoS and have described a high-level overview of the SE process. We also provided an introduction to an optical systems model that can, itself, be considered an SoS model. We provided some simple equations for determining the spatial resolution of a circular or square well-designed optical system with or without atmospheric turbulence. We also showed how to determine the projected pixel size in object space for an arbitrary pixelated sensor. Through the integrated case study, we were also able to see some specific considerations involved in understanding high-level attributes of a UAV optical system and how some integral SE concepts relate to the technical understanding and development of the optical system. In the next chapter, we will present some concepts related to model-based SE.

References

Barrett, H.H. and K.J. Myers. 2004. *Foundations of Image Science*. Hoboken, NJ: John Wiley & Sons, Inc.

Biberman, L.M. and S. Nudelman. 1999. *Photoelectronic Imaging Devices*, Volume 2: Devices and Their Evaluation. New York: Plenum Press.

Blanchard, S. and W.J. Fabrycky. 2011. *Systems Engineering and Analysis*, 5th edn. Englewood Cliffs, Upper Saddle River, NJ: Pearson Education, Inc.

Boehm, B. and W.J. Hansen. 2000. Spiral development experience, principles, and refinements. *Spiral Development Workshop*, University of Southern California, Los Angeles, CA, February 9, 2000. http://www.sei.cmu.edu/reports/00sr008.pdf (accessed March 30, 2014).

California Department of Fish and Wildlife. Marine sportfish identification: Other fishes. http://www.dfg.ca.gov/marine/mspcont7.asp#tomcod (accessed March 30, 2014).

California Grunion, obtained from the California Department of Fish and Wildlife, http://www.dfg. ca.gov/marine/mspcont7.asp \l "tomcod" http://www.dfg.ca.gov/marine/mspcont7.asp#tomcod (accessed November 18, 2014).

Evans, S. 1996. *The Deer Hunter's Almanac*. New York: The Hearst Corporation.

Firesmith, D. 2010. Profiling Systems using the Defining Characteristics of Systems of Systems. Pittsburg, Pennsylvania: Software Engineering Institute, Carnegie Melon University, p. 9.

Gansler, J.S., W. Lucyshyn, and A. Spiers. 2008. Using spiral development to reduce acquisition cycle times, UMD-AM-08-128. College Park, MD: Center for Public Policy and Private Enterprise. http://www.dtic.mil/dtic/tr/fulltext/u2/a494266.pdf (accessed April 21, 2014).

Harris Corporation. 2012. Systems engineering manual, S-401-001, Revision 47, Revised November 12, 2012. New York: Harris Corporation, Government Communications Systems Division.

Lorenz, E. 1972. Predictability: Does the flap of a butterfly's wings in Brazil set off a tornado in Texas? *139th Meeting of the American Association for the Advancement of Science*, Boston, MA, December 29, 1972. http://eaps4.mit.edu/research/Lorenz/Butterfly_1972.pdf (accessed March 30, 2014).

Motorola White Paper. 2012. Video surveillance trade-offs, a question of balance: Finding the right combination of image quality, frame rate, and bandwidth. http://www.motorola.com/web/Business/_ Documents/static%20files/VideoSurveillance_WP_3_keywords.pdf (accessed March 30, 2014).

National Aeronautics and Space Administration Goddard Institute for Space Studies. 2012. Earth's energy budget remained out of balance despite unusually low solar activity. http://www.giss. nasa.gov/research/news/20120130b/ (accessed November 18, 2014).

Office of the Deputy Under Secretary of Defense (ODUSD) for Acquisition and Technology Systems and Software Engineering. 2008. Systems engineering guide for systems of systems, Version 1.0. Washington, DC: ODUSD(A&T)SSE. http://www.acq.osd.mil/se/docs/SE-Guide-for-SoS.pdf (accessed March 30, 2014).

Pedrotti, F.L. and L.S. Pedrotti. 1987. *Introduction to Optics*. Englewood Cliffs, NJ: Prentice-Hall, Inc.

Quirrenbach, A. 1999. Observing through the turbulent atmosphere. In: *Principles of Long Baseline Stellar Interferometry*, P.R. Lawson, Ed. Pasadena, CA: JPL Publications.

Rausch, R., D. Benevelli, and S. Mort. 2007. Economic history of the North American fur trade, 1670 to 1870. In: *Encyclopedia of Earth*, C.J. Cleveland, Ed. Washington, DC: Environmental Information Coalition, National Council for Science and the Environment.

Robbins, P. (editor). 2007. Encyclopedia of Environment and Society. London, GB: SAGE Publishing.

Roggemann, M.C. and B. Welsh. 1996. *Imaging through Atmospheric Turbulence*. Boca Raton, FL: CRC Press.

Saundry, P. and EH.Net. 2012. Economic history of the North American fur trade, 1670 to 1870. http://www.eoearth.org/view/article/151941 (accessed March 30, 2014).

Sherif, J.S. and T.-L. Tran. 1994. An overview of the quality development (QFD) technique. Software Assurance Group, Jet Propulsion Laboratory, Pasadena, CA, http://trs-new.jpl.nasa.gov/ dspace/bitstream/2014/33621/1/94-1598.pdf (accessed March 30, 2014).

The Next Generation in One System Technology AAI Corporation. 2009. http://www.aaicorp.com/ pdfs/ugcs41709a.pdf (accessed March 30, 2014).

USAF. 1969. Military standard system engineering management. http://www.everyspec.com/MIL-STD/ MIL-STD-0300-0499/download.php?spec=MIL-STD-499A.010375.PDF (accessed March 30, 2014).

U.S. Customs and Border Protection—Border Security. 2012. U.S. border patrol total apprehensions by southwest border sectors fiscal year 1960–2012. http://www.hsdl.org/?view&did=734433 (accessed March 30, 2014).

U.S. Customs and Border Protection: Office of Air and Marine. 2012. UAS on leading edge in homeland security. http://www.dtic.mil/ndia/2012targets/TKostelnik.pdf (accessed March 30, 2014).

U.S. Department of Transportation, Federal Highway Administration. 2007. Testing programs for transportation management systems: A technical handbook, FHWA-HOP-07-088. Washington, DC: U.S. Department of Transportation, Federal Highway Administration, May 10, 2007. http:// ops.fhwa.dot.gov/publications/tptms/handbook/tptmshandbook.pdf (accessed March 30, 2014).

4

Model-Based Systems Engineering

> A good engineer thinks in reverse and asks himself about the stylistic consequences of the components and systems he proposes.
>
> **—Helmut Jahn, Architect of the Sony Center (Helmut 2013)**

4.1 Overview of MBSE

Systems engineering (SE) was started to identify and analyze systems requirements and determine if these requirements had been followed and properly implemented throughout the systems development life cycle (SDLC). Hence, systems requirements can be considered one of the most crucial elements of the SE processes. At the core of the SE process, it is essential that the unique requirements obtained from the systems requirements generation process are used to drive the system's design and development processes. To ensure that the systems requirements are clearly communicated throughout the life cycle, model-based systems engineering (MBSE) has been introduced to support the systems requirements process (INCOSE Systems Engineering 2007). A team of contractors that may or may not be managed by a prime contractor often develops large, complex systems. For large teams, unambiguous requirements are essential for successful integration of the constituent subsystems. In SE, a good system is one that has its requirements clearly established and effectively communicated to the development team. In other words, the different parties responsible for designing and developing a system must clearly comprehend the requirements and ensure that the requirements are complete, measurable, and testable. A system that has good requirements will not only have fewer problems and surprises but will also save cost in the long term. MBSE tools are a tremendous aid in capturing, documenting, and effectively communicating requirements to stakeholders and the development team. They are also instrumental as analytical tools and are useful in capturing and communicating design and test artifacts.

Presently, a commercial or government entity that takes MBSE into consideration will usually use models to architect their enterprise and developmental systems. According to the Institute of Electrical and Electronics Engineers (IEEE 1990), "A model is an approximation, representation, or idealization of selected aspects of the structure, behavior, operation, or other characteristics of a real world process, concept or system." Models are also applicable, developed, and used at multiple levels in a given systems development effort. Typically, there are three categories of models used in developing a system: operational models, system models, and component models.

According to the IEEE Standard 1471, which describes the architecture of a software-intensive system, "A model usually offers different views in order to serve different purposes. A view is a representation of a system from the perspective of related concerns or issues" (IEEE 2000). According to the International Council on Systems

Engineering (INCOSE) article, "MBSE is the formalized application of modeling to support systems requirements, design, analysis, verification and validation activities beginning in the conceptual design phase and continuing throughout development and later lifecycle phases" (INCOSE Systems Engineering 2007).

In the next section, an introduction and description of the MBSE process is given. In addition, an explanation of several MBSE methodologies, tools, and paradigms is provided. Lastly, we describe the importance of MBSE to the modern-day SE discipline.

4.1.1 Introduction to MBSE

From the practical viewpoint, MBSE acts as an integrative vehicle that starts from the conceptual design phase of a project and carries the system's requirements, evolving design attributes, analytical capabilities/results, and verification and validation throughout all the SE life-cycle phases. In the system's life cycle, all phases are unique and the output of each phase acts as an input to the next phase, and together all the phases represent a precedence model. The idea behind MBSE is to model the enterprise architecture and provide visualization tools, requirements management tools, and analytical tools for essential systems activities throughout the system's life cycle. When MBSE methods are applied and executed meticulously and diligently throughout the life cycle of a systems development effort, the likelihood of success is increased dramatically.

MBSE facilitates clear communication between stakeholders and the development team. To supervise, evaluate, and observe complex as well as simple systems, MBSE provides organized and rational systems modeling tools. As a result, MBSE ensures consistency, correctness, and completeness of data and facilitates information flow for decision making throughout the system. Lastly, MBSE not only captures the requirements and aids in the performance analysis and verification and validation of a project, but it also leverages and reuses these steps for other systems. Today, MBSE philosophies and methodologies are central to many top-performing SE organizations. MBSE has taken SE processes to a higher level ensuring correctness, completeness, and aiding in integrating activities throughout an organization. MBSE, as the name implies, means engineering a system using MBSE methods and tools.

From a conceptual standpoint, the systems requirements of a project/program/system service can be visualized as a tree structure. The prime node, or node from which all branches originate, is the "A" specification (A spec), or the system specification. From the prime node, any number of branches can stem depending on the system being modeled. Each subsequent node, or child node, of the prime would represent another specification. For example, given a tree diagram used to model the systems requirements of a weapons system, the "A" spec would outline the systems requirements and present a clear operational view of the project. Two branches coming from the prime node could be two "B" specs, or development specifications. One of the "B" specs would contain the hardware specifications of the system, while the other "B" spec would contain the software specifications. There could, of course, be more than two "B" specs stemming from the "A" spec. For instance, there could be multiple hardware subsystems and multiple software configuration items, each with their own "B" spec. It is important to note that regardless of the complexity of a system, or its modeled tree diagram, there is only ever one "A" spec, while there may be many other specifications at any given level.

Traditionally, systems requirements were written first, and then models were developed based on the requirements documents. This traditional document-oriented process of developing the model using only requirements has created numerous problems in terms of originality, completeness, and efficiency. According to INCOSE, "MBSE is expected to replace the document-centric approach that has been practiced by systems engineers in the past and to influence the future practice of systems engineering by being fully integrated into the definition of systems engineering processes" (ECSS 2012). Therefore, we can model the systems requirements from the beginning by applying new MBSE methodologies and techniques. Implementing these new MBSE techniques helps decrease requirements errors and subsequent analytical, design, or testing complications. MBSE is expected to provide good communications, reduce risk, and provide a better understanding of the elements of the system. For instance, in the 14th IEEE International Requirements Engineering Conference, an article about the MBSE application in quantitative requirements analysis was published. In this article, an idea is shown to connect quantitative requirements analysis to MBSE. This idea uses and combines emerging ideas from modern-day developmental environments. We present a few of these emergent ideas to illustrate the impact of MBSE methods on the modern SE environment.

"Requirements engineering" results in clearly characterized requirements (both functional and nonfunctional). These resulting requirements produce constraints that help narrow design options and highlight the early assessment, investigation, and confirmation of developmental planning activities. MBSE methods and tools facilitate the requirements engineering processes by providing an integrative, actionable, instantiation of the requirements through enterprise modeling, requirements management, and business process modeling tools. MBSE methods and tools can also directly link to design and test articles. For instance, rapid prototyping systems are made possible by MBSE methods and tools wherein changes in requirements directly and immediately affect a design or test element. For example, by making a change (or potential change) in top-level requirements, the effect is immediately observable in the particular design element, hardware-in-the-loop (HiTL) or software-in-the-loop (SiTL) testing element.

"Design by shopping emphasizes revealing the space of options available from which to choose (without presuming that all selection criteria have previously been elicited), and provides means to make understandable the range of choices and their ramifications" (Feather et al. 2006).

"Model-based systems engineering emphasizes the goal of utilizing a formal representation of all aspects of system design, from development through operations, and provides powerful tool suites that support the practical application of these principles" (Feather et al. 2006).

An MBSE application to a complex, vital, multi-entity system: In a 2010 IEEE Systems Conference, one of the papers on MBSE applications was to capture a disaster management system's (DMS) conduct and structure. The idea was to begin with the design framework required for identifying and confining the DMS. The authors state that it is important to recognize that "the management of disasters as complex adaptive systems (CAS), whose performance cannot be improved through the isolated optimization of their constituents." MBSE methods could then be used to capture the system's conduct along with the causal structure for the activities. The Systems Modeling Language (SysML), which is a domain-independent language, can be used to put the MBSE into operation. The holistic application of MBSE methods is expected to bring a new standard into DMS and is intended to decrease their overall complexity. Application of the holistic approach begins with initially addressing the DMS behavioral aspects and subsequently the use of the DMS. Afterward, the MBSE focuses on physical structural aspects that could possibly lead to the observed behavior. The model can be applied to either the desired behavior (ideal DMS) or the actual behavior (existing DMS). Once instantiated, assessments and benchmarking studies can be conducted on the DMS. The authors conclude, "By enforcing traceability between the DMS subsystems, and by leveraging artifacts such as viewpoints to develop perspectives that are tailored to specific stakeholders and users, it is anticipated that a MBSE-based holistic approach such as the one presented in this paper will promote proper use of the scientific method for the design, verification, validation and improvement of the disaster management systems" (Soyler 2010).

From a 2011 IEEE Aerospace Conference, one paper described the application of MBSE for future complex systems. According to this paper, building a complex system requires joint effort by the collaborating team and disciplines including various modeling processes, software tools, and methodologies. The nature of MBSE is to facilitate a steady, logical, interoperable, and developed model during the life cycle of the system. On the other hand, currently there exists no comprehensive modeling language that shows all system characteristics (especially in the case of systems of systems [SoS]) throughout the systems development life cycle (Bajaj et al. 2011). In the world of SE, it is likely that generally large and complex systems, with their subsystems of different scale, must not only coexist but need to work seamlessly together. MBSE is created to maximize the compatibility between subsystems. According to INCOSE, "Applying MBSE is expected to provide significant benefits over the document centric approach by enhancing productivity and quality, reducing risk, and providing improved communications among the system development team" (ECSS 2012). Therefore, applying the MBSE method is becoming an essential aspect of modern-day SE practices. According to an IEEE paper, the teams who are working together shall "speak" a similar language and shall work on a similar "matter" to build the more complex and interdisciplinary SoS. The system model is the "matter," so general, inclusive, adaptable, and friendly modeling languages are required for effective communications. The MBSE paradigm is advancing in the SE discipline, so it is likely to become a regular practice in the near future. "As an emerging paradigm for the systems of the 21st century, it seems useful to overview the current state of the art concerning the developing standards, the embryonic formalisms, the available modeling languages, the methodologies, and the major applications" (Ramos 2012).

Conventionally, documents have been characterized as the principal medium to write down the requirements throughout the system's life cycle, but after applying the new MBSE methodologies, we saw a dramatic reduction in the ambiguity of the requirements as the requirements become more clear, accurate, and complete. This notion is illustrated in an IEEE article that appeared in an Avionics Systems Conference in 1998. According to this paper, documents like system specifications and trade-off studies are expected to be the most important products for systems engineers. Explaining the requirements to software and hardware designers by using these documents seems like adding unneeded complexity. By applying MBSE, no ambiguity exists in requirements specification documents. Therefore, the model built after applying MBSE methods completely describes the functional conduct, input/output (I/O), and necessary aspects of the system's physical architecture. In addition, the performance and resource requirements also give a combined, reliable, and traceable design. An example was provided for the development of an Automotive Personal Assistance System (APAS), using the Global Positioning System (GPS) capabilities, identical to General Motors OnStar System. The APAS model contains important artifacts from the complete SE development life cycle, from originating requirements through system validation. Also, there exists no need to incur the high cost of prototype development because their model correctness can be verified by fully executing their behavioral model. Therefore, applying an MBSE methodology will support increasing the product quality while decreasing the overall development cost (Fisher 1998).

As a result, SE projects that use MBSE methodologies and paradigms, along with sound SE principles in their project, will likely enhance communications among team members, decrease errors and development time, and positively impact their chances of success.

In implementing MBSE methods, several factors such as the project scope, the functional decomposition of the system, and the allocation of requirements are important considerations in the SDLC. The new model-based approaches bring accurate and clear methods to define the functions of the system. Applications using the MBSE approach have been proven to increase communications between the development team, primary stakeholders, and other interested parties. MBSE methods facilitate integration, fluid workflow, organization, control, capitalization, and usability. Therefore, it becomes important to master MBSE concepts and to apply them throughout the life cycle of a project. In the next section, we provide some useful MBSE-related terminology and present some useful MBSE tools. In particular, we focus on representative MBSE tools that are useful for optical SE projects. The following are examples:

- No Magic, Inc. *MagicDraw*™ is an emerging Unified Modeling Language (UML)/ SySML modeling tool with requirements management, enterprise architecture, and other useful plug-ins.
- *ENVI*™ (Environment for Visualizing Images) is a well-developed software tool used in image processing applications and as an analytical tool for geospatial imagery.
- MATLAB®, *Simulink*®, and *Mathematica*™ are analytical, scientific/engineering/ technical modeling, and rapid prototyping tools.

Note that these modeling tools are meant to provide a representative example and to provide a point of reference. There are many other excellent tools available on the commercial market. The previously mentioned tools serve as illustrative examples in the next section.

4.1.2 MBSE Tools

Before we start with the description of some useful MBSE tools, it is important to recall that MBSE tools act as an aid to provide clear communication between stakeholders and the development team. To gain better understanding of the MBSE tools, and how they are applied in SE, it is important to understand the following definitions from the *Systems Engineering Guidebook* written by *James Martin*, which explain the concepts of methodology, process, and tools (Martin 1996):

> A process (P) is a series of logical tasks carried out to accomplish a specific objective. A process describes *"what"* is to be done, without telling *"how"* each task is performed. The structure of a process provides several levels of aggregation to allow analysis and definition to be completed at various levels of detail, to support different decision-making needs. A method (M) has various techniques that specify *"how"* each task is performed, in this regard, the words "method," "technique," "practice," and "procedure,"

are frequently used interchangeably. At any level, process tasks are carried out using methods. Each method is also a process itself, with a series of tasks to be performed for that specific method. In other words, the *"how"* at one level of abstraction becomes the *"what"* at the next, lower level. A tool (T) is an instrument applied to a specific method to increase the effectiveness of the task, but only if it is applied in a correct manner, and by someone with the right skills and training. The use of a tool should be to assist the accomplishment of the *"how."* In a wider sense, a tool boosts the *"what"* and the *"how."* The majority of the tools used to support Systems Engineering are computer or software-based tools, which are known as Computer Aided Engineering (CAE) tools.

Martin (1996)

As the MBSE paradigm is a relatively new advancement in the SE world, many tools related to MBSE use state-of-the-art and cutting edge technology. For example, many MBSE tools use the SysML, which is a very effective and popular modeling language. "The Modelling language is just the language, and must be combined with a methodology to be useful" (INCOSE UK 2012). Today, SysML is regarded as one of the most powerful and widely used languages for modeling systems. SysML was built in partnership between the Object Management Group (OMG) and INCOSE and is based on the language known as UML, which has gone through many iterations and is now known as UML 2.0 (OMG 2006). In a system model, SysML targets three major classes of diagrams. The three classes that SysML supports are structure diagrams, behavior diagrams, and requirement diagrams. The structure diagrams describe a systems structure and its parameters. The behavior diagrams describe how a system operates or functions. The requirement diagrams describe the goals of a system from a design and performance perspective.

As an example of the utility of MBSE methods for space systems, an MBSE challenge project was launched to develop an imaginary FireSat satellite system to assess the appropriateness of the SysML for defining space systems. The hypothetical nature of the project prohibited the realization of the system and thus could not serve as an example to prove the effectiveness of the tool in that specific application. The study did, however, serve as a template for the standard CubeSat model, which was used in the Radio Aurora Explorer (RAX) mission, built by the Michigan Exploration Lab (MXL) and SRI International (Spangelo 2012).

If an SE project uses tools like MagicDraw (built by No Magic, Inc.), various fundamental SE concepts and best practices are already inherent in the tool. The MagicDraw tool can use the basic components of the SysML modeling language. Various MagicDraw elements such as blocks, properties, and constraints are used. Parametric models and diagrams can readily be implemented. One significant benefit of using MagicDraw is that by using constraint blocks, MagicDraw can be linked to MATLAB and Simulink functions and scripts, thereby providing the ability to directly link requirements to executable code. Another benefit is the integration of the MATLAB and Simulink existing models into bigger MagicDraw models, which can increase the speed of systems development. As an example, an integrated mechatronic system was achieved by building the links between SysML and MATLAB/Simulink models (Qamar 2009).

Another well-known and highly practical tool that can be used to build analytical models for complex SE problems is MATLAB. MATLAB is one of the most prevalent computational packages among engineers and scientists. It has the capability of providing complicated graphics in an easy manner. The simple plots are fairly easy to model, but specialized plots are possible through the use of "handles." Handles present the means to entirely control and customize graphical objects in MATLAB. After understanding the handle-based system, users can generate plots according to prescribed requirements and quality standards generally set by professionals. Various instances are developed to demonstrate this concept (Green 2007). MATLAB is a powerful prototyping tool that has come a long way since its inception. MATLAB comes embedded with numerous toolboxes that increase the power and usefulness of the tool while remaining relatively simple. For example, to solve parallel computing problems, MATLAB has a parallel computing toolbox add-on that makes it easy for users to write the parallel computing native functions. Similarly, for the mathematics, statistics, and optimization problems, MATLAB offers a sequence of toolboxes such as partial differential equations, statistics, curve fitting, optimization, global optimization, neural networks, and model-based calibration to solve these complex problems in an easy manner. In control system design and analysis, MATLAB provides control system, system identification, fuzzy logic, robust control, model predictive control, and aerospace toolboxes. To deal with signal processing and communications problems, MATLAB provides signal processing, digital signal processing, communications systems, radio frequency, phased-array systems, and wavelet toolboxes. Similarly, in image processing and computer vision applications, MATLAB provides a sequence of toolboxes including image processing, computer vision systems, and image acquisition and mapping toolboxes. MATLAB contributes in the test and measurement area, as well, by providing data acquisition, instrumental control, and vehicle network toolboxes. MATLAB has modeling capabilities in the areas such as computational finance and computational biology by offering financial, econometrics, data feed, database, financial instruments, trading, and bioinformatics toolboxes.

In the next section, we will discuss the building blocks of optical systems, in which various elements of optical systems are analyzed. Additionally, the next section also discusses how the various MBSE tools are used throughout the optical system's life cycle. These tools are primarily responsible for relating requirements of the various building blocks of the optical system and for analytical and simulation purposes.

4.1.3 Transition to Optical Systems Building Blocks

Modeling the components of an optical system can be difficult due to its complex nature. Understanding the optical components often requires a multidisciplinary approach and expertise in various fields. The visible sensor is operated in a different band of wavelengths than the infrared (IR) sensor. With visible light, we deal primarily with reflected light, whereas IR light may be both reflected and emitted (in the mid-wave IR) or emitted radiation (long-wave IR). The visible light interacts differently with the atmosphere than radiation in the IR part of the electromagnetic spectrum. Dominating noise terms, spatial resolution capabilities, thermal shielding requirements, optical coatings, and even signal processing requirements can differ substantially between the visible and IR detection regimes. Thus, to build the complete system, all of these optical system building blocks must work together to generate a common output.

Consequently, to connect the optical system's requirements from the architectural level down to a component level, it becomes essential to employ MBSE tools. Different optical system components such as the IR sensors, visible sensors, data storage, image processors, or optical modulators interact with each other. The output obtained from one component is used as the input to other components. These different components perform different and distinctive tasks but work in concert to generate the final system's output. This stresses the significance of using MBSE tools to connect the requirements of the different optical system components and build integrated system models used to analyze and simulate the optical system's performance. As an example, all optical systems requirements can be divided into functional and nonfunctional requirements. The interfaces definitions between the functional decomposition of the requirements are also critical and can be captured by MBSE tools. For complex systems like optical systems, it is essential to manage requirements by using MBSE tools such as MagicDraw. MagicDraw uses NoMagic's Cameo and Requirements Management Software. It can also interface directly to the Dynamic Object-Oriented Requirements System (DOORS), a well-known and widely used requirements management software suite from IBM. It provides MBSE support for general-purpose traditional requirements management software as well as provides a variety of functionality with different plug-ins. The following sections detail optical building blocks, which introduce the fundamentals of optical SE, and a case study, which demonstrates these optical methods and MBSE principles.

4.2 Optical Systems Building Block: The Basic Optical System

We have looked at electromagnetic light and its propagation characteristics in previous chapters and now want to characterize how light interacts with some basic optical devices such as lenses mirrors, beam splitters, turning mirrors, collimators, corner cubes, filters, and telescopes. We will treat more specialized optical elements such as acousto-optical devices, liquid crystals, and special devices elsewhere in the text and, instead, focus on basic optical elements here in this chapter. According to Saleh and Teich, different approaches to characterize electromagnetic phenomena in decreasing amounts of rigor are (1) quantum optics, (2) electromagnetic optics, (3) wave optics, and (4) ray optics also known as geometrical optics (Saleh and Teich 1991). Quantum optics describes virtually all interactions of light with matter and presents the most complete and rigorous theory. Electromagnetic optics describes the interaction of light with matter in a classical sense, and wave optics, also known as scalar diffraction theory, is a theoretical simplification that is useful in Fourier optics or studying the propagation of optical light through atmospheric turbulence. In this section, we focus on the simplest form of describing the interaction of light with materials and devices—the geometrical optics approach, also known as ray optics.

This simple description of the propagation of light and its interaction with matter does not describe diffraction effects and also assumes that light is monochromatic. Nevertheless, the geometrical optics approach is useful in many applications including describing the propagation of limiting "rays" of light such as paraxial rays that indicate the relative size of the image on a detector element. As an example, a typical purpose of the optics between the telescope entrance pupil and the optical detector element is often to place the spatial Fourier transform of the electromagnetic wave in the optical system's entrance pupil completely on the imaging system's detector element in as efficient way as possible. As will be seen in this chapter, lenses, turning mirrors, beam expanders, collimators, beam splitters, neutral density filters, and other basic optical elements are used to accomplish this task. Geometrical optics methods are useful in such an application as well as other applications where diffraction theory does not play a significant role. The advantage of the geometrical optics approach is that it provides relatively simple mathematics to describe the propagation of light and the interaction of light with optical devices. An excellent text on a concise overview of geometrical optics is found in Saleh and Teich's book on photonics (Saleh and

Teich 1991). More comprehensive information can be found in standard modern optics texts such as Guenther's book on *Modern Optics* (Guenther 1990). In this section, we present some fundamental geometrical optics principles that need to be understood by the optical systems engineer and working professional working on EO/IR systems.

4.2.1 Basic Optical Principles

Most objects are observable to us because they reflect, refract, transmit, or radiate light in the visible part of the electromagnetic spectrum. Our eyes are keenly tuned to receive light in the visible part of the electromagnetic spectrum having a peak transmission at the yellow color of our sun. What we visually perceive around us is the superposition of the electromagnetic fields in the visible part of the electromagnetic spectrum that are reflected, transmitted, refracted, or radiated into our eyes field of view. When dealing with the propagation of light through a chain of optical elements, frequently some of the more elegant and mathematically rigorous methods are not required, and a simpler and a mathematically more tractable approach such as geometrical optics suffices. An example would be in determining the size of an image on an imaging system's detector element. Another example would be in estimating how light refracts through a series of optical elements. If accounting for diffraction effects is not important to a particular application, and we are dealing with larger objects and distances as compared to the wavelength of light under consideration, then a geometrical optics approach provides a reasonably accurate and tractable solution for our needs. In this section, we will develop a geometrical optics-based interpretation of refraction, reflection, and the interaction of light with mirrors, lenses, optical coatings, prisms, and filters. We present a matrix-based ray tracing method and provide insight into some commercial software that is available for this purpose. We end with a discussion on telescopes and provide a useful model in describing their functionality. By understanding these methods, the reader should be able to model basic optical elements found in a collection of basic optical components.

4.2.1.1 Refraction

An incident wave on a surface may also be refracted. Refraction occurs when light propagating in one medium encounters another medium that permits transmission of the light. At the interface boundary, the transmitted light leaves the boundary at a different angle. For example, a train of plane waves of light incident on a plane surface separated by two transparent substances such as air and glass is split: part of the incident light is reflected at the surface, and part passes into the medium and is refracted (e.g., changes direction).

The directions of the incident, reflected, and refracted light are specified in terms of the angle they make with the surface normal at the point where the incident light meets the substance boundary. The following observations can be made by looking at the interaction of the incident, reflected, and refracted light at the point of incidence on the boundary surface:

1. There is one plane that contains the rays of light from the incident, reflected, refracted, and surface normal vector.
2. The angle of incidence is equal to the angle of reflection at all wavelengths.
3. For monochromatic light, the ratio of the sine of the angle of incidence to the sine of the angle of refraction is a constant.

These three observations are quite useful and form the basis for ray tracing methods that allow us to describe the path of an optical ray through a collection of optical elements. The mathematical articulation of item (3) is given by the law of refraction:

$$C = \frac{\sin(\theta_1)}{\sin(\theta_2)},$$ (4.1)

where
C is a constant
the angles θ_1 and θ_2 are the angles of incidence and refraction, respectively, as shown in Figure 4.1

The left side of Figure 4.1 shows the incident ray of light propagating in Medium 1 with a refractive index of n. The refractive index is the ratio of the speed of light in a vacuum to the speed of light in a particular material:

$$n = \frac{c}{v}.$$ (4.2)

The index of refraction is a function of temperature and, as will be seen in a later chapter, the variations in the index of refraction are the primary contributing factor in atmospheric turbulence effects. For now, we note that as propagating rays encounter other materials, the velocity of light changes in that material and Equation 4.2 defines the index of refraction of the material. Notice that the numerator and denominator in Equation 4.2 have the same units, and so the index of refraction is unitless.

When light passes from one material with one index of refraction to a different material with a different index of refraction, the light is refracted, or bent. Snell's law determines the amount that the light is bent:

$$n_1 \sin(\theta_1) = n_2 \sin(\theta_2).$$ (4.3)

Here
n_1 is the index of refraction in Material (or Medium) 1
n_2 is the index of refraction in Material 2
the angles θ_1 and θ_2 are the angles that the incident/refracted light make with the material boundary normal vector, respectively

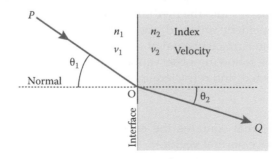

FIGURE 4.1
Graphical rendering of Snell's law. (Image courtesy of Cristan. 2006. Wikimedia Commons. http://commons.wikimedia.org/wiki/File:Snells_law.svg, accessed December 30, 2014.)

This phenomenon is well known by spear fishermen or birds hunting fish. If Material/ Medium 1 is air and Material/Medium 2 is water, and the point of observation makes an angle θ_1 with the normal to the water line (shown vertically in Figure 4.1), then the "apparent" position of the fish would be on the line obtained by extending the incident ray through the boundary at the same angle. However, the actual position of the fish would be along the trajectory of the refracted ray. Not accounting for this difference in angles would likely send the fisherman or bird home empty handed and hungry.

Another interesting phenomenon is the case of total internal reflection. Total internal reflection occurs when the medium that contains the incident light has an index of refraction greater than the medium that has the refracted light. Additionally, the angle of incidence of the incoming ray must be equal to or greater than a critical angle determined by

$$\theta_C = \sin^{-1}\left(\frac{n_2}{n_1}\right),$$ (4.4)

where

θ_C is the critical angle

n_1 and n_2 are the indices of refraction of the first medium (medium with incident ray) and second medium (medium with refracted ray), respectively, as depicted in Figure 4.2

The case for total internal reflection is shown at the right side of Figure 4.2: where the angle of the refracted ray θ_2 is 90°. Equation 4.4 becomes apparent when substituting θ_2 equal to 90° into Snell's law. Physically, if an observer were to look along the incident ray under conditions for total internal reflection, the observer would not see anything in Medium 2—Medium 2 would appear opaque to the observer. It is interesting to note that the critical angle condition given in Equation 4.4 is only possible for boundaries where the index of refraction of the material containing the incident ray is greater than that of the refracted ray. If this is not the case, as is depicted on the left of Figure 4.2, there is no possibility of total internal reflection and rays for any valid angle θ_1 are refracted into Medium 2. Also, if the incident ray has an angle θ_1 that is less than the critical angle, the refracted ray is transmitted into Medium 2 with an angle $\theta_2 < 90°$. We now look at some properties of reflected light.

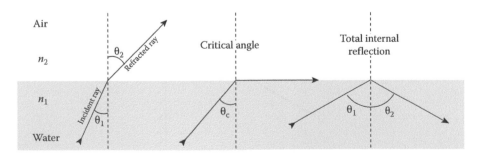

FIGURE 4.2
Total internal reflection. (Image courtesy of Josell7. 2012. Wikipedia. http://en.wikipedia.org/wiki/File:RefractionReflextion.svg, accessed December 30, 2014.)

4.2.1.2 Reflection

During daytime, the majority of what we see is due to reflected light. Commonly, the reflected light is a diffuse reflection having the property that it is scattered in all directions. Diffuse reflection occurs when the reflecting surface is rough when compared to the optical wavelength undergoing the reflection.

Another type of reflection is specular reflection. In this type of reflection, incoming light rays are reflected in a given direction. Specular reflection occurs when the reflecting surface is smooth compared to the size of the wavelength of the incoming light. The following are examples:

Reflection from a blotter is diffuse.

Reflection from a mirror is specular.

There are also cases in which there are both diffuse and specular reflections. Part of a material may be "optically rough" and cause the incident light to undergo diffuse reflection, whereas parts of the material might be highly polished and optically smooth thereby producing a specular reflection component to the reflected light. These reflecting components are additive at the electromagnetic field level. In other words, given an observation point *P* some distance away from a reflecting surface, the observed electromagnetic field would be the sum of all the reflected components (due to both diffuse and specular reflections) that are incident at point *P*.

Figure 4.3 represents light reflected from a boundary such as a mirror. When light is reflected, there is no change in medium; thus, there is no change in index of refraction. By Snell's law, the incident angle is equal to the reflected angle.

In Figure 4.3, the incident light is shown at the top of the figure making angle θ_i with the normal vector to the surface boundary of the mirror. It is reflected back as illustrated by the departing ray at the bottom of the figure making angle θ_r with the surface normal vector. Snell's law still holds with both the medium for the incident and reflected ray being

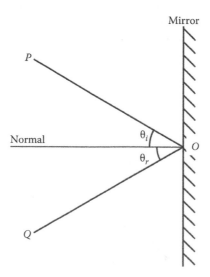

FIGURE 4.3
Illustration of a reflected light ray from a flat surface (e.g., mirror). (Image courtesy of Johan Arvelius. 2005. Wikipedia. http://en.wikipedia.org/wiki/File:Reflection_angles.svg, accessed December 30, 2014.)

the same and having the same index of refraction. Looking at Equation 4.3 with equal indices of refraction, the incident ray angle θ_1 and the exiting ray angle θ_2 must be the same.

4.2.1.3 Mirrors

Mirrors are highly reflective surfaces that use the principle of reflection. The geometrical optics approach to understanding the interaction of light rays with the optical surface of the mirror has already been presented in Snell's law. The task in modeling the interaction of light with a mirror surface is simply applying Snell's law at each surface. Mirrors come in three types, convex, planar, and concave, according to the radius of curvature of the optical surface. Assuming that light is traveling from left to right, a concave mirror has its focal point to the left of the mirror surface at some distance F. A planar mirror has its focal point at an infinite distance from the optical surface, and a convex mirror has a virtual focal point to the right of the optical surface. An example of the focal point for a concave mirror is shown in Figure 4.4.

The focal length F is given as half the radius of curvature R of the concave mirror:

$$F = \frac{R}{2}. \tag{4.5}$$

Light rays arriving at the concave mirror from infinity (parallel rays) will be focused onto the focal point as shown in Figure 4.4. At this point, we need to choose a sign convention used when dealing with distances in relationship to mirrors and lenses. There are various conventions, but the one we will use is the same for spherical mirrors and for lenses. In building our optical model, we assume that light always comes in from the object side and is called the incoming light if it is on the object side of a particular surface (mirror or lens). In the figure, we show the incoming light coming in from left to right corresponding to a source object on the left. For a mirror (or lens) surface, the radius of curvature is the distance from the vertex V to the center of curvature of the mirror (or lens) surface. If the center of curvature lies to the left of the mirror, as is the case for a concave mirror, then

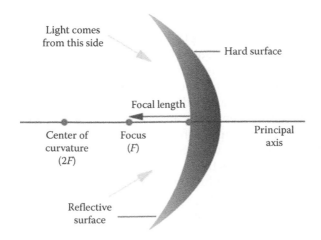

FIGURE 4.4
Reflection from a concave mirror. (Image courtesy of Cronholm144. 2007. Wikimedia Commons. http://commons.wikimedia.org/wiki/File:Concave_mirror.svg, accessed December 30, 2014.)

the radius of curvature R is positive. Also, according to Equation 4.5, the corresponding focal length is positive. If the radius of curvature of a mirror is to the right of the mirror, as in a convex mirror, then the radius of curvature is negative as is the focal length. If s_1 is the distance from the vertex V of the mirror to the object, and s_2 is the distance from the mirror vertex to the image, then for both convex and concave mirrors, a useful sign convention is that the object and image distances are positive if located to the left of the vertex V and negative if located to the right of vertex V. Another way to say this is the object and image distances are positive if the spherical mirror makes a real image (in front of the mirror) and the object and image distances are negative if they make virtual images (behind the mirror) for either concave or convex mirrors. With these assumptions, a mirror equation that is valid for all spherical mirror types can be given that relates the radius of curvature of the mirror surface to the object and image distances:

$$\frac{1}{s_1} + \frac{1}{s_2} = \frac{2}{R}. \tag{4.6}$$

With regard to the sign convention of vertical dimensions such as the object and image height, the adopted convention is that distances are positive if they are above the optical axis and negative if they are below the optical axis. With this convention, we can define a magnification for the mirror given by

$$m_m = -\frac{h_i}{h_o} = \frac{s_2}{s_1}, \tag{4.7}$$

where h_i and h_o are the image height and object height, respectively, in meters. The subscript m is used to indicate the magnification due to a mirror element. Notice that for a planar mirror, both s_1 and s_2 are positive and infinite, so the mirror magnification is one. The magnification and mirror equations are very useful in understanding the interaction of light rays with mirrors. We will discuss telescope systems that consist of mirrors and lenses later on in this section.

4.2.1.3.1 Types of Mirrors

In addition to characterizing mirrors as planar or spherical, there are characteristics of mirrors that need to be understood. One such characteristic is the mirror type. Predominantly, there are two types of mirrors, *front-side mirrors* and *back-side mirrors*. Front-side mirrors are often used in optical systems applications since they reflect light from the front surface. This type of mirror requires special handling since the optical reflective coating is on the front surface of the mirror element and can easily be damaged. In field experiments, or operational environments, these types of mirrors need to be protected against the elements and incidental handling and are usually contained within an environmental conditioning chamber. The second-type mirror is the *back-side mirror*. This type of mirror is often used in day-to-day applications such as in bathroom mirrors. For back-side mirrors, the optical coating is put on the rear side of a mounting surface such as glass. The glass itself acts as a protective cover, and so these mirrors are quite sturdy and practical for day-to-day use. Notice that although the back-side mirror is very rugged, it introduces an extra material interface by the addition of the glass mounting surface. An optical ray first undergoes a reflection at the air glass interface that is on the order of 4% of the incident optical power. Consequently, 96% of the light is transmitted into the glass and then propagates to the reflective surface at the back of the mirror. At the back end of the mirror, the light is reflected back through the glass where it again encounters the glass/air interface. At this interface, some light is again reflected and the remaining light is transmitted into the air. Notice that in this case, Snell's law must be applied to three surfaces (the original air/glass interface, the glass/polished surface interface, and then the glass/air interface), and so the angle of incidence and the angle of reflection are not exactly the same. Also, the phase of the electromagnetic wave undergoes a phase shift due to the extra optical path difference between the incident ray at the air/glass interface and the emerging ray. Because of these differences, front-side mirrors are typically preferred in optical applications that require high accuracy.

4.2.1.4 Lenses

Lenses are optical elements that are of primary importance in building an optical system. As we learned earlier, lenses take the optical spatial Fourier transform of the incident irradiance thereby producing the spectrum of the object in the lens's focal plane. Another lens placed in the focal plane of the first lens would produce an inverted image of the object in its focal plane. In this section, we present an overview of basic optical principles needed to understand how lenses work. Before we do this, we first want to develop a useful model for analyzing a general-purpose optical lens system and also define some terms that are used to understand the propagation of light through these optical elements using the geometrical optics approximation. We start with Figure 4.5, which presents an example of how to model a general lens system. More optical elements could be present, but the approach for determining important optical parameters that are essential in describing the optical system is the same as presented in this example.

In Figure 4.5, consider one dimension of an object that is shown at the left of the figure. For any series of optical elements, there is a limiting aperture that acts as the *aperture stop* of the lens system (shown at the center of the figure). The aperture stop is the object in the chain of optical elements that limits the size of the angle U_2 through the optical system (e.g., it determines the maximum value of U_2). The *marginal ray* is the ray from

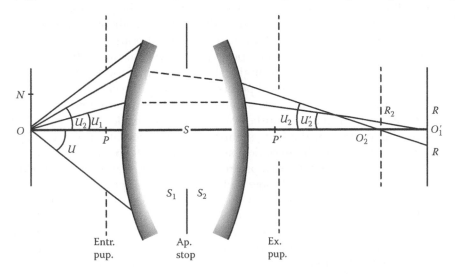

FIGURE 4.5
Representative example of a general lens system. (Image courtesy of Blleininger. 2010. Wikimedia Commons. http://commons.wikimedia.org/wiki/File:ABERR1.svg, accessed December 30, 2014.)

the object that skims the edge of the aperture stop and just barely makes it through the optical system. Conversely, any ray that goes through the center of the aperture stop is called a *principal ray* (or *chief ray*). One important point to consider is that for any ray in a given medium, light will continue linearly in that medium if there is no change of the index of refraction. Consequently, in a constant index of refraction medium, a ray of light will continue in a straight path until it encounters an interface. At the interface, Snell's law can be used to determine the change in direction of the ray of light. As will be seen later, this simple fact can be used to trace a collection of rays emanating from an object through an optical system to determine the image. Snell's law is a fundamental principle for many ray tracing algorithms used in geometrical optics modeling software. The *entrance pupil* is the image of the aperture stop when viewed from the object side, whereas the *exit pupil* is the image of the aperture stop when viewed from the image side. It is interesting to note that the principal ray will also appear to pass through the center of the entrance pupil and the exit pupil. For well-designed telescope systems, the entrance pupil is also the location of the primary telescope mirror. Hence, the diameter of the telescope mirror (e.g., an 8.5 in. telescope) is also the diameter of the optical system's entrance pupil. In general, light interacting with an optical system that consists of an arbitrary number of optical elements can be reduced to a simple model of refraction at so-called principal planes. Figure 4.5 represents a simple two-lens example model showing only the exit pupil (Ex. Pup.) on the right side of the figure.

Any combination of simple optical elements like lenses and mirrors can be modeled by introducing two fictitious planes known as principal planes. Two of these planes are shown as the entrance pupil (called H_1 for brevity's sake) and the exit pupil (called H_2 hereafter) in Figure 4.5. As will be seen in the following ray tracing section, a single transfer matrix can model an arbitrary number of optical elements between these principal planes. A property of the principal planes is that a ray leaving the rear principal plane (H_2) appears to be at the same height from the optical axis as a ray viewed from the front of the first principal plane (H_1). As a consequence, all the optical elements between the principal planes can be reduced to a set of single refractions that occur at points on the principal planes. The intersections of the principal planes with the optical axis are the principal points P and P'. If the medium surrounding the lens system is the same (e.g., air or water), then the location of the principal points is also the location of the nodal points. Nodal points have the property that if a ray is aimed at a nodal point at a particular angle, that ray appears to leave the other nodal point at the same angle. The collection of the three types of points—principal points, nodal points, and focal points—makes up the set of *cardinal points* of an optical system. For an ideal, rotationally symmetric optical system, knowing the location of these points on the optical axis is sufficient to determine basic imaging properties like the image size, the location of the image plane, and the orientation of the image. There are also other important stops in an optical system such as field stops that limit the field of view of the optical system. Other stops are used to control aberrations, for mounting purposes, and/or to reduce glare. We now look at some useful properties of a thin lens.

A lens is an optical device that interacts with the incident light, converging, diverging, or collimating the incident beam. Figure 4.6 shows a positive (converging) lens.

There are two spherical surfaces associated with the lens, and we refer to the radius of curvature of the first surface encountered by the light as R_1 and the second surface encountered by the light as R_2. We illustrate in the previous model that incoming light is to the left of the lens and outgoing light from the lens is to the right of the lens. Our sign convention for lenses is similar to that of the mirror in all respects except for the radius of curvature. If the center of curvature for the surface is to the left of the surface vertex, as in the case of

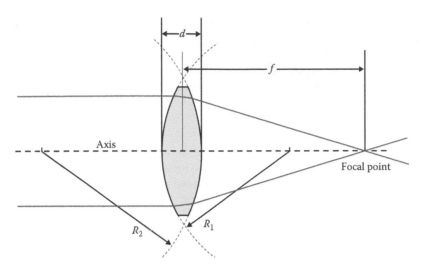

FIGURE 4.6
Positive (converging) lens. (Image courtesy of DrBob. 2006. Wikimedia Commons. http://commons.wikimedia.org/wiki/File:Lens1.svg, accessed December 30, 2014.)

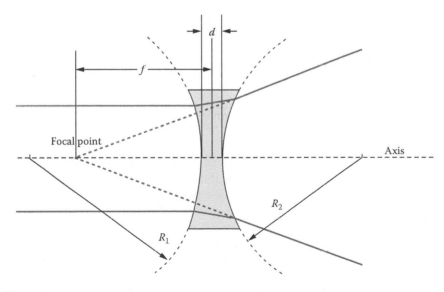

FIGURE 4.7
Negative (diverging) lens. (Image courtesy of DrBob. 2006. Wikimedia Commons. http://commons.wikimedia.org/wiki/File:Lens1b.svg, accessed December 30, 2014.)

a diverging lens, then the radius of curvature is negative. This occurs if the first surface is concave. Figure 4.7 shows a negative (diverging) lens. On the other hand, if the center of curvature is to the right of the surface vertex, then the radius of curvature and the focal length are positive.

In Figure 4.7, the radius of curvature R_1 for the first surface (left-most surface) is negative, whereas the radius of curvature R_2 for the second surface is positive. The remaining sign conventions are the same as for the case of the spherical mirror. Namely, if the object is to the left of the lens, it is a real object and the distance to the object is positive. If the object lies to the right of the lens, it is a virtual object and the distance from the lens to the object is negative. If the image is a real image (e.g., formed by a real object and to the right of the lens as in a converging lens), then the image distance is positive. If the image is a virtual image (formed to the left of the lens from a real object as would be the case for a diverging lens), then the image distance is negative. The vertical distances, just as in the case for the spherical mirrors, are positive if they are above the optical axis and negative if they are below the optical axis.

4.2.1.4.1 Thin Lenses

The simplest kind of lens is a thin lens. A lens can be classified as a thin lens if its thickness along the optical axis is small compared with the focal length of the lens. To determine the focal length, f', the previous sign convention is used along with the so-called lens maker's equation:

$$\frac{1}{f'} = \left(\frac{n_1}{n_2} - 1\right)\left(\frac{1}{R_1} - \frac{1}{R_2}\right),$$ (4.8)

where
  n_1 is the index of refraction of the lens material
  n_2 is the index of refraction of the surrounding medium

Notice that for a converging lens like the one shown in Figure 4.6, the radius of curvature R_1 of the first surface is positive and the radius of curvature of the second surface R_2 is negative. For this case, the shape factor K_s is positive and is given by

$$K_s = \left(\frac{1}{R_1} - \frac{1}{R_2}\right).$$ (4.9)

The shape factor is a useful quantity. If the lens is in air, then the index of refraction of the surrounding medium n_2 will be 1 and the index of refraction of the lens itself (n_1) will be >1, and so the shape factor then determines whether the lens is a converging lens or a diverging lens. In air, if the shape factor K_s is positive, the lens will be a converging lens. If the shape factor K_s is negative when the lens is in air, then the lens will be a diverging lens. If the surrounding medium is not air, then whether or not the lens is a converging or diverging lens depends on the relative nature of the material surrounding the lens, the material of the lens itself, and the shape factor. In general, the *power* of a lens can be used to determine whether or not a lens will diverge or converge light. The power of a lens is given by

$$\text{Power} = \frac{1}{f'},$$ (4.10)

where f' is the focal length of the lens. If the power of the lens is negative, then the lens will diverge light; if the focal length is positive, it will converge light. The lens maker's equation in Equation 4.8 can be used to determine a wide range of practical lenses. If the first surface is flat (Plano), then radius of curvature is located at infinity and the first term of the shape factor goes to zero. The sign of the shape factor is then determined by the radius of curvature of the second surface of the thin lens. Using our sign convention, a

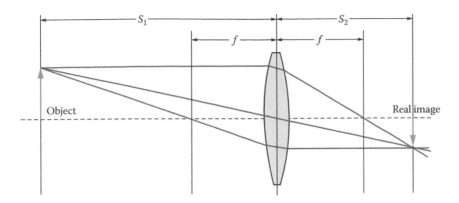

FIGURE 4.8
Imaging model for a simple thin lens. (Image courtesy of DrBob. 2006. Wikipedia. http://en.wikipedia.org/wiki/File:Lens3.svg, accessed December 30, 2014.)

wide range of useful lenses can be specified using the lens maker's equation such as biconvex, plano-convex, plano-concave, convex–concave (meniscus), and biconcave. Regarding meniscus lenses, they come as positive and negative meniscus lenses depending on the relative radius of curvature of the surfaces. If the first surface has a larger radius of curvature than the second surface, the shape factor shows that this will be a negative meniscus lens. Conversely, if the radius of curvature for the first surface is smaller than the second surface, the result is a positive meniscus lens.

We now want to look at imaging properties of thin lenses. We define a first focal length (front focal length) shown as f in Figure 4.8. The object space is to the left of the thin lens, and the image space is to the right of the lens. We see that for a thin lens, there is one principal plane that vertically bisects the lens and separates the object space from the image space.

The entrance pupil, exit pupil, aperture stop, and field stop, as well as the nodal points and principal plane points, all lie in this plane. In the image space, f' is the focal length given by the lens maker's equation in Equation 4.8. For imaging applications, we define the distance from the thin lens to the object as S_1 and the distance from the thin lens to its corresponding image as S_2. If a real object is in the object space as shown, then the biconvex lens will form a real image in the image space as shown. The image will be inverted and will appear in the image plane. The image plane can be found by projecting a parallel ray from the top of the object through the lens. The lens will focus the ray through the focal point f. By projecting another ray leaving the top of the object through the nodal point at the center of the lens on the optical axis, the intersection of the two projected rays will then determine the image plane. Besides helping determine the image plane (e.g., the plane where the optical image is formed), the focal length is useful in a variety of applications. If optical rays are projected through the first focal point as shown in Figure 4.9, the rays leave the thin converging lens in parallel lines and are *collimated*. This principle can be used quite effectively in expanding or contracting the dimensions of a beam of light.

An interesting application is when two lenses are combined as follows. Beams of parallel rays arrive at a lens L_1 and are focused through its focal point. A second lens L_2 has its front focal point at the same position as the back focal point for the first lens L_1. When the beam emerges from L_2, it is once again collimated but with a different diameter that is a simple function of the ratio of the focal lengths of the lenses. If f_1 is the back focal

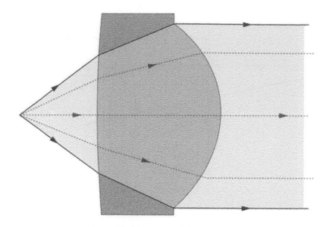

FIGURE 4.9
Generating parallel rays from a thin lens. (Image courtesy of RCarboni. 2009. Wikipedia. http://en.wikipedia. org/wiki/File:CollimatingLensSVG.svg, accessed December 30, 2014.)

length of the first lens and f_2 is the front focal length of the second lens, then the output diameter of the ray bundle leaving the second lens is related to the diameter of the incoming ray bundle by

$$D_{out} = D_{in} \cdot \frac{f_2}{f_1}. \tag{4.11}$$

If the focal length of the second lens is larger than that of the first lens, then the two-lens system is a *beam expander*. Another important parameter related to the optical throughput of a lens (or imaging system) is its *f*-number. The *f*-number is given by

$$N = \frac{f}{\#} = \frac{f}{D}, \tag{4.12}$$

where
  D is the diameter of the lens (e.g., the diameter of the entrance pupil of the imaging system)
  f is the focal length of the lens

For the case of a wide angle, wherein the focal length is small and the diameter of the lens is large, the *f*-number is very small. Conversely, if the diameter is very small and the focal length is very large, then the *f*-number is very large. This metric is counterintuitive, and so another metric called the numeric aperture is often stated. The numeric aperture is given by

$$NA_i = n \sin\left(\arctan\left(\frac{D}{2f} \right) \right) \cong n\frac{D}{2f}, \tag{4.13}$$

where n is the index of refraction of the medium surrounding the lens. The subscript i indicates that the numeric aperture is viewed from the image side. By accounting for the magnification m of the optical system, the numeric aperture as viewed from the object side is given by

$$NA_o = \left(\frac{2(m-1)}{nm}N\right)^{-1},$$ (4.14)

where
  N is the f-number given by Equation 4.12
  n is the index of refraction of the surrounding medium (1 for air)
  m is the lateral magnification (assumed negative)

We now want to extend some of these useful results to thick lenses.

4.2.1.4.2 Thick Lenses

Other than thin lenses, all real lenses are considered thick lenses and have additional complexity. Recall that for a thin lens, there was only one principal plane that served as the entrance pupil, exit pupil, and aperture stop of the optical system. Figure 4.10 shows a model for a thick lens. Thick lenses have two principal planes whose locations depend on the radius of curvature of the front and back surfaces. The focal points are illustrated by f and f'. The remaining cardinal points are shown as C_1 and C_2.

Note that the nodal points and principal points coincide at the principal plane for a general optical system. The lens maker's formula determines the optical power of the thick lens as

$$\phi = \frac{1}{f'} = \frac{(n_L - n_1)}{n_1 R_1} - \frac{(n_L - n_2)}{n_1 R_2} + \frac{(n_L - n_2)(n_L - n_1)}{n_1 n_L} \frac{t}{R_1 R_2},$$ (4.15)

where
  n_L is the index of refraction of the thick lens
  n_1 is the index of refraction of the medium that interfaces on the first surface (assumed to the left of the thick lens in Figure 4.10)
  n_2 is the index of refraction of the medium on the object side (right) of the thick lens
  t is the thickness of the lens along the optical axis
  R_1 and R_2 are the respective radii of curvature as before

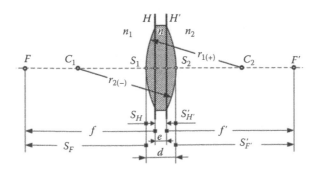

FIGURE 4.10
Model of a thick lens. (Image courtesy of Tamasflex. 2011. Wikipedia. http://en.wikipedia.org/wiki/File:ThickLens.png, accessed December 30, 2014.)

If the thick lens is in air, as is often the case, Equation 4.15 reduces to

$$\phi_{air} = \frac{1}{f'} = (n_L - 1)\left[\frac{1}{R_1} - \frac{1}{R_2} + \frac{(n_L - 1)}{n_L}\frac{t}{R_1 R_2}\right]. \tag{4.16}$$

Notice that if the thickness t is reduced to zero, we have the thin lens result. There is a useful relationship to relate the image and object distances to the front and back focal lengths. If the distance from the object to the first principal plane of the lens is given by S and the distance from the lens' second principal plane to the image is given by S', then the relationship between the front and back focal lengths and the object and image distances is given by

$$\frac{n_3}{f'} = \frac{n_1}{f} = \frac{n_1}{S} + \frac{n_3}{S'}, \tag{4.17}$$

where
   n_1 is the index of refraction in the object space
   n_3 is the index of refraction in the image space

Not only is the general shape and thickness of a lens important but also the amount of light that makes it through our optical system. We provide a summary of some of the notable results relating to mirrors and lenses in Table 4.1.

In the next section, we look at antireflection coatings to maximize the amount of light that reaches the detector.

4.2.1.5 Antireflective Coating

When setting up optical imaging systems that are designed to work with low-light levels, the treatment of the surfaces of the optical components becomes important. As we stated earlier, if we leave a surface untreated, a certain portion of the light is reflected from the surface. In glass, this is roughly 4% of the optical power at each glass interface. The more untreated optical elements we introduce in the optical chain of the imaging system, the more loss of optical power there is on the detector. A method for reducing the reflection from an optical surface is to coat it with an *antireflection coating*. An antireflection coating, as the name implies, reduces the amount of reflected light by causing destructive interference between the reflected light on the surface of the coating, and the rays of light that are reflected back from the glass interface are transmitted through the air/film interface.

The film thickness is designed to be equal to one-quarter of the wavelength of the incident light, so a down and back round trip path of the ray (from the air/film boundary to the film/glass boundary and back) is $\lambda/2$ m long. The two reflected rays are roughly of equal magnitude and 180° out of phase and so destructively interfere with each other—no power is carried away in the reflected direction. Consequently, a high percentage of the optical power in the incident ray is transmitted into the glass.

4.2.2 Achromatic Principle

The index of refraction is not constant with changes in wavelength. The use of lenses to focus a broadband image may result in a poor image due to differences in focal length for the various wavelengths. Variations due to color (wavelength) in the index of refraction must be corrected. Figure 4.11 shows the differences between a normal lens (shown at the right of the figure) and an achromatic lens (shown at the left of the figure).

TABLE 4.1

Summary of Notable Equations

Image formation	S = object distance from principal plane (always positive for real object)
	S' = image distance from principal plane (positive, real image; negative, virtual image)
	f = front focal length
	f' = back focal length
	R = radius of curvature
	h_i = image height
	h_o = object height
	D = diameter of entrance pupil
	t = thick lens thickness at optical axis
	Spherical mirror (in air) $\dfrac{1}{S}+\dfrac{1}{S'}=\dfrac{2}{R}$
	Plane mirror (spherical mirror for $R \rightarrow \infty$) $S = -S'$
	Spherical refracting surface $\dfrac{n_1}{S}+\dfrac{n_2}{S'}=\dfrac{(n_2-n_1)}{R}$
	Thick lens $\dfrac{n_3}{f'}=\dfrac{n_1}{f}=\dfrac{n_1}{S}+\dfrac{n_3}{S'}$
Thin lens	$\dfrac{1}{f'}=\left(\dfrac{n_1}{n_2}-1\right)\left(\dfrac{1}{R_1}-\dfrac{1}{R_2}\right)$
Thick lens optical power	$\phi=\dfrac{1}{f'}=\dfrac{(n_L-n_1)}{n_1R_1}-\dfrac{(n_L-n_2)}{n_1R_2}+\dfrac{(n_L-n_2)(n_L-n_1)}{n_1n_L}\dfrac{t}{R_1R_2}$
Lateral magnification m	$m=-S'/S$
Magnitude of lateral magnification $|m|$	$|m|=h_i/h_o$
Angular magnification	$m_\gamma=\dfrac{\gamma_i}{\gamma_o}$
F/# (f-number)	$N=\dfrac{f}{\#}=\dfrac{f}{D}$
Numerical aperture (image side) NA_i	$NA_i=n\sin\left(\arctan\left(\dfrac{D}{2f}\right)\right)\cong n\dfrac{D}{2f}$
Optical invariant	$m\cdot m_\gamma=\dfrac{n_o}{n_i}$

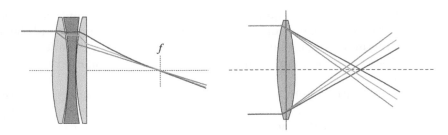

FIGURE 4.11

(See color insert.) Comparison of normal lens and achromatic lens. (Achromat reprinted with permission from Egmason, May 8, 2010. http://en.wikipedia.org/wiki/File:Apochromat.svg, accessed November 20, 2014; Normal lens reprinted with permission from Bob, October 9, 2004. http://en.wikipedia.org/wiki/File:Lens6a. png, accessed November 20, 2014.)

Achromatic lenses are designed to focus the different colors of light onto the same focal point. This is done by adding a second material of a different index of refraction and designing the thickness and curvature of the interface, to focus the wavelengths onto the same focal point.

4.2.3 Ray Tracing

Ray tracing is a tool used to describe optical systems, where each optical component is expressed in a matrix form. This is a powerful methodology that traces rays through a host of optical components using the geometrical optics principles discussed earlier. The input ray to the optical system is defined by its height and also by the angle that it makes to the horizontal plane at the point of interest. This ray can be propagated through the chain of optical elements by repeated application of Snell's law when the ray encounters an optical surface. In the paraxial limit, the propagation of light rays can be described by a series of matrices that characterizes the interaction of the light ray when it encounters a particular surface or propagates through a particular medium. The new height and angle of the ray can be expressed as the matrix multiplication of the old height and angle of the ray and the matrix representing the particular surface or propagation distance. The following equation defines the relationship between the height of the original ray from the optical axis and the angle it makes to the horizontal surface and a particular propagation or surface encounter represented by the matrix element:

$$\begin{bmatrix} h_{out} \\ \theta_{out} \end{bmatrix} = \begin{bmatrix} A & B \\ C & D \end{bmatrix} \begin{bmatrix} h_{in} \\ \theta_{in} \end{bmatrix}, \tag{4.18}$$

where the A through D elements of the matrix are uniquely defined dependent on the type of interaction the ray encounters. For instance, for a thin lens with unit magnification, the A, B, C, and D parameters are given by

$$\begin{pmatrix} 1 & 0 \\ -\dfrac{1}{f} & 1 \end{pmatrix}. \tag{4.19}$$

By substituting Equation 4.19 into Equation 4.18, we see that, at the point where the light ray intersects the thin lens, the output height is equal to the input height and the output angle is equal to the input angle minus the input height divided by the focal length of the lens. Table 4.2 shows some common matrices for the ray propagation principles discussed previously.

The utility of the matrix method becomes apparent when tracing a ray of light through a collection of optical elements using simple transfer matrices. By multiplying all the matrices of our optical system, we end up with one matrix that describes the whole optical system and relates the input ray and input angle to the output ray and output angle. We can illustrate this concept by deriving the $ABCD$ matrix for a thick lens. For a thick lens, the optical ray in a medium with refractive index n_1 first encounters a refractive surface. The optical ray then propagates through the thick lens of width t in a medium that has an index of refraction n_L. The optical ray then encounters

TABLE 4.2

ABCD Matrices for Simple Optical Surfaces and Components

Optical Element (or Operation)	ABCD Matrix	Remarks
Free space propagation or index of refraction is constant	$\begin{pmatrix} 1 & d \\ 0 & 1 \end{pmatrix}$	d = distance.
Refraction at a flat interface	$\begin{pmatrix} 1 & 0 \\ 0 & \dfrac{n_1}{n_2} \end{pmatrix}$	n_1 = first refractive index, n_2 = second refractive index.
Refraction from a curved boundary	$\begin{pmatrix} 1 & 0 \\ \dfrac{n_1 - n_2}{R \cdot n_2} & \dfrac{n_1}{n_2} \end{pmatrix}$	R = radius of curvature. $R > 0$ for convex surface. $R < 0$ for concave surface. n_1 = first refractive index. n_2 = second refractive index.
Reflection from flat surface (e.g., flat mirror)	$\begin{pmatrix} 1 & 0 \\ 0 & 1 \end{pmatrix}$	Identity matrix.
Reflection from a curved surface (e.g., curved mirror)	$\begin{pmatrix} 1 & 0 \\ -\dfrac{2}{R} & 1 \end{pmatrix}$	R = radius of curvature. $R > 0$ (convex surface). $R < 0$ (concave surface).
Thin lens	$\begin{pmatrix} m & 0 \\ -\dfrac{1}{f} & m \end{pmatrix}$	f = back focal length of lens. $f > 0$, for convex lens. $f < 0$, for concave lens. Valid for thin lens model only. The factor m is the lateral magnification of the optical system.

another refractive surface that separates the thick lens from a medium with refractive index n_3. The resulting $ABCD$ matrix M_{tl} is given by

$$M_{tl} = \begin{pmatrix} 1 & 0 \\ \dfrac{n_L - n_3}{n_3 R_2} & \dfrac{n_L}{n_3} \end{pmatrix} \begin{pmatrix} 1 & t \\ 0 & 1 \end{pmatrix} \begin{pmatrix} 1 & 0 \\ \dfrac{n_1 - n_L}{n_L R_1} & \dfrac{n_1}{n_L} \end{pmatrix} = \begin{pmatrix} A & B \\ C & D \end{pmatrix}, \qquad (4.20)$$

where R_1 and R_2 are the respective radii of curvature for the first and second surface of the thick lens. Notice the way the matrices are "stacked" is from right to left. The first operation on the incident ray is shown as the right-most matrix. Subsequent operations are ordered from right to left. In general, an $ABCD$ matrix M that consists of a sequence of N operations, each having a 2×2 matrix formulation, is given by

$$M = M_N \cdot M_{N-1} \rightleftharpoons M_2 \cdot M_1. \qquad (4.21)$$

In this manner, rays can be propagated through a complicated set of optical elements, and the resulting output ray position and angle can be determined by a single matrix multiplication consisting of the reduced $ABCD$ matrix from the sequence of 2×2 matrices. The $ABCD$ matrix method applies to a wide variety of optical operations and optical devices such as turning mirrors, corner cubes, beam splitters, and prisms. *Turning mirrors* are used to change the direction of propagation of the light. A common use is to extend the focal length of an imaging system by folding the optical path length so that a larger focal length fits into a given small space. *Corner cubes* are simply two, 90° reflections from mirrored surfaces (canted at 45° to the incident light) that reflect the light back in the direction that it came. A corner cube was placed on the moon so that lasers fired from the earth to the moon would have their concentrated laser pulses reflected back toward the earth. In that manner, lasers can be used to measure the time of flight of an optical pulse from the earth to the moon and back.

4.2.4 Optical Filters

Optical filters are useful to select certain wavelength to be processed in an optical system. A *narrow-band filter* passes light in a certain frequency band (e.g., 50 nm wide and centered at a particular wavelength). A "red" filter passes red light but rejects other wavelengths. Looking through a red filter, the transmitted light all appears red. Filters also come in *high-pass, low-pass,* or *notch* varieties based on whether they pass high-frequency light, low-frequency light, or light in a particular frequency range. *Polarizers/polarization filters* are designed to select/remove polarization components from the transmitted imagery. *Beam splitters* split the optical power in different directions. For instance, a 50/50 beam splitter reflects half the optical power in one direction and transmits half the optical power through a partially reflecting mirror in another direction. Different ratios of reflected to transmitted power can be specified. *Dichroic beam splitters* can pass light in a certain wavelength range and reflect light in another wavelength range. *Neutral density filters*, polarization components, and *optical gratings* can all be used to adjust the optical power through a chain of optical elements. *Optical gratings* and *prisms* can be used to separate different colors or reflect a particular color at a determinable angle. All these optical components can be used to control light, as it propagates through an optical system. We will look at some of these components in closer detail in later sections. We will also take a closer look at optical aberrations that arise in our optical system due to imperfections in the optical components, effects of the turbulent atmosphere, and even the random nature of the light itself.

4.2.5 Optical Design Software

In the previous sections, we discussed the ray tracing technique used for modeling the optical systems by developing the matrices and where understanding of the matrices calculation is clear-cut and simple. But using the ray tracing method for 3D structures of an optical system to mark out bundles of rays can be a complex and tedious task. There exist software packages that are useful in designing and analyzing the optical systems. Several of these software packages are licensed, and some packages are free. What follows is a list of optical design and analysis software that can be helpful for ray tracing, designing, and analysis of an optical system:

- Advanced Systems Analysis Program (ASAP)
- Code V
- dbOptic
- LightTrans Virtual Lab 5
- Optics Software for Layout and Optimization (OSLO)
- HEXAGON

- Optix
- FRED Optical Engineering Software
- Zemax

In general, commercially available software is better documented and maintained and less error prone than freeware. For example, "Zemax optical design and analysis software" has been available on the commercial market for a long time, and they have a refined and excellent product (Radiant Zemax 2014). These optical design and analysis software suites simplify the process of designing an optical system and have become fundamental tools for the optical designer. The specific tool used, however, is determined by the needs of the organization and hopefully determined by a requirement-driven SE process.

In the next section, we will use our integrated case study to apply some of the methods learned so far. We will demonstrate how the MBSE tools are applied to the optical system. As part of the case study, the SE concept of the "analytical hierarchy process" (AHP) is explained.

4.3 Integrated Case Study: MBSE, MATLAB®, and Rapid Prototyping at FIT (Basic Optical Components)

The objective of this case study is to present how MBSE tools may be used in a working environment and show how they are an integral aspect of the modern SE-oriented development environment. We also use this section to provide an example of how the optical building blocks material is used in the development of the optical system. The AHP is presented as a tool for conducting a trade-off study for selecting evaluation software for a critical component for our scenarios unmanned aerial vehicle (UAV) optical system, an optical modulator.

As part of the Department of Homeland Security UAV contract that fantastic imaging technologies (FITs) have recently won, FIT has selected to design and develop the basic optical components of the optical system mounted on a UAV. In this application, the optical modulator is used to modulate a beam of light before it reaches the optical detectors to help reduce background noise and help in signal detection. As part of these discussions, the technical aspects of the basic optics systems needed to be understood before embarking on the optical modulator design. The company is also applying MBSE methods to help manage the requirements, design, and testing aspects of the system. For this project, FIT uses the MagicDraw tool with its requirements management plug-in, to manage the requirements of the system, and Zemax™ as the primary optical design and analysis software tool. For their high-speed, atmospheric turbulence compensation (ATC) applications, FIT has developed a rapid prototyping system that implements their patented MATLAB-based ATC algorithms.

A number of brainstorming meetings occurred between the systems engineers at FIT to discuss the basic optical system and the optical modulator. An important topic of the discussion between the systems engineers and the optical designers was the technical requirements for the basic optical system for the imaging sensors for the DHS UAV project. Karl Ben, the senior systems engineer, called a meeting with one of his junior systems engineers, Jennifer O. (Jen) to discuss the basic optical system. We join the two as they begin their discussions.

(Tuesday morning, bright and early.)

Karl: "Hi Jen, How's life been treating you?"

Jen: "I am doing good and how are you?"

Karl: "Great! Thanks for asking! So, how is that radiometric study coming along?"

Jen: "Pretty good! We are wrapping up some of the final details but we are practically certain that we can get away with a passive imaging system."

Karl: "That is good news. When do you think the study will be finished?"

Jen: "Fairly soon. We need to have a better understanding of the basic optical system so that we can put that into our models. We are coming along pretty well on that front also."

Karl: "Great, tell me about it."

Jen: "We were looking at the Raytheon MTS-B-20 gimbal pod and we now know that it has a 20 inch (0.508 m) primary mirror for both the visible and infrared parts of the spectrum. We have identified representative sensors for both the visible and infrared parts of the spectrum that we can use as placeholders to give us a feel for the imaging scenario."

Karl: "That's good. We are can use these in our current MBSE models and then use the actual sensor parameters once the trade study completes. That way, we can get started with a soft design and see what issues we run into."

Jen: "Right. That brings up an issue though. We haven't chosen our component-level MBSE tool yet. I was wanting to do a trade-off study to select the right tool but I have been tied up with the radiometric model, and basic optical systems analysis and so haven't gotten a chance to do that yet."

Karl: "Well, you still have the weekend right?! Just kidding! The radiometric study has priority, followed by understanding the optical system. Let's continue with that."

Jen: "Hah, good one! Ok, the sensors we are using for both the visible and imaging sensors both have a 1 cm^2 active area with 100 percent fill factor. The read-out electronics are on the back of the sensor elements. They each have 2048 × 2048 pixels. With some relay optics, we can get an effective focal length that ranges between (0.2 and 2 m). We are using a nominal 1 m effective focal length to start."

Karl: "What does the Raytheon gimbal provide?"

Jen: "It provides the primary telescope mirrors, the pointing and tracking mechanisms, servo-control mechanisms, drivers, gimbal command software, and a PC interface. The rest is up to us."

Karl: "OK, what about the transmittance of the optics?"

Jen: "We are going to minimize the number of optical surface in our relay optics and want to keep the transmittance through all the optical components greater than 95%. The optical engineers are running ZeMax to work on some preliminary designs."

Karl: "Ok, what are we using for the Fried parameter in our analysis?"

Jen: "Well, at visible wavelengths, the atmospheric coherence length often ranges between 8 and 12 cm for air-to-ground applications. We decided to use 10 cm for a nominal value of r_o. For the infrared spectrum, ro scales as the ratio of the center wavelength at the IR wavelength to that of the visible wavelength with the result raised to the 6/5th. So, the atmospheric coherence length can be found by multiplying our 10 cm nominal ro by $(\lambda_{ir}/\lambda_{vis})^{6/5}$."

Karl: "Sounds reasonable. We may want to do a worst-case–best-case analysis, along with the nominal case once we have all the necessary parameters. What are we using for the separation distance between the sensor and our intended human object."

Jen: "The nominal UAV altitude above ground-level is 15,000 ft (4,572 m). Maximum altitude is 50,000 ft (15,240 m), and minimum altitude is 13,123 ft (4,000 m). With the human object standing at the fence line, that gives a nominal sensor-to-human standoff distance of 9543 m."

Karl: "OK, again, we should probably look at the best and worst cases to go along with the nominal case."

Jen: "Sure, no problem."

Karl: "What are the transverse and lateral magnifications of the basic optical system?"

Jen: "We still have to determine that during the actual design process. For now, we are going with the variable ranges of 1 to 25 for both longitudinal and transverse magnifications for both the visible and IR optics."

Karl: "What mean wavelengths are you using for the visible and IR sensors?"

Jen: "For the visible, we are using 500 nm. We are using 9.5 μm for the infrared mean wavelength based on the peak emission wavelength for our human object."

Karl: "Good. We will have to talk about sources soon. Sounds like you have a good handle on the basic optical properties. I should be able to determine a good number of the attributes and capabilities from the information you just provided. I think that is enough for now. Maybe you can get started on trade-study that you mentioned earlier."

Jen: "I was hoping you'd ask!"

Karl: "I learned that you are studying Systems Engineering at that great local university. How is the class and coursework going?"

Jen: "I like every bit of the class. I am studying Systems Engineering and its principles and gaining a lot of knowledge. Most importantly, the knowledge I am getting can be directly applied to several projects that I have been working on at my job."

Karl: "Make sure that when you do that trade study, we can use the resulting tool to help model the optical modulator that we have been discussing for background suppression both for this and future applications."

Jen: "Sure, I will make that a goal of the study."

Karl: "Oh, and make sure that you use a formal, repeatable, and justifiable analytical methodology so that we can defend the results."

Jen: "Oh, what a coincidence! In the system engineering class, we just finished studying the Analytical Hierarchy Process or AHP for short. AHP is an analytic method used to analyze alternatives and aid in decision-making. I think it's perfect for these types of problems."

Karl: "That's great! I would like you to consider at least three MBSE tools and determine which would be best in analyzing the optical modulator, integrating with the rest of our MBSE tools, and providing the best features for our future work."

Jen: "I understand. I would like to assess MATLAB, ENVI, and Mathematica, as they are all prominent analytical tools, and some may even be useful in rapid prototyping applications"

Karl: "I want you to specifically focus on development time, cost, maintenance and support, and also ease of use."

Jen: "Alright. I'd like to get started right away."

Karl: "Let me know as soon as you're done."

As Jen started the trade-off study, her mind drifted back to what she learned about the AHP method in her SE class. She recalled that the AHP method was well suited for choosing the best alternative given a set of (possibly qualitative) criteria. Dr. Thomas Saaty developed the AHP methodology in the 1970s. It has been applied in many disciplines, such as social sciences, engineering/science, national-level models, and private- and public-sector models. In implementing the AHP methodology, the first step is to choose a particular goal for the analysis. In this case, the goal is to select the best long-term analytical tool used for analyzing key optical components such as an optical modulator. In order to use AHP, one must understand matrix manipulation and be familiar with the calculation of eigenvalues and eigenvectors. Two matrices are created with dimensions equal to the number of decision criteria that are to be evaluated. The first matrix is known as the criteria versus criteria (CvC) matrix, which represents the previously mentioned decision criteria. The second type of matrix is called the alternative versus alternative (AvA) matrix, of which, one is created for every criterion. Four simple steps are followed in using the AHP method to select alternatives:

1. Establish decision criteria (e.g., price, development effort, maintenance, quality) and alternatives (e.g., Job A, Job B).
2. Collect pair-wise importance/preference data for paired criteria and for each alternative.
3. Synthesize and determine the relative priority of each criterion using the AHP method.
4. Aggregate all priorities and compute weights of all alternatives; select the best alternative.

In these steps, CvC matrices are generated for each alternative based on a pair-wise comparison and the *decision consistency* or perhaps more accurately the *decision inconsistency* is determined. Eigenvalues and eigenvectors help to quantify this aspect.

Figure 4.12 shows a block diagram for the decision-making process in this scenario. The top level of the block diagram shows the goal of the analysis. The middle row shows the criteria that will be used in the analysis. The third level shows the alternatives available for comparison. Jen noted that she was given the criteria for evaluating the MBSE tools:

- Maintenance support
- Ease of use
- Development time
- Cost

She was also given a list of alternatives:

- MATLAB
- ENVI
- Mathematica

Jen then proceeded with her analysis. Table 4.3 illustrates the CvC comparison matrix to pick from the different MBSE tools. Table 4.3 has four rows and four columns, where every row and every column correspond to a chosen criterion. When the comparison of the same criteria is performed, then its matching cell is allocated a weight factor of 1 as shown. An arbitrary, with appropriate granularity, is chosen in making the pair-wise comparisons. The results of the preferences in Table 4.3 can be understood as follows: Cost is

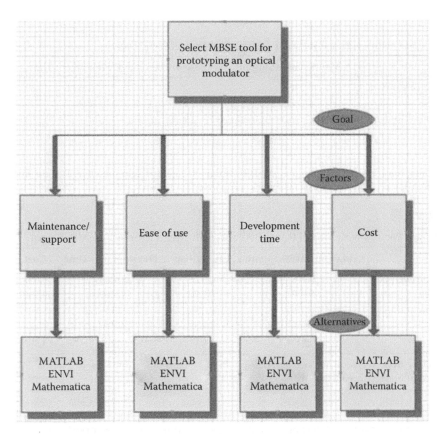

FIGURE 4.12
Normal lens. (Reprinted with permission from Dr Bob, October 9, 2004, http://en.wikipedia.org/wiki/File:Lens6a.png, accessed November 20, 2014).

TABLE 4.3

AHP Weight Factors Used to Select an MBSE Tool

	Maintenance/Support	Ease of Use	Development Time	Cost
Maintenance/support	1	2	1/3	1/5
Ease of use	1/2	1	3	2
Development time	3	1/3	1	1/3
Cost	5	1/2	3	1

Source: Data from Technical Report on AHP by Bradley Barteotti, submitted to Florida Institute of Technology (Barteotti 2014).

preferred five times more than maintenance/support and three times more than development time but only half as much as ease of use. Notice that the comparisons are made on a pair-wise basis, from row to column. The diagonals are all one because any criteria must be preferred the same to itself.

Table 4.4 is identical to Table 4.3, with the addition of a "total" row, which includes the sum of the given column. This step is in preparation of normalizing the cell entries so that each matrix entry is listed in terms of a percentage and not its actual value.

In Table 4.5, each column was divided by its total, normalizing it. This allows each column entry to be relatively compared to the other column entries so that the magnitude

TABLE 4.4

AHP Weight Factors with a Total Row Used to Select an MBSE Tool

	Maintenance/Support	Ease of Use	Development Time	Cost
Maintenance/support	1	2	1/3	1/5
Ease of use	1/2	1	3	2
Development time	3	1/3	1	1/3
Cost	5	1/2	3	1
Total	9.5	3.83	7.33	3.53

Source: Data from Technical Report on AHP by Bradley Barteotti, submitted to Florida Institute of Technology (Barteotti 2014).

TABLE 4.5

AHP Weight Factors with a Total Row Used and Row Mean to Select an MBSE Tool

	Maintenance/Support	Ease of Use	Development Time	Cost	Mean
Maintenance/support	0.105	0.522	0.045	0.057	0.182
Ease of use	0.053	0.261	0.409	0.566	0.322
Development time	0.316	0.087	0.136	0.094	0.158
Cost	0.526	0.130	0.409	0.283	0.337
Total	9.5	3.83	7.33	3.53	

Source: Data from Technical Report on AHP by Bradley Barteotti, submitted to Florida Institute of Technology (Barteotti 2014).

TABLE 4.6

AHP Weight Factors of the Alternatives with respect to Maintenance/Support

Maintenance/Support	MATLAB®	ENVI	Mathematica	Mean
MATLAB	1	4	3	0.623
ENVI	1/4	1	1/2	0.137
Mathematica	1/3	2	1	0.239
Total	1.583	7	4.5	

Source: Data from Technical Report on AHP by Bradley Barteotti, submitted to Florida Institute of Technology (Barteotti 2014).

of any individual entry does not dominate over equally valid but smaller magnitude cell values. The mean of each row was then calculated. A new column was added to illustrate this. As will be seen, the mean column is the eigenvector in this analysis.

The next step in the AHP is to create one AvA matrix for every decision criterion that indicates how each alternative meets the stated criterion. In this case, four matrices are created, representing maintenance/support, ease of use, development time, and cost. Table 4.6 shows the matrix created for maintenance/support. These matrices are also normalized in the same fashion as the CvC matrix was normalized. Additional rows and column were added to show the means and totals, respectively.

Table 4.7 shows the repeated process for ease of use criterion.

Next, development time was considered, and a corresponding AvA matrix was generated. Table 4.8 shows the result.

Finally, the AvA matrix for cost was generated and normalized as for the previous matrices. The results are shown in Table 4.9.

TABLE 4.7

AHP Weight Factors of the Alternatives with respect to Ease of Use

Ease of Use	MATLAB	ENVI	Mathematica	Mean
MATLAB	1	5	3	0.671
ENVI	1/5	1	1/3	0.114
Mathematica	1/3	3	1	0.287
Total	1.533	9	4.333	

Source: Data from Technical Report on AHP by Bradley Barteotti, submitted to Florida Institute of Technology (Barteotti 2014).

TABLE 4.8

AHP Weight Factors of the Alternatives with respect to Development Time

Dev. Time	MATLAB	ENVI	Mathematica	Mean
MATLAB	1	3	7	0.872
ENVI	1/3	1	3	0.340
Mathematica	1/7	1/3	1	0.120

Source: Data from Technical Report on AHP by Bradley Barteotti, submitted to Florida Institute of Technology (Barteotti 2014).

TABLE 4.9

AHP Weight Factors of the Alternatives with respect to Cost

Cost	MATLAB	ENVI	Mathematica	Mean
MATLAB	1	6	3	0.718
ENVI	1/6	1	3	0.305
Mathematica	1/3	1/3	1	0.160
Total	1.500	7.333	7	

Source: Data from Technical Report on AHP by Bradley Barteotti, submitted to Florida Institute of Technology (Barteotti 2014).

TABLE 4.10

AHP Alternatives versus Criteria

	Maintenance/Support	Ease of Use	Dev. Time	Cost
MATLAB	0.623	0.671	0.872	0.718
ENVI	0.137	0.114	0.340	0.305
Mathematica	0.239	0.287	0.120	0.160

Source: Data from Technical Report on AHP by Bradley Barteotti, submitted to Florida Institute of Technology (Barteotti 2014).

The next step in the AHP is to construct a matrix that summarizes the row means of each AvA with its decision criteria. Table 4.10 shows the summary. Each row represents a different alternative. Each column of the matrix represents the row mean of the AvA for each respective decision criterion.

The final step in obtaining a ranked preference order for each alternative is to multiply, by matrix multiplication, the alternative versus criteria matrix shown in Table 4.10 by the mean column (eigenvector) obtained in Table 4.5. The results are shown in Table 4.11.

TABLE 4.11

AHP Alternatives versus Criteria into Row Mean Vector

	Maintenance/Support	Ease of Use	Dev. Time	Cost		Mean		
MATLAB	0.623	0.671	0.872	0.718		0.182		0.710
ENVI	0.137	0.114	0.340	0.305	×	0.322	=	0.219
Mathematica	0.239	0.287	0.120	0.160		0.158		0.209
						0.337		

Source: Data from Technical Report on AHP by Bradley Barteotti, submitted to Florida Institute of Technology (Barteotti 2014).

TABLE 4.12

Alternatives Ranking

MATLAB	1
ENVI	2
Mathematica	3

Source: Data from Technical Report on AHP by Bradley Barteotti, submitted to Florida Institute of Technology (Barteotti 2014).

The resultant vector provides the ranked order of alternatives based on the relative ranking of the specified criteria. The highest number is the best choice. In this case, the MATLAB MBSE tool clearly scores higher than the other MBSE tools. Table 4.12 presents the resultant rankings of the three MBSE alternative tools.

Although the AHP analysis is complete, there is an additional step that will reveal the consistency of the rankings given to the criteria. The idea is to determine a consistency index that provides an indication for how consistent (or inconsistent) the weighting factors were determined. An example is as follows: say that you prefer ease of use three times to cost and prefer cost three times as much to development time. This implies that you prefer ease of use to development nine times as much (3 × 3). If, on the other hand, during the pair-wise comparison step, you assigned a value of 7 to the ease of use to development time matrix entry, then you would have an inherent "inconsistency" in assigning the weights. The consistency index captures this aspect of the analysis. If all of the pair-wise comparisons were perfectly consistent, when the relative weights were assigned, then Saaty showed that the eigenvalue estimates of the CvC matrix would exactly equal the number of criteria used in the AHP analysis (in this case 4). If any decision inconsistency were present, the eigenvalue estimates would deviate from 4. Determining the eigenvalue estimates can be done for any of the matrices using their row means, but we will illustrate using the original CvC matrix in Table 4.3 and its eigenvector shown as the mean in Table 4.5. Multiplying the original CvC matrix by the eigenvector produces a 4 × 1 vector. This vector can be divided element-wise by the original eigenvector to estimate the eigenvalues. The result is shown in the last row of Table 4.13.

Subtracting the number of criteria (4 in this case) from each element of the result and dividing each element by the number of criteria minus 1 (3 in this case) produce the consistency index. Dividing the consistency index by a statistical correction factor known as the ACI that is associated with the table of randomly generated pair-wise comparisons (equals 0.9 for 4 criteria) produces the consistency ratio. If the consistency ratio is larger than 10% (which it is in this case), then the decision maker should revisit the preference

TABLE 4.13

Decision Consistency

	Maintenance/ Support	Ease of Use	Development Time	Cost	Mean		Lambda *P*	Lambda
Maintenance/support	1	2	1/3	1/5	0.182		0.947	5.195
Ease of use	1/2	1	3	2	0.322	=	1.563	4.851
Development time	3	1/3	1	1/3	0.158		0.925	5.841
Cost	5	1/2	3	1	0.337		1.885	5.589
Total	9.5	3.83	7.33	3.53				

Source: Data from Technical Report on AHP by Bradley Barteotti, submitted to Florida Institute of Technology (Barteotti 2014).

weights because there is a large degree of inconsistency present in assigning the preference weights to the pair-wise comparisons. (Jen wonders what could have gone wrong.) (The next day....)

Karl: "Hi Jen, it's good to see you again. Were you able to finish the analysis on which MBSE tool we should use to evaluate the optical modulator?"

Jen: "Hi Karl, yes I was able to apply the Analytical Hierarchy Process to our problem but I ran into a problem."

Karl: "What's wrong?"

Jen: "Well, I was able to conclude that MATLAB is the best MBSE tool from applying the AHP. However, when I calculated the consistency ratio, I found that I had a large degree of inconsistencies present when the weights were assigned. This means we can't really trust the results until we straighten out the inconsistent rankings."

Karl: "Well, at least we have a metric that let's us know when we are off-base. How did you determine the weights?"

Jen: "I asked different people in the organization who had a stake in a criteria to make an estimate. For instance, I asked those that were going to do the analysis to rate the importance of 'ease of use' to the other criteria. I asked Joe in finance, the importance of cost, and Roger in software to evaluate the development time, and the lead service tech on the maintenance aspects."

Karl: "Did you ask them to be consistent in their relative rankings and with each other?"

Jen: "Oh, no! I didn't!"

Karl: "Well, there is your problem. None of them were enforcing consistency and so these inconsistent weightings crept into your analysis. I would get them all together, explain the AHP method, and ask them to revise their rankings while considering the overall consistency. I think you will get better results next time."

Jen: "Thanks Karl, I'll get right on it!"

Karl: "Ok, good luck! Let me know when you finish."

Jen conducted a second AHP; this time gathering all persons giving input together to discuss the criteria ranking. She was able to achieve a consistency ratio of 0.0023. MATLAB was still the MBSE tool of choice.

In this scenario, Jen used AHP to analyze which MBSE tool would best meet the organizational needs for an analytical tool for analyzing optical components like an optical modulator for the UAV project and future applications. This scenario illustrates the usefulness of AHP as an alternative and decision-making tool. One unique features of the AHP method is its ability to accept "fuzzy" inputs. Simply put, AHP can be used to mathematically analyze attributes such as "ease of use," which was used in this example. Another example of "fuzzy" input would be "works well." Another unique feature is its ability to provide a consistency check. Due to its unique features and its relative ease of use, AHP is an essential and robust SE method for considering alternatives and making decisions.

4.A Appendix: Acronyms

AHP   Analytical hierarchy process
APAS   Automotive Personal Assistance System
ASAP   Advanced Systems Analysis Program
CAE   Computer-aided engineering
CAS   Complex adaptive systems
DMS   Disaster management systems
DOORS   Dynamic Object-Oriented Requirements System
ENVI   Environment for Visualizing Images
FIT   Fantastic imaging technologies
GUI   Graphical user interface
IEEE   Institute of Electrical and Electronics Engineers
INCOSE   International Council on Systems Engineering
MBSE   Model-based systems engineering
MXL   Michigan Exploration Lab
OMG   Object Management Group
OSLO   Optics Software for Layout and Optimization
RAX   Radio Aurora Explorer
SDLC   Systems development life cycle
SE   Systems engineering
SoS   Systems of systems
SysML   Systems Modeling Language
UAV   Unmanned aerial vehicle
UML   Unified Modeling Language

References

Bajaj, M. et al. 2011. SLIM: Collaborative model-based systems engineering workspace for next-generation complex systems. Paper presented at *Aerospace Conference, 2011 IEEE*, InterCAX LLC, Atlanta, GA.

Barteotti, B. 2014. *Technical Report on AHP*. Submitted to Florida Institute of Technology, Melbourne, FL.

Cornford, S.L. et al. 2006. Fusing quantitative requirements analysis with model-based systems engineering requirements engineering. Paper presented at *14th IEEE International Conference*, Jet Propulsion Lab, California Institute of Technology, Pasadena, CA, September 11–15, 2006.

European Cooperation for Space Standardization (ECSS). October 1, 2012. *Space Engineering—Software Engineering Handbook*, European Space Agency (ASA) Requirements and Standards Division, Noordwijk, the Netherlands. ECSS-E-HB-40A PR-Draft1.

Fisher, G.H. 1998. Model-based systems engineering of automotive systems. Paper presented at *Proceedings of the AIAA/IEEE/SAE Digital Avionics Systems Conference, 17th DASC*, Vitech Corporation, Vienna, VA.

Green, R.A. July 2007. Getting a handle on MATLAB graphics. *Potentials IEEE*, 26(4): 31–37.

Guenther, R. 1990. *Modern Optics*. New York: John Wiley & Sons.

Helmut, J. 2013. BrainyQuote.com, Xplore Inc. "Hemult Jahn at BrainyQuote.com." http://www.brainyquote.com/quotes/quotes/h/helmutjahn325250.html (accessed April 2, 2014).

IEEE Std 610.12-1990. September 28, 1990. *IEEE Standard Glossary of Software Engineering Terms*. New York: The Institute of Electrical and Electronic Engineers.

IEEE Std 1471-2000. October 9, 2000. *IEEE Recommended Practice for Architectural Description of Software-Intensive Systems*. New York: The Institute of Electrical and Electronic Engineers.

INCOSE Technical Operations. 2007. *Systems Engineering Vision 2020*, version 2.03. Seattle, WA: International Council on Systems Engineering, Seattle, WA, INCOSE-TP-2004-004-02.

INCOSE UK. January 2012. Chapter Z9 Model Based Systems Engineering. Obtained from https://incoseonline.org.uk/Documents/zGuides/Z9_model_based_WEB.pdf (accessed November 20, 2014).

Martin, J.N. 1996. *Systems Engineering Guidebook: A Process for Developing Systems and Products*. Boca Raton, FL: CRC Press.

Object Management Group. May 2006. *OMG Systems Modeling Specification*. Needham, MA: Object Management Group.

Qamar, A. 2009. Designing mechatronic systems, a model-based perspective, an attempt to achieve SysML-MATLAB/Simulink model integration. Paper presented at *2009 IEEE/ASME International Conference on Advanced Intelligent Mechatronics*, Singapore, July 14–17, 2009.

Radiant Zemax 13. Release 2—The industry standard, powerful optical and illumination design software. http://www.radiantzemax.com/rz/news (accessed July 21, 2014).

Ramos, A.L. 2012. Model-based systems engineering: An emerging approach for modern systems. *IEEE Transactions on Systems, Man, and Cybernetics, Part C: Applications and Reviews*, 42(1): 101–111.

Saleh, B.E.H. and M.C. Teich. 1991. *Fundamentals of Photonics*. New York: John Wiley & Sons.

Soyler, A. 2010. A model-based systems engineering approach to capturing disaster management systems. Paper presented at *Fourth Annual IEEE Systems Conference*, Department of Industrial Engineering and Management Systems, University of Central Florida, Orlando, FL.

Spangelo, S.C. 2012. Applying Model Based Systems Engineering (MBSE) to a standard CubeSat. Paper presented at *2012 IEEE Aerospace Conference*, University of Michigan, Ann Arbor, MI.

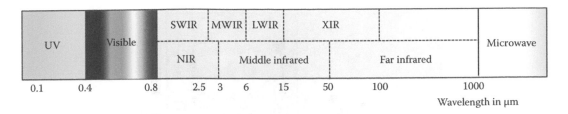

FIGURE 1.10
Expanded view of the visible and IR parts of the electromagnetic spectrum.

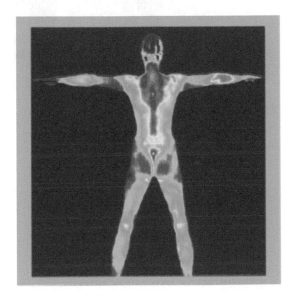

FIGURE 1.12
IR image of human body. (Reprinted from NASA, 2014, http://commons.wikimedia.org/w/index.php?title=File: Atmospheric_window_EN.svg&page=1. With permission.)

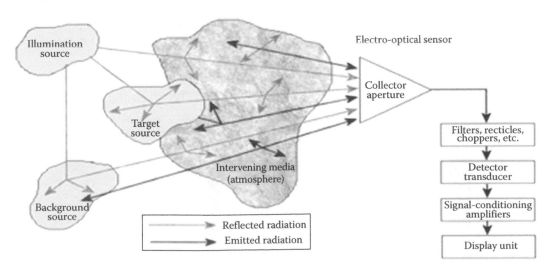

FIGURE 3.8
General-purpose physical SoS model for analyzing an optical system.

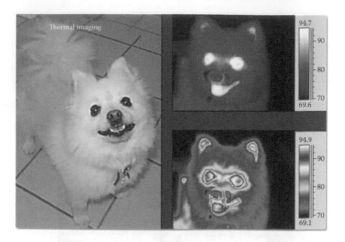

FIGURE 3.9
IR images of a dog. (Reprinted from NASA. With permission.)

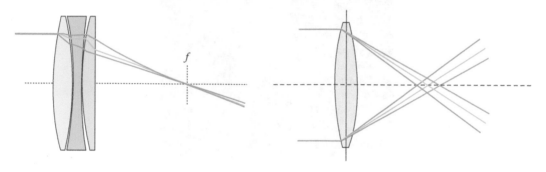

FIGURE 4.11
Comparison of normal lens and achromatic lens. (Achromat reprinted with permission from Egmason, May 8, 2010. http://en.wikipedia.org/wiki/File:Apochromat.svg, accessed November 20, 2014; Normal lens reprinted with permission from Bob, October 9, 2004. http://en.wikipedia.org/wiki/File:Lens6a.png, accessed November 20, 2014.)

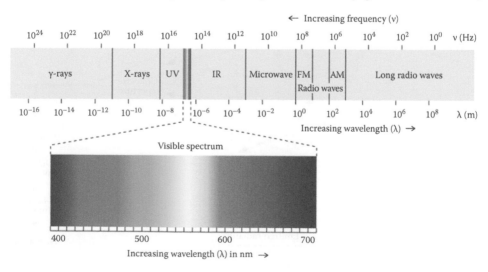

FIGURE 5.4
Spectrum of visible light. (Reprinted with permission from Ronan, P., EM spectrum, http://en.wikipedia.org/wiki/File:EM_spectrum.svg, accessed April 1, 2014, 2007.)

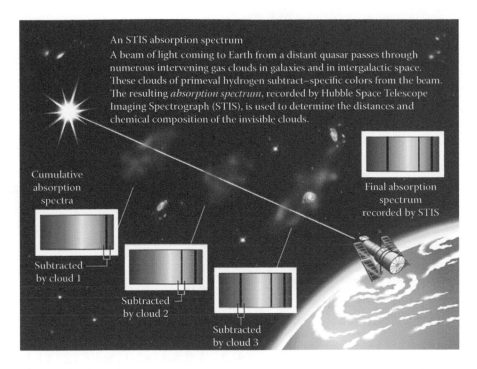

FIGURE 6.8
Hubble Space Telescope Imaging Spectrograph. (Obtained from NASA/STSci: http://hubblesite.org/newscenter/newsdesk/archive/releases/1998/41/image/r, 1998, accessed November 26, 2014.)

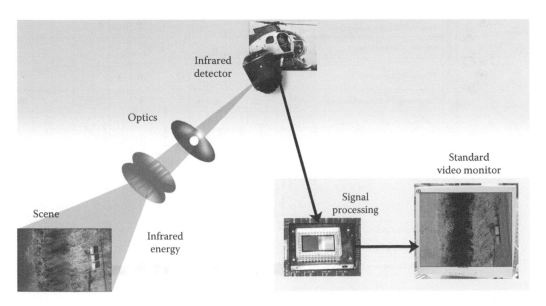

FIGURE 6.16
Basic infrared imaging detection system. (From Richard, Q., 2014, accessed April 19, 2014.)

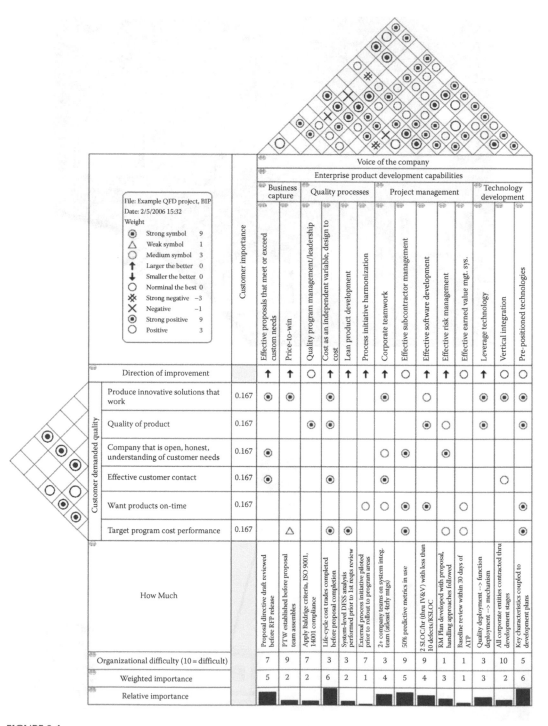

FIGURE 9.4
House of quality. (Obtained from Cask05, http://en.wikipedia.org/wiki/File:A1_House_of_Quality.png, 2006, accessed November 29, 2014.)

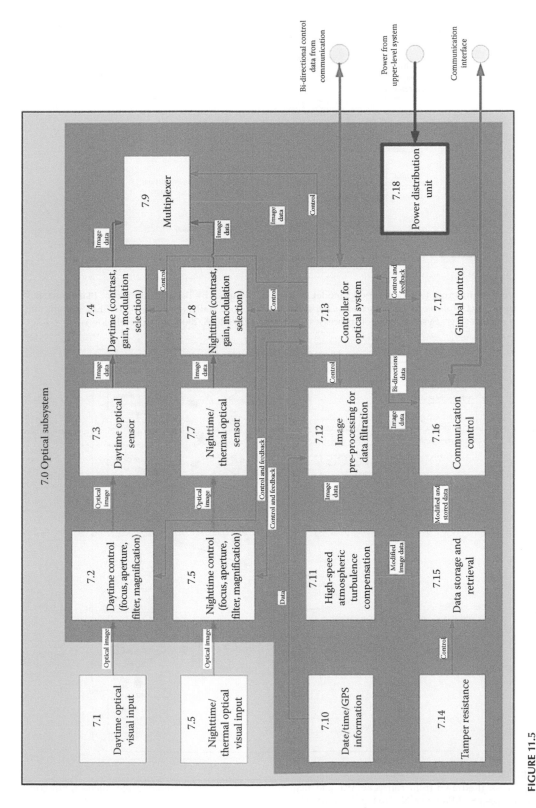

FIGURE 11.5
Optical system FBD. (Courtesy of Dean Smith.)

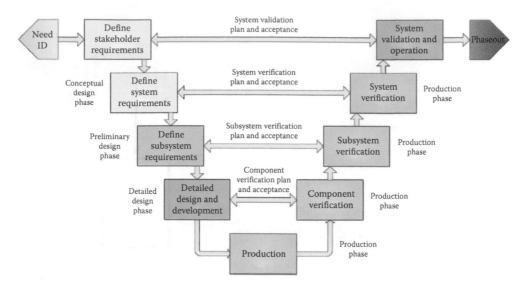

FIGURE 13.22

FIT systems engineering technical process model (V-Model). (Adapted by Brian Zamito from Blanchard, B.S. and Fabrycky, W.J., *Systems Engineering and Analysis*, 5th edn., Prentice Hall, New York, 2011; IEEE, Systems and software engineering—Systems lifecycle processes, IEEE STD 15288-2008, 2008. With permission.)

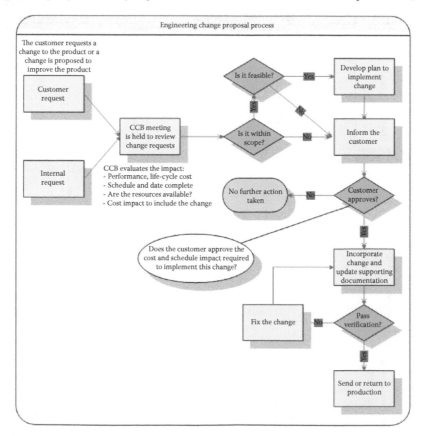

FIGURE 15.3

ECP process. (Redrawn from Fabrycky, W.J. and Blanchard, B.S., *Systems Engineering and Analysis*, 5th edn., Pearson, Upper Saddle River, NJ, 2011, p. 147, ISBN 13: 978-0-13-221735-4.)

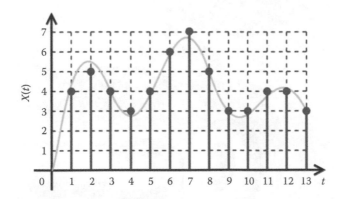

FIGURE 15.19
A/D Conversion. (Obtained from Wdwd. 2010b. Wikimedia Commons. http://commons.wikimedia.org/wiki/File:Digital.signal.svg, accessed December 7, 2014.)

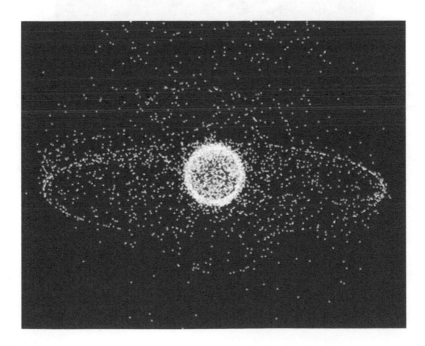

FIGURE 16.5
Space debris plot by NASA. (Image courtesy of NASA Orbital Debris Program Office, Johnson Space Center. 2014. http://orbitaldebris.jsc.nasa.gov/photogallery/beehives.html, accessed December 8, 2014.)

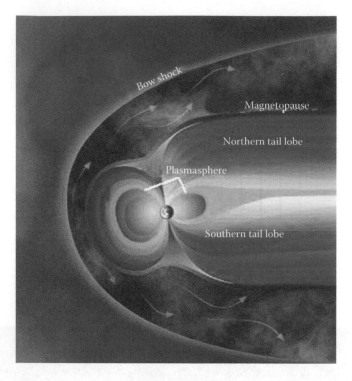

FIGURE 16.16
Representation of the earth's magnetosphere. (Image courtesy of Dennis Gallagher. 1999. Wikimedia Commons.
http://commons.wikimedia.org/wiki/File:Magnetosphere_Levels.jpg, accessed December 8, 2014.)

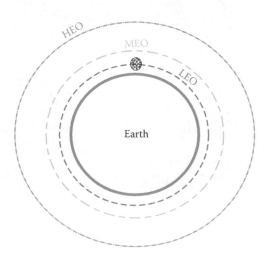

FIGURE 16.18
Orbital zones. (Created by and courtesy of Marty Grove.)

Section II

Application of Systems Engineering Tools, Methods, and Techniques to Optical Systems

5

Problem Definition

His name stands magnificently over the portal of classical physics, and we can say this of him; by his birth James Clerk Maxwell belongs to Edinburgh, by his personality he belongs to Cambridge, by his work he belongs to the whole world.

—Max Karl Ernst Ludwig Planck

In systems engineering, a critically important task for a successful system, service, product, or process (SSPP) is to properly define the problem that needs to be solved. This chapter explores systems engineering topics such as stakeholder identification; the determination of stakeholder's needs and wants; stakeholder requirements generation and some associated problems; generating a concept of operations (CONOPS); defining the mission, its goals, and objectives; and establishing the overall scope of the systems engineering development effort. We begin this section with some fundamental systems engineering principles and methods that are useful in carrying out the problem definition process.

5.1 Systems Engineering Principles and Methods for Problem Definition

Systems engineering is a science, process, or discipline that deals with the system as a whole. Regardless of the complexity, the main emphasis of the systems engineering discipline is to bring systems into being in the most cost-effective and optimal manner as possible. Processes have been developed to assist stakeholders and the development team with the system definition, design, development, production, support, and disposal of functional, reliable, and trustworthy systems that meet established functional and nonfunctional requirements.

In a very complex and large system, it is impossible to be the technical master of all aspects of the system. Therefore, the systems engineering processes must integrate and unify complex information and provide interaction mechanisms for merging activities and products from many related teams and departments. To efficiently accomplish these tasks, systems engineering processes must be focused on the capability of balancing technical and organizational requirements in multifaceted systems.

It is stated in the *NASA Systems Engineering Handbook* that "Systems engineering is about trade-offs and compromises, about generalists rather than specialists, and also about looking at the big picture, not only ensuring that they meet requirements, but that they get the right design" (NASA 2007, pp. 10–15). The goal of systems engineering principles, techniques, and methods is to provide an accurate, complete, functional, reliable, trustworthy, and cost-effective SSPP to the system's stakeholders.

The systems engineering discipline deals with all phases of the systems life cycle. The systems development life-cycle (SDLC) phases are presented sequentially as follows:

- Conceptual design phase
- Preliminary design phase
- Detailed design and development phase
- Production, construction, or manufacturing phase
- System operations and support phase
- System disposal and retirement phase

These phases are sometimes aggregated and bundled in various ways by different authors and organizations. For instance, the initial four phases are sometimes called research development, test, and evaluation (RDT&E). These same four phases are also called the *acquisition phase*, whereas the last two phases are called the *utilization phase* (Blanchard and Fabrycky 2011). The first three phases are often called the system *design phases*. In any case, the aforementioned SDLC phases are broadly accepted and have been in use in one form or another for a very long time.

Systems engineering is most effective when it is initiated as early as possible in the SDLC. The beginning of the systems life cycle is when the problem definition processes take place (e.g., stakeholder's needs definition/requirements, feasibility analysis, and conceptual design). Throughout these phases, the systems engineering process breaks the systems development into even more detailed design activities. The system-level design is accomplished during the conceptual design phase and ends with the baselined system segment specification (A specification). The system is decomposed into constituent subsystems, and these are designed in the preliminary design phase. Interfaces between the decomposed elements (subsystems in this case) are defined and documented in interface control documents (ICDs). The outputs of this phase are the collection of development specifications (B specifications) that capture the requirements for each of the subsystems, the ICDs, preliminary lower-level specifications, and the integrated subsystem design itself. The subsystems are then further decomposed during the detailed design and development phase into the lowest-level design element, the components/assemblies/pieces. At the detailed design level, the requirements are captured in the product specification (C specification) along with process requirements (D specification) and material requirements (E specifications). The requirements documents, ICDs, functional block diagrams, functional flow diagrams, model-based systems engineering (MBSE) products, prototypes developed in the detailed design phase, and the subsystem and component-level designs form the basis for proceeding on to production. After production, first the components are assembled and tested, then the subsystems, and then the system itself.

Systems engineering does not only handle the system, service, or product but also the processes involved in the systems development effort. Benjamin S. Blanchard (Blanchard 1998, pp. 45–50) states that "systems are composed of components which are operating parts of a system consisting of an input, process and output; attributes which are the properties or discernible manifestations of the components of a system, and relationships that are the links between components and attributes" (Blanchard 1998, pp. 45–50). It is very important that during the design phase, all the components and their interactions are clearly defined. Feasibility and trade-off studies may need to be accomplished to define all the necessary components and interactions. Failing to determine (or making false assumptions about) the system/subsystem/component interactions can result in cost overruns and schedule delays for the project.

The systems engineering discipline emphasizes three fundamental areas in the systems engineering approach to developing systems: system management principles, system engineering processes, and system engineering tools/technologies. These fundamental areas are mutually supportive and provide a rich framework for optimal systems development.

In recent years, due to rapid changes in technology, the developmental environment has been very volatile, highly interactive, and complicated. As a result, there have been notable increases in design and life-cycle costs and an overall decrease in system effectiveness. These increases in cost occur along with rising technical problems and schedule delays.

Often, these problems can be linked to poorly elicited, defined, and/or documented requirements. One way to address requirements issues was developed by Volere and described by Robertson (2006). Robertson states that the Volere requirements process (and the associated techniques and templates) is a great method to follow when designing a system, covering a detailed approach between the project blastoff stages, and delivering the requirements specifications to the design team. This process captures essential requirements characteristics and attributes needed by the development team and answers questions dealing with subsystem and component interactions and product operation. It is emphasized in the *NASA Systems Engineering Handbook* that "The most important role and responsibility of a systems engineer is ensuring the system technically fulfills defined needs and requirements and that a proper system engineering approach is followed" (NASA 2007, pp. 10–15). Standardized approaches like the Volere process are beneficial in fulfilling this role. We now focus on the beginning and, perhaps, the most important of the SDLC processes, problem definition.

It is very important to take enough time to properly define the problem before any design work begins to avoid any design failures later in the project. A "design it now, fix it later" approach causes cost overruns and schedule delays. Often in many system development environments, there is a "must do now" mentality. There seems to never be enough time to thoroughly plan up front; however, there is always time to do it again and fix the ensuing problems often at greater expense than if it were done right the first time. Problem definition contains the system requirements generation process and results in the baseline system requirements document for the subsequent subsystem and eventual component-level designs. A careful definition of the problem to be solved will result in a clear understanding when embarking on design activities and will increase the chances of designing a successful system that meets stakeholder's needs and fulfills their expectations.

This chapter focuses on a vital part of a systems development effort, the problem definition process. In the upcoming sections, we will discuss the identification of stakeholders, stakeholder's needs, stakeholder requirements, and some important systems engineering tools and methods like the quality function deployment (QFD) and the CONOPS. The project's scope, goals, and objectives of the systems engineering process will also be presented and examined. These processes and activities will be tailored to optical systems applications. The problem definition stage is the most important stage of systems engineering as it helps properly define the system that needs to be designed. Omitting the problem definition phase is not an option. It is like starting a design without knowing what is to be built.

5.1.1 Stakeholder Identification

A stakeholder is defined as an organization, individual, or entity that has some stake in the SSPP and is ultimately affected by the SSPP in some way. From here on, for brevity's sake, we use the term project to represent the SSPP in question. Where necessary, we will use the appropriate term for emphasis and clarity. An example of the stakeholders for a racing team would include the driver, support staff, investors, promoters, banks, insurance companies, legal team, marketing team, owners of the team, competitors, raceway officials, regulating agencies, and the fans. To build a stakeholder engagement strategy, the following critical questions should be asked:

- How will the stakeholders be affected by the project?
- To what degree will each stakeholder be affected?
- What will be the influence of each stakeholder on the project?

The first step is to identify the full complement of stakeholders. There is a likelihood that the needs of the stakeholders in the system will differ and even conflict. Once a full list of stakeholders is identified and their unique requirements are understood, it is the responsibility of the systems engineers to define, analyze, synthesize, and integrate all these varied requirements into a system that fulfills the needs, wants, and desires of each stakeholder.

Identification of the stakeholders is crucial to getting the answers to some basic questions. First, we ask the sponsor of the project questions along the following lines of reasoning:

- Who are the real stakeholders?
- Who can answer design questions?
- Are there any stakeholders ignored?
- What are the hidden decisions and priorities about the project? (You may have to search for these or approach these indirectly.)
- Who can decide?
- Who is the originator of this requirement?
- Who will be affected if this is done?

"Stakeholders" are not just the individuals or organizations that fund the project but also the customers of the project. On the other hand, all stakeholders are not customers. An example of the types of stakeholders that may not be customers for the product or service can include representatives from a customer, user, competitors, governmental regulators, community interfaces, or external, health, and environmental concerns.

There are two convenient categories or types of stakeholders: primary and secondary stakeholders. Primary stakeholders are the individuals or organizations directly affected by the outcomes of a project. Examples of primary stakeholders are project originators, funding agencies and/or investors, the development team, users of the system, maintainers of the system, organizations or individuals that interface with the system, competitors, distributors, marketing and business development, administration, insurance, legal, support services for the system, promoters, and evaluator of the system in its current evolution. Secondary stakeholders have inputs and requirements for the system and may have future benefit from offshoot technologies or subsequent evolutions of the system.

From a future business case perspective, the requirements of the primary stakeholder should be considered, where possible, with an open architecture perspective (i.e., being adaptable for future projects). This will make the system more marketable and standardized. The more standard the system is, the more supportable and adaptable it is, leading to a quicker design cycle and less cost in evolving the system.

For example, if an optical system is produced for and funded by the Department of Homeland Security (DHS), it may benefit from having adaptable features to meet the needs of other future stakeholders, for example, the FBI. The FBI may want to fund this system or make a variation on the system produced for the DHS. The needs of the FBI will not be completely the same as the needs of the DHS, but the system is designed with flexibility and adaptability in mind so that it can readily be evolved for the secondary stakeholders. It must be noted that the flexible design cannot occur at additional expense to the primary stakeholder. If, for instance, an 8 in. diameter telescope aperture would satisfy the requirements for the DHS primary stakeholder, but there is a known requirement for a 20 in. optical diameter for a future Department of Defense (DOD) stakeholder, the optical system cannot be overdesigned with the 20 in. optical diameter and charged to the DHS primary stakeholder. Doing so would likely result in a perturbed stakeholder and possible legal action. Instead, the optical system should be designed for an 8 in. optic but with a flexible interface that could be adapted to the 20 in. optic with additional DOD funding. Good modern-day design principles such as open architecture principles, modular design methods, enterprise architecture (EA) methods, and MBSE are essential elements in designing for flexibility and adaptability.

Secondary stakeholders are individuals or organizations who are not directly affected by the immediate project outcomes, but they may have a future interest in the project or possible evolution of the project. Examples are downstream customers or sponsors, organizations, or individuals that can adapt the system to their purposes, along with future investors.

In some cases, a negative stakeholder is added to the classification that is explained earlier. Members of the group are called key stakeholders. Phil Rabinowitz explains key stakeholders in an Internet website from Kansas University's Community Tool Box as follows:

> Key *stakeholders*, who might belong to either or neither of the first two groups, are those who can have a positive or negative effect on an effort, or who are important within or to an organization, agency, or institution engaged in an effort. The director of an organization might be an obvious key stakeholder, but so might the line staff—those who work directly with participants—who carry out the work of the effort. If they don't believe in what they're doing or don't do it well, it might as well not have begun. Other examples of key stakeholders might be funders, elected or appointed government officials, heads of businesses, or clergy and other community figures who wield a significant amount of influence in the success of the project.
>
> **Rabinowitz (2013)**

An example of a negative stakeholder is an employee who thinks that their tasks will be replaced by the system that is being designed. Negative stakeholders may also include competitors, opposing political figures, or even dueling coworkers. Allowing these detractors to influence the project requirements can be costly and present real danger for the project's ultimate success. These negative stakeholders should be identified early to avoid the risk of superfluous requirements with associated added costs and schedule overruns.

As mentioned, the primary stakeholders, and their needs, cover a range of system requirements. It is easy to understand that the concerns and priorities of multiple customers may be different. These customers may think that they are the only system stakeholder, but the customer is not always the individual or organization who will ultimately use the product, so we have another stakeholder type called users. The users are the most critical stakeholders. They are the technical operators who can explain the exact system needs, from the perspective of their use of the product or system.

For example, while designing an optics system for high-powered binoculars for military target acquisition, the user preference might be to have oversized protruding buttons on the unit that allow operation when wearing military gloves. In some cases, there are multiple types of users for a given system. The same binocular system may be designed for coalition forces, and the language used for the labels may be a requirement.

Customers and users are mentioned previously, but what about other types and classification of stakeholders? There are many methodologies to classify and identify stakeholders. One example from Sharp states, "One of the methodologies suggests the split into three categories; those internal to the project team, external to the team but internal to the business and those external to both the team and business" (Sharp et al. 1999, pp. 387–391). This methodology is a good way of classifying and prioritizing the stakeholders, and it may be an applicable guideline for future projects. Table 5.1 provides some guidelines and suggestions for possible stakeholders. It breaks out the various stakeholder types and provides a description of their individual roles.

Table 5.1 is intended to provide a glimpse into possible stakeholders and is by no means a comprehensive list. Most of the stakeholders described in the table are pretty clear and apparent.

TABLE 5.1

Stakeholder Descriptions

Stakeholder Type	Description
Consultants and subject-matter experts	Internal or external to the project, can be directly involved in the systems development or to an adjacent system.
Management	Engineering management internal to your organization, program and product manager, and board members.
Inspectors	Government or private; safety, technical, and other inspectors.
Legal	Lawyers, police, local and federal law enforcement agencies.
Negative stakeholders	Those who have no interest in your product and may suffer negative consequences from the implementation.
Industry standard managers	Professional and government organizations that may set standards for the product type.
Public opinion	General public who may use or be influenced by the system.
Special-interest groups	Representatives from special-interest groups who may be directly or indirectly affected by the system.
Cultural interests	Representatives from different cultures who may be directly or indirectly affected by the system.
Adjacent systems	If the system under development is a subsystem to a larger one, it would be wise to contact these stakeholders for input on features and function.
Previous users	Interface with previous users of past systems to discuss deficiencies of past systems.
Customer service representatives	Usually internal to the organization, individuals who deal with customer service on past delivered systems can provide input on deficiencies of past systems.
Maintenance and field service technicians	Usually internal to the organization, these stakeholders have input on maintainability.

Sources: Data adapted from NASA, *NASA Systems Engineering Handbook*, National Aeronautics and Space Administration (NASA), Washington, DC, 2007; Blanchard, B., *Systems Engineering and Analysis*, Prentice Hall, Upper Saddle River, NJ, 1998, pp. 45–50.

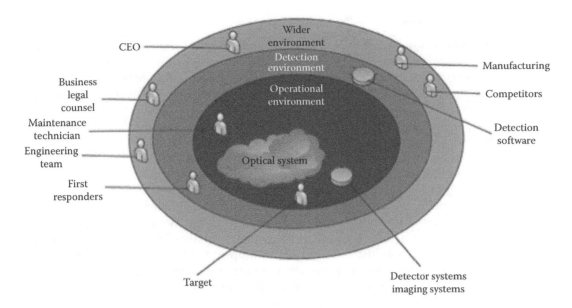

FIGURE 5.1
Optical system stakeholder onion diagram. (Adapted from Robertson, S., *Mastering the Requirements Process*, Addison-Wesley, Boston, MA, 2006.)

One way to depict the stakeholders of a system is to use an onion diagram. It is easy to show the stakeholders of a system and the interaction between these stakeholders by using this tool. Figure 5.1 is an example of an onion diagram for a generic optical system. It is critical to define the relationships of the stakeholders with the project, to define requirements, and to prioritize the requirements. At the center of the diagram of Figure 5.1, the "cloud" depicts the system under design; this is our generic optical system:

> The darkest ring from there is the operational work environment for the system. Here, we see a target (what the optical system is intended to detect), some of the actual subsystems of the optical system, and a maintenance technician. The next ring of the onion is the containing business. These stakeholders are not actual users, but they use the data or services the system provides to perform some other function. Lastly, we have our wider environment ring containing stakeholders who have some direct influence on the optical system. Here we list employees of the optical system company such as the CEO, legal counsel and the engineering team along with industry competitors.

> **Rabinowitz (2013)**

Briefly, the stakeholder identification process of a system is very important because we are able to get a better view of who has inputs to the system, who is affected by the system, whose requirements influence the system, and how the stakeholders interact with the system. This assists the systems engineer to prioritizing stakeholder requirements and provides insight as to how the stakeholders view the system. Identifying stakeholders is the starting point to defining stakeholder's needs and subsequently stakeholder requirements.

5.1.2 Stakeholder Needs

Once the stakeholders are identified, the needs and wants of the stakeholders must be determined. The stakeholder identification process can result in answers to many complicated questions but can also be the source of even more questions. Stakeholder's inputs need to be properly captured, documented, and used to generate a proper set of requirements for the system.

It is important to remember that there can be a difference between what stakeholders want and what they need due to differences in perspective and technical proficiency. Nevertheless, the needs of stakeholders may vary because their concerns or interests may vary. Each stakeholder may have different concerns such as economic, cultural, ideological, and ethical. It is important to examine conflicting concerns, as priorities and constraints of individual stakeholders or subgroups may be substantially different.

There are many different ways of producing a product or a system. Money and time are always a constraint that defines the system. The system "quality" (i.e., the ability to meet stakeholder's needs, its reliability, and operational supportability) is also an important factor. The stakeholder's needs and requirements are often dynamic, and as they are more educated in the system features and capabilities, the stakeholder's needs, wants, and concerns may change over time. As a result of evolving stakeholder's needs and perceptions, the identification of stakeholder's needs is an iterative and complicated, but crucial, stage in developing a better system, and it must be performed very carefully.

There are a variety of useful methods in gaining insight into stakeholder thinking and motivations. Some of these techniques include interviews, focus group questionnaires, requirements workshops, brainstorming, use-case model development, business modeling, and reviewing customer requirement specifications. No matter which one of these methods or combination of methods is used, it is fundamentally important to the system development effort's success to understand and capture what motivates and drives the stakeholders.

One of the most widely used methods in understanding stakeholder's needs, wants, motivations, and drivers is the interviewing technique. The interview should try to provide answers to questions such as what needs to be done, what constraints are there? why do the stakeholders want to solve this problem, are there any hidden agendas or motivations, what are the conceptual trade-offs for the project, what do they think is the best way to solve the problem, what outstanding issues are there, and what concerns do the stakeholders have.

On the other hand, knowing the background, expertise, and "stake" the stakeholder has in the project before interviewing, by conducting background research and/or organizing a preparatory meeting with the stakeholder (if possible), can make all the difference in directing the right questions. For example, if designing an optical system for border protection purposes, asking a local politician what range the optical system should detect a human target would be an inappropriate question. Asking this same stakeholder "What is your view of the system?" and "What do you expect it to do?" may be a very general but appropriate questions, allowing the stakeholders to share their attitudes, insights, and perspectives on the potential optical system.

Educating and involving the stakeholders and users on the system assumptions and design concepts early in the program conceptual design phase and capturing their input can create a positive impact during the deployment stages. Defining the critical questions that will help the team understand the stakeholder's real perspectives should be completed before they are interviewed. Using open-ended questions and avoiding provoking questions make the interviewing process more efficient. Stakeholders and project members often meet to discuss the CONOPS and learn more about the project. We describe the CONOPS in a later section, but in summary, it is a tool that summarizes the characteristics of a suggested system from an operational point of view. It is helpful for the stakeholder to review the CONOPS in order to have a good understanding of the system and help aid the interview.

In the optical system example for border protection, discussed earlier, if interviewing an optical expert, it may be a good idea to give this individual a good high-level description of the system as this may shape their answers to certain specific optics questions that could be asked. Aside from a one-on-one meeting, focus groups could also be held with stakeholders to promote the flow of ideas between individuals. The interaction between individuals accelerates the process of understanding the needs of stakeholders. However, sometimes a person with a very strong opinion could sway other stakeholders. The focus group leader should ensure that a range of views are presented and considered and that everyone in the focus group has a voice.

For a more efficient process, it is best to organize a focus group meeting that has defined groups, a defined agenda, and a systematically organized session before the meeting. To facilitate this kind of meeting, it is convenient to develop a stakeholder list with their functions.

Sometimes, it is necessary to organize the stakeholder in terms of their priorities, goals, end use, or some common "look and feel" criteria. Summarizing these in table form helps the team understand the stakeholders in a better way. Table 5.2 is an example of a stakeholder analysis table for an optics system.

Table 5.2 first lists the type of stakeholder, their name, and their priority before digging into their actual system requirement input. Prioritizing stakeholders and placing higher-priority stakeholders at the top can aid in executing the important needs first. Table 5.2 is an abbreviated view of a project. It may be derived from an excerpt of a larger one for an optical system. The features or functions listed along the top help the team establish design priorities.

TABLE 5.2

Stakeholder Analysis Template

Stakeholder Identifier	Name	Priority	Goals	Look and Feel	Range Performance
Client	Bob Smith	High	Marketable product	Modern, easy to use	Detection of a target 3000 m
Customer	Carl Jackson	High	System that performs specified functions	Well built, rugged, long lasting	Detection of a target 1500 m
User 1	Mike Barron	High	Easy to use, reasonably priced	Easy to use, reliable	Detection of a target 200 m
Local politician	Jesse Bloomfield	Medium	Marketable as project for re-election	Works well for users	No opinion
Optics consultant 1	James Dalberth	Medium	Future consulting work	Most advanced features implemented	Detection of a target 5000 m
Design team	Thomas McCurdy	Medium	Well-rounded product, marketable	Easy to use, meets size and weight requirements	Detection of a target 5000 m

Source: Data adapted from Rabinowitz, P., Community tool box, http://ctb.ku.edu/en/tablecontents/chapter7_section8_main.aspx, accessed April 1, 2014, 2013.

In addition, the project team needs to be aware of the interaction and relationship between stakeholders and record this in order to understand and balance different views and needs of the stakeholders. Even though stakeholders have already been prioritized above, this can assist in further prioritizing the information they provide.

(Sharp et al. 1999, pp. 387–391)

One of the other ways to understand stakeholder's needs is the use-case method. A use-case method describes the actions accomplished by a system for a particular application (use). It gives a noticeable consequence of value to an actor. In a use-case sense, an actor is something within the system that has a role and can be a person or an external system. A use-case method defines a set of actions that the system will execute for different scenarios. Using these scenarios is one way to stimulate vision of the end use by envisioning the specific execution steps of a use case. Some of the meaningful questions in determining use cases include the following:

- What are the actor's priorities and responsibilities?
- Will the actor generate, store, alter, and take out data in the system?
- Will the actor require notifying the system about unexpected outside changes?
- Does the actor require being educated about definite occurrences in the system?

Sometimes, questionnaires are used to define stakeholder's necessities. It includes a series of questions that are carefully selected for understanding each stakeholder's identity and needs.

Identification of stakeholder's needs is one of the most important processes of the problem definition phase, but it can be a huge undertaking. Identifying the required resources and associated timeline for identifying stakeholder's needs is essential. The results of the interviews should be carefully managed and made available for future use by the project team and may also be useful for downstream teams working similar efforts with the same or similar stakeholders.

5.1.3 Stakeholder Requirements

In the previous section, we discussed the importance of a stakeholder, the tools used to identify stakeholders, and their needs, wants, motivations, and drivers. The next step is to transform their needs and wants into a clear set of stakeholder requirements. In order to identify the stakeholder requirements, effective communications between the stakeholders and the development team are essential. "Stakeholder requirements play major roles in systems engineering as they form the basis of system requirements activities, system validation, and stakeholder acceptance, and act as a reference for integration and verification activities, and serve as a means of communication between the technical staff, management, finance department, and the stakeholder community" (Sharp et al. 1999, pp. 387–391).

Even after working through this process, the needs and wants of the customer stakeholder (or client, hereafter referred to as customer in this section) can still get muddled in with the rest of the needs and wants of other stakeholders, especially as the design and even manufacturing process evolves (Yang 2003; Pacheco 2008, pp. 472–477). To mitigate this, Yang identified the QFD process, which can be used as a planning tool to convert customer needs and wants into requirements and track these requirements through the development process (Yang 2003).

The QFD process is a formal method that consists of making a matrix diagram that correlates stakeholder requirements with a company's engineering parameters of a system. It is a very useful tool for the stakeholders to see all this information together in one table. Putting stakeholder requirements in a table using the QFD methodology facilitates requirements discussions. It increases the quality of interchange between stakeholders and promotes effective stakeholder interactions. According to Cox, a QFD is used as follows: "It can be used in many different design phases and you do not have to use QFD fully. QFD can be used in different ways depending on the project phase where it is used. Utilizing this QFD technique, being constantly mindful of how requirements creation and design decisions and changes will affect the customer's satisfaction of the product, can assist in driving towards a faster product development cycle, reduction in redesigns later in the process, improved overall quality, and customer satisfaction" (Cox 2001, pp. 245–259).

Additionally, to achieve the benefits of the QFD, a multifunctional team is needed. According to Christiano, "In order to carry out such a process with the customer, heavily involved with producing input, a cross-functional team comprised of marketing, design, quality and production team all lead by an experienced moderator are needed" (Cristiano 2001, pp. 81–95).

The QFD methodology (Herzwurm 2003, pp. 1–2; Desai 2008) has many types of diagrams and processes that can be followed to reach the same goal of extracting the most useful information and learning the priority of this information, from the customer for implementation through the design process. According to Francheschini, "However, at the heart of the process is the House of Quality Diagram assisting in preventing three common occurrences that have haunted design teams for years:

1. The customer was not properly represented by the design of the product.
2. The customer's input was lost during the design process.
3. Different interpretations were made among the several design sub-teams within the project" (Francheschini 2001, pp. 21–33).

The QFD makes an attempt of answering the customer's "whats" with "how" the system will be designed to meet the specific stakeholder requirements. The house of quality diagram goes even further to take into account the design's "how" interdependencies, and even competitors take on similar objectives (Francheschini 2001, pp. 21–33).

While creating this diagram, the customer's "whats" need to be identified, which may require many different design's "hows" to satisfy each of them, often achieved by surveys, interviews, forums, and informal interactions as mentioned in the section earlier.

In the QFD method, the results of these "whats" are listed on the left side of the house of quality diagram. In the next step, the "hows" are created by the design team to address each of the "whats" the stakeholder has identified. A maximum or minimum rating helps the design team form a view of stakeholder priorities and helps establish the trade space between the stakeholder requirements. The QFD diagram consists of a relationship matrix that emphasizes how well the "hows" correspond to the "whats."

In a common example, the QFD generates a relationship table between the stakeholder requirements and how each requirement will be satisfied. The QFD matrix shows the relative importance of each requirement and how these requirements impact the technical responses. This matrix is used to display the strength of the relationship between the row requirements ("whats") and the column technical responses ("hows") through the weights at the intersection points of the matrix. The house of quality diagram also takes into account benchmarking and strategic planning for each stakeholder's need.

Benchmarking is when something is compared to a standard, an industry leader, or the "best of the best." This aspect of the QFD can be used for benchmarking requirements and/or technical response attributes or as a comparison tool with competitors. The QFD diagram also shows the technical performance measures that correspond to each technical response. Figure 5.2 depicts an abridged house of quality diagram for a simple optical system.

Figure 5.2 consists of a diagram that shows a matrix between what the customer's needs are with a ranking and how they will be satisfied. According to Yang, "This can be used to identify synergies (depicted by a plus sign) between 'how's' and where compromises (depicted by a minus sign) between them need to be made to choose one over the other." Herzwurm (2003, pp. 1–2) states, "Re-visiting this diagram throughout the system development process can help ensure the customer 'voice' is being heard at all times" (Herzwurm 2003, pp. 1–2).

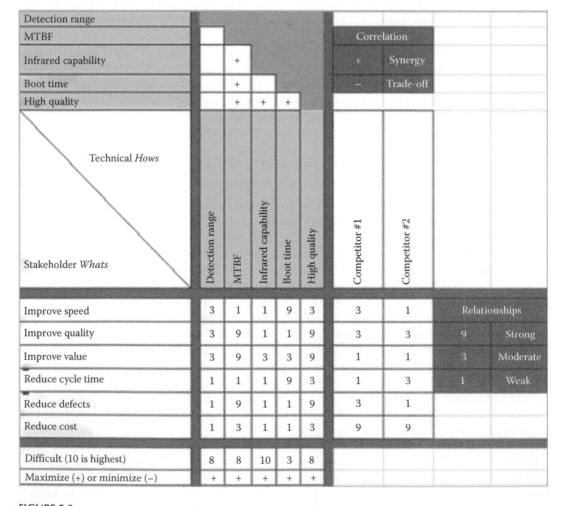

FIGURE 5.2
House of quality diagram. (Adapted from Francheschini, F., *Advanced Quality Function Deployment*, CRC Press, 2001, pp. 21–33.)

The QFD and the house of quality diagram help the stakeholders and design team to understand and prioritize the stakeholder's needs in an effective way. These tools help the design team communicate to the stakeholders how each of their needs/requirements will be addressed and how they compare to the industry's benchmark. The QFD method and corresponding house of quality are very important because they provide a good visualization mechanism to talk about requirements and they serve as an excellent stimulus for stakeholder and developer communications.

5.1.4 Concept of Operations

Creating or obtaining a high-level graphical view of a system is a way to communicate with those that are unfamiliar with the system. It describes what the system actually does, using a graphical view. This provides a "big picture" and can be used to learn about the functions, operations, and the support environment of the system. In terms of a system, the CONOPS is the most useful if there is a high-level graphic, along with a one-page brief story or representation of what the system actually does from the viewpoint of each key stakeholder. A CONOPS diagram allows system engineers to easily represent what their system must do, allowing them to focus their attention on users, responsibilities, constraints, and goals and objectives of the system.

A CONOPS diagram does not focus on details and the subparts or subsystems; it is an integrated view of a system. According to Firesmith, "In some cases, fairly rare, a customer may have a CONOPS diagram already created for their system and they can pass this diagram off to the design team for a good feel on how the system should operate. More commonly, the diagram needs to be created by the design team with input from the customer on how they expect the system to operate" (Firesmith 2008).

CONOPS diagrams are used to provide high-level use cases and identify major components in order to educate the stakeholders on the system being designed. It is then the job of the design team to further educate the system architect(s) on technological or product-related limitations. Identifying these high-level limitations early on can help shape the CONOPS and prevent issues later on in the development.

The most important aim of a CONOPS is to get concurrence from the stakeholders on their tasks, how the system will work, the surroundings, system potential, and processes that the system will maintain. This is best accomplished by presenting the information from the viewpoint of each stakeholder. It is very important to form a CONOPS that is easy to understand by each stakeholder and the design team.

Sometimes, the CONOPS diagram represents the top-level system's use cases and scenarios in a visual format. To be effective, very complex systems may require very complicated CONOPS. In some cases, for very complicated systems, CONOPS diagrams can go beyond one diagram to outline multiple use cases. The well-known mnemonic KISS (keep it simple, stupid) principle should be applied while designing a CONOPS. It only needs to be able to describe enough so that each stakeholder can understand and learn about the "big picture" easily. Keeping the diagram on one page has many advantages and can prevent confusion.

It is very challenging to create a CONOPS diagram because a good CONOPS should reflect the viewpoint of each stakeholder, and the stakeholders should understand it easily. Boardman states,

> By the way, the architects are not experienced about the system and the technical issues of the system. In some cases, multiple agencies within a customer's organization may be utilizing the system all with similar but slightly differing needs from it. Capturing

these needs in the stakeholder diagram and explaining to each of these architect(s) early on, why one implementation may be better than another, or why one feature was added or deleted based on another agency's input, is extremely important.

Boardman (2008)

Figure 5.3 is an example of a CONOPS diagram for an optical detection system meant to be deployed on the U.S.–Mexico border.

Figure 5.3 depicts a long-range monitoring station at the U.S.–Mexico border with long-haul communications back to a command center sending video and imagery data. This pictorially tells the story of what is going on with the system. From this CONOPS design, it is easy for the stakeholders of the optical system to realize interrelationship of the system components, characteristics, and functions that the stakeholder has requested. Once agreed, the design team must move forward to create a proposed system that acts within the expected confines illustrated in the detailed interactions shown in the CONOPS.

An example in real terms is taken from the U.S. Military. According to a diagram from the Internet on the Operational View 1 (OV-1), "The United States Department of Defense utilizes a Department of Defense Architecture Framework (DoDAF) that defines different Operational Views (OVs). Included in these views is the High-Level Operational Concept Graphic (OV-1). This OV-1 is a formalized version of the CONOPs diagrams listed above and has been used as a standard for managing CONOPs diagrams across many industries" (U.S. Department of Defense 2013).

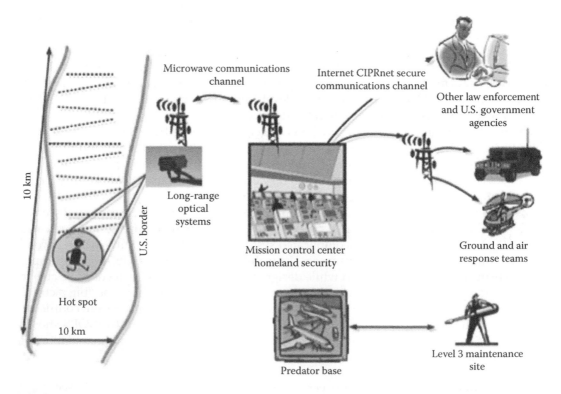

FIGURE 5.3
Optical system CONOPS diagram.

To summarize, the CONOPS should help the stakeholders realize what is going on with the system. Designing a CONOPS graphic helps form an agreement between stakeholders, architects, and designers. The design teams and architects should be aware of each stakeholder, and they must focus on forming a CONOPS diagram that contains every major aspect of the project. They also must avoid waiting for a prepared CONOPS diagram by stakeholders, and they should never postpone preparing and maintaining the CONOPS throughout the project.

It is the system project leader's responsibility to identify any CONOP changes and ensure that each new team member is on the same page in terms of what the system is supposed to do. Barnhart states, "When working through the design process, it is always a good idea to weigh each design decision against the CONOPS and ensure it is assisting in meeting the system goals. It can also assist in the system drifting away from its core values" (Barnhart 2010, pp. 668–673).

Another critical project task is identifying and managing the scope of the project. Requirements changes can be viewed as progressive discovery or sometimes scope creep. It is important to address the project's scope to manage the program effectively. This will be discussed in the next section.

5.1.5 Project's Scope

According to Carmichael, "Scope can be defined as setting boundaries across a project to properly define the extent of work to be undertaken" (Carmichael 2003). The project's scope is directly related to the work that must be accomplished to produce a system or product. The stakeholders and design team need to agree on the project's scope early in the project. The project's scope is a critical project artifact since it provides clarity and focus for project's deliverables and outlines the conditions for the expected work. Robertson stated,

> A good project scope statement should comprise the following items; project scope justification (how and why your project came to be), project scope objectives (what the project is trying to achieve), product scope description (results that your project will produce), product acceptance criteria (standards required to satisfy the customer's quality expectations and gain acceptance of the final product), project constraints (technological, physical, constraints of the project, and so forth), project assumptions (the statements that you believe to be true for accomplishing the project).
>
> **Robertson (2006)**

A properly scoped project can prevent cost and schedule overruns and keeps the design team focused on what they should be working on. Scope management, for a project lead, is a tool to defer extra internal or external requests that may be made on the system. A good project scope statement also provides a road map, which explains how to decide when there is a valid engineering change request that may be eligible for additional compensation system development team. Using good scope management techniques can be firepower for cost growth and schedule creep on the project. It makes it easier to manage the project when new requirements are added, rather than absorbing the impact and implementing them without the benefit of good technical integration and analysis.

Badiru stated that "In fact, scope can be broken out into five key steps across a project; (1) Scope Planning, (2) Scope Definition, (3) Work Breakdown Structure (WBS) creation, (4) Scope Verification, and (5) Scope Control" (Badiru 2009). After working through these steps, the project lead can expect to have created the following deliverables: scope statement(s), the WBS, a process for verification and acceptance of the system, and a process for change requests to the system scope.

5.1.5.1 Scope Planning

The first step in the scope development and maintenance process is scope planning, which consists of a set of activities used to determine the project's scope. Planning works as a data collection and information gathering process, and it includes activities like creating templates and forms based on company practices for scope creation, assessing preliminary scope statements from the stakeholders, and reviewing the project charter (if available).

5.1.5.2 Scope Definition

Once these items have been established, the second step in the process is scope definition, which kicks off with the design team performing an analysis on the project's needs based on CONOPS and use cases. Throughout this process, stakeholder requirements and the business needs should be examined carefully, and all of the information gathered should be used to perform the scope definition. One of the most important parts of a scope definition process is preparing a work breakdown structure (WBS). This is a spreadsheet or graphically formatted document that encompasses all of the major activities for the system development (and eventually the subsystems and component development of the system), maintains comprehensiveness, and provides some structure to how the system elements may work together. The WBS should have related cost estimates and completion times for each WBS element to control schedule and cost overruns and aid in impact analysis.

5.1.5.3 Scope Verification

The next step is scope verification, which aligns the scope as defined by the development team to what the stakeholders are actually expecting. This step maintains the focus and the attention of the development team on the stakeholder's needs and ensures that the stakeholders and development team have a common understanding with regard to what must be done. The result of this process is a common and unique scope for the project.

5.1.5.4 Scope Control

The final step in scope definition is scope control for the system. Anyone can define a scope; it is controlling it that tends to be extremely challenging over the life of a complex system development. During the design phases, there may be some stakeholders, contractors, and managers wanting changes that are out of the project's scope.

The scope should include a guideline that shows what actions are taken against change requests. Identifying the project road map maintains objectivity and minimizes the problems that may occur between stakeholders and the design team. Regarding an out-of-scope request, Enos states that "An internal request such as this would be considered out of scope and would definitely have some cost, schedule, resource and possibly product quality implications for the customer. Each of these items needs to be weighed by the organization for impact to the project and decided on" (Enos 2007, pp. 57–73).

The project's scope statement process is vital for projects in terms of managing the project. Scope issues should be clarified before the design phase begins so that cost or schedule targets can be established and impacts assessed. Scope creation and management across a project can lead to more consistent results and prevent disagreement between the design team and the stakeholder. Using the aforementioned tools can clearly assist in defining boundaries, constraints, assumptions, acceptance criteria, and justifications for changes. They also provide the framework to set goals and objectives.

5.1.6 Goals and Objectives

Setting goals and objectives for a project can produce many positive outputs for both an engineering organization and the project itself. This process encourages all stakeholders to work toward the same purpose, and this forms team synergy. Setting strong goals can produce measures of success and, at the completion of the project, can give an organization a clear understanding of what can be improved upon for the next project. The act of setting good and achievable goals provides common challenges that unite the efforts of the team and that helps in evaluating progress and outcomes.

Goals also create a basis for setting objectives. They can boost morale and the productivity of a project team, empowering individuals to be more effective in their project work. Setting strong goals can keep stakeholders on track when harvesting information from them. They keep discussions on track and avoid wasting time on insignificant details. Goals can keep the stakeholders on track with a mutual understanding of the project goals (Poynting and Thomson 1906, pp. 230–231).

Enos further stated, "Measuring each goal can help the project team set effective goals for the project and guide the development team to success from the start" (Enos 2007, pp. 57–73). Nevertheless, the priorities of the goals may vary. Some goals are immediate; however, some of them can also be classified as short-term or long-term goals. This categorization is useful when scheduling plans are prepared.

Setting the goals is a basis for setting the objectives. Objectives support the goals by indicating the outcomes of activities that must be accomplished for achieving the goals. Objectives also should be feasible, realistic, measurable, straightforward, clearly defined, and purposeful. The objectives should also be captured in a process plan.

In summary, setting the goals and objectives of a project provides guidelines for the team, unites the team, and identifies milestones that need to be accomplished during the SDLC. Goals and objectives serve as common reference points for all stakeholders so that they can measure the performance or evaluate the outcomes.

5.1.7 Transition to Optical Systems Building Blocks

Just as the problem definition phase lies at the beginning of the system engineering development process, optical sources lie at the beginning of the electromagnetic propagation process. In the next section, we discuss the conceptual and mathematical framework needed to understand optical sources. This conceptual framework can be used to develop mathematical models for optical systems and gain insight into the nature and properties of optical sources.

5.2 Optical Systems Building Blocks: Optical Sources

In this section, we focus on electro-optical (EO) and infrared (IR) sources. Our intent is to provide an overview of the topic and develop some physics-based mathematical applications and examples that are useful for analyzing, characterizing, and simulating optical sources in both the visible and IR parts of the electromagnetic spectrum. We limit ourselves to describing the source characteristics and focus on radiometric propagation analysis in another section. In this section, we accomplish the following:

- We start with some fundamental properties of light and describe the parts of the electromagnetic spectrum that are useful for optical imaging. We describe some fundamental differences between the visible and IR parts of the spectrum and then describe some source analytical methods.

- Predominantly useful in the IR part of the spectrum, the reader will come to understand blackbody-type sources and their beneficial uses in calibration and optical systems test. We distinguish between blackbody-type and gray-body-type sources.

- Radiation characteristics of some IR sources ranging from rockets to personal vehicles are then discussed as real-world applications.

- We end this section presenting some use cases showing the basic application of the methods in this section for some practical cases such as the sun, molten iron, and human sources.

When reading this chapter, the reader must take note that the text is approached from the perspective of an EO and IR systems engineer. Much of the theory is well documented in the technical world.

5.2.1 Visible and IR Parts of the Electromagnetic Spectrum

The electromagnetic spectrum may be partitioned into different categories as a function of wavelength, frequency, or wavenumber. Radiation within a category shares some common attribute or characteristic. For instance, visible light is a category for optical radiation that can be observed by the human eye. Similarly, the IR part of the electromagnetic spectrum cannot be seen by the naked eye but can be measured as heat energy. Different sources, detectors, and technologies may be involved in generating and detecting radiation in separate categories. In this section, we focus our attention on optical sources in the visible and IR parts of the electromagnetic spectrum.

As commonly understood, the human eye is sensitive to light that lies in the visible part of the electromagnetic spectrum roughly between 400 and 700 nm in wavelength as shown in Figure 5.4. This spectrum has various colors that we know and can demonstrate using a prism or diffraction grating.

A similar partition is defined as the IR part of the electromagnetic spectrum. The IR part of the spectrum lies around 0.7 μm to 1 mm in wavelength as shown in Figure 5.4. However, the practical parts of the IR are broken down into the short-wave IR (SWIR), mid-wave IR (MWIR), and the long-wave IR (LWIR) as shown in Figure 5.5. Radiation in the IR part of the spectrum requires detectors that can respond to changes in temperature and may require cooling for the best performance. The IR spectrum can be measured by using

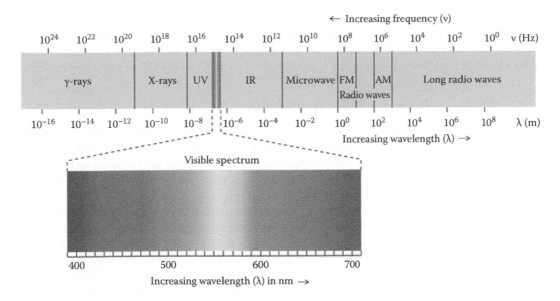

FIGURE 5.4
(See color insert.) Spectrum of visible light. (Reprinted with permission from Ronan, P., EM spectrum, http://en.wikipedia.org/wiki/File:EM_spectrum.svg, accessed April 1, 2014, 2007.)

	Visible		Infrared		Microwave	
	SWIR		MWIR		LWIR	
	Near infrared		Middle infrared	Far infrared		Extreme infrared

| 0.0 | 1 | 1.5 | 2 | 3 | 4 | 6 | 8 | 10 | 15 | 20 | 30 | Wavelength, µ |
| | 10,000 | | 5,000 | | 2,500 | | | 1,000 | | 500 | | Wavenumber, cm |

FIGURE 5.5
IR spectrum.

a device like a monochromator to split into narrow bands of light. A thermal energy meter can measure the energy of the radiation present in the given narrow wavelength range or spectral region.

Although visible light only includes a small part of the electromagnetic spectrum, it is the only part of the spectrum that the human eye can see. The sun is a dominant source of visible light. However, the sun also emits energy in other parts of the electromagnetic spectrum. The vast majority of observed light that can be seen during the day is emitted by the sun, the greatest source of white light, which also emits other frequencies of radiation outside of the visible range.

Visible light also comes from artificial sources, such as charged gases (fluorescent lights) or glowing metal filaments (tungsten devices), light-emitting diodes (LEDs), and lasers. Some of these light sources emit energy in both the visible and IR spectrums.

IR radiation lies in a part of the electromagnetic spectrum where the human eye cannot see. Instead, in this thermal region of the electromagnetic spectrum, the energy can be measured and felt as heat. The sources in this part of the spectrum radiate thermally, and the radiation at the peak wavelength is a function of the source average temperature.

As the average temperature increases, the peak shifts to shorter wavelengths. The radiation pattern as a function of wavelength can often be modeled by the blackbody radiation curve discussed later on in this chapter. An example of thermal radiation is that of the human body wherein radiation in the form of heat is emitted in the IR part of the spectrum.

5.2.1.1 IR Sources: Thermal and Selective Radiators

If a given source is a thermal radiator such as a heated solid or liquid, then the curve of the radiation distribution is continuous with a single maximum. Thermal radiators are the most prevalent sources from the standpoint of an IR designer and will include typical thermal radiators such as personnel, cars/trucks, terrain, and aerodynamically heated surfaces.

On the other hand, if a source is a selective radiator such as a flame or electrical discharge in gas, then the associated flux will be concentrated in narrow, spectral intervals with multiple local maxima and minima. This would be seen in examples like the stream of hot gases from a jet engine's exhaust, gas discharge lamps (such as sodium or mercury vapor light bulbs used for lighting), or the shock-excited layer surrounding a reentry vehicle. In a line spectrum, the intervals appear to be extremely narrow with sharply defined lines. This is typical of an emitting gas. A band spectrum consists of bands of narrow lines.

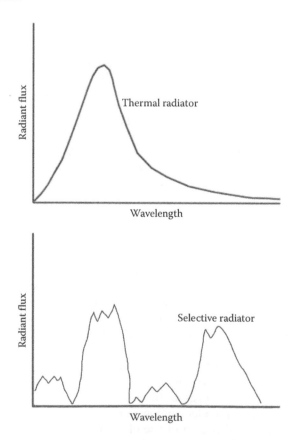

FIGURE 5.6
Thermal and selective radiators.

Figure 5.6 shows a typical spectral response for representative thermal and selective radiators. The radiant flux is plotted versus wavelength. The two spectral responses for both types of radiators look significantly different and behave as explained earlier.

5.2.2 Absorption and Emission Spectra

For atoms and molecules, energy is absorbed and emitted in very narrow bands. Left to their own devices, electrons in the atoms and molecules seek to find the lowest energy state. If radiation at the right energy level interacts with the atom or molecule, then the electrons are kicked into higher energy states and try to return to their lower energy states by either giving off heat or transitioning from a higher energy state to a lower one and giving of a photon at a wavelength corresponding to the band-gap energy (the difference between the energy between the two states). The absorption and emission interactions are based on quantum mechanical rules. The presence of atoms, molecules, and aerosols in the atmosphere serves to absorb some of the optical radiation that is passing through it. The absorption occurs at distinct bands based on the absorbing atom, molecule, or aerosol. Atoms, molecules, and aerosols in the atmosphere also scatter light and generally reduce the optical flux that is passing through the atmosphere. In Figure 5.7, we can see how the IR emission of a Bunsen flame behaves. By using a library of known IR emissions, operators can identify elements and objects by the shape of its emission graph.

5.2.3 Thermal Radiation in the IR Spectrum

Thermal radiation can be classified as electromagnetic radiation that emanates from an object's surface and is related to its temperature. It stems from the movement of charged

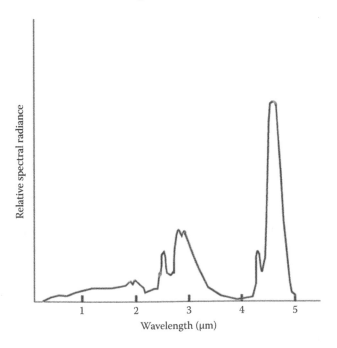

FIGURE 5.7
Representative IR emission from a Bunsen flame. (Adapted from Plyler, E.K., *J. Res. Natl. Bur. Stand.*, 40, 113, 1948.)

particles within the object generating heat that is converted into electromagnetic radiation. Thermal radiation is generated when heat from the movement of charged particles within atoms is converted to electromagnetic radiation. The emitted electromagnetic spatial power density (radiant emittance) of the thermal radiation varies as a function of wavelength and temperature and, for a genuine blackbody, is given by Planck's law of radiation. Wien's law gives the wavelength corresponding to the peak radiant emittance for a blackbody radiator as a function of temperature, and the Stefan–Boltzmann law gives the total radiant emittance of the object as a function of temperature.

5.2.4 Planck's Law

Planck's law describes the blackbody radiation radiant emittance distribution as a function of wavelength and temperature. The term c_1 is the first radiation constant, and c_2 is the second radiation constant. Note that these constant terms are simply weighted arithmetic combinations of the speed of light, Plank's constant, and the Boltzmann constant. The following equation illustrates Plank's law:

$$W_\lambda = \left(\frac{c_1}{\lambda^5} \right) \left(\frac{1}{e^{(c_2/\lambda T)} - 1} \right),$$ (5.1)

where (in SI units)
W_λ is the spectral radiant emittance, $W/(m^2 \cdot \mu m)$
λ is the wavelength, μm
T is the absolute temperature, K
$c_1 = 2\pi h c_2 = $ first radiation constant $= 3.7411 \times 10^8 \ W \cdot \mu m^4/m^2$
$c_2 = ch/k = $ second radiation constant $= 1.4388 \times 10^4 \ \mu m \ K$
h is the Planck's constant $= 6.6256 \pm 0.0005 \times 10^{-34} \ W \cdot s^2$
c is the speed of light $= 2.99792458 \times 10^8 \ m/s$ (exact)
k is the Boltzmann constant $= 1.3806503 \times 10^{-23} \ W \cdot s/K$

Figure 5.8 shows several notable characteristics of the radiation from a blackbody, which is evident in these curves. Notice that the total radiant emittance is actually the integrated spectral radiant emittance shown in the figure. As the temperature rises, the peak wavelength of the spectral radiant emittance moves to the left. Also, the higher the temperature becomes, the higher the spectral radiant emittance, regardless of the wavelength.

The electromagnetic wave has properties that can be described both from a particle perspective and from a wave perspective—the so-called duality of light. James Clerk Maxwell found that phenomena such as the refraction of light from lenses and prisms, diffraction effects, scattering, absorption, and reflection can be explained by the wave properties of the electromagnetic field. On the other hand, Newton and Planck described light in terms of discrete units. These light quanta, or photons, traveled in a linear direction and had discrete amounts of energy that were proportional to their radiation frequency:

$$Q = h\nu,$$ (5.2)

where
Q is the radiant energy
$h = 6.626 \times 10^{-23} \ J \cdot s$ (Planck's constant)
ν is the frequency of photon

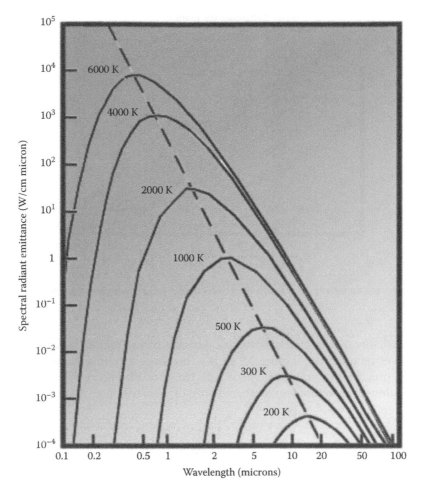

FIGURE 5.8
Spectral radiant emittance of a blackbody by temperature.

Planck's equation explains the photoelectric effect, which is the absorption of photons upon certain metal surfaces causing the emission of electrons.

5.2.5 Stefan–Boltzmann Law

In 1879, Stefan found from his experimental measurements the same conclusion that Boltzmann reached in 1884 by applying thermodynamic relationships. They introduced what is known as the Stefan–Boltzmann law that states that the radiant emittance from a blackbody is proportional to its temperature, raised to the fourth power. The Stefan–Boltzmann law provides the total radiant emittance from the blackbody (W) in watts/cm^2 over its entire surface area and all its wavelengths. The proportionality constant is the Stefan–Boltzmann constant (σ), and the temperature (T) is expressed in kelvins (K). The radiant emittance is given by

$$W = \sigma T^4, \tag{5.3}$$

where
$\sigma = 5.6697 \times 10^{-8}$ W/m^2 K^4 (Stefan–Boltzmann constant)
T is the temperature in kelvins

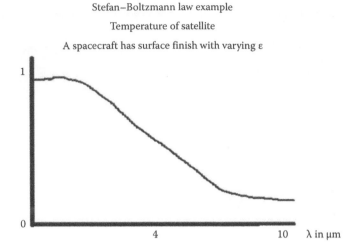

FIGURE 5.9
Representative varying emissivity as a function of wavelength for man-made object. (Courtesy of Maryam Abdirad.)

From Equation 5.3, it can be seen that hotter blackbodies have a higher radiant emittance than those with lower temperature. Figure 5.9 shows the emissivity (y-axis) versus wavelength (x-axis) for a representative spherical spacecraft. Underneath Figure 5.9 is a quick example of how to calculate the satellite temperature using the Stefan–Boltzmann law.

Power in = power out	H = solar constant
$HA_1\alpha = A_S\sigma\varepsilon T^4$	A_1 = projected area of satellite
$T^4 = \dfrac{H}{\sigma}\cdot\dfrac{A_1}{A_S}\cdot\dfrac{\alpha}{\varepsilon}$	A_S = surface area of satellite
For a sphere, $\dfrac{A_1}{A_S} = \dfrac{\pi R^2}{4\pi R^2} = \dfrac{1}{4}$	
Assume $\dfrac{\alpha}{\varepsilon} = 0.5$	α = absorptance in solar region
$T^4 = \dfrac{1350}{5.67\times10^{-8}}\cdot\dfrac{1}{4}\cdot\dfrac{1}{2} = 2.976\times10^9$	ε = emissivity in equilibrium temperature region
$T = 233.6\,\mathrm{K}$	$\sigma = 5.67\times10^{-8}\,\mathrm{W/m^2\,K^4}$

5.2.6 Kirchhoff's Law of Radiative Transfer

Introduced in 1860, Kirchhoff's law stated that *good absorbers are also good radiators*. He also proposed the term *blackbody*, a body that absorbs all of the incident radiant energy and that must also be the most efficient radiator, as his law asserts. This would make the blackbody the ultimate thermal radiator and set a standard to compare with other sources.

Kirchhoff's law relates the total emitted radiation over all wavelengths and the entire object surface to the total absorbed radiation under the same conditions. Kirchhoff found that for a fixed temperature, the radiation emitted from a source divided by the radiation emitted from a blackbody source at the same temperature is constant for all sources. Consequently, the emissivity is defined as

$$\varepsilon = \frac{M}{M_b},$$ (5.4)

where
 ε is the emissivity
 M is the emittance of an object
 M_b is the emittance of a blackbody at the same temperature

The limiting value of emissivity for a true blackbody is 1 and that of a perfect reflector is 0.
   Through Kirchhoff's studies of radiation transfer, Kirchhoff found that the ratio of radiant emittance to absorptivity is a constant for all materials at a given temperature and that this ratio is equal to the radiant emittance of a blackbody at the same given temperature.
   Kirchhoff's law can be stated as

$$\frac{W'}{\alpha} = W - M_b,$$ (5.5)

where
 W' is the radiant emittance of non-blackbody
 W is the radiant emittance of a blackbody
 α is the absorptivity

5.2.7 Emissivity

Incident radiant energy is processed on a surface by a combination of three processes. The incident energy may be absorbed, reflected, and/or transmitted. In his work on thermal radiation, Kirchhoff proposed a definition of an ultimate radiant energy absorber: the *blackbody*.
   For the blackbody, all incident radiant energy on the body is absorbed. No part of the radiant energy is reflected or transmitted. The beauty of the blackbody reference is twofold:

* Kirchhoff and successors were able to use the blackbody model to describe and formulate properties of the ultimate radiant energy *source* by virtue of his law that *good absorbers are also good emitters.*
* The blackbody (and its properties) placed the upper limit on both sources and surfaces. It defines unity for the analysis of emissivity.

To consider the emissive properties of a source other than the *pure* blackbody, a factor is added to the blackbody formula, and that factor is called emissivity. The emissivity is determined by dividing the radiant emittance of the source W' by the radiant emittance W of a perfect blackbody radiator at the same temperature:

$$\varepsilon = \frac{W'}{W}.$$ (5.6)

The resultant number will have a value of zero for a nonradiating source and one for a blackbody source.

To take into consideration material type, surface finish, wavelength, and temperature of a non-blackbody source, the spectral emissivity, ε_s, is used:

$$\varepsilon = \left(\frac{1}{\sigma T^4}\right)\int_0^\infty \varepsilon(\lambda)W_\lambda \, d\lambda. \tag{5.7}$$

Kirchhoff's law can be rewritten in the following form:

$$W' = \alpha W. \tag{5.8}$$

In combination with the Stefan–Boltzmann law, this can be reduced to

$$\frac{\varepsilon\sigma T^4}{\alpha} = \sigma T^4. \tag{5.9}$$

Canceling the common terms on both sides, Equation 5.9 can be simplified down to

$$\varepsilon = \alpha. \tag{5.10}$$

This means that the emissivity of any material at a given temperature, T, is numerically equal to its absorptivity at that temperature. Kirchhoff's findings show that good absorbers are good emitters.

For convenience, the relationship is rewritten in terms of easier to measure reflectance:

$$\varepsilon = (1-\rho). \tag{5.11}$$

5.2.8 Emissivity of Common Materials

Table 5.3 provides an example of how the emissivity of given materials will vary with temperature and surface finish. These values should be used primarily as a guideline for relative or differential temperature measurements. When high accuracy is necessary, the exact emissivity of a particular material should be determined.

A practical use of this information comes from the emissivity of the surface of materials and Kirchhoff's law. An airplane on the ground, in the sun, will have a lower internal temperature if painted with a lacquered paint. The emissivity is much higher for the lacquer compared to aluminum. This means the lacquer will absorb more radiant energy than the aluminum. And, thanks to Kirchhoff's law, the lacquer will radiate away more radiant energy as well.

5.2.9 Sources That Approximate Blackbody Radiators

A blackbody is a thermal radiator, which can serve as a basis of calibration for IR detectors. The reader must understand that a perfect blackbody is a theoretical concept of a body that absorbs all electromagnetic radiation and reflects none; however, it has the ability to radiate any given wavelength. Blackbody radiation is also known as cavity radiation. Kirchhoff established conditions that must be met in order to properly construct this thermal radiator.

TABLE 5.3

Emissivity of Common Materials

Nonmetals		Metals	
Material	Emissivity	Material	Emissivity
Asbestos		Alloys	
Board	0.96	80-Ni, 20-CR, oxidized	0.87
Cement	0.96	Aluminum	
Cement, red	0.67	Unoxidized	0.03
Cement, white	0.65	Highly polished	0.09
Cloth	0.90	Roughly polished	0.18
Paper	0.93	Commercial sheet	0.09
Slate	0.97	Bismuth	
Asphalt, pavement	0.93	Bright	0.34
Asphalt, tar paper	0.93	Unoxidized	0.06
Concrete		Brass	
Rough	0.94	Matte	0.07
Tiles, natural	0.63–0.62	Burnished to brown color	0.40
Brown	0.87–0.83	Unoxidized	0.04
Black	0.94–0.91	Carbon	
Cotton cloth	0.77	Unoxidized	0.81
Granite	0.45	Candle soot	0.95
Gravel	0.28	Graphitized	0.76
Ice		Chromium	0.08
Smooth	0.97	Chromium	0.26
Rough	0.98	Cobalt, unoxidized	0.13
Marble		Copper	
White	0.95	Matte	0.22
Smooth, white	0.56	Polished	0.03
Polished gray	0.75	Rough	0.74
Sand	0.76	Gold	
Sawdust	0.75	Polished	0.02
Snow		Iron	
Fine particles	0.82	Oxidized	0.74
Granular	0.89	Unoxidized	0.05
Soot		Mercury	0.12
Acetylene	0.97	Silver	
Camphor	0.94	Plate (0.0005 on Ni)	0.06–0.07
Candle	0.95	Polished	0.02
Coal	0.95		

To establish a baseline for these conditions, Kirchhoff compared the radiation of a black-body to the radiation of an isothermal enclosure through an opening that is greatly smaller than the diameter of the enclosure. An isothermal process is that one where its temperature remains constant. If the isothermal condition is met, then the radiation through the opening slit approaches blackbody radiation. The relative size of the slit in an isothermal enclosure to the diameter of the enclosure affects the accuracy of the blackbody approximation. The technical performance measure for this source is the effective emissivity.

The emissivity characterizes how closely the radiation spectrum of a real heated body corresponds to that of a blackbody. Reviewing a concept from a previous chapter, the emissivity of a material is a measure of the ability of its surface to emit energy by radiation, and in theory, its range could vary from zero up to one. The ideal blackbody radiator would have an effective emissivity of one, while the perfect reflector would have an emissivity of zero. Depending on how the spectral emissivity varies, we can identify three types:

1. Blackbody: The perfect radiator, its emissivity is one for all wavelengths.
2. Gray body: Its emissivity is a constant fraction of what the corresponding blackbody would radiate at the same temperature. The emissivity is constant and less than one ($\varepsilon < 1$). Gray bodies are independent of wavelength, and their spectral shapes are the same as a blackbody.
3. Selective radiator: The emissivity ε is an explicit function of λ.

A conceptual example of these types of radiators is shown in Figure 5.10.

Notice that for both the blackbody and the gray body, the spectral emissivity is constant across the wavelengths, whereas the spectral emissivity can vary for a selective radiator. Often, this difference can be used as a discriminator when evaluating the spectral response from an unknown object.

Leslie's cube, a hollow metal cube that contains hot water, is an experiment that demonstrates that some materials are better at absorbing radiation than others. The four faces of the cube have different finishes, for example, matt white, matt black, dull aluminum, and polished aluminum. If we place our hand or a radiation thermometer near the surface of each side, we can notice that the black side radiates maximum energy and the polished metal surface the minimum (Poynting and Thomson 1906, pp. 230–231). A Leslie's cube is shown in Figure 5.11.

A typical example of effective emissivity is shown in Figure 5.12. The graph shows the comparison of the curves of different effective emissivity values. The emissivity ranges from a surface emissivity of 0.65 (bottom curve) to 0.95 (top curve), as seen by the three

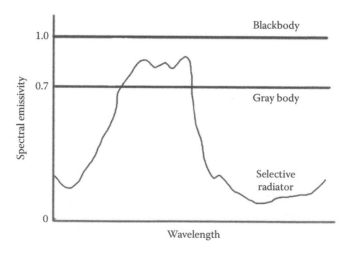

FIGURE 5.10
Spectral emissivity curves for various types of radiators. (Courtesy of Maryam Abdirad.)

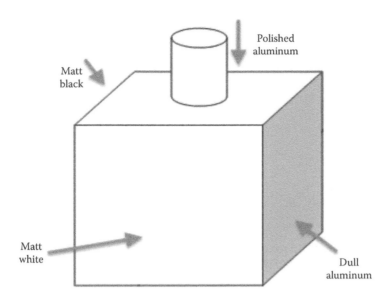

FIGURE 5.11
Leslie's cube.

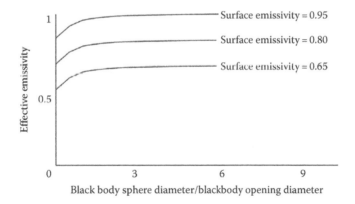

FIGURE 5.12
Graph demonstrating effective emissivity change. (Courtesy of Maryury Diaz.)

lines in the graph. As the aperture of the blackbody simulator becomes smaller in relation to the diameter of the spherical blackbody type, the effective emissivity increases, and the blackbody-type becomes more like a true blackbody. The emissivity of metals increases with increasing temperature, oxidation, and surface roughness. On the other hand, for nonmetals, including ceramics and organic materials including dielectrics, the emissivity decreases slowly with temperature.

Andre Gouffé, who strived to construct precise blackbodies using an assumption that the walls of the isothermal enclosure are *diffuse reflectors*, performed further study of the effective emissivity of a blackbody. A diffuse reflector is a surface whose reflected brightness is the same in all directions (De Vos 1954). An example of a diffuse reflector is illustrated in Figure 5.13.

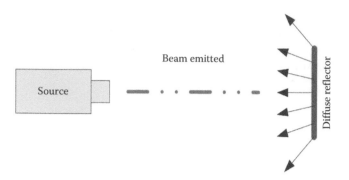

FIGURE 5.13
Diffuse reflector example.

Using this assumption (Sparrow and Johnson 1962), Gouffé formulated an equation to find the effective emissivity of an enclosure:

$$\varepsilon' = \frac{\varepsilon(1+K)}{\left(\varepsilon\left(1-(A/S)\right)+(A/S)\right)}, \qquad (5.12)$$

where
  ε' is the effective emissivity of enclosure
  ε is the emissivity of enclosure walls
  A is the area of enclosure opening
  $K = (1-\varepsilon)(A/S - A/S_o)$
  S is the surface area of enclosure including opening
  S_o is the measurement of the surface area of a sphere with diameter being from the opening to the most distant wall of the enclosure

An important concept is that changing the size of the enclosure can modify the effective emissivity of a given surface enclosure. A typical result is that the enclosure with the greatest surface area has the greatest effective emissivity.

Some of the calculation assumptions earlier require further discussion. Gouffé made the previous assumption that the walls of the enclosure were *perfectly diffuse reflectors*. De Vos found similar results to Gouffé when making the assumption that these walls may be diffuse or nondiffuse reflectors (De Vos 1954). One key difference is that De Vos determined that as the walls become less diffuse, the effective emissivity of the enclosure decreases.

Individuals including Gouffé, Sparrow, Kelly, Johnson, Moore, Campanaro, and Ricolfi, among others, have performed several related studies (Sparrow and Johnson 1962; Kelly and Moore 1965; Campanaro and Ricolfi 1967) on emissivity. These studies have debated the accuracy that results from the calculations for effective emissivity for spherical enclosures based on error introduced by assumptions made by various researchers. According to the formulas put forth by Gouffé, the cylindrical and conical enclosures have an error of up to 2%. This type of accuracy is often good enough for calculations of effective emissivity.

To further refine the accuracy of blackbody source emissivity, one must switch to using the Monte Carlo method developed by V.I. Sapritsky and A.V. Prokhorov

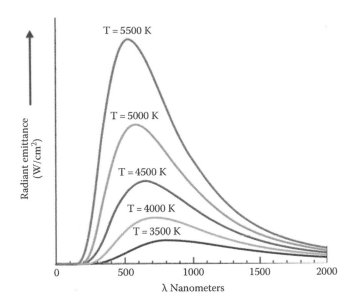

FIGURE 5.14

Blackbody emission brightness as a function of maximum wavelength. (Reprinted with permission from 4C, Wiens law, http://en.wikipedia.org/wiki/File:Wiens_law.svg, accessed April 1, 2014, 2006.)

(Sapritsky and Prokhorov 1992). This method has become the national standard for calculating the effective emissivity of a blackbody radiation source due to its low relative error of 0.0001%.

The spectral emittance from a blackbody is determined by Planck's equation (Equation 5.1).

An important parameter of a blackbody source is the wavelength where the spectral emittance is the maximum value; this is the wavelength where the most power is emitted, as shown in Figure 5.14. The physicist Wilhelm Wien established Wien's law, Equation 5.13, that indicates that the hotter a blackbody becomes, the shorter its peak wavelength associated with the spatial power density (emittance—watts/m^2) becomes:

$$\lambda_{max} = \frac{b}{T} \tag{5.13}$$

where
  $b = 2898 \ \mu m \ K$
  T is the temperature of the blackbody in K

5.2.10 Tools for Radiation Calculations

Radiation calculations are often much too tedious to be accomplished by hand. Today, mostly all radiation calculations are done using some type of computer application or program. Alternatively, programs such as MATLAB® can be used to precisely model electromagnetic interactions with matter and simulate radiometric propagation. FEMLAB is an excellent multiphysics tool that has finite element–based simulation capabilities for multiphysics applications that include electromagnetic propagation modeling, along with a variety of other useful physical modeling capabilities.

5.2.11 Scale Choice in Plotting

The plotting scale choice affects the shape of the blackbody spectral distribution curve. Figure 5.15 shows the combinations of linear–linear (first chart) and log–log (second chart) scales on a set of blackbody curves. Note that the curves on each respective chart have a similar shape, but the Wien function for the log–log scale chart is a sloped line instead of an exponential curve as seen in the linear–linear scale.

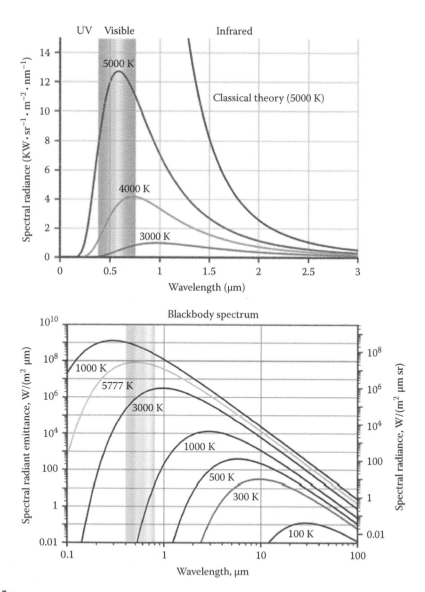

FIGURE 5.15
Relative blackbody spectral distribution curves. (Reprinted with permission from Kule, D., Black-body radiation, http://en.wikipedia.org/wiki/Black-body_radiation, accessed April 1, 2014, 2010; Sch, BlackbodySpectrum loglog, http://commons.wikimedia.org/wiki/File:BlackbodySpectrum_loglog_150dpi_de.png, accessed April 1, 2014, 2006.)

5.2.12 Radiation Efficiency

The thermal radiation efficiency is given in the following equation:

$$\frac{W_\lambda}{W} = \left(\frac{c_1}{\lambda^5}\right)\left(\frac{1}{e^{((c_2/\lambda T)-1)}}\right)\left(\frac{1}{\sigma T^4}\right), \tag{5.14}$$

where the efficiency is defined as the radiant emittance at a given wavelength divided by the total radiant emittance given by the Stefan–Boltzmann law. Image detection in the IR is highly dependent on the background temperature. When the target temperature is close to the background temperature, the contrast becomes so low that target detection can be very difficult.

The maximized value of radiation contrast (C_{rad}) can be obtained by taking the partial derivative of the Planck function with respect to temperature. This is shown in the following equation:

$$C_{rad} = \frac{\partial W_\lambda}{\partial T} = W_\lambda \frac{\psi}{T}, \tag{5.15}$$

where ψ is a value from a table provided by Pivovonsky et al. (1961).

For a given blackbody temperature–wavelength combination, the maximum rate of change of the spectral radiant emittance occurs at a wavelength that obeys

$$\lambda_c T = 2411, \tag{5.16}$$

where the "c" in the subscript of the wavelength denotes change.

5.2.13 Making Blackbody Sources

The preceding section describes some of the characteristics of a laboratory blackbody-type radiation source. The studies of V.I. Sapritsky and A.V. Prokhorov previously mentioned have developed a Monte Carlo–based methodology to produce high-quality blackbody-type enclosures with an accuracy approaching a near-perfect blackbody radiator (Sapritsky and Prokhorov 1992; Hanssen 2004). For the purpose of simulating the IR part of the electromagnetic spectrum, a typical blackbody-type source can be used to generate a range of responses. As an example, a blackbody-type source with a half-inch opening can generate responses corresponding to a temperature range from 400 to 1300 K. When developing this calibration tool, the accuracy depends on the enclosure configuration (spherical, conical, cylindrical), isothermal generation, enclosure wall emissivity, and the diameter of the cavity as compared to the exit slit length. The accuracy of the simulated blackbody also depends on the choice of the core, which is the large mass of material within the enclosure. A uniform heating method must be implemented to ensure that an isothermal condition is approximated.

By varying the features of the core, the emissivity of the enclosure walls can be modified. Different surface finishes and varying temperatures affect emissivity. The surface condition and oxidation of treated surfaces have a significant effect on the emissivity. Figure 5.16 is

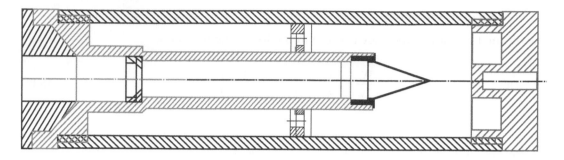

FIGURE 5.16
Fixed-point blackbody crucible with cylindro-conical cavity. (Reprinted with permission from Hanssen, L., Monte Carlo modeling of effective emissivities of blackbody radiators, Physics Laboratory—Optical Technology Division, National Institute of Standards and Technology, Gaithersburg, MD, accessed May 30, 2007, 2006.)

an example from the National Institute of Standards and Technology (NIST) showing the typical construction of a conical blackbody radiation source (Prokhorov et al. 2010).

Measuring temperature of any blackbody-type source is typically done with a spectral radiance and irradiance primary scale as prescribed by Gibson at the NIST (Gibson 2010). A picture of the standard setup for a spectral radiance calibration is shown in Figure 5.17. A calibrated lamp is placed at a separation distance "d" from a diffuser plate surface. The orientation of the lamp placement is chosen so that the diffuser plate receives perpendicular illumination. Consequently, the illuminated plate acts like a uniform radiation source. The radiation from the plate provides the illumination source for the radiance input optics shown at the bottom of Figure 5.17. Pisulla reported that if the distance, x, from the diffuser plate to the radiance input optics is too large, then the input optic will not be uniformly illuminated by the plate and the calibration procedure will be invalid (Pisulla et al. 2009, pp. 516–527).

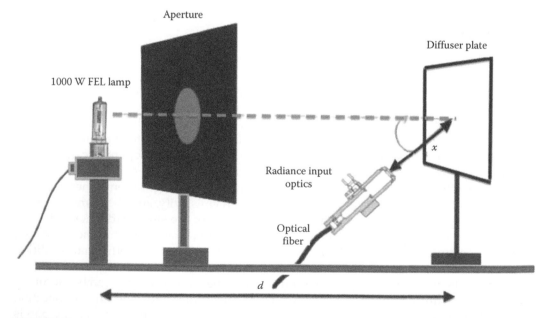

FIGURE 5.17
Standard spectral radiance calibration setup.

More inexpensive temperature devices are often implemented due to the fact that extremely accurate temperature readings are typically not required. Simple thermocouple calibrations to match temperature controller set points are often sufficient. Standard stainless steel blackbody types can be manufactured in a laboratory environment.

Various calculations can be implemented to determine the changes in radiant emittance at an assumed temperature and effective emissivity. For example, using the equations of Stefan–Boltzmann, we can determine the radiant emittance at 800 K with an effective emissivity of 0.99 by using Equation 5.15:

$$W = \varepsilon\sigma T^4, \tag{5.17}$$

where
  W is the radiant emittance (W/cm^2)
  ε is the emissivity of enclosure walls
  T is the temperature (K)
  $\sigma = 5.6697 \pm 0.0029 \times 10^{-12}$ W/(cm^2 K^4)

By differentiating Equation 5.16, we determine the percent change in radiant emittance at a given temperature (K). The result is shown in Equation 5.17:

$$\frac{dW}{dT} = 4\varepsilon\sigma T^3. \tag{5.18}$$

By multiplying Equation 5.16 by the area of the surface of the source, we determine the optical power radiated by the source:

$$P = A\varepsilon\sigma T^3, \tag{5.19}$$

where
  P is the power radiated (W)
  ε is the emissivity of enclosure walls
  T is the temperature (K)
  $\sigma = 5.6697 \pm 0.0029 \times 10^{-12}$ W/(cm^2 K^4)

The proper construction of a blackbody type requires that the temperature be managed within fractional degrees (De Vos 1954). Gouffé has addressed the effects of limiting aperture on emissivity (Kelly and Moore 1965). However, one must note that these aperture affects do not apply to thermally isolated types.

Proper ways to measure apertures may include a toolmaker's microscope or projection techniques onto a calibrated screen using a method known as optical comparison. Fowler reported on the technique of optical comparison that yields an error of 0.04% in the worst-case scenarios and is a standard of the NIST (Fowler and Gyula 1995, pp. 277–283). Note that this technique is very useful when measuring nonsymmetrical holes. The general standard for aperture size is to be >0.01 in., due to the fact that dust may have greater effects on apertures of this size. More advanced techniques including use of a radiometer may be implemented to determine the size of the small apertures. When striving for higher-precision measurements, one must realize that a blackbody is a Lambertian source. A Lambertian source has the notable characteristic that the source radiance obeys Lambert's cosine law.

FIGURE 5.18
Leitenberger's IR calibration device datasheet. (Reprinted with permission from Larason, T.C. and Houston, M., *Spectroradiometric Detector Measurements: Ultraviolet, Visible, and Near-Infrared Detectors for Spectral Power*, NIST Special Publication 250-41, National Institute of Standards and Technology, Gaithersburg, MD, 2008, http://www.nist.gov/calibrations/upload/sp250-41a.pdf, accessed April 1, 2014.)

Lambert's cosine law stated by Fowler says that "Radiance of certain idealized optical sources is directly proportional to the cosine of the angle—with respect to the direction of maximum radiance—from which the source is viewed" (National Telecommunications and Information Administration 1996). Therefore, the viewing angle must be noted, and corrections for the viewing angles must be accounted for in the calculations. Just as a final point of clarification for blackbody types, one must mention the variation of size and aperture that are available today. Some blackbodies can range from mere ounces to over 300 lb, and the respective aperture may range from fractions of an inch to several inches. There are also a variety of proprietary configurations with little detailed information.

New technologies can have calibration capabilities that range to the 1/100th of a degree like Leitenberger's IR calibration device. As an example, a picture of Leitenberger's calibration device is seen in Figure 5.18.

Some of the key technical performance measures for Leitenberger's IR calibration device that should be pointed out are that the temperature range for the precision reference thermometer is from −50°C to 199.99°C and works with different temperature controllers. The device has a 0.01°C resolution and produces an emissivity >0.95.

5.2.14 Sources, Standards, and NIST

Measurements for sources of radiant energy are benchmarked to the NIST. Physical measurements and local calibrations will reference this standard. In applications for government including the Department of Defense, the NIST is typically the guiding measurement standard. Since 2006, the NIST has standards for radiant temperature measurements known as the ITS-90 scale. This scale specifies the precise calibration points such as the melting point of gallium or the freezing point of tin. These points are used for calibration purposes. Note that the typical understanding is that the calibration points are 10 times more precise than the calibrations that are derived from using these points (Johnson et al. 1994, pp. 731–736).

There are many standards that the NIST has established for the use of calibrating IR devices. The most important to this text is the standard for IR blackbody spectral characterization. In this standard, the NIST uses fixed-point blackbody sources, Monte Carlo calculation programs for emissivity, and a medium-resolution Fourier transform spectrometer for spectral radiance measurements with errors <0.1% (Hanssen 2006). All NIST-certified IR blackbody spectral characterization measurements are performed at the medium background IR facility (MBIR). The facility houses the large-area blackbody (LABB) source, MBIR absolute cryogenic radiometer (MBIR ACR), and the thermal-IR transfer radiometer (TXR). The LABB is a blackbody source capable of measurements between 180 and 350 K and has an estimated cavity emissivity of 0.999 from 1 to 14 μm (Johnson and Rice 2006c). A pictorial representation of the low-background IR facility (LBIR) provided by the NIST is shown in Figure 5.19.

The MBIR ACR is a radiometer with an active cavity that is used for the absolute measurement of broadband radiance (Johnson and Rice 2006a). A pictorial representation provided by the NIST is shown in Figure 5.20.

Finally, the TXR is used to take radiance measurements in the IR range. Johnson and Rice (2006b) reported that its accuracy has been calculated to an error of 0.2%. A pictorial representation provided by the NIST is shown in Figure 5.21.

Some standards for which the NIST uses the MBIR include the following: thermal radiation, thermal radiation lamps, and spectral radiance lamps as calibrated against blackbody types.

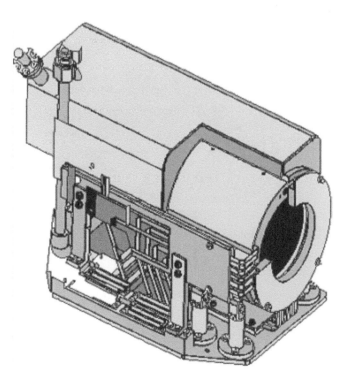

FIGURE 5.19
LBIR facility. (Reprinted with permission from NIST 2010a.)

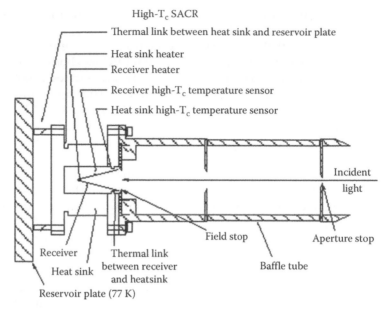

FIGURE 5.20
Schematic of the MBIR ACR. (Reprinted with permission from NIST, High-T$_c$ SACR, http://www.physics.nist.gov/TechAct.98/Div844/div844h.html, accessed April 1, 2014, 2002.)

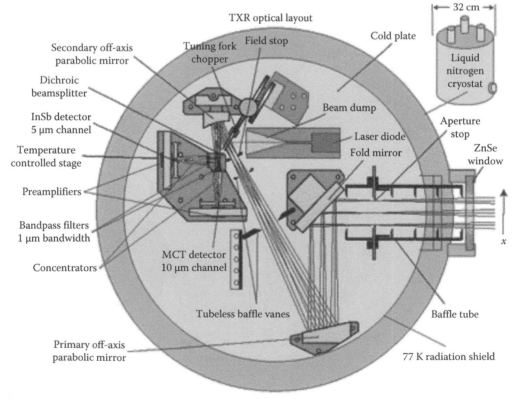

FIGURE 5.21
Optical schematic of the TXR. (Reprinted with permission from NIST 2010b.)

5.2.14.1 Optical Sources

In this section, we present an overview of available optical sources. We present the common sources in the ultraviolet (UV), visible, and IR parts of the electromagnetic spectrum along with a short description of their properties. We separate the sources according to what part of the electromagnetic spectrum that they belong. We start with UV sources, then visible, and end up with IR sources.

5.2.14.2 Sources of Ultraviolet Radiation

The UV part of the electromagnetic spectrum is divided into four distinct categories: (a) long-wave UV (UVA) that covers 315–400 nm, (b) medium-wave UV (UVB) that covers 280–315 nm, (c) short-wave UV (UVC) that covers 200–280 nm, and (d) vacuum UV (VUV) that covers 100–200 nm. Applications include UVA, bonding and curing materials, nondestructive testing; UVB, bonding and curing, sunlight aging studies; UVC, germicide, sterilization, rapid surface curing; and VUV, microelectronics in vacuum. Delivery mechanisms include direct illumination and light guides such as optical fiber.

5.2.14.3 Light-Emitting Diodes

UV LED sources produce optical power in the UV spectrum that is narrowband and can be tuned. These sources are durable as compared to standard UV lamps and can last longer than 20,000 h. Representative wavelengths range across the UV spectrum, and a variety of center wavelengths are commercially available (e.g., 265, 280, 310, 365, 385, 395, 400, 405, and tunable sources are available). Power densities of 30 kW/m^2 are achievable. Some disadvantages include the following: power densities achieved by conventional arc lamps are confined to spot sizes on order of a millimeter, achievable power is about 20% of conventional arc lamps, and some curing methods require additional wavelengths (Lumen Dynamics 2014).

5.2.14.4 UV Lamps

UV lamps put out optical power in the range of 250–600 nm using gas-discharge mechanisms. Common gases used are mercury, argon, deuterium, and xenon. Fluorescent UV lamps produce short-wave illumination at 185 and 253.7 nm using mercury, with 253.7 nm being the dominantly produced wavelength. The lower wavelength is often filtered out by impurities in the quartz tube containing the mercury. Short-wave UV lamps are used in sterilization applications such as food processing, laboratories, and also for disinfecting water supply. Other applications for UV lamps include tanning beds, urine detection, forgery detection, and crime scene inspection. Disadvantages include toxic materials such as mercury in their construction, relatively long warm-up times (minutes), and lower durability than LEDs.

5.2.14.5 UV/Visible/IR Lasers

Since laser sources span the UV, visible, and IR parts of the electromagnetic spectrum, we combine the discussion on laser sources and present it only once, with applicability to the entire optical spectrum. The word laser stands for light amplification by stimulated emission of radiation. Lasers have high radiance, directionality, are coherent and are easily modulated, making them good candidates for sources in the UV, visible, and IR parts of the electromagnetic spectrum. Today, lasers are the primary sources used by industry.

TABLE 5.4

Type of Lasers (Classified according to the Material)

Type of Laser	Examples	Characteristics
Solid state	Glass	Not semiconductor.
	Crystal	
		High-power applications.
		Pulsed-mode applications.
Gas	Copper vapor	Cheap, simple technology.
	CO_2	
	Molecular gas	
		Used in the visible and invisible parts of the spectrum.
		Low-power applications.
		CO_2 lasers emit energy in the far IR.
Liquid	Dye	Used in the generation of ultrashort pulses.
		Variety of wavelengths in the visible part of the spectrum.
Semiconductor	Aluminum gallium arsenide	Overview of different geometries.
	Indium phosphide	
		Optical data transmission systems (CD and DVD).

They are in common use in the medical, commercial, industrial, and military sectors and are even prevalent in the entertainment industry and academia. Lasers emit a continuous beam or pulses of light that are nearly monochromatic and in phase. Lasers are highly directional and can be focused to small spots producing high spatial power densities. The wavelength of light that is radiated is highly dependent on the energy band-gap properties of the laser material. There are different types of lasers; they can be classified by the laser medium material, which can be solid, gas, liquid, or a semiconductor as shown in Table 5.4.

Common for all lasers, regardless in what part of the spectrum they emit, is the requirement for the laser material, a laser cavity, and mirrors that serve as the mechanisms for stimulating and amplifying the laser emissions. The first stimulated emission was observed in the microwave part of the electromagnetic spectrum—the microwave amplification by stimulated emission of radiation (MASER). The choice of a type of laser depends on the application. UV lasers are used for dermatology, oncology, high-precision surgery, laser spectroscopy, mass spectroscopy, light radar (LIDAR), micromachining, and more. Visible lasers are used for cutting, alignment, inspection, printing, communications, entertainment, and host of scientific, military, medical, industrial, and commercial applications. IR lasers applications include illumination sources, especially in the MWIR part of the spectrum, standoff explosive detection, IR spectroscopy, spectral imaging, thermal vision, industrial process control, medical diagnostics, hazardous chemical detection, and military thermal countermeasures.

5.2.14.6 Visible Light Sources

Just as in the UV part of the spectrum, a variety of visible light sources are commercially available including incandescent lamps, fluorescent lamps, LEDs, and lasers. Indoors, the predominant types of visible light sources are incandescent lamps and fluorescent lamps. Outdoors, the sun is the predominant natural light source during the day and starlight, moonlight, and fires in remote areas at night. Depending on location, natural light can be observed from sources such as the aurora borealis (northern lights), bioluminescence from some marine life (e.g., fire worms, parchment worm), and fireflies, meteors entering the atmosphere comets, volcanoes, lightning, cosmic rays, clouds, and such. A variety of man-made light sources exist (too many to list) including nuclear and high-energy sources, direct chemical light sources, electric powered, combustion, and other various sources (Lighting 2014). Commercial sources come in a wide variety from broadband, narrowband, quasi-monochromatic, and monochromatic. Sources can be coherent, partially coherent, or incoherent either spatially or temporally. Often, using a broadband source and a monochromator can produce a narrowband source. The line width of the spectral source is then determined by the size of the exit slit on the monochromator. Tunable, narrowband sources exist throughout the visible spectrum. Fire is an exothermic chemical reaction associated with combustion that produces energy in the UV, visible, and IR parts of the spectrum. For observable light in the visible spectrum, the temperatures can range from 525°C (red) to 1500°C (brilliant white).

5.2.14.7 IR Sources

In order to emit IR radiation, an object must have an absolute temperature >0 K. A variety of IR sources exist, including blackbody radiators, tungsten lamps, and silicon carbide sources, among others. When implementing optical systems testing or alignment, the sources available for use may include commercial sources with spectrometers, solar simulators, or calibration devices. These sources are usually a cheap alternative to full NIST-certified measurements. The most common sources are discussed in the following sections.

5.2.14.8 Nernst Glower

The German physicist Walther Nernst developed this source in 1897. These devices consist of a tube of rare-earth oxides and produce radiation similar to a blackbody. The typical wavelength range of operation is 2–14 μm. The Nernst glower has an emissivity on the order of 0.6 for this range. It also has the unique design characteristic of being a nonconductor when cold and a conductor when heated above 400°C. This source is useful in IR spectroscopy. A representation of the Nernst glower is shown in Figure 5.22.

Advantages of the Nernst glower over the Globar are that the Nernst glower operates in air, uses less power, and has long life. The Nernst glower is better for shorter-wavelength ranges than the Globar. Disadvantage is that it requires preheating.

EVZ-066 Nernstlampe

FIGURE 5.22
The Nernst glower. (Reprinted with permission from Wikipedia, Nernst lamp, http://en.wikipedia.org/wiki/Nernst_lamp, accessed April 1, 2014, 2013.)

5.2.14.9 Globar

Globar, a silicon carbide heated radiator, is used with spectrometers and is easily heated to operate typically around 1200–1400 K with an emissivity of 0.8 on average. For its operation, no preheating is needed. Globar is more stable than the Nernst glower but must be cooled and requires more power than the Nernst glower.

5.2.14.10 Carbon Arc

The carbon arc lamp creates a higher radiance and intensity from the interaction between two carbon electrodes. The electrodes are touched together to spark the arc, and then they are brought apart to make a small gap. The electricity vaporizes the carbon and produces an intensive white light. The carbon arc lamp is ideal for use in solar simulators because it operates in the region of 5800–6000 K. Water cooling of the electrical contacts is needed to prevent arcing.

5.2.14.11 Tungsten Lamp

The tungsten lamp is used for generating calibrated radiation in the near-IR part of the electromagnetic spectrum. For an example, the spectral outputs for tungsten lamps have a variable relative spectral power, based on the wavelength that they are propagating. Tungsten lamps radiate thermally from approximately a lower wavelength of 300 nm to an upper wavelength of 1400 nm. As such, they put out a continuous spectrum from the UV to the near IR. At approximately 900 nm, the spectral radiant emittance of the lamp's output is maximized. The tungsten lamp provides incandescent illumination by means of passing an electrical current through a small filament of tungsten material. Some common examples are the conventional light bulb or lamp. Tungsten is chosen as the base material since it is efficient as an optical radiator.

5.2.14.12 Sun

The sun provides a means for measuring irradiance, which must be error corrected for atmospheric turbulence. This source, therefore, provides an unpredictable means for emitting IR. The amount of solar irradiance that is received on the Earth's surface is dependent on various factors such as the presence of dust, particles, molecules, aerosols, haze, and clouds. The altitude of the observer and the relative angle that the sun makes with the observer are also important. Even though the sun's maximum wavelength is in the visible spectrum, it emits more energy in the IR region.

Table 5.5 shows a representative set of common types of IR radiation sources. It shows the material type, the radiation source, and the emitted wavelength.

5.2.15 Conventional Target Characteristics

Besides the IR sources mentioned earlier, there are a variety of objects/targets that provide their own thermal radiation that can be detected by IR systems. We now present a short description of some of the more useful conventional targets.

TABLE 5.5

Types of IR Sources

Type	Material	Radiation Source Example	Wavelength (μm)
	Tungsten	IR bulb	1–2.5
Thermal	Silicon carbide	Globar	1–50
Radiation	Ceramic	Nernst's glower	1–50
	Metal	Sheath heater	4–10
	Carbon	Carbon arc lamp	2–25
Cold	Mercury	Mercury lamp	0.8–25
Radiation	Xenon	Xenon lamp	
Stimulated	Carbon dioxide	CO_2 lamp	9–11
Emission	Lead compounds	PbSnTe laser	6–7

5.2.15.1 Aircraft: Turbojet

One of the first objects to be analyzed includes the turbojet engine. This source is a useful object due to the fact that it emits a large amount of radiant energy. This radiant energy is the result of the high-temperature combustion process. The classified nature of turbojet engines used in military applications cannot be discussed; however, the declassified commercial or civilian turbojet will be referenced. Various Internet sources for tutorials on jet propulsion and the evolution and design of the turbojet engine are available. Figure 5.23 provides a rudimentary understanding of the parts associated with a jet engine.

Using Internet search engines, various aircraft and missile specifications can be found, along with public releases of studies done by major corporations such as Lockheed Martin and Raytheon, on their continuing studies.

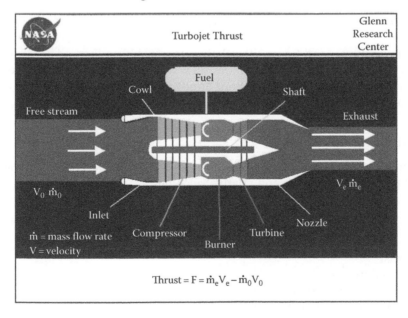

FIGURE 5.23
Basic turbojet diagram. (Reprinted with permission from NASA, Turbojet thrust, http://www.grc.nasa.gov/WWW/K-12/airplane/turbth.html, accessed April 1, 2014, 2008.)

Understanding the basic mechanics from the sources previously, the reader should note that the dominant sources of IR radiation from the turbojet engine would be from the metallic tailpipe and the hot exhaust gas stream, which is known as the plume. Due to the fact that the tailpipe is a long cylindrical shape with an aperture letting hot gases out, the tailpipe emulates a blackbody-type source. The exhaust gas temperature (EGT) is typically measured by thermocouple devices, and the effective emissivity is often assumed to be 0.9 for the turbojet tailpipe. The constant emissivity that is less than one, independent of wavelength, makes the turbojet engine similar to a gray-body device.

The combustion process yields the elements of water vapor and also carbon dioxide. These elements affect the propagation of spectral radiance as specified at 4.4 and 2.8 µm. This is valuable information for detection purposes. The following equation can be used to determine the temperature of gas after it has progressed through the exhaust nozzle and expanded:

$$T_2 = T_1 \left(\frac{P_2}{P_1} \right)^{\gamma - 1/\gamma}, \tag{5.20}$$

where
  T_2 is the temperature of the gas after it has expanded thru the exhaust nozzle (K)
  T_1 is the EGT (K)
  P_2 is the pressure of gas after it has expanded (atmospheres)
  P_1 is the pressure of gas in the tailpipe (atmospheres)
  γ is the ratio of gas heat to volume (specific heat for constant pressure and volume)

The following equation is used for determining EGT when the assumption is that the gas pressures are only expanding to ambient pressure:

$$T_2 = 0.85 \times T_1. \tag{5.21}$$

The actual atmospheric conditions with relative atomic, molecular, and aerosol concentration, as well as temperature of the gas molecules, will affect the propagation of radiation from the turbojet. More information about atmospheric effects is discussed in a subsequent Chapter 6.

5.2.15.2 Aircraft: Turbofan Engine

As we begin to study the turbofan engine, the following equation presents how to determine thrust given the mass flow and relative velocities:

$$F = m(V_e - V), \tag{5.22}$$

where
  F is the thrust
  m is the mass flow
  V_e is the velocity rearward of exhaust
  V is the velocity forward of exhaust

The thrust of a turbofan engine is derived from the turbojet portion and the huge mass of air from the fan. Figure 5.24 gives a depiction of the turbofan's internal workings. Typically, the radiation from a turbofan is less than that of a turbojet.

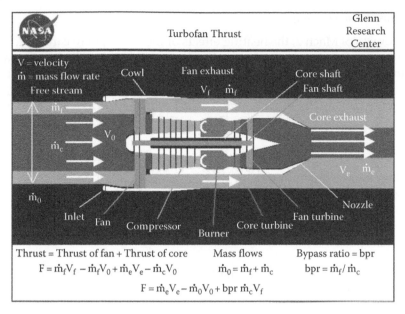

FIGURE 5.24
Basic turbofan diagram. (Reprinted with permission from NASA, Turbofan thrust, http://www.grc.nasa.gov/WWW/k-12/airplane/turbfan.html, accessed April 1, 2014, 2010.)

5.2.15.3 Aircraft: Afterburning

To give a quick definition of what afterburning is, the reader has to be familiar with the operational environment of a typical jet engine. Usually, one-third of the available oxygen is burned in the combustion chambers of the engine itself, and the other two-thirds is burned in the tailpipe. This tailpipe burn is known as the afterburn process, which aids in increasing the airplane's thrust. With the increase in the pressure ratio across the exhaust nozzle, the plume temperature decreases. For example, at Mach 3, the EGT is one-half of subsonic temperatures. Even though this temperature is decreasing, the engineer must realize that the radiation effects associated with the plane body heating increase as speed increases.

5.2.15.4 Aircraft: Ramjet

A ramjet lacks the compressor, of conventional turboprops, in front of the combustion chambers. The only compression comes from the moving ramjet engine. At Mach 2, the compression of the engine is a factor of 7; however, the engine must work to these speeds before the ramjet engine becomes effective. A combination of the turbojet and ramjet seems to be a good combination for maximum power at variable speeds. The ramjet has no moving parts, and EGT lowers at higher speed, and the same calculations can be used for the ramjet as for the turbojet.

5.2.15.5 Rockets

A rocket operates on its own fuel and oxidizer and, therefore, requires no atmospheric air. It, therefore, can operate in a vacuum such as space. The standard rocket engine setup consists of a propellant, combustion chamber, and exhaust nozzle. Rosenberg's calculations on a rocket propelled with liquid O_2 and kerosene depict a continuous radiation from the exhaust cone (Rosenberg et al. 1962). The principal source of radiation for the rocket is considered to be the plume.

5.2.15.6 Atmospheric Heating

At speeds in excess of Mach 2, the heating effects of the atmosphere can be clearly seen by an IR detector. Any point on a moving body where air stops is called a stagnation point. The following equation can calculate the temperature of these points:

$$T_S = T_O\left(1 + r\left(\frac{\gamma - 1}{2}\right)M^2\right), \tag{5.23}$$

where

T_S is the stagnation temperature (K)
T_O is the static temperature (K) at point of stagnation
r is the factor of recovery
$\gamma = 1.4$, the ratio of the specific heats of air at constant pressure and volume
M is the Mach speed

Surfaces that are good radiators show smaller evidence of heating because the majority of the heat is radiated away.

5.2.15.7 Humans

The skin on a human body averages an emissivity of approximately 0.98 and 0.99 at >4 μm wavelengths. Emissivity is not affected by skin color. The average male with a skin temperature of 32°C (305 K) has a radiant intensity of 93.5 W/sr, and clothing reduces this radiant intensity (Hudson 1969). Skin temperatures can reach as low as 0°C depending on environmental conditions.

5.2.15.8 Ground-Based Vehicles

Typical paint on surface vehicles has an emissivity of 0.85. Factors such as corrosion, dust, and deterioration increase the emissivity. Vehicle designers in military applications have well understood that the radiance in a vehicle is predominately the result of exhaust pipes, and therefore, the vehicle designers have made it a point to hide these emitters under the vehicle's body.

5.2.15.9 Stellar Objects

The apparent visual magnitude of stars is the illuminance that the star provides when the observational point of the star is outside the Earth's atmosphere. Bright stars are detected using the visible or the near-IR spectrum. Spectral radiance is evaluated in only a small number of visible stars. Only 19 have a spectral irradiance $>10^{-12}$ W/cm^2/μm. In Figure 5.25, we can see the blackbody radiation curves for various stars. These profiles were estimated from temperature and visual data (Ramsey 1962). The peaks of the respective curves were determined using Wien's displacement law.

5.2.16 Background Radiation and Clutter

When detecting objects using IR systems, it becomes very likely that there will be objects that interfere with the detection process. Some background radiation sources that can potentially interfere with the detection process include the Earth, sky, outer space, stars, and planets.

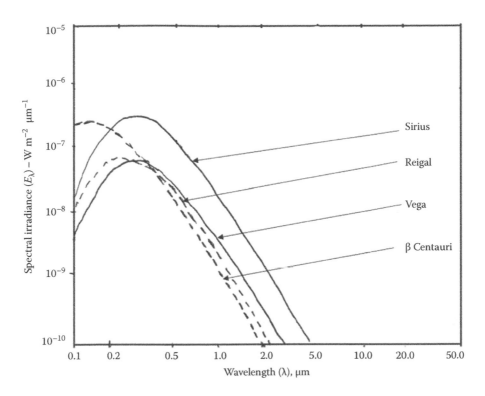

FIGURE 5.25
Calculated spectral irradiance from the brightness stars outside of the Earth's atmosphere. (Adapted from Ramsey, R.C., *Appl. Opt.*, 1(4), 465, 1962.)

Another problem is objects, other than the intended target, that radiate at the same (or near) wavelengths as the intended objects. These are known as clutter objects. Optical filtering and modulation methods can sometimes be used to mitigate some background and clutter effects.

5.2.16.1 Earth

The Earth is a source of thermal emissions, along with the surface radiation and reflected and scattered sunlight. The maximum spectral radiance of the sun is at a wavelength of about 0.5 μm. The maximum spectral radiance of the Earth is at a wavelength of 10 μm. Regardless of the terrain, whether it be snow, grass, soil, or white sand, the radiance curves hold true. The curve depicted in Figure 5.26 is the average spectral radiance of the Earth during daytime. Regardless of the type of terrain, the spectral radiance curves for other materials have a somewhat similar shape to the soil curve in Figure 5.26. The soil response is similar to a blackbody response at 35°C.

5.2.16.2 Atmosphere and Beyond

When observing the sky, the molecular structure of the atmosphere and aerosols need to be considered. The water vapor, carbon dioxide, and ozone all play an important role in affecting the emissivity of the atmosphere. We have to know the temperature of the atmosphere and elevation angles of our line of sight in order to determine the radiance of the sky. At low elevation, we see a long path through the atmosphere, and at higher elevation

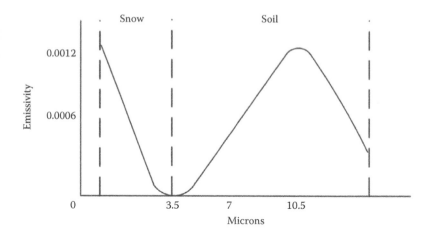

FIGURE 5.26
Representative spectral radiance of the Earth during daytime (soil at 32°C greater than 3.5 μ) and snow (descending curve from 2 to 3.5 μ). (Courtesy of Maryury Diaz.)

angles, the path is shorter and emissivity becomes low. At elevation angles that are 0°, the sky spectral radiation looks similar to that of the Earth in that it approaches a blackbody curve that peaks around a wavelength of 10 μm. However, at even a few degrees of elevation, absorption bands in the atmosphere strongly attenuate the signal between 7 and 15 μm and again at 16–20 μm.

Deep space, on the other hand, has no atmosphere. The background radiation of deep space is on the order of 3 K. Background radiation in the near-Earth environment can be substantially higher and can swing radically (e.g., 175–393 K are representative examples).

5.2.17 Emissivity of Common Materials

The emissivity of common materials has been measured and can be found in many reference books. Table 5.7 shows the emissivity of various common metallic materials. The interesting thing to note is that the surface condition is also a factor in the emission data.

Table 5.6 shows that for polished metals, the emissivity is low. However, the emissivity increases with temperature and may increase considerably with the formation of an oxide layer on the object surface.

5.2.18 Common Nonmetallic Material Emissivities

Nonmetallic materials have fairly high emissivity values. This often makes the detection of such items a very achievable measurement from distances including from space satellites. Table 5.7 shows the common nonmetallic materials found in common use in construction and industry. It shows by material type that many of these are commonly measured at values >0.9.

Table 5.7 shows that the emissivity of nonmetallic surfaces is usually >0.8 at room temperature. However, the emissivity decreases as the temperature increases.

TABLE 5.6

Emissivity of Metals and Other Oxides

Metals and Other Oxides	Condition	Emissivity
Aluminum	Polished sheet	0.05
	Sheet as received	0.09
	Anodized sheet, chromatic acid process	0.55
	Vacuum deposited	0.04
Brass	Highly polished	0.03
	Rubbed with 80-grit emery	0.2
	Oxidized	0.61
Copper	Highly polished	0.02
	Heavily oxidized	0.78
Gold	Highly polished	0.21
Iron	Cast polished	0.21
	Cast oxidized	0.64
	Sheet, heavily rusted	0.69
Nickel	Electroplated, polished	0.05
	Electroplated, not polished	0.11
	Oxidized	0.37
Silver	Polished	0.03
Stainless steel	Type 18-8, buffed	0.16
	Type 18-8, oxidized	0.85
Steel	Polished	0.07
	Oxidized	0.79
Tin	Commercial tin-plated sheet iron	0.07

Source: Data adapted from Infrared Information and Analysis Center (Ann Arbor, MI), *The Infrared and Electro-Optical Systems Handbook. Sources of Radiation*, Vol. 1, Society of Photo Optical, Bellingham, WA, http://oai.dtic.mil/oai/oai?verb=getRecord&metadataPrefix=html&identifier=ADA364020, accessed April 1, 2014, 1993.

The fact that every object emits radiation can be used in a number of useful applications. Depending on the target that needs to be detected, there are many specific applications that may generally be classified by the military. However, practical nonclassified objects abound. The radiation characteristics of these targets can be analyzed by the application of conventional radiation laws, such as Planck and Wien's laws.

5.2.19 Use Cases

For each of the following cases, the radiant emittance, the total power radiated from the object, and the wavelength at which it emits most strongly can be calculated. The formulas used are as follows:

- Radiant emittance: We will use the Stefan–Boltzmann law, Equation 5.16.
- Power radiated: We will use Equation 5.18.
- Maximum wavelength: We will use Wien's law, Equation 5.12.

TABLE 5.7

Emissivity of Nonmetallic Materials

Nonmetallic Materials	Emissivity	
Brick	Red common	0.93
Carbon	Candle soot	0.95
	Graphite, filed surface	0.98
Concrete		0.92
Glass	Polished plate	0.94
Lacquer	White	0.92
	Matte black	0.97
Oil, lubricant (thin film of nickel base)	Nickel base alone	0.05
	Oil film: 1, 2, 0.005 in.	0.27, 0.46, 0.72
	Thick coating	0.82
Paint, oil	Average of 16 colors	0.94
Paper	White bond	0.93
Plaster	Rough coat	0.91
Sand		0.90
Human skin		0.98
Soil	Dry	0.92
	Saturated with water	0.95
Water	Distilled	0.96
	Ice, smooth	0.96
	Frost crystals	0.98
	Snow	0.90
Wood	Planed oak	0.90

Source: Data adapted from Infrared Information and Analysis Center (Ann Arbor, MI), *The Infrared and Electro-Optical Systems Handbook. Sources of Radiation*, Vol. 1, Society of Photo Optical, Bellingham, WA, http://oai.dtic.mil/oai/oai?verb=getRecord& metadataPrefix=html&identifier=ADA364020, accessed April 1, 2014, 1993.

5.2.19.1 Use Case 1: The Sun

The sun's diameter is approximately 1.392×10^9 m, and its surface temperature is about 5778 K. Stars are considered to be perfect blackbodies; therefore, the epsilon value for this case is 1.

The radiant emittance is

$$W = 5.67 \times 10^{-8} \times 5778^4 = 63,196,526.546 \text{ W/m}^2.$$

The power radiated is

$$P = AesT^4 = AW = 4p\left(\frac{1.392 \times 10^9}{2}\right)^2 \times 5.67 \times 10^{-8} \times 0.5778^4 = 3.85 \times 10^{26} \text{ W}.$$

The maximum wavelength is

$$\lambda_{max} = \frac{2898 \,\mu\text{m K}}{5778 \,\text{K}} = 5.02 \times 10^{-7} \text{ m} = 502 \text{ nm}.$$

The sun's wavelength is in the visible part of the electromagnetic spectrum. The color of the sun in space is actually white; however, the sun appears yellow to those on the Earth's surface because of the absorption and scattering of light. When observing the temperature of stars, the hotter stars look blue. Cooler stars are apparently more orange and reddish in color. It is interesting to note that the solar constant in space is around 0.140 W/cm², and on the surface, this value is approximately 0.09 W/cm².

5.2.19.2 Use Case 2: Iron Furnace

In this case, we use a circular pool of molten iron in a furnace diameter of 3 m and temperature of 1800 K. The emissivity for a molten iron surface is 0.42.
The radiant emittance is

$$W = 0.42 \times 5.67 \times 10^{-8} \times 1800^4 = 249{,}989.85 \, \text{W}/\text{m}^2 = 250 \, \text{KW}/\text{m}^2.$$

The power radiated is

$$P = AesT^4 = pr^4W = p1.5^2 \times 0.42 \times 5.67 \times 10^{-8} \times 1800^4 = 1.76 \, \text{MW}.$$

The maximum wavelength is

$$I_{max} = \frac{2898 \, \mu\text{m K}}{1800 \, \text{K}} = 1.61 \times 10^{-6} \, \text{m} = 1610 \, \text{nm}.$$

The maximum wavelength is in the IR region, but it is still observable in the visible part of the spectrum.

5.2.19.3 Use Case 3: Person in the Desert

For the case of a person in the desert, assume a temperature of 305 K, 0.42 m² estimated viewable cross section seen by a border patrol, and airborne IR system at 25° to the viewable object. According to Table 5.7, the emissivity of the human skin is 0.98.

The radiant emittance is

$$W = 0.98 \times 5.67 \times 10^{-8} \times 305^4 = 480.8 \, \text{W}/\text{m}^2.$$

The power radiated is

$$P = Ae\sigma T^4 = 0.42 \times 0.98 \times 5.68 \times 10^{-8} \times 305^4 = 201.95 \, W.$$

The maximum wavelength is

$$\lambda_{max} = \frac{2898 \, \mu\text{m K}}{305 \, \text{K}} = 9.501^{-6} \, \text{m} = 9.5 \, \mu\text{m}.$$

The maximum emission wavelength of a representative human is around 9.5 μm, and it is located in the long-wave part of the IR spectrum. The human body does not give out any practical visible radiation, but human traffic can be detected during the night by using an IR system.

5.2.20 Practical Detection Applications

Nowadays, for example, EO and IR systems are being used in border patrol applications in order to track human traffic. EO systems are used for high-resolution spatial imaging during daytime, and IR systems are often used at night. Human bodies are a good IR source; the typical temperature of a human body is around 305 K. The peak spectral radiant emittance for humans is at 9.5 μm, and so detectors that are optimized for detection at/ or around this wavelength are often used.

5.2.21 Transition to a Dialog to Develop Stakeholders, Requirements, and Scope

In the next section, we will look at a typical scenario that may occur in a general company at the beginning of a project. Just as in previous chapters, we use a fictitious company called Fantastic Imaging Technologies (FIT) to illustrate the concepts discussed in the previous sections in a modeled real-world scenario. The intent is to show how the systems engineering principles developed in Section 5.1 shape the application of the technical material presented in the optical building block material in Section 5.2. We will demonstrate how the QFD method can be used in the problem definition phase to capture and understand stakeholder's needs. We will also show how the analytical methods in Section 5.2 can be used to capture CONOPS-related physical constraints and aid in the general understanding of the performance characteristics of the optical system.

5.3 Integrated Case Study: Introduction

In this integrated case study, we find FIT engaged in a variety of activities stemming from their recent DHS contract for equipping unmanned aerial vehicles (UAVs) with a state-of-the-art imaging system for 24/7 border patrol applications. As a result of recent meetings, FIT is conducting a radiometric study to determine whether there is sufficient detectable signal present to go with a passive imaging approach or an active imaging (e.g., laser-based imaging approach will be required). The optical design team, led by Tom Phelps, analyzed the Raytheon MTS-B-20 gimbal and is preparing a feasibility-level basic optics design to integrate the contracted FIT imaging systems into 21 government-furnished UAVs. The purpose of the feasibility-level optics design is to aid in radiometric throughput analysis and to provide preliminary optical systems parameters for spatial resolution, field of view, and scanning analysis and modeling. Garry Blair, FIT's chief information officer, is working with Arlene G. from business development, Jean H. from operations management, Steven E. (Steve) from IT and technical support, Marie C. from requirements management and configuration management, and Ginny R., the newly appointed program manager for the DHS effort, on initiating artifacts for the EA. The systems engineering team, lead by Karl Ben, the senior systems engineer, and Jennifer O. (Jen), one of his systems engineers, just completed a study selecting some MBSE tools. Dr. Bill Smith, the CEO, opens the weekly staff meeting.

Bill: "Good morning everybody. There has been lots of activity this week. I am looking forward to hearing what you have to say. Karl, how is that radiometric study coming?"

Karl: "We have made great progress! Tom has done a great job getting his team to put together a preliminary optics model for the visible and IR optics systems using the Raytheon MTS-B-20 gimbal system that is on the DHS UAVs. We think …"

Bill: "Wait. Sorry to interrupt, but how can we do a preliminary optics model if we don't even know if we are going to use an active or passive imaging system?"

Karl (smiles): "Ah, great question! If our radiometric study indicates that we need an active imaging system, the power delivery optics will be a separate system from the receiver optics. We are currently modeling the relay optics for the visible and infrared bands. I better let Tom tell you about it, since he has the details."

Tom: "Thanks Karl. The model we are building will be in MATLAB and Simulink® so that it can integrate with our adaptive optics simulation and compensation tools. We are linking the block diagrams in Simulink to the block diagrams that will be used in our MBSE tools. That way, when we make changes to the block diagrams during the design process, the results will be immediately observable through the executable Simulink models. At this point, we just want to implement a simple, expandable optics model to give us an idea about the basic performance parameters of the optical system and to allow us to run some *what-if* scenarios. The modular design methods we are using will allow us to adapt to the real, optimized design once we get that far."

Bill: "That sounds good! Wilford at DHS will be happy to hear that. He fully supports the model-based engineering approach and will probably want to contribute with some of the modeling parameters. By the way, for those who haven't met her, I'd like to introduce Ginny R., our program manager for this effort. Ginny came over to us from DRS, where she was running a joint DHS and DOD contract for ground-based spectral imaging for counterterrorism. She personally knows Wilford at DHS and had great success in her projects over at DRS. We are lucky to have her! She also brought with her Miss Arlene G., a feisty business development manager with leads into DHS and several of our targeted secondary stakeholders. Welcome to you both!"

Ginny: "I am very happy to be here." (Arlene waves from the opposite side of the round table.)

Bill: "Ginny, you have lead on this effort. Let me know what resources you need and keep me informed of what I need to know. I think I don't need to remind everyone that this is our top-priority program."

Ginny: "You got it Bill. You have an excellent team here, and I have already started working with many of them. For now, we need to complete that radiometric propagation analysis. Tom, how is that coming along?"

Tom: "We should have something for you next time we meet. As I said earlier, we are finishing up the basic optics model. We are implementing a standard zoom optics configuration, but there is a compact, dual field-of-view telescope approach that works in the long-wave IR as well as in the visible. This method looks like it will have some advantages in size, weight, and power, as well as having some reliability benefits. We will conduct a trade study as part of the design process. However, for now, we are modeling our conventional approach to dynamically adjusting the instantaneous field of view (IFOV)."

Bill: "Is that all you need to complete the radiometric propagation study?"

Tom: "Well, almost. Once we complete the basic optics model, we need to link it to our detector models, atmospheric turbulence simulator models, atmospheric turbulence compensation models (if we need them), and the source models. Sounds like a lot, but we are almost there. We are using a generic 2048 × 2048 pixel detector model with a 100 percent fill factor for a placeholder for the actual detectors (both visible and IR). The atmospheric turbulence simulation and compensation models already exist and can be easily accessed and integrated into our MBSE environment. We were just finishing up the source modeling just before we came to this meeting."

Bill: "OK. How are we modeling the source?"

Tom: "My team, and Jen from systems, was looking at the infrared part of the system, since the nighttime imaging system is the one that may or may not have an issue with available optical power. The daytime system will have plenty of signals since the energy from the sun and the local environment will provide plenty of power for our detectors to respond. For the IR system, we used a gray body thermal model to determine the optical power at the source. We know that the mean temperature of a healthy human is 98.6°F (or about 310 K) and the emissivity of skin is about 0.98. However, some studies have shown that equivalent skin radiating temperature in a warm climate is on the order of 32°C or 305 K. In cold climates, the skin temperature can get down to 0°C or 273 K. In the colder months, at night, the desert can get to this temperature, so we will have to run some worst-case scenarios once the models are integrated. For now, we are going with the 305 K for nominal conditions. So, using the Stefan–Boltzmann law, we get a radiant emittance of 480.8 W/m². Using an estimated cross section of a human of 0.42 m², we can determine the optical power at the source to be 201.95 W radiating in the direction of our infrared sensor on the UAV."

Bill: "How did you get that scattering cross section? I've heard that a human is closer to 2 m²."

Tom: "That is true. I have seen numbers that range from 1.86 to 2 m² for a human radiating in all directions. However, when accounting for the directionality and angle that the human makes with our airborne sensor, the actual scattering cross section is much lower than that. Our conservative estimates put it at 0.42 m²."

Bill: "That's a good point. So how does this source optical power translate to power on our detector?"

Tom: "That will require us to finish our radiometric study. We should have that for you next time we meet. We believe that this is sufficient for our detectors and this CONOPS based on previous experience for some of our commercial customers. However, we want to complete the study and make sure."

Bill: "Well, that sounds great! You all are coming along nicely. Thanks for all the hard work and let's keep it up. Since you mentioned the CONOPS, let's make sure that we have our problem definition well in hand. I will be talking with Wilford later on in the week, and I want to give him an update. Ginny, how are we doing with our problem definition artifacts?"

Ginny: "Pretty well! The technical team has done a great job in getting the technical details for this product from our primary stakeholder. Arlene is working with Garry, Jen, and Tom to develop the OV-1 and some of the basic technical and systems artifacts for our enterprise architecture. Arlene can then use this when she starts to probe the interest of some of our secondary stakeholders."

Bill: "Make sure that we have a cross-disciplinary team that includes manufacturing and support input when we start developing some of these products for our EA."

Ginny: "Good idea. Based on what we have learned so far, we generated the following OV-1 for this program. (Ginny punches up Figure 5.27, her Department of

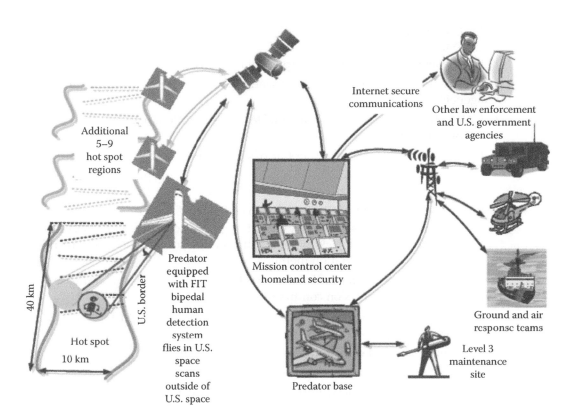

FIGURE 5.27
UAV optics system CONOPS diagram.

Defense Architectural Framework 2.0 OV-1 artifact on her MagicDraw MBSE tool, and projects it onto the screen.)

I think this CONOPS diagram captures the major actors and the overall DHS border patrol mission. Arlene, Garry, Jen, and Tom put this together after harvesting information from our stakeholders. Notice that we replaced the microwave towers for long-haul communications with a direct satellite link. During our stakeholder interviews, we found that DHS had access to this link, and it is their preferred option. We can make the following observations from the CONOPS diagram (OV-1):

- UAV optics systems monitor up to 10 hot spots along the border with the hot spots changing over time.
- Each UAV optics system is directed to a location and asked to send back imagery data.
- Each UAV optics system scans a 10 km × 10 km area for targets.
- UAV optics system detects an object of interest and sends video information back to the DHS headquarters via satellite communications (on screen at MCC within 5 seconds of candidate detection).
- MCC is in a secure, remote location.
- MCC processes the data for validity as being bipedal movement and an actual target of interest.

- MCC notifies appropriate partner agencies (e.g., border patrol agents, DEA, U.S. Marshals, FBI, CIA, DIA, law enforcement, Interpol, international cooperating agencies) of location of targets of interest for intercept/interdiction.
- MCC sends image products, reports, and EA artifacts to interested partner agencies.
- UAV optics system is maintainable at UAV base maintenance shop for repair or routine maintenance.
- UAV can be operated from either line of site or from a mission control center.
- UAV optics system can be controlled by, and send images to, the mobile command center units that can assume control locally from mobile command centers in special trucks or deployed portable command centers.
- Situational reports (as necessary) and image products (within 5 seconds of acquisition) are sent back to MCC from mobile command centers using long-haul satellite exfiltration methods.
- Mobile command center units have a real-time, direct, secure communications link with the mission command center.
- Images, control signals, and communications are encrypted with 128-bit encryption methods.
- UAVs have onboard data storage for mission critical video.
- UAVs have tamper protection for stored data.
- UAVs cannot communicate with or control other UAVs."

Bill: "Pretty good! We will have to check with Wilford on the actual number of hot spots and see if they are completely funded for 17 UAVs with 5 spares. That means 22 of our optics systems. We need to make sure they have the funding for this and that the funds are committed. Also, let's make sure on the type of UAV that they are using. I thought they were running MQ-9s. Finally, I think that they have a direct satellite communications link with some of their special customers. Right now, we are showing only a tower link. Let's check on this and update the OV-1 if needed."

Ginny: "You got it Bill. Arlene and the crew also did some great work in harvesting information about the stakeholders themselves."

(Ginny puts Table 5.8 up on the screen.)

Ginny: "As you can see, this table lists the major stakeholders, their positions, and our rankings of how important these stakeholders are to our current effort. We want to make sure that our high-priority stakeholders get a correspondingly high amount of support. That doesn't mean that we are not paying attention to the lower-priority stakeholders, but we are making sure that our high-priority stakeholders get the service and attention they deserve. To see these relationships better, Arlene and Jen came up with this onion diagram of our stakeholders (Figure 5.28)."

Ginny: "This is a visual representation of our current stakeholder list and how they relate to our effort. Notice that we have the operational environment central to

TABLE 5.8

UAV Optics System Stakeholder Matrix

ID	Category	Function	Employer	Name	Priority
1	Interfacing technology	Technical specialist	Homeland Security, U.S. Customs and Border Protection	Glen H.	Medium
2	Maintenance operator	Maintenance technician	Homeland Security, U.S. Customs and Border Protection	TBD	Medium
3	Normal operator	Dispatch authority	Homeland Security, U.S. Customs and Border Protection	Kyle, Rebecca, Andy	Medium
4	Normal operator	System operator	Homeland Security, U.S. Customs and Border Protection	Kyle N.	Medium
5	Operational support	Maintenance technician	Homeland Security, U.S. Customs and Border Protection	TBD	Low
6	Client	N/A	Homeland Security, U.S. Customs and Border Protection	Wilford Erasmus	High
7	Functional beneficiary	Patrols	Homeland Security, U.S. Customs and Border Protection	L.T. Daniels, L.T. Chang	Medium
8	Functional beneficiary	Liaison	Mexican military	Colonel Eduardo Torrez	Low
9	Functional beneficiary	Liaison	Interpol	Mike King	Low
10	Functional beneficiary	Liaison	DIA	Alex Smith	Low
11	Functional beneficiary	Liaison	U.S. Marshal	Scott Holmes	Medium
12	Functional beneficiary	Liaison	ATF	Ted Gunderson	Low
13	Functional beneficiary	Liaison	DEA	TBD	Medium
14	Functional beneficiary	Liaison	FBI	Corey Goon	Medium
15	Functional beneficiary	Liaison	CIA	TBD	Medium
16	Customer	N/A	Homeland Security, U.S. Customs and Border Protection	Wilford Erasmus	High
17	External consultants	Landowners, United States	N/A	Various	Low
18	External consultants	Landowners, Mexico	N/A	Various	Low
19	External consultants	U.S. congressmen	U.S. Congress	Various	Low
20	Negative stakeholders	Intruder	N/A	Various	Medium
21	Negative stakeholders	Competitors	Various	Various	Medium
22	Negative stakeholders	Immigration activists	Various	Various	Low
23	Core team members	CEO	FIT	Bill Smith	High
24	Core team members	Project engineer	FIT	Marie C.	High

(Continued)

TABLE 5.8 (*Continued*)

UAV Optics System Stakeholder Matrix

ID	Category	Function	Employer	Name	Priority
25	Core team members	Technical lead	FIT	Gerry Blair	Medium
26	Core team members	Development team	FIT	Gerry Blair, Marie C., Phil K., Amanda R., Ron S., Tom Phelps	Medium
27	Consultants	Imaging specialist	Optics Now Consulting	Tom Phelps	High

Sources: Data adapted from NASA, *NASA Systems Engineering Handbook*, National Aeronautics and Space Administration (NASA), Washington, DC, 2007; Campanaro, P. and Ricolfi, T., *J. Opt. Soc. Am.*, 57(1), 48, 1967.

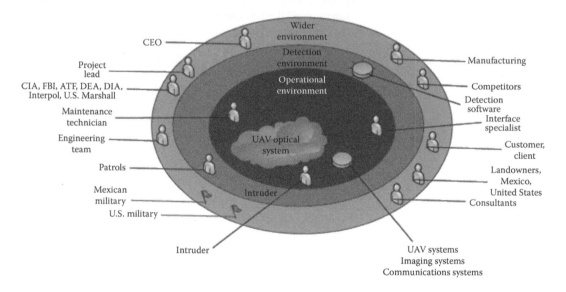

FIGURE 5.28
UAV optics system stakeholder onion diagram. (Adapted from Robertson, S., *Mastering the Requirements Process*, Addison-Wesley, Boston, MA, 2006.)

the effort. The stakeholders that most strongly relate to this environment are listed there. We also are including negative stakeholders in order to look at our approach and methods from their perspectives. They obviously want to avoid being detected, so we may need to take steps to ensure we are ready for what they do. For example, I know of a pair of professors at the university here that are looking at game theory as a decision aid for determining sensor deployments in just this kind of an application."

Bill: "Game theory! Wow, Ginny can you bring them in? This might be useful."

Ginny: "I'll arrange it, I'm sure they'd love to talk with us. Towards the outer part of the onion diagram, are shown the stakeholders that are more distant from the direct operational environment. This is a nice visual representation of our stakeholders. Next, Arlene and the systems wizards generated this stakeholder input table."

TABLE 5.9

UAV Optics System Stakeholder Input Table

Stakeholder's Needs (What)	Stakeholder Rank	Stakeholder Comments	Characteristics (How)	Technical Response
Operate from a UAV	(5) Very important	Must be from this platform.	Vibration and shock isolators	
Discern bipedal human traffic	(5) Very important	This is the mission.	Optimized proprietary algorithms	TBD km daytime target recognition TBD km nighttime target recognition
Daytime operations	(5) Very important	We have 24/7 requirement.	Visible sensor system	
Nighttime operations	(5) Very important	We have 24/7 requirement.	IR sensor system	
Real-time imagery communications	(5) Very important	5 s or less.	Image processing/ communications onboard buffering for live stream	5 s or less from capture to communications
Continuous operation	(5) Very important	24/7 We may stagger UAVs, but we need to meet our operational availability requirement.	Component selection	MTBF of TBD
Ease of maintenance	(5) Very important	We will also want to do simple maintenance.	LRU packaging	
Maintenance support	(5) Very important	Yes. For harder things. We would like on-site support, training for our maintenance techs, at our depot, and a 5-year warranty.	On-site and depot product support	On-site training 5-year warranty
Technical support	(5) Very important	Yes. Training, service, maintenance, technical questions for 3 years with optional renewals.	Help desk	24/7 access to help desk 3-year contract with optional renewal
Information assurance	(4) Somewhat important	Yes, will be levied on us in the future by our stakeholders.	Tamper protection	
User-selectable day/night mode	(4) Somewhat important	We want to be able to use either sensor at our choice. You could have an automatic feature but not at the expense of being able to select the sensors ourselves.	User interface	
User-selectable imagery mode	(4) Somewhat important		User interface	
Long-haul communications capable	(4) Somewhat important	This is a must. You could downlink to a local site and use Internet protocol to send to command center. Not sure you would have this available everywhere. You must get the images/data back to the MCC.	Satellite communications with encryption	Will use Ku because the C-band is saturated

(Continued)

TABLE 5.9 (*Continued*)

UAV Optics System Stakeholder Input Table

Stakeholder's Needs (What)	Stakeholder Rank	Stakeholder Comments	Characteristics (How)	Technical Response
Ease of installation	(4) Somewhat important	This is important. Our people will likely remove/install various optical packages.	LRU packaging	Mean time to replace = TBD
Software upgrades	(4) Somewhat important	Yes. This is important. We want to be able to bring in new algorithms as we find them/have them developed.	User-programmable SW	
Spares	(4) Somewhat important	We need to meet our availability requirement. Apart from that, we don't care.	Logistics planning	Spares per TBD
Onboard image storage	(3) Important	Don't care whether it is onboard or not as long as we have access to images.	How many bytes to hold flight imagery?	Refer to real-time imagery communications
Ease of stored image retrieval	(3) Important	This would be good. However, with training, it could be more complicated. We have smart people.	LRU interface (only matters if we have onboard storage)	Refer to real-time imagery communications
Documentation	(3) Important	We need enough to know what we need to do and troubleshoot and maintain our systems.	Hard-copy and electronic manuals	User manual, Level 1 maintenance manual, Level 2 maintenance manual

Source: Data adapted from Sapritsky, V.I. and Prokhorov, A.V., *Appl. Opt.*, 34(25), 5645, 1992.

(Ginny flips Table 5.9 on the screen.)

Ginny: "This tables shows what our stakeholders are interested in and, where possible, what metrics we are using to ensure we meet their needs. We want to make sure that we are solving the right problem and that our stakeholders get what they want. I periodically review this to make sure that we are staying on focus and addressing their needs and doable wants as much as possible. Once we had a good idea of the stakeholder needs/wants, we prepared a quality function deployment (QFD) diagram. That captures their high-level wants/needs."

(Ginny punches up Figure 5.29 on her computer and links it to the screen.)

Bill: "Ah, the good old QFD! This will be important for my talk with Wilford and for the stakeholder presentation we have in two weeks. Very nice!"

Ginny: "Thanks Bill. It was Arlene, the systems folks, and Tom and his engineering team that came up with it. Notice that the QFD gives a relative priority of each of the customer requirements. This QFD was aimed at the folks from DHS, so we are using the specific term "customer" in this case. Based on our discussions with them, we came up with the relative weights shown on the QFD. Bill, you may want to get them to split these out a little better during your presentation to them. Either that or confirm our estimates of their ratings. Right now, there

Row #	Relative weight (%)	Customer importance	Maximum relationship	Customer requirements (whats)	1 Vibration and shock isolators	2 Optimized target detection algorithms	3 Visible sensor system	4 Infrared sensor system	5 Proprietary image processing	6 Component selection	7 Product packaging	8 Product support infrastructure	9 Communications subsystem	10 Tamper protection and detection	11 Operator and user interface	Row #
1	9	9	9	Operate from a UAV	●	○	▽	▽	●	●	▽	▽	●	▽	▽	1
2	9	9	9	Bipedal human traffic detection	▽	●	●	●	●	○	▽	▽	▽	▽	▽	2
3	9	9	9	Daytime operation	▽	●	●	▽	●	○	▽	▽	▽	▽	▽	3
4	9	9	9	Nighttime operation	▽	●	▽	●	●	○	▽	▽	▽	▽	▽	4
5	9	9	9	Real-time imagery communications	▽	●	▽	▽	●	○	▽	▽	●	○	○	5
6	9	9	9	Continous operation	▽	●	○	○	●	●	▽	○	●	▽	▽	6
7	9	9	9	Ease of maintenance	▽	▽	○	○	▽	●	●	○	○	▽	○	7
8	9	9	9	Maintenance support	▽	▽	▽	▽	▽	○	○	●	▽	▽	▽	8
9	9	9	9	Technical support	▽	▽	▽	▽	▽	○	▽	●	▽	▽	▽	9
10	3	3	9	Information assurance	▽	●	○	○	●	○	▽	▽	●	●	▽	10
11	3	3	9	Day/night mode user selection	▽	●	○	○	●	▽	▽	▽	○	▽	●	11
12	3	3	9	Imagery mode user selection	▽	●	○	○	●	▽	▽	▽	○	▽	●	12
13	3	3	9	Long-haul communications capability	▽	○	▽	▽	●	○	▽	▽	●	●	▽	13
14	3	3	9	Ease of installation	▽	▽	▽	▽	▽	●	●	●	▽	▽	▽	14
15	3	3	9	Software upgrades	▽	▽	▽	▽	▽	▽	▽	●	▽	▽	●	15
16	3	3	9	Spare parts availability	▽	▽	○	○	▽	●	▽	●	▽	▽	▽	16
Relationships				Max relationship	9	9	9	9	9	9	9	9	9	9	9	
Strong ●				Technical importance rating	171	547	300	300	618	476	212	347	388	165	206	
Moderate ○				Relative weight (%)	5	15	8	8	17	13	6	9	10	4	6	
Weak ▽																

FIGURE 5.29
UAV optics system QFD base diagram. (Courtesy of Maryury Diaz.)

are too many of them that are the same to help us effectively distinguish their relative importance. Our technical team came up with the technical responses to their requirements shown as the column entries on the QFD. At the bottom of the QFD, we show an overall technical importance rating of each technical response. The team then evaluated the correlations of the technical responses with each other to show dependencies between our technical responses."

(Ginny produces Figure 5.30 on the overhead.)

Ginny: "The '+' signs indicate a positive correlation between the technical responses, and blanks show no correlation between the technical responses. For instance, the intersection of the 'infrared sensor system' and the 'optimized target detection algorithms' shows a positive correlation."

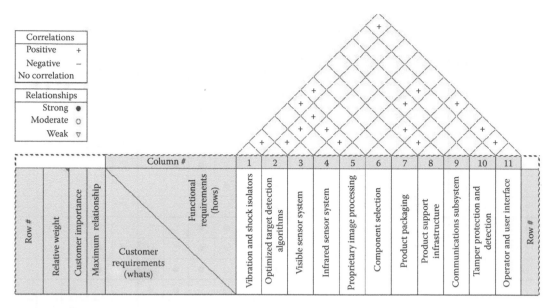

FIGURE 5.30
UAV optics system QFD "roof." (Courtesy of Maryury Diaz.)

Bill: "Ginny, this is great information. Before we move on to our next project, I would like for you all to understand the scope of this project, its importance. As you know, this is our first contract with the DHS, and this represents a major opportunity for us to get involved in providing our systems to the government and help protect our borders. Please always keep in mind that our systems will be used on the front lines and that lives may be at stake. Our mission is to *provide the world's best and most reliable imaging systems for our stakeholders and provide the best support and service for those systems throughout their operational life cycle.* This means we need to understand current and upcoming related technologies and push the envelope when it comes to integrating new capabilities. We do not cut corners, and we do not settle for second best. The scope of this project is *to provide the best possible imaging systems, image analysis software, and image products to DHS for their UAV fleet; integrate these systems and products into their UAV fleet and mission command center/mobile command centers; and provide the best service and support for these imaging systems and products throughout their operational lifetime.* We fully intend to leverage this effort to help other secondary stakeholders in the future. I would like to leave you and the senior team with the following goals and challenges:

- Give this customer a world-class product and a world-class experience throughout this development effort and over its life cycle.
- Leverage this effort into 3 major contracts with secondary stakeholders by the end of this contract.
- Generate an overseas product line within two years of contract completion.

- Achieve 15% of the U.S. aerial high-resolution imaging market within 5 years of completing this contract.
- Expand this product line to ground systems, underwater systems, and space systems with 10 years of contract completion.

I will leave it up to the executive staff to work with you to generate measureable objectives for each of these goals that will take us there. Overall, great job everyone! Keep up the good work! Let's move on to the next project."

5.A Appendix: Acronyms

°C	Degrees Celsius
°F	Degrees Fahrenheit
C-band	Portion of the microwave region, electromagnetic spectrum
CD	Compact disk
CEO	Chief executive officer
CIA	Central Intelligence Agency
CO_2	Carbon dioxide
CONOP	Concept of operations
Cr	Chromium
DHS	Department of Homeland Security
DODAF	Department of Defense Architectural Framework
DVD	Digital video disk
EGT	Exhaust gas temperature
EO/IR	Electro-optical and infrared
FBI	Federal Bureau of Investigation
FIT	Fantastic Imaging Technologies
GPPP	General-purpose parallel processing
HSBP	Homeland Security Border Patrol
IR	Infrared
JSE	Junior systems engineer
K	Kelvin
KISS	Keep it simple, stupid
Ku	Portion of the microwave region, electromagnetic spectrum
LABB	Large-area blackbody
LASER	Light amplification by stimulated emission of radiation
LRU	Line-replaceable unit
LSE	Lead systems engineer
MASER	Microwave amplification by stimulated emission of radiation
MBIR ACR	Medium background infrared absolute cryogenic radiometer
MCC	Mission control center
MQ-1	Predator UAV
NASA	National Air and Space Agency
Ni	Nickel

NIST	National Institute of Standards and Technology
OV	Operational views
QFD	Quality function deployment
RFI	Request for information
RFP	Request for proposal
SI	International System of Units (metric system)
SW	Software
T	Temperature
TBD	To be determined
TXR	Thermal-infrared transfer radiometer
UAV	Unmanned aerial vehicle
U.S.	United States
UV	Ultraviolet
WBS	Work breakdown structure

References

4C. 2006. Wien's Law of Radiation. http://en.wikipedia.org/wiki/File:Wiens_law.svg (accessed November 22, 2014).

Badiru, A. 2009. *STEP Project Management*. Boca Raton, FL: CRC Press.

Barnhart, E. 2010. Integrated mission models and simulation through the entire program life-cycle. *Military Communications Conference, 2010—MILCOM 2010*, San Jose, CA, October 31 to November 3, 2010, pp. 668–673.

Blanchard, B.S. 1998. *System Engineering Management*, 2nd Ed. Hoboken, NJ: John Wiley and Sons.

Blanchard. B.S. and W.J. Fabrycky. 2011. *Systems Engineering and Analysis*, 4th Ed. Upper Saddle River, NJ: Pearson Prentice Hall.

Boardman, J. 2008. *Systems Thinking*. Boca Raton, FL: CRC Press.

Campanaro, P. and T. Ricolfi. 1967. New determination of the total normal emissivity of cylindrical and conical cavities (1917–1983). *Journal of the Optical Society of America*, 57(1): 48–50.

Carmichael, D. 2003. *Project Management Framework*. Boca Raton, FL: Taylor & Francis.

Cox, C. 2001. *Manufacturing Handbook of Best Practices*. Boca Raton, FL: CRC Press.

Cristiano, J. 2001. Key factors in the successful application of quality function deployment (QFD). *IEEE Transactions on Engineering Management*, 48(1): 81–95.

Desai, A. 2008. Engineering course design based on quality function deployment (QFD) principles. *38th Annual, Frontiers in Education Conference, 2008—FIE 2008*. Saratoga Springs, NY, pp. T2G-17–T2G-21.

De Vos, J.C. 1954. Evaluations of the quality of a blackbody. *Physica*, 20(7–12): 669–689.

Enos, D. 2007. *Performance Improvement: Making It Happen*, 2nd edn. Boca Raton, FL: Auerbach Publications.

Firesmith, D. 2008. *The Method of Framework for Engineering System Architectures*. Boca Raton, FL: Auerbach Publications.

Fowler, J.B. and D. Gyula. 1995. High accuracy measurement of aperture area relative to a standard known aperture. *Journal of Research of the National Institute of Standards and Technology*, 100: 277–283. http://www.nist.gov/calibrations/upload/100-3-95.pdf (accessed November 22, 2014).

Francheschini, F. 2001. *Advanced Quality Function Deployment*. Boca Raton, FL: CRC Press.

Gibson, C.E. 2010. Facility for spectroradiometric calibrations (FASCAL). Physics Laboratory—Optical Technology Division, National Institute of Standards and Technology, Gaithersburg, MD. http://www.nist.gov/pml/div685/grp01/spectroradiometry_fascal.cfm (accessed April 1, 2014).

Hanssen, L. 2004. Infrared blackbody spectral characterization. Physics Laboratory—Optical Technology Division, National Institute of Standards and Technology, Gaithersburg, MD. http://www.nist.gov/publication-portal.cfm?authorLastName=hanssen&dateTo=1/1/2007& page=3 (accessed November 22, 2014).

Hanssen, L. 2006. Monte Carlo modeling of effective emissivities of blackbody radiators. Physics Laboratory—Optical Technology Division, National Institute of Standards and Technology, Gaithersburg, MD (accessed May 30, 2007).

Herzwurm, G. 2003. QFD for customer-focused requirements engineering. *Proceedings of 11th IEEE International Requirements Engineering Conference*, Washington, DC, September 8–12, 2003, pp. 330–338.

Hudson, R.D. 1969. *Infrared System Engineering*. New York: John Wiley & Sons.

Infrared Information and Analysis Center (Ann Arbor, MI). 1993. *The Infrared and Electro-Optical Systems Handbook. Sources of Radiation*. Vol. 1. Bellingham, WA: Society of Photo Optical. http://oai.dtic.mil/oai/oai?verb=getRecord&metadataPrefix=html&identifier=ADA364020 (accessed April 1, 2014).

Johnson, B.C., G. Machin, C. Gibson, and R.L. Rusby. 1994. Intercomparison of the ITS-90 radiance temperature scales of the National Physical Laboratory (U.K.) and the National Institute of Standards and Technology. *Journal of Research of the National Institute of Standards and Technology*, 99: 731–736.

Johnson, C. and J. Rice. 2006a. Active cavity absolute radiometer based on high-temperature superconductors for MBIR calibrations. Physics Laboratory—Optical Technology Division, National Institute of Standards and Technology, Gaithersburg, MD.

Johnson, C. and J. Rice. 2006b. Thermal infrared transfer radiometer (TXR). Physics Laboratory—Optical Technology Division, National Institute of Standards and Technology, Gaithersburg, MD. http://citeseerx.ist.psu.edu/viewdoc/download?doi=10.1.1.169.7456&rep=rep1&type=pdf (accessed April 25, 2014).

Johnson, C. and J. Rice. 2006c. Large-area blackbody source (LABB). Physics Laboratory—Optical Technology Division, National Institute of Standards and Technology.

Kelly, F.J. and D.G. Moore. 1965. A test of analytical expressions for the thermal emissivity of shallow cylindrical cavities. *Applied Optics*, 4(1): 31–40.

Kule, D. 2010. Black-body radiation. http://en.wikipedia.org/wiki/Black-body_radiation (accessed April 1, 2014).

Larason, T.C. and M. Houston. 2008. *Spectroradiometric Detector Measurements: Ultraviolet, Visible, and Near-Infrared Detectors for Spectral Power*. NIST Special Publication. Gaithersburg, MD: National Institute of Standards and Technology. http://www.nist.gov/calibrations/upload/sp250-41a.pdf (accessed April 1, 2014).

Lighting. 2014. http://en.wikipedia.org/wiki/List_of_light_sources (accessed July 22, 2014).

Lumen Dynamics. 2014. UV product literature. http://www.ldgi.com/technology-learning-center/led/ (accessed July 22, 2014).

NASA. 2007. *NASA Systems Engineering Handbook*. Washington, DC: National Aeronautics and Space Administration (NASA).

NASA. 2008. Turbojet thrust. http://www.grc.nasa.gov/WWW/K-12/airplane/turbth.html (accessed April 1, 2014).

NASA. 2010. Turbofan thrust. http://www.grc.nasa.gov/WWW/k-12/airplane/turbfan.html (accessed April 1, 2014).

National Telecommunications and Information Administration. 1996. Department of Commerce. Telecommunications: Glossary of Telecommunication Terms. Federal Standard 1037C. August 7, 1996. http://www.everyspec.com/FED-STD/download.php?spec=FED-STD-1037C.004685.pdf (accessed April 1, 2014).

NIST. 2002. High-T_C SACR. http://www.physics.nist.gov/TechAct.98/Div844/div844h.html (accessed April 1, 2014).

NIST. 2010a. Low background infrared (LBIR) facility. http://www.nist.gov/pml/div685/grp04/absolute_facility.cfm (accessed April 1, 2014).

NIST. 2010b. TXR optical layout. http://www.nist.gov/pml/div685/grp04/transfer_txr.cfm (accessed April 1, 2014).

Pacheco, C. 2008. Stakeholder identification methods in software requirement: Empirical findings derived from a systematic review. *The Third International Conference on Software Engineering Advances—ICSEA '08*, Sliema, Malta, October 26–31, 2008, pp. 472–477.

Pisulla, D., G. Seckmeyer, R.R. Cordero, M. Blumthaler, B. Schallhart, A. Webb, R. Kift et al. 2009. Comparison of atmospheric spectral radiance measurements from five independently calibrated systems. http://www.ndsc.ncep.noaa.gov/UVSpect_web/SkyRadiance/Radiance_final.pdf (accessed April 25, 2014).

Pivovonsky, M., M. Nagel, and S. Ballard, 1961. *Tables of Black body Radiation Functions*. MacMillan Monographs in Applied Optics. London, U.K.: MacMillan.

Plyler, E.K. 1948. Infrared radiation from a Bunsen flame. *Journal of Research of the National Bureau of Standards*, 40: 113.

Poynting, J. and J. Thomson. 1906. *A Textbook of Physics*. London, U.K.: Charles Griffin & Company.

Prokhorov, A.V., L.M. Hanssen, and S.N. Mekhontsev. 2010. Calculation of the radiation characteristics of blackbody radiation sources. In: *Radiometric Temperature Measurements: I. Fundamentals*, Z.M. Zhang, B.K. Tsai, and G. Machin, (eds.), Oxford, U.K.: Academic Press, 42: 181–240.

Rabinowitz, P. 2013. Community tool box. http://ctb.ku.edu/en/tablecontents/chapter7_section8_main.aspx (accessed April 1, 2014).

Ramsey, R.C. 1962. Spectral irradiance from stars and planets, above the atmosphere from 0.1 to 100.0 microns. *Applied Optics*, 1(4): 465–471.

Robertson, S. 2006. *Mastering the Requirements Process*. Boston, MA: Addison-Wesley.

Ronan, P. 2007. EM spectrum. http://en.wikipedia.org/wiki/File:EM_spectrum.svg (accessed April 1, 2014).

Rosenberg, N.W., W.M. Hamilton, and D.J. Lovell. 1962. Rocket exhaust radiation measurements in the upper atmosphere. *Applied Optics*, 1: 115.

Sapritsky, V.I. and A.V. Prokhorov. 1992. Calculation of the effective emissivities of specular-diffuse cavities by the Monte Carlo method. *Applied Optics*, 34(25): 5645–5652.

Sch. 2006. BlackbodySpectrum loglog. http://commons.wikimedia.org/wiki/File:Blackbody Spectrum_loglog_150dpi_de.png (accessed April 1, 2014).

Sharp, H., A. Finkelstein, and G. Galal. 1999. Stakeholder identification in the requirements engineering process. http://eprints.ucl.ac.uk/744/1/1.7_stake.pdf (accessed April 1, 2014).

Sparrow, E.M. and V.K. Johnson. 1962. Absorption and emission characteristics of diffuse spherical enclosures. *Journal of Heat Transfer*, 84(2): 188–189.

U.S. Department of Defense. 2013. OV-1: High level operational concept graphic. http://dodcio.defense.gov/dodaf20/dodaf20_ov1.aspx (accessed April 1, 2014).

Wikipedia. 2013. Nernst lamp. http://en.wikipedia.org/wiki/Nernst_lamp (accessed April 1, 2014).

Yang, K. 2003. *Design for Six Sigma for Service*. New York, NY: McGraw Hill.

6

Feasibility Studies, Trade Studies, and Alternative Analysis

> Difficult to see. Always in motion is the future.
>
> —**Yoda,** *Star Wars Episode V: The Empire Strikes Back*

This section presents a survey of the types of analyses that systems engineers and design teams perform in support of designing new systems, products, or services. Feasibility studies, trade studies, and alternative analyses are highlighted in this chapter. These are the integral parts of the systems engineering (SE) approach to undertaking complex, costly, and risky projects through the systems development life cycle (SDLC).

Validating requirements with feasibility studies provides the design team with the information needed to select viable approaches, alternatives, or methods to fulfill them. Through feasibility studies, the design team can determine if an approach to a requirement is practical and realistic, where trade studies and analyses of alternatives focus on ensuring that the system attributes meet stakeholder's needs/requirements. Alternative analysis provides a rigorous, objective, and justifiable methodology for deciding among mutually exclusive technical approaches. This is an iterative process and is executed until the best choice is identified and therefore accepted and implemented. Trade studies provide a similar structured decision-making process for optimizing the system configuration and choosing the best allocation of resources used in the development of the system under design. A common attribute of the methods in this section is that they respond to and are guided by stakeholder's needs, stakeholder input, and the consideration of all aspects of the systems life cycle.

It is recommended that we know whether the system is feasible and optimally configured well in advance of actually producing the system. For example, one of the first questions of feasibility for an optical system is concerned with the actual detection and delivery of the image data. If the optical system cannot provide images and/or video to the stakeholder(s), the system cannot perform its intended mission. Additionally, when a design team ignores the SE approach and dives into the detailed design of system components right away, they may succeed at designing an excellent communication subsystem. However, without the benefit of a feasibility study, the result may be subtle and expensive errors upon integrating the subsystems, such as the communication system not being able to read and convert the image data format into a signal or the overall data transmission rate of the system being severely throttled at the signal processing subsystem and communication subsystems interface. Although implementing these analyses in the systems life cycle can take some time and money up front, it is a small fraction of the cost and schedule delay that it can prevent.

Several preparatory discussions are in order to develop the concepts behind these beneficial analyses. First, the concept of feasibility is discussed immediately following the introduction, and then the inputs, outputs, methods, and goals of feasibility studies are elicited in the section afterward. Afterward, trade studies and alternative analysis are presented as fundamental decision-making processes in the design (and other) phases of the SDLC. Next, the analytic hierarchy process (AHP) is discussed as a method that is adaptable for use in both analysis of alternatives (AoA) and trade studies. Then the discussion about the use of sensitivity analysis in an AHP alternative analysis method is extended to a general discussion about the link between feasibility and risk. Afterward, integrated product teams (IPTs) are discussed along with how such teams might relate to the purpose of feasibility and trade studies. Finally, a case study demonstrates how an SE team might tackle critical design decisions using the tools presented in this chapter.

6.1 Understanding What Is Feasible

Feasibility and risk go hand in hand. The more risk involved, the less feasible the goal. Less risk can also mean a more achievable goal. Whenever we assess the feasibility of a task, we are actually performing risk analysis. Feasibility, or risk, essentially defines the objectives that can be achieved within the constraints and restrictions that have been identified by the stakeholder.

6.1.1 Risk Categories

Stakeholder constraints and restrictions are often proportional to their tolerance for risk. For example, a risk-averse stakeholder would often choose a well-established and mature technology over a potentially better-performing technology that has not been proven or completely tested. A physical example would be choosing a well-established and tested signal processing board in a sensor application that provides the capability of detecting a required target versus a new signal processing board that just hit the market that *could* detect the required target but also other targets of interest (TOIs). While tolerance for risk plays a key role in establishing constraints and restrictions, the identified risks can often be classified into the following useful categories: (1) technical, (2) economical/financial, (3) organizational, (4) operational, (5) motivational, and (6) managerial/political.

Technical risk is based on the known limits of the technology that is to be applied to the product. For example, if a product is designed using existing components and processes, the risk is lower than a product that uses newly developed or unproven components and processes.

Economical/financial risk is based on the estimated cost that is required to develop, manufacture, and perhaps support a product. Cost risk can be related to time, materials, labor, safety, occupational, and other identified risks. Economical/financial risk can also draw a direct correlation to the technical risk. Products that require higher technical risk may also have greater economical/financial risk.

Organizational and operational risks can sometimes be related. According to a member of a traditionally risk-averse industry—the banking industry—organizational risk is "The risk of loss resulting from inadequate or failed internal processes, people and systems, or from both, internal or external events" (Black Sea Trade and Development Bank 2014). From an SE perspective, organizational and operational risks may not always be apparent or identified in the initial stakeholder requirements; however, these risks should become apparent after the risk assessment and quantification process has been completed. Risk assessment and quantification can be performed using tools such as failure mode and effects analysis (FMEA), work breakdown structures, and conducting a life-cycle cost analysis. Reliability engineering, safety engineering, and quality engineering use FMEAs to determine, document, and mitigate system points of failure. Motivational and schedule risks are used to determine if a project is feasible based on the support for the project (e.g., willingness/commitment to do the project) and the time (or estimated time) that it will take to complete it.

Managerial/political risks are often difficult to quantify and are sometimes quite nebulous. For example, something that is technically achievable may not be organizationally or politically correct. Just because the product can be specified, designed, and produced does not mean that it will be successful or accepted. Examples are projects that have severe environmental impact or are harmful to society. While these projects may be effective and achieve their end goal, the organizational and political risks may be too great for their development and production to be supported.

6.1.2 Uses of Feasibility Studies

The definition of feasibility studies varies between sources; however, there are a few common defining characteristics. According to Blanchard and Fabrycky (2011), feasibility studies are used to determine whether the proposed systems concept is physically feasible and realizable within the required schedule based on the resources that are available. Therefore, based on their specific stakeholder requirements, the feasibility study can specifically respond to each stakeholder requirement and identify whether or not their needs, wants, and constraints are realizable within the given schedule and with the provided resources. Later in the SE process, refined feasibility studies might respond to system, subsystem, or component-level requirements.

Feasibility studies can also be used to demonstrate that a potential solution to a requirement is achievable (design feasibility). Feasibility cannot be determined unless it can be demonstrated. A feasibility study should be capable of producing a measureable assessment of whether or not a potential concept is feasible.

The feasibility study can be used to realize alternative system designs or technological approaches. If a systems concept, components, or operational functions are not well defined before feasibility studies are conducted, a feasibility study can be used by the development team so that alternative conceptual designs, operational functions, system components, and technology choices can be identified. Trade-off studies can then be accomplished to choose between the identified alternatives.

6.1.3 Application of the Feasibility Study

Let us assume that we have a stakeholder that needs a visible light optical system (VLOS). The stakeholder's need for the VLOS is translated into a set of stakeholder requirements, which are in the form of a series of short, measurable statements. These stakeholder requirements are then used as inputs to the system requirements analysis process. Based on the stakeholder requirements, the systems engineer might consider the following hardware and software for producing high-quality images: aberration correcting components such as wavefront sensors and deformable mirrors, image processing hardware and software, techniques targeted at adaptive optics (AO), and/or atmospheric turbulence compensation (ATC) methods. The life-cycle considerations of these proposed technical approaches also need to be considered. Relying on bigger, better, and more precise mirrors, lenses, and sensors will likely identify maintainability, predictability, supportability, disposability, and component supply issues. However, reliance on immature image processing algorithms that require specific optical data may identify other issues that can impact the development schedule and technology integration risks. A new software user interface may also raise end-user human factor issues.

A feasibility study may focus on the technical aspects of a set of requirements, such as establishing the required spatial resolution for an optical system for an unmanned aerial vehicle (UAV) platform. The output of the study could identify some high-risk technical findings that imply a higher probability of schedule and cost risk. A similar feasibility study might have been applied to several different design concepts. These may include different platforms for the optical system or even using human eyes on the ground to provide the target information.

The findings from the initial feasibility studies will continue to propagate directly and indirectly through the project's life cycle. As the conceptual design phase progresses, feasibility studies conducted on stakeholder requirements both aid the system's conceptual design process and mature the system requirements.

The initial stakeholder feasibility studies are extremely important in ensuring that the project goals are within the stakeholder's parameters and risk tolerance. They continue to have indirect influence on studies and analyses such as refined feasibility studies, alternative analysis, and trade studies by providing the baseline feasibility information.

6.1.4 Need for Feasibility Studies

So, why are feasibility studies needed? The simple answer is that systems engineers use this method to determine whether or not some aspect of the system is doable. Let us consider a system engineer working on a new system. Before the stakeholder will be willing to invest significant time and resources into the detailed design and manufacturing plan, it is the system engineer's responsibility to prove that the systems development effort is doable.

A new system can mean several things; all or some of a system's proposed components might need to be developed. It could also mean that all of the components could already exist and could be arranged into a new configuration that fulfills the stakeholder's needs. For example, it may be found that a commercial off-the-shelf (COTS) image processing algorithm, a COTS laptop, and a COTS camera could be combined to meet a stakeholder's surveillance and facial recognition needs. This concept allows the systems engineer to reshape existing technologies, ideas, and products to form a new system. Proven technology can provide a higher degree of feasibility and a lower degree of risk.

On the other hand, implementing a new system could mean the difficult task of creating a completely original idea based on an untested concepts and configurations. Developing a system this way is very risky in most aspects: schedule, cost, technology risk, operational, organizational, and even political. This type of project is usually only feasible if the stakeholder's motivation is high, the stakeholder needs it, resources support it, and there is no other cost-effective way to achieve the results.

Figure 6.1 shows some high-level, sequential activities that occur throughout the SDLC. This figure uses an information system as an example and focuses on macroscopic activities through the life cycle. Due to their importance and propagating effects throughout the systems life cycle, feasibility studies need to be initiated in the early stages of the systems life cycle. Figure 6.1 shows feasibility studies squarely planted in the systems concept development activity. In reality, they occur in one form or another throughout the SDLC. The estimates of costs and risks attached to requirements are important outcomes from

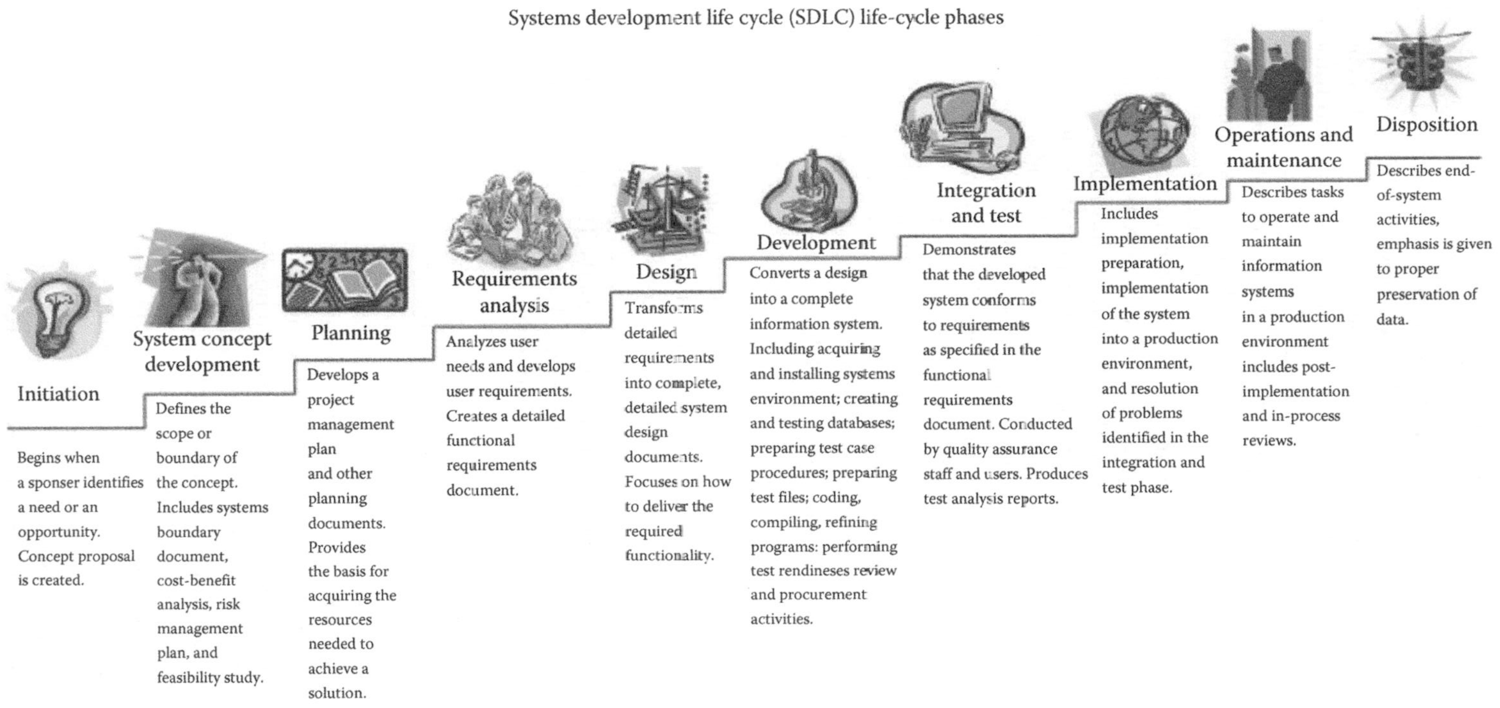

FIGURE 6.1
Systems development lifecycle (SDLC), Lifecycle Phases. (Obtained from U.S. Department of Justice, Systems development lifecycle (SDLC), lifecycle phases, 2003, accessed November 24, 2014.)

the initial feasibility studies, and they need to be developed in enough detail for the stakeholder to logically determine whether the project should proceed or stop.

If it is determined that the benefits of developing/fielding a better optical system outweigh the costs (usually evaluated at milestone events like the System Requirements Review, Preliminary Design Review, Critical Design Review, Functional Configuration Audit, or Physical Configuration Audit), then the systems development is continued.

It is critical to conduct risk assessment early rather than later to avoid unnecessary cost. The longer it takes to identify the risk, the greater the possibility of missing scheduled deadlines and increasing the cost of the project. Early feasibility studies also allow time to identify multiple solutions in the trade space to select the optimum solution for the stakeholder's requirements. Cost, schedule, and technical risk estimates associated with each individual systems concept can be primary drivers in comparing and choosing between competing systems concepts. At the very least, the results of feasibility studies produce the initial *pros* and *cons* that are fed into the trade-off analysis and AoA.

Communicating the results of feasibility studies can also provide useful feedback to the primary stakeholders regarding their own perceived needs and wants before the AoA is concluded. The possibility that a function or design feature deemed necessary by the stakeholder may in reality not address the problem the system is intended to address (Blanchard and Fabrycky 2011). If the development team properly addresses the scope of the stakeholder's problem, feasibility studies and functional analyses can uncover logical inconsistencies between true operational needs and stated needs.

A system might not be physically or technologically possible whatsoever, and only a future scientific breakthrough would enable it. Ultimately, a system is feasible because it can be designed and implemented to achieve the system requirements and satisfy the project goals.

The responsibility of performing and validating a feasibility study may fall upon the system procurer or the system provider depending on the situation. An intentionally nameless city provides a case study where the responsibility and consequences of an inadequate feasibility study fell upon the public-sector procuring agent. The goal of the city was to increase revenues by increasing tourist traffic during the early 2000s. The system that the city chose to meet this financial requirement was a ferry system. The thought was that creating a ferry system would draw more tourists to the city and, even with the new expense of the ferry system, would provide increased revenues for the city. Before performing any research or feasibility studies on whether or not a ferry would attract tourists, city personnel forged ahead, procured the ferry, and ordered the work on the infrastructure that was needed to support the system. After the ferry system was installed and operational, it failed to produce the intended financial results. Additionally, no sensitivity analysis was conducted on how the financial success of the ferry would be dependent on fuel prices. Fuel prices rose, and the ferry was docked. In this situation, the stakeholder names a technical solution as a need, even though it goes against the actual operational need (Blanchard and Fabrycky 2011).

Whether or not the system procurer participates in the conceptual design, the stakeholder may demand specific methods be used to meet their needs. For example, a robust SE conceptual design process should identify multiple potential system solutions. Stakeholder feedback is critical in narrowing down the choices to the system's design, capabilities, and functionality to meet their needs. The SE team should present multiple potential options for the systems development, along with a recommended preliminary system configuration. If potential alternatives exist, the risk of one approach failing is mitigated by other outlets. Additionally, this reduces the chance that a better configuration is excluded from consideration.

The set of alternatives proposed to the stakeholder must be reduced to those that are both feasible and credible. This means showing a low risk of failing technical, cost, and schedule requirements. The recommendation should be presented along with a justification of how the conclusion was made. Citing technical performance measures (TPM), other metrics, and the results of trade studies and AoA may accomplish the justification. Based on the overall results of the feasibility studies and an alternative analysis, a case may be made for, or against, previously demanded system functions. On the other end of the spectrum, a case may be made that identifies additional and unforeseen capabilities that should be incorporated as operational requirements.

Feasibility studies, AoA, and trade studies ultimately add specificity and focus to any type of proposed systems concept. The system may be a nationwide telecommunications network, a highway system and all of its supporting infrastructure and maintenance operations, a single orbiting telescope, or a factory that produces consumer products. Very different systems in form, function, and scope, however, all need to be analyzed for feasibility and the associated risk.

6.1.5 Establishing the Feasibility of Requirements

The earlier section described the inputs, outputs, and the general merits of the feasibility study. This section addresses the details of carrying out a feasibility study, by which the feasibility of requirements is established. Some requirements may need a short and simple assessment for feasibility. More complicated requirements that contain a large degree of uncertainty may require extended testing, research, and in some cases subject matter experts to establish feasibility. No matter the nature of the requirement, technical, economic, organizational, schedule, risk, political, and operational factors must be properly considered to anticipate and respond to the possible obstacles and problems in the systems development effort. In addition, at the beginning of the project's life cycle, the results of feasibility studies provide insights into how the system may meet stakeholder's needs.

Feasibility studies must point to how the system will function. This makes sense, as the systems engineer cannot claim a requirement that can be met without demonstrating the method by which it can be met. Depending on the progress of design, several alternative designs with defined operational functions may already be available. The job of the feasibility study in this situation may be to determine how well the system will function. "Design Dependent Parameters (DDP) are attributes and/or characteristics that are inherent to the design, for which predicted or estimated measures are required, or desired (e.g. design life, weight, reliability, predictability, maintainability, and others). The feasibility analysis is then focused largely on determining values for cost and effectiveness measures and benchmarks in terms of DDP for each design being considered" (Blanchard and Fabrycky 2011).

Note the subtle shift of the term feasibility study to the term feasibility analysis. Some sources will use these interchangeably or just generally lump feasibility concerns into the *design* activities (Blanchard and Fabrycky 2011). The use of the word analysis instead of study generally indicates that numerical results are generated for physical design attributes. Some of these numbers may be derived statistics such as availability, flexibility, and supportability. The complements to derived statistics are more physical and performance-oriented metrics, commonly referred to as figures of merit (FOMs), such as weight and standoff distance that may be derived from expected DDP values.

The feasibility analysis focuses on system effectiveness, cost, and a spectrum of risk parameters for conceptually developed designs. If the term feasibility study is used, it might inclusively refer to a more holistic role in design synthesis beginning with customer operational requirements, brainstorming, research, and consulting in-house experts. No matter how it is semantically stated, the conceptual design process needs to come up with at least one design that can feasibly address stakeholder's needs to move to preliminary design. There should be compelling numerical or categorical results that support feasibility arguments. All requirements must be evaluated, and several design concepts should be analyzed.

The methodology by which a feasibility analysis produces these metrics is dependent on the individual situation. Models of system functions based on physical laws may be employed to assess basic achievability. Technical experts, journals, and organizational lesson–learned databases might be consulted to address technological feasibility. Technological feasibility analyses may cite benchmark TPMs for technology that will likely become system components, and a broader assessment may be made about the technology's maturity and adaptability to the proposed system. A categorical assessment of technology's maturity may contribute to a probabilistic model of schedule, cost, and technical risk. The same type of holistic feasibility to quantitative risk factor translation may be performed to estimate organizational risk in the pursuit of requirements where experience and flexibility count. A design team and their organization should realistically look at the risk of working on completely unfamiliar functionalities. Economic feasibility may involve referencing economic databases to estimate costs of labor and materials. If a particular component is critical for a design, suppliers should be analyzed and an economic model to assess market sensitivity may be warranted.

Papers published by the International Council on Systems Engineering (INCOSE) community increasingly emphasize simulation and modeling for determining system feasibility before it is designed and built in detail. Furthermore, simulation can apply to all aspects of the systems life cycle. Powerful tools such as Simulink® or MATLAB® can perform numerical simulation and directly handle and analyze signal output from electronic components to garner performance measures. System integration can be modeled in terms of component interface interactions, such as passing data and the effects of control and feedback. A Monte Carlo simulation may be called upon to estimate the effects of potential schedule and cost failures by drawing random results from the individually varying activities of design, prototyping, testing, training, and manufacturing. Queuing simulation may be used to assess the maintenance turnaround time to check if the maintenance concept is sufficient for the stakeholder's support needs. Later on in the project's life cycle, these models may be extremely important for alternative analysis and trade studies.

No matter how they are garnered, DDP and other results of feasibility analyses are extremely useful for comparison as long as they are translated into common FOMs that correspond to operational requirements. A design team may be able to declare a potential design infeasible if an estimated FOM is below its requirements-driven threshold value. However, such decisions may not be so clear-cut during the initial conceptual design, so usually multiple candidate designs or configurations press onward for alternative analysis.

FOMs and other derived system life-cycle metrics can enable design comparison in a more focused and justifiable way. Complicated systems and conflicting operational needs can make distinguishing the best system extremely difficult. No one particular system may be the best in all categories assessed. For decision making, trade studies and AoAs are called upon to complement feasibility studies. The metrics derived from the feasibility studies and analyses act as vital input into the process. Of course, numbers are not everything. Stakeholder input, operational priorities, and decisions, aided by a numerically supported recommendation, are important factors to consider.

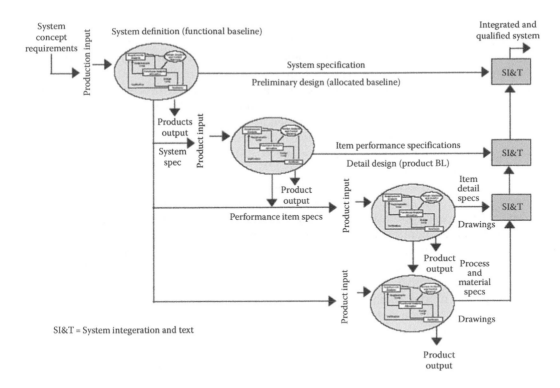

FIGURE 6.2
Specification and levels of development. (Obtained from Systems Engineering Fundamentals, Defense Acquisition University, http://en.wikipedia.org/wiki/File:Specification_and_Levels_of_Development.jpg, 2007, accessed November 25, 2014.)

Much has been discussed about some of the internal considerations and processes of analyzing designs for feasibility. During the conceptual phase, there may be no design yet to analyze. At such a stage, the feasibility study must identify functionalities, operating aspects, components, and possible configurations of the system. This is a high-level synthesis activity that is directly bridging the stakeholder's needs and operational requirements to a feasible design.

Figure 6.2 captures the overall structure of the SE development process. Multiple phases and subphases of the systems development process are shown in the figure, and they take the form of circuits between the system definition (SD) circle, the preliminary design (PD) circle, and the detailed design (DD) circle. The initial stakeholder requirements are part of the systems concept requirements activities and are used to develop the functional baseline (e.g., the system specification aka A spec) in the top-left circle. Feasibility studies play prominently in the evaluation of the stakeholder requirements and also the requirements produced as major outputs of the circles in Figure 6.2. Feasibility studies also have a dominant role in forming and evaluating potential designs and design alternatives. Specifically, the feasibility studies will output DDPs, provide models to translate the DDPs into FOMs, and provide benchmarks and necessary component interface relationship information that will be used for testing and integration. As stated earlier, this information is the basis for narrowing the alternative solutions within a particular circle. As the system moves into later phases with more detailed requirements captured in the PD and DD circles, all these activities continue to operate in the safe space established by feasibility studies.

A single feasibility study does not alone determine all elements of a system design. All of the feasibility studies in response to all requirements, stakeholder input, and other concurrent design processes produce knowledge that the systems engineer synthesizes to increment the system design. The system definition box of Figure 6.2, in which feasibility analyses play a prominent part, represents the important processes and activities that must be accomplished in order to increment the system design from the conceptual design to preliminary design stage. Altogether, feasibility studies must point to a consistent set of system features. Furthermore, feasibility studies should point to several possible self-consistent systems concepts and designs. Feasibility studies should also lay down the foundation for comparing those systems. The comparison of multiple systems and the optimization of the system that survives the AoA is where trade studies take hold.

The highest level of attainment for a feasibility study is demonstration. Demonstration in proxy of the real system is the driving purpose behind estimation, simulation, and modeling activities mentioned earlier. Only for a truly trivial requirement, feasibility may be demonstrated with a risk analysis that shows trivial risk by means of low chance of failure, low impact of failure, and/or many alternatives in the case of failure. For less certain requirements that imply highly technical applications, a computer simulation, prototype testing, physical model, mathematical models, and/or industry and economic data may be used to show that concept and technology can be combined to produce results within the range expected by the stakeholder. The feasibility study that commands the most confidence is the one that shows the requirement or functionality successfully implemented and operating in the environment the stakeholder intends. Section 50.2.10.2 of Mil-STD-882B cites demonstrated COTS technology as a standard for a low risk factor (Department of Defense 1984).

An example of the best kind of feasibility demonstration is as follows. Imagine a news organization, CBS News, as the stakeholder. CBS is interested in getting real-time video of breaking stories, and they want to exploit civilian remote-controlled aircraft to hover overhead to do so. Many types of feasibility studies could analyze their requirements of having robust, easy-to-use, and good-image-quality-producing equipment on a journalism budget. The best and quickest feasibility demonstration would point to the existence of robust and ruggedized UAV optical systems in addition to available COTS video processing and data transmitting hardware and software. The company employing the system engineer who uses this strategy would have a competitive advantage in the bidding process. They can come to the table quickly with a feasible and credible system. Credible system designs, by definition, use only technology and capabilities that have been created and applied previously (Shishko, et al. 1995). While it is not always the case, a COTS approach generally signals good value and reliability to the stakeholder. In the next section, we will examine trade studies.

6.1.6 Conducting Trade Studies

Multiple technical solutions may be available at the system or subsystem level, and understandably, the primary stakeholder only wants to pay for the development of just one system. Trade studies are used to figure out the best possible approach. The function of a trade study is to optimize a particular aspect of the system design given a well-defined set of technical choices and parameters (Blanchard and Fabrycky 2011). The inputs of the trade study are the DDPs of competing technical solutions or just one widely variable solution. These inputs can consist of models of those technical solutions, the stakeholder's priority-ranked operational requirements, systems life-cycle considerations, and other data. The output of a trade study is a decision on which depends the technical route to follow, based on the desired characteristics of the system, within the alternatives being studied. The trade study will also document the justification of the decision. The goal of this section is to explain the process that provides the trade study its justifying power; a little more detail on those inputs and outputs, when and how many trade studies are conducted, what the sensitivity analyses are, and how alternatives are analyzed are related to trade studies.

AoA and trade studies can be roughly distinguished as being concerned with big and little decisions, respectively. AoA usually refers to the decision-making process during conceptual design to pick the optimal design from categorically different approaches, that is, a ground-based telescope versus a space-based telescope. In that situation, the systems may have different DDP even though they could be built and deployed in various numbers to address similar operational requirements. A trade study may focus on how to best allocate changing variables within a relatively stable configuration. The variables being manipulated relate back to operational functions required and ranked by the stakeholder criteria. An example could be varying the aperture size of an optical system, which related back to two highly important concerns of the stakeholders: imaging performance versus system cost. A trade study conducted on the light collecting function of the optical system will reveal negatively correlated trade-off of cost and the ability to collect light.

The term alternative analysis is rarely used to describe anything outside the conceptual design phase; the Joint Capabilities Integration and Development System (JCIDS) requires an AoA to be performed even before potential suppliers have concluded their conceptual design phases. This makes sense as big decisions that impact configuration greatly should not be happening after significant design work has been performed in the conceptual design phase. Making big changes essentially throws away design work and therefore becomes more and more costly later in the systems life cycle. The many smaller decisions and optimizations that need to be made throughout the acquisition phases are referred to as trade studies (Blanchard and Fabrycky 2011). However, the semantics are not rigorous from every source. The decision to be made between clearly defined and non-sliding-scale subsystem configurations and components may invoke the usage of the term alternative analysis. There is an additional factor that sets AoA apart. The stakeholder may conduct an AoA from a procurement perspective when deciding between multiple competing systems providers. In fact, the U.S. Government demands that agencies operating under the JCIDS must perform an AoA early on in the acquisition process (Department of Defense 2009). As provider of systems, National Aeronautics and Space Administration (NASA) endorses the use of trade studies before making system decisions as part of a robust SE process (Shishko, et al. 1995).

The fact that system providers and stakeholders, private- or public-sector agencies, conduct alternative analyses and trade studies suggests a great deal about the justifying value of this robust SE decision-making process. Figure 6.3 helps to explain how resultant AoA and trade-study decisions are well supported, accepted by stakeholders and program management alike, and lead to stable designs. The overall structure of the decision-making process develops rating criteria based on system goals and rates system solutions according to their performance against FOM corresponding to those goals. In addition to corresponding to the immediate design concerns that deal with performance and the current design phase cost and schedule, FOM should be drawn from the entire procurement and utilization systems life cycle. A broad base of evaluation criteria should address performance trade-offs from system conception to system retirement and everything in between.

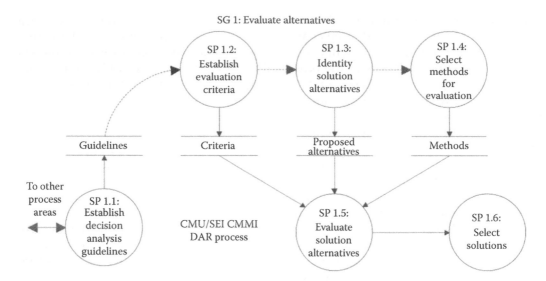

FIGURE 6.3
Decision analysis and resolution (DAR) activity. (Courtesy of Peter Hartman.)

The decision-making process that begins with the system provider creating over-all decision-making guidelines with stakeholder input is represented by circle SP 1.1 in Figure 6.3. Including stakeholder input means that any decision reached was done so in a consensus-building manner. These guidelines help establish credible comparison criteria (SP 1.2). The sources of these criteria are from the inputs mentioned earlier: DDP, important benchmarks, and stakeholder operating function priorities. All alternative solutions (SP 1.3) that have been identified are entered into consideration. Generally, the decision guidelines (SP 1.1) allow alternative solutions into consideration. The reasoning used is that it is better to consider designs, which turn out to be infeasible, than accidentally exclude the best design (Blanchard and Fabrycky 2011).

The designs and solutions under consideration as alternatives can refer to multiple levels and stages of the system under development: competing design concepts, alternative con-figurations, subsystems, system components, and different levels of resource allocations that can fulfill the same operational requirement. Before these different solutions can be evaluated against each other, a common system of measurement is needed. Even if all DDPs do not match, different solutions can be translated into FOM, which are essentially metrics specified in requirements. Any models developed during feasibility studies may be extremely helpful as methods of converting DDPs into FOMs. For use in trade studies, the parameters of models and simulations can be varied to show the effect of incremental changes in design. This quickly shows an estimate of the impact on FOMs without having to build a model system first.

Once FOMs are produced for every category considered, they can then be compared apples to apples within each category. Then the difficult proposition of comparing design alternatives when no design alternative leads in every category is next. Afterward, the cross-category comparisons need to be made. This is why the ranking of categories by the stakeholder is important for the final evaluation and decision process.

Cost and risk are also included as important considerations, but due to the uncapped nature of certain cost and risk potentials, these parameters could potentially drown out all the others. It may help to leave these as separate details in the analysis, even as the effects of smaller technical details are lumped together for ease of presentation to the stakeholder or program management. For significant decisions, the stakeholder should be involved and informed by a recommendation. The recommendation should also come with a sensitivity analysis to show the relative risk of FOMs changing as the system model moves to a real system. If options are close to each other in their overall ranking, but one option's FOMs have a large variance according to the component model or SOS interactions, then it is clear that option is much riskier than the other and may therefore change the overall solution preference previously assumed (Haskins 2006). The stakeholder ultimately makes the decision of exactly how scalable their cost and schedule criteria are for individual operational function trade-offs and should these criteria come into significant conflict.

Stakeholders care about FOMs such as the mean time between failures (MTBF), which is a derived figure for reliability. Reliability and maintainability are categorically different FOM metrics. In order for good, comparative evaluation of solutions to exist, these FOMs and operational performance metrics should be ranked to build consensus in the decision-making process. For example, a system designer brings the results of an AoA to the stakeholder's attention. The stakeholder's need is to stare deep into space to find human extinction event–sized objects. The alternative optical system platforms to choose between are a satellite versus a telescope on the ground. Since the atmosphere will present challenges to image spatial resolution for aperture sizes larger than the atmospheric coherence length, there will be significant functional differences between the space- and earthbound optical systems for equivalent aperture sizes. Although the system may integrate atmospheric compensation techniques, they might not achieve exactly the same image quality of images offered by the satellite platform. They must rank their concern of cost with image-quality FOMs. Furthermore, the atmospheric compensation could be pursued by lower-cost but lower-maturity technologies. The stakeholder must compare the cost difference of mature versus developing technologies with their tolerance for system risk, in which the schedule risk of new technology will play a factor. Finally, once the stakeholder and system provider have settled on a method, there will still be a need for trade studies. If a satellite platform is chosen, the stakeholder and the provider should both agree that maintainability is a much lower priority than reliability—unless space shuttles become cheap and abundant.

The timing and frequency of trade studies have not yet been discussed. Trade studies occur throughout the acquisition cycle. Trade-off studies are best begun during the tail end of functional analysis (Blanchard and Fabrycky 2011). This is when functions are broken down to the next level of detail. In general, every new level of system detail will mean a new functional analysis, requirements allocation, and trade studies. During every segment of development, many alternative mechanisms may be clarified, and trade-off studies will be needed before the evaluation of alternative components and configurations can be concluded.

An example of functional analysis leading to trade-off studies is the breakdown of the image processing function that advanced real-time video systems require. Two conceptual objects handle the flow of optical data below the image processing function level: software and hardware. Several software applications and types of hardware may be available. A model may be needed to decide what combination of software and hardware determines the most cost-effective way to reach a required 30 Hz frame rate FOM. In this case, it is easy to trace the translation of the operational requirement into an FOM frame rate. The speed requirement was passed from the optical system level to the image processing subsystem and then allocated between the software, hardware, and other subsystems that impact the speed of image delivery.

Hardware and software represent an allocation of specific resources to accomplish a function. A model involving the interaction of hardware, software, and subsystem data interfaces would be needed to predict the impact of the design decision on the overall system's ability to push the optical data to video. There will be a certain amount of uncertainty with the validity of such a model until it is built. Trade studies, therefore, also produce sensitivity analyses. A delicate balance of value is delivered to the customer when their ranked priorities are estimated and adjusted for optimal trade-offs. However, if a small software bug could cause the system to perform below the 30 Hz requirement, then the SE team should perform a trade study to report this sensitivity. A risk management reaction might emphasize choosing more mature and robust software or extra testing for the selected software. In this case, the sensitivity analysis did its job of catching cost, schedule, and technical risks that might have been buried by the notion that the requirement had been fulfilled.

6.1.7 Evaluating Alternatives: The Analytic Hierarchy Process

Quality function deployment (QFD), factor scoring, and the AHP comprise a short list of excellent processes used for alternative analysis. All of these processes involve the categorical ranking of FOMs, but their evaluating models differ. Evaluation modeling can be either analytical, relying on deterministic mathematical relationships and algorithm mechanics, or simulation based, relying on random processes, transaction analysis, and/or the mathematics of probability to describe the outcome of events and decision through iterative cycles. This section focuses on the AHP method, which is typically a deterministic model.

The AHP is a well-established and a widely recognized method. This is an important point since the SE team needs a credible process they can show the stakeholder, almost as much as they need good results. Dr. Thomas Saaty published *Decision Making for Leaders: The Analytical Hierarchy Process for Decisions in a Complex World* in 1982 to relay this methodology (Saaty 2012), and the AHP has since been widely used and cited to address complex decision problems (Haas and Meixner 2011).

A typical situation where AHP might be used comes at the end of the conceptual design, though it can be adapted for use in trade studies of smaller scope. The inputs are the alternative designs, of course, stakeholder rankings of operational requirements and the associated FOMs, and other results of the feasibility studies conducted on those designs. The results of feasibility studies and previous trade studies are excellent sources of information and metrics considering the "-ilities," risk, cost, and previously developed models.

Even though the point of the AHP is to pick something and go with it, the design analyses, sensitivity analysis, and the documentation generated during alternative analysis are highly valuable for a systems development team. As explained later on, the eigenvector analysis of AHP matrices might literally point to a theoretical optimal allocation of resources in n-dimensional space. Although it might not be a strictly valid reality, it is an analytical method that will provide design insight. In a more result-oriented perspective, evaluation of all feasible alternative approaches to design early in the life cycle should help to promote greater design maturity earlier (Blanchard and Fabrycky 2011). Furthermore, the documentation of decent alternatives may be revisited later in the project and prove to be highly usable for the purposes of risk mitigation. In the event that method "A" fails, the SE team already has a good idea that method "B" will work. Also, the process of evaluating model outlines metrics and testing criteria for later verification and validation. Identifying ways to measure and test designs reduces uncertainty and life-cycle risk. This is an easy point to see from the stakeholder's perspective. A stakeholder probably would not be interested in acquiring a complex system that has not been tested. On the other hand, they would probably be thrilled to hear that they will be getting the best of multiple designs and will receive tangible proof.

Part of the reason why the AHP is credible to the stakeholder is that the stakeholder's preferences drive the alternative selection process. The stakeholder reports preferences as input for the AHP, and then those exact same preferences are reported back with their implication for design choices, cost, and risk. The AHP in particular encourages the stakeholder to use a simple pair-wise ranking in the decision-making process to determine their relative preferences between a set of mutually agreed upon criteria.

For example, the stakeholder interested in searching deep space for asteroids might propose *optical performance, reliability, maintainability, supportability,* and *availability* as a set of categories that are important for the proposed system. They may generally decide that optical performance is more important than reliability: optical performance ability over reliability. Then they move on to the next pair-wise comparison: optical performance over availability. Then they decide: reliability over availability. A completely ordered ranking system can be constructed from this pair-wise comparison process in a relatively straight-forward manner.

First, create a table with a row and a column for each of the categories; an example of this is shown alongside Table 6.2. Next, compare the cross column and row category and assign each box with greater, less than, or the same. Now count the "greater" preferences for each row and then order them. A category whose row contains five "greater" has a higher priority than a row with three "greater." This is a good method to rank all criteria in order. However, it may not necessarily communicate that reliability is three times more important than maintainability in the stakeholder's view. It is, however, a start.

Consider a ranking order as follows: performance over reliability, reliability over maintainability, and maintainability over performance. This is a highly illogical ordering, at least on the surface appearance. Such rankings like previously shown entail many implications for consistency analysis, transitivity (the logical consistency of all possible relationships between a set of objects), independence versus dependence, and why cost and risk were not named as general categories for the AHP ranking system. First, the AHP uses linear matrix methods. A fundamental assumption in the mathematical analysis is that the variables are independent.

The previously shown ordering may be a simple logical error, or it may be communicating that not all of these variables are independent. The order of preference of availability could be changing as a trade-off with performance and reliability is being made. In that case, availability and one of those variables are dependent and the system may be nonlinear. As far as analytical models of alternative analysis go, "The most important concern in the selection of criteria or attributes is that they be independent of each other" (Blanchard and Fabrycky 2011).

The comparison of a satellite-based telescope and a ground-based telescope is an easy example to see how the order of preference could be dependent on alternatives. For a satellite, maintainability may not be important at all to the stakeholder. It is not cost-effective currently to repair the system in orbit, so there is little impact if there are no access panels for the repair personnel. For the same reason that maintainability is not important, a satellite needs to be very reliable and correspondingly have a very low failure rate. The opposite case may be true for a ground-based system. The observatory crew has much better access to maintain the telescope as needed. Additionally, reliability is likely to be traded against cost in some aspects of the system since the system is accessible to be repaired and downtime may be afforded.

For the satellite versus observatory case of alternatives, it would be very difficult to get the AHP criterion comparison matrix to work. There would need to be two criterion comparison matrices produced to accommodate both cases. However, this leads to a logical error in comparison. A criterion comparison matrix is used to produce a column of numbers corresponding roughly to the eigenvector of the matrix. For now, imagine that this is a variable, x, that pops out of that algorithmic equation. With two criterion matrices, two eigenvectors are produced: x and x'. Glancing down to Table 6.7, the logic flow stops as only one column (C1–C4) can accept x or x' but not both. We are halted tantalizingly close to the alternative ranking, which is the right side of the matrix equation in Table 6.7. There is a methodology that can be used to fix this logical error, however. The dependent categories—maintainability and reliability—can be grouped together to form one independent variable. The stakeholder may be able to assign a ranking to this synthesized "ability to continue to operate" category.

The problems of inconsistency and dependent variables need to be guarded against in the AHP. The AHP uses two simple checks for this. During the ranking of criteria, the AHP demands transitivity, which excludes nonlogical sequencing. A simple demonstration of transitivity is that $1 < 2 < 3$ should never be $1 < 3 < 2$ (Blanchard and Fabrycky 2011). The AHP defines matrices where there are more relationships for each member entry than purely adjacent order as shown by 1, 2, and 3. In this case, the logical relationship includes both ordering and the scale of ordering. In order for a matrix to maintain transitivity, both the ordering and relative magnitude of two entries must be consistent and reproducible if they are to be compared by transforming them into other entries. That is, if $A < B < C$ (ordering), $B = 2A$ (relative scale), and $C = 2B$, then four of A should be translatable into one C. There are many other senses of the definition that could be used based on the mathematical space being referenced, but simple ordering and an implied set of algebraic equations are enough to describe transitivity in the AHP. To learn more about ideally transitive matrices, reference an introductory text on group theory.

The second check is the consistency check, which results in a consistency ratio (CR). The CR test will fail if transitivity fails. In that case, the SE team performing the AHP should go back to the stakeholder or source of the original preference ratings. More subtle errors are also caught by the consistency check. Optical performance, reliability, and maintainability may pass the transitivity gate, but they may fail relative consistence. This would result if optical performance is three times more important than reliability and nine times more important than maintainability and reliability is only two times more important than maintainability. The logical implication should have been that reliability is three times more important than maintainability. Such a discrepancy may or may not cause the consistency check to fail. Some inconsistency is allowed as long as the relationship is approximately linear in the trade space considered.

Returning to the optical AHP example, assume that the stakeholder ranked a large number of "-ilities." However, the SE team chose to include optical performance, reliability, availability, and supportability only. There are several factors that influenced their decision to limit the number of categories.

The first factor is that decisions become more difficult and time consuming as more variables are involved. Less important parameters should be excluded to ensure that the AHP is not too complex to use in actual decision making but includes enough factors to make it useful. Arthur Felix contributed a paper to a 2004 INCOSE conference that extensively analyzed trade studies in practice. The recommendation Felix produces is that "between three and nine criteria should be involved" (Felix 2004, June 20–24). This idea is also supported by psychologist George Miller's research, which found that the human brain is capable of handling only so many conceptual objects and discrete figures at a time (Miller 1956). The magic number turns out to be a magic range of 7 ± 2 objects. Interestingly, this research heavily influenced the choice of the seven-digit phone number.

The second factor is more obvious; only factors that are actually affected by the alternative choices should be included. A more subtle decision must be made about factors that are equally impacted by whatever technical choice is made. Categories that all alternatives influence equally should be excluded since there is no difference to decide upon (Felix 2004, June 20–24). This does not mean that impacted FOM should not be tracked in some form. If all alternatives result in the same predictability, but the predictability is unacceptably low for all choices, that fact needs to be documented.

Cost and risk should not be included in the decision analysis. Cost and risk should be applied to the results of the initial AHP ranking of the technological solutions (Felix 2004, June 20–24). The data analysis produced during the first stages of the AHP can feed directly into the cost and risk models. An important contribution to the risk model is the sensitivity analysis that the AHP produces during the evaluation stage. The quality of the cost and risk assessment may actually benefit from not being included in the AHP decision matrices. The technical solution evaluation also benefits since there is more focus on aspects that determine the real value of the solution to the project: performance, predictability, maintainability, and so forth. Generally, when one tries to decide the value of a product, the cost is asked for, although the cost is the only determinant of the actual value. Felix recommends a course of action that any smart value shopper knows. Allowing cost to get involved up front allows the possibility that the purchaser will lose sight of effectiveness.

A fourth consideration for selecting ranked criteria involves the stage of the project. Depending on the progress and the system level of the project, the general category of maintainability could turn into several things. If alternatives being compared are at very small level, such as figuring out quality of capacitors to use, the failure rate statistic might be an appropriate category. However, an argument could be made that the entire MTBF FOM should be used for the entire system. The reliability of capacitors used everywhere may impact the system's reliability in a global fashion. It is important to see how such a decision would affect the entire system, especially in comparison with the MTBF TPM. It might also make sense, for technical comparison purposes, to designate the criteria at the functional or component level. This would make sense if there were a unit dedicated to checking and self-correcting a system operation.

Much thought goes into selecting the proper criteria to use as categories in AHP. This is quite sensible since the bridge is being made between the subjective to the objective. With proper criteria selected, an AHP can be created from it. Before populating a matrix, the preference order needs to be converted into numbers. The AHP criterion comparison is quasi-normalized in the sense that there is a maximum and minimum range of importance. This type of standardization is fairly common anywhere AHP is used. In the following examples, 9 represents maximum importance and 1 represents minimum importance. For example, optical performance is far more important than supportability for our user, so performance is 9 to supportability's 1. The ratio is 9/1. When comparing supportability to performance, the ratio is 1/9. Any criterion compared to itself is of equal performance and is therefore 1/1. This information is summarized in Table 6.1.

Now that the weights have been assigned to the criteria, they can be entered into a matrix that expresses the relative ranks of criterion preferences against one another. The row criterion entries are typically compared (preferred) relative to the column

TABLE 6.1

AHP Criterion Ranking Definitions

Intensity of Importance	Definition	Explanation
1	Equal importance	Choices are equally weighted.
3	Moderate importance	One choice is slightly preferable to another.
5	Strong importance	One choice is strongly preferred over another.
7	Very strong importance	One choice is preferred very strongly over another.
9	Extreme importance	One choice favored over another without repute.
2, 4, 6, 8	Intermediate values	Choose when the weighting is between values.
Reciprocals	Apply to opposite relation	If choice A favored to some level relative to choice B, then choice B has reciprocal relationship to choice A.

TABLE 6.2

Example AHP Three-Criterion Pair-Wise Ranking Matrix

	Criterion #1	Criterion #2	Criterion #3
Criterion #1	Criterion #1 vs. Criterion #1 ranking	Criterion #1 vs. Criterion #2 ranking	Criterion #1 vs. Criterion #3 ranking
Criterion #2	Criterion #2 vs. Criterion #1 ranking	Criterion #2 vs. Criterion #2 ranking	Criterion #2 vs. Criterion #3 ranking
Criterion #3	Criterion #3 vs. Criterion #1 ranking	Criterion #3 vs. Criterion #2 ranking	Criterion #3 vs. Criterion #3 ranking

TABLE 6.3

Example AHP Three-Criterion Comparison Matrix

	Criterion #1	Criterion #2	Criterion #3
Criterion #1	1	3	7
Criterion #2	1/3	1	3
Criterion #3	1/7	1/3	1

criterion entries. The diagonal of the matrix will always have values of 1 since a particular criterion is compared to itself (Table 6.2).

With numbers filled in, Table 6.3 now looks like a square N-by-N matrix. In order to interpret the priorities, the matrix boxes are referenced first by row and then by columns. Starting from row 2, which represents Criterion #2, go across the row to find the number three. Tracing the column up, the label Criterion #3 is found. In the AHP format, this means that the criterion in row 2 is three times more important than Criterion 3. Looking at row 3 and column 2, the matrix reads that Criterion 3 is one-third as important as Criterion 2. Cell (3,2) is the reciprocal of (2,3) and vice versa.

These criterion ranking matrices are mathematical models of preference. They must be square matrices, although they can grow in size as more criteria are added. Shortly, a second type of matrix will be constructed to show how the solutions relate to each other in response to a criterion. Those are the two types of foundational matrices that AHP uses: one that models how criteria relate to each other in importance and a second that gives information on how the technical solutions relate to each other with respect to each criterion. In order to tell what solution is best, the information in both matrices needs to be synthesized.

Given the synthesized preference information, which solution has the best overall ranking across criteria when those categories are given their proper weight? In other words, which solution represents the best allocation of criteria? The ability of solutions to satisfy separate criteria cannot be compared with other solutions' ability to satisfy criteria across categorically different criteria. However, those criteria become numerically comparable when the criteria–criteria ranking information is added. This allows a solution's criterion score to be mapped to a common magnitude across criteria. This common magnitude can be summed for each solution and then compared directly.

An eigenvector analysis allows a square matrix relationship to be summarized with a column vector that captures the relative preferences of the criteria. These "summary" vectors can be produced for the criteria–criteria and solution–solution matrices. As will be seen in the following, the relative magnitude of the preferred solution is determined when the criteria–criteria eigenvector right multiplies the matrix formed from the solution–solution eigenvectors for each criterion.

An eigenvector is not really a "summary" vector of course, but for AHP, it is useful to think of it as such. In AHP, the eigenvector is considered to be a preference vector. An eigenvector is loosely defined as a vector, when left multiplied by a matrix, does not change direction. For this reason, an eigenvector is sometimes called the direction of the matrix. The repeated use of these vectors in the AHP suggests that finding the best solution is analogous to a geometry problem. In many ways it is. A column vector with three rows can be represented as (x, y, z) coordinates. In the same way, an eigenvector from an N-by-N criteria–criteria matrix is an N-dimensional vector.

Looking down at Figure 6.4, the column to the very right is an estimate of the eigenvector. Now imagine that the matrix at the very left is the optical criteria–criteria matrix. The second row has large numbers, and so it means that it is an important category like optical performance. The top row has very small fractional numbers. This represents the relatively low-ranked supportability comparing unimportantly to other criteria. Going across the rows to find the eigenvector, the second row, which corresponds to optical performance, is the largest number. Just above it in the supportability column, there is the smallest number in the eigenvector. If the eigenvector were plotted in a 4D space against the optical performance, supportability, availability, and reliability axes, it would be found that the vector has longer components along the most important criterion axes. For the less important supportability, the vector would barely nudge in that direction.

A					A					A^2					X
1	1/7	1/2	1/3		1	1/7	1/2	1/3		4	.4968	2.381	1.345		0.0718
7	1	5	3		7	1	5	3		33	4	19.5	10.83		0.5877
2	1/5	1	1/2	×	2	1/5	1	1/2	=	6.9	.8524	4	2.267		0.1224
3	1/3	2	1		3	1/3	2	1		12.33	1.495	7.167	4		02182

FIGURE 6.4
Squared (A-by-A) criteria matrix and incremental eigenvector. (Courtesy of Greg Legters.)

The eigenvector graphically represents that, and it represents other subtle factors. The relative magnitude of importance, between 9 and 1, is reflected in it. The additional relationships between various one-by-one criterion comparisons are also weighted in the eigenvector. For this application, it truly is a summary vector for the mapping of criteria to a preferential value. The solution–solution matrix is mapping a relative preference of the proposed solutions when regarded from the perspective of a given criterion. Similar to the criteria–criteria matrix eigenvector, the solution–solution matrix eigenvector can be regarded as "directions in M-dimensional space (where M represents the number of alternative solutions)" of the solution preferences with respect to the particular criterion.

There are several methods that can be used to calculate or estimate the eigenvector of a square matrix. The analytical solutions can be found using the following eigenvector equation (Strang 1993):

$$Ax = \lambda x, \tag{6.1}$$

where
  x is an eigenvector of matrix A
  λ is a scalar called the eigenvalue

For properly constructed matrices for use in the AHP, normally only one eigenvector and the largest positive eigenvalue per matrix constructed need be considered. Notice how x is on both sides of the equation and λ pops into the place of matrix A. Every time that the vector is multiplied by the matrix, λ multiplies to a higher power. This property of the eigenvalue has important implications for one of the techniques that will be discussed in the following.

Notice in the following equation, replacing matrix A with A multiplied by itself 10 times implies that lambda must multiply itself by 10 times:

$$A^{10}x = \lambda^{10}x. \tag{6.2}$$

If A has a second eigenvector x_2 and a second corresponding eigenvalue λ_2, the same equation can be written as shown in the following equation:

$$A^{10}x_2 = \lambda_2^{10}x_2. \tag{6.3}$$

With the same matrix A, if $\lambda_1 > \lambda_2$, then after successive self-multiplication, the value of $\lambda_1^{10}x$ will be much greater than $\lambda_2^{10}x_2$. Essentially, the eigenvector with the largest eigenvalue will dominate the matrix at higher powers. Not surprisingly, the first technique to find the eigenvector is multiplying the matrix by itself over and over again (Haas and Meixner 2011).

Normalizing the eigenvectors will be extremely important for the AHP. Look down at Table 6.7. All columns on the left side of the equation are eigenvectors from the solution–solution matrices of interest. Notice how these are all approximately normalized columns, since they are comprised of fractional numbers that sum to about one when each term in the column is squared individually and added. Now imagine that the coworker assigned to figuring out the column, represented by Criterion #4, reliability, forgot to normalize it. Instead of [0.0755 0.2291 0.6954], the column is [75 229 695]. This is a factor of 1000 times greater than the other criterion columns. Notice how column 4, when right multiplied by the C1–C4 column, is only multiplied by the bottom number in the C1–C4 column. This means that criterion that should have a weighting of 0.2179 has an effective weight of 218, when the other criteria are <1. Since solution #3 happened to do quite well with regard to Criterion #4, reliability, the overall ranking represented by the right side of the equation in Table 6.7 will change. With this scaling error, solution #3 will be ranked higher than solution #2 because the number in its row will increase proportionally more than the number in the row corresponding to solution #2. The correct ranking, according to Table 6.7, is that solution # 2 is better than #3 overall when the columns have been properly normalized. This example of false rankings illustrates why normalization is functionally important for the AHP.

So, to summarize and recap, the first technique to find eigenvectors involves multiplying the matrix by itself over and over until the columns converge to a nearly fixed vector direction. The columns are averaged to produce one column, and then that column is pseudo-normalized by dividing the component numbers by their sum. If the column average of the matrix is [2 1], then it will be divided by (2 + 1) to produce [2/3 1/3]. Due to the iterative matrix self-multiplication, programs, such as MATLAB, that have robust algorithms that guard against truncation errors best handle this method.

The second way to estimate a pseudonormalized eigenvector is much quicker, but it comes at the expense of some accuracy. The second technique involves the normalization of columns and then averaging rows to get a normalized eigenvector (Haas and Meixner 2011). Essentially, all columns of the matrix are treated as representative eigenvectors of the matrix. Each column of the matrix is pseudonormalized, so the numbers in each column sum to one. Then the columns themselves are averaged to produce an average column, which should also sum to one. This column is the estimated eigenvector of the matrix. For the criteria–criteria matrix in the AHP, this may be a decent assumption as long as the stakeholder rankings exhibit transitivity, independence, and consistency, as was discussed in the opening paragraphs of this section.

These two eigenvector estimation techniques will now be applied to the criteria–criteria ranking matrix. Table 6.4 represents the stakeholder's priority and relative importance information. The system provider SE team transformed stakeholder rankings and statements into this matrix.

The next step is to get the summary vector of the relative priorities for each criterion. The first technique of matrix multiplication will be used to estimate this matrix's eigenvector. Since the matrix self-multiplication method can be iterated infinitely, engineering judgment should be used to determine when to stop. Figure 6.4 contains the estimation of stopping at the second power. To get the final column of [0.0718 0.5877 0.1224 0.2182], all of the rows of the matrix can be summed. This creates four sums that make up the row entries in a single column. Then each entry in the column can be divided by the sum of all entries in the matrix, which in this case results to 114.6. The operation of summing the first row and normalizing the entry is (4 + 0.4968 + 2.381 + 1.345)/114.6.

There are a number of similar algorithms that could be stated to pseudonormalize the square matrix on the right side of Figure 6.4. The overall objective is to express the

TABLE 6.4

Example AHP Four-Criterion Comparison Matrix

	S	O	A	R
Supportability	1	1/7	1/2	1/3
Optical performance	7	1	5	3
Availability	2	1/5	1	1/2
Reliability	3	1/3	2	1

eigenvector in a percentage form. If A^4 is calculated, then averaging the column of the resulting A^4 matrix to produce one column and then pseudonormalizing the resulting column give an eigenvector estimate of [0.0722, 0.5872, 0.1228, and 0.2179]. Even though A squared and A to the fourth power may look nothing alike, the resulting eigenvector estimates do not change much. This indicates that the value is converging. When A^8 is used, the same [0.0722, 0.5872, 0.1228, and 0.2179] is produced. Although the answers are fractionally different, at this level of significance, the result functionally has not changed.

Table 6.5 results when the second method is applied to the matrix shown in Table 6.4. The algorithm is somewhat more rigid in this case since the columns of Table 6.4 criteria–criteria matrix are normalized with respect to themselves. The first column of [1 7 2 3] is divided by its own total of 14. This becomes column [0.0769 0.5385 0.1538 0.2308] in Table 6.5. Each column is normalized with respect to itself since it is not assumed that all columns are converged on the same direction. Therefore, an equal weight is desired for all four-column directions suggested by the matrix in Table 6.4. Pseudonormalization with respect to each column produces the square matrix in Table 6.5. The averages of those columns produce the single column on the right side of Table 6.5. That is the estimate of the eigenvector. Notice how it is similar in magnitude to the column vector produced by the first method. However, it is not the same. Generally, this is a less precise method.

The same exact process of constructing a priority matrix and then estimating the eigenvector can be used for solution–solution ranking matrices. Table 6.6 shows the results for solution–solution ranking matrices. There are four matrices since there are four criteria. For Criterion #4, reliability, all three potential solutions are ranked relative to each other. The numeric entries of the matrix are determined using similar rules to those used in translating stakeholder priorities to a fixed scale. Looking at row 3, column 1 of the reliability matrix, there is a 7 entered. This means that solution 3 is preferred seven times over solution 1 when it comes to reliability. Whatever the rationale, solution 3 is preferred seven times to solution 1 when it comes to reliability.

There are many ways to handle solution–solution ranking, and a discussion is in order about which solution-specific measurements should correspond to an overall criterion. We start first with a discussion on the use of eigenvectors. Table 6.6 overall shows that each matrix produces an eigenvector that is essentially a summary ranking vector. The vector produced from the solution–solution matrix in response to the reliability criteria will rank the technical solutions as if the only criterion to consider is reliability.

Solutions 1, 2, and 3 correspond to 7.55%, 22.9%, and 69.54%, respectively, in the reliability criterion matrix. This is essentially saying that solution 3 has the highest score for satisfying the stakeholder's reliability needs. The geometric eigenvector interpretation might suggest that the maximum length of the satisfaction vector is maximized in the direction of [0.0755 0.2291 0.6954] as those entries correspond to the technology axes. This interpretation is, of course, context dependent.

However, reliability is not the only criterion being considered. There are four. These matrices do not contain their relative importance. Therefore, their eigenvector ranking

TABLE 6.5

Example Four-Criterion Matrix with Normalized Columns and Eigenvector

0.0769	0.0852	0.0588	0.0690	0.0725
0.5385	0.5966	0.5882	0.6207	0.586
0.1538	0.1193	0.1176	0.1034	0.1236
0.2308	0.1989	0.2353	0.2069	0.218

TABLE 6.6

Example Three-Solution/Four-Criterion Evaluation Matrices and Eigenvectors

Criteria 1	Sol #1	Sol #2	Sol #3	C1 EV	Criteria 2	Sol #1	Sol #2	Sol #3	C2 EV
Sol #1	1	1/4	3	0.2051	Sol #1	1	1/3	5	0.2654
Sol #2	4	1	8	0.7166	Sol #2	3	1	9	0.6716
Sol #3	1/3	1/8	1	0.0783	Sol #3	1/5	1/9	1	0.0630
Criteria 3	Sol #1	Sol #2	Sol #3	C3 EV	Criteria 4	Sol #1	Sol #2	Sol #3	C4 EV
Sol #1	1	1/2	1/5	0.1220	Sol #1	1	1/4	1/7	0.0755
Sol #2	2	1	1/3	0.2297	Sol #2	4	1	1/4	0.2291
Sol #3	5	3	1	0.6483	Sol #3	7	4	1	0.6954

result should be pseudonormalized, so the externally reported importance is not exaggerated due to internal ranking numbers. However, the criteria–criteria ranking eigenvector, produced from the A^4 eigenvector convergence estimation process, does contain the relative importance of these criteria. If each criterion isolated solution–solution eigenvector is multiplied by the eigenvector from the criteria–criteria matrix, then the solution–solution vectors will be scaled in terms of relative importance.

Table 6.7 shows how this multiplication is performed in matrix notation. Note how the criteria-specific solution–solution eigenvectors are grouped into a 3 × 4 matrix and multiplied by a 4 × 1 column criteria–criteria eigenvector to produce a final 3 × 1 column on the far right. In general, the grouped solution–solution eigenvectors form an $M \times N$ matrix, where M is the number of solutions and N is the number of criteria.

The column on the extreme right suggests that solution 2 is the best since solution 0.5242 is greater than any other "satisfaction" magnitude in the far right column. Once again, the alternative interpretation is that in order to maximize the stakeholder satisfaction, 52% of resource allocation should go to solution 2, 20% of resource allocations should go to solution 1, and 28% should go to solution 3. That interpretation does not work if the solutions cannot be mixed or synthesized. In the later mutually exclusive case, the highest ranking wins!

The AHP is not complete once the solution ranking is obtained. Before the results can be presented for use in the decision-making process, the results should be checked for logical consistency, inaccuracy due to estimation methods, and a possible deviation in the overall ranking if assumptions are changed.

A sensitivity analysis generally estimates the risk of making an "other than optimal" decision based on the variability of the input parameters. The assumptions in the matrices may vary widely, be subject to change, or just potentially be wrong. The sensitivity analysis checks what the variance of the final solution rank outcome is based on the variability of the inputs. If the sensitivity analysis reports that the ranking has a very high probability of changing, and the solutions correspond to a critical function, then there is potentially

TABLE 6.7

Example AHP Solution Ranking

	Criteria 1	Criteria 2	Criteria 3	Criteria 4		C1–C4		Sol #1/#2/#3
Sol #1	0.2051	0.2654	0.1220	0.0755		0.0722		0.2021
Sol #2	0.7166	0.6716	0.2297	0.2291	×	0.5872	=	0.5242
Sol #3	0.0783	0.0630	0.6483	0.6954		0.1228		0.2738
						0.2179		

a great deal of risk in pressing forward with a solution choice. The mitigation strategy to deal with that risk, either through more investigation to remove uncertainty or through searching for more alternative solutions, needs to be provided by the development team. On the other hand, if a low probability of change is reported by the sensitivity analysis, then the AHP either produced a robust analysis, or one solution was significantly better than the alternatives. Both scenarios imply trustworthy results.

A consistency analysis checks for logical errors as discussed in the opening paragraphs of this section. If the consistency analysis fails, then the stakeholder ranking and solution ranking matrices may need to be checked for fundamental assumptions. The consistency analysis will also catch excessive errors associated with the mathematical estimation assumptions used in the AHP. Essentially, the consistency analysis is heavily based on seeing if the estimated eigenvectors are close to the real eigenvectors of the matrix. Both estimation error and structural problems with a matrix's internal ranking system will show up in the form of an eigenvector deviation.

Before discussing the calculation mechanics, the mathematical basis of the test is useful to understand. Recalling the property of eigenvectors, when a matrix multiplies its own eigenvector, it should return that eigenvector in the same direction times a scalar multiplier, λ. Essentially, the consistency check plugs the estimated eigenvector back into the eigenvector equation and checks to see how far it deviates from a scalar multiplication. Any deviation says that the estimated eigenvector is in a slightly different direction than the actual eigenvector of the matrix. The following equation demonstrates this:

$$Ax_{est} = x_{mut}. \tag{6.4}$$

A consistency analysis begins by plugging in each estimated eigenvector, x_{est}, and the matrix from which it was derived into this equation. This should be done for the criteria–criteria matrix and all of the solution–solution matrices. Unlike the exact eigenvalue equation in Equation 6.1, the matrix by column multiplication in Equation 6.4 is unlikely to return the same column vector multiplied by a scalar. Instead, the direction and the magnitude should simultaneously change unless the exact eigenvector was found. The returned column vector is x_{mut}.

Essentially, each row of the estimated column has been changed by an unknown factor. A way to find the unknown change of magnitude for each entry of x_{mut} is to divide every entry of x_{mut} by the corresponding entry from x_{est}. All of these factors can be added up and divided to get λ_{avg}. Looking back at Equation 6.1, λ_{avg} estimates the would-be eigenvalue. According to Saaty (1987) and mathematical proofs, λ_{avg} will exceed the order of the matrix unless a perfect eigenvector is chosen. If a logically inconsistent 4×4 matrix is assessed, it should result in $\lambda_{avg} > 4$. An example calculation on how to derive the unknown factors using the criteria–criteria matrix and its estimated eigenvalue is shown in Table 6.8.

TABLE 6.8

Criterion Ranking Matrix, Consistency Calculation

	Crit #1	Crit #2	Crit #3	Crit #4		CEV		Result		Result/CEV		λ
Crit #1	1	1/7	1/2	1/3		0.0722		0.2901		0.2901/0.0722		4.0183
Crit #2	7	1	5	3	×	0.5872	=	2.3603	$\lambda=$	2.3603/0.5872	=	4.0196
Crit #3	2	1/5	1	1/2		0.1228		0.4936		0.4936/0.1228		4.0195
Crit #4	3	1/3	2	1		0.2179		0.8758		0.8758/0.2179		4.0194

The true deviation is not represented by λ_{avg} but by its difference from the expected value, n. The expected value is the order of the matrix. For a 4×4 matrix, use $n = 4$. For a 3×3, use $n = 3$ and so on. A consistency index (CI) can summarize this deviation, and it is calculated with the following equation:

$$CI = \frac{(\lambda_{avg} - n)}{n - 1}. \tag{6.5}$$

In order for the CI to have meaning, it needs a standard for comparison. Borrowing concepts from statistics and from empirical evidence (Saaty 1987) developed a standard of comparison to outline consistency tolerances. A CR must be calculated to compare with Saaty's standard. Equation 6.6 shows how to compute the CR using the CI. The average consistency index (ACI) is a standard scaling factor developed by Saaty for matrices of different orders. The ACI must be looked up. See the following equation:

$$CR = \frac{CI}{ACI}. \tag{6.6}$$

Using the criteria–criteria scenario in Table 6.8, the CR can be calculated by first calculating the average value of λ (e.g., $\lambda_{avg} = 4.0192$). Plugging λ_{avg} and $n = 4$ into Equation 6.5 yields a CI of 0.0064. In order to get the CR, the ACI as indicated in Equation 6.6 must divide the CI. For $n = 4$, the tabulated value is 0.90. The calculated CR is 0.0071. Using the information about the solution–solution matrices in Table 6.6, the other CRs can be calculated at 0.0158, 0.0251, 0.0032, and 0.0659.

Notice how the CRs are all below 0.10 or 10%. That is Saaty's standard. If the CR is less than 10%, then the eigenvector solution for that matrix is sufficiently consistent. Saaty's reason for setting a CR threshold of 0.10: "Although the mind is primarily concerned with constructing a consistent decision, it must allow a modicum of inconsistency in order to admit new information, giving rise to change in the old judgments. However, inconsistency is less important than consistency by one order of magnitude (the 10% tolerance range)" (Saaty 1987). The example matrices thus far are well within the range of what is considered consistent.

Does the calculation method used to estimate the eigenvalue matter? The final CR will change slightly if different methods are used. However, it should not radically alter the result of the consistency test. The derived value of $CR = 0.0071$ came from an eigenvector estimated using the matrix self-multiplication method. Specifically, data were used from the A^4 criteria–criteria matrix. If the normalized column average method is used on the data with an eigenvector of [0.0725 0.586 0.1236 0.281], the CR is 0.0055. This represents approximately a 30% change, which is a significant difference between methods. However, in comparison to Saaty's 10% rule, it barely changes. Some further insight can be gained by looking at the variability behind the calculated λ_{avg} of 4.015. Whereas the individual multiplying factors were very consistent in the far right of Table 6.8, there is more spread in the factor data [4.0000, 4.0358, 4.0057, 4.0183] calculated with the column method.

A more interesting case is presented in Table 6.9. This 4×4 matrix is purposefully inconsistent. Essentially, Saaty's test will be tested to see if it can catch the inconsistent matrix. The matrix is inconsistent because the rankings between several pairs do not translate across other pairs. Note how Criterion #3 is twice as important as Criterion #1 and Criterion #4 is thrice as important as Criterion #1. There are many other contradictions to be found, but hopefully the consistency test will indicate to what extent the data are inconsistent.

TABLE 6.9

Inconsistent 4 × 4 Criterion Ranking Matrix, Eigenvector and λ

	Crit #1	Crit #2	Crit #3	Crit #4		CEV		Result		Result/CEV		λ
Crit #1	1	1/9	1/2	1/3		0.0533		0.2547		0.2547/0.0533		4.7819
Crit #2	9	1	2	7	×	0.6036	=	2.8862	λ =	2.8862/0.6036	=	4.7819
Crit #3	2	1/2	1	1/5		0.1198		0.5728		0.5728/0.1198		4.7819
Crit #4	3	1/7	5	1		0.2234		1.0683		1.0683/0.2234		4.7819

The eigenvector of the data [0.0533 0.6036 ...] is produced by the matrix self-multiplication method. When the vector is rerun through the matrix using Equation 6.4, an x_{est} of [0.2547 2.8862 ...] is the output. Using the algorithm outlined to get the implied multiplication factors, the right side of the table is produced. Plugging λ_{avg} of 4.7819 and $n = 4$ into Equation 4.5 gives $CI = 0.2606$. With an ACI of 0.90, using Equation 6.6 yields a CR of 0.2896. Saaty's rule has successfully pegged this matrix as suspect. Using the column pseudonormalization method produces an estimated eigenvector of [0.0570 0.5564 0.1399 0.2467]. Using the same process, the CR is calculated at 0.3047. Looking at the differences in the eigenvectors, there is a significant variation in the result between the methods, but it is well below the sensitivity of Saaty's method. If the team performing the *AHP* method produces such a matrix, it will then be their job to look for one of the many logical errors previously discussed that could result in a consistency test failure.

6.1.8 Connection between Feasibility and Risk

Logically speaking, feasibility and risk are complementary terms. As the feasibility of a requirement goes up, the risk goes down. As things get riskier, the project seems less feasible. The connection between feasibility and risk is inversely proportional, and in many ways, there is a sliding scale. In the section on feasibility studies, it was stated that feasibility studies measure the extent of feasibility. In effect, a feasibility study asks if the project is likely to succeed given schedule, technology, and budget constraints.

Feasibility studies attempt to establish a robust start to the project in order to reduce risk. This involves not only forecasting technical viability but also documenting and responding to potential problems down the road. For example, a great deal of risk is associated with manufacturability, and so a feasibility study determines, ahead of time, if a model or other method can demonstrate whether or not a design approach is likely to have manufacturing problems. A certain argument can be made that the more time and effort are spent on feasibility analyses and trade-off studies, the less risky the project will be. If it is discovered that there is no way for a system provider to meet the requirements as a result of the feasibility study, then this is in some ways a victory. The overwhelming probability of an event or effort where time, money, and resources are wasted is avoided.

Potential problems and alternative approaches that are documented during feasibility studies feed directly into risk analyses. The various technological approaches studied during feasibility studies will become risk mitigation factors. If an alternative approach may be taken, then in the chance event that a current approach does not work, it will not critically hinder the project since there is another viable alternative available. Consider, for example, a portable surveillance equipment project that looks at wavelength and phase diversity technology to achieve ATC. They may initially decide to go with phase diversity because the technology is more mature. However, if they later find that the hardware to support image processing will be too heavy and too expensive, all is not lost since they still have the opportunity to pursue the second route (wavelength diversity) that was documented. The results of the initial feasibility study will also flow into the risk analysis and provide helpful information about the likelihood of failure, the severity of the failure, and the identification of models that may provide useful quantitative predictions and spreads of outcomes.

6.1.9 Transition to Optical Systems Building Blocks: Propagation of Radiation

A fundamental consideration of an optical system is that it must be able to take in enough optical power to produce an image or detectable signal on the detector. In this section, we take a closer look at radiometric aspects of the propagating electromagnetic wave to help us understand and model the relative signal levels between the source and the optical imaging system.

6.2 Optical Systems Building Block: Optical Radiation and Its Propagation

In this section, we discuss concepts and methods for propagating radiated source properties to a distant detection plane for the purposes of determining the amount of signal received by the optical system. Concepts such as absorption as well as common instruments, which measure the spectrum, are highlighted in this section. Lastly, the infrared (IR) and visible spectrums are highlighted.

In order to determine what the level of confidence is in the proposed optical system described herein, a method known as "decision engineering" (DE) will be used. DE is an emerging discipline that integrates a variety of contemporary engineering practices for the purposes of creating decisions. Some of the tools include traditional requirements analysis methods, use case and scenario planning, quality engineering/assurance, security, optimization, and design methods. "During the decision execution phase, outputs produced during the design phase can be used in a number of ways; monitoring approaches like business dashboards and assumption based planning are used to track the outcome of a decision and to trigger re-planning as appropriate" (Pratt and Zangari 2008).

One view of how some of these elements are combined to create the DE framework is shown in Figure 6.5.

Even though those methods for supporting decisions have been studied and developed for decades, these methods often find less traction than the traditional spreadsheet approach in decision making. "Decision engineering seeks to bridge this gap, creating a critical mass of users of a common methodology and language for the core entities included in a decision, such as assumptions, external values, facts, data, and conclusions. If a pattern from previous industries holds, such a methodology will also facilitate technology adoption, by clarifying common maturity models and road maps that can be shared from one organization to another" (Pratt and Zangari 2008).

Parnell et al. (2011) discuss the following methodologies in the book titled *Decision Making in Systems Engineering and Management*. "The lead systems engineer is responsible for guiding the team and ensuring that the decision making process is based on a proven approach method." Generally, when a complex system is being evaluated, a method known as "value-focused thinking" (VFT) is practical. VFT focuses on the values that people care about. For example, Sections 6.2.1 through 6.2.7 are based on measurements that can be

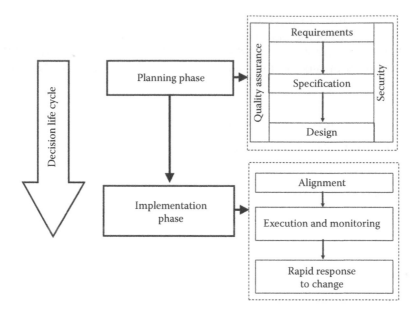

FIGURE 6.5
Elements of decision engineering. (Obtained from Prazan via: http://en.wikipedia.org/wiki/File:DEFramework. png, 2008, accessed November 26, 2014.)

quantified through various numerical and modeling analysis methods. The values that are retrieved from these measurements and models can be used by the VFT method since they are tangible metrics that can be used to determine the risks and uncertainties of the design. These values can also be used to determine actual and potential consequences of actions and inactions (Parnell et al. 2011).

6.2.1 Radiometry

Radiometry is defined as a scientific field interested in the study of the measurement or detection of the electromagnetic spectrum (McGraw-Hill Radiometry 2006). More simply, radiant energy is the energy transmitted by electromagnetic waves, and radiometry is the measurement of this energy. This portion of the chapter focuses on the measurement of the power (or flux) of electromagnetic radiation in the ultraviolet (UV), visible, and IR parts of the spectrum. We are interested in understanding source characteristics in relation to background radiation, the interaction of the radiation with its environment, and how it propagates and is detected. Figure 6.6 shows the source, target, channel, and the optical detection system.

When considering detection objects (e.g., targets), there are two broad source categories that are quite useful: point sources and extended sources. A point source has physical dimensions that are smaller than the projected pixel size in object space. An extended source has physical dimensions that are larger than the projected pixel size in object space.

In Figure 6.7, the concept of the steradian measurement unit is presented. A steradian (sr) is in the International System of Units (SI), a measurement of a solid angle on a sphere. A solid angle is the ratio of the area of a portion of a sphere that is r meters away from a point source to the square of the radius r of that sphere. A complete sphere is equal to 4π sr or about 12.57 sr. Notice that the area of the sphere is $4\pi r^2$ and that the steradians for the

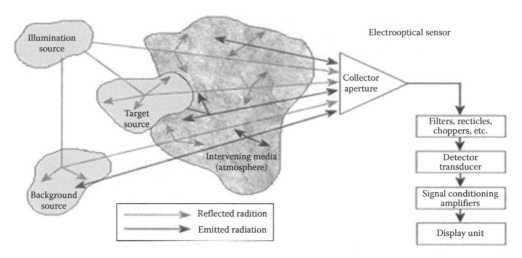

FIGURE 6.6
General detection model for optical system.

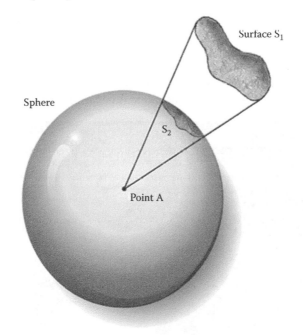

FIGURE 6.7
Geometry for determining solid angle in steradian. (Courtesy of John Green.)

sphere would be the area of the full sphere divided by r^2. In the field of radiometry, sr units are often used in the calculation of radiant intensity or how brightly an unresolved object shines in watts (W)/sr.

If we consider a point source that is radiating into a hemisphere, the associated steradian calculation would produce 2π sr. The following equation gives the solid angle that results from a point source radiating into a cone with half angle θ:

$$\Omega = 2\pi(1 - \cos\theta). \tag{6.7}$$

If θ is 90°, the result is the hemisphere of 2π sr. Using DE to evaluate the radiant intensity metric, the geometry for determining the solid angle in the radiant intensity metric is presented in Figure 6.7. Equation 6.7 supports the DE philosophy that creating a visual language (e.g., mathematical model) can provide better communications between stakeholders and the development team and promotes the broader use of analytical techniques and technical methods (Pratt and Zangari 2008).

6.2.2 Absorption

Electromagnetic absorption occurs in several ways, for instance, absorption is produced when a light source that produces a continuous spectrum (all the wavelengths of the spectrum are included) passes through a gas or a liquid medium (The Penguin Radiometer 2009). At the other end of this medium, the spectrum coming out has gaps and cuts, pieces of the spectrum absorbed by the medium or scattered out of the field of view of the detection system. Recall from an earlier chapter that when an electromagnetic wave propagates through a medium that contains atoms, aerosols, and molecules, the energy of the electromagnetic wave at a particular wavelength can be absorbed, scattered, or transmitted. A material's absorption spectrum shows the fraction of electromagnetic radiation that is absorbed by the material over a range of wavelengths relative to its incident radiation. Chemical elements have absorption lines at unique wavelengths that correspond to energy atomic orbital energy differences. The absorption lines given in this fashion are often used in determining the material properties of stars or other remote gaseous objects. In our case, we are considering the atmosphere to be our medium and the radiation source (e.g., star light) our target. Figure 6.8 is another illustration of the concept of absorption

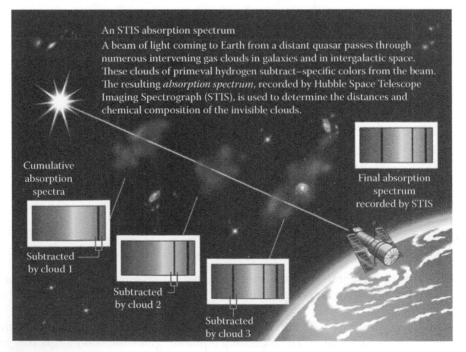

FIGURE 6.8
(See color insert.) Hubble Space Telescope Imaging Spectrograph. (Obtained from NASA/STSci: http://hubblesite. org/newscenter/newsdesk/archive/releases/1998/41/image/r, 1998, accessed November 26, 2014.)

using the Hubble Space Telescope observing a star at a great distance away. As you can see, the gaps in the spectrum are evidence of absorption. Assuming that there is no intervening atmosphere or medium between the star and the Hubble detector, the absorption lines detected are from absorption processes within the star itself providing a fingerprint of the star's composition. However, if there is dark matter in between the star and the sensor, the absorption lines could be due to the dark matter instead of the star.

Using DE to evaluate absorption, existing functional designs and working models are used to provide "reasoning structures." "Reasoning structures are used in complex decisions where the design aspect of decision engineering draws from other representative technologies (Cokely and Kelley 2009). In this case, the Hubble Space Telescope has been used to indicate that gaps in the spectrum can be directly related to Absorption, and in turn related to the composition of the star."

6.2.3 Gaseous Absorption Spectra

The rotational spectrum of a molecule is the result of changes in rotational energy. Likewise, the vibrational spectrum is the result of changes in the vibrational energy of a molecule (Figure 6.9).

The interaction within molecules is very complex. This may include vibrational overtone bands, additive or subtractive combination bands, and rotational frequency effects on the higher vibrational frequency. The speed of this vibration is proportional to the absolute temperature of a gas. The absorption of radiation and the resulting temperature changes can interfere with our ability to see temperature or radiation from an object. Figure 6.10 shows the wavelength (in microns) of the vibration–rotation band of carbon dioxide.

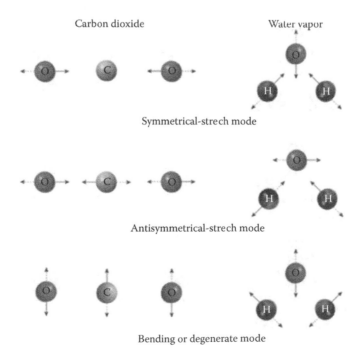

FIGURE 6.9
Vibrational modes of gases. (Courtesy of John Green.)

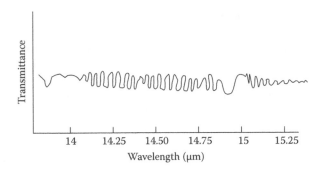

FIGURE 6.10

Fifteen micron vibration–rotation band of carbon dioxide. (Adapted from Hudson, R.D., *Infrared System Engineering*, Wiley & Sons, New York, 1961.)

6.2.4 Liquid and Gas Absorption Spectra

Liquids and solids have a much greater density than gases. Vibrational–rotational spectrum effects broaden and then disappear as the density increases. There is no associated shift in wavelength of the absorption band as you move from a gaseous to a liquid state. Rotational spectra evident in a gaseous state are not observed at all in a liquid state. As further temperature decrease continues, the absorption band narrows and shifts slightly toward longer wavelengths. Figure 6.11 gives a comparison of transmittance for the gas and liquid states of hydrogen chloride.

Figure 6.12 demonstrates the combined system effect on optical power. As you can see, the final detected power on the optical sensor is the product of spectral properties of the source, the propagation distance, the transmittance of the atmosphere or intervening

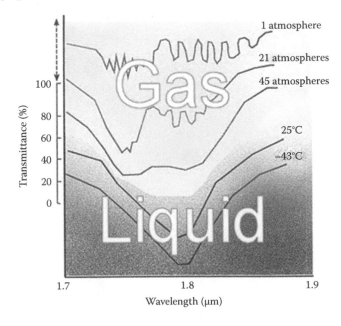

FIGURE 6.11

Relative 1.75 μ overtone bands for hydrogen chloride. (Adapted from Hudson, R.D., *Infrared System Engineering*, Wiley & Sons, New York, 1961.)

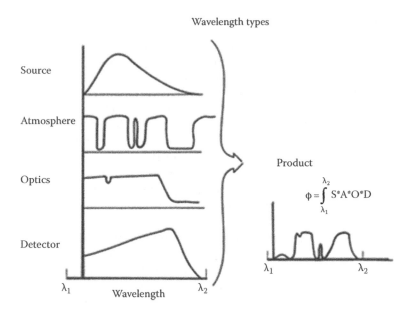

FIGURE 6.12
Combined source, channel, and system effects on optical power as a function of wavelength. (Courtesy of John Green.)

medium, the transmittance of the optics, and the spectral responsivity of the detector itself. For a radiometric detector, the detector material responds to the total power that falls on the active area of the detector (e.g., the power integrated over all wavelengths that is incident on the detector).

6.2.5 Radiometer

A radiometer is a device for measuring radiant flux over a broad spectral interval (broadband measurements). These devices are typically used in the measurement of radiation in visible, IR, and near-UV spectrum. The basic elements of a radiometer can be seen in Figure 6.13. A radiometer works by collecting some of the radiant flux from the source by the optics and focuses it on the detector element, which produces an electrical signal that is proportional to the incident flux. The typical radiometric quantity of interest is the flux (optical power) or irradiance on the detector (optical spatial power density).

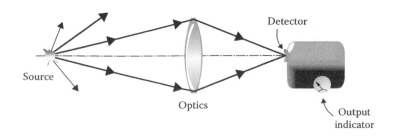

FIGURE 6.13
Basic radiometer elements. (Courtesy of John Green.)

6.2.6 Spectroradiometer

A spectroradiometer is a device for measuring the radiant flux within a small spectral interval (narrowband measurements). If there is no intervening medium, the spectroradiometer measures the spectral distribution (i.e., the variation in flux as a function of wavelength) of the radiant flux. It consists of two parts—a wavelength selection part such as a monochromator (which provides radiant flux over a narrow band of wavelengths) and a radiometer (which measures this flux). Within the monochromator, a prism or diffraction grating disperses the flux from the source into a spectrum. Then, a small portion of flux passes through the exit slit of the monochromator to the radiometer. The monochromator has a feature for "tuning in" different narrow bands of light around a center wavelength. These narrow bands of light can be tuned over a range of wavelengths permitting the determination of spectral properties of the source such as the source radiance (Figure 6.14).

Another important consideration of imaging is reflectance. Reflectance refers to the amount of radiation reflected by a surface as compared to the total amount of radiation incident on the surface (McGraw-Hill Radiometry 2006). As shown in Figure 6.15, the higher the reflectance ratio, the greater the reflected return of energy. In the visible part of the electromagnetic spectrum during daytime, the sun is the dominant source of electromagnetic radiation. Snow is an excellent emitter of visible radiation, but its reflectance is low. Clouds, on the other hand, are excellent reflectors of visible radiation. By calculating the reflectance ratio, a spaceborne optical sensor can discriminate between "white" clouds and "white" snow. The horizontal axis in Figure 6.15 is the optical depth of the material through which the light propagates.

6.2.7 Radiation in the Visible and Infrared Parts of the Electromagnetic Spectrum

This section discusses several useful techniques for detecting visible and IR radiation. The visible part of the electromagnetic part of the spectrum is dominated by reflected light. This is why it becomes dark as the sun sinks below the horizon. For low-temperature objects such as objects that have temperatures at or below our body temperature, there

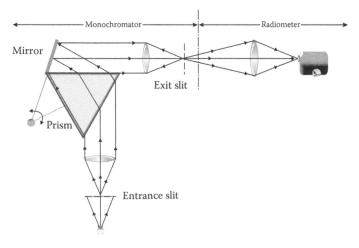

FIGURE 6.14
Spectroradiometer elements.

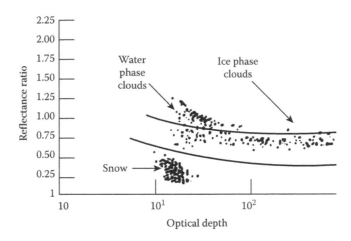

FIGURE 6.15
Reflectance ratio example with snow and clouds. (Courtesy of John Green.)

is not enough energy emitted in the visible band of frequencies to detect with our eyes. The IR spectrum, on the other hand, contains more detectable, emitted energy. The IR spectrum is consequently a particularly useful portion of the electromagnetic spectrum with regard to surveillance. Sensors that operate in this range can detect targets in adverse weather conditions where visible sensing technology would be unable to collect information on a target. The IR spectrum is beyond the visible spectrum; this segment of the electromagnetic spectrum ranges from 0.75 μm to 1 mm. There are different detection techniques involved for different parts of the IR spectrum. Table 6.10 summarizes the different techniques correlated to the portion of the spectrum.

Table 6.10 is helpful in identifying useful portions of the IR spectrum that can be exploited by IR detection systems. Sometimes, optical detection systems are mounted on manned aerial vehicle and UAV. Most of these airborne vehicles include a detection system capable of providing both visible and IR images. Some of these types of imaging systems

TABLE 6.10

IR Radiation Spectrum

Division Name	Wavelength (μm)	Characteristics
NIR	0.75–1.4	The closest part of the IR spectrum to visible light. Shorter wavelength than other parts of the IR leads to better spatial resolution.
SWIR or NIR	1.4–2.5	Spectral region for long-distance telecommunications.
MWIR or MIR	3.3–8	In guided missile technology, such as surface-to-air seekers (DESIDC 1990). This part of the IR spectrum has both reflective and emissive properties and has spatial resolution better than LWIR but worse than SWIR.
LWIR	8–15	Good for detection of passively radiating sources like humans. Works in pitch-dark, nonreflecting environments. Spatial resolution is worse than all other bands of the IR except the far IR (FIR).
FIR	15–1000	Closer to the microwave region of electromagnetic spectrum (NASA 2007). Useful in spectroscopy, plasma physics diagnostics, explosive detection, and chemical warfare detection. Less spatial resolution than other IR regions because of longer wavelengths.

include detectors in the visible range, the near-IR (NIR) region, the mid-wave IR (MWIR), and the long-wave IR (LWIR) part of the electromagnetic spectrum. Each of these regions may have its own detector and matched optics and can be considered a separate optical imaging system. The combined systems provide an integrated image detection capability. Lasers are also used at times to provide an optical source in dark conditions. The visible sensor provides high-resolution daytime images, where the IR systems provide lower-resolution thermal systems for nighttime or foul weather conditions. The MWIR (night vision) can work with reflected light (either moon, star, or laser in pitch darkness). The LWIR works strictly on emitted radiation.

Let us provide a quick example and consider the targets that a typical surveillance UAV will encounter. We will analyze the IR radiation coming from them and present some of the issues that must be considered in the overall optical imaging scenario. Our main target will be a human; in order to analyze the imaging scenario, we must determine which one of the following definitions will fit best:

- Point source—If the spatial resolution of the imaging system is larger than the maximum spatial extent of the object
- Extended source—If the spatial resolution of the imaging system is smaller than the maximum spatial extent of the object

As discussed in earlier chapters, if the optical system is within the earth's atmosphere, then the effects of atmospheric turbulence must be included in the spatial resolution estimate of the optical imaging system (unless ATC or AO methods are employed). The pixel size of the detectors projected onto the target has to be smaller than the spatial resolution of the optical system (either diffraction-limited spatial resolution for ATC/AO systems or atmospheric turbulence–limited spatial resolution for non-ATC/AO systems). For well-designed optical systems, the projected pixel sizes are usually matched to the spatial resolution capabilities of the optical system. For example, if the optical system has an ATC/AO system, then the projected pixels of the detector are designed to be smaller than the classical diffraction-limited resolution of the optical system. If there is no ATC/AO system present, then the projected pixels are matched to the expected spatial resolution of the atmospheric turbulence–limited optical system. A practical rule of thumb would be to see how many pixels are covering the target in a given linear direction. If 10 resolution cells (pixels) or more can fit in a given direction, then we can assume that the target feature is recognizable in that direction. For instance, if we wanted to identify an individual's eye color, then by this rule of thumb, we would want to have 10 resolution cells (projected pixels) across the 1 cm width of the eye equating to a desired spatial resolution of 1 mm for the optical system. This is a conservative estimate, and image features may be recognizable with fewer resolution cells. The number 10 should, therefore, be considered as a guideline.

An IR imaging device is a device that converts an invisible IR image into a visible image. The radiation available for imaging will, in this case, be emitted by targets (thermal radiation) or, in the case of the MWIR, can have both reflected and emitted components. Figure 6.16 is a high-level diagram showing the main components in an IR imaging system.

As mentioned earlier, it is common that airborne surveillance systems provide a combination of visible and IR systems into a single integrated package. There is typically a visible detection system that detects light in the visible part of the electromagnetic spectrum,

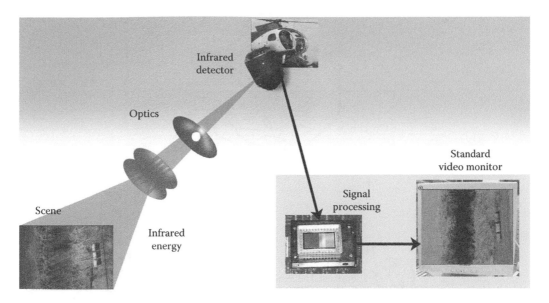

FIGURE 6.16
(See color insert.) Basic infrared imaging detection system. (From Richard, Q., 2014, accessed April 19, 2014.)

a thermal sensing system that detects emitted radiation in the form of heat, and a night vision system that works with reflected light and that typically provides slightly better spatial resolution than the LWIR imaging system. The visible optical system works if there is enough reflected light from the TOI such as in moonlight or when illuminated by a laser. The night vision subsystem amplifies the detected electrons from the low-light visible or IR radiation and displays the result on a viewing screen.

A key aspect to consider is that the optics and detector have to be able to capture sufficient radiation to detect the subject or target. Therefore, we need to make sure that we take the radiometry of the source into account. In a radiometric sense, we are interested in first determining a relevant radiometric quantity such as the source radiance (N) in flux per object area per solid angle or $W/(m^2 \, sr)$, radiant emittance (W) in flux emitted or reflected from the source per source unit area, or intensity that is flux per solid angle or W/sr. Any one of these relationships will do; however, the intensity is one of the easiest to work with. If we know the intensity of the source, we can then determine the optical power that is collected by the imaging system by first multiplying the source intensity by the solid angle that is formed by the collecting aperture of the imaging system. A good approximation of the subtended solid angle is given by dividing the area of the collecting aperture of the imaging system (e.g., the entrance pupil for well-designed, telescopic imaging systems) by the entrance pupil to source/target distance squared (in meters). This result must also be multiplied by the atmospheric transmittance for broadband detectors or the spectral atmospheric transmittance corresponding to the wavelengths for spectrally sensitive detectors such as spectroradiometers. Once the optical power is determined in the entrance pupil of the optical system, it is then the function of the optics to relay this power onto the active detector surface. The power in the entrance pupil is then multiplied by the transmittance of the optics resulting in the optical power that is deposited onto the detector surface. Many useful detection metrics can then be determined such as the detector output signal level, the detector output signal-to-noise ratio, and the detector detectivity.

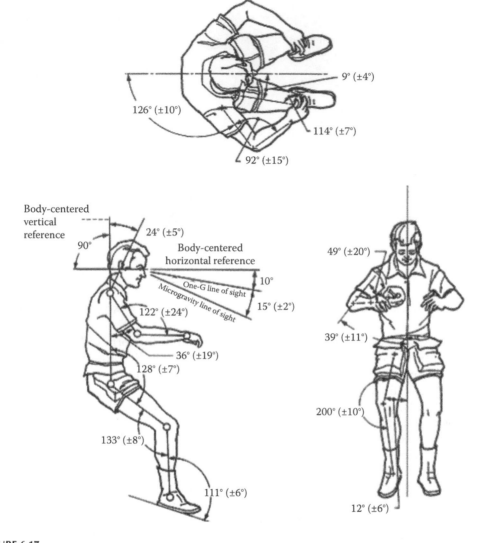

FIGURE 6.17
Target cross-sectional area from different aspects. (From Griffin, B.N., *The Influence of Zero-g and Acceleration on the Human Factors of Spacecraft Design*. JSC-14581, NASA-JSC, 8–78, 1978.)

The specific case of an airborne imager presents a unique situation in which the area of a human can change depending on the perspective of the sensor. If the sensor is directly above the TOI, the area that needs to be considered as a source is considerably smaller than the source area if the imaging system is looking at the target at an angle. For illustration purposes, let us look at the different areas of a shape from different perspectives (Figure 6.17).

The images shown were from a NASA Skylab study (Griffin 1978) that evaluated natural body positions (shown by the angles and standard deviations in the image). For our purposes, we note the difference in cross-sectional area from a frontal and top-down view. Note that this difference would be even more pronounced if the person were standing. Let us look at the surface area of our average human: assuming that our target is about

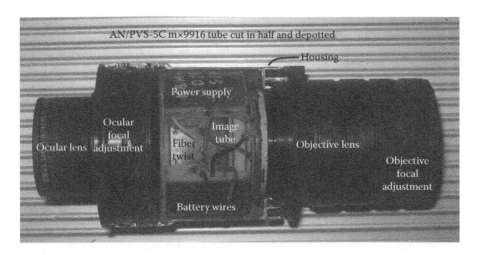

FIGURE 6.18
AN/PVS-5C night vision system. (Courtesy of David Kitson. http://en.wikipedia.org/wiki/File:AN-PVS-5C-Cut_image.jpg, 2010, accessed November 26, 2014.)

170 cm tall and about 50–60 cm wide, our total area from a horizontal perspective will be about 2 m² based on average humans. However, if we consider an angle of 25°, we have a surface area of about 74 cm × 60 cm or 0.445 m² for our optical system. These adjustments need to be made when determining the source radiometric quantities.

In our introduction, we mentioned that the IR optical detection systems mounted in a UAV need to be able to detect different parts of the IR spectrum (e.g., short-wave IR [SWIR], MWIR, and LWIR). We mentioned that a night vision system is typically available to gather reflected low light levels and amplify the results so that an observable image is generated. Figure 6.18 demonstrates how night vision sensing works.

The model that best describes a night vision imaging system is a photon model. Imagine that under low-light-level conditions, a source photon is detected by the night vision system. In one configuration, the detected photon causes an electron to be generated from a photomultiplier tube initial detection surface. This electron is amplified through several amplification stages where a single electron generates multiple electrons (say 10) at each stage. At the first stage, the single electron generates 10 electrons. At the second stage, the 10 electrons each generate 10 electrons, and so we have 100 electrons; the third stage leads to 1000 electrons and so forth for each stage. After six stages, this scenario would produce one million electrons detected. The low light levels would easily be detected after the amplification stages. Notice that the night vision system still requires some reflected photons to be present. In pitch-blackness, these night vision systems do not work. Instead, thermal systems are required that detect emitted light.

For thermal optical detection systems, the radiation of an object changes according to its temperature. We will calculate that the radiation emitted by the target is enough to be detected by our imaging system and then use this to create a thermal image. A thermal image of a human is shown in Figure 6.19.

At this point, we have discussed what IR radiation is and how it plays an important part in the surveillance detection systems, especially airborne imaging sensors. We also saw that since human bodies have a relatively fixed, stable temperature, the IR radiation emanating from them can be easily identified by typical surveillance equipment. We also

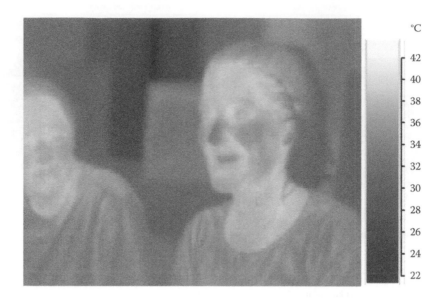

FIGURE 6.19
Thermal image of human girl. (Courtesy of Masgatotkaca from http://en.wikipedia.org/wiki/File:Ir_girl.png, 2008, accessed November 24, 2014.)

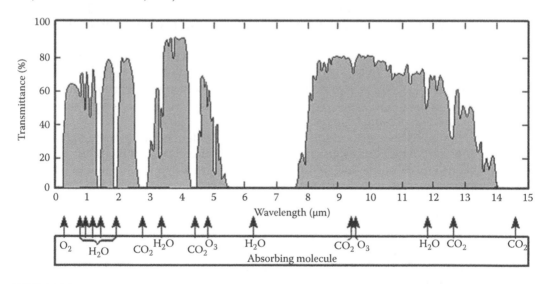

FIGURE 6.20
Atmospheric transmittance as a function of wavelength. (Courtesy of U.S. Navy, http://web.archive.org/web/20010913094914/http:/ewhdbks.mugu.navy.mil/transmit.gif, 2010, accessed November 26, 2014.)

mentioned how the propagation of radiation through the earth's atmosphere is affected by reflection, scattering, and absorption. In Figure 6.20, we see the effect that the atmosphere has on optical radiation.

At the lower end in the UV part of the electromagnetic spectrum, the atmosphere is opaque and strongly absorbs UV radiation. You can also see strong molecular absorption due to water (H_2O), carbon dioxide (CO_2), and oxygen (O_2) as an example. In these regions, radiation is strongly absorbed, and so building long-range detectors or sources in these regions is not advisable or practical.

The next section presents an integrated case study that focuses on how feasibility analysis and optical propagation analysis would be used in a fictitious working environment. The intent is to demonstrate how fundamental SE principles and methods such as feasibility analysis work together with technical analysis in understanding, analyzing, and characterizing optical systems.

6.3 Integrated Case Study: Establishing Technical Feasibility through Optical Propagation Analysis

In this section, we continue with our integrated case study and show how the SE concepts of feasibility analysis, trade studies, and alternative analysis are used in conjunction with the optical technical principles, learned in the previous sections, in a simulated realistic technical setting. In this case study, we see FIT coming to grips with the need of a feasibility study to show that their concept will work for their primary stakeholder, the Department of Homeland Security (DHS), U.S. Customs and Border Protection. In preparation for their first formal kickoff meeting with their customer, FIT is diligently conducting feasibility analyses, trade studies, and alternative analysis, including optical propagation analysis and spatial resolution analyses, to show that their conceptual, optical detection scenario is feasible. The UAV platform that will house FIT's optical detection system is shown in Figure 6.21.

(The kickoff meeting was proposed to begin at 10:00 a.m. Monday morning.)

In this integrated case study scenario, we have Dr. Bill Smith, FIT's chief executive officer (CEO); Tom Phelps, FIT's chief technology officer (CTO) and optics expert; Karl Ben,

FIGURE 6.21
Reaper unmanned aerial vehicle (UAV). (Obtained from USAF Photograph Archives via: http://en.wikipedia.org/wiki/File:MQ-9_Reaper_in_flight_ (2007).jpg, 2007, accessed November 24, 2014.)

FIT's senior systems engineer; Ginny R., FIT's UAV optical system program manager; Ron S., FIT's systems engineer; Phil K., FIT's software engineer; George B., FIT's hardware specialist; Amanda R., FIT's mechanical engineer; Wilford Erasmus, chief of operations and procurement, U.S. Customs and Border Protection, DHS (primary stakeholder); and Dr. Glen H., DHS technology specialist.

6.3.1 Part 1: An Unexpected Meeting

Karl (stopped by Ron's office to alert him about the impending meeting): "Ron, I would like for you to attend the kickoff meeting with our stakeholder on the predator drone UAV (unmanned aerial vehicle) project. There may be some questions on some of the analyses that were done, and Tom and you can provide the details."

Karl: "By the way, please tell me that the optical propagation analysis has been completed. I need to get the results to Bill prior to the meeting with DHS."

Ron: "Yes, Tom's group completed them over the weekend. The results show that we can reliably get sufficient signal to detect a human. The optics group now has a basic optical model that links the source, through the atmospheric models, through their basic sensor optical model, and onto a generic detector. They also did some initial spatial resolution studies using the software, but you'll have to talk to Tom about those. I'm sure Tom will be more than happy to talk about it. It looks like we won't need an active imaging system and can get by with a passive imaging approach."

Karl: "That is great news! I will tell Bill. That will put a smile on his face. He hasn't had his first cup of coffee yet!"

Ron: "What's the scope of the meeting with DHS?"

Karl: "As you know, our stakeholder is looking to install an enhanced optical imaging system onboard the UAV to monitor the U.S.-Mexico border for illegal border crossings and have selected FIT to build, integrate, and support the optical system for their UAV fleet. They've done their internal feasibility studies, trade studies, and analysis of alternatives already, and they want to compare their results with ours as part of the project kickoff meeting."

Ron: "Will decision engineering process come into play at some point?"

Karl: "You know, I've heard of decision engineering, but I don't really have too much exposure to it. Why would you suggest decision engineering processes for this project?"

Ron: "The methodology is based on a number of best engineering practices for making decisions and promotes a structured approach to problem solving."

Karl: "That's interesting. Let's see how this meeting goes, and if you feel we need to apply it, then open the discussion on this topic."

Ron: "OK. I will look over the documents you gave me and put some notes together for discussion. The more information I have, the better, so can you tell me a little of what we want to accomplish in the meeting?"

Karl: "Sure! The stakeholder had supplied an aggressive schedule requirement, with some operational requirements that imply the need for some fairly advanced technology. Because of the technology and schedule constraints, this could be a fairly risky project to undertake. The goal of this meeting is to go over some of those stakeholder requirements and documentation with the stakeholder and ask questions that will aid our feasibility studies and get us connected to their technical team. Over the coming weeks, we'll need to perform a feasibility analysis and trade-off studies to see if this optical system is practical within the time and cost targets that they are looking for. Getting a feel for operational priorities during the meeting is also important, since it will help us to design a successful system for our stakeholder. It'll be important when we conduct alternatives analyses and trade studies."

Ron: "I hope this meeting will conclude with sufficient information that will help us to get a clearer understanding of the values that drives the alternatives instead of the traditional method of just defining the alternatives first."

Karl: "Well all right then! That all sounds good! OK, let's head to the meeting room."

(Karl and Ron arrive at the meeting room.)

Karl: "Good morning Tom."

Tom: "Morning Karl. I'm told that our meeting today will have to be brief, due to the customer's travel schedule."

Tom: "Dr. Smith, the CEO, will be down in just a minute with two representatives from DHS. We will have a short introduction and then transition quickly to the program kickoff discussion."

(Other attendees arrive at the meeting.)

Bill: "Good morning and welcome to you all. For those who don't know me, my name is Bill Smith. I am the CEO for FIT. As this is our first face-to-face meeting, I'd like to go around the table and have each of you introduce yourself to the rest of the room, starting with Tom."

Tom: "Thanks Bill. Good morning! My name is Tom Phelps, and I am the chief technology officer for FIT. Karl?"

Karl: "Good morning everyone. My name is Karl Ben. I am the senior systems engineer at FIT. Ron?"

Ron: "My name is Ron, and I am a systems engineer here at FIT. Thank you."

Jen: "Hi, I'm Jen, and I am also a systems engineer here at FIT."

Ginny: "I'm Ginny, the program manager for this effort. Hi Wilford, good to see you again!"

(Wilford waves.)

Wilford: "My name is Wilford Erasmus, and I am the chief of operations and procurement for the U.S. Customs and Border Protection, Border Patrol. We're under the auspices of the Department of Homeland Security."

Glen: "My name is Dr. Glen H. I have been asked by Mr. Erasmus to act as a technology liaison between DHS and FIT on this program. I will answer any technical questions that you may have regarding the requirements of the optical system planned for the predators while I'm here today. I can answer any specific optical system operational questions that you may have."

Bill: "OK. Thank you everyone for those introductions. To that end, I'd like to cede the floor to Tom and Ginny, who have been looking through the signed contract we received from DHS."

Karl: "Mr. Erasmus, would you please elaborate on the contract: What other stakeholders would you like us to include as part of this effort? What is the current status of funds on the contract and are you considering any options in the future? Is there an expected price point per system?"

Wilford: "No problem, and call me Wilford! The contract comes from DHS. The contract is cost plus incentive fee (CPIF). If the contract is delivered on time and within budget, the contractor receives a 10% incentive fee. If the contract comes in either over budget or over schedule, the contractor incentive fee is reduced to 5%. The total contract award will be for 22 optical imaging systems for our current fleet, with an additional 8 spares, or 30 total systems. The first unit must be integrated into our fleet not later than 2 years from contract award, 8 units between 2 to 3 years, and the remaining 21 units by year 4. Each optical system unit, complete with analytical software, and mission control center/ mobile command center image/data distribution formatting capabilities will be developed/produced at a not-to-exceed cost of $4M per unit. That brings the development portion of the contract to $120M. Life-cycle service, support, and preplanned product improvement (PPPI) is 8% per annum, adjusted for inflation of the overall contract award, or $9.6M per year plus inflation adjustments. Expected period of performance is 10 years after delivery of the last units, with an option to extend to 25 years. Based on performance, there are also options in year 4 to develop ground-based optical systems and helicopter-based optical systems as well as some possibilities for additional systems for some of our silent friends.

   As far as stakeholders go, the information collected by the UAV will be used primarily by border patrol agents, but we will also share our border image and video database with other agencies: Central Intelligence Agency, Interpol, local law enforcement, you name it. The actual operators of the drones are based in our mission control center at Schriever AFB in Colorado Springs, Colorado. If you look in the contract, there are some usability requirements to that end and some human factor information such as the typical skill level and training of the operators."

Tom: "Mr. Erasmus, are there other borders that the system will be monitoring or just the U.S.-Mexico border? What is the typical climate and terrain in the monitored area? What is the typical population expected to be in the monitored areas, and why is this border being monitored?"

Wilford: "For now, the target is the U.S.-Mexico border. We want to monitor ten variable "hot spots" along the border. These are all away from population centers with variations in terrain indicated by the different landscapes. The hot spots will likely change over time. We should be prepared to deal with all terrain types along the border. The reason why border patrol is monitoring hot spots is to detect human foot traffic for immigration, customs, and homeland security. For the most part, the current systems in place can detect vehicles coming from a mile away. Detecting living creatures and categorizing them as human traffic versus critters is the challenge, and doing it day or night is the driving need for this system".

Karl: "Dr. Glen H., I think I read in the Request for Proposal (RFP) and the subsequent contract you issued to us that you have ten predators and plan to get more based on need. I also saw that target detection should be provided 24 × 7 days with a 98% operational availability. I did not see where the UAV airstrips were relative to the hot spots. But you said the ten hot spots could move, and there is predator transit time between the airstrips and the hot spots to be considered, not to mention maintenance time and MTBF to be factored in. Bottom line, we will need to propose procurement of additional predators to compensate for all the unavailable times. How would you like us to account for all these variables?"

Glen: "Call me Glen! Good catch! At the time of the RFP, we had 10 UAVs in the fleet. Since then, we have added 7 more to our inventory, and we will have the remaining 5 units by year's end. You should document all of these variables in your feasibility study. Off the top of my head, I know our UAVs fly out of Fort Huachuca, Arizona, and a naval air station at Corpus Christi, Texas. There are other bases we could fly out of with the say-so of DHS. In general, there's no U.S.-Mexico border area that is more than 300 miles away from a base capable of housing predator drones. The MQ-9's that we have cruise at roughly 200 miles per hour. We set a maximum round trip at 3,600 miles or 14 hours of flying time. For a given mission, we use the lower limit to guard against equipment loss. For that reason, we also want to fly them high and well away from the border. My memory is pretty good, but you should check the location, standoff, and predator specs. I would think you might create a model of a system that can satisfy those criteria, randomize the hot-spot locations and compute the distribution using a Monte Carlo method, and model the transit times to the hot spots accordingly. That should give you an estimate on the most likely number of predators needed to satisfy the mission."

Bill: "That reminds me, we may have an interesting way to determine the 'hot-spot' locations using game theory. We have been talking with some professors from a local university that have been working on this, and they have some promising results."

Wilford: "Really? I know the air force, army, and navy have shown some interest in decision aids. I think this will tie in very well with some synergistic efforts that we have and may address a need we have for this program as well. I am very interested, keep me posted."

Bill: "We certainly will!"

Wilford: "OK. Anyone have a question they need answered now?"

Ron: "Mr. Erasmus, can you rank the importance of the operational parameters and characteristics on a scale of one to nine on behalf of your organization?"

Wilford: "On a scale of one to nine, being able to classify humans and get the operator's attention has to be a nine. Operational availability is also pretty important. We can't enforce border security if we don't know what is going on. I'd call it a seven. The schedule is a six. Normally, there is some wiggle room, but there is enormous political pressure on us from above to start collecting data about border crossings ASAP. Usability is about a five. All the bells and whistles you can provide won't matter if our operators can't consistently and effectively use them. Cost is maybe a three. We'd be willing to go for a somewhat more costly system as long as it is justifiable in giving us features that we want beyond what we need. Flexibility and adaptability come to mind. Finally, safely operating the UAV such as preventing it from crashing and being shot down is only as important as it affects operational availability and cost. Maybe that's a two. We've been operating our UAVs for a while now, and losing our birds doesn't appear to be a problem in this area. Ordinary maintenance concerns are far more palpable. In fact, I'd say maintainability is about a four".

Wilford: "Dr. Smith, I apologize, but we need to wrap up this meeting. I have a plane to catch back to Washington, and Dr. Glen H. is heading out to Texas to observe a mobile border patrol control center in operation".

Bill: "Excellent. Thank you everyone for your time. FIT team, I would like for us to meet Wednesday morning, 9 a.m., to discuss the status of our feasibility analysis."

(Everyone filters out of the conference room.)

Karl: "Ron, good job in the meeting. Both Tom and I forgot to put stakeholder priorities on our agendas. That would have set us back several days waiting on Mr. Erasmus to get back to us with that information. Tom and I are getting together to start brainstorming over the contract details for a while. Later, you and I can discuss some of the material presented during the kickoff meeting and any notes you collected."

Ron: "Great! I'd like that."

6.3.2 Part 2: Lunch and Learn

Karl: "Hi Ron. You have the system designed and built yet?"

Ron: "Oh, sure! I'm still trying to figure out how to get started!"

Karl: "Ron, can you assist me with the contract details a little bit?"

Ron: "That would be fantastic!"

Karl: "Good! Let's go down to the cafeteria and grab lunch. We'll stop back and grab your notes and my laptop on our way to the discussion room."

(After a tasty lunch, Karl and Ron settle in the discussion room.)

Karl: "Well, we need to incorporate the information we learned today as soon as possible. If you recall from the meeting, the border patrol is looking to equip UAV aircraft with an advanced optical system that they can use to spot 'bipedal targets' trying to illegally cross the US-Mexico border. DHS has included a high-level statement of work (SOW), which is basically an itemized list of the work and deliverables that they need accomplished as part of this effort. The contract from DHS is asking for us to respond to each requirement with a feasibility study. It states that we need supporting documentation for the design approach we take, so that means we'll be doing an analysis of alternatives (or AoA) and getting Dr. Glen H.'s input and weigh in. It's also asking for trade-off study that supports our system configuration."

Ron: "What is a bipedal target?"

Karl: "Human beings. This presumably implies illegal aliens crossing the border. They don't care about wild animals (dogs, goats, horses, whatever else). They also don't want to detect weather events, haboobs (that's a dust storm, by the way) or any other noise or clutter."

Karl: "We have to develop an optical system that can be installed as the payload of a UAV aircraft (specifically, the MQ-9 version). There needs to be enough of these systems and spares to allow DHS to monitor ten 'hot spots' or target areas of land at a time, 24 hours a day, 7 days a week, with 98% operational availability. The hot spots can change over time, and the complete fleet must be sufficient to cover the time required for the UAVs to travel from their bases to the hot spots, all while accounting for the MTBF and maintenance characteristics of the optical and UAV subsystems. The optical systems must store the images captured for a period of time, and the images must also be streamed live via satellite to the border patrol command center. We're mainly concerned with an optical system that can fit into that Raytheon MTS-B-20 gimbal pod under the nose there."

Ron: "We can design a new pod if we need to, right?"

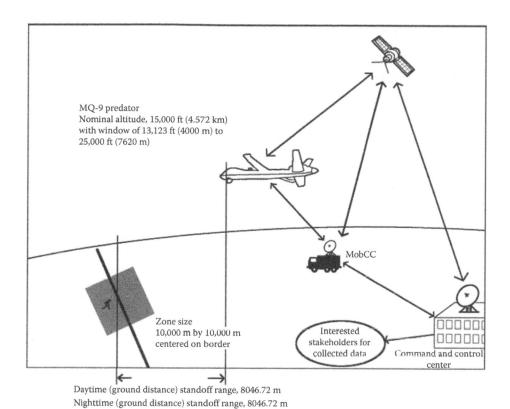

MQ-9 predator
Nominal altitude, 15,000 ft (4.572 km)
with window of 13,123 ft (4000 m) to
25,000 ft (7620 m)

MobCC

Zone size
10,000 m by 10,000 m
centered on border

Interested
stakeholders for
collected data

Command and control
center

Daytime (ground distance) standoff range, 8046.72 m
Nighttime (ground distance) standoff range, 8046.72 m

FIGURE 6.22
Revised basic CONOPs diagram.

Karl: "Yes and no. I say no because you heard the stakeholder describe the schedule situation. Even if they were willing to pay for a new mount that would allow a giant aperture size and crystal clear pictures, they simply can't swing the schedule. It's not feasible for that reason."

Ron: "I guess the theme is that we should do our best to work with what is available. From the simplified CONOPS, Figure 6.22, I see that we need to send our visuals through the existing satellite and mobile command link ups. I'm guessing there's a bandwidth limit, but there's no way to know how much information we need to send until we know about our detector."

Karl: "Yes, that was one of the things that I talked with Tom about. One of our previous projects developed a 2048 × 2048 detector, detector control system, and some relatively reusable software. We're going to make a systems engineering assumption and use that for our initial feasibility studies. It's a good baseline to start from. It's already well-developed technology, so there's almost no schedule risk and no need for the stakeholder to pay for developing it. That will help us manage costs if the detector passes the initial feasibility checks. Anyway, the number of pixels leads directly to the pipeline bandwidth question that you brought up, and it leads to the distances on the CONOPS. We need to project the pixel size out from the maximum and nominal height to the target. Then we'll figure out what we can and cannot see based on the detector itself. We also need to look at the spatial resolution capability of the optics with and without our atmospheric turbulence compensation or adaptive optics systems."

Ron: "But for that, we need to know the instantaneous field of view (IFOV). We'd probably want to relate the IFOV determined from the detector pixels when they're projected with the height and worst-case and nominal standoff distances. We'd make the pixel dimensions the size we need and work backward to get the IFOV so that our unzoomed picture is at a sensible resolution for its operating distances. We'd also have to consider the workable effective focal length, but that can be our starting point. There are a couple performance parameters there: magnification, detector pixels, and distance from the target. We do a trade-off study to vary them for the best mix for our operational criteria: detecting human traffic and usability. For example, I doubt our operators would be very happy or productive if they have to play around with the zoom system."

Karl: "That's a good idea for a trade study. That might give us a quantitative justification for how we configure the optical subsystem. However, you mentioned something that made me realize we're going about this backward. We should start with the requirements first, and then figure out the projected resolution cell size we need, and then figure out if we can do it before worrying about how we're going to put our mirrors, prisms, lenses, and detector together."

Ron: "I'm starting to feel like we are losing site of the requirements, and I also feel like we need to take a step back."

Karl: "If we go back to the requirements, it doesn't say anything about how many pixels we have to put on a target; it just says we need to recognize it as a person rather than a coyote or some other critter. How many pixels do we need to do that?"

Karl: "Actually, Tom and I were just talking about that. Some human factors studies suggest that an operator viewing a screen needs 10 or more well-defined pixels on a target to determine whether it moves like a human or not. I'm assuming that if we're going to write a detection algorithm that alerts the operator that something is moving like a human, we need to have 10 or more resolution cells on the target. If we translate that into Δx, the symbol optical engineers at FIT use for projected resolution cell edge sizes, we'd need something in the range of 0.1 and 0.2 meters."

Ron: "So we need at least that for our longitudinal and transverse cell detector cells. The standoff distance between the target and UAV could be as much as 18.616 kilometers according to the worst-case scenario and 10.995 km for the nominal scenario with the new relaxed requirements from today's meeting. What's the largest aperture size in that pod thing?"

Karl: "It's about 0.5 meters."

Ron: "With those numbers, we should probably check to see whether we're going to have sufficient spatial resolution to be able to classify targets as human or not. We might need atmospheric turbulence compensation (ATC) to give us the spatial resolution that we need."

Karl: "Why don't you plug the new numbers into the new models that Tom's group and Jen developed and see what we get?"

Ron: "OK, I'll get on it after this meeting. Let's run down on our spatial resolution feasibility study strategies. For human detection, we want to determine if we can get a Δx of 0.2 meters at 18.616 km, the worst-case scenario, by checking it against the diffraction limit, the coherence length limit, and our estimated projected pixel size."

Karl: "Don't forget we have a nighttime requirement. We'll have to use infrared for that, but the process is basically the same."

Ron: "Then there's availability. I guess our optical system really won't affect the number of predators needed by the stakeholder, since it's mostly a travel and position game."

Karl: "Yes and no. We should still do the feasibility study there. If we don't do it, we may not have enough sensors to reliably meet the DHS mission. Remember Bill's speech at last week's meeting? We need to make sure that what we deliver fully satisfies their needs. Also, we need to write our own system requirements, so our subsystem failure rate and maintenance concept don't significantly degrade the operational availability of the whole system. I think there's plenty of field data available for the type of drone we're working on, so we'll be able to calculate what we need for our overall optical system. Also, we need to make sure that we consider COTS solutions if they are feasible and available."

Ron: "Based on the ranking that Mr. Erasmus gave us, we should consider a fast-turn maintenance concept. COTS, if I'm not mistaken, can really help here in terms of reliability and maintainability. There is a trade-off analysis that needs to be made though to consider if a preexisting solution meets the very fast maintenance concept. If the system is built with modules that can be replaced in the field, or field replaceable units (FRUs), those can be swapped out fast and sent to a workshop after the fact for diagnosis. We can show our maintenance concept, and MTBF system requirement will be feasible for meeting the availability requirement."

Karl: "OK, I have a feeling that when we meet with the CEO next, we'll be put on an integrated product team (IPT), and we'll be able to consult a communications guy for the bandwidth issue we have. I'm not sure if our satellite link-up will allow us to stream 2048 (pixels) times 2048 (pixels) times 3 (colors) times 12 bits (quantization per pixel) times thirty (image frames per second). That's a rough estimate of the data we would have to transmit to get real-time video with the full field of view. I'm assuming we'll end up compressing and sending image segments or something like that. We'll also have to consider all of the options—including COTS video compression and image management software."

Ron: "OK. For usability, it would probably be best to quantify that early and show it's feasible somehow."

Karl: "That one is difficult to handle. My initial plan is to look at what software predator drone operators have to use, and then borrow the metrics they used to show that software met usability requirements for the ground station (GS) skill level of the operators. Essentially, we'll have to write requirements for all of the functions we add: compression, ATC, zoom, edge detection mode, etc., won't significantly change the existing interface and negatively impact the metrics. Once again, if we don't get too carried away with our requirements, we may be able to use a COTS solution."

Ron: "I can't think of any other operational requirements off the top of my head from the meeting. I'd need to see the actual contract document to work with you further."

Karl: "I'll e-mail that to you after we finish here. Anyway, after we handle our feasibility studies for our optical system, we'll have a better idea whether the project is doable. We've already been discussing a few alternative approaches we might take and some trade-offs between system characteristics. Our next steps after that will be to decide on which of the possible technical approaches we might take. For that, you already helped out by getting a rough estimate of the stakeholder's ranking of operational characteristics and attributes during the meeting."

Ron: "That reminds me of the analytic hierarchy process (AHP) that Arlene and Jen did last week. You know, where you make a stakeholder criteria matrix and then make technical solution matrices in response to each criterion. It's also a good time to talk about decision engineering and how we can actually design some of the decisions that we make during the process."

Karl: "I was about to get to that. The AHP tool is a great tool. I am sure we will be using that often. We can also use your suggestion for decision engineering. We'll have to figure out which stakeholder criteria matter the most for our technical evaluations, choose the pair-wise preferences, and implement the AHP. Don't forget the consistency ratio as a check on the consistency of your assigned ratings for criteria–criteria and alternative–alternative matrices."

Ron: "Also, how do you do your sensitivity analysis?"

Karl: "I use the Monte Carlo approach, where I can randomly vary the stakeholder criteria ranking and options to see if the outcome changes. That models what happens if the stakeholder's opinion changes. Using the same approach, I can vary the solution–solution ranking, usually by varying the figure of merits (FOM). I don't do that randomly of course. Well, you can do it randomly, but I mean in an ordered way. Basically I nudge the parameters used for comparison in a physically authentic way and find the variance of the statistical distribution if it's applicable. There are a lot of ways to estimate FOM variability. We have lots of internal data available from past FIT projects, so we have real data whose distribution we can borrow. Otherwise, we can take a mathematical model that outputs the FOM and perform an error propagation analysis and inputs realistic measurement errors."

6.3.3 Part 3: The Trade-Offs Begin

(At a working lunch meeting the next day.)

Ron: "Gary, we have a problem. I did a calculation of the required spatial resolution. Visible is working just fine; however, infrared may have an issue."

Karl: "What? I'm sorry Ron, but I haven't gotten down to doing the studies just yet. Can you explain?"

Ron: "The pixel size is too large with 9.5 micrometers radiation at 10.995 km. It's larger than 0.25 m."

Karl: "That's not too far from the 10 segment rule of thumb. That is fairly close to the required 0.2 m. It may actually be sufficient where it is now for a human to reliably classify the target as human or not. Remember, the 10-segment rule is a guideline. I am not too concerned at this point. Have you run the worst-case scenario yet?"

Ron: "I was just getting to that next. Let me punch the numbers into the model and see what we get. (Ron's fingers fly on his notebook keypad.) At 18.616 km standoff range, we get about 0.43 m with the AO or ATC system."

Karl: "That is getting a little harder to accept. That only gives us about 3–4 resolution cells on target. What options do we have?"

Ron: "Changing the pod will not make the stakeholder happy. That will cause a very long redesign delay and put a lot more risk into the production phase. So we have to fly lower, which, to my knowledge, doesn't affect anything. Why do they want that altitude?"

Karl: "I'm guessing for political reasons, they don't want to be seen, heard, or shot down. In terms of being heard and being shot down, I'm pretty sure the UAV could fly miles closer and a mile lower without any problem. As far as visibility, it probably wouldn't matter at night."

Ron: "I am not sure I understand. Can you explain please?"

Karl: "Let's perform an AoA using the AHP and try to convince our stakeholder that our way is the best. We have three technical solutions: a new gimbal, fly lower and closer, or do nothing."

Ron: "OK, I'll create a spreadsheet and put this together."

Karl: "Excellent."

6.3.4 Part 4: Changes

(Wednesday, FIT team meets again.)

Bill: "I got a memo yesterday that the operational parameters for the UAV border patrol detection scenario have changed. The flight floor and standoff range have been lowered for night operations. Mr. Erasmus said that the change resulted from an exchange between our technical teams. Good work!"

Bill: "So far, it seems like the feasibility studies Karl and Ron are conducting have gone quite well. I'd like to know how this affects our radiometric propagation analysis. Tom?"

Tom: "I'd be glad to tell you about it; Ron, Jen, and our team worked pretty hard getting an integrated radiometric model ready. We can now rapidly do 'what-if' scenarios like this one, when changes occur. We decided to run the worst-case scenario first. If it works for that, then changes like the one you mentioned today (e.g., ones that increase the signal and relax the constraints) will work also. In any case, we first updated our source model. The human cross section of 2 m^2 seemed to us to be too large for the expected population (adults and kids), but this number is used regularly in the literature, and so we decided to keep it but make some parameterized adjustments. Next, the emissivity of 0.98 is for a naked human body. I don't think this will be the case for the folks we are interested in. We accounted for a good portion of the body being clothed (80%–90%) and reduced the effective emissivity to 0.9. This gives an effective radiant emittance of 441.6 Watts/m^2. Given that this radiant emittance is a broadband result and that our optics system has a narrowband filter for clutter reduction that is centered on the peak wavelength of 9.5 µm, we had to develop a spectral model of the source radiant emittance. The source radiant emittance in the pass band of the narrowband filter is 35 Watts/m^2. This result assumes that we are looking straight on to the target, but the sensor is on a UAV platform and is at angle to the normal vector to the target. The worst-case imaging scenario determined the angle. This angle is also parameterized, so we can see the change in the radiometry as a function of angle variation in our models. In any case, we used the new operational changes and analyzed the worst-case imaging scenario. At a nominal cruising altitude of 4572 meters above local ground, 8046.72 meters back from the fence line on the U.S. side for alert, safety, standoff mode, and a target 10 km back from the fence line in Mexico (furthest point away, for minimum signal), we get a viewing angle of 15 degrees from the target normal vector. Assuming a diffuse scattering model for the human, the radiant emittance drops off as the cosine of the viewing angle, and so we get a spectral radiant emittance of 33.8 Watts/m^2 in the direction of our sensor. The atmospheric scattering and absorption models provide an atmospheric transmittance of 80% over the wavelengths in the pass band of the narrowband filter, and so these effects reduce the signal to 27 Watts/m^2. At the source, the scattering cross section of the human was 2 m^2 as previously mentioned, and we already accounted for the view angle. With our assumption of a diffuse scattering surface, we can determine an effective intensity from the radiant emittance and the cross-sectional area. This ended up being 17.22 Watts/sr. The steradian associated with the imaging system's entrance pupil for the longest standoff range of 18,616 m is 5.66×10^{-10} sr. Therefore, the optical power collected by the imaging system is 9.75 nWatts. The optical components themselves introduce another 2% loss in getting this power from the entrance pupil to the detector. The optical power on the detector for one human is 9.56 nWatts, well within the feasible range of detection for conventional detectors available on the market today. As previously said, relaxing the safety margin and moving closer to our fence line just increases this signal on our detector."

Bill: "Great job! I want to also run the scenarios for best case and nominal case."

Tom: "I'll see to it."

Bill: "What about the spatial resolution of the imaging system?"

Tom: "Well, that's where it gets interesting. The projected pixels provide adequate spatial sampling. Even for our worst-case scenario, we get a projected pixel size that is less than the required 20 cm resolution cell for our classification rule of thumb (e.g., telling that we are viewing a human vs. a coyote). However, for the classical diffraction-limit case, like we would get using our adaptive optics system, we only get 3 to 4 resolution cells across the long dimension. This was with the original flight envelope. Bill, what is the worst-case human-to-sensor standoff range with the new operational flight profile?"

Bill: "They were talking about reducing the flight deck of the UAV to 4000 m and moving the UAV at night to 2000 meters back from the fence line. The UAV, of course, could fly all the way up to the fence line, but in worst-case protected mode, they would fly 2000 meters back and no lower than 4000 meters. Also, their CONOPS would scan 5 km deep into Mexico and 5 km deep into the U.S. The fence line would sit in the middle of the scan zone."

(Ron punches in the new numbers into the integrated optics model.)

Ron: "That works! That gets us 0.184 m for the resolution cell."

Bill: "Great, however, I think the original CONOPS may still be OK for classification, but it is below our rule of thumb of having 10 resolution cells across our target. In the case of looking through the atmosphere with no ATC/AO, the target will definitely look like a point source, and we can't tell what it is. It looks like we will need the AO or ATC system."

Bill: "Great job everyone. Tom, why don't you call Dr. Glenn and talk about it. I think they were leaning towards the AO/ATC system anyways; it's what set us apart from our competitors. Tell them we can adapt to the new CONOPS if they don't want to take the slight degradation at the extreme ranges. Let me know what you find out."

Tom: "Roger!"

Bill: "By the way, I'd like to introduce George B., Phil K., and Amanda R. George is our hardware guy. He should be able to contribute on any image processing and communication hardware subsystem issues. Phil is an up and coming software engineer. Amanda works in our optical science lab, and she'll be able to fill in the gaps as far as adaptive optics, optical materials, and optical system maintenance concerns. I have to rush off to put out some fires, and unfortunately, these aren't the fun kind of fires that burn marshmallows at the camp grounds."

(The CEO leaves the conference room.)

Phil: "Hey Karl, you were right. There's no way the satellite uplink can handle all the raw data. We'll have to figure out some allocation of the following functionalities: onboard storage, onboard image processing, interesting activity detection within the IFOV, and image data segment selection for real-time transmitting. That's a trade study waiting to happen, but from the introduction, it sounds like we may be adding AO/ATC to that list."

Karl: "That is correct. We're considering the trade-offs of placing an AO versus ATC on the UAV. We'll be looking at things like weight, cost, hardware and software, integration risks, and the impacts on maintainability and availability."

Amanda: "Well, I am most familiar with the deformable mirror end of adaptive optics. My lab has assembled and tested a wide variety of them. They work extremely well, but their control systems can get sort of bulky. After a while, you have to calibrate them. Also, they can be tricky to manufacture. It's definitely possible to put them in airplanes. The Hubble survived being shot into space. There's sort of a challenge here that the Hubble didn't have. NASA could design the satellite around the optical system. We're going to be designing to fit into a pod. There's definitely going to be some maintainability, weight, cost, and schedule risk issues with this if we go with the traditional AO system. I think Phil is familiar with software methods for ATC. Phil?"

Phil: "Yes, there are several different techniques at our disposal. A lot of them are really slow and only used in astronomy and postprocessing applications. I'm thinking mainly of phase diversity (PD), though I have heard of a wavelength diversity (WD) technique. I'll talk about PD first. This method has already been developed and used before; proven technique. There's not much risk in terms of getting it to work, and it probably wouldn't be expensive to implement, except in terms of computing power, where it would be considered expensive. I suspect we'd need to greatly increase our designs for processing power to support this function. Either that or have the processing done at the MCC, or mobile command center. The bad news is that PD doubles the amount of data that needs to be sent out of the predator since you now need two simultaneously captured images (an in-focus and slightly out-of-focus image pair). More bad news is that PD requires special hardware to capture those simultaneous images. Even more bad news is that traditional ATC methods with video would be difficult to implement because of the extended processing time. The good news is that we are working on that in Tom's group. Now WD is almost the opposite. There's a guy here in Florida who just published on this, and it seems like it would be really doable. Whatever source code there may be, it hasn't been released yet. So I'd have to work with an optical engineer like Amanda or a physicist and produce the software after we get the licensing rights. It's a bit risky schedule-wise, but it's still far cheaper than AO stuff that Amanda does in her lab. Also, it's inherently more efficient.

It can be implemented on a parallel processing board, and it doesn't double the required data or need any special hardware."

Tom: "We'd need to make sure our optical system collects multiple wavelengths and stores the data in a specific fashion. Honestly, this isn't a major challenge. The adjustments to the optics end would be pretty minor. There are even some detectors that simultaneously collect that multiple wavelength information at each pixel (an RGB digital camera, for instance)."

Amanda: "Thanks Tom. Anyway, WD seems like it could be fast enough to deliver real-time video if it performed on image segments of maybe 100 by 100 pixels, but definitely not 2048 × 2048."

George: "With a parallel processing architecture dedicated to our optical system, I think we could get the throughput on the itty-bitty image segments and still have computing and bus resources left over to do memory housekeeping. I'd have to see the algorithm."

Ron: "I'm sorry I can't add much to this conversation. It sounds, though, that WD is a potential candidate for providing an ATC capability without adding much weight or cost. Also, hypothetically, if our ATC system were broken, it wouldn't affect the conventional operations of the system. The system would just revert to our typical, high-end, conventional imaging system."

George: "Basically, based on Ron's point, we beef up the onboard processor. We can then update the software anytime later, and the system can perform ATC."

Karl: "This approach will offer flexibility and adaptability for future needs."

Tom: "FIT can produce a trade study where we add a feature with practically no schedule risk as long as we asterisk that it is for a future operational function only. It's just a straight exchange for capability and money."

Karl: "That's assuming we haven't overlooked something. I can definitely do a cost estimate from our company's database on the development of new software technology and pass along a report to Dr. Glen H."

Tom: "OK! Everyone please forward to Karl what he needs from your end to do this."

6.4 Summary

Feasibility studies, trade-off studies, alternative analyses, and the AHP are important for the decisions that affect the form, function, cost, and riskiness of the system. The most important inputs to these studies come from an environment supportive of robust SE practices that span the whole systems life cycle, a multidisciplinary team, research, data, stakeholder's needs, and of course funding. An environment established in integrated product and process development principles is one way to ensure this type of environment.

Feasibility studies are launched early on in the SE process. They determine the basic feasibility of requirements, and they also assess the extensiveness of feasibility in terms of cost, schedule, and technical risk. This includes gathering information about proposed system design ahead of time. This may include DDPs, models, economic research, alternative types of technology, and preliminary configurations. Feasibility studies are a high-level, synthetic activity, and oftentimes creative solutions are derived from the process of determining feasibility. Any team assigned to feasibility analyses is expected to be in communication with other segments of the program and stakeholders as needed.

Trade studies and alternative analyses are structured methods for making design decisions. AoAs usually have a wider scope of categorically different technological paths, whereas trade studies are required during functional analysis when further refinement and definition of the system are being developed. General guidelines show how these decision-making processes are formulated and conducted. The criteria important to the primary stakeholder are set as comparison criteria, and an evaluation of the different alternatives is made in relation to those criteria. The criteria may include the ability of the system to meet technical performance, cost goals, and project's life-cycle activities and attributes such as maintenance, reliability, and producibility.

The AHP is a very particular process for performing alternative analysis. The AHP translates stakeholder operational requirement rankings and other criteria into matrices that quantify those rankings. Like the structured process described for AoA and trade studies, the rules for comparison and evaluation are stated ahead of time. The performance and life-cycle impacts of technology decisions are quantified in terms of FOM that are important to the criteria chosen by the stakeholder. Then the possible solutions are ranked per each criterion in a matrix according to their performance with regard to the particular criteria. An eigenvector matrix manipulation process multiplies the technology ranking vectors by a criterion ranking vector. Although a ranked list of solutions is produced at this point, a sensitivity analysis and consistency analysis should be performed before the results can be deemed acceptable for presentation to the stakeholder as a recommendation.

6.A Appendix: Acronyms

ACI	Average consistency index
AHP	Analytic hierarchy process
AoA	Analysis of alternatives
ATC	Atmospheric turbulence compensation
AT&T	American Telephone and Telegraph
BS	Bachelor of Science
CEO	Chief executive officer
CI	Consistency index
CMMI	Capability Maturity Model Integration
CMU	Carnegie-Mellon University
CONOPS	Concept of operations
COTS	Commercial off the shelf
CR	Consistency ratio
CTO	Chief technology officer
DAR	Decision analysis and resolution
DHS	Department of Homeland Security
EV	Eigenvector
FIT	
FIT (context dependent)	Fantastic Imaging Technologies
	Florida Institute of Technology
	Fashion Institute of Technology
FOM	Figures of merit
INCOSE	International Council on Systems Engineering

IPPD	Integrated product and process development
IPT	Integrated product team
JCIDS	Joint Capabilities Integration and Development System
MTBF	Mean time between failures
NASA	National Aeronautics and Space Administration
RFP	Request for Proposal
SE	Systems engineering
SEI	Software Engineering Institute
SG	Specific goal
SOW	Statement of work
SP	Specific Practice
TPM	Technical performance measures
UAV	Unmanned aerial vehicle
U.S.	United States
VLOS	Visible light optical system

References

Black Sea Trade and Development Bank. 2014. Operational risk management policy. http://www.bstdb.org/about-us/key-documents/Operational_Risk_Management_policy.pdf (accessed April 12, 2014).

Blanchard, B.S. and W.J. Fabrycky. 2011. *Systems Engineering and Analysis*, 5th edn. Pearson Education, Inc. Upper Saddle River, NJ.

Cokely, E.T. and C.M. Kelly. February 2009. *Cognitive abilities and superior decision making under risk: a protocol analysis and process model evaluation*, in Judgment and Decision Making, 4(1): 20–33.

Defense Acquisition University. 2007. Specification and Levels of Development. http://en.wikipedia.org/wiki/File:Specification_and_Levels_of_Development.jpg (accessed November 26, 2014).

Department of Defense. 1984. Military standard. System Safety Program Requirements. MIL-STD-882A MIL-STD-882B.

Department of Defense. 2009. Operation of the Joint Capabilities Integration and Development System. Chairman of the Joint Chiefs of Staff Instruction. CJCSI-3170.01-H.

Defense Scientific Information and Documentation Centre (DESIDC). 1990. Guided missiles. Delhi, India: Defense Scientific Information and Documentation Centre (DESIDC) http://drdo.gov.in/drdo/data/Guided%20Missiles.pdf (accessed April 17, 2014).

Felix, A. 2004. Standard approach to trade studies. A technical paper from *INCOSE 14th Annual International Symposium Proceeding*, Toulouse, France, June 20–24, 2004.

Griffin, B.N. 1978. *The Influence of Zero-g and Acceleration on the Human Factors of Spacecraft Design*. JSC-14581, NASA-JSC, 8-78.

Haas, R. and O. Meixner. 2011. An illustrated guide to the *analytic hierarchy process*. https://alaskafisheries.noaa.gov/sustainablefisheries/sslmc/july-06/ahptutorial.pdf (accessed May 2014).

Haskins, C. 2006. *INCOSE Systems Engineering Handbook: A Guide for System Life Cycle Processes and Activities*. San Diego, CA: International Council on System Engineering.

Hudson, R.D. 1961. *Infrared System Engineering*. New York: John Wiley & Sons.

Kitson, D. 2010. AN/PVS-5C Night Vision System. http://en.wikipedia.org/wiki/File:AN-PVS-5C-Cut_image.jpg (accessed November 26, 2014).

McGraw-Hill Radiometry. 2006. *McGraw-Hill Concise Encyclopedia of Science and Technology*. http://www.credoreference.com.portal.lib.fit.edu/entry/conscitech/radiometry (accessed April 6, 2014).

Miller, G. 1956. The magical number seven, plus or minus two, and some limits on capacity for processing information. *The Psychological Review*, 63: 81–97.

NASA/STSci. 1998. Hubble Space Telescope Imaging Spectrograph. http://hubblesite.org/newscenter/newsdesk/archive/releases/1998/41/image/r (accessed November 26, 2014).

NASA. 2007. The electromagnetic spectrum. The infrared. http://science.hq.nasa.gov/kids/imagers/ems/infrared.html (accessed April 17, 2014).

Parnell, G.S., P.J. Driscoll, and D.L. Henderson. 2010. *Decision Making in Systems Engineering and Management*, 2nd Ed. Hoboken, NJ: John Wiley & Sons.

Pratt, L. and M. Zangari. 2008. Decision Engineering White Paper: Overcoming the Decision Complexity Ceiling through Design. Denver, CO: Quantellia Inc.

Prazan. 2008. Elements of Decision Engineering. http://en.wikipedia.org/wiki/File:DEFramework.png (accessed November 26, 2014).

Saaty, T. 1987. Its priority and probability: Risk analysis.

Saaty, T.L. 2012. *Decision Making for Leaders: The Analytical Hierarchy Process for Decisions in a Complex World*. Pittsburg, CA: University of Pittsburgh.

Shishko, R. et al., 1995. *NASA: Systems Engineering Handbook*. Washington, DC: National Aeronautics and Space Administration.

Strang, G. 1993. *Introduction to Linear Algebra*, 2nd edn. Wellesley, MA: Wellesley Cambridge Press.

The Penguin Radiometer. 2009. *The Penguin Dictionary of Physics*. http://www.credoreference.com.portal.lib.fit.edu/entry/pendphys/radiometer (accessed April 6, 2014).

US Navy. 2006. Atmospheric transmittance as a function of wavelength. http://web.archive.org/web/20010913094914/http:/ewhdbks.mugu.navy.mil/transmit.gif

7

Systems and Requirements

> Simultaneous contrast is not just a curious optical phenomenon - it is the very heart of painting.
>
> —Josef Albers

It is all about the requirements! This chapter addresses what some consider as the most critical aspect of systems engineering (SE) and the heart of the systems development effort—the requirements. We present the preeminent role that requirements have throughout the systems development life cycle (viz., the conceptual design phase, preliminary design phase, detailed design and development phase, production/manufacturing/construction phase, operations and support phase, and system retirement and disposal phase).

The SE discipline is extremely relevant to governmental, commercial, industrial, and academic sectors and equally benefits large and small endeavors. From designing a satellite system with a high-end payload, to running a small, high-tech, consulting business, to involving students in long-term research projects in universities (Arrasmith 2007), properly tailored and correctly applied SE principles, methods, and techniques are practical, beneficial, and cost effective in today's complex, high-paced, international, and multidisciplinary world. Systems requirements define the necessary characteristics and/or attributes of a system and subsystem and component/assembly/part. They define the systems functional performance as well as nonfunctional aspects of the system (e.g., reliability, producibility, availability, and other so-called ilities). In essence, the requirements are what the systems development project is all about (Table 7.1).

The International Council of Systems Engineering (INCOSE) provides the following definition for SE, "an interdisciplinary approach and means to enable the realization of successful systems" (INCOSE 2004). SE involves defining stakeholder needs and required systems functionality and "documenting requirements, and then proceeding with design synthesis and system validation while considering the complete problem" (Valerdi 2012). On a more macroscopic level, the definition of a system is a set of elements or services joined together to meet the stakeholders' desired expectations. A system may possess a vast range of components/elements to meet the objectives required in which SE provides the resources and guidance to ensure that these objectives are met to reach the overall goals throughout its life cycle. In all these definitions, understanding what the stakeholder needs and wants, and developing a set of requirements that captures these needs and approved wants, is the activity that defines the overall systems development effort. If this activity is not done correctly, then the wrong system will be built!

Once the system developer and stakeholder identify the systems requirements for functional operations in customer language, the system developer and its team conduct further analysis to identify functions systematically to ensure that all system elements are analyzed and hierarchically arrange functions so that key functions are documented. An in-depth analysis of the stakeholder requirements should be accomplished to ensure

TABLE 7.1

Examples of Nonfunctional Requirements

Reliability requirements
Quality requirements
Producibility requirements
Maintainability requirements
Look-and-feel requirements
Security requirements
Manufacturability requirements
Aesthetic requirements

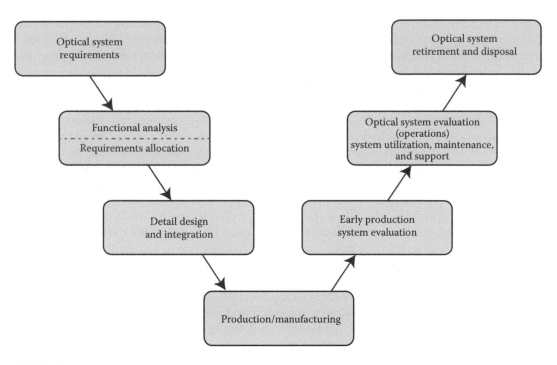

FIGURE 7.1
Optical systems development process. (Courtesy of Robert Mulligan. With permission.)

that they are easily understood, and agreed upon, by all parties, prior to finalizing a base-line set of requirements. Figure 7.1 shows the optical system development process and explains how systems requirements initiate the optical systems development process. It begins with optical systems requirements. Each requirement is examined thoroughly to understand how feasible they are in terms of technical issues, cost, risk, and schedule.

The requirements must be assessed to determine whether or not they are feasible within the stakeholder's constraints. In the case that a particular requirement cannot be satisfied, it may become necessary to develop an alternative approach, or revise, or eliminate the requirement, with stakeholder concurrence. When these types of problems are not cor-rected or taken into consideration at initial system development, issues are likely to arise later in the product's life cycle that could significantly reduce the project's success and certainly change life-cycle cost. Thus, it is essential that life-cycle considerations be part of a system alternative study.

The overall impact of the project's life-cycle execution occurs up front by identifying clear requirements. In subsequent sections, we will be discussing systems requirements based on how they are generated, what they are telling the system developer to actually accomplish, what is a good requirement compared to a poorly written one, the case for a requirements document, and how to manage changes in requirements, along with a number of other concepts that will provide an appreciation for importance of requirements to the SE discipline.

7.1 Requirements Generation Process

A requirement is a characteristic and/or attribute of a system (or a subsystem, process, product, or service) that is fundamental to the system's intended purpose. That is, without that requirement, the system is undefined or ill defined. A requirement is usually defined during the initial phase of the life cycle prior to design, but it can at times be defined during design. A requirement is typically written as a "shall" statement, rather than a suggestion. It is a demand that the system developer must uphold and constitutes a contractual agreement; thus, the system developer cannot simply change it at a later time. Statements in the requirements document with a "should" instead of "shall" are interpreted as a recommendation. These requirements are defined with a great amount of thought, with input provided by many stakeholders, before they are finalized. This is because they may not always be realistic and are sometimes infeasible. Stakeholders may dictate some requirements, but they may not always have an idea on the feasibility in terms of what can actually be done technically, or within a given budget or schedule. In the next section, we will discuss the requirements generation process and the differences between user requirements and systems requirements along with identifying correct and incorrect requirements.

7.1.1 Determining the "Whats!"

The first point about understanding a requirement is that "a requirement deals with what the system should do and not how the system should do it" (Maiden 2012a). Consequently, the system developer should be told "what" to do and not "how" to do it. The following example shows a poor requirements statement.

- The optical system shall interoperate with the existing computer database located in the command center using a Java-based protocol.

This requirement breaks the rule for explaining "what" the requirement needs to do and instead defines how the design team is required to make the optical system interoperate with the computer database. This creates a constraint on the system developer that would prohibit them from thinking of the best approach to satisfy the requirements. Not only would they potentially force an ineffective solution, but this suboptimal approach may also drive the development cost higher and/or increase the schedule due to downstream rework requirements. A more appropriate requirement would read as the following:

- The optical system shall interoperate with the existing computer database in the command center.

This less constrained requirement allows the system developer to focus on what the system must do, while enabling their development team to provide a creative solution. Well-written requirements also allow the system developer to propose a system more likely of meeting their expected cost and schedule targets.

Two types of requirements that should be derived at the beginning of the SE process are stakeholder requirements and systems requirements (Maiden 2008, 90–100). A stakeholder requirement comes from "a user or other type of stakeholder and expresses a property of the domain, or business process, that the introduction of a new system will bring about" (Maiden 2008, 90–100). Essentially, a stakeholder requirement is a necessary system component to satisfy the terms of the contract. Two good examples of stakeholder requirements are as follows:

- A user of the optical system shall be able to search for files within its internal storage.
- A user shall be able to add, edit, or delete files within the optical systems internal storage.

These two requirements explain what the stakeholder may do rather than what the system can do.

A systems requirement "expresses a desirable system property that, when implemented in the domain or business process, will lead to the achievement of at least one user requirement" (Maiden 2008, 90–100). Notice the shift in perspective—from what operational need that needs to be satisfied to what the system must do to satisfy that need. The following two examples show good systems requirements based on the previous stakeholder requirements:

- The optical system shall have the ability to have files retrieved from its internal storage device.
- The optical system shall allow for the adding, editing, and deleting of files from its internal storage device.

Even though these two types of requirements seem similar, they are in actuality much different. The first example is written from the point of view of the user, whereas the second requirement is written from the perspective of what the system must do to satisfy the user's need. The SE process clearly delineates this transition with a set of processes to transition from a user focus to a systems focus. Initially, there exists a particular need that a set of stakeholders needs to satisfy. The SE process begins with identifying the relevant set of stakeholders and understanding their needs, wants, and motives in the problem definition process. Once understood, the needs are articulated in a set of concise, short statements known as stakeholder requirements. As previously stated, these requirements capture the intent of what needs to be accomplished and holistically describes the attributes of success or failure of the project. In other words, if the stakeholder requirements are met, the system, service, product, or process (SSPP) is considered a success. On the other hand, if the stakeholder requirements are not met, the project fails. The distinction of the relative importance of the individual requirements is often used in the terminology of the requirement itself, by the use of the word "shall" or "should." If a requirement is considered highly important, the term "shall" is used. If the requirement is considered highly desired, but not mandatory, then the word "should" is used in the particular requirements statement. Notice that in the previous requirements examples, all of them made use of the word "shall" and so are considered mandatory requirements. By mandatory, we mean that a given effort is considered to have failed if the mandatory requirement has not been met. Failure to meet a requirement that is stated with a "should" does not necessarily mean that the project has failed. For instance, the requirement, "The crate to ship the optical sensor system should be painted blue with the words, 'Woo hoo, we're done!' displayed prominently on all six sides of the container," likely would not be considered a project failure, if violated.

Once the stakeholder requirements have been fully defined, it is the purpose of the conceptual design phase to use these requirements to generate conceptual solutions to the stakeholder requirements, choose the best solution from the available options, and translate the stakeholder requirements to systems requirements through the conceptual design process. The systems requirements are captured, stored, and maintained in a requirements specification known as the systems specification (A spec). The collection of requirements in the systems specification is "baselined" (i.e., enter formal configuration control) at the major milestone event, the system segment review (SSR) that marks the formal end of the conceptual design phase and the start of the preliminary design phase of the systems development effort. In the modern-day SE environment, instead of (or in addition to) the traditional document-based system, the collection of requirements (stakeholder requirements, systems requirements, and other requirements generated later on in the development effort) is contained in model-based systems engineering (MBSE) tools, which are current, immediately available to the development team, and integrated with other MBSE tools like enterprise architecture (EA) tools and rapid prototyping tools. When properly used, these MBSE tools facilitate team integration, reduce requirements and other errors, and streamline the development process.

Requirements may be separated into two categories: functional and nonfunctional requirements (Baca 2007). Functional requirements state what the system must do. This involves actions as to what the system must carry out to meet the needs of the stakeholder. On the other hand, nonfunctional requirements are the components that your system must have (Baca 2007). Usually, we consider nonfunctional requirements as qualities that make the product attractive, usable, or fast. They illustrate the values of the functional requirements (Baca 2007). For example, a functional requirement may state that a flashlight shall have a range of 30 ft, whereas a nonfunctional requirement would say the flashlight shall be operational for a period of 72 h of continuous operation. According to the author Claudia Baca, "A unique difference between functional and non-functional requirements is that the non-functional requirements are entities that fulfill the functional requirements" (Baca 2007).

It is helpful to link requirements to goals established early on in the development process (Ramasubramaniam and Venkatachar 2007). As previously discussed, requirements are considered binding contracts that must be achieved by the developed system. Therefore, it is critical to establish the feasibility of the requirements, along with inherent risks, early on in the systems development effort. Notice the inverse relationship between feasibility and risk. The greater a requirement's feasibility, the less inherent risk there is in achieving the requirement.

7.1.2 The "ilities!"

Nonfunctional requirements are commonly referred to as the "ilities." By "ilities," we mean design considerations such as maintainability, reliability, and supportability. Figure 7.2 shows a collection of "ilities" to consider when designing a system.

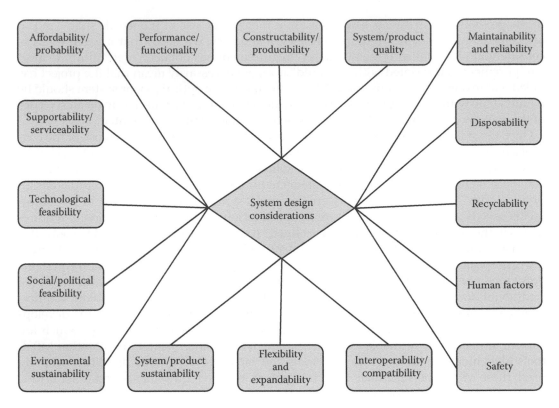

FIGURE 7.2
The "ilities" and their relation to system design. (Courtesy of Robert Mulligan. With permission.)

These "ilities" are quite useful and essential in eliciting holistic systems requirements that cover nonfunctional aspects of the system over its entire life cycle. We will describe several of the commonly used "ilities." In doing so, we can better understand their importance and their connection to the design process.

One of the most significant nonfunctional requirements that often impact system performance and ultimately stakeholder satisfaction is the reliability of the system. Reliability is "the ability of a system to perform and maintain its function in routine circumstances, as well as hostile or unexpected circumstances" (Blanchard and Fabrycky 2011). This is the system's ability to run as expected by successfully satisfying the given requirements during system operations. Use cases and operational scenarios are used in the development process to analyze the performance space in systems design. Probability-based reliability models are typically generated in concert with functional block diagrams and functional flow diagrams as part of the functional analysis activity. The reliability models are then used to predict reliability-related performance parameters such as mean time between failure (MTBF), mean time to failure (MTTF), and the system, subsystem, or component failure rate, to name a few. These parameters are often used as technical performance parameters in the requirements themselves. We will take a more detailed look at methods for predicting systems reliability in later chapters.

Another useful category of nonfunctional requirements are usability requirements. Usability is "the ease of use and learnability of a human-made object" (Blanchard and Fabrycky 2011). This simply means that the system can be used with ease; basically, how comfortable can the end user be or how quickly can the end user come up to speed and perform tasks when operating the system. Critical considerations such as training requirements, documentation, human factors, safety, ergonomics, and required skill level are critical for this category.

Maintainability relates to the ease with which maintenance of the system can be performed in accordance with prescribed requirements. A whole host of considerations must be addressed in generating requirements for this category: What levels of repair should there be? How many spares are needed? How much time will it take to repair systems? What is the preventative maintenance schedule? What equipment is needed? Questions such as these are just the tip of the iceberg. Careful analysis of these questions spawns maintainability-related requirements. If this "ility" is not done well, the system may become unusable after a failure, potentially impacting the mission and adversely affecting the stakeholder's satisfaction, the reputation of the maintainers, and perhaps potential future business. One can also see a connection between maintainability and reliability if a system requires extensive and costly maintenance frequently. The ability to reliably repair and replace components of the system in the least amount of time, at the least cost and with a minimum amount of resources, is the goal of maintainability analyses. System developers should generate maintainability requirements as part of the requirements generation process to ensure that the ultimate design of the system can be properly maintained. For example, designing an engine of a car so that the engine must be removed in order to replace a fuel filter is an example of a lack of proper consideration of maintainability.

Affordability deals with the life-cycle cost of the entire system. For instance, costs of maintenance activities and failure rates greatly impact the overall system costs. Improving the overall systems reliability may increase the initial cost of a system by building in redundancies, using high-quality components, and designing in access points for repair, but reduce the cost of the system over its entire life cycle by providing a more reliable and easily maintained system. Reliability models can be used to predict the impact of potential design changes on the overall MTBF of the system. More failures means more repairs or preventative maintenance will be required if the system must also be maintained as part of the contracted deliverables. Cost savings by using a cheaper part in the design could quickly become lost or exceeded in the operational phase due to increased maintainability requirements. Whether or not a system component should be repaired or replaced must be carefully evaluated to see which option is less expensive. In a similar fashion, each "ility" is considered from the perspective of what nonfunctional requirements need to be generated to fully describe the needed system attributes throughout its life cycle.

After considering the "ilities" and generating a holistic set of nonfunctional requirements, the combination of functional and nonfunctional requirements is used to drive the design process. In the conceptual design phase, the objective is to start from the stakeholder requirements and generate the systems requirements. The design process translates the "whats" (i.e., the stakeholder requirements) to the "hows" (the system level design captured by the systems requirements). In essence, the higher-level stakeholder requirements are broken down into more detailed, lower-level systems requirements that fully describe needed attributes of the system. As part of the conceptual design process, technical performance measures (TPMs), also known as measures of performance (MOP), are established. TPMs help system developers and stakeholders define how well the system meets its high-level requirements. TPMs capture essential technical aspects related to the systems performance. As part of the design process, particular design choices may force trade-offs between different TPMs. For this reason, it is important to prioritize the TPMs with the stakeholders. TPMs are often categorized into design-dependent parameters (DDPs) where trade studies occur. An example could be the spatial resolution provided by a particular optical imaging system. As the name implies, the DDP is a function of the design. If the optical components or their relative arrangement changes as a function of a design changes, the DDP also changes. There can be a number of DDPs identified. For instance, for an optical system, the effective focal length, the diameter of the entrance pupil, the size of the pixels, and the transverse and lateral magnification of the optical system are all representative examples of DDPs. As part of the design activity itself, trade studies, alternative analysis, feasibility analysis, functional analyses, and requirements allocation are conducted to find the best way of implementing the stated requirements.

7.1.3 Techniques for Writing Good Requirements

Writing good requirements is both an art and a science. By this statement, we mean that there are methods and techniques that can be learned that aid in writing good requirements; however, there is a component to writing good requirements that is related to skill and experience. We provide, as previously stated, that the requirements must be well written in a way that they are clear to stakeholders and the development team. They must be complete in the sense that nothing is left out. The collection of requirements at the stakeholder level must completely describe what is expected for project success. At the system level, the requirements must completely describe the system functional and nonfunctional aspects, not only at delivery but also over its entire life cycle. When writing requirements, it is often useful to begin with a step-by-step format that addresses requirements from users and then organizes them into an understandable state that is clear to all of the parties involved.

Before the requirements are written, there are usually stakeholders who fulfill the following roles: author, publisher, reviewer, and implementer (Hull et al. 2005). The author is the originator of the requirements and adds any changes needed. The publisher dispenses and archives the requirements document. The reviewer evaluates the requirements as well as recommends changes to the document. Finally, the implementer scrutinizes the requirements, presents any proposed changes, and ultimately implements the requirements. These stakeholders are all required to ensure that each written requirement satisfies the following criteria:

- Describes what and not how
- Has one purpose or idea, not two or more
- Is unique and not duplicated
- Classifies every statement in a requirement by importance, type, and urgency
- Provides traceability of the requirement so that we can understand when it is written, reviewed, satisfied, and then qualified
- Reviewed in terms of completeness by either performance information, quantification, test, rationale, or in some other method

Key requirements, usually abbreviated as "KURs (Key User Requirements) or KPIs (Key Performance Indicators), are small subsets of requirements taken from the whole set of requirement that capture the essence of the system" (Hull et al. 2005). Two important questions should be considered when looking at KURs: At the stakeholder level, "Will I buy the product (or am I interested in the product), if the solution does not provide me this capability?" (Hull et al. 2005). At the system level, "Will I still buy it, if the system didn't do this?" (Hull et al. 2005). Of course, the answers should be an automatic no! Key requirements are essential and should never be optional.

Key performance parameters (KPPs) are critical performance attributes of a system for fielding/developing an effective capability. KPPs are typically military in nature and have a threshold value for minimum acceptable performance associated with low-to-moderate risk and an objective performance value associated with higher risk. KPPs usually have a threshold for representing the minimum acceptable value achievable at low-to-moderate cost, schedule, and performance risk. There is also a target objective that achieves the desired operational goal, but at greater risk. A measure of effectiveness (MOE) is "A criterion used to assess changes in system behavior, capability, or operational environment that is tied to measuring the attainment of an end state, achievement of an objective, or creation of an effect" (Military Dictionary 2014). Table 7.2 provides several definitions about the differences of key requirements.

To capture the true essence of the stakeholders needs, requirements must be clear and concise. Both the system developer and stakeholders must agree and understand the systems requirements. Establishing a consistent and comprehensive set of requirements attributes is often beneficial in establishing the requirements and effectively communicating them to the development team.

For example, the following is a representative example of requirements categories that are useful for defining requirements attributes and effective communications: requirements identification, intrinsic characteristics, priority, source, ownership, context, satisfaction argument, verification argument, approval authority, process support, and more (Hull et al. 2005). Each of these categories helps bring together the expectations for how requirements are to be handled throughout the system life cycle. They provide useful information about the current state of the requirement and can readily be connected to the requirements and tracked in conventional requirements management tools. The life-cycle phase attribute defines in what part of the project life cycle the requirement is assigned and managed and plays a prominent role. The priority level attributes distribute the requirements

TABLE 7.2

Definitions for Metrics Related to Requirements, Performance, and Effectiveness

Term	Definition
Key performance parameters (KPPs)	Performance attributes of a system considered critical to the development of an effective military capability (DAU Acquipedia [KPP] 2014).
Key user requirements (KURs)	These requirements are a small subset abstracted from the whole that capture the essence of the system (Hull et al. 2005).
Key performance indicators (KPIs)	Agreed and quantifiable measures related to critical success factors for an organization or effort (DAU Acquipedia [KPI] 2014).
Measure(s) of performance (MOP)	System-particular performance parameters such as speed, payload, range, time on station, frequency, or other distinctly quantifiable performance features (DAU Acquipedia [MOP] 2014).
Technical performance measures (TPMs)	Typically exhibited at the system level, TPMs are measures related to technical performance aspects and attributes of the design itself.
Measure of effectiveness (MOE)	A criterion used to assess changes in system behavior, capability, or operational environment that is tied to measuring the attainment of an end state, achievement of an objective, or creation of an effect (Military Dictionary 2014).

into separate key categories such as mandatory, optional, or desirable to aid in focusing the development efforts. INCOSE uses the must, should, could, and would categories for priority levels (Tronstad 1996, 64). During requirements analysis, the estimated cost category is one commonly overlooked attribute. This is because the basis of the cost estimates is usually developed during the pursuit portion of the program and is accomplished to determine the expected cost of the program. After the requirements have been allocated to the subsystems, the engineering staff reviews them and determines the development and design cost. Due to the extra time and effort required to make associations of individual requirements to design, development, test, and support levels, the cost estimates are rarely assigned to specific requirements and are usually provided at a higher and more aggregated level. However, requirements with high risk to the program's success must be managed and gathered to provide insight to potential cost drivers.

When writing requirements, an important aspect is ambiguity (or better, the lack of it). To reduce costly misunderstandings, stakeholders and the development team alike must correctly interpret requirements. The following useful guidelines for writing requirements are provided by Tronstad (1996, 64). A typical stakeholder requirement may be written as follows:

- "The 'stakeholder' shall be able to 'ability'."

With constraints, it would look more like the following:

- "The 'stakeholder' shall be able to 'ability' within 'performance' of 'event' while 'operational condition'."

A well-written requirement using the stated guidance would be as follows:

- The radio operator shall be able to transmit audio from the radio within 3 s of his sensor being triggered while in severe rainstorm conditions.

This mnemonic is useful in establishing a consistent flow to the requirements and ensuring that the requirement is complete. Requirements provide guidance to the designers/developers as to what the system must do under expected operating conditions. They also provide guidance to the testers that demonstrate that the developed system met the stated requirements.

7.1.4 Importance of the Requirements Document

During the system design effort, the written requirements are traditionally contained in a single document. Nowadays, MBSE tools also act as a repository for the requirements. Whether a model-based system, a document-based system, or both are used, it is essential to have a common location that contains the system's contractual expectations for the engineers and stakeholders. As previously stated, this is similar to a binding contract between stakeholders and the development team and provides a thorough understanding of what is required (Burns 2010). Apart from directing the design activity itself, a major purpose for the requirements document is monitoring the requirements for test compliance during the end of your project (Maiden 2013, 16–17). The responsibility of the system developer is to provide proof of compliance for each requirement levied upon the program by the stakeholders. This becomes the testing team's task to interpret the requirements and develop test plans and test procedures that verify and validate the requirements through analysis, demonstration, testing, and inspection.

With regard to the requirements document, it should be formatted to be easily viewable and quickly accessed by the development team. This document is a working document and is typically well used, so it is important that it is written like a reference document with numbered sections. The requirements document should begin with a consistent outline of the contents. This outline should be holistic and may be reused, but tailored, for each program. The following are some essential considerations before embarking on writing a requirements document:

- Ensure that you are well prepared and have an established outline of a requirements document to capture all topics.
- Realize that the requirements document will go through much iteration before being signed off by the stakeholders.
- Think about the system from all perspectives, and interview the stakeholders and end users to have a good understanding of the high-level stakeholder needs.
- Embrace the operations of the end user, and consider what changes to the system may make it more functional for them, without straying too far from the high-level requirements for the system.
- Write the document in a clear and concise manner that is easy for the readers to interpret.
- Remain focused on the structure of the requirements document, and ensure that each section is complete and stays on topic.

A requirements document explains everything that the system must do in a single document (Maiden 2012b, 8–9). Through testing, one can take a product, go through each requirement, and determine that the system fulfills each one. Along with the requirements document, system developers are also responsible for creating a requirements verification traceability matrix (RVTM) that accurately confirms the verification of each requirement to the stakeholder through the associated plans and test procedures. Test reports are used to document the test results. In summary, each system development effort is based on the interpretation of requirements documents and the development team's ability to carefully create design documents that meet those requirements, while not expanding the scope of the system beyond what was agreed upon with the stakeholders. A good rule of thumb is to keep the requirements simple and say only "what" the system must do, not "how" it must be done.

7.1.5 Managing Requirements and Change: Model-Based Systems Engineering Requirements Tools

In a sufficiently complex development effort, it is possible to have a large number of requirements. Having hundreds of requirements or more is not unusual. Requirements management tools allow multiple people and organizations to work on the requirements and make edits, track changes, and provide a configuration control capability should the need arise to revert to previous versions (Bangert 2010, 24). Today, there are a variety of MBSE tools that provide integrated capabilities to the stakeholders and development team. Excellent examples are EA tools that facilitate the integration of multiple entities on complex systems development efforts. These tools have inherent architectural frameworks such as the Department of Defense Architecture Framework (DODAF) and a set of tools to rapidly build enterprise models and architectural framework artifacts and for unifying the efforts of a diverse set of stakeholders.

Cost is a primary factor to consider when selecting a requirements management tool. Most engineering requirements tools are expensive and are charged per license. A recent study had shown that 60% of 30 vendors of engineering requirements tools had costs above $1000 per license (Carrillo de Gea et al. 2011). Depending upon the options needed by the development team, the requirements management tools can be very powerful or simple. These tools provide templates and reports to communicate requirements volatility and status information as well as help keep requirements manageable and organized. The tools also perform requirements traceability, provide the status of the verification/validation of each requirement, provide analytical tools and graphs, and capture associated requirements attribute data such as implementation status, testing status, and defect rates. These data are often used for presentations to the stakeholders, management, or the development staff. These tools are used extensively in program quality control and configuration control functions to ensure that there is a historical database of updates and a record of changes. Requirements management tools can synchronize changes across platforms, and even across organizations, and integrate with EA tools and other tools to provide a powerful suite of capabilities to the system developer and stakeholder. INCOSE maintains a list of useful requirements management tools along with links to the vendors (INCOSE 2010).

Dynamic Object-Oriented Requirements System (DOORS) and MagicDraw are two commonly used MBSE tools. DOORS is a requirements management tool that allows for the optimization of requirements and provides a means for communicating and authenticating the requirements throughout an organization (Feinman 2007, 22). One of the great characteristics of DOORS is that it is easily manageable and can be used to capture, track, and review changes to the requirements as the requirements evolve through the systems development effort. DOORS provides individual user accounts with configurable permissions, so that some users will only be able to view the document (or sections of the document), while others are permitted to make changes. For multiple users, DOORS allows access for editing, configuration control, and analysis, which can be tracked so that changes are attributable, and not lost during the requirements generation process (Rubinstein 2003). During the integration and test phase of the program, DOORS is used to track the implementation and success rate of the requirements by tracing and tying requirements to the test plans and procedures. Since there are often multiple instances where requirements are linked to external sources, you can also associate them with that external information.

There are scalability benefits for converting the requirements document to the DOORS format in a similar structure as outlined in the requirements document. Links between multiple requirements level and from requirements to test are easily established. Once implemented in DOORS or some similar requirements management tool, adding, deleting, and modifying complete sections, as the scope of the system design changes, become easier than in the document base system. For example, if a system meets the needs of the current stakeholder, but must be modified for the needs of international customers, the modified set of requirements can be easily added to new sections of the DOORS database as a part of the overall systems requirements. These can be kept separate from the individual stakeholders. As requirements change, the impact of the change on other requirements and on test can be dynamically assessed and rapidly implemented.

MagicDraw is a modeling tool that integrates information from the DOORS (and other) requirements management tool with team collaboration support (MagicDraw 2014). MagicDraw is object oriented and is specifically geared toward business/software analysts, programmers, and quality assurance engineers. It allows the user to break a project into more manageable parts that can be worked on by separate parts of the team. MagicDraw is useful in collaborative efforts where multiple team members need to simultaneously view and update project artifacts. It also gives the user the ability to export reports in various formats such as PowerPoint, XML, and HTML. As a whole, MagicDraw is a tool that is useful throughout the development effort. MagicDraw also has a number of plug-ins to customize the tool and provide enhanced capabilities. MagicDraw with its plug-ins can be used for a variety of MBSE applications to include requirements management, EA modeling, business process modeling, systems modeling, system decomposition, and analysis to name a few.

7.1.6 Transition to Optical Systems Building Block: Optical Modulation

Requirements drive the system design. When a proper SE development environment is established and maintained, the requirements themselves become the forcing function for the design activity. When dealing with systems in general, requirements at any level will inspire feasibility analysis and trade studies to determine the best technical response to achieving the requirements. For instance, a stakeholder-level requirement to "see" a particular target at some standoff range in a given operational scenario will inspire feasibility and trade studies to determine the basic optical system needed to satisfy the driving requirement. As part of the basic optical system, optical modulation methods may be needed to remove clutter and pull a weak signal out of a noisy environment. In the next section, we will discuss optical modulation techniques that are part of the optical system itself and that are needed for detecting the signal of interest (SOI) or the target of interest (TOI).

7.2 Optical Systems Building Block: Optical Modulation

An optical modulator is a device that is placed in the optical path of an imaging system to aid in such applications as target tracking, target characterization, direction finding, and background suppression (Diggers 2003). Optical modulators, reticles, or choppers are often mechanical devices that permit the separation of a target attribute (such as single or multiple target detection, target size determination, target range, speed, heading, and/or target-type identification) from surrounding clutter. The inclusion of a reticle in the optical system often makes the difference between success and failure of the optical systems intended purpose. Reticles often complement software-based signal and image processing methods and are sometimes the critical optical element in an optical imaging or optical tracking system. The optical modulator can be a rotating mechanical device that has sections of transparent and opaque materials. The incoming optical radiation is "modulated" by the particular pattern on the reticle, and the irradiance on the detector is then processed to find the target attribute of interest. Optical modulators can also be stationary and use electrical or acoustic properties to modulate incoming light as in the case of electro-optic modulators and acousto-optic modulators. In this section, we emphasize reticle patterns and the connection to target attributes of interest. We reserve discussions on particular devices such as electro-optic modulators and acousto-optic modulators for later.

Some of the more common uses for reticles include

- Signal conversion (DC to AC)
- Background suppression
- Target attribute patterns

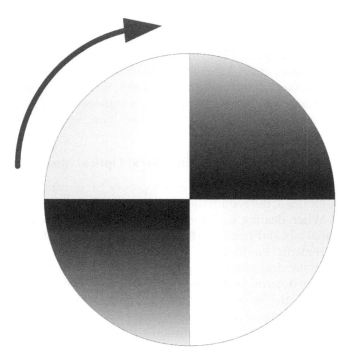

FIGURE 7.3
Simple optical modulator. (Courtesy of Dong Wang. With permission.)

As a simple example, the optical modulator shown in Figure 7.3 has a series of alternating transparent and opaque triangular segments on a disk that rotates. The reticle is placed in the image plane where it rotates with a rotational frequency f_r.

The detector behind the optical modulator can be either a single-element detector—where the output of the detector measures the total irradiance transmitted through the rotating reticle—or a multielement detector like an imaging camera. The type of detector used depends on the purpose of the optical system. Notice that the regions behind the opaque fan blades (shown in black in Figure 7.3) are not irradiated but the transparent sections (shown in white in Figure 7.3) are irradiated. Different areas of the detector are either revealed or concealed as the reticle rotates. If the image of a target object has dimensions smaller than the opaque portion of the reticle, it is possible to temporarily miss the target when the opaque blades rotate between the target image and the detector itself. The target becomes detectable, given sufficient signal strength relative to noise considerations, when the opaque blade rotates out of the way and the transparent area exposes the target irradiance to the detector.

7.2.1 Early History of Optical Modulation

A physical example of the optical modulator shown in Figure 7.3 is given by looking at a moving object through a common household fan (one with triangular blades). As the blade of the fan passes in front of the observing eye, the last position of the object is retained by the observer's brain. In reality, the object continues to move, and when the opaque blade no longer obscures the scene, the target appears to instantly jump to the new location. If the target was a stationary pen light (simulating the point source object above) in a dark room, the pen light would appear to blink on and off through the fan. Although this effect was likely noticed in ages past, the optical modulator has come on the scene relatively recently. We provide the following timeline of key events in the historical development of optical modulators.

1928: A. H. Pfund (Pfund 1928) initially used the optical modulator technique when the flux illuminating a thermopile detector was "chopped" using a pendulum.

1934: H. A. Dahl filed a patent to find targets by their heat radiation.

1940: Reticles were used in WWII by Germany in their tracking and guidance systems.

1950: H. L. Clark (Clark 1950) presents a V-2 rocket guidance technique that seeks the sun.

1956: A reticle-based infrared (IR) tracking system is used on sidewinder missile.

1959: G. F. Aroyan develops spatial filtering technique.

1966: L. M. Biberman develops refinement to reticle calculations.

1984: Terrain contour matching is used by Tomahawk missile's guidance system.

2001: International Space Station uses laser reflectometry–based tracking system to dock spacecraft.

Not much unclassified literature is published regarding reticle-based targeting and guidance research due to the sensitive military nature of this topic. Nevertheless, some fundamental principles of optical modulation are readily available and in the public domain for nonmilitary applications, and these are further detailed later.

7.2.2 Optical Filtering

Optical filtering is often a critical component of a well-designed optical system and is application dependent. There are a variety of optical filters that we will discuss that are application dependent. Some optical filters are hardware based while others can be implemented in software. Filters can be used by themselves or in combination with other filters. Optical modulators, although strictly speaking not optical filters, have similar properties as optical filters in that they accomplish some of the same functions (e.g., background suppression and target detection). Optical filtering occurs predominantly in the frequency domain of the SOI. For instance, a *spatial filter* can be used to clean up aberrations in optics or remove higher-order modes in a laser beam due to variations in the laser's gain medium. The spatial filter can be a tiny pinhole aperture that is placed at the focal point of a lens. The lens takes the 2D Fourier transform of the light it receives. In the case of the laser, the lens focuses the 2D Fourier transform of the laser beam onto the pinhole, and, if the pinhole is small enough, only the lowest mode of the laser beam (TEM_{oo}) is transmitted through the clear aperture of the pinhole. Similarly, for optical aberrations, the lens focuses the 2D Fourier transform of the aberrated light onto the pinhole. The central portion of the optical beam has the least aberrations and is transmitted by the clear aperture of the pinhole. Another lens that placed a focal length behind the pinhole performs another 2D Fourier transform of the spatially filtered light resulting in reduced aberrations and a "cleaned up" beam. One limitation of using a spatial filter is that the smaller the pinhole gets, the more light is rejected and the optical power through the optical system is decreased. If the incoming irradiance levels are low to begin with, then the spatially filtered signal may be too low to be detected, and the detector noise may wash out the SOI. Another common optical filter is the narrowband filter.

In many applications, the region of interest is confined to a narrow band of frequencies. For instance, if a signal is radiating at a wavelength of 500 nm, then from

$$\lambda v = c, \tag{7.1}$$

we know that the optical frequency for the SOI is at 6×10^{14} Hz. If we want to isolate this wavelength and only transmit light at this center frequency and a narrow band of wavelengths surrounding it, then we can use a narrowband optical filter, or "notch" filter, to achieve this purpose. As an example, let us assume that we have an optical filter that is centered at 500 nm and has a 50 nm window measured at full width at half maximum (FWHM) of the filter. If we differentiate Equation 7.1 with respect to the frequency v and take the magnitude, we get

$$|\Delta v| = \frac{c}{\lambda^2} |\Delta \lambda| \tag{7.2}$$

where the bandwidth, Δv, is seen to be 6×10^{13} Hz that is 10% of the original optical frequency. If this were an ideal optical filter, then signals between 5.4×10^{14} Hz and 6.6×10^{14} Hz would be passed by the filter, and all other frequencies would be rejected. Since many noise types are spread over all frequencies, this filter effectively reduces system noise and suppresses background noise and clutter. In reality, optical filters are not perfect, and so some unwanted noise, clutter, and/or background signals pass through the filter but at a much-reduced level. The notch filter is very important since many detectors respond to a very broad band of light (e.g., over the entire visible range and beyond). If we can isolate the spectral response of our SOI, then we can develop a set of *matched filters* for our SOI. The detector then only "sees" light in the frequency range of interest and rejects the unwanted light. This approach works well if the SOIs are known ahead of time or can be predicted. Sometimes, the SOI is not known ahead of time, but it is known that the signal has a narrowband feature. In this case, tunable filters can be useful. Some devices such as acousto-optic tunable filters (AOTF) and electro-optic tunable filters (EOTF) can be used to tune the filters to different center frequencies. Other kinds of filters such as *low-pass filters* pass frequencies below a certain cutoff frequency and reject frequencies above the cutoff frequency. *High-pass filters* accomplish the reverse in passing frequencies above the cutoff frequency and rejecting signals below the cutoff frequency. We will discuss these types of filters in more detail in the signal processing section of this text. Other types of filters are polarization filters that select certain polarization states of the incoming light and neutral density filters that reduce the overall brightness of the light but are not wavelength selective. Neutral density filters are useful if the detector is too sensitive, and the irradiance on the sensor would either damage or blind the sensor.

One issue that occurs for thermal sensors that optical filtering cannot correct is the condition of "washout." Washout occurs when the irradiance due to the background is comparable to the thermal radiation of the target. A visible light analogy might be trying to find a white rabbit on a field of snow. Optical filtering will not help this situation. In thermal imaging applications, IR sensors will typically experience this situation at least twice a day. Figure 7.4 illustrates the practical considerations governing washout. In Figure 7.4a, we have a color image of a Humvee truck. This corresponds to a normal image captured by an optical imaging system. An optical system can form a color image by using a color scheme such as a red, green, and blue (RGB) color scheme. Some sensor elements, like those that are in many modern digital cameras, provide these three-color components as part of their detector design. For instance, a single pixel may consist of active elements that respond to radiation centered

(a) (b)

(c) (d)

FIGURE 7.4
Representative detected images with (a) normal visual image, (b) grayscale image, (c) optimal condition, and (d) washout condition.

on the RGB center wavelengths. Other detectors may have broadband sensors that require narrowband filters as discussed earlier to isolate particular wavelengths like the RGB color components to form color imagery (Holst 2006). On the top right, we see a grayscale image formed, for example, by combining weighted RGB components of a color image.

Forming the grayscale image may be useful if communication bandwidth limitations prevent sending the entire tricolor image over a given communications channel. Instead, a single image can be made out of the color components, and only the single grayscale image (instead of all three color components of the image) can be sent over the data link substantially reducing the amount of transmitted data. Figure 7.4c shows an IR image wherein the Humvee is much hotter than the surrounding background. Figure 7.4d shows the conceptual idea of "washout." The background radiation is largely comparable to that of the TOI, and so the TOI is difficult to distinguish from the background. This condition occurs at least twice a day for many different types of objects. For instance, in the case of the Humvee, at night if the engine of the Humvee is off, the metal on the Humvee will cool overnight and be colder than the surrounding environment leading to a negative contrast condition where the Humvee would appear blacker than the surrounding terrain. If the Humvee was operational, then the heat from the engine would heat the metal of the Humvee, and a thermal imaging camera at night would produce an image similar to Figure 7.4c—a positive contrast scenario. For the case where the Humvee has not been in use all night, when the sun emerges, the Humvee will gain heat faster than the surrounding environment. At some point, the Humvee will heat up to the point where the temperature of the Humvee is similar to its surroundings—this is the first condition of washout. Through the remainder of the day, the Humvee continues to absorb solar radiation and is hotter than the surroundings. As the sun sets, the Humvee begins to cool. Since the Humvee cools faster than the surrounding environment, at some point, the Humvee's temperature will once again be equivalent to the average temperature of the surrounding environment—the second washout condition. Neither broadband optical modulators nor filtering methods help if the radiation of the source is not distinguishable in some manner from that of the background or clutter. That being said, some background suppression methods, active detection methods, and/or spectral methods (discussed later) may sometimes prove useful in distinguishing the target from background or clutter.

7.2.3 Background and Clutter Suppression

When attempting to distinguish a TOI from a surrounding scene or a cluttered environment, the relative spectral content and distribution of the target spectrum to the background spectrum and clutter spectrum is of critical importance. A distinction between background and clutter is in order. A background signal is when the sensor is collecting data from a scene that has no target or clutter objects present. A clutter object is something in the scene that may possibly be confused with the target. A clutter object has some property that is similar to the TOI—as in the so-called "rat in the rocks" scenario. As an illustration, let us say we have an imaging system that is trained on a brown rocky cliff. There are a variety of gray rocks of different sizes and shapes similar to scale as our TOI (a gray rodent) present in the scene. If we wanted to find the little gray rodent that was wandering about the rocky cliff with our imaging system, the observer might have some difficulty in distinguishing the rodent from among the clutter (gray stones). If we swept the hillside and removed the gray rocks (clutter) and also removed the gray rat (target), we would then be left with the background scene (brown cliff). If we now released the gray rat on the brown surface, the observer would have an easier time finding the gray rodent on the brown surface since the clutter has been removed. Since we often work in the spatial frequency domain, the background and clutter will have their own spatial frequency content and structure. If frequency content significantly overlaps that of the target, in observable parts of the spectrum, then filtering out the background and clutter becomes challenging. Since different materials have different spectral content, placing filters in select spectral locations can aid in discriminating TOIs from background and clutter. If the clutter or background spectral contribution lies outside the dominant spectral features of the TOI, optical modulation and filtering methods can be used to reject or suppress the background/clutter and dramatically improve the imaging systems' ability to detect the TOI. For instance, the detected irradiance from sunlight scattering from clouds can sometimes be 4–5 orders of magnitude larger that of a distant target. Optical modulators can be used to suppress the cloud background and detect the target in this case.

Schmieder and Weathersby categorized clutter into three regions, namely, low, moderate, and high. Their experiments resulted in the definition of the signal-to-clutter ratio as given by (Schmieder and Weathersby 1983)

$$\text{SCR} = \frac{|\,\text{max tgt value} - \text{background mean}\,|}{\text{rms (clutter)}} \tag{7.3}$$

where
  SCR is the signal-to-clutter ratio
  max tgt value is the maximum signal level of the target
  background mean is the mean value of the background signal level
  rms is the root mean square

7.2.4 Chopping Frequency Equation

At this point, we need to describe the rotational effects of the optical modulator. Since in many applications, the TOI is much smaller than the instantaneous field of view (IFOV) of the imaging system, the signal coming out the back end of the detector appears "chopped." The bright and tiny target is "on" in the clear areas of the optical modulator and is "off" while opaque fin is in between the target and the detector. The output appears to be a modulated signal of "on" and "off" states, and the TOI is readily distinguished from a larger, relatively unchanging background. Of fundamental importance is the "chopping frequency" that is given by

$$f_c = nf_r. \tag{7.4}$$

In Equation 7.4, n represents the number of clear and dark segment pairs. As a reference, in Figure 7.3, there are two pairs of clear and opaque triangular pairs in this optical modulator. The chopping frequency, f_c, is the product of the number of clear/opaque segment pairs in the optical modulator and the reticle rotation frequency f_r. In the next section, we look at a simple optical modulator and step through the detection process. The discussion will outline the approach for understanding the functionality of a candidate optical modulator.

7.2.5 Simple Reticle System

In this section, we provide an example of a simple reticle to show how reticles can be used to suppress background information and also to show how the signal coming from a detector can be used to discriminate a TOI from the background. We start with an assumption (typically true) that the TOI is much smaller than the dimensions of the opaque fan blades shown in Figure 7.5. The optical system detector output is represented at the bottom right of Figure 7.5. The target is assumed to be at some distant point away from the imaging system "optics" (entrance pupil aperture and relay optics to get the light collected in the entrance pupil to the detector). The optical modulator is placed directly in front of the detector element in the image plane. The optical modulator rotates with chopping frequency f_c. A converging lens often focuses the "chopped" light onto a single detector element. By single detector element, we mean that the output of the detector itself provides only temporal information and not spatial information of the target. In essence, the detector does not have any picture elements (pixels) and consists only of a surface material that

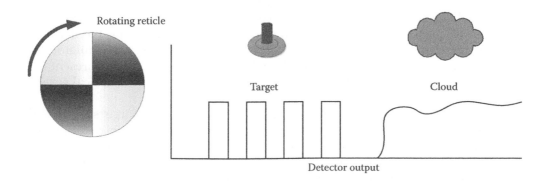

Detector output

FIGURE 7.5
Spatial filtering using a rotating reticle. (Adapted from Hudson, R.D., *Infrared System Engineering*, John Wiley & Sons, Inc., New York, 1969; Courtesy of Dong Wang. With permission.)

responds to the total input irradiance across the detector element. The resulting irradiance on the detector will be that of a spatially modulated scene as determined by the IFOV of the optical system and the optical modulator. This optical modulation concept works in equally well in the visible, IR, and other parts of the electromagnetic spectrum. The detector could be responsive to the visible part of the electromagnetic spectrum by simply switching out the active detector with one that is sensitive to wavelengths in the visible part of the electromagnetic spectrum. On the middle and right side of Figure 7.5 are detector outputs corresponding to different representative types of targets and backgrounds. At the left side of Figure 7.5, we see the reticle that was used in the optical system. In this case, it happens to be the same reticle that was shown in Figure 7.3.

Let us first assume that there is no target present in the IFOV of the optical system. Let us further assume that the optical system is looking at a uniform background such as a cloudless sky with no clutter. In this case, the transparent areas of the reticle would transmit equal amounts of light regardless of the relative position of the opaque fans. The output of the detector would be constant over time corresponding to a fixed, background optical radiation level. The output expected from the detector would be constant at some very low voltage or current level. If the background happened to be a very large cloud, then the detector output would be mostly constant with some slight ripples due to slight surface variations of the cloud itself. This case is shown at the right of Figure 7.5. Let us now consider a small (point source–like target), very bright (with respect to the background), constant brightness target that is located within the IFOV say near the top of the optical modulator shown at the left of Figure 7.5. As the optical modulator rotates, the small object (assumed much smaller than the width of the fan blades in the figure) would produce a square wave pattern as shown in the central parts of Figure 7.5. The signal out of the detector would equal the constant background-only case mentioned previously when the opaque blades of the reticle obscure the target. When the target is in the transparent fan area, the full irradiance from the object—assumed much brighter than the background—falls on the detector. A good mental image might be a very bright little, point source–like dot that is near the top of the rotating reticle in Figure 7.5. Notice that the period of the square wave would decrease as the tiny target object moved toward the center of the reticle. Near the center, the detector output would oscillate rapidly between the fully obscured and fully transparent states and suddenly equal the target-free background illuminated state when the target is obscured by the opaque center. If we slightly altere the reticle in Figure 7.3 by making a tiny portion of the center transparent instead of opaque, then the target that is fully centered in the IFOV of the optical system would produce a constant detector output that is much higher than the background irradiance on the detector. By calibrating the detector to the prevalent background levels, an optical system with this optical modulator and a single-element detector could determine if a point source–like target was present in the scene, how close to the center of the optical axis the target is, or if the optical system is boresighted on the target.

If we now let the target have some finite spatial extent but have the target still be smaller than the opaque blade, then a new discriminating observable can be introduced. When the target is fully behind the opaque blade, the detected radiation is the same as that of the background radiation (assuming no observable changes in the background radiation have occurred). As the extended object (assumed square for simplicity) starts to emerge from behind the opaque blade, the irradiance detected starts to increase. The more the target emerges from behind the opaque fan blade, the more the irradiance increases until the entire object is in the clear aperture part of the reticle. The reticle continues to turn, and the irradiance remains constant until the front edge of the target starts to become obscured by the next opaque fan blade of the reticle. The irradiance decreases until it once again reaches the background level irradiance when the target is fully obscured by the opaque blade. The temporal shape of the detector output looks like a trapezoid! The indication of the positive and negative slope on the rising and falling irradiance levels shows that the object has some finite area. For a point source object, the transition from fully obscured to fully observed is immediate, and the temporal output pattern of the detector is more like a pulse of square waves instead of trapezoids. Note that if the extended object size is exactly equal to that of the width of an opaque fan blade that is near the top of the reticle, then the pattern from the detector looks somewhat like a triangular pulse train. If this extended object moves toward the center of the reticle, then parts of the object become increasingly exposed to the transparent fan resulting in an increase in the average irradiance with a superimposed triangular modulation. If the extended object was circular with uniform irradiance and centered on the reticle, then the detector output would be constant corresponding to a higher average irradiance level than the background irradiance level with no modulation in evidence. Nonuniformities in the target irradiance and object detail would also lead to modulation effects that potentially could result in ambiguities (e.g., are we looking at an extended object at the center of the reticle with object detail, or are we looking at cloud features from partially viewed clouds with small inhomogeneity?). To reduce ambiguity, this reticle is often restricted for use with targets that can be modeled as point sources.

Let us look at the situation if both target and cloud were present in the IFOV of the optical imaging system. As previously stated, the irradiance of the cloud can be 10^4–10^5 times stronger than that of the target. Let us say further that the cloud covers only part of the reticle and several segments of the transparent fan blades view the cloud. We know that there will be an increase in the irradiance on the detector because multiple transparent segments are exposed to the cloud. The amount of chopping is reduced in the detector output signal, and the output signal due to the cloud component would look like relatively flat as shown in the right part of Figure 7.5. The target would produce the nicely modulated square wave pattern shown in the center portion of Figure 7.5. The combined target and cloud image would be a slightly modulated and largely flat signal. If a narrowband filter is centered about the chopping frequency, the cloudy background is suppressed, and the target pattern can be extracted. As the target moves further away from the optical system, the modulation that the target makes becomes smaller and smaller. When the target modulation is on the order of the modulation caused by the background scene, then this is known as the background-limited condition. Background-limited photon (BLIP) detectors operate under this condition. In early IR systems, the active detector element was sometimes in the 2–2.5 μm wavelength range. This part of the electromagnetic spectrum is sensitive to reflected solar radiation from clouds. Many newer IR detection systems have their active elements in the 3.2–4.8 μm wavelength region, which is less susceptible to sunlit reflections from clouds.

7.2.6 Optical Modulation and Coding

Optical communications involves carrying information over distances using light with either visual or electronic devices. Typically, a source such as a laser is used to send an encoded and modulated signal through a noisy channel. Figure 7.6 shows this process. On the left side of Figure 7.6, U represents the user information (e.g., the message) that is being sent through the noisy optical channel. The message is first broken into bits of

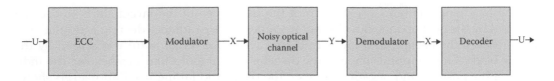

FIGURE 7.6
The optimal communications channel. (Image was obtained from NASA and was placed in the public domain.)

code and tagged with error-correcting code (ECC) as shown in the first block of Figure 7.6. The ECC is optically modulated and aggregated into recognizable symbols by the optical modulator block (second block of Figure 7.6). The modulated and encoded message shown as X in Figure 7.6 is transmitted through the noisy optical channel producing the noisy encoded and modulated message Y shown to the right of the "noisy optical channel" block in Figure 7.6. The noisy, modulated, and encoded message is then demodulated to produce the message in terms of the encoded symbols X (shown to the right of the demodulator block in Figure 7.6). Finally, the symbols are decoded by the decoder block and produce the recreated message U.

The inclusion of the ECC and symbol structure increases the probability of the correct message being retrieved at the receiver and is part of many modern-day communications systems.

7.2.7 Reticle Applications

In the previous section, we described a simple reticle that can be used to suppress cloud clutter in order to detect the presence of a small bright target. We found that with the right interpretation of the output signals from the detector, useful target positioning attributes can readily be determined. There are many different types of reticles having a variety of patterns. We will look at a few of these and describe their critical functional characteristics and their applications. For instance, in military applications, some representative examples of optical modulators are to aid in target detection, tracking, pointing and steering applications, target recognition, satellite and missile navigation, and surveillance. Reticles are used in industry for applications such as quality assurance testing, ranging, "sights" on firearms, and tracking. Medical examples include endoscopy, contrast enhancement in microscopes, and noise suppression.

7.2.8 Reticle Considerations

It must be understood that many reticles work best with distant, small targets that can be well approximated as point sources. Some reticles applied to extended objects can suffer from ambiguity issues with background and/or clutter signals. For uniformly spaced optical modulators, the modulated output signal from the detector has the highest modulation when the size of the target equals the spacing dimensions of the reticle pattern. The size of the target, the relative impact of interference from the reticle's rotational frequency, and the magnitude and spectral distribution of the clutter/background dramatically affect the functionality of the reticle and must be considered in the optical system design process. Optical modulators can have a variety of patterns, some of them very finely spaced and complex. The manufacturing methods of reticles require precision, care, and careful handling. Aroyan (1959) and Biberman (1966) have developed mathematical algorithms for processing complex reticle information. Whitney developed a reticle to suppress background signals without requiring a filter (Whitney 1961).

7.2.9 Reticles for Directional Information

Reticles can be used to find out directional information about the target. Some example applications for these reticles include reticles for missile guidance systems, star/sun seeking trackers that are useful for navigational purposes, and trackers for fire control systems. Target direction information can be obtained by applying different types of reticles. Irradiance due to the TOI, clutter, and background is incident on a reticle, and directional details can be determined from the resulting spatially modulated detector output. The type of information that is extracted depends on the reticle pattern employed. We will present several functional details for optical modulators ranging with modulation patterns ranging from the very simple to the complex.

7.2.10 Rotating Reticles

There are two types of reticles in common use, fixed reticles, and rotating reticles. In this section, we describe rotating reticles. One of the simplest optical modulators is the two-sector or two-segment rotating reticle. This optical modulator has one-half side that is transparent, and the other side is opaque. Figure 7.7 shows this optical modulator. At the top right of Figure 7.7, the white half of this rotating modulator represents the transparent side, and the black side represents the opaque side. There is a point source–like target shown as a "dot" on the top optical modulator in the transparent section of the reticle that makes an angle of δ_1 degrees from the left horizontal axis. The reticle is rotating in the clockwise direction with rotational frequency f_r. We assume that the target is stationary. While the opaque side of the reticle is moving toward the target (e.g., the rotation of the reticle is less than δ_1 degrees), the detector "sees" the target and provides a constant output signal until the reticle has reached the target (rotated exactly δ_1 degrees). At this point, reticle obscures the target from the detector and the detector output immediately goes to zero. The first pulse at the top-left part of Figure 7.7 shows this.

Notice that the target could lie anywhere along the dashed line shown at the top-right part of Figure 7.7, and the corresponding detector output, shown on the top-left part of

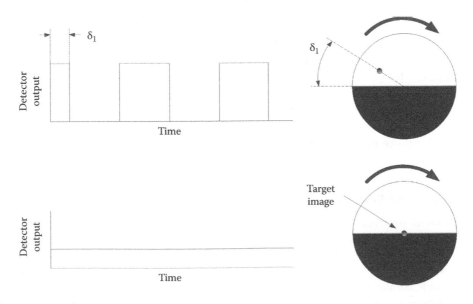

FIGURE 7.7
Directional information using a two-segment reticle. (Adapted from Hudson, R.D., *Infrared System Engineering*, John Wiley & Sons, Inc., New York, 1969; Carbonara et al., 1960.)

Figure 7.7, would be the same as shown for the first pulse. In essence, the angular position of the target given by δ_1 can be determined from the length of the first modulated signal from the detector (e.g., the first left-most pulse at the top-left part of Figure 7.7). To emphasize this result, let us imagine that the target moved to the right side of top-right reticle in Figure 7.7. In this case, the angular position of the target would be given by a different angle δ_2 (not shown). If the reticle's starting position is the same as the previous case and if the rotational frequency of the reticle is the same as before, then it would take the reticle longer to reach the new target position at δ_2 than it would to reach the previous target position at δ_1. Just as before, the detector output would show the presence of the target (e.g., be "on") until the reticle reached the target angular position δ_2 whereupon it would go to zero when covered by the opaque side (e.g., be "off").

In this case, the length of the first pulse shown at the top left of Figure 7.7 would be longer than the pulse in the first scenario. In essence, the angular position of the point source target determines the length of the first pulse. Since this reticle provides angular information, it cannot determine how close the target is to the center of the reticle, only the angular position of the target with respect to the reticle starting point—assumed here as the left horizontal axis. The subsequent modulated detector output does not provide any additional information. It is the length of the first pulse or the relative shift of the subsequent pulses from the reticle starting point that determines the angular position information.

Notice the reticle example shown at the bottom of Figure 7.7. Here, the target is at the center of the reticle. If the center of the reticle is clear, then the signal output from the detector will always be "on" regardless of the angular position of the reticle. The detector output would be constant as shown at the bottom left of Figure 7.7. If the center of the reticle were opaque, then the detector output would be zero, instead of the constant output. As can be seen, this reticle provides angular directional information and can also be used to tell whether or not the reticle is boresighted on the target.

A nice technique to prevent loss of carrier phase information (e.g., detector output "on" and "off") at the zero pointing condition is to use a double-modulation technique. Figure 7.8 shows an example of this method. As long as the target remains within the field of view, the carrier information is available with this method.

Placing a two-segment reticle in front of a reticle that has a large amount of segments (as shown on the right side of Figure 7.8 as Reticle 2) creates a high-frequency carrier without regard to the target placement in the field of view. The field of view is determined by the opening that contains Reticle 1. Since the amplitude is proportional to the target's irradiance, it can be used for automatic gain control purposes or for detecting the presence of the target. The more densely segmented second reticle helps with regard to background suppression.

7.2.11 Background Rejection

Reticle patterns produce a variety of interesting effects. For instance, the reticle pattern shown in Figure 7.9 accomplishes several things. The semitransparent section shown at the bottom of Figure 7.9 produces angular direction information just like the two-segment reticle previously discussed. It also ensures that there is a break in the modulation every time the reticle rotates through half a turn. The fanned pattern at the top helps with background suppression. As previously discussed, a narrowband filter centered on the chopping frequency may also be required. If a small target is exactly centered in the reticle, the opening produces a constant strong signal that is unmodulated.

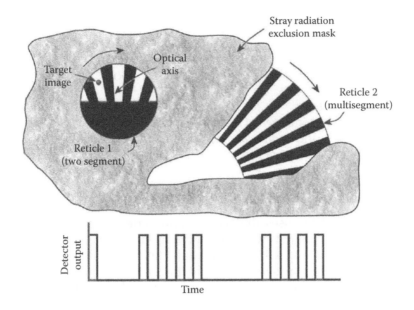

FIGURE 7.8
Double-modulation carrier loss prevention with zero pointing error. (Adapted from Hudson, R.D., *Infrared System Engineering*, John Wiley & Sons, Inc., New York, 1969; Robert, A. and Deslaudes, J., Tracking devices, U.S. Patent No. 2.975.289, March 14, 1961; Chitayet, A.K., Light modulation system, U.S. Patent No. 3.024.699, March 13, 1962.)

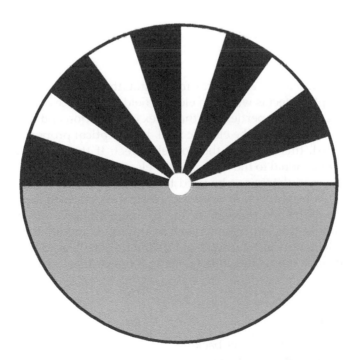

FIGURE 7.9
Sky background rejection reticle. (Courtesy of Dong Wang. With permission.)

The output of this reticle is characterized by

$$f_c = Knf_r \tag{7.5}$$

where

f_c is the chopping frequency
K is the reciprocal of the total area of the reticle occupied by the target sensing portion
n is the number of clear and opaque segments in the target sensing portion of the reticle
f_r is the reticle rotational frequency

7.2.12 Acousto-Optic Modulator

An acousto-optic modulator uses the acousto-optic effect to diffract electromagnetic waves. The acousto-optic effect is where sound is used to control light. For an acousto-optic modulator, sound is injected into a medium, such as a Bragg cell, and alters the medium so that it changes the properties of light that passes through it. The acoustic wave perturbs the refractive index of the medium and in turn changes the propagation direction of the light. One way this can be accomplished is through the use of a piezoelectric transducer. A piezoelectric transducer is based on the piezoelectric effect, wherein charge is produced in a medium subjected to mechanical pressure. This transducer can be added to a material like glass. As a result, the electric signal oscillates forcing the piezoelectric transducer to vibrate. The vibrations, in turn, create sound waves inside the glass material. In the acousto-optic modulator, these sound waves modulate the refractive index causing the electromagnetic wave that is passing through it to diffract at an angle related to the velocity of the sound. Figure 7.10 shows the angular diffraction. There is also a shift in the frequency of the transmitted light.

7.2.13 Electro-Optic Modulator

An electro-optic modulator is an optical modulator that is based on the electro-optic effect. The electro-optic effect is where an electric field can induce changes in a medium's optical properties. Optical properties like the index of refraction and the material absorption can change as a function of the electric field. If the optical property changes linearly with the electric field, this is called the Pockels effect. If the optical properties of the material change proportional to the square of the electric field, then this is called the Kerr effect. Electro-optic modulators can be used to change or control the phase of the electromagnetic wave, its brightness, or even its polarization state. In anisotropic material, the x and y linear polarization states of the electromagnetic wave can be separately controlled leading to interesting applications such as amplitude modulators, optical switches (like a laser Q-switch), optical retarders, and diffracting devices (Saleh and Teich 1991). Putting an electro-optic modulator in between two crossed polarizers can make amplitude modulators.

7.2.14 Two-Color Reticles

Two-color reticles are used to reject background signals as can be seen by the following example. Imagine a jet aircraft against a background of reflected sunlight as shown in Figure 7.11. The sun's temperature is approximately 5900 K. Table 7.3 provides some

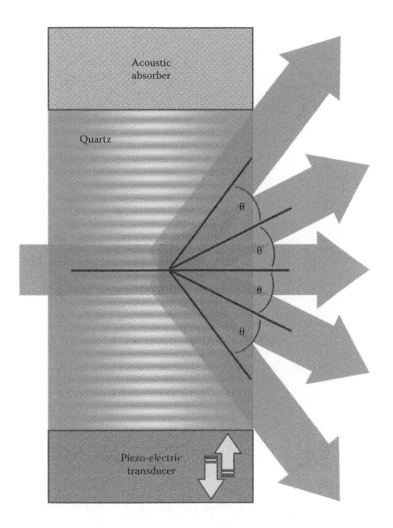

FIGURE 7.10
The Acousto-optic Modulator. (Image courtesy of J. S. Lundeen. 2005. Wikipedia. http://en.wikipedia.org/wiki/File:Acousto-optic_Modulator.png, accessed November 28, 2014.)

convenient blackbody relationships. According to Table 7.3, the median wavelength of the sun's spectral distribution is 0.69 μm. Therefore, half the radiant flux occurs on either side of this wavelength.

By using a two-filter system, a high-pass filter transmits all wavelengths above 0.69 μm, while a low-pass filter transmits all wavelengths below 0.69 μm. Even though we generally use clear and opaque sections of a reticle, these two filters may replace them (e.g., the high-pass filter material replaces the clear sections of the reticle, and the low-pass filter material replaces the opaque sections or vice versa). If this is used in a simple rotating reticle system along with a detector responding equally with all wavelengths, the reticle functions as a neutral filter. Reflected sunlight is equally transmitted through both the high-pass and low-pass sections, and so there is no modulation present.

For a jet aircraft approximating an 800 K blackbody, less than 10^{-6} percent of the flux is composed of wavelengths less than 0.69 μm. Therefore, using such a reticle system, one could theoretically reject solar radiation completely; however, in real practice, complete rejection is very difficult to achieve.

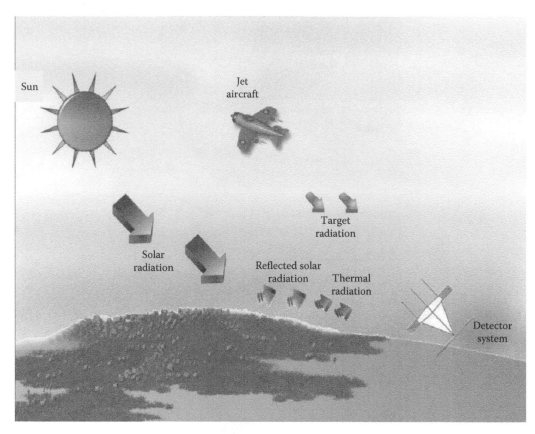

FIGURE 7.11
Background rejection example scenario.

TABLE 7.3

Convenient Blackbody Relationships for λ in μ, T in K, and W in W cm^{-3}

Wavelength Type	Relevant Equation	Optical Power in Band
Peak	$\lambda_m T = 2898$	0.25 W lies between 0 and λ_m.
		0.75 W lies between λ_m and ∞.
Half-power	$\lambda' T = 1780$	0.04 W lies between 0 and λ'.
	$\lambda'' T = 5270$	0.67 W lies between λ' and λ''.
		0.29 W lies between λ'' and ∞.
Median	$\lambda''' T = 4110$	0.50 W lies between 0 and λ'''.

There are a few difficulties in this solution when dealing with variations in the spectral distribution of the source and atmospheric transmittance unpredictability. This makes the construction of an effective two-color reticle system very difficult. However, under certain controlled conditions, such as detection of specific aircraft types against a static high altitude background, a two-color reticle system could be useful.

7.2.15 Guide Star Systems and Reticles

Illuminated reticle eyepieces are used with guide star systems on large telescopes, using positioning controls to place a reticle crossline pattern at the desired position in the field

FIGURE 7.12
Keck Observatory using a guide star system. (Obtained from Keck Observatory home page, image link: http://www.keckobservatory.org/gallery, accessed June 14, 2014.)

of view. One example of a guide star system was deployed on the Keck Observatory (Kinoshita 2003). This is shown in Figure 7.12. When studying distant objects, this guide star system at the Keck Observatory allows scientists to remove atmospheric turbulence effects, thereby greatly increasing the spatial resolution of Keck's imaging system. The guide star system works by projecting an artificial star into the sky with the use of a powerful laser. The guide star acts as a point source, and the returned light at the Keck receivers can analyze this light for optical aberrations and correct the atmospheric aberrations in real time.

Sufficiently bright natural stars are not available in all parts of the sky. Thus, the technique of the "guide star" reduces this dependency (Kinoshita 2003). An increase from 1% to over 80% sky coverage is produced from scientists using the guide star system.

7.2.16 Target Tracking

Various types of weapons targeting systems use reticle systems for target identification and tracking such as riflescopes, which use reticle eyepieces. The sidewinder missile is one of the most influential military optical target tracking systems. Figure 7.13 shows the missile's main sensor located at the tip of the missile.

The main sensor employs an IR detector and then homes in on a target's exhaust gases, thus directing the missile to the target's engine. Many upgrades to the sidewinder's sensors have taken place since its commissioning in 1956. One of the first versions of sidewinder, the AIM-9L, used a frequency-modulated reticle to "reduce the effect of a target's increasing size with decreasing range on the seeker error signal output" (Puckett 1959). The next major upgrade to the sidewinder came when the AIM-9R was released. Its targeting sensor was upgraded to include an optical device that allowed the missile to see a difference between the background and the target. Recently, the missile was upgraded again to include a mid-wave IR focal plane array (Zimmerman and O'Donnell 2007).

FIGURE 7.13
AIM-9L missile. (Image courtesy of Sr. Airman. Theodore J. Koniares. 1982. Wikipedia. http://en.wikipedia.
org/wiki/AIM-9_Sidewinder#mediaviewer/File:AIM-9L_DF-ST-82-10199.jpg, accessed January 2, 2015.)

7.2.17 Optical Modulation Summary

Optical modulation involves the use of a reticle device that has a specific pattern of opaque
and transparent segments. The reticle or the optical axis is rotated, causing a chopping
effect on the incoming flux. It is an effective mechanism for background suppression and
target detection and tracking.

The reticle is a selective modulator that is most effective against point source targets.
The highest efficiency in target detection is obtained when the transparent segments of
the reticle match the target size.

7.3 Integrated Case Study: Systems Requirements and the Need for Optical Modulation

This section of this document will focus on an integrated case study that applies the subject matter learned earlier in this chapter to a fictitious optical systems development effort. The topics
that will be discussed are systems requirements and optical modulation. FIT has recently met
with representatives from their primary stakeholder, the Department of Homeland Security
(DHS), U.S. Customs and Border Protection, for a kickoff meeting to develop high-resolution
imaging systems that are to be installed on DHS unmanned aerial vehicles (UAVs) for border
patrol applications. FIT has conducted some feasibility studies and trade studies for some critical aspects of the optical imaging system determining that an adaptive optics system or a high-
speed, atmospheric turbulence compensation (ATC) system is required for this application.

In this integrated case study, we have Dr. Bill Smith, FIT's CEO; Karl Ben, FIT's senior
systems engineer; Tom Phelps FIT's chief technology officer (CTO) and head of the optics
team; Ginny R., FIT's UAV program manager, Jennifer O. (Jen), FIT systems engineer; Phil K.,

software engineering; George B., hardware specialist; Amanda R., optical engineer and adaptive optics specialist; and Marie C., requirements management and configuration control.

In this part of the integrated case study, the FIT team is working to develop the systems requirements, but there seems to be a snag. We join the FIT team in the middle of an integrated product team (IPT) meeting.

Ginny: "OK, I know that our adaptive optics system (AO) or atmospheric turbulence compensation (ATC) system is going to provide us awesome spatial resolution, but it seems like we are on the marginal side of performance of the extreme operational ranges of the sensor. I don't like asking our stakeholder to change their operational profiles. There has got to be something that we can do."

Tom: "Ginny, even with the AO, or ATC system, we are limited by the size of the collection aperture. We will be capturing the images with spatial resolutions better than anything conventionally available, but it still won't give us the spatial resolution in the IR to get 10 resolution cells across our target. If we use the original conservative target to sensor standoff range of 18.616 km, we get a resolution cell of about 43 cm. That puts less than 5 resolution cells across the target in the longwise direction. Across the arms, you may only get 1 or 2 resolution cells. That's a very blocky image. At the relaxed distance of 8628 meters, ..."

Ginny: "Tom, sorry for interrupting, but I don't want to entertain relaxing the requirements unless we absolutely have to. It sounds like we may be able to make out a human target with our infrared system even at the worst-case operating conditions, but it is right on the hairy edge."

Tom: "That's right. There is nothing we can do to get increased spatial resolution except decreasing the wavelength, which will affect the signal level we get from the target, or increasing the diameter of the entrance pupil, which requires a redesign of the Raytheon pod."

Amanda: "Perhaps we are going about this the wrong way. We are wracking our brains trying to figure out an effective way to increase our spatial resolution for a marginal use case scenario—for example, the extreme operational range case. The way I see it, the requirement is not to get more spatial resolution (that's a 'how') but rather to provide a capability to detect humans trying to illegally cross the border at this maximum range (that's a 'what'). We don't necessarily have to resolve the target to do that."

Ginny: "What do you mean?"

Amanda: "Well, there may be some things that we can do with optical modulators that doesn't require the target to be resolved. In fact, some of these optical modulation methods work best if the target is a point source. We may be able to expand the field of view to ensure that the target is a point source and possibly use some optical modulation techniques to tell us that 'something interesting' is detected by a lit up pixel, sort of like a tip and queue for the high-resolution system."

Jen: "What's the point? If we increase the instantaneous field of view (IFOV), and consequently the pixel size on the target so that the target looks like a point source, and we use the optical system in a tip-and-queue mode, won't we still have to verify what is in the pixel? Doesn't that get us back to a high spatial resolution system?"

Amanda: "Jen, sort of, but in a safe way. If we have a tip-and-queue mode, the UAV can fly with its current conservative mission profile, with a wider IFOV to look for potential points of interest. Once they find something interesting, a potential target of interest (TOI) from the tip-and-queue system, they can survey the area in between them to see if it is clear and then fly closer, zoom in, and see what is really there. This then would give a spatial resolution cell size of 25 cm, and we get over 8 resolution cells on target. This likely will be enough to tell what we are looking at. Better yet, this ensures the UAV's safety and detects the target at 10,000 meters back from the fence line. Also, by working with a slightly lower wavelength, we may also be able to pick up the full 10 resolution cells or at least get real close to it."

Ginny: "That sounds promising! I don't understand what you mean by an optical modulation method though."

Tom: "That is an old technology that is used a lot in IR tracking systems such as missiles. They are great in suppressing background information and pulling weak signals out of the background or for a variety of other applications."

Ginny: "But didn't you say in the meeting with the customer that we had enough signal from the source? Why do we need the optical modulator?"

Tom: "That's right, we have enough signal, and so we won't be using the optical modulator for that. Do you remember when Wilford was talking about other critters? He was concerned about other things setting off the sensor and giving false alarms. If we use an optical modulator, we can sort of tune the sensor response to a human and reduce the number of false alarms."

Ginny: "Really? How does it work?"

Amanda: "In this case, I was thinking of an optical modulator that has triangular slices of equal sizes, sort of like an equally sliced pizza. Every other slice would have a low-pass filter that has a cutoff frequency matched to the median value of a human's spectral radiance distribution. According to Table 7.3, that is 13.5 μm. The other slices in the optical modulator are high-pass filter windows that are also centered at this median value. We place this optical modulator in the optical system so that the entire IFOV is contained in one of the slices. That way, the entire image plane alternately sees either the low-pass view or the high-pass view. Since the system is centered on the human's median wavelength, half the available optical power from the human goes through the low-pass filter elements, and half goes through the high-pass system. In effect, there would be no modulation present if we were looking at a human (or something that has similar surface area and body temperature). If another bright target is out there at a different temperature, like a fire, the median optical power would be a different wavelength, and a different amount of optical power would pass through the low-pass and high-pass filters, thereby modulating the received brightness of the pixel."

Tom: "It's a great idea. However, we might want to use a slightly different median frequency since the effective temperature of a human can vary from the nominal value."

Amanda: "That's true. The body temperature can go down to 0 degrees Celsius in extremely cold climates. I don't think we will be experiencing that in this scenario, however, but we should make some calculations. We also need to account for the atmospheric transmission window since it drops off rapidly at around 13 μm. That will skew the power optical power distribution. We have to account for that and pick a good center wavelength corresponding to the human target. Once we do that, the spectral radiant emittance curve is wide enough so that the expected human temperature variations shouldn't introduce too much of a modulation effect."

Ginny: "Sounds like a design consideration. I like the idea! This way, the UAV can fly in its current, safe, operational profile and use a wide field of view to search the scene for targets of interest. If a human is in the search field, a pixel will light up, and the brightness of that pixel will stay relatively constant. The UAV can then check the intervening landscape to make sure things are safe, fly closer, and turn on the high-resolution imaging system to confirm."

Jen: "How does this affect the complexity and maintenance aspects of the system?"

Amanda: "As Tom said, this technology has been around for quite a long time and is already in use in other operational systems. It will be low risk to include in the current system. The modulator will occasionally need to be cleaned, or exchanged, but we can consider this in the design process and supportability analysis."

Phil: "Are we going to have to provide some code to analyze the modulation for the user?"

Ginny: "Yes. We are offering a complete turnkey system. If we have an optical element that is providing critical information to the operator, the operator will need to select the capability, in this case the optical modulator, and understand the results."

Phil: "I will put a feature stub into the image processing module."

Ginny: "All right. We will need some systems requirements that capture these discussions. Something on the optical system's capability to discriminate between human and other targets, and something on supportability, and maintenance of the system."

Amanda: "I'll write something up."

Jen: "What about the daytime system? Do you need the optical modulator for that?"

Amanda: "Actually, we have plenty of spatial resolution for the daytime system. The wavelength of 500 nm is about a factor of 4 smaller than the IR system, and so spatial resolution cell size will also be 4 times smaller than the IR system. That gives us about a 2.2 cm resolution, which is a factor of 9 better than what we need for the worst-case scenario. The problem with the daytime imaging system is not the spatial resolution or that we are getting enough of a signal from the target; rather, we are getting too much signal from everything. During the daytime, the sunlight is reflecting off of humans, rocks, dirt, bushes, everything in the IFOV. How do we find the humans in all that stuff? We definitely have sufficient signal and spatial resolution from the daytime imaging system, but knowing what to point at is the real problem. Maybe we can use the optical modulator idea here too?"

Tom: "Go on."

Amanda: "Well, the sun's median radiance is in the visible wavelength regime, right about 0.69 μm right?"

Tom: "Yep."

Amanda: "So all the reflected energy will peak in the visible part of the electromagnetic spectrum and will barely contribute to the long-wave IR signal at the median wavelength peak of the human spectral radiance curve of 13.5 μm. If there is enough separation between the peak of the human and the peak of the background thermal radiation, we can use the same optical modulator idea to discriminate the human target from the background. We could use the IR system as a tip and queue for the visible system too."

Tom: "I'm not too sure that will work. During daytime, the dirt and sand have a spectral emittance curve that peaks at roughly the same wavelength that humans do. The radiance from the background is actually a clutter signal at roughly the same temperature as the human temperature. We can't use the previous optical modulator to discriminate the background."

Amanda: "Hmmm, well it was an idea."

Tom: "Wait a minute though. Let's not scrap this idea yet. We may be able to use another optical modulator for this. This clutter radiation will be distributed throughout the scene, whereas the spectral radiance of the human is localized to a 2 m^2 area. Let's look up the daytime spectral radiance of the common materials in the scene and use the worst-case (largest spectral radiance) material present in the scene to see what the contribution of the clutter radiation turns out to be. If it is substantially less than the spectral radiance of a human for the same surface area, then the human will look like a bright point source in the scene. We can use the typical optical modulator that has alternating opaque and clear pizza slices as the triangular elements of the optical modulator. I think I remember seeing a figure of something along those lines in that optical systems engineering book. Ron, you have a copy of that from our library don't you?"

Ron: "I sure do! I happen to have it with me too. Here you go."

(Tom flips through the pages.)

Tom: "Here it is. It looks like the brightest background source is soil which looks almost like a blackbody. This chart is a bit coarse, but it looks like the background peaks at 10 μm and has a worst-case spectral radiance of 1.1×10^{-3} watts/(cm^2 sr) at 9.5 μm and 32 degrees C. That gives a spectral radiance for the clutter signal of about 22 watts/(m^2 sr) or a clutter radiant emittance of 69.1 watts/m^2 if we assume a diffuse scattering surface. Remember from our radiometric models that the human radiant emittance was 33.8 watts/m^2. I'm not sure that would give us enough of a modulation signal compared to variations in the background scene. We will have to run some analyses."

Ginny: "Wait a minute. What's the requirement for this? I can see why we need to do something for the nighttime system since we are getting questionable spatial resolution performance of our conceptual system at the extreme operational ranges. However, for the visible sensor, we are providing a system that has sufficient spatial resolution to make out the targets. Figuring out where to point is out of scope for this project. Also, the customer may already have some automatic target recognition (ATR) algorithms, or they may just go with the operator looking where he or she wants to look. Are we potentially adding a bell and whistle here?"

Karl: "Ginny you are right! If we can come up with something, we may provide a future upgrade, but that wasn't in the scope of the current effort. Good catch! I am having lunch with Bill and I will give him an update. He'll probably want us to run this by Wilford's team."

Tom: "OK, I will have my team run the optical models to firm up the numbers. Amanda will get you the preliminary systems requirements statements by close of business tonight. Anything else? OK, good job everyone. Let's get together tomorrow at 10 a.m. to look at the systems requirements."

(10 a.m. the next day.)

Karl: "Welcome back everyone. Before we jump into the requirements, let's set the focus of the meeting by providing a short overview of what we are trying to accomplish. The primary mission, for the UAV's optical system, is to provide effective and efficient bipedal human detection (walking) and analysis to a mission control center (MCC) for the border patrol. The system must also reliably transmit the captured video/images from the UAV back to a fixed MCC and/or to a mobile command center (MoCC) within 5 seconds of image/video capture. Remote control of the optical system and an image analysis capability must also be provided for both the MCC and the MoCC. Image products must be in a format compatible with MCC data/image distribution system. A total of 30 optical systems of which 8 are spares are planned for production. All optical systems, software, and hardware will be supported and maintained for 10 years after fielding a system, with an optional extension to 25 years. Preplanned product improvements are expected as part of the support. Marie, can you bring up the current operational requirements in our requirements management tool?"

Marie: "Sure!" (Marie brings up Table 7.4 on the overhead.)

Marie: "Table 7.4 is a table of initial requirements using four columns. The first column explains their identity. The second column details the operational requirements that were derived from column four which defines the wants/needs of the stakeholder. The third column simply shows the importance level of the listed requirements ranging from a 1 to 5 scale with 1 being least important to 5 being the most important."

Karl: "OK, this is a good start. We will have to add quite a few more requirements to this as we get further into the conceptual design."

Ron: "Like what?"

Karl: "Well, we need to state our own operational availability requirement. Currently, the way the requirement is written, we have to meet the UAV's operational availability requirement. We are not responsible for the UAV requirement. We need to figure out what our operational availability needs to be to support their mission and state the requirement that way. Something like 'The optical system operational availability shall be 99%.' Also, where are the testing requirements, logistics requirements, reliability requirements such as mean time between failure? We've got a good start, we now need to keep going and fill in the details. Marie, do we have the A spec template in our requirements management tool?"

Marie: "Yes, we do. I have given all the team members access to a working copy. I am maintaining the official configuration controlled version."

Karl: "Excellent! Let's use the A spec template as a guide when we are fleshing out our conceptual design. That way, we make sure that all the proper categories are addressed in developing our concept. As we go along, let's make sure that each requirement is feasible, that we have a qualification strategy for the requirement, and that we identify any technical performance measures and metrics, key user requirements, and key performance parameters. We need to also link the requirements to any design and test artifacts. We can track this with special categories in our requirements management tool. Also, let's make sure we use our MBSE tools and update them as necessary."

TABLE 7.4

Operational Requirements

ID	Operational Requirements	Stakeholder Rank	Stakeholder Wants/Needs
Mission requirements			
OR-1	The optical system shall operate from a UAV flying at a nominal altitude of 4.572 km, with an operating altitude window of 4.000–7.620 km.	(5) Very important	UAV operation
OR-2	The optical system shall be capable of discerning human bipedal traffic within the operating altitude window of the UAV.	(5) Very important	Human bipedal traffic detection
OR-3	The optical system shall be capable of operating in daylight.	(5) Very important	Daytime operations
OR-4	The optical system shall be capable of operating at nighttime.	(5) Very important	Nighttime operations
OR-12	The optical system shall transmit imagery that is compatible with the current long-haul communications in the UAV and line of sight communications in the UAV.	(4) Somewhat important	Long-haul communications
OR-5	The optical system shall transmit imagery to the MCC or MoCC in 5 s or less after image/video acquisition, with said imagery being captured at up to 30 Hz.	(5) Very important	Real-time imagery
OR-6	The optical system shall meet the operational availability requirement of the UAV of 98%.	(5) Very important	Continuous operation
OR-9	The optical system image storage medium shall employ information assurance.	(4) Somewhat important	Information assurance
Operator/user interface requirements			
OR-10	The optical system shall permit the operator to select day/night modes of operation independently of each other.	(4) Somewhat important	Day/night mode
OR-11	The optical system shall permit the operator to choose ATC—or not—when viewing imagery.	(4) Somewhat important	ATC control
OR-7	The optical system shall permit the user to perform simple maintenance.	(5) Very important	Ease of maintenance
OR-13	The optical system shall permit the user to perform software upgrades.	(4) Somewhat important	Upgradable software
OR-15	The optical system image storage product shall permit the user to retrieve stored imagery.	(3) Important	Stored image retrieval
Support requirements			
OR-8	FIT shall provide maintenance support for the optical system over the projected life cycle of the product.	(5) Very important	Maintenance support
OR-8a	The maintenance support shall include training, documentation, service, and preplanned software upgrades.	(5) Very important	Maintenance support
OR-16	FIT shall provide technical support for the optical system over the projected life cycle of the product.	(5) Very important	Technical support
OR-16a	The technical support shall include training, documentation, and access to technical support staff.	(5) Very important	Technical support
OR-14	FIT shall provide adequate spares to meet mission requirements over the projected life cycle of the product.	(4) Somewhat important	Spares availability

References

Aroyan, G.F. 1959. The technique of spatial filtering. *Proceedings of the Institute of Radio Engineers*, 47: 1561.

Arrasmith, W.W. 2007. A systems engineering entrepreneurship approach to complex, multi-disciplinary university projects. *National Conference of the American Society of Engineering Education*, Honolulu, HI, June 23–27, 2007.

Baca, C. 2007. *Project Management for Mere Mortals*. Boston, MA: Addison-Wesley.

Bangert, M. 2010. Managing customer specific requirements. *Quality*, 49(9): 24.

Biberman, L.M. 1966. *Reticles in Electro-Optical Devices*. New York: Pergamon.

Blanchard, S.B. and W.J. Fabrycky. 2011. *Systems Engineering and Analysis*, 5th edn. Boston, MA: Prentice Hall.

Burns, M. 2010. How to document requirements. *CA Magazine*, June–July, 143 ed., p. 13.

Carbonara, V. E., et al. 1960. Star Tracking System. U.S. Patent No. 2.947.872, August 2, 1960.

Carrillo de Gea, J.M., J. Nicolas, J.L.F. Aleman, A. Toval, C. Ebert, and A. Vizcaino. 2011. Requirements engineering tools. *Software*, 28(4): 86–97.

Chitayet, A.K. 1962. Light modulation system. U.S. Patent No. 3.024.699, March 13, 1962.

Clark, H.I. 1950. Sun follower for V.2 rockets. *Electronics*, 23: 71.

DAU Acquipedia (KPI). 2014. Measure(s) of performance. https://dap.dau.mil/glossary/pages/2237.aspx (accessed July 28, 2014).

DAU Acquipedia (KPP). 2014. Key performance parameter (KPP). https://dap.dau.mil/acquipedia/Pages/ArticleDetails.aspx?aid=7de557a6-2408-4092-8171-23a82d2c16d6 (accessed July 28, 2014).

DAU Acquipedia (MOP). 2014. Measure(s) of performance. https://dap.dau.mil/glossary/pages/2237.aspx (accessed July 28, 2014).

Driggers, R.G. 2003. *Encyclopedia of Optical Engineering*, Vol. 3. Boca Raton, FL: CRC Press.

Feinman, J. 2007. Telelogic opening DOORS for business software projects. *Software Development Times*, 171: 22.

Holst, G.C. 2006. *Electro-Optical Imaging System Performance*. Winter Park, CO: JDC Publishing.

Hudson, R.D. 1969. *Infrared System Engineering*. New York: John Wiley & Sons, Inc.

Hull, E., K. Jackson, and J. Dick. 2005. *Requirements Engineering*. London, U.K.: Springer.

INCOSE. 2004. International council on systems engineering homepage. http://www.incose.org/practice/whatissystemseng.aspx (accessed April 23, 2014).

INCOSE. 2010. INCOSE requirements management tools survey. http://www.incose.org/productspubs/products/rmsurvey.aspx?&session-id=b0d230059a2b6b3432b0ba46443858c3 (accessed July 28, 2014).

Kinoshita, L.K. 2003. W. M. Keck Observatory, October 2003. http://www.keckobservatory.org/article.php?id=46 (accessed June 13, 2007).

MagicDraw. 2014. MagicDraw Homepage, http://www.nomagic.com/ (accessed November 28, 2014).

Maiden, N. 2008. User requirements and system requirements. *Software*, 25(2): 90–100.

Maiden, N. 2012a. Exactly how are requirements written? *Software (IEEE)*, 29(1): 26–27.

Maiden, N. 2012b. Framing requirements work as learning. *Software (IEEE)*, 29(3): 8–9.

Maiden, N. 2013. Monitoring our requirements. *Software (IEEE)*, 30(1): 16–17.

Military Dictionary. 2014. Official US DoD definition of measure of effectiveness (accessed June 14, 2014).

Pfund, A.H. 1928. Resonance radiometry. *Science*, 69: 71–72.

Puckett, A.E. 1959. *Guided Missile Engineering*. New York: McGraw-Hill.

Ramasubramaniam, K.S. and R. Venkatachar. 2007. *Goal-Aligned Requirements Generation*. Berlin, Germany: Springer-Verlag, pp. 245–254.

Robert, A. and J. Deslaudes. 1961. Tracking devices. U.S. Patent No. 2.975.289, March 14, 1961.

Rubinstein, D. 2003. Telelogic opens DOORS to new development. *Software Development Times*, 82: 5.

Saleh, B.E.A. and M.C. Teich. 1991. *Fundamentals of Photonics*. New York: John Wiley & Sons.

Schmieder, D. and M. Weathersby. 1983. Detection performance in clutter with variable resolution. *IEEE Transactions on Aerospace and Electronic Systems*, AES-19(4): 622–630.

Tronstad, Y.D. 1996. Requirements drive system engineering. *Electronic Engineering Times*, p. 64.

Valerdi, R. 2012. What is systems engineering? Industrial Engineer, IE 44, 2.

Whitney, T.R. 1961 (February 21). Scanning discs for radiant energy responsive tracking mechanisms. U.S. Patent No. 2.972.276.

Zimmerman, M. and J. O'Donnell. 2007. *The Evolution and Effects of the Smart Weapons System*. Pittsburgh, PA: University of Pittsburgh. http://fie.engrng.pitt.edu/eng12/Author/final/7065.pdf (accessed June 13, 2007).

8

Maintenance and Support Planning

An ounce of prevention is worth a pound of cure.

—**Benjamin Franklin**

When developing a system, one of the biggest challenges is to optimize the balance between performance, quality, and cost in a dynamic, international, market. Reliability is an important factor when designing any system, and it is directly linked to the system's quality and performance. The system that is being developed needs to be reliable enough to work over the time period specified by the stakeholders. It is also not enough to only design a system up to its launch. For example, a system that meets the stakeholder's functional requirements but is only available 10% of the time is not very useful or practical. As John Hsu and Satoshi Nagano state, "the success of aerospace programs is directly linked to the successful application of the systems engineering (SE) invoked from program initiation through program closure" (Hsu and Nagano 2011). Support for a system does not end until its termination, and until then that system needs to be efficiently and effectively maintained.

Maintenance and support planning is a fundamental activity that comes quite natural to systems engineers. According to Kamrani and Azimi, systems engineers optimize the cost and performance of a system by viewing the system as a whole, across its entire life cycle (Kamrani and Azimi 2010). Consequently, maintenance and support planning is a natural response to systems thinking and life-cycle planning. To be competitive on the global market, developers push for better performance, lower costs, and shorter development time frames when developing systems. However, effectively and efficiently developing systems are becoming increasingly more challenging as time goes on. Technology is advancing rapidly and systems are becoming more complex, while expected delivery times and available resources are shrinking. With regard to operational systems, mechanical parts can fail, code can contain defects, and components may not work correctly together. These problems will need to be fixed, and as such, the system will need to be effectively and efficiently maintained and supported.

One of the major responsibilities of the systems engineering team is to plan early on in the systems development cycle for proper maintenance, service, and support. A standard way this is accomplished is by analyzing and modeling the system in three related areas: system's reliability, maintainability, and availability (SRMA).

SRMA parameters, often included as requirements, are an important part of system design because they describe how well the system functions. Everything from the vehicles we drive to the homes we live in can be thought of as systems, and the ongoing functionality of these systems is directly related to their reliability, maintainability, and availability (Dhillon 2006). Consideration of maintenance and support aspects must be established early on in the development of the system since maintainability requirements can have a significant impact on the system design. Also, failure to properly consider these requirements can severely degrade operations in the operations and support phase and drive up the overall system's life-cycle cost.

8.1 Introduction to Maintenance and Support Planning

The first section of this chapter will discuss the systems engineering process of maintenance and support planning. The major elements to be discussed include the importance of the early consideration of maintenance and support elements; the generation of reliability, maintainability, and availability models; and finally covering some SRMA methods throughout the life cycle of the system. The second section will cover optical detectors, the various types of optical detectors, and their performance and comparisons. The third and final section of this chapter is an integrated case study that uses a fictitious company, Fantastic Imaging Technologies (FIT), engaged in an optical systems development effort to illustrate and apply the principles, methods, and techniques presented in this chapter.

8.1.1 Importance of Early Consideration of Maintenance and Support Elements

Improvements to the reliability and maintainability of a system are ways to improve quality, improve stakeholder satisfaction, reduce large costs associated with inventories, and maintain lean production systems (Madu 2005). By making the system easily maintainable, it becomes easier to repair, thus preventing parts from failing and needing replacements. Complementing this, improving the reliability of the system allows the system to last longer without needing to be repaired. Together, maintainability and reliability methods work hand in hand and greatly improve the quality of the system. Improvements in reliability and maintainability will lead to lower support and maintenance costs because of, on average, longer-lasting products, fewer required repairs, less parts and spares, and faster maintenance.

Maintenance programs exist for preventive maintenance actions and to correct unplanned events that affect the proper function of the system. In some cases, maintenance is not planned in advance, but becomes necessary by unforeseen failures. For both preventive and corrective maintenance cases, detailed procedures explaining how to maintain the system should be produced. These procedures should also be tested, and technicians should be trained to carry out the maintenance actions. Unless the system is under warranty, maintenance costs are typically the responsibility of the system user.

System maintenance costs also include the production, acquisition, and storage of spare parts, which can often be the largest cost of system ownership (Wessels 2010). Producing too many spare parts is costly. Not only has the company wasted money by having unused parts, but the storage space used by these spare parts might be better used for other systems. This used inventory space can hold back other projects, as the lack of storage space can hold back production and cause a failure to meet demand. Also, a shortage of spare parts can result in other costly issues. Not having the spare parts on hand will cause delays in the repair of the system. The part in question must be either ordered or, in the worst case, manufactured. If the part must be manufactured and the manufacturing setup previously used to create that part was reallocated to another project, even more issues, delays, and costs arise. If no spare manufacturing machinery is available, the company would either have to purchase more machinery, outsource the production, or stop current production to reallocate machinery to produce the needed parts. While all of these scenarios cost the company time and money, the last scenario can result in lost opportunity costs for other systems. These costs can become quite large and have a negative impact on a company's position within the market (Wessels 2010).

Without the early consideration of maintenance and support, companies run the risk of falling into a strategy of trying to sustain the system. To reduce systems reliability issues,

SRMA analysis should be conducted at an early stage of development where changes inflict the least amount of cost (Birolini 2010). Early consideration of maintenance and support is necessary to properly plan for their inclusion throughout the system life cycle.

8.1.2 SRMA

During the early stages in development, it is important to understand what expectations the stakeholders have with regard to the maintenance and support aspects of the system. These expectations can generally be found in the stakeholder requirements and are typically identifiable as nonfunctional requirements in one of three systems categories/focus areas, system reliability, maintainability, and availability (SRMA). SRMA are systems focus areas that relate to how a system accomplishes its designated goal over a specific period of time. When designing a system, it is important to keep SRMA in mind through each design stage, not just as a step to be completed at a point in the development process. SRMA models should be developed as part of the functional analysis process and updated as more information about the system becomes available and the design evolves and matures. The functional decomposition of the system through functional block diagrams (FBDs) and functional flow diagrams (FFBDs) is an excellent place to start developing reliability models. Once developed and depending on the level of decomposition, these models can be used to predict the reliability of the system, subsystems, or even components. Further, these models can be validated and refined during the operational phase of the system. In the next section, we present some of the concepts needed for understanding reliability models.

8.1.2.1 Reliability

Reliability is the probability of an object performing its intended function over a specific period of time (Birolini 2010). The reliability of a system is very important since it directly reflects the system's ability to work and directly relates to the maintenance, support, and operational availability of the system. With the reliability of a system, models can be generated to identify possible modes of failure that the system can undergo (Wessels 2010). In order to find the reliability of a system, the components of that system must be analyzed considering the system's specified operating conditions. This includes factors such as environmental conditions, operating temperatures, and vibrations from external forces. It is also important to consider the conditions the system will face during storage and transport. Depending on the circumstances, a system may spend a significant amount of time in storage or transport. For example, a system designed to function in a desert should still be designed for humidity while in storage on site. Also if a system is being delivered in a truck, the vibrations sustained during transport can cause damage to internal components if not properly secured. Once the normal operating range and usage conditions are understood and defined, the reliability of the system, subsystem, and components can then be mathematically determined through the application of probability methods.

In reliability analysis, the probability of failure is the probability that something will fail over a given period of time. Probability is the chance that an event will occur, and the event, in this case, is the item failing over the said period. An example would be if the probability of failure for a system was 30% over a run time of 3 weeks, then that system has a 30% chance of failing during that time period. Notice that the time period is an intrinsic part of the "event" and must be inherently defined. To explain, let us begin with the time to failure (τ). Given that an object's time to failure has a probability density function $f(\tau)$, the probability $F(t)$ that a failure occurs within the time interval $(0, t)$, where t is the amount of time being analyzed, is given by

$$F(t) = \int_0^t f(\tau)d\tau. \tag{8.1}$$

Similarly, the reliability function $R(t)$ is the probability that no failure will occur in the time interval $(0, t)$. It can be written as

$$R(t) = 1 - F(t). \tag{8.2}$$

Using $R(t)$, we can find the instantaneous failure rate $\lambda(t)$. This is the number of failures over a given period of time by

$$\lambda(t) = \frac{-dR(t)/dt}{R(t)}. \tag{8.3}$$

If the system is brand new, time will equal zero, and $R(0) = 1$. $R(t)$ can be determined by substituting the dummy variable x (for t) into expression (8.3) and integrating both sides with respect to time over the interval $(0, t)$

$$R(t) = e^{-\int_0^t \lambda(x)dx}. \tag{8.4}$$

In many applications, it can be assumed that the instantaneous failure rate remains constant over time, such that $\lambda(t) = \lambda$. In this circumstance, $R(t)$ can then be given as

$$R(t) = e^{-\lambda t}. \tag{8.5}$$

The reliability of a system is not commonly expressed with its failure rate λ. Generally, it is expressed as the *mean time between failure* (MTBF), which instead gives failure expressed as a mean time (Wessels 2010). For a constant failure rate, the MTBF is the inverse of the failure rate and can be represented as

$$\text{MTBF} = \lambda^{-1}. \tag{8.6}$$

$R(t)$ can then be rewritten in terms of the MTBF as

$$R(t) = e^{-t/\text{MTBF}}. \tag{8.7}$$

Notice that when time equals the MTBF, the probability that the system will work is approximately 37%. Another characteristic of reliability is the *mean time to failure* (MTTF), which is the average time, measured from the current point in time, in which there is no failure (Birolini 2010). It is also the time difference between the MTBF and the current time and can be written as

$$\text{MTTF} = \text{MTBF} - t. \tag{8.8}$$

MTTF is typically used with nonrepairable replaceable parts, while MTBF is typically used for repairable parts (Bazovsky 2004).

Obtaining the reliability of each piece of a system does not necessarily mean the reliability of the entire system has been found. While the reliability of each component is important, it is also important to look at how each piece interacts with the rest of the system.

8.1.2.1.1 Reliability Block Diagram

A *reliability block diagram* (RBD) is a mathematical model designed to "relate the individual 'block' reliability to the reliabilities of its constituent blocks or elements" (Blanchard 2004). Essentially, the reliability of a system is modeled by modeling the way its components interact, specifically how the input and output of each part are handled within the system. Different components may interact in one of two ways: in series or in parallel. Both of these ways are shown in Figure 8.1.

In this diagram, Subsystem C is a statistically independent component that is in series with the Subsystem B. In a reliability sense, a series connection is one in which the elements in question do not act as a redundancy for the other subsystem. For example, if two light bulbs are placed in series, then a cut in the wire to the first light bulb causes both lights to go out. The second light was not a redundancy for the first light, so, in a reliability sense, the components are arranged in series with respect to each other. For systems in series, the reliability is the product of each of the individual element reliabilities. For instance, for two subsystems, the total system reliability is given by

$$R_T = R_A \times R_B,$$ (8.9)

where R_A and R_B are the reliabilities of each one of the subsystems in series.

For systems in series, the resulting total system reliability is smaller than each of the individual reliabilities of the subsystems.

The second way in which subsystems can interact is in parallel, as seen in Figure 8.1. In this diagram, Subsystem B is composed of three entities set up in parallel that receive input from Subsystem A. Subsystems are considered to be in parallel in a reliability sense if one subsystem acts as a redundant subsystem or fail-safe for the other. Revisiting the light bulb example from earlier, if two light bulbs are in parallel, breaking either one of the bulbs would not cause the other one to go out. The lights effectively act as a redundancy for each

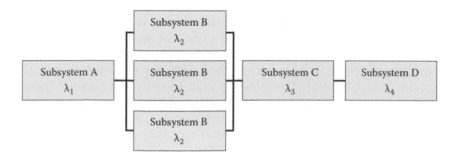

FIGURE 8.1
RBD. (Obtained from Wyatts, Reliability block diagram, http://en.wikipedia.org/wiki/File:Reliability_block_diagram.png, accessed April 3, 2014, 2005b.)

other. In this case, the connection between the two bulbs is a parallel connection. The total reliability of a system with independent and identically distributed elements in parallel is

$$R_T = 1 - \left[(1 - R_1)(1 - R_2)...(1 - R_n)\right], \tag{8.10}$$

where
  R_T is the total reliability of the system
  R_n is the reliability of the nth element in parallel
  R_1 and R_2 are the reliabilities of the first and second components, respectively

If the elements were identically distributed, and independent, the element reliability R_1, R_2, through R_n would all be the same.

It is important to note that while elements (e.g., subsystems or components) interact in series or parallel, the entire system may be a combination of series and parallel components. In this case, the system must be simplified part by part. Sets of reliability blocks that are in series are multiplied, whereas parallel loops are reduced using Equation 8.10. Some complex, interconnected elements can be reduced to determine the overall system reliability by repetitive applications of the series and parallel reliability equations earlier. Advanced reliability methods are required for statistically dependent elements and complex reliability interconnections.

An RBD is useful in that it shows the interconnections of the reliability elements and can be used to determine the total reliability of a system. It can also be used to identify weak areas of the design and to identify redundancy points that may be required. It can also be used as a basis for a reliability-centered maintenance program.

8.1.2.1.2 Failure Mode, Effects, and Criticality Analysis Model

With the RBD of a system, it is possible to find the weak links in the system. One method of doing this is through *failure modes, effects, and criticality analysis* (FMECA) model or a *failure modes and effects analysis* (FMEA) model. Márquez states that this method "includes the necessary steps for examining all ways in which a system failure can occur, the potential effects of failure on system performance and safety, and the seriousness of those effects" (Márquez 2009). An FMECA can be applied either functionally or physically. Fabrycky and Blanchard state, "the FMECA needs to address both the product and the process" (Fabrycky and Blanchard 2011). In short, not only must the system itself be addressed but also the developing, storing, shipping, and handling of the system. Conducting an FMECA is an eight-step process, seen in Figure 8.2.

The first step in an FMECA is to define the system and its environment in which it operates. Then, the second step is to define ground rules and declare any necessary assumptions about the system. The third step is to construct block diagrams of the system to clearly define the system and its boundaries. The fourth step is to identify how the system can fail (e.g., the failure modes). The fifth step is to identify the cause of these failures. This step also includes identifying the effects of a failure, how it will affect the system elements, and other parts of the system. The sixth step is to assign detection methods for the failure modes and establish some compensating provisions. The seventh step is to assign severity rankings based on a classification of the failure effects. This is the last step in the commonly used FMEA model. The eighth and final step is to determine the criticality part of the FMECA. In this step, the severity or consequence of the failure is mapped into the range [0,1], and the probability of the failure occurring is determined. Often, the probability of detecting the failure is also established. These normalized consequence,

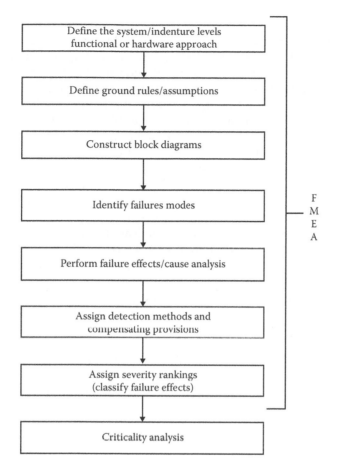

FIGURE 8.2
Typical FMECA flow. (From Department of the Army, TM 5-698-4, Failure Modes, Effects and Criticality Analyses (FMECA) for Command, Control, Communications, Computer, Intelligence, Surveillance, and Reconnaissance (C4ISR) Facilities, September 29, 2006.)

probability of failure, and probability of detect rankings are multiplied together resulting in the *risk priority number* (RPN) of each failure mode (Blanchard 2004). As part of the final step, critical system failure modes are identified. This includes finding failure modes with high RPNs, finding their cause(s), and making recommendations for improvement (Fabrycky and Blanchard 2011). So, if a particular system element has a high failure probability of occurrence, a high severity rating, and a large probability of not being detected, that system element should be analyzed to determine how its criticality can be reduced. Extreme combinations such as very high consequence and very low probability of occurrence or very high consequence and extremely low probability of detect should be separately considered.

An FMECA model is useful in determining what failure mode elements of the system are most critical. While this is a very useful tool for systems engineers, the FMECA model is limited in that it does not capture the chain of events or conditions that lead up to the failure. Typically, a fault tree analysis (FTA) is accomplished for this purpose. As will be seen in the next section, FTAs complement FMEA/FMECAs and are essential in understanding a system's failure modes.

8.1.2.1.3 Fault Tree Analysis

Another commonly used model to analyze the reliability of a system is the *FTA*. Harms-Ringdahl defines an FTA as "a graphical representation of logical combinations of causes of a defined undesired event or state" (Harms-Ringdahl 2001). In short, an FTA is used to determine the different ways a specific mode of failure can occur. Unlike an FMECA, the FTA can describe the events that lead up to a particular failure and better handles human and environmental influences on a system's failure (Birolini 2010). An FTA is a series of events connected with AND/OR gates, as seen in Figure 8.3.

The FTA begins with a top-level event. This event needs to be specific. Because it analyzes a specific mode of failure, simply stating "something fails" would not help. Next, the causes of the event need to be determined. These causes are referred to as intermediate events. Subsequently, the causes of the intermediate events are determined, and this cycle repeats until it reaches the lowest level failure event, known as a basic event. In the case that breaking down an event becomes too complicated, it is labeled as an undeveloped event and can be analyzed with a separate FTA. Once the fault tree is constructed, the probability of occurrence of the top-level event is established. This is accomplished by determining the probabilities of all the other events on the tree and consolidating those probabilities based on logic (e.g., "AND" and "OR" gates).

The FTA and FMEA/FMECA complement each other. Once the RBD of a system is developed, an FMEA/FMECA is established to determine possible problem areas of the system. From that point, an FTA is used to gain more insight into the events that cause the specific modes of failure. With these methods, the systems engineers can determine critical failure aspects of the system that directly affect the overall reliability of the system. Perfectly reliable systems are not possible in this realm of existence. Consequently, it is important to focus on how one can prepare for failure through maintenance.

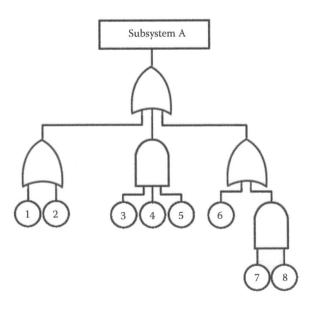

FIGURE 8.3
Fault tree. (Obtained from Wyatts, Fault tree, http://en.wikipedia.org/wiki/File:Fault_tree.png, accessed April 3, 2014, 2005a.)

8.1.2.2 Maintainability

Maintainability can be considered a complement to reliability. Reliability determines the odds of a system or component working, whereas the maintainability relates to how the risk of system failure is reduced (preventive maintenance) and how it will be fixed if it fails. *Maintainability* is a design characteristic of a system and is the ability of a system to undergo maintenance and repair (Fabrycky and Blanchard 2011). Maintenance can be split into two types, the first being corrective maintenance and the second being preventive maintenance. When designing a system, preparations should be done for both types of maintenance.

Corrective maintenance is an unscheduled repair resulting from a failure in the system (Dhillon 2006). Despite it resulting from an unexpected failure, the steps taken with corrective maintenance follow the same pattern. This pattern is the corrective maintenance cycle, which can be seen in Figure 8.4.

Corrective maintenance begins with the detection of a problem, which is typically when the failure occurs. After the failure is found, the system is prepared for maintenance. Once maintenance begins, the problem component is located and isolated. After that, the component in question is disassembled or removed from the system. At this point, the maintenance can go one of two paths. Either the part is replaced with a spare part or the part is repaired to working order. After that, the part is reassembled into the system. Systems that are mostly software can provide a challenge for corrective maintenance as it is not simple to disassemble or remove software from the system. With software, the corrective maintenance usually involves finding and fixing the error in the software, rebuilding the software, and installing the corrected software (often referred to as "patching"). The system is then adjusted or reset back to working order. Finally, the system is tested to confirm the maintenance was successful. For systems that are mostly software, a set of "regression" tests are performed to confirm that the changes to the software have not adversely affected other parts of the software.

Preventive maintenance is a type of scheduled maintenance that is periodically done to prevent part failure and keep the system in its operational state (Dhillon 2006). This can be done through various means, including periodic inspection and

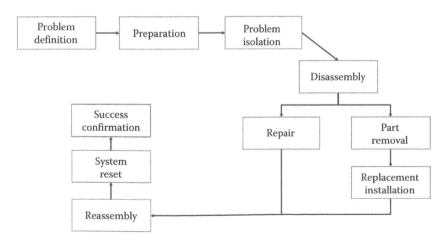

FIGURE 8.4
Corrective maintenance cycle. (Courtesy of Yassine Rayad.)

replacing a part after a predetermined run time. While this type of maintenance might prevent failure, there are factors that limit the amount of preventive maintenance. One factor is cost, due to the replacement parts and labor. Another factor is time, as if the *mean time between maintenance* (MTBM) is too short or the maintenance takes too long, the system would not be available for operations as expected. Other factors include testing, transportation, and facilities used for maintenance, which can become costly if repair cannot be done on site. Both preventive and corrective maintenance are affected by these factors.

For systems that are mostly software, preventive maintenance may include such tasks as making sure that hard disks and other storage media are not out of space. For many systems that must run 24 h, 7 days a week with a high reliability, parts of the system may be redundant. A regular preventive maintenance task may be to perform a failover or switchover on the redundant system periodically to make sure the system is in working order. Another task might be to periodically stop and restart computers as a preventive measure to avoid system crashes due to undetected memory leaks or other problems that only show up after many months or years.

Preparing for only one type of maintenance is not a good idea. Pure preventive maintenance leaves the system vulnerable should it suddenly experience failure. Pure corrective maintenance can get costly and can cause the efficiency of the system to lower as parts become more unreliable over time. As such, a mixture of both preventive and corrective maintenance is needed. Finding the time it takes for a system or component to undergo maintenance can give an idea of when to perform preventive maintenance, as well as an idea of how long corrective maintenance takes.

8.1.2.2.1 Maintenance Time

One of the easiest ways to measure maintainability is through time. One example is the *mean corrective maintenance time* (MCMT) (M_{ct}), also known as *mean time to repair* (MTTR). M_{ct} is the average time taken for corrective maintenance. Let us say that we have n maintenance facilities and each have an average corrective maintenance time of M_{cts}; then the average corrective maintenance time for all of the service centers is just the total average of the individual corrective maintenance times:

$$M_{ct,mean} = \frac{\sum_{x=1}^{n} M_{cts}}{n}.$$ (8.11)

At each service center, the MCMT at that service center M_{cts} is given by

$$M_{cts} = \frac{\sum_{i=1}^{l} \lambda_i M_{cti}}{\sum_{i=1}^{l} \lambda_i}$$ (8.12)

where
    λ_i is the rate of failure for the item undergoing maintenance
    M_{cti} is the average corrective maintenance time for the item
    l is the total number of corrective maintenance events at the service center

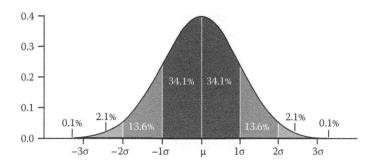

FIGURE 8.5
Normal curve showing standard deviation. (Author Petter Strandmark (aka Mwtoews) based on figure by Jeremy Kemp [2005], Standard deviation diagram, http://en.wikipedia.org/wiki/File:Standard_deviation_diagram.svg, accessed November 28, 2014, 2007.)

Maintenance times typically have one of three distributions. The normal distribution, seen in Figure 8.5, is used for simple tasks with little variation and a fixed time. The exponential distribution is used for parts that can simply be replaced. The third distribution, log-normal distribution, corresponds to maintenance tasks that vary in time and practice.

The standard deviation (σ) of a normal distribution used to make Figure 8.5 can be found as

$$\sigma = \sqrt{\frac{\sum_{i=1}^{l}\left(M_{cti} - M_{cts}\right)^2}{l-1}}. \tag{8.13}$$

With these data, it is possible to determine the MCMT for the service center along with the expected standard deviation in the MCMT for expected tasks. With supporting financial data, these estimates are relatable to expected maintenance costs.

Another measure of maintainability is *mean preventive maintenance time* (MPMT) (M_{pt}). M_{pt} is the average time spent conducting preventive maintenance activities according to a preventive maintenance frequency (f_{pt}) over m maintenance cycles. If we use the subscript s to designate a particular service center (e.g., $s = [1, 2, \ldots, nth]$ for the 1st, 2nd, ..., nth service center), then the MPMT at service center s is given by

$$M_{pts} = \frac{\sum_{i=1}^{m} f_{pti} \times M_{pti}}{\sum_{i=1}^{m} f_{pti}} \tag{8.14}$$

where
   f_{pti} is the preventive maintenance frequency for the ith system element
   M_{pti} is the corresponding MPMT for the ith system element
   m is the total number of preventive maintenance events

With this, the mean time interval between preventive maintenance at the service center can be estimated. Other common measures of maintainability include the *median active corrective maintenance time* (M_{act}) and the *median active preventive maintenance time* (M_{apt}).

M_{acts} is the time needed at the service center, such that half of the corrective maintenance times are equal to or below this value. The M_{acts} is given by

$$M_{acts} = \text{antilog} \frac{\sum_{i=1}^{l} \log M_{cti}}{l}. \qquad (8.15)$$

When viewing this on a normal distribution curve as shown in Figure 8.5, this time is the mean time corresponding to the time associated with the peak of the Gaussian curve. In other distributions, this is not always the case. For the log-normal distribution, the median is the same as the geometric mean (Fabrycky and Blanchard 2011). M_{apts} is essentially the same thing for preventive maintenance times at the service center and can be calculated as

$$M_{apts} = \text{antilog} \frac{\sum_{i=1}^{m} \log M_{pti}}{m}. \qquad (8.16)$$

However, as stated earlier, it is important to plan for both preventive and corrective maintenance. This can be done using the *mean active maintenance time* (MAMT) (M_{amt}). M_{amt} is the average time to perform both M_{ct} and M_{pt}. If we are considering maintenance activities at service center s, then we obtain

$$M_{amts} = \frac{\lambda_s \times M_{acts} + f_{pts} \times M_{pts}}{\lambda_s + f_{pts}} \qquad (8.17)$$

where
    λ_s is the average failure rate at service center s
    f_{pts} is the average preventive maintenance frequency

The *maximum active corrective maintenance time* (M_{max}) is the maintenance time below which a specific percentage of all corrective maintenance should be completed. For service center s, the maximum active corrective maintenance time is given by

$$M_{max\,s} = \text{antilog}\left(\overline{\log M_{acts}} + Z\sigma_{\log M_{cti}}\right), \qquad (8.18)$$

where
    the bar over $\log M_{cts}$ signifies the mean of the logarithms of the M_{cti} terms
    the $\sigma_{\log M_{cti}}$ term is the standard deviation obtained using the logarithm of the M_{cti} terms
      (Fabrycky and Blanchard 2011)

Note that if there is only one service center, then Equation 8.11 can be disregarded and the s subscript dropped from all of the aforementioned equations.

An important consideration is that these formulas do not take into consideration *logistics delay time* (LDT) or *administrative delay time* (ADT). LDT is the amount of time wasted during maintenance when parts or tools for maintenance are being procured (Blanchard 2004). ADT is the time wasted during maintenance for any administrative means like paperwork or lunch breaks (Fabrycky and Blanchard 2011). What does include these values is the *maintenance downtime* (MDT), which is the total amount of time taken up by maintenance and delay times (Dhillon 2006).

Other factors involving maintenance time are the labor hours and labor costs for maintenance, in which each action of preventive and corrective maintenance cost money. The labor, spare parts, and maintenance equipment are all factors that cost the system operators. Furthermore, the higher the MDT, the less the system will be active and performing its job. Generating a good maintenance plan is key to lowering these costs to acceptable levels.

8.1.2.2.2 Reliability-Centered Maintenance

One way to generate a maintenance plan is through *reliability-centered maintenance* (RCM). Linneberg states, "RCM addresses elements and systems with a predictable failure pattern, where degradation cannot effectively be measured before failure" (Linneberg 2012). If the degradation of a component or subsystem is noticeable and measurable, preventive maintenance could be performed once the degradation reaches a certain threshold. But if the degradation of a part is not easy to detect, it would be difficult to determine the proper time for preventive maintenance. By focusing on a predictable failure pattern, this can be circumvented, and preventive maintenance can be planned for the failing component or subsystem. In some cases, it may not be economically feasible to repair the system. In order to determine when this is the case, another type of maintenance plan should be used.

Many systems use data collected by the system to analyze information that is gathered by the system to make an RCM determination of preventive or corrective maintenance. For example, a rail system in a cold climate may use weather sensors to detect when the conditions are likely to produce ice. The ice can get in between the pieces of the switch mechanism and cause the switch to malfunction. The system will then automatically turn on a switch heater or snow blower to clear the switch. Further, the system may then automatically cycle through the switches and periodically request the movement of each switch to proactively detect a switch failure due to the weather.

8.1.2.2.3 Level of Repair Analysis

A *level of repair analysis* (LORA) is a type of analysis done to determine whether or not it costs too much to bother repairing a system (Fabrycky and Blanchard 2011). In some cases, it may cost too much to perform corrective or preventive maintenance. For example, having to take apart the system's exterior, remove a subassembly, disassemble the subassembly, and then replace a part may cost too much. In these cases, it may be better to just dispose of the faulty part or subassembly and replace it with a fresh one.

It was mentioned that the repair of a system could cost too much. This cost could either be in terms of money or time. Time can be a very important factor for a system. If maintenance takes too long, the amount of time a system is operational decreases. This would give the system a lower availability than desired.

8.1.2.3 Availability

The availability of a system does not give any information about the system that has not already been discussed. However, it presents that information in an easy to understand way for both the system engineers and stakeholders. *Availability* is the ratio of the system's uptime to its total time (uptime plus downtime) (Birolini 2010). There are three forms of availability, each one taking into consideration a less ideal state. We assume one service center and drop the *s* subscripts from the previous equations for clarity.

The first form of availability is the *inherent availability*. This is the probability that a system is operating as desired at any given time while under an ideal support environment (Fabrycky and Blanchard 2011). An ideal support environment is one in which any needed supplies, tools, spares, or personnel are readily available. It is important to note that this does not include preventive maintenance actions, ADT, or LDT. The inherent availability (A_i) of a system can be expressed in terms of MTBF and MCMT and can be written as

$$A_i = \frac{\text{MTBF}}{\text{MTBF} + M_{ct\,\text{Mean}}}. \qquad (8.19)$$

The second form of availability is known as *achieved availability*. It is the probability that a system will, at any given time, operate as desired when used under set conditions while in an ideal support environment (Fabrycky and Blanchard 2011). While this definition is nearly identical to inherent availability, the difference is in the set of factors that are taken into account when calculating achieved availability. This form of availability does take into consideration the preventive maintenance of a system. It is calculated from the MTBM and MAMT of a system but still excludes the LDT and ADT of a system. Achieved availability (A_a) can be expressed as

$$A_a = \frac{\text{MTBM}}{\text{MTBM} + M_{amt}}. \qquad (8.20)$$

The third form of availability is the *operational availability*. This one covers a more realistic state and is defined as the probability that a system will operate correctly while in a real run time environment and when run under those same set conditions as in achieved availability (Fabrycky and Blanchard 2011). The key phrase here is the environment, as this form of availability does take LDT and ADT into its calculations. It can be found from the MTBM and MDT of a system. Operational availability (A_O) can be written as

$$A_O = \frac{\text{MTBM}}{\text{MTBM} + \text{MDT}}. \qquad (8.21)$$

When describing the availability of a system, operational availability is most commonly used and is usually expressed as a percentage. For example, the availability may be expressed as 99.9%, 99.99%, or 99.999%. Respectively, these are referred to as "3 nines," "4 nines," and "5 nines" availability. However, this term may not be as appropriate to use as a figure of merit when determining design requirements for particular pieces like equipment suppliers who do not have control over the operational environment in which their equipment is expected to function. In each case, it is important to define what exactly is meant by "availability" and how to apply it.

Now that SRMA concepts have been defined, it is necessary to show how to use these concepts in practice. Reliability, maintainability, and availability all work together during systems development. In fact, SRMA should be used as early as possible in a system's development life cycle.

8.1.3 SRMA Methods throughout the Systems Development Life Cycle

As previously mentioned, SRMA is an important part of development and should be considered early on in the life cycle of a system. However, this is not just something that is done once early on. SRMA should be a constant factor through all stages when designing a system. In fact, the SRMA should be considered even after the system is launched. In this section, methods for determining and defining SRMA throughout the system life cycle will be discussed. There are six major phases in a typical life cycle of a system: conceptual design, preliminary design, detailed design and development, production/construction, operational use and system support, and, finally, system disposal and retirement.

8.1.3.1 SRMA in Conceptual Design

The first stage of the systems development life cycle is the conceptual design phase. During this stage, there are several items to consider in regard to SRMA. This includes determining the operational requirements. The operational requirements will form the basis for determining if the system is performing in a "satisfactory" state. It is this state that the engineers will be designing the system around and determine the required reliability and maintainability of the system. Some key parameters to be defined from the system requirements include the A_O, MTBM, and MDT. With this in hand, a basic maintenance and support cycle should be defined. Allocating the reliability and maintainability measures and metrics through the functional decompositions of the system can do this. The requirements for maintainability and reliability are then transformed into lower-level design criteria (Fabrycky and Blanchard 2011). These lower-level requirements act as the basis for defining support factors throughout the life cycle of the system. It is also important to take into consideration pre-planned product improvements (PPPI) from an early point in the system design.

PPPI is a type of maintenance and support strategy that takes into consideration performance improvements regularly throughout the system's life cycle. The idea behind PPPI is to begin planning early on for the incorporation of needed improvements to the system throughout its life cycle. For example, the initial design of the product would allow for growth in order to meet the changes in the products operating environment. PPPI can also be used to decrease the costs of a system across its life cycle (*Acquisition Strategy Guide* 1999). For example, you can prepare to potentially use different components for repairing the system in case they become more financially advantageous (e.g., new supplier agreements) than the current parts. Once these steps have been completed, the validity of a design should be estimated. One way to do this is to perform modeling or simulations. An early FTA model can be used to check if there are any faults during the conceptual design.

8.1.3.2 SRMA in Preliminary Design

In preliminary design, a more in-depth functional analysis of the design is performed. System analysis leads to the identification of alternative functions and subfunctions. Once there is a high-level functional design, the reliability of individual potential components can be calculated and compared. Once reliability for each component is determined, an RBD can then be made. Allocation of various system requirements is done in this stage. This includes the allocation of performance factors, effectiveness requirements, and system support requirements. All of these have an impact on the expected availability of the system. During this time, an FMECA can be developed using the parts selected for the system to determine any possible areas of failure. The operational requirements should be used to define factors of the operating environment that could have impacts on the reliability of the components in the system. For example, if the system is expected to operate at high altitudes, certain parts may have a lower MTBF than at sea level. These factors need to be considered when selecting and comparing components to become part of the system. Also, a more refined FTA can be performed at this time. As the parts and functions of a system develop, more detailed information can be used to create more complicated

FTAs and FMEAs/FMECAs. As the development on the design continues, it will eventually enter the detailed design and development phase.

8.1.3.3 SRMA in Detailed Design and Development

During the detailed design phase, subsystems are broken down further into components. Prototypes may be built during this stage where possible. Prototypes are great tools, as their performance can be measured to determine the overall reliability and maintainability of the current design. During this stage, more FMEAs/FMECAs and FTAs are developed as the system's design changes with each prototype or design alteration. These models can give an idea of support factors such as the ease of access or ease of repair based on the parts and possible failure modes. Other benefits of prototypes include those that affect supply. For example, the availability of parts can be better estimated. During this phase, the system's design is tested and evaluated to see if it still meets its operational requirements. As actual physical components are acquired, stress testing can be performed on them. How a particular part reacts under environmental stresses can have a large impact on the overall reliability and availability of the system. If a part is proven to fail under certain durations of stresses, that may lead to further definition of preventive maintenance actions to take in regard to that part, as well as possibly completely replacing the part for a more reliable one.

Finally, the requirements for the system's retirement should be taken into consideration. As systems become obsolete and are no longer capable of being maintained, they must be discarded. The disposal of a system should also have little negative impact on the natural environment. By developing the retirement plan before system completion, the disposal of the system can become easy, quick, and at low cost (Fabrycky and Blanchard 2011). Once the design is considered complete, it is time to begin the production phase.

8.1.3.4 SRMA in Production

The production phase of a system's life cycle is just that: the beginning of production. In this phase, parts must be ordered and manufacturing setups must be created. During this phase, it is important to initiate the logistic support structure for the system. When the system is turned over to the stakeholder, the support and service activities commence. Specifically, maintenance depots will have to be established and equipped. People will have to be trained in the methods of systematic and preventive maintenance for the system. Spare part allocation should be considered as well, as the lack of parts can put a hold on the system's usability. Once the system is sold as a product, the final phase of the system's life cycle begins.

8.1.3.5 SRMA in Operations and Support

The operations and support phase of the system life cycle is where systems are delivered to the stakeholder (user) and put into operational use. During operations, system status is monitored, and preventive maintenance will be performed as determined in earlier phases. The logistics support structures defined in previous phases are necessary for maintaining the system. Because these have already been put into place, corrective maintenance should be supported and capable. These support structures should also be analyzed and refined to better handle system performance, as the system grows older. For example, more spare parts may be required as the older parts become less reliable. Field data on reliability metrics such as system/subsystem/component MTTF can be gathered and fed back to the design team to update their SRMA (and other) models. Finally, the retirement conditions defined previously in the systems design and development phases come into play. Once the system is deemed fit for retirement, the system is disposed of in the proper, predefined manner.

While we may have covered the methods of using SRMA throughout a system's life cycle, a more detailed understanding of the system in question is necessary to develop that system's reliability and maintainability factors. In order to get a good understanding of what determines these factors, let us look at optical detectors as a representative example.

8.1.4 Optical Detectors and Associated Maintenance and Support Concepts

Even though all the topics discussed throughout the chapter so far can be applied to any system, subsystem, component, assembly, or part, we want to illustrate these concepts using optical systems examples. An optical system will have various maintenance and support requirements in connection with its critical systems. One of the main critical systems of an optical system is the optical detector itself. The optical detector must function correctly in order for the system to operate as intended. In the next section, we will provide an introduction to optical detectors and illustrate some performance characteristics and useful performance metrics.

8.2 Fundamentals of Optical Detectors

An *optical detector* is a device that "views" objects by absorbing the electromagnetic radiation emitted from objects. The detector then outputs an electrical signal, which typically is proportional to the object's irradiance. *Irradiance* is the amount of radiant energy emitted from an object that is applied to another object's surface (Smith 2007). Optical detectors are typically designed to work over a narrow range of wavelength. For optical detectors, these wavelength ranges typically fall either into the visible or infrared parts of the electromagnetic spectrum.

The way optical detectors work is that when the material used for detecting is struck by light, a change occurs (Jones 1959). For example, this change could be an alteration in the electrical resistance of the material or the absorption of energy from the light, expelling electrons currently in the material and creating a voltage. Whatever the event, it will persist as long as the light is still in contact with the material. Only after the light is removed will this effect stop. Either way, these changes do not affect only the detector material, but alter the voltage or current in the circuitry connected to it. Specifically, what happens is either the voltage changes while the current remains the same, or the opposite. These changes are used to detect not just the presence of the light but also its brightness.

Optical detectors that function within the infrared portion of the spectrum are used as a transducer of the radiant energy of the electromagnetic waves. What this means is that the infrared detector takes the radiant energy and then converts it into another type of energy. This could be the generation of an electrical current, a chemical change blackening a photographic plate or even a mechanical change in the physical properties of the detector.

Some examples of the detectors used in the 0.2–15 µm region can be seen in Figure 8.6. The arrows represent the ranges of each detector where its response is atleast 20% of its maximum value. The image in the middle is the transmittance curve of the atmosphere,

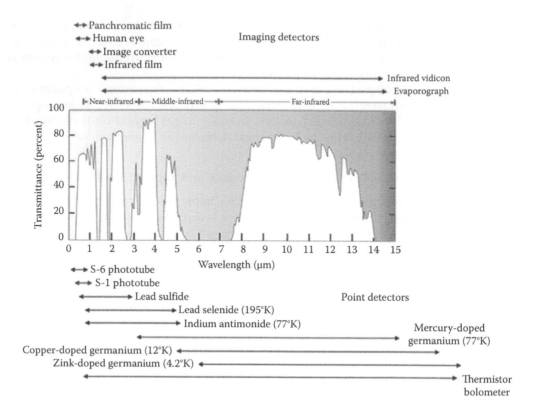

FIGURE 8.6
Representative detectors for the 0.2–15 μm region.

which shows the atmospheric windows for these wavelengths. As can be seen, most of the detectors in Figure 8.6 only cover a specific range of the electromagnetic spectrum.

The detectors shown in Figure 8.6 are split into two basic groups. These two groups consist of *imaging detectors* and *point* or *elemental detectors*. An imaging detector works like a standard camera, it takes in the entire image in one instance to create a picture. The point detector, however, creates an image over time by sequentially scanning its target. Specifically, the point detector responds to the average irradiance over its active area, which, in an imaging application, is usually a small, often square, picture element (pixel) that spatially samples the object scene. Interestingly enough, an imaging detector acts as a set of point detectors, each one responding to the irradiance at one pixel.

8.2.1 Types of Optical Detectors

Optical detectors are grouped by the physical means used in detection. These groups can be further split into subgroups depending on how the detection is accomplished. The first group in question is the *thermal detector*. A thermal detector is a type of detector that detects incident radiation using a temperature-dependent material property such as electrical resistance. Essentially, the materials used in a thermal detector respond to temperature. For optical detection, a thermal detector absorbs the heat from an object's incident radiation. This causes some change in the detector, such as an increase in electrical resistance or some change in the absorption property of the material. Despite being able to be used as an optical detector, only the heat radiating from an object affects thermal

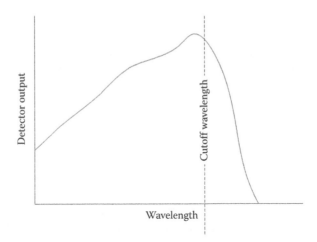

FIGURE 8.7
Relative detector output as a function of wavelength for photon detectors. (Courtesy of Terry D. Cox.)

detectors, and so thermal detectors can function independently from the electromagnetic spectrum, unlike the other group of optical detectors.

The second group of optical detectors is *photon detectors*. In photon detectors, the energy from light reacts with the detector's sensing material (Hudson 1969). The incident photons of the light react with the electrons of this material. This reaction causes the material to generate free electrons. Unlike thermal detectors, photon detectors are highly dependent on the electromagnetic spectrum. Essentially, a photon detector's wavelength response has a wavelength cutoff. If the wavelength of the oncoming photons is longer than the cutoff wavelength, the energy produced by the photons will be too small to produce a free electron. The end result is that the response of the photon detector dramatically drops to zero for wavelengths above the cutoff wavelength. Cooling should be used to reduce detector noise and get the most out of these detectors (Crouch 1965).

Figure 8.7 depicts the relative shape of the output curves for photon detectors as a function of wavelength. The output of the photon detectors increases as a function of increasing wavelength. The output drops to zero after reaching the cutoff wavelength. In contrast, a thermal detector has a constant response as a function of wavelength. As previously stated, thermal detectors respond to temperature variations and are not sensitive to changes in wavelength. In short, these two detector types respond differently to incident radiation. Thermal detectors respond to the amount of heat they absorb from the target. Photon detectors, on the other hand, respond to the electromagnetic radiation itself.

8.2.2 Thermal Detectors

As mentioned previously, thermal detectors rely on the heat absorbed from an object's incident radiation to produce a physical change in the detector material. This response is independent of wavelength, and only the amount of energy absorbed matters. Due to this independence of wavelength, thermal detectors are popular in systems where cooling is not required or not preferred (although there are some cases where cooling a thermal detector is preferable, such as some types of bolometers).

One drawback to thermal detectors is their low time constant, typically being around a few milliseconds. As such thermal detectors are rarely used in situations where a large data-sampling rate is necessary, such as searching applications. There are a variety of implementation mechanisms for thermal detectors. Some thermal detectors, such as a mercury thermometer, provide visual feedback on the amount of heat absorbed. Thermal detectors that are useful for detecting optical radiation include bolometers, Golay cells, thermocouples, calorimeters, and pyroelectric detectors.

8.2.2.1 Bolometers

Bolometers are a type of thermal detector whose electrical resistance depends on its temperature (De Waard and Wormser 1959). Bolometers can be made from a variety of temperature-sensitive materials, such as thermistors, carbon-based resistors, superconductors, or even germanium (Hudson 1969). The heat capacity of a bolometer depends on the size of the temperature-reactive material. This will affect the detector's time constant. For all thermal detectors, this material needs to be coated black to better absorb the heat of the incident radiation. Unlike most thermal detectors, some bolometers need to be cooled to be effective.

One example of this is the carbon resistor bolometer (Hudson 1969). Made from a carbon wafer, this type of bolometer is cooled with a cryogenic liquid to reduce its temperature and increase its sensitivity. It is also necessary to isolate the electrical connections thermally to avoid errors. The blackened carbon wafer is faced toward the incident radiation, while the cryogenic liquid cools the other side. This reduces the amount of electrical heating the carbon wafer experiences, causing the change in voltage from the material to be very small. A precise instrument is required to correctly measure the change in voltage as the temperature changes.

While bolometers may be useful in detecting temperature effects, there are some cases in which using one is not desirable. One example is in remote detection applications. If the intent is to detect the temperature of a remotely located object, the local environment surrounding the bolometer is an important consideration. Take for example a blizzard. The cold air and snow surrounding the bolometer's detecting material would throw off its ability to correctly read the temperature of the remote object. In this case, using a detector that is designed with its detector material protected against the environment would be a better option. One such detector is a Golay cell.

8.2.2.2 Golay Cell

Golay cells are thermal detectors whose heat-sensitive material is surrounded by gas (Hudson 1969). Golay cells operate at room temperature and are one of the earliest types of infrared detectors to do so. A Golay cell's temperature-sensitive material is located in a chamber filled with gas on both ends. This material, typically a blackened membrane, absorbs the incident radiation on one end of the chamber. As the material is heated, the temperature of the gas surrounding it rises as well due to convection heating. On the other side of the chamber lies a temperature-sensitive flexible mirror, which is heated by the gas, causing it to alter. These alterations are then sensed and measured by an optical system, which translates them into changes in temperature. Interestingly enough, the amount of noise that a Golay cell receives is typically only from the convection with the gas (Putley 1973). Because of this, Golay cells can be very accurate.

However, while accurate, Golay cells can be rather delicate. As it requires a mirror and detector material to be suspended inside a closed chamber, external force and vibrations affect the measurements or even may damage the device. Also, if the housing gets covered or somehow loses its level of transparency, readings could become inaccurate. For some operations a simpler detector, such as a thermocouple, may be required.

8.2.2.3 Thermocouples and Thermopiles

A *thermocouple* is the combination of two wires, whose wires have a different level of thermoelectric power (Hudson 1969). Typically, a thermocouple is used to measure temperature via physical contact. But when a thermocouple is blackened or is touching something blackened, it can be used to detect incident radiation. A thermopile is essentially a collection of thermocouples, joined in series. This increases the responsivity and time constant of the detection.

A thermocouple's time constant (τ) can be obtained with the thermal capacity of its junction (C) and the rate of energy loss (*Delta*), and can be expressed as

$$\tau = \frac{C}{Delta}. \tag{8.22}$$

Typically, τ ranges from 4 to 50 ms. This depends on the detecting material's thickness and type as well as the thickness and type of the blackened radiation-absorbing material coating the detection material. The blackened coating adds to the thermal capacitance of the detecting material. If a sealed housing is used, conductive paths to get the detected signal out of the housing can also affect the time constant. For example, a large enough thermal conductance will lead to a decrease in the thermocouple's time constant.

For some operations, a time constant is not that much of a big deal. When you need to simply measure temperature over a brief period of time, it may be hard to get an accurate reading from a thermocouple. In this case, a calorimeter would be a much better option.

8.2.2.4 Calorimeters

Calorimeters are commonly used to measure the pulses, or short bursts, of thermal energy. Different types of calorimeters have been developed in order to measure the amount of energy found in an optical pulse. One such example is the black radiation detector, whose results can be used to determine the responsivity of other detectors (Hudson 1969). These measurements, however, only determine the exact amount of energy in a pulse. The ability to define the shape of this pulse is not typically within a calorimeter's capabilities, as they typically do not respond fast enough. This may be due to the size of the standard calorimeter. Because the pulses calorimeters detect tend to be very small, the calorimeter's temperature-sensitive material must also be small. This is to allow the absorbed energy to be rapidly distributed throughout the device. Another requirement for the calorimeter is that it must be thermally isolated from its surroundings. This is to prevent the loss of the heat the calorimeter just absorbed, which would reduce its accuracy. Calorimeters used for optical detection tend to use blackbody absorbers with low thermal mass. By placing the calorimeter next to the absorber, the change in temperature can then be determined. With this information, the amount of energy obtained in an optical pulse can be found. While determining the energy of a single pulse is useful, sometimes an application demands that thermal radiation is detected over a longer signal. For this purpose, consider using a pyroelectric detector.

8.2.2.5 Pyroelectric Detectors

The pyroelectric detector detects heat. An electrical signal is produced through a three-stage process. First, the thermal radiation interacts with the detector surface by changing its temperature. Second, the changing temperature induces a corresponding change in the electrode charge density producing a current. Third, the detector current is amplified by signal processing electronics producing a signal that is proportional to the input heat and the area of the detector. This current (i_p) can be found using the pyroelectric constant (p), the area of the elements of the detector (A_s), and the change in temperature (dT_p) with respect to the change in time (dt) and can be expressed as

$$i_p = p \times A_s \times \frac{dT_p}{dt}. \tag{8.23}$$

With all these different types of thermal detectors, it can be difficult to choose which one would work best for your system. The choice of which type of thermal detector is chosen depends on the nature of the thermal detection application. In some cases, thermal detectors do not provide sufficient sensitivity. In these cases, photon detectors can be considered as a possible alternative.

8.2.3 Photon Detectors

Typically, photon detectors have detectivity about one or two orders of magnitude greater than their thermal counterparts. *Detectivity* is a measure of detector performance and will be discussed later in Section 8.2.4. The direct interaction, between the photons and electrons of the detector material, results in a very short response time. Despite this, many photon detectors typically would not work well unless they are cooled to cryogenic temperatures. The photon that enters the detector must have sufficient energy to exceed the cutoff energy of the detector. This is accomplished by ensuring that the illuminating wavelength is shorter than the cutoff wavelength.

Like thermal detectors, photon detectors may be grouped according to the fundamental physical response of the detector. Some fundamental detector types are listed:

- *Photoelectric*: These detectors are rather simple. When photons hit the light-sensitive detector material, they release energy into electrons on the material's surface, which could be enough energy to release those electrons (Hudson 1969). These free electrons create a current and are then collected in an external circuit.

- *Photoconductive*: A material is considered photoconductive when its resistance decreases while being struck by light (Cashman 1959). The free electrons from the light cause the electrical conductivity of the light-sensitive material to change with respect to the light's irradiance. Photoconductive detectors are typically made from semiconductor materials.

- *Photovoltaic*: These detectors consist of a semiconductor with a p–n junction (Hudson 1969). This type of junction is a region where an n-type material and p-type material join together in a particular host material. In this type of detector, photons striking the detector material create electron–hole pairs that can be separated with an applied electric field. This charge separation generates a photo-voltage (e.g., a voltage that is proportional to the incident photons).

- *Photomagnetic*: These detectors are also made from semiconductors. This time, however, the detection occurs when the surface of the semiconductor generates electron–hole pairs when in struck by photons (Hudson 1969). The electrons end up diffusing into the material in order to balance the electrons in the material. The electron–hole pairs can then be separated by a magnetic field.

For imaging applications, two different types of photon detectors are commonly used to convert the incident electromagnetic radiation into an electrical signal that is stored and saved as picture data. These detector types are known as the *charge-coupled device* (CCD) and the *complementary metal-oxide semiconductor* (CMOS).

8.2.3.1 Charge-Coupled Device

A CCD is an electrical device often used in imaging applications. In an imaging configuration, the CCD consists of an array of pixels that can store and transfer charge through associated readout electronics. Like all other photon detectors, this type of detector can respond to incident light; however, this type of detector can also respond to inputs in the form of an electrical charge. The output of the CCD is an electronic signal. The typical substrate used to fabricate CCDs is p-type, and it has a thin surface of n-type material. The configuration is shown in Figure 8.8. On top of the n-type substrate is an insulating material made of silicon dioxide. Finally, electrodes, also called gates, are placed on the insulator. The gates can be metallic or made from heavily doped polycrystalline silicon and form a conducting layer.

A common use of the CCD detector is in imaging applications. Each pixel generates charge from being struck by photons and this charge can be read out using *readout electronics* either on the sides of the detector or on the back of the detector. Sometimes, an application may require that each pixel is separately read out. In this case, a set of separate point detectors like in a CMOS detector can be used.

8.2.3.2 Complementary Metal-Oxide Semiconductor Detector

A CMOS detector is another common form of imaging detector. Unlike the CCD, each pixel in the CMOS is capable of converting its charge to a voltage. A useful attribute is that the CMOS detector has all the circuitry it needs to convert each pixel output to a digital signal. This makes it an array of point imaging detectors. This is not without its drawbacks, however, since the added circuitry sometimes displaces the active detection area as well as increases the complexity of the detection system. The detector uniformity is also relatively low since each pixel has independent circuitry from the other pixels.

Knowing the basic features of a detector goes a long way in determining which detector is appropriate for a given application. However, usually, more detailed information is

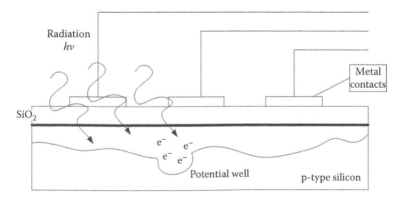

FIGURE 8.8
Diagram of a CCD. (Courtesy of Terry D. Cox.)

required to ultimately select the detector. In the next section, we present some basic detector performance metrics that can be used to evaluate, analyze, or predict detector performance. We focus only on basic metrics here and look at more detailed metrics in the next chapter.

8.2.4 Detector Performance

In order to determine the performance of an optical detector, some key terminology must first be defined and explained. One of the easiest ways of describing the performance of an optical detector is its responsivity. The responsivity of a detector is the ratio of its output signal to input power (Jones 1959). Many infrared detectors are used with a chopper. As we found in an earlier chapter, a chopper can be used to modulate either the phase or amplitude of the input irradiance to the detector. In order to reduce some of the detector noise, cryogenic cooling is typically required (Crouch 1965). The responsivity (R) of a detector is a function derived from its signal voltage (V_s), the input irradiance (H), and the area of the detecting material exposed to the irradiance (A_d) and can be expressed as

$$R = \frac{V_s}{H \times A_d}.$$ (8.24)

The more responsive the detector, the more voltage it will emit. The response time of a detector is characterized by the time it takes, after an incident change in radiation, for the detector output to reach 63% of its final value. This is known as its *response time constant*. If a rotating optical modulator is used (chopper), then the detector's responsivity is affected by the chopping frequency. For a majority of detectors, the responsivity, as a function of chopping frequency f, follows an exponential law and relates to the responsivity at zero frequency (R_0), and the time constant (τ), in the following manner:

$$R_f = \frac{R_0}{\left(1 + 4 \times \pi^2 \times f^2 \times \tau^2\right)^{1/2}}.$$ (8.25)

As the chopping frequency of the detector increases, its responsivity decreases. Responsivity is a convenient parameter for comparing various detectors. However, it gives no indication of the minimum detectable radiant flux. The minimum amount of noise in the output can be related to another parameter that is useful in describing detector performance. The *noise equivalent power* (NEP) introduces the noise element to detector analysis. The NEP indicates the amount of optical power on the detector's active area that produces a detector output signal equivalent to the detector noise. Given that the output signal is difficult to measure in this case (since the detector output signal-to-noise ratio is 1), this measurement is often accomplished at higher signal levels. This can be expressed with the same parameters as responsivity, with the addition of the voltage output produced from noise (V_n) and can be calculated as

$$\text{NEP} = \frac{H \times A_d \times V_n}{V_s}.$$ (8.26)

Bear in mind that these calculations are used with the assumption that the signal output of the detector is linearly related to its input. For signal-to-noise ratios of 10^3 or less, this assumption is usually valid. It is also necessary to state the electrical bandwidth of the circuit used to measure the noise since the NEP is a function of the electrical circuit's bandwidth.

A similar measurement used in describing optical system detector performance is the *noise equivalent irradiance* (NEI). NEI is the spatial optical power density (irradiance) that is required to produce a detector output signal that is equivalent to the detector noise:

$$\text{NEI} = \frac{H \times V_n}{V_s}. \tag{8.27}$$

When comparing detectors, the detector that produces the largest output response per input power has the highest responsivity. However, the detector that can respond to the lowest signal levels has the lowest NEP. Since it is rare to rate a situation as better with a lower value, the reciprocal of the NEP is used and known as *detectivity* (Nudelman 1962). This is strictly used for convenience sake, as by using the reciprocal of the NEP calculation, the better detector will be the one with the higher detectivity. Detectivity (D) can be measured as

$$D = \frac{1}{\text{NEP}}. \tag{8.28}$$

The conditions of measurement must be standardized for an accurate comparison of detectors being measured by different labs or being analyzed by an engineering team. One immediately notices a problem with the current detectivity metric. First, a larger detector area produces a larger output signal. This makes it difficult to directly compare detectors that have different size active areas. Second, the NEP is a function of the chopping frequency, and so detectors that are run with different chopping frequencies will have different detectivities. Third, the NEP is also a function of the electrical circuit bandwidth, and so the detectivities will be different for detectors with different electrical bandwidths. Factors like the operating environment (vibrations, temperature, and noise) and the wavelength of the illuminating wavelength also affect the NEP and so also the detectivity. With regard to temperature, a common way of making measurements easily achievable and readily reproducible is to make the measurements at temperatures such as those of normal ambient conditions (300 K), solidified carbon dioxide (195 K), liquid nitrogen (77 K), liquid hydrogen (20 K), and liquid helium (4.2 K).

A number of theoretical and experimental studies have indicated that detectivity varies inversely as the square root of the area of the detector and can be seen as

$$D A_d^{1/2} = \text{constant}. \tag{8.29}$$

Noting this fact, a more comparable and standardized metric was developed—D^* (dee-star). The quantity D^* is the detectivity referenced to an electrical bandwidth of 1 Hz and a detector area of 1 cm². This uses the chopping frequency (Δ_f) and can be written as

$$D^* = \frac{\left(A_d \Delta_f \right)^{1/2}}{\text{NEP}}. \tag{8.30}$$

Notice that the D^* metric is useful when the task is comparing detectors with each other. However, to understand the performance of a given detector, the detectivity D is often best. When measuring NEP, NEI, or D^*, the use of a calibrated blackbody source is often convenient. This permits the precise specification of the radiant flux or irradiance on the detector. To further define the measurement conditions, it is helpful to use D^* and specify the temperature of the blackbody and the chopping frequency. For example, D^*(300 K, 600) indicates a value of D^* measured with a 300 K blackbody at a chopping frequency of 600 Hz. If the detector response is desired at a particular mean wavelength, then D^* can be written as D^*(9.5 μm, 600), where in this case, the first number in parentheses indicates the wavelength at which the measurement was made while the second still indicates the chopping frequency. And finally, when the wavelength is at a maximum in the spectral bandwidth of the detector, the temperature is 300 K, and the chopping frequency is 600 Hz, the normalized detectivity is given by D^*(peak, 300 K, 840).

These numbers form a basic set of measurements that give insight into detector performance. Keep in mind, however, that while one detector may outperform the other, its usage may not meet the requirements of the system. As such it is necessary to keep the type of optical detector in mind while looking at comparisons.

8.2.5 Detector Comparison

When comparing detectors, thermal detectors are typically shown separately from photon detectors. The detectors in each group are further classified based on their operating temperature. An ideal thermal detector is one that responds equally to all wavelengths. In this case, the D^* value at any wavelength is ideally the same for equivalent input power. Differences between the expected flat spectral responses are deviations from the ideal.

Different photon detectors have varying detectivities and have different spectral ranges and peak wavelengths at different parts of the electromagnetic spectrum. While some detectors may have better ranges than others, they also have varying levels of detectivity over certain ranges. For example, a thermocouple covers a wide range of wavelengths but has a low detectivity. On the other hand, an ideal photovoltaic detector has a very high detectivity around a particular wavelength, but drops sharply at higher wavelengths. In the next section, we calculate some of the useful detector parameters as background material for the unmanned aerial vehicle (UAV) integrated case study that will be presented shortly.

8.2.6 Application to Integrated Case Study

The primary stakeholder requirement of the UAV project of FIT is human traffic monitoring during day and night. For daytime operations, simple imaging detectors, such as a camera, can give the maximum spatial resolution. For nighttime application, however, these are not possible to use due to the lack of light. In this case, infrared detectors can be used, as human bodies are very good sources of infrared radiation. In previous chapters, we used an effective temperature of 305 K (or about 31.9°C or 89.3°F) considering a clothed human in a nighttime, cool environment. In keeping with the discussion of standardization, here we use a human source temperature of 310.2 K (or 37°C or 98.6°F). Since there is a temperature difference between the human and its surroundings, there is a temperature gradient and heat will be exchanged. In order to find out the detector characteristics, it is first necessary to start with the source radiation characteristics.

As discussed earlier, Planck's law can be used to approximate the radiation properties of humans. This provides an upper limit for a human's radiation characteristics. For a

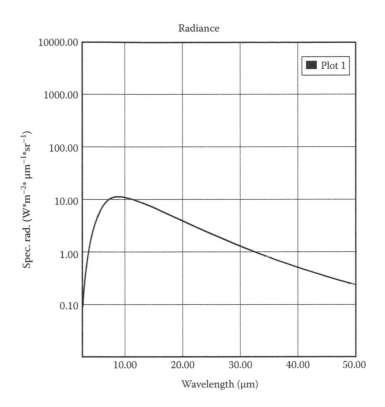

FIGURE 8.9
Blackbody radiation curve of human body.

typical human body (assumed naked and at nominal temperature), this curve is shown in Figure 8.9. The curve shows that the significant portion of radiation occurs in the infrared part of the electromagnetic spectrum having its peak in the vicinity of 9.35 µm.

Often, an effective temperature of 305 K is used to account for nighttime outdoor temperature and clothing effects resulting in a peak radiation wavelength of 9500 nm (9.5 µm). By consulting graphs of infrared detectivity versus wavelength for various materials (readily available in the literature and Internet), a detector with maximum detectivity in the region of human body radiation is the HgCdTe detector. VENUS LW, manufactured by SOFRADIR, is a typical example of an HgCdTe detector operating in the 7.7–9.5 µm spectral range. For perfect radiators, the power radiated (*P*) is given by the Stefan–Boltzmann law multiplied by the area of the detector

$$P = \sigma \times A \times T^4, \tag{8.31}$$

where
 σ is the Stefan–Boltzmann constant
 A is the surface area of the radiator
 T is the temperature of the radiator

For objects that are not perfect emitters or perfect absorbers, the emissivity (ε) of the object needs to be included:

$$P = \varepsilon A \sigma T^4. \tag{8.32}$$

The emissivity of the human body is 0.98. This makes the power emitted by the human body at 310 K equal to 1026 W considering that a commonly given value of human area is 2 m^2. The radiant emittance is equal to 513.2 W/m^2. The spectral radiance at the peak wavelength of 9.35 μm is 11.5 W/(m^2 sr μm). By integrating the region under the curve in Figure 8.9 from 8.75 to 9.35 μm (the spectral range of narrowband filter for peak atmospheric transmission), the band radiance is 6.9 W/(m^2 sr). Using the area of human area of 2 m^2 and neglecting look angles, the source intensity is 13.8 W/sr.

Irradiance at the entrance pupil (H) is calculated using the distance between the target and entrance pupil (r) and the intensity of the signal (J). This can be calculated as

$$H = \frac{J}{r^2}. \tag{8.33}$$

Assuming a worst-case standoff distance between target and object of 18,616 m, the irradiance is about 39.8 nW/m^2.

Taking into account the effect of the atmosphere in the region of the electromagnetic spectrum from 8.75 to 9.35 μm is given roughly by Figure 8.6 to be 80% (i.e., only roughly 20% of the irradiance is absorbed or scattered by the atmosphere), the actual irradiance at the entrance pupil of the imaging system is approximately given by 0.8 × 39.8 nW/m^2 or about 31.9 nW/m^2.

The flux at the entrance pupil (ϕ) is determined from the irradiance (H) and the area of the entrance pupil (A_d) and can be found as

$$\phi = H \times A_d. \tag{8.34}$$

Assuming a circular aperture with radius of the entrance pupil equal to 20 in. (0.508 m), the area of the optical system's entrance pupil is 0.81 m^2. Therefore, the flux is equal to 25.83 nW. The mean temperature of the detector is assumed to be conditioned in the range of 20°C–35°C, and the responsivity of the detector (found from a particular HgCdTe detector specification sheet) is 8.14 × 10^2 V/W at 35°C. Therefore, the signal voltage at the output of the detector is equal to 21.03 μV. If the noise levels of the detector are comparatively low, this is a workable detector output signal that will need to be amplified and digitized (more on noise and signal processing in a later chapter).

Now that some basic detector capabilities and metrics have been discussed, the connection to SRMA is a bit clearer. Each type of detector has its own reliability and maintainability issues. Some detectors may require cooling; others may require shock and vibration isolation or shielding against the local environmental conditions. Just picking one model that fulfills the stakeholder's optical requirements does not mean it will fulfill the maintainability and reliability constraints. In order to give a more real-life example, let us look back at the fictitious company FIT in our integrated case study.

8.3 Integrated Case Study: Maintenance and Support in Context of the Enterprise

In this section, we see FIT in the midst of the conceptual phase, evaluating the utility of some optical detectors. A decision needs to be made soon on the type of detector. The optical design team wants to update their optical systems model-based systems engineering tool to include detector modeling. The systems engineering team wants to develop reliability, maintainability, and availability models for the optical system.

The characters in this integrated case study are Dr. Bill Smith, FIT CEO; Karl Ben, FIT senior systems engineer; Tom Phelps, FIT chief technical officer and optics expert; Ginny R., FIT UAV program manager; Jennifer O. (Jen) FIT systems engineer; Ron S., new hire systems engineer; and Rodney B, FIT field service.

(We find Jen in the office becoming edgy, waiting for Ron to arrive.)

It is 2:00 PM in the office and Jennifer has been working heavily on an FTA for the current optical detection setup. Tired and fed up with the results, she notices the new employee, Ron, entering the office. Jennifer quickly collects her notes and makes a beeline toward Ron right before he can sit down.

Jen: "Hey Ron! Can I get a minute from you?"

Ron: "Oh! Goodness, you startled me Jennifer. Yeah, I just came back from lunch and...."

(Jennifer interrupts Ron, trying to quickly get to the point.)

Jen: "Great! Great, I'm glad you're ready to go now that you're back from your break."

Ron: "Well yes actually, I just turned in some paperwork to Karl about the current detector study I was working."

Jen: "Oh how wonderful! So, have you made any progress on which detector type we plan to use on this project?"

Ron: "Ah well, a little bit, really Tom's group has been helping me a lot. I mean I've read the papers and I am beginning to get it, but I'm no expert."

Jen: "Well, that makes you the best candidate in our section to help me come up with the selection criteria."

(Ron looks around with that deer-in-the-headlights look.)

Ron: "Selection criteria? We just narrowed in on the type of detector to use; isn't this way too early in the development cycle to select the actual detector?"

Jen: "Well, yes. But it isn't too early to start thinking about what makes a good detector for our application. I don't think you want to wait until the detailed design and development phase to decide what criteria your detector needs to satisfy. It will be too late then. What if you can't meet all of your criteria? If you find a problem with the choice of your detector so late in the game that can seriously impact the program."

Ron: "Ah, I get it. Besides, if we start looking at the detector criteria right now, we are building our case for justifying our selection. We can use the criteria for conducting detector trade studies to select the particular detector, and we have the rationale to justify the choice. I like it! Should we use the AHP? That seems real popular around here."

Jen: "I don't think we need to do a full blown AHP at this time. I think a weighted factor scoring method will be quicker and work just as well. AHP is great when you have decisions that involve relative preferences. For example, how much do you prefer the reliability of the system to maintainability aspects of the system? In this case, we don't have quite the same type of considerations, or at least we may not have enough information at this stage to make those kinds of comparisons."

Ron: "What do you mean?"

Jen: "In this case, we are weighing a bunch of technical factors that will be used to evaluate, and ultimately select, the detector used for this project—things like, size, weight, power, number of pixels, size of pixels, responsivity, noise, and so forth. In some cases, it doesn't make much sense to ask how much do you prefer one factor to another. For instance, how much do you prefer responsivity (bigger is better) to pixel size (smaller is better) is a difficult question to answer because you need both factors to be successful. If your responsivity is low, then you don't get high enough output signals from your detector, and you fail to see the target. On the other hand, if the pixel size is too large, you may not have enough spatial resolution in your optical system, and you fail to see the target. In essence, you need both responsivity and the spatial resolution of the optical system to be sufficient to permit detecting the target in the expected operating environment. These are more like binary decisions, you either have enough to meet the human detection requirement or you don't. With regards to the factor itself though, a detector with a higher responsivity has more potential to detect a lower-level irradiance, and a detector with smaller size pixels has the potential to provide better spatial resolution assuming other technical factors permit it."

Ron: "What do you mean?"

Jen: "Well, just because you add more pixels, that doesn't mean your spatial resolution will go up. Remember the meeting the other day? If you have an atmospheric turbulence compensation system, then the best resolution you can get is the classical diffraction limit, and this depends on the size of the optical system's entrance pupil (usually the telescope primary mirror). The optical system has to be designed to have a projected pixel size smaller than the classical diffraction limit; however, any pixel sizes that are smaller than that don't give you any additional advantage. In fact, if they are too small, they reduce your potential instantaneous field of view. The point is that the interaction of the technical factors is sometimes more complicated than a pair-wise comparison method like the AHP is capable of clearly describing."

Ron: "So why factor scoring?"

Jen: "Weighted factor scoring! In this case, we have a bunch of technical factors that we know are important but don't really know how important they are with respect to each other. So, initially, you set all the weights equal to each other so that each factor is considered of equal importance. Say that, currently, we only use three factors such as responsivity, weight, and size. I know there are more but I am just using this as an example. We would initially weight each one equally 1/3, 1/3, and 1/3—the weights sum to 100%. Let's say that in evaluating these three factors, we find that most available sensors barely fit into the sensor housing allocated for space on the UAV platform. Let's also assume that our studies show that we have decent conservative signal estimate with about a 10% margin and we plenty of weight margin (i.e., the weight of the detector is substantially lower than the allocated weight). In this case, we might change the relative priority of the weights to 10% for the weight category, 50% for the size category, and 40 percent for the responsivity category. We could use a 1 to 10 ranking scheme, where a 1 would be worst and 10 would be best in each evaluation category, and evaluate each candidate detector against the rating. Additionally, you could include some binary or threshold conditions for each category that would eliminate a candidate from consideration if the condition were violated. For example, if a candidate detector's size was too big to fit in the environmental conditioning unit, or drew too much power, or had pixels that didn't provide an acceptable field of view or adequate spatial sampling for the optical system design, the detector would be eliminated from consideration."

Ron: "I get it! All of the categories are important, and their relative importance is adjusted by the weighting factors. If the technical factor isn't inherently good enough to meet the optical system requirements, then it is eliminated from consideration. What if the optical systems design changes and the new design accommodates some of the detectors that were previously rejected? Hey! I just had an idea. What if we built a tool that captures the constraints and selection criteria for the detector? We could integrate it with the analytical metrics that are being developed over in the optics and imaging group."

Jen: "Ron, that's a great idea. The optics folks are working on a detector model as an upgrade to their model-based systems engineering tool that they used in the staff meeting the other day for the radiometric propagation analysis. That model will be an accessible artifact on our enterprise architecture. We can pull the design factors into our own MBSE tool that links these factors to the optical systems requirements. That way, when we evaluate potential candidate detectors, we will be using the current optical systems design and optical systems metrics in evaluating the detectors. If the design changes, the model will sort through all the detectors that are in our system and automatically recheck the constraints. If it passes, the candidate detector will automatically be reevaluated against our current set of criteria and weights. Also, if we are running "what-if" scenarios, bad assumptions will trip a constraint in the decision logic and show what requirements are violated. We also get a built-in justification for the detector choice, and the design itself. If we are careful in how we build it, this feature may be reusable in other projects too."

Ron: "This approach also fits in with the rapid-prototyping philosophy that we have here at FIT."

Jen: "Let's bring Amanda over here to go over their planned technical metrics and see what they think. Let's also bring in Phil K. from software since he, or someone in software, may have to do some programming. We also need Mr. Garry Blair to get the enterprise architecture perspective. It will be up to systems to make sure this MBSE tool is reusable across projects, and so we will have to carefully consider how the links are made to the stakeholder and systems requirements. Before we get too far on this, let me talk with Karl."

Ron: "Ok, sounds good."

Jen: "Did you say earlier that you finished the detector-type study?"

Ron: "Pretty much. We have a few more things to look at, but we have a pretty good idea of what detector type to use."

Jen: "That's great! I have 3 days to get a preliminary high-level failure modes and effects analysis completed and I need to know the detector type to make any more progress. As you know, the different detector types have different failure modes."

Ron: "Ah, yes. Well, we eliminated thermal detectors because we have to scan through a relatively wide area and rather quickly too. We were thinking about using a thermal point detector to view a wide area but that would mean we would have a second optical system acting as a tip and queue for a higher resolution imaging camera. In the end, we chose a photon detector as the detector type. They are leaning towards HgCdTe detector material based on some promising

D^*, responsivity, and noise equivalent power numbers they saw. They haven't decided yet on whether they are going to use a CCD or CMOS configuration; that is still being looked at."

Jen: "That certainly helps! I can get enough information to help with the FMEA. What values of D^*, responsivity, and NEP are they using?"

Ron: "Let me check my notes... ah, here it is. This is preliminary since they only looked at a few detectors so far, but one interesting HgCdTe detector used the photovoltaic detection mechanism. It had a noise equivalent power (NEP) of 2×10^{-12} W and a detectivity of 5×10^{11}. The D^* peak was at 10.6 µm. It had a D^* value of about 3×10^{10} cm Hz$^{1/2}$/W at 9.5 µm and at a temperature of 77 K. Its responsivity was 4×10^4 V/W."

Jen: "The temperature is 77 K? Cryogenic cooling is doable, but it adds complexity to the optical system. I wouldn't want to use it unless we are forced to. Also, wouldn't that increase maintenance costs? I assume you would have to have some scheduled maintenance in order to replace the cryogenic fluid and maintain the cooling system."

Ron: "Right, and we will need to check the reliability of the cooling system too, since if that fails, the detector won't work right. We should also look at using preventive maintenance actions if the reliability of the cooling system isn't sufficiently high enough."

Jen: "Well Ron, let's not get ahead of ourselves. We first have to determine the actual detector and see if there is a requirement to cool it. Putting in stubs for our reliability model for the cooling system is probably a good idea at this point."

Ron: "Right, well it looks like I've found my next project to work on. Also, in order to avoid having our detector being outdated, perhaps, we can plan for adapting newer models into our system as they develop or become cheaper to purchase."

Jen: "Right! Preplanned product improvement, I like that. Good, it looks like we have a preliminary set of technical factors that we can start within the detector evaluation process. We will have to determine constraint, threshold, and target values and then use the MBSE models to link to the design and requirements. I'll go talk to Karl about our approach. Great job Ron!"

Ron: "Thanks Jen, talk to you in a bit. I'll give Amanda and Phil a heads-up so that they aren't caught off-guard."

(Early the next morning, Rodney B. from Field Service stops by for a discussion with Jen.)

Rodney: "Hi, Jen. Just wondering if you had time to discuss the maintenance plan for the new UAV project?"

Jen: "Sure, Rodney, the data we found on available UAV reliability data show they have a pretty high failure rate. You are the one that will live with this thing in the field, what do you recommend?"

Rodney: "Well, based on what I have seen before, I think we should concentrate on some kind of preventive maintenance program. After all, if we can manage to fix things before they break, then we can keep the optical system functioning, and that will make everybody happy!"

Jen: "They will for sure! Except for Ginny, she will be worried about the cost of replacing parts when it isn't necessary. Hmmm, I guess we will need to find a balance between too much maintenance and too little. OK, let us get something together and then we will pass it by you for a review. Thanks for stopping by."

Rodney: "No problem, I will look forward to reviewing what you come up with."

(Later that day, Jennifer sees Ron in the break room.)

Jen: "Hi Ron, didn't you help Karl with the initial reliability model?"

Ron: "I sure did. It wasn't as difficult as I thought it would be, but I don't know if I could do it by myself yet. You don't have more of that for me do you?"

Jen: "No, don't worry; I just would like your help on something else. I am working on the maintenance plan and was talking with Rodney from Field Service, and he suggested that we consider the maintenance aspects for the optical system. I have seen a way to do this by using the reliability analysis in what is called reliability-centered maintenance."

Ron: "Oh, I get it. You can use the failure rates we found to determine a plan to do preventive maintenance?"

Jen: "Yes, that's it exactly! And we can also structure the plan to incorporate the reliability data from the fielded UAV optical systems as we get it and feed it back into the plan to fine-tune it. Could you gather up the failure rates on the top-level blocks of the optical system and stop by my desk later so we can go over them?"

Ron: "Sure, see you later."

(Later that day, Ron stops by Jennifer's desk.)

Ron: "Hi, Jen. I have the data you asked for. OK, if we look at the top level and consider the UAV as a functional block and the optical system as a functional block, then we found that the UAV had a MTBF of 288 hours and the optical system had an MTBF of 10 years."

Jen: "Why did you bring me the UAV data? We are only responsible for the optical system part of it. The UAV maintenance is Homeland Security's concern."

Ron: "That's true. But there is a chance that, depending on what breaks, they may have to mess with the gimbal or optical sensor housing. We may need to do some system calibration and testing if they are messing with any of our systems or if the power goes out on some of our systems. We can also take advantage of the downtime to tweak our system and make sure everything is running optimally. You know, check the cryogenic system if we end up with one, and make sure the system is boresighted and in tip-top shape. The UAV failure rate data may also affect the number of spares we need on hand and the logistics for getting them there."

Jen: "Those are good points. We will have to consider all that in our planning. We are interested in the failure rate. That is the lambda value. Do you have the lambda values?"

Ron: "Yes, the lambda for the UAV is 12 failures per year, and for the optical system, lambda is 0.1 failures per year."

Jen: "Wow! There is a big difference in those numbers. I guess that makes sense since the UAV has a lot of mechanical parts that can wear out from friction, vibration, rough landings, and so on, and the optics is mostly sealed inside the housing.

Jen: "I will write the preventive maintenance plan to include the possibility of monthly corrective maintenance to include test and calibration of the UAV optical systems and preventative maintenance on the optical system perhaps semiannually.

We probably want to do the maintenance when the optical system reliability falls below 70%. I, also, will write in a section that we gather data on the fielded UAV optical system and feed it back into the calculations to adjust the failure rate estimates and update the preventive maintenance plan accordingly. Thanks for your help."

(Later the next week, Rodney stops by Jennifer's office.)

Rodney: "Jennifer, here are my review comments on the maintenance plan inputs. Glad you showed that too me. I had to go back and try to get more funds for field service, and I was able to do that based on the data you and Ron provided. A year from now, I wouldn't have been able to cover those costs, so you saved my bacon by finding that now!"

References

Bazovsky, I. 2004. *Reliability and Theory Practice*. Mineola, NY: Dover Publications.

Birolini, A. 2010. *Reliability Engineering: Theory and Practice*. Berlin, Germany: Springer-Verlag.

Blanchard, B.S. 2004. *Logistics Engineering and Management*, 6th edn. Upper Saddle River, NJ: Pearson Education.

Cashman, R. 1959. Film-type infrared photoconductors. *Proceedings of the Institute of Radio Engineers*, 47(9): 1471–1475. doi:10.1109/JRPROC.1959.287039.

Crouch, J. 1965. Cryogenic cooling for infrared. *Electro-Technology*, 75: 96–100.

Department of the Army, TM 5-698-4. Failure Modes, Effects and Criticality Analyses (FMECA) for Command, Control, Communications, Computer, Intelligence, Surveillance, and Reconnaissance (C4ISR) Facilities. September 29, 2006.

De Waard, R. and E.M. Wormser. 1959. Description and properties of various thermal detectors. *Proceedings of the Institute of Radio Engineers*, 47(9): 1508–1513. doi:10.1109/JRPROC.1959.287049.

Dhillon, B.S. 2006. *Maintainability, Maintenance, and Reliability for Engineers*. Boca Raton, FL: CRC Press.

Fabrycky, W.J. and B.S. Blanchard. 2011. *Systems Engineering and Analysis*, 5th edn. Upper Saddle River, NJ: Pearson Education.

Harms-Ringdahl, L. 2001. *Safety Analysis*. London, U.K.: Taylor & Francis.

Hudson, R.D. 1969. *Infrared System Engineering*. New York: John Wiley & Sons.

Hsu, J. C., and S. Nagano. 2011. Introduction: Systems Engineering, *Journal of Aircraft*, 48(3): 737.

Jones, R.C. 1959. Phenomenological description of the response and detecting ability of radiation detectors. *Proceedings of the Institute of Radio Engineers*, 47(9): 1495–1502. doi:10.1109/JRPROC.1959.287047.

Kamrani, A.K. and M. Azimi. 2010. *Systems Engineering Tools and Methods*. Boca Raton, FL: CRC Press.

Linneberg, P. 2012. Reliability based inspection and reliability centered maintenance. In *Bridge Maintenance, Safety, Management, Resilience and Sustainability*, F. Biondini and D.M. Frangopol (eds.). pp. 2112–2119. Leiden, the Netherlands: CRC Press.

Madu, C.N. 2005. Strategic value of reliability and maintainability management. *International Journal of Quality & Reliability Management*, 22(2): 317–328.

Márquez, A.C. 2009. The maintenance management framework: A practical view to maintenance management. *Journal of Quality in Maintenance Engineering*, 15(2): 167–178.

McDaniel, N.A. 1999. *Acquisition Strategy Guide*, 4th edn. Fort Belvoir, VA: Defense Systems Management College Press.

Nudelman, S. 1962. The detectivity of infrared photodetectors. *Applied Optics*, 1(5): 627–636.

Putley, E. 1973. Modern infrared detectors. *Physics in Technology*, 4: 202–222.

Smith, W. 2007. *Modern Optical Engineering: The Design of Optical Systems*, 4th edn. New York: The McGraw-Hill Companies, Inc.

Strandmark, P (aka Mwtoews). 2007. Standard deviation diagram. http://en.wikipedia.org/wiki/File: Standard_deviation_diagram.svg (based on figure by Jeremy Kemp [2005], accessed November 28, 2014).

Wessels, W.R. 2010. *Practical Reliability Engineering and Analysis for System Design and Life-Cycle Sustainment*. Boca Raton, FL: CRC Press.

Wyatts. 2005a. Fault tree. http://en.wikipedia.org/wiki/File:Fault_tree.png (accessed April 3, 2014).

Wyatts. 2005b. Reliability block diagram. http://en.wikipedia.org/wiki/File:Reliability_block_diagram.png (accessed April 3, 2014).

9

Technical Performance Measures and Metrics

> Measurement is the first step that leads to control and eventually to improvement. If you can't measure something, you can't understand it. If you can't understand it, you can't control it. If you can't control it, you can't improve it.
>
> —H. James Harrington

9.1 Introduction to Technical Performance Measures and Metrics

In the field of systems engineering (SE), the vital aspect that connects stakeholder wants and needs to dynamic, possibly complicated systems/services/products/processes is requirements. Seen as one of the most important concepts of SE, requirements allow system characteristics and attributes to be defined without constricting the manner in which they are achieved. However important requirements may be, the act of translating them into quantifiable, measurable tasks is what forms the basis of the systems development effort. Technical performance measures and metrics (TPMs&Ms) play the pivotal role in defining the essential measurable attributes and characteristics of requirements in the requirements generation process and are used to verify and validate the requirements when the system and its elements undergo testing. Once established, TPMs&Ms can be used to establish the performance envelope for the system or system elements. TPMs&Ms can be separated into technical performance measures (TPMs) and their associated metrics. For example, if a requirement stated that the system must be capable of reaching an altitude of 3 km with respect to sea level, the TPM for this requirement would be altitude in kilometer with respect to sea level, and the technical performance metric would be 3 km. If the prototype developed during the detailed design and development phase failed to reach this altitude, then this requirement is clearly not met and the system is not meeting its performance requirement in this area. It is back to the drawing board! As is illustrated by this example, TPMs&Ms can trigger flags such as the need to have change reviews, the need for increased funding, or an indication to shift focus. Often, in the literature, the term TPM is used and the associated metric is implied. We will adopt this convention and include the technical performance metric as needed.

As another example, perhaps the TPM is a threshold for the ambient temperature for which a ruggedized laptop computer must be able to operate without failing. This TPM will follow and govern the design team throughout the SE life cycle. It may drive specific environmental exposure tests, or it may focus the engineering team's attention on heat transfer rates through specific metal types for rapid dissipation during peak performance or stress. In short, TPMs not only allow for a design team to trace system requirements throughout the course of a project life, but they allow for designers to evaluate requirements with respect to each other. A prioritization factor is involved in this process, which allows for a responsible design while still adhering to stakeholder requirements. Through the use of tools such as the quality function deployment (QFD) models and house of quality (HOQ) matrices, TPMs give an engineering team the ability to track satisfactory progress and identify areas of concern that may have otherwise gone unnoticed.

Dr. Bahill, author of *Technical Performance Measures*, defines TPMs as "tools that show how well a system is satisfying its requirements or meeting its goals," saying they are "functions of time" that "provide assessments of the product and the process through design, implementation, and test" (Bahill 2004–2009). Similarly, Guerra, author of *Technical Performance Measures Module*, states TPMs are "measures of the system technical performance ... based on the driving requirements or technical parameters of high risk or significance" (Guerra 2008). Finally, Professor Blanchard, author of *Systems Engineering and Analysis*, says TPMs "are quantitative values (estimated, predicted, and/or measured) that describe system performance" (Blanchard and Fabrycky 1990). Perhaps the most comprehensive definition is that "TPMs measure attributes of a system element within the system to determine how well the system or system element is satisfying specified requirements" (Roedler and Jones 2005). In summary, TPMs establish measures related to goals and requirements in order to establish measureable performance criteria to demonstrate progress toward achieving the goals and satisfying the requirements.

Historically, as with other aspects of SE, the idea of developing TPMs might be called to question. The rigorous procedure of measuring performance requires collecting, analyzing, and reporting various aspects of the system's behavior, which is additional work to an already complex design process. However, the science of measuring performance parameters enables system stakeholders to hold managers responsible for project performance and accountable for successful progress. According to *Technical Measurement: A Collaborative Project of PSM, INCOSE, and Industry*, the attributes of technical measures allow for "the continuing verification of the degree of anticipated and actual achievement of technical parameters" (Roedler and Jones 2005). They go on further by defining a TPM template as follows: (a) achieved to date, (b) current estimate, (c) milestone, (d) planned value (target), (e) planned performance profile (analysis model), (f) tolerance band (decision criteria), (g) threshold, and (h) variance(s) (Roedler and Jones 2005). All in all, TPMs can be used to "judge the value" a vendor creates for a stakeholder while also providing managers with the data necessary to improve performance (Behn 2003).

Figure 9.1 depicts a model highlighting the relationship between speed, quality, flexibility, dependability, and cost, five commonly used parameters (Lichiello and Turnock 1999). The figure also demonstrates the internal and external relationship to the system benefits (as seen within the box) when performance measures are properly used (Tangen 2003). As seen, different requirements are categorized within the context of these five parameters and transition to performance measures. Figure 9.1 demonstrates that although these five parameters are independent of each other, they all stem from the need for high total productivity.

Perhaps one of the most famous telescopes, NASA's Hubble Space Telescope (HST), serves as a perfect example of how not developing TPMs for a system can result in failure. The telescope, at almost 12.8 m long, was $2.5 billion over budget when it was launched in 1990 (Kasunic 2011). Within a few weeks of its launch, the scientists and engineers monitoring the telescope's performance had discovered a flaw in the optical system. The first image was collected after supposed perfect focus was achieved, but issues with the primary mirror showed an aberration: it had been ground to the wrong shape, which resulted in the out-of-focus images.

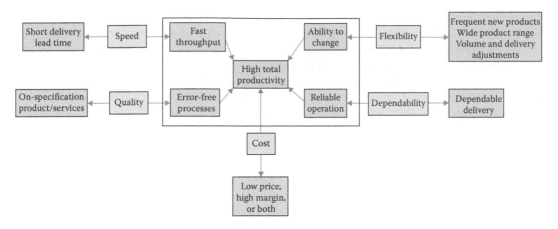

FIGURE 9.1
Interaction of performance measures.

NASA created an investigatory commission to determine how the error occurred and propagated. Lew Allen, former Jet Propulsion Lab director, led the commission, which found that the reflective null corrector (RNC), a device used to identify any defects on the surface of a mirror, had been incorrectly built. PerkinElmer, the company that was contracted to test the mirrors, had used a conventional device for the job. Before completion of the project, however, a specialized corrector was used that was assembled incorrectly in such a way that it was still precise, but not accurate. Inconsistent results occurred between the standard device and the custom device, one of which produced perfect results and the other which was in error. In the end, the presence of spherical aberrations in the mirror was dismissed in favor of the incorrect device's results, which is a significant error in any testing methodology. The following are excerpts from the NASA official report on the projects failings:

- "... there was a surprising lack of optical experts with experience in the manufacture of large telescopes during the fabrication phase."
- "The quality assurance people at Perkin-Elmer ... were not optical experts and, therefore, were not able to distinguish the presence of inconsistent data results from the optical tests."
- "Fabrication of the HST mirror was the responsibility of the Optical Operations Division of Perkin-Elmer, which did not include optical design scientists and which did not use the skills ... which were available to Perkin-Elmer."
- "Perkin-Elmer line management did not review or supervise their Optical Operations Division adequately. In fact, the management structure provided a strong block against communication between the people actually doing the job and higher-level experts both within and outside Perkin-Elmer."
- "The Optical Operations Division at Perkin-Elmer operated in a 'closed door' environment which permitted discrepant data to be discounted without review."
- "The Perkin-Elmer Technical Advisory Group did not probe at all deeply into the optical manufacturing process .... This is particularly surprising since the members were aware of the history ... where spherical aberration was known to be a common problem" (Allen et al. 1990).

It was clearly noted in the commission report that any concerns with the data produced by the RNC were held at the internal level. In the aftermath of this incident, PerkinElmer, to avoid a lawsuit, paid a $25 million head off. Faced with this financial responsibility for the Hubble's crippling optical issue, PerkinElmer sold their previously highly successful optics division. Had greater credence been given to TPMs provided by the conventional devices been considered, the errors may have been discovered earlier and avoided.

Measuring performance and its role in the SE process is vital and serves as the focus of this chapter. First, we will delve into the components of TPMs and how they allow for responsible design in the SE design process. Then, a detailed introduction of the science of detectors and noise with relation to optical systems is provided. Finally, a fictional case study highlighting the importance of using TPMs throughout the product life cycle will be applied to optical detectors.

9.1.1 Role of Technical Performance Measures in the Systems Engineering Process

TPMs are connected with many parts of the SE development process. Figure 9.2 shows TPMs as part of the repetitive systems analysis and control block.

It can be seen in Figure 9.2 that the SE process is quite detailed. In the beginning, input is required from the stakeholders with respect to missions, constraints, and measures of effectiveness (MOE). This is accomplished through interviews with stakeholders to help identify and prioritize items of importance and determines which items warrant monitoring throughout the systems development and, possibly, the system's life span. The next step involves conducting a requirements analysis, which evaluates all inputs from the stakeholders for the purpose of producing top-level, systems requirements. From there, functional analysis/requirements allocation and synthesis studies are conducted. In generating the requirements, TPMs are defined/identified to establish the performance criteria of the requirements. The TPMs can then be used as quantifiable requirements attributes by the engineering team. Having been specified, the TPMs can be used as performance yardsticks to evaluate potential designs or results or serve as target values for feasibility studies. In short, the top-level purpose of TPMs is to track progress through measurement and evaluation, which is ultimately used to gauge the progress toward satisfying stakeholder's needs and agreed wants.

9.1.2 Types of Measurements

In a general view of technical measurement, there are four different terms used to describe performance. Two of these measures, TPMs and key performance parameters (KPPs), have already been discussed. Two additional measures, which are frequently used in demonstrating and judging the overall system performance, are MOE and measures of performance (MOP).

MOEs focus on performance, suitability, and affordability, but more prevalently defined by Lockheed Martin Senior Program Manager Garry Roedler as "the 'operational' measures of success that are closely related to the achievement of the mission or operational objective, being evaluated, in the intended operational environment under a specified set of conditions, i.e. how well the solution achieves the intended purpose" (Roedler and Jones 2005). From the view of the customer/stakeholder, MOEs are the benchmark under which the project is judged with respect to the goals and needs of the mission. In some cases, ones of a technical nature, there is some room for change as long as the stakeholder agrees to the change.

MOPs are related to MOEs and can be described as "the measures that characterize physical or functional attributes relating to the system operation, measured or estimated

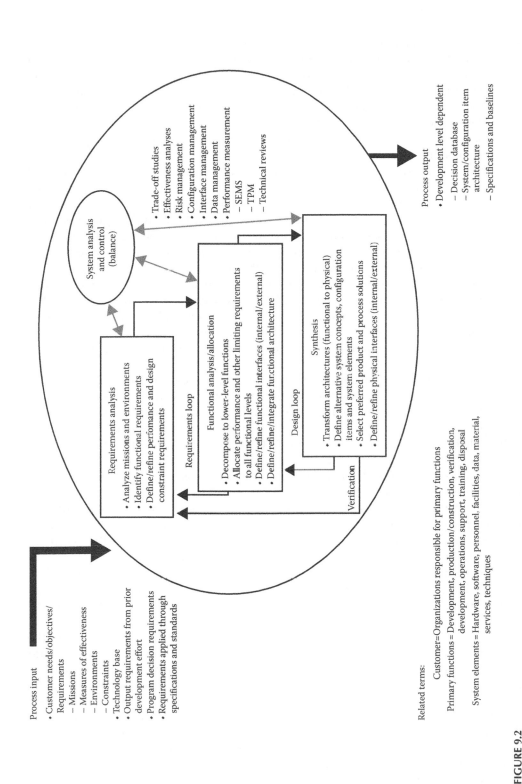

FIGURE 9.2

Systems engineering process. (Obtained from Defense Acquisition University, Systems engineering fundamentals supplemental text, http://www.acqnotes.com/Attachments/DAU%20Systems%20Engineering%20Fundamentals.pdf, p. 31, 2001, accessed November 29, 2014.)

under specified testing and/or operational environmental conditions" (Roedler and Jones 2005). As MOEs set the standard for the goals of the mission, MOPs judge whether or not the system has the ability to meet those requirements. They are used when a system has components critical to achieving the system's operational objectives. MOPs also examine a delivered systems ability to meet the original system level requirements. MOEs are closely related to system requirements or KPPs in that they consist of quantitative features such as speed, mass, or range. Like all technical measurements, MOEs are used as benchmarks throughout the development process, and when contrasted with MOPs, they can present the system engineers or designers with insight into developmental issues or areas of concern.

Some of the most important elements of the TPM process are KPPs. They are described as "a critical subset of the performance parameters representing those capabilities and characteristics so significant that failure to meet the threshold value of performance can be cause for the concept or system selected to be reevaluated or the project to be reassessed or terminated" (Roedler and Jones 2005). Approximately five KPPs should be given because the more KPPs a program has, the more opportunities for budget and scheduling delays. Because TPMs involve "predicting the future value of the higher level end product under development," it is noted that performing beyond the standard, set forth by a KPP, is considered a desirable result (Hagan 2009). Used to characterize factors such as supportability, interoperability, and operational performance, KPPs allow for a perspective as to whether the customer's needs and approved wants are being satisfied and also give the designing team a summary of the overall progress of the development effort. Unlike MOEs and MOPs, KPPs contain two parts: a threshold value and an objective value. Stakeholders can evaluate decisions and design choices by using KPPs.

The Lockheed Martin press release, issued July 22, 2002, concerning the success of their F-22 Raptor program, serves as an excellent example of how KPPs and, by extension, TPMs are used to demonstrate a compliant and successful delivery of the stakeholder's technical needs. The press release describes in great detail how the company managed to meet *and* exceed the stakeholders needs stating that the KPPs including a "perfectly balanced blend of three technologies: stealth, super-cruise speed, and advanced integrated avionics" were achieved (Polidore 2002). It should be noted that the press release neglects to mention how Lockheed ensured their goals were met.

TPMs, MOEs, MOPs, and KPPs are used to evaluate the project's progress with the goal of determining whether or not it was successful. All of these measurements have an integrated relationship with one another, which is worth noting. Even though they each are associated with separate areas of the project, they are used in tandem with one another. If the intent is to consider the project itself and ensure its performance is maximized, then MOPs are considered, whereas a completion of the wants/needs of the stakeholder is where MOEs come into play. For this purpose, it is vital to understand the difference between validation and verification. In the process of verification, the essential question being asked is, "Did we build it to the design specifications?" In contrast, the process of validation begs the question: "Did we build what the stakeholder wanted, and does it operate in a real-world environment?" In the lifetime of a system, verification is used during component, subsystem, and system test, and validation is when the competed system is tested under operational conditions. It is interesting to note the relative priority of MOEs, MOPs, KPPs, and TPMs. MOEs drive the development of KPPs and MOPs, TPMs are derived from MOPs, and KPPs play an important role in determining MOPs. The established connection among these technical measures is outlined in Figure 9.3. As the systems development advances, the scope narrows and technical insight increases. As time goes on, these different measures help provide focus to the systems development effort.

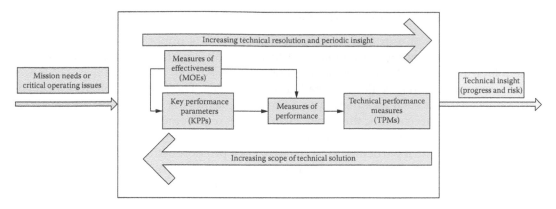

FIGURE 9.3
Relationships of the technical measures.

Consider this simple example (Roedler and Jones 2005):

- MOE: Operational time of data system used to log highly sensitive data.
- MOP: The system downtime shall not exceed 5% of yearly uptime.
- TPM: System redundancy, fault tolerance, and failure KPP.

A pure comparison among the different kinds of technical performances is possible through uniform definitions (Mitchell 2010). When evaluating the correlation between availability and reliability measures, a scenario often encountered becomes apparent: if redundant equipment is not needed for normal operation, but becomes due to be serviced for maintenance, how should this situation be managed? If the spare is taken out of commission for maintenance, the system is not immediately affected, provided the primary equipment does not fail. The availability of the system will not change as long as the primary remains functioning; however, if the redundancy has been removed from the equation due to maintenance, the reliability of the system has to be downgraded. If the system under discussion is related to life support, lessening the probability of reliability may be unacceptable. In this situation, the backup equipment must be replaced with identical equipment for the servicing or maintenance to occur. A requirement for a life-support system may mirror the following: "the system shall have a mean time to failure (MTTF) of XX years." The requirement would be determined based upon the stakeholder's needs and could take into account a factor such as how long the system operates in a functional time period. If the usage is noncontinuous, the exposure period to failure may potentially be reduced. The MTTF could be determined by a combination of quantitative techniques such as failure mode, effects, and criticality analysis (FMECA) and fault tree analysis (FTA). The FMECA could be performed on the system components such as circuit boards or mechanical connectors, and then a quantitative fault tree could use undesired events driven from the FMECA to perform a top-down evaluation of failure modes in relation to undesired outcomes. Software tools such as Reliability Workbench can handle extensively complex scenarios and perform the mathematics required to generate MTTF rates. This depiction of TPMs and the need for properly written requirements are further explored in the following section, which describes, in detail, this fundamental connection of TPMs to requirements.

9.1.3 Connection of Technical Performance Measures to Requirements

During the infancy of a project, in its proposal stage, it is important to create a set of TPMs as a way of judging the standard set forth by adhering to the wants/needs of the stakeholder. TPMs must be important, relevant, and measurable. Because of this, TPMs are closely associated with specific and detailed requirements. The TPMs are often associated with project aspects such as adhering to a timetable, achieving stakeholder approval, or remaining within an economic threshold. Even though requirements are directly related to TPMs, the objective of the systems engineer should be to limit the combination of requirements associated with a particular TPM. This practice promotes efficiency and helps avoid confusion throughout the life cycle. TPMs can be any number of system characteristics, such as reliability, cost, or throughput. When it comes to assigning TPMs, the complexity of the system, the stakeholders, and resource availability are all factors that must be considered. After all is said and done, a prioritization scheme can be assigned and appropriately evaluated for each TPM (Oakes 2004–2005).

The importance of assigning weight factors to inputs or characteristics of a system cannot be understated. Whether the system under evaluation is a commodities portfolio or a submarine, all TPMs are not created equal. For example, the Kalman filter, which in one application is a commonly used mathematical algorithm for probabilistic location determination in guidance and navigation systems, is an example of a system where prioritization factors are of high value. The Kalman filter is especially adept to using a variety of different inputs, in an iterative process, to determine the validity of location. For example, inputs to a navigation system may include a GPS system, dead-reckoning techniques, and RFID sensors. If the RFID sensors are surveyed to an exact latitude and longitude, any input of RFID-related variables to the system should be weighted with a higher degree of confidence than dead-reckoning variables, which are a function of speed and time. Dead reckoning may become highly inaccurate over extended periods of measurement due to the fact that speed and time do not necessarily accurately predict location unless the movement is in a completely straight line and velocity estimates are precise. Therefore, the designer of the system needs to weigh the RFID inputs with much higher mathematical importance than the dead-reckoning inputs and calculations. While all of the inputs to the Kalman filter are important for the system to function as a whole, certain elements have higher priority weights. This example highlights the need for careful thought and domain expertise when assigning the relative importance of system-related TPMs.

One of the most common ways to create, prioritize, and track TPMs is by the use of a QFD and the associated HOQ. The QFD references the stakeholder needs/wants with respect to requirements priorities. This is a very effective way of establishing vital aspects of a systems development effort and presents a means for documenting stakeholder preferences. Each of the stakeholder needs/approved wants is listed as row entries in the QFD and is typically prioritized by the stakeholder. This gives an indication to the development team as to the relative importance of the stakeholder needs/wants. The example HOQ in Figure 9.4 shows a variety of useful attributes of the QFD.

Note that in this example, all of the stakeholder needs/wants are of equal weight. The triangles to the left of the stakeholder needs/approved wants indicate the correlation of the respective needs/wants (e.g., if one of the needs/wants increases, what is the effect on the others/wants?). The column entries on the QFD are typically the response of the developer to the stakeholder needs/wants. The triangular "roof" of the HOQ indicates correlations between the technical responses. The matrix intersection of the stakeholder needs/wants with the technical responses often indicates the relative importance of the technical

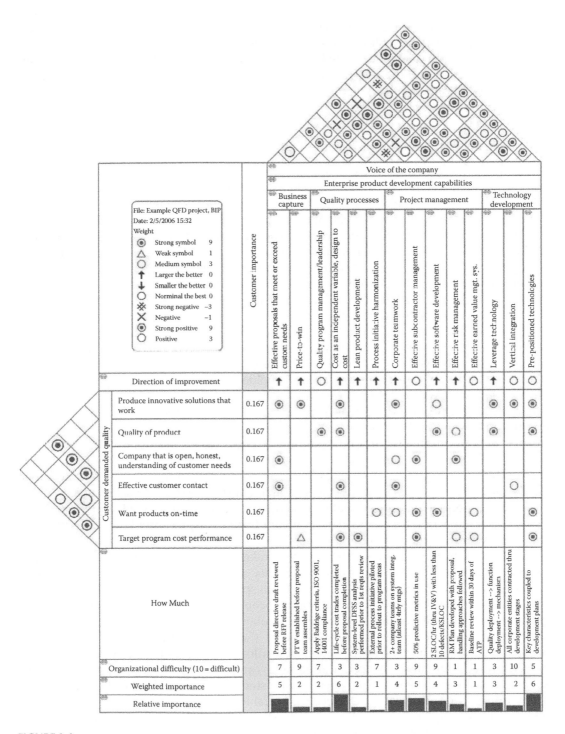

FIGURE 9.4
(See color insert.) House of quality. (Obtained from Cask05, http://en.wikipedia.org/wiki/File:A1_House_of_Quality.png, 2006, accessed November 29, 2014.)

response to the stakeholder needs/wants based on a common rating scheme (posted at the top left of the example in Figure 9.4). The bottom of the HOQ relates to the TPM identified for the technical response column and provides useful information such as the organizational difficulty in implementing the technical response, the weighted importance of the technical response, and the relative importance of the technical responses to each other. Different useful categories can be presented in the QFD based on the target audience and the purpose of the QFD.

9.1.4 Systems Engineering Approach to Optical Systems

When the systems engineers are defining the stakeholder requirements, they must first determine the difference between the stakeholders' needs and wants. Determining this difference may be difficult because the stakeholder may not be able to recognize one from the other. The stakeholder may feel that they "need" a Bavarian Motor Works (BMW), but their budget supports a more modest vehicle. The reality is that the stakeholder "wants" the BMW but "needs" reliable transportation. In this case, the vendor clearly lays out the pricing options for the stakeholder, and the stakeholder satisfies his or her "need" at another dealership. In this case, the "want" was not realistic and the perceived need was not met. On the other hand, another stakeholder with adequate financial means may walk into the same BMW dealership with a need for a reliable vehicle and a "want" of a convertible BMW with blue exterior and red interior, and the stakeholder has the budgetary means to see both the needs and wants met. In this case, the stakeholder leaves the BMW dealership with a new blue convertible BMW and has his or her needs and approved wants satisfied. Both needs and wants are important. The stakeholder's needs and approved wants eventually develop into stakeholder requirements.

In designing an optical system, everything from the viewing distance, the field of view and the quality of the image, right down to the environment that the system will be used in, need to be accounted for. In understanding the operating conditions and the tasks the system is expected to accomplish, one can establish performance requirements. By handing off the performance requirements to the development team, fulfilling the customer needs and approved wants will be easier, simpler, and more directly achievable.

A specification table is traditionally created to provide an organized summary of the vital factors impacting a systems development effort. It displays the key factors used to shape the system without delving too deeply into the design details. It demonstrates the means by which the engineers can fulfill the stakeholder requirements. Organizing the key factors into one table allows for accountability and traceability. Imagine, for example, a high-powered telescope set to be launched into space. In order to meet the field of view and spatial resolution requirements, it might become necessary to use a large number of lenses. However, as the number of optical components increases, so does the cost, weight, and size of the optical payload. At some point, the size, weight, and cost budgets may become exceeded. Using the specification table allows engineers to clearly see when they exceed preset boundaries. A specification table can also be used to identify redundant specifications. Often, redundant specifications stem from multiple stakeholders having a buy-in on an original concept. Specification tables can also be used to find inconsistent and/or incomplete specifications. Inconsistent specifications must be identified and resolved to prevent design conflicts later on in the development. Incomplete specifications leave gaps in the design process leading to ambiguity and potential errors in the design itself. By summarizing the important specification factors in one table, they are on display for all involved to see, understand, reference, and evaluate.

TABLE 9.1

Trade Table Subsection Example

KPP	Digital Camera	Space Telescope	Binoculars
Availability	Now	Delay	Soon
Size	Small	Large	Small
Cost	Low	High	Medium
Weight	Low	High	Low
Power usage	Medium	High	Low
FOV	Large	Small	Large
Reliability	Medium	High	High
Image storage	Yes	No	No
COTS	Yes	No	Yes

TPMs also play a role in trade studies. One use of trade studies is to determine a technology's ability to meet a system-/subsystem-/component-level requirement. An example of a trade study table can be seen in Table 9.1. By comparing and contrasting the differing characteristics, the advantages and disadvantages of individual technologies become clear.

When applying a trade table to an optical system, as with any systems requirements development effort, the needs and wants of the stakeholders must be considered. Ask: what can this option (technology) do for the stakeholder? A stakeholder may want a compact optical system with a low weight. But they may not know whether they need high reliability or not. Does the environment in which the optical system will operate require high reliability? Does the optical system mission require high reliability? These are the types of questions a trade table brings to the surface. They also demonstrate the trade-off between technologies. Including TPMs in a trade study can often help a stakeholder determine the best technical option or approach.

The preliminary design of the subsystems not only identifies the requirements of each individual subsystem, but also defines how those subsystems interface with one another. Both the hardware and software components must be considered, evaluated, and defined through requirements. Also, in defining the way the subsystems interact, KPPs must be considered, such as requiring a downtime no greater than 5%. How will your power source function in a power surge? Are all the subsystems protected? Will they shut down for a time greater than 5%? What redundancies will prevent the system from staying down for longer than 5%? How will the system track the 5% downtime?

As the subsystems begin to be designed, a separate process begins simultaneously: developing the test plan. The test plan outlines the testing methodology for the requirements. By testing each requirement, a system can demonstrate that the requirements have been satisfied. Often, test plans consist of several different types of tests such as black box, white box, and beta testing. Testing could be as complex as demonstrating 95% reliability for the duration of the systems lifetime or as simple as proving that 95% of first-time users between the ages of 15 and 25 years old, being able to turn on the device within the first 5 min of picking it up.

9.1.5 Transition to Optical Systems Building Blocks

As with any engineering system, an optical system's TPMs&Ms provide the critical specifications and insights into the optical system's performance. Failing to properly specify the TPMs will introduce uncertainty into the requirements generation process and adversely affect the overall optical systems design. Examples of some TPMs in a general optical system include minimum and maximum detection range, the instantaneous field of view (IFOV), the spatial and angular resolution of the optical system, the optical system's tangential and longitudinal magnification, the effective focal length, image quality metrics, and detector signal and noise performance measures and metrics. These TPMs are often interdependent in the sense that as some of these TPMs change, they affect other TPMs. For instance, for a general optical system, changes in the detection range affect the signal-to-noise characteristics on the detector. On the other hand, some TPMs may remain unaffected by changes in other TPMs. For instance, changes in range do not affect the angular IFOV as long as the optical system's transverse and longitudinal magnifications remain the same. In the context of optical systems design, the TPMs must be optimized for a specified concept of operations (CONOPS). Optical systems modeling and simulation methods coupled with multiparameter optimization methods help to determine the performance bounds of the optical system and facilitate exploratory modeling and facilitate agile SE methods. In this section, we focus on the TPMs of the detector itself. Specifically, we look at analytically modeling detector noise and evaluating and analyzing the critical characteristics and attributes of the detector. We also present mathematical relationships for establishing the performance limits on optical detectors and discuss basic optical systems testing.

9.2 Optical Systems Building Block: Detector Noise, Characteristics, Performance Limits, and Testing

9.2.1 Introduction to Detector Noise

Noise can be considered as unwanted fluctuations in an electrical signal of a system. While noise is common throughout optical systems, the focus of this chapter is on electro-optical and infrared (EO/IR) optical systems. In terms of EO/IR systems, noise is considered either internal or external to a detector.

External sources of noise that impact EO/IR systems are created either by natural means or by man-made devices. Some examples of natural noise sources include background radiation, scattering of light into the IFOV of the optical system, natural clutter signals, multipath effects, shock and vibration coupling effects from natural sources such as tectonic movement or sea swell, and natural events such as lightning, storms, and local environmental conditions (e.g., temperature, pressure, humidity, altitude, atmospheric effects, rain, snow, dust, and winds). The optical photons themselves are a source of noise in a statistical optics-formulated detection scenario. Man-made noise sources include induced vibrations and shock into the optical system (e.g., motor/generator vibrational coupling or shock and jitter introduced into the optical system by human or machine handling of the optical system), man-made clutter objects such as camouflage in the visible part of the electromagnetic spectrum, and thermal or photonic man-made objects in the infrared part of the electromagnetic spectrum.

In contrast to these external sources of noise, internal noise sources are generated within the optical system itself. For instance, for an IR optical system, the temperature of the components themselves will cause radiation in the IR part of the electromagnetic spectrum. Detector cooling methods can be used to decrease this component of the noise. Vibration and shock coupling to optical components from internal motors introduce jitter and impulsive noise that must be filtered and minimized for proper optical system performance. In this section, we present a mathematical description of the various noise terms associated with EO/IR detectors. We start with a statistical description of noise.

9.2.2 Statistical Description of Noise

Even if all the external and internal noise effects are minimized, the very nature of the photons themselves introduces noise into an optical system. In this section, we present some fundamental statistical concepts needed to properly characterize detector noise. At the heart of the statistical analysis lies the probability density function (PDF). If the PDF is known or if it can be estimated, then the powerful framework of statistical analysis can be adopted in describing and analyzing the detector noise characteristics. Generally, there are two types of PDF models, continuous and discrete. Discrete PDF models typically occur when there are isolated events such as low-light detection scenarios such as photon-counting cameras. In this case, a discrete Poisson PDF is appropriate. In the other extreme such as high-light-level imaging scenarios where there is an abundance of independent photon events, the continuous Gaussian PDF is often appropriate. Since many applications involve high light levels, we start with the Gaussian distribution. Figure 9.5 shows a Gaussian PDF.

The mathematical expression for a single variable Gaussian PDF is given by

$$p(x) = \frac{1}{\sigma\sqrt{2\pi}} e^{-\left(\frac{x-\mu}{2\sigma}\right)^2}, \tag{9.1}$$

where
σ is the population standard deviation
μ is the population mean

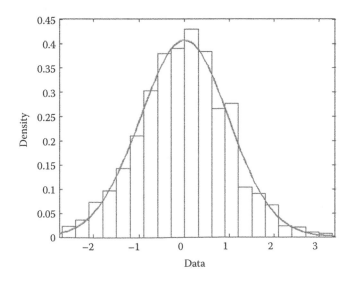

FIGURE 9.5
PDF for a Gaussian distribution.

If x represents an optical statistical parameter of interest, then the probability of occurrence of that statistical parameter is given by $p(x)$. The mean of the population and standard deviation can be calculated from measured data. The estimate of the population mean m can be determined from the commonly known sample mean

$$m = \frac{\sum_{i=1}^{N} x_i}{N},$$
(9.2)

where N is defined as the total number of elements in the data set. The parameter x_i is the ith sample element, and the sample variance s^2 is determined by using

$$s^2 = \frac{\sum_{i=1}^{N} (x_i - m)^2}{(N-1)}.$$
(9.3)

As the quantity of data increases, m approaches μ and s^2 approaches σ^2. As an example, we have included a MATLAB® script found in Appendix 9.A that generates a noisy signal using a pure sine wave and a random noise signal with a Gaussian distribution as shown in Figure 9.6.

In Figure 9.6, the amplitude of the sinusoid was set to 0.5, and the standard deviation of the Gaussian noise was set to 1. A histogram plot of the zero mean sine wave and the random Gaussian noise input would show a normal distribution, with 99% of the sampled values within three standard deviations of the mean. Essentially, with noise that follows a normal distribution, the filtering and signal processing methodology designed to reproduce the original signal is well established and has a high success rate of producing quality original signals, with the majority of the Gaussian noise removed.

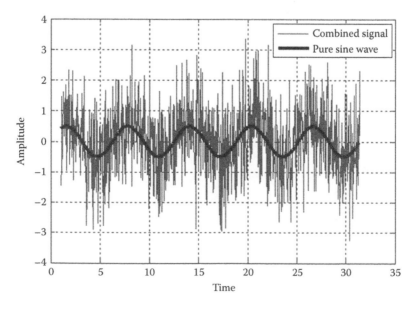

FIGURE 9.6
Noisy sine wave signal.

9.2.3 System Noise and Figures of Merit

Figures of merit (FOMs) also play a critical role in describing detector performance. FOMs are established for devices such as detectors to determine their usefulness and utility. The concept of FOM was introduced in previous chapters. In this section, we illustrate the impact that system noise has on EO/IR detector FOM.

At the system level, some of the more notable FOMs in characterizing the performance of an EO/IR system are the signal transfer function (SiTF), noise-equivalent differential temperature (NEDT), noise-equivalent power (NEP), signal-to-noise ratio (SNR), noise-equivalent irradiance (NEI), responsivity, and detectivity.

- SiTF is the factor that converts an input temperature differential to the detector signal response or the video brightness input to the detector voltage or current response. This function is usually measured as volts/degree, amperes/degree, or counts/degree in an IR optical system or volts/amperes/counts per watt in an EO system. This function is important in finding the NEDT for IR systems.

- NEDT is the input that produces an SNR of one, taken as the smallest signal that the system can differentiate. The NEDT is used in defining the optical system's dynamic range.

- NEP is the ratio of the input power to the output SNR.

- SNR is the ratio of a signal (e.g., voltage, current, and counts from an object of interest) to a noise quantity (e.g., total noise, background-limited noise, and photon noise) at a particular point of reference in the optical system (at the detector input, at the detector output, in the entrance pupil, and so forth).

- NEI is the NEP divided by the active area of the detector.

- Responsivity is the ratio of the detector response (either voltage or current) to the detector input power.

- Detectivity is the inverse of the NEP.

Understanding these FOM provides essential insights into the performance of an optical system. For example, establishing the NEI would provide insights to what signal levels would be required for detecting a potential object of interest. If the irradiance level of the object of interest is below the noise floor, then additional filtering or modulation methods would have to be employed to pull the signal out of the noise floor. Figure 9.7 shows various FOMs throughout the imaging channel from source to display.

In determining the overall imaging system's performance, FOMs related to optical system characteristics such as the ambient noise, environmental factors, the overall range, the quality of the detector itself, the complexity and filtering ability of the signal processing circuitry, and some visual representation of the noise as a function of time or magnitude are depicted in Figure 9.7.

9.2.4 Three-Dimensional Noise Model

A typical 3D noise model includes two spatial dimensions as well as time. Three-dimensional noise models can also have the third dimension be a function of frequency, wavelength as in spectral imaging systems, or even a third spatial dimension as in 3D imaging applications. In many instances, a 2D spatial (EO) or thermal (IR) imaging

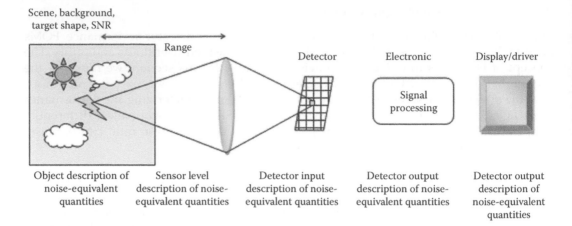

FIGURE 9.7
Top-level IR systems FOMs.

camera that takes sequential image frames over time as in a video camera serves as a representative 3D optical system for understanding 3D noise. In order to understand the modeling of 3D noise, there are several noise components that commonly occur in optical systems. These noise components are listed in Table 9.2, and they constitute what is generally known as the 3D noise model (Holst 1998).

An illustration of the dimensions of the representative 3D noise model can be seen in Figure 9.8.

As the name suggests, there are three dimensions that come into play. The first lies on the y-axis and measures the number of pixels in the y direction. The second is on the x-axis and simply measures the number of pixels in the x direction. Finally, the n-axis lays out the number of frames in the video (time).

An image can be run through a simple program, as seen in Appendix 9.B, to model the effects of the noise components on an image. Temporal noise terms can be determined by evaluating sequential image frames for temporally related noise components like random noise, rain, streaking, and flicker. The effect of this can be seen in Figure 9.9.

It is important to understand that some noise components are measured over time, and so cannot be determined using a single image frame. For example, fixed pattern noise is associated with noise patterns in digital imaging sensors and typically occurs during extended sampling periods and is characterized by an easily recognizable mix of bright- and dark-color pixel arrays. Similarly, column and row noises follow distinct horizontal

TABLE 9.2

3D Noise Model Components

3D Noise Components	Description
σ_{TVH}	Random 3D noise
σ_{VH}	Spatial noise fixed in time, also known as *fixed-pattern noise*
σ_{TH}	Variations in columns over time, also known as *rain*
σ_{TV}	Variations in rows over time, also known as *streaking*
σ_V	Row noise fixed in time, also known as *horizontal lines* or *banding*
σ_H	Column noise fixed in time, also known as *vertical lines*
σ_T	Variations in frame intensity over time, also known as *flicker*

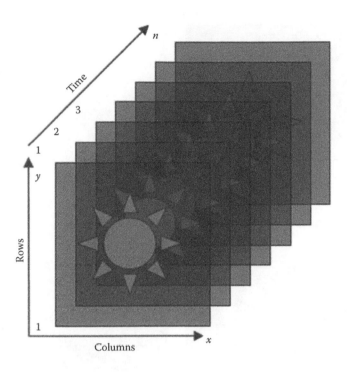

FIGURE 9.8
Noise cube diagram.

or vertical patterns and are a result of imbalanced sampling properties on the detector itself. Digital signal processing techniques are used to remove the respective fixed pattern, row or column noise. A clean reference image is shown in Figure 9.9a.

9.2.5 Sources of Noise

The types of noise that are present in a given EO/IR imaging scenario depends on the type of detector and its operating environment. Optical system observables and components such as the photon flux, the detector, preamplifier, and analog-to-digital (A/D) conversion circuitry are examples of sources of optical system noise. A noise that overwhelms any other type of noise in an optical system is considered the limiting noise. An illustration of limiting noise performance is presented in Figure 9.10.

In Figure 9.10, further to the left, the signal of interest (SOI) is the less noise it experiences. Even the arriving photons from our SOI inherently have random fluctuations and are therefore a source of noise. Noise is often displayed in terms of voltage, current, or electrical power for optical detectors. Using the concept of PDFs as discussed earlier, the variance of noise can be described using the following equation. A random variable (v) is used to describe the noise voltage fluctuation over time:

$$\overline{(\Delta v)^2} = \overline{(v - \overline{v})^2} = \frac{1}{\Delta t} \int_0^{\Delta t} (v - \overline{v})^2 \, dt, \tag{9.4}$$

where Δt is the time interval under consideration and the statistical assumption is that the temporal average is the same as the ensemble average. By assuming that the

FIGURE 9.9
Examples of image noise. (a) Ideal image, (b) example of fixed-pattern noise, (c) example of column noise, and (d) example of row noise.

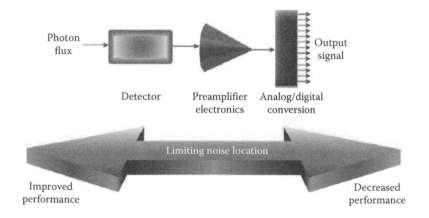

FIGURE 9.10
Limiting noise location performance.

average voltage is equal to zero, the mean square voltage fluctuation equation can be determined, as seen in Equation 9.5. When ($\Delta t = 0$), the autocorrelation of the temporal voltage signal reduces to the average power in the signal (assuming unit resistance) (Bradshaw 1963):

$$\mathcal{R}\left(\Delta t = 0\right) = \int\limits_{-\infty}^{+\infty} v(t)v(t)dt = \int\limits_{-\infty}^{+\infty} v^2(t)dt. \tag{9.5}$$

Equation 9.5 is useful since the average power is an observable quantity even if the underlying random noise voltage signal has zero mean.

9.2.6 Types of Noise

In this section, we expand on some of the more common noise terms that occur in EO/IR imaging systems. As will be seen, some of these noise terms are dependent on temperature, and so cooling techniques may be employed to reduce these noise terms. Other noise terms may be dependent on the optical system's IFOV, and so shielding methods against stray signals may be effective. Pre- and/or postdetection optical modulation methods and filtering techniques can often be used to drastically reduce optical system noise and discriminate low signals from other noise effects. We focus on common noise terms associated with the optical detector itself, specifically

- Johnson–Nyquist thermal noise (Johnson noise)
- Shot noise
- $1/f$ noise
- Generation and recombination (GR) noise
- Popcorn noise
- Radiation or photon noise
- Quantization noise

In the upcoming sections, we look at each of these terms separately and provide some basic information as to the source of the noise term, a mathematical description where practical and any aliases associated with the noise term. We start off with a description of the temperature-dependent Johnson–Nyquist thermal noise term.

9.2.6.1 Johnson–Nyquist Thermal Noise (Johnson Noise)

In 1928, the discovery of noise in a resistor resulted in the development of a mean square voltage equation that described the so-called Johnson noise. Johnson noise is a type of thermal noise that is produced by conductive materials that have microcurrents that are functions of temperature. These conductive materials allow for the movement of electrons, which interact with the detector material. The movement of electrons is what creates a very small current. While the total effect of all of these currents is negligible over time, in short time periods, these current fluctuations cause Johnson noise. Johnson noise depends on the temperature of the detector material and the resistance associated with the detector electrical circuit model. Johnson noise is independent of the specific electrical circuit frequency; however, Johnson noise is a function of the electrical circuit bandwidth. With Johnson noise, current fluctuations in equilibrium occur without any applied voltage and without any average current flowing. These concepts are presented as follows:

$$\left(v^2\right)^{1/2} = v_j = \sqrt{4kTR\Delta f}, \tag{9.6}$$

where
 k is Boltzmann's constant
 T is the detector material temperature in Kelvin
 R is the detector electrical circuit equivalent resistance in ohms
 Δf is the electrical bandwidth of the detector

Figure 9.11 shows that Johnson noise is not a function of any particular frequency but that it is, instead, a function of the electrical system's bandwidth.

As seen in Figure 9.12, Johnson noise can be modeled by an ideal voltage source that generates a noise voltage at the location shown by the voltmeter. The model shown is a Thevenin's equivalent circuit model of the Johnson noise source associated with the detector.

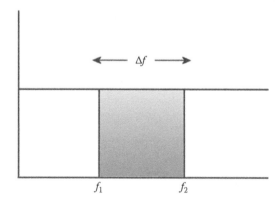

FIGURE 9.11
Johnson noise independent of frequency.

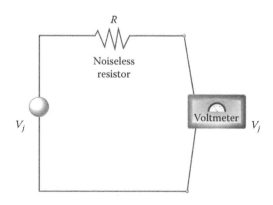

FIGURE 9.12
Thevenin equivalent resistor noise.

In considering the impedance of the detector, Johnson noise is only a function of the resistive aspects of the detector and therefore is not a function of its inductance or capacitance.

A Norton's equivalent circuit of the detector can also be used in a current generator model. The resistive part of an ideal Norton's equivalent detector circuit is shown in Figure 9.13, where i_j is equal to the Johnson's mean squared noise current obtained by dividing the mean squared noise voltage in Equation 9.6 by the detector resistance.

Equation 9.7 is associated with the Norton equivalent noise current i_j and is found by

$$i_j = \frac{\sqrt{4kTR\Delta f}}{R},$$ (9.7)

where R in the denominator is the resistive element associated with the detector impedance. Figure 9.14 shows a Johnson noise circuit with a constantly fluctuating voltage across a resistor–capacitor combination.

Johnson noise is considered "white noise" in the sense that its power spectrum is flat. Since this is the case and as previously stated, Johnson noise is clearly not a function of frequency. However, since the Johnson noise spectrum is flat, an increase in the electrical systems bandwidth would also increase the Johnson noise. Cooling the detector would have the effect of reducing the Johnson noise (Bradshaw 1963).

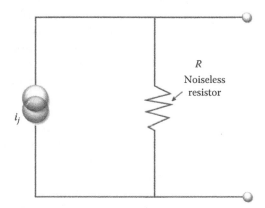

FIGURE 9.13
Norton equivalent circuit. (Courtesy of Christopher Cox. With permission.)

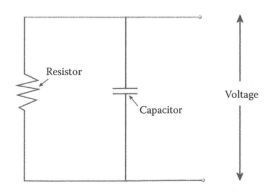

FIGURE 9.14
Johnson noise circuit. (Modified and courtesy of Christopher Cox. With permission.)

Another thermal noise term is present in circuits that have capacitors. Thermal noise in a simple resistive–capacitive (*RC*) circuit is known as *kTC* noise and is a function of the capacitance and the temperature *T* as shown in the following:

$$v^2 = \frac{kT}{C},$$
(9.8)

where
    C in the denominator is the circuit capacitance
    k is the Boltzmann constant
    T is the temperature

Cooling such detectors reduces the *kTC* noise just as in the case of Johnson noise. Increasing the electrical circuit capacitance also reduces the *kTC* noise but adversely affects the time constant of the circuit. In the next section, we look at another noise source that also is a function of bandwidth (Bradshaw 1963).

9.2.6.2 Shot Noise

Like Johnson noise, shot noise is associated with the flow of electrons. An example of shot noise is a vacuum tube, wherein the arrival rate of the electrons is random and consequently is a source of noise at the collecting electrode (Hudson 1969). Discovered in 1918 by Schottky, shot noise is prevalent in electronic and telecommunication devices. A difference between shot noise and Johnson noise is that shot noise has a direct current (DC) flow, and the individual electrons arrive randomly at their destination. In contrast, in Johnson noise, the random motion of the electrons sums to zero, and only variations in the random motion of the electrons over short periods give rise to the Johnson noise. An example of a device that exhibits shot noise is the temperature-limited diode.

Similar to Johnson noise, shot noise has a flat spectrum and is therefore not a function of any individual frequency. Since it reflects a counting process, a Poisson PDF can be used to calculate statistical noise parameters. Similar to the arrival rate of electrons on the collecting plate of a vacuum tube, photons arriving on a photon-counting detector experience shot noise (or photon noise) due to the random nature of the arrival of the photons on the detector. In the following equation, the variable *I* indicates the brightness that is incident on the detector surface:

$$\Delta I^2 = \left(I - I\right)^2.$$
(9.9)

This average brightness term gives rise to an average current due to the emission of electrons from the detector. This current can be described by Bradshaw (1963)

$$\overline{i^2} = 2eI_{DC}\Delta f,$$
(9.10)

where
    the bar over the term on the left side is the time average
    e is the charge on an electron
    I_{DC} is the applied DC current
    Δf is the detector electrical circuit bandwidth

Shot noise is the limiting noise in many high-end optical systems wherein the random arrival rate of the photons—photon noise—introduces a variance term as shown in Equation 9.9. This effect can be seen in photomultiplier tubes and avalanche photo diodes.

9.2.6.3 1/f Noise

One of the more dominant noise terms is the so-called 1/f noise. This noise term occurs at lower electrical circuit frequencies and is, as the name suggests, inversely proportional to the frequency itself. As the frequency becomes larger, this type of noise decreases and is often smaller than other noise terms at frequencies above a few hundred hertz. Expressions for the mean squared noise voltage and/or current are complex. However, the equation for the power spectrum is relatively simple since it is directly proportional to the DC (current raised to a value α near 2) and is inversely proportional to the frequency (raised to a parameter β that ranges from 0.8 to 1.5). Other common names for 1/f noise are contact noise (for carbon resistors), modulation noise (in semiconductors), flicker noise (in vacuum tubes), and excess noise in electrical circuits since it exceeds shot noise at low frequencies.

9.2.6.4 Generation and Recombination Noise

GR noise results from differences in the rate that charge carriers are generated and recombined in the detector material. GR noise occurs in semiconductors and in typical photodetectors and is observable in the intermediate electrical circuit frequency ranges between the 1/f noise–dominated and the Johnson noise–dominated spectral regions. GR noise is flat across the spectrum up to a cutoff frequency that is roughly equal to the inverse of the carrier lifetime. For higher frequencies, the GR noise power spectrum decreases at roughly 6 dB per decade.

9.2.6.5 Popcorn Noise

Popcorn noise can result from a variety of conditions such as surface contamination in semiconductors and defects in manufacturing processes. A distinctive feature is a rapid change in the noise voltage level that can last for many milliseconds (Markov and Kruglyakov 1960). Popcorn noise can also arise during the manufacturing process when high-energy ions are implanted in a material (Pruett and Petritz 1959). Popcorn noise can result from the trapping or release of charge carriers at interfaces and can occur over a short or long duration. The name derives from this type of noise giving off a particular popping sound when hooked up to an acoustic speaker. Other names for popcorn noise are impulse noise, burst noise, bistable noise, and random telegraph signal noise.

Figure 9.15 shows several of the important noise terms and where they dominate in the spectrum. Figure 9.15 shows the total noise power as a function of frequency for the effects of 1/f noise, GR noise, and Johnson noise for a photoconductor.

The plot in Figure 9.15 shows the behavior of the total detector noise voltage as a function of frequency as represented by

$$v_{total,rms} = \sqrt{v_{1,rms}^2 + v_{2,rms}^2 + \cdots + v_{m,rms}^2}. \tag{9.11}$$

In Equation 9.11, each noise term is assumed independent from the other, and so the total noise voltage of the optical device is given by the root of the sum of the squared noise terms of the individual noise components.

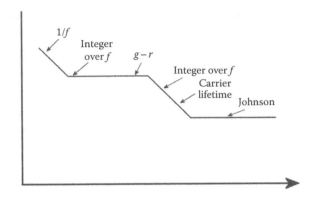

FIGURE 9.15
Prominent photoconductor noise types by frequency. (Modified and courtesy of Christopher Cox.)

9.2.6.6 Radiation or Photon Noise

The limiting form of noise in all optical detectors is photon noise. This noise would exist even if all other forms of noise were eliminated. Photon noise, or radiation noise, is produced when the detector interacts with arriving photons. The arrival rate of the photons itself is random, and so the incident flux has fluctuations that result in an associated noise voltage at the output of the detector. This type of noise is negligible in many commercial applications but can be observed in high-end, cooled optical detection equipment. In high-end optical equipment, such as photon-counting cameras, the arrival rate of the photons can be considered a counting process and so the Poisson distribution is often readily applicable. Consequently, the variance of the noise associated with the signal is equivalent to the mean signal level. In a photon-counting camera, the detection of a photon can be considered a photo event (e.g., the photon is detected and an electron is generated by the detector material). In the case of no other noises besides photon noise, the mean signal level output by the detector is then proportional to the average number of photo events. Consequently, the SNR of a photon-counting camera scales as the square root of the average number of photo events. A graphical illustration is given in Figure 9.16 wherein the mean number of detected photo events is 100. The standard deviation about the mean is equal to the square root of the mean in a Poisson process, and so the standard deviation is seen to be 10 photo events.

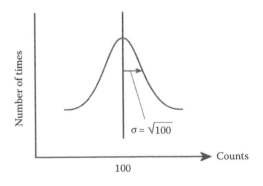

FIGURE 9.16
Distribution of counts around mean.

This observable variation in the arrival of photons would manifest even if all other sources of noise were minimized or eliminated, and so photon noise is the ultimate limit on detector performance.

9.2.6.7 Quantization Noise

Quantization noise is associated with the signal conversion process when converting from a continuous signal to a discrete time signal such as in A/D conversion. Rounding and truncation errors occur because discrete time and digital signal are just approximations of the true, continuous signal (Bradshaw 1963). Rounding errors occur when a continuous value is represented by the nearest quantized level. For example, if the individual quantization levels are uniformly distributed across the continuous signal amplitude (e.g., the voltage step size Δv for each "bin" of the quantizer is the same), then any continuous signal that falls into a particular bin of the quantizer returns the same digital output. For instance, a voltage range given by Δv may define a particular level of the quantizer, and then any voltage between $-\Delta v/2$ and $\Delta v/2$ would produce the same digital output signal from the A/D converter. Figure 9.17 illustrates this concept.

In Figure 9.17, any continuous voltage signal in the interval Δv would produce the same digital output signal. Equation 9.12 provides the magnitude of the maximum possible rounding error produced by an A/D converter:

$$E_R \leq \frac{2^m}{2},$$

(9.12)

where m is the number of bits in the A/D converter (Oppenheim and Schafer 1975).

Truncation errors occur when an equation that requires an infinite sum is approximated with a finite number of terms. For instance, in a sinusoidal series expansion of square pulse, an infinite number of terms are required in the series to fully describe the corners of the object. Truncation error occurs when only a finite set of elements in the series is used to approximate the corner. Equation 9.13 provides the magnitude of the maximum error associated with truncation errors in A/D converters:

$$E_T \leq 2^{-m}.$$

(9.13)

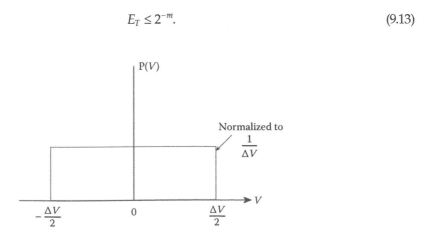

FIGURE 9.17
Uniform distribution for probability of voltage within one hit.

9.2.7 Equivalent Noise Bandwidth

In many of the previous equations, we used the term electrical bandwidth (e.g., Equation 9.6 describing Johnson–Nyquist noise). A mathematical definition and description is in order. Throughout this book and consistent with Hudson (1969), we adopt the following definition for equivalent noise bandwidth:

$$\Delta f = \frac{1}{G(f_{max})} \int_0^\infty G(f)\,df,$$ (9.14)

where

$G(f_{max})$ is the maximum value of the power gain

$G(f)$ is the power gain of the electrical circuit as a function of frequency

The power gain can be obtained by squaring the voltage gain as a function of frequency. The noise-equivalent bandwidth Δf can be found by evaluating Equation 9.14 and then assuming a rectangle with the same height as $G(f_{max})$ and a width that produces the same area under the curve as the power gain (Table 9.3).

9.2.8 Detector Figures of Merit and Performance Characteristics

A vital aspect of a thriving optical systems development business is the presence of, or access to, high-quality, optical testing facilities and equipment as well as sound methods for quantifying the characteristics of optical detectors. This is essential for any organization that wishes to design and integrate optical detectors in any of their projects. By understanding and creating a high-quality test environment with access to a wide variety of detectors and test equipment, effective equipment calibration and optical test measurements can be made. The following set of sections describes various topics including detector performance measures and FOMs, equipment required for measurements, laboratory equipment procedures, and the characteristics of some detectors.

TABLE 9.3

Noise Bandwidth for 3 dB

Number of Poles in Filter	Δf
1	$\left(\frac{\pi}{2}\right)\Delta f_{adB}$
2	$\left(\frac{\pi}{2}\right)^{1/4}\Delta f_{adB}$
3	$\left(\frac{\pi}{2}\right)^{1/6}\Delta f_{adB}$
4	$\left(\frac{\pi}{2}\right)^{1/8}\Delta f_{adB}$
5	$\left(\frac{\pi}{2}\right)^{1/10}\Delta f_{adB}$

9.2.8.1 Detector Figures of Merit

Detector performance is best determined by FOMs associated with important detector characteristics. This allows both a prediction and evaluation of performance by an engineer before a particular component is integrated into a system. Assuming perfect conditions such as performance under noiseless conditions, though idealized, often provides useful performance bounds. In the perfect world, a detector is assumed to be unaffected by external or internal noise. This is where background-limited photon detectors, or BLIP detectors, come into play. BLIP detectors are photoconductive and photovoltaic detectors that are basically unaffected by optical system noise, and their performance is only limited by the background radiation itself.

We will present some results for BLIP detectors later on in this chapter. First, we provide brief descriptions of important FOMs associated with optical detectors.

The first and most general FOM is referred to as the spectral response of the detector. This FOM simply describes the response of a detector as a function of wavelength or optical frequency. In other words, the spectral response of an optical detector refers to how the detector responds to input radiation that is spectrally separated and distinguishable. An example would be a tunable monochromator that is placed in front of a radiometer so that only a narrow band of light falls on the active area of the detector. The detector output current is measured as the monochromator is tuned over wavelengths that produce observable detector output. The resulting curve is the spectral response.

The next FOM is responsivity, which relates optical power that is incident on the detector surface to the output signal. For infrared detectors, a common method is to use a calibrated blackbody source to provide a fixed irradiance on the detector with an active area A. The associated temperature for blackbody reference is set at a desired temperature such as 500 K, and the detector output voltage or current is measured. The ratio of the detector input power to the output voltage or current is known as the responsivity. The responsivity has units of amps/watt or volts/watt and is shown as

$$R(T,f) = \frac{v_s}{p} = \frac{v_s}{HA}, \tag{9.15}$$

where v_s is the root mean square (RMS) signal voltage at the output of the detector with active area A and frequency f in response to incident radiation of RMS power p and irradiance H modulated at frequency f from a blackbody of temperature T. Similarly, $R(\lambda, f)$ represents the spectral responsivity measured at modulation frequency f in response to narrow-band radiation centered about wavelength λ.

A related FOM to the responsivity of a detector is its uniformity. Uniformity is a consistency measure that is important in imaging applications. For instance, in a charge-coupled device (CCD), the uniformity FOM would measure the pixel-to-pixel variations in the detector response for a given input power. The uniformity can be measured either broadband or spectrally and can also be measured over time. Consistent differences between pixel responsivities can be accounted for and compensated during image postprocessing.

One of the more commonly used FOMs is the so-called SNR. The SNR provides a relative measurement of signal strength to that of the strength of noise. There are quite a few variations in this FOM that are application dependent and equally valid. For instance, the SNR can be calculated or estimated at different physical locations in the imaging scenario (e.g., at the entrance pupil, at the input of the detector, at the output of the detector, and at the input of the display device). The SNR can also be calculated within a spectral band or it can be a broadband quantity. With regard to the noise term(s), the SNR can include all practical noise terms or just a few noise terms as would be the case for establishing performance bounds. Consequently, it is proper and practical to provide the specific details regarding which quantities are being evaluated. For instance, one might say the detector output voltage SNR having defined the noise terms or the BLIP power SNR at the input of the detector. Statements such as "We need to calculate the SNR" are ambiguous and ill defined. The SNR FOM is useful in radiometric calculations and provides an indicator whether or not there is sufficient signal present in the imaging scenario to register a detection event. A useful rule of thumb is that a particular SNR of 10 or greater provides sufficient signal strength for most applications. For instance, if the average signal level on a pixel in an imaging camera is 10 times larger than the average total noise of that pixel, then there is typically enough signal present to provide confidence that the detector would "see" the detection event and would be able to discriminate the event from the total noise at the pixel location. It must be noted that the choice of a particular SNR of 10 is just a rule of thumb and that detections can be made with lower SNR. Signals can even be less than the noise and still be reliably detected with optical modulation methods, hardware-based optical filters, and postprocessing filtering methods.

A related FOM to the detector output voltage SNR is the NEP. The NEP is the ratio of the RMS power incident on the detector surface to the voltage SNR at the output of the detector. The incident power on the detector can be determined by multiplying the detector irradiance by the detector's active response area. This FOM can be determined at high signal levels where the discrimination between the output signal and noise is more readily observable. The inverse of the NEP is the detectivity FOM. The lower the NEP, the higher the detectivity will become. The NEP may be reduced by increasing the voltage (or current) SNR at the output of the detector. The detector output voltage (or current) SNR can be increased by minimizing the contribution of noise terms at the output of the detector.

The NEI FOM is readily determined from the NEP by dividing the NEP by the active detector area. As the name suggests, this FOM provides the irradiance level on the detector that would produce an output signal comparable to the noise. The NEI is therefore a good metric for determining lower limits on radiometric quantities associated with the imaging scenario. For example, projecting the NEI to the target space would provide an indicator of the minimum radiant emittance of the source required to produce a signal on the detector equivalent to the noise floor of the detector. In subsequently analyzing the radiometric properties of the source, predictions of whether or not a particular source would produce a detection event are possible.

Another important FOM is D^*. This FOM is related to the detectivity but normalizes out the effects of variations in detector size (e.g., the active area of the detector) and electrical bandwidth of the detector and so provides more of an "apples-to-apples" comparison of individual detector *detectivities*. For the D^* FOM, the assumption is that the detector noise-equivalent bandwidth or electrical bandwidth Δf is 1 Hz and the detector area is 1 cm^2, and so this FOM provides an excellent way to compare the performance of detectors with different areas when used in circuits having different electrical bandwidths. The NEP is related to D^* in

$$D^* = \frac{\left(A_D \Delta f\right)^{1/2}}{NEP},$$

(9.16)

where
  A_D is the detector area in cm^2
  Δf is the noise-equivalent electrical bandwidth in Hz

Since NEP is a FOM that is a function of the active detector area and the electrical bandwidth, the FOM of D^* removes these sensitivities from the detectivity equation and allows for direct comparisons between different detectors. The detectivity, however, provides a better FOM for an individual detector where the active detector area and electrical bandwidth are already established.

Another important FOM is D^{**} (dee double star) that effectively normalizes the IFOV associated with the detection measurement to a solid angle of π steradians (effectively a hemisphere). This FOM is useful in comparing detector performance and is given by

$$D^{**} = D^* \left(\frac{\Omega}{\pi}\right)^{1/2}, \tag{9.17}$$

where the omega term on the left side of the numerator is calculated as the equivalent solid angle for a hemisphere surround. Evaluating this equivalent solid angle in terms of the half-angle θ measured from the detector to the limiting aperture of the optical system, the functional relationship between D^{**} and D^* is given by

$$D^{**} = D^* \sin(\theta). \tag{9.18}$$

If there is no limiting aperture, then the angle θ is $90°$, and the equivalent solid angle is a hemisphere. In this case, any physical aperture does not limit the input irradiance on the detector in the optical system, and the value of D^{**} is the same as D^*.

The response time of a detector is characterized by its response time constant, the time that it takes for the detector output to reach 63% of its final value following an initiating change in the detectors input power. The time constant τ FOM is given by

$$\tau = \frac{1}{2\pi f_\tau}, \tag{9.19}$$

where f_τ is the frequency at which the *responsivity* of the detector is 0.707 of its low-frequency value. Some interesting trades begin to emerge. For instance, in order to increase the detector response time, the electrical bandwidth needs to be increased. However, increasing the electrical bandwidth also increases the detector noise. In the next section, we present some basic components needed to measure these FOMs and to analyze the actual performance of a given optical detector.

9.2.8.2 Basic Equipment for Measuring Detector Figures of Merit

In the previous section, we defined some common FOMs that are useful in determining the performance characteristics of a given detector and also for comparing detectors with each other. In this section, we describe some basic equipment needed to measure these FOMs. The basic test set for EO/IR detectors usually measures the following detector characteristics: signal and noise voltages, electrical system time constant, the detector frequency response, and detector resistance. Figure 9.18 shows a block diagram for the basic configuration and components for an EO/IR detector characterization system. This basic block diagram accommodates IR detectors and includes the following components: a light source (blackbody source for IR detectors), photodetection circuitry, light choppers and their associated control circuits, preamps and amplifiers, shutters, potentiometers, and test equipment such as oscilloscopes and frequency generators.

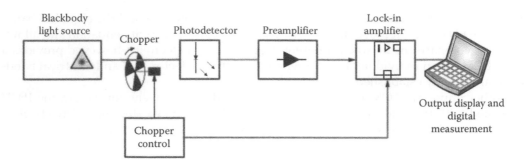

FIGURE 9.18

Basic test configuration of EO/IR detectors. (Obtained from Chris Cox, With permission, 2014; Adapted from Biezl, http://en.wikipedia.org/wiki/File:Lockin_amplifier_experimental_setup.svg, 2010, accessed November 29, 2014.)

The device under test (DUT) is shown to the right of the chopper and includes any of the beam-shaping optics (e.g., filters, limiting apertures, collimators, beam expanders, shutters, polarizers, mounts, potentiometers, vibration isolators, and environmental conditioning equipment to properly conduct the test). We discuss optical testing in more detail in a later chapter but present a preliminary overview of the essential basic components of an optical test set in the next section.

9.2.8.2.1 *Calibrated Light Sources and Controllers*

In order to precisely determine the output of a detector, the power that is placed on the detector needs to be controlled. Calibrated sources serve this purpose. Calibrated sources and calibrated detectors can be obtained from the National Institute of Standards and Technology (NIST). Both calibrated sources and calibrated detectors are available in broadband and spectrally calibrated varieties. For instance, for IR sources, NIST can provide a calibrated blackbody source that precisely generates source radiometric quantities, such as radiance or radiant emittance at a specific temperature and/or tunable over a select temperature range that is traceable back to a calibration standards maintained at NIST. With this source, the precise power incident on the detector can be determined through analysis. Alternatively, a NIST calibrated detector will precisely determine the radiometric quantities incident on its detector surface and is also traceable back to calibration procedures conducted at NIST.

To control the amount of light that falls on a detector, beam-forming and beam-shaping optics are used. Optical beam expanders and collimators are used to produce parallel beams of light with a desired diameter. These beams of light can be matched to the detector surface area to precisely control the amount of power that is incident on the detector surface. An illustration of a collimator is shown in Figure 9.19.

A beam expander, as the name implies, can change the width of the column of light produced by the collimator. Placing a matched converging lens after a diverging lens can make a simple beam expander. The first lens diverges the beam, whereas the matched second lens collimates the beam with a larger beam diameter. Of course, the reverse process is also true in that a smaller diameter beam can be made with the combination of a converging lens and subsequent matched diverging lens.

9.2.8.2.2 *Controlling the Amount of Light with Limiting Apertures*

For the purposes of controlling and decreasing the irradiance on a specific detector, limiting apertures are used (Pedrotti and Pedrotti 1993). Various sizes of apertures are available that range from sizes much smaller than 1 mm (on order of tens of microns or even smaller

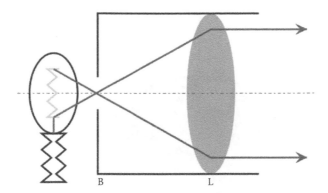

FIGURE 9.19
Collimator. (Obtained from Krishnavedala, http://en.wikipedia.org/wiki/File:Collimator.svg, 2012, accessed November 29, 2014.)

with cooling) to apertures on the order of inches or larger. These apertures are used to limit the optical power incident on the detector and also to precisely control the amount of optical power that illuminates the detector. If the detector is sensitive to the IR part of the electromagnetic spectrum, the limiting aperture may have to be cooled to prevent it from being a thermal source that is detected by the IR camera.

9.2.8.2.3 Chopper Controlling Circuitry

The next component of interest in the basic optical test configuration is the optical chopper control circuitry (Hudson 1969). It is often difficult to establish reliable and repeatable rotation rates over the general operating ranges of these choppers. The goal with this component is to create specific types of waveforms (square or triangular to give some examples). It is in fact difficult to achieve a constant waveform of the chopped pulse, maintain a constant chopped frequency, and also to eliminate the transient noise spikes that the motor and speed controller generate. Dependent on the application, expected rotational frequencies could range from a few Hz for thermal detectors to rotational frequencies in excess of 10 kHz for photon detectors.

Chopping geometry (χ) can be defined as the ratio of the angles subtended by the aperture (θ_α) and by the tooth-slot pair (θ_t):

$$\chi = \frac{\theta_\alpha}{\theta_t}. \tag{9.20}$$

Fourier analysis techniques indicate that a square waveform consists of a fundamental as well as odd-order harmonics. It is known that a limiting aperture is typically circular. Therefore, a good way to approximate a square wave is to cut straight-sided slots into the rim of a disk. The geometry involved is shown in Figure 9.20.

In Figure 9.20, D_α is the diameter of the source-limiting aperture, and D_c is the diameter of chopped disk. When the value of θ_α decreases, we know that the value of D_α will get close to zero and subsequently a square waveform will be produced. The irradiance on the detector is specified in terms of the RMS value of its fundamental component. Ranging from 0.28 to 0.45 peak-to-peak value, the RMS conversion factor is used with relation to the component's RMS value. This of course is dependent upon the chopper type design. Table 9.4 displays each conversion factor as it relates to chopper geometry.

As can be seen from Table 9.4, the RMS value of the fundamental component is 0.45 times the peak-to-peak amplitude of the square wave. The following equation describes the irradiance on the detector when the chopping pattern resembles a square wave (Hudson 1969):

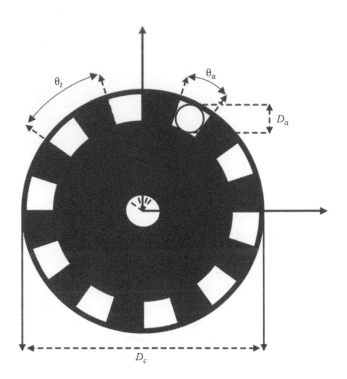

FIGURE 9.20
Square wave chopper geometry. (Courtesy of Christopher Cox. With permission.)

TABLE 9.4

Values of RMS Conversion Factor for Various Chopper Geometries

χ	Conversion Factor (RMS)
0 (square)	0.450
0.05	0.448
0.08	0.445
0.10	0.442
0.15	0.433
0.20	0.421
0.25	0.405
0.30	0.386
0.40	0.340
0.50 (triangular)	0.286

$$H = 0.45 \frac{\sigma T^4 A_s}{\pi d^2}. \tag{9.21}$$

where
    T is the temperature of the blackbody source measured in Kelvin
    A_s is the limiting source aperture that is measured in cm^2
    d is the measured distance between the limiting aperture and the detector element plane
    σ is Boltzmann's constant

When θ_α equal to $\theta_t/2$ and χ is 0.5, a triangular waveform is produced. The RMS value of the fundamental component of a triangular wave is 0.28 times the peak-to-peak amplitude. The equation for χ can be rewritten as follows:

$$\chi = \frac{\eta D_\alpha}{\pi(D_c - D_\alpha)},$$ (9.22)

where η is the total number of teeth on the chopper. In most instances, the assumption is that a square wave is used in calculating the irradiance on the detector. The error in this assumption can be calculated from the percentage difference in the conversion factors in Table 9.4. For instance, the error between a square wave assumption and a χ value of 0.05 is less than ½%. An equation to create a chopper with a maximum number of teeth in which the RMS conversion factor differs from the true square wave by 1% or less is given by (Hudson 1969)

$$\eta = 0.251 \frac{(D_c - D_\alpha)}{D_\alpha}.$$ (9.23)

We can illustrate this concept by using this equation in an example. If given a chopper with a limiting aperture diameter of 0.2 cm and a chopper diameter of 20 cm, then in order to maintain an error less than 1% between the conversion factor of a pure square wave and the conversion factor for the chopper geometry in question, the chopper must have less than 24 teeth.

The potential high speed of the blade of the chopper also poses a safety hazard, and safety enclosures are necessary to prevent serious injury. Variable speed choppers can provide chopping frequencies from a few Hz to the kHz range. Variable speed motors can be used to generate chopping frequencies up to 50 kHz and more. Wave analyzers with automatic frequency control provide relatively the best solution for a stable laboratory chopping frequency.

9.2.8.2.4 Frequency Meter

This component of the basic detector test setup is used in combination with the chopper to determine its rotational frequency. Several motor speed controllers include frequency-metering features. The chopper frequency can vary over the course of different tests and different detectors from a few Hz to very high frequencies such as 50 kHz. The frequency meter is used to monitor the chopper's rotational frequency and helps ensure that no errors are introduced in the test set by uncertainties in the chopper's rotational frequency.

9.2.8.2.5 Shutter

Shutters are used to "turn off" the source so that detector noise properties can be measured. For instance, for IR systems, a shutter would close the opening aperture and block the light from the blackbody source from reaching the detector. The detector can then be used to measure ambient conditions, or if shielded from the environment, the detector provides the noise associated with the detector itself. If the detector is an IR detector, then the detector shield that isolates the detector from its local environment must be cooled, so that the shield itself does not emit an unwanted signal.

9.2.8.2.6 *Mounting Hardware*

At the heart of the performance characterization effort is the DUT itself, the detector that is being evaluated. It is practical to be able to adjust the location and orientation of the detector in its test stand to help with alignment and positioning. By having the ability to precisely adjust the detector's position and orientation, source to sensor separation distances can be reliably determined providing for repeatability in the experiment and consistency in the irradiance levels seen at the detector. Calibrated optical mounting hardware serves this purpose.

9.2.8.2.7 *Noise Isolation and Signal Amplification Component (Preamplifier)*

The preamplifier is a critical test component. The purpose of the preamplifier is to amplify the low-signal-level outputs of the detector. A key consideration in selecting a preamplifier is that the noise introduced by the preamplifier and the rest of the characterization circuit must be much lower than that of the detector itself. Other considerations include low impedance so that low impedance connection hardware can be used to connect the preamplifier with the rest of the readout circuitry.

9.2.8.2.8 *Shielded Enclosure*

A shielded enclosure is necessary to protect the detector as well as other testing components. In order to further ensure that ambient light does not penetrate the enclosure, the inside is painted black. In certain instances, radio frequency (RF) shielding is required, which is accomplished by constructing a metal box or pipe that surrounds the components that require shielding. The inside of the pipe often has baffling that is also painted black to help shield against unwanted light. If RF shielding is required, this box would have metal gaskets and would provide for necessary filtering of the internal components. The leads that enter and exit the enclosure would also be shielded (Hudson 1969).

9.2.8.2.9 *Audio Oscillator and Calibrated Attenuator*

This is a calibration device for generating calibrated voltages in the audio range. These voltages have a known amplitude and frequency and are used in determining the gain between the detector and the output device. Calibrated attenuators introduce a known signal loss into the circuitry. This can be done with voltage dividing circuitry, but care needs to be taken not to adversely affect the operating point of the detector.

9.2.8.2.10 *Precision Potentiometer*

The purpose of precision potentiometer is to work in conjunction with a platinum resistance thermometer or thermocouple to give an accurate measurement of the blackbody cavity temperature as well as the temperature of the compartment of the detector. The potentiometer itself is a way of measuring and/or introducing variable voltage using variable resistance in dependence with a physical slider. The resistor itself is composed of a very thin wire that is tightly wound around an insulating form. It is important to note that a potentiometer does have physical limitations (wearing of the sliding unit) as well as an intrinsic possibility of randomly affecting the output signal (Wheeler et al. 1996).

9.2.8.2.11 *Amplifier*

Since the preamplifier signal is too low for direct application to the final readout device, an amplifier is used. The amplifiers that may be used include those in the audio range due to the fact that most chopping frequencies are in this range. An important consideration is impedance matching when introducing the amplifier (or any other circuit component) into the output detection circuitry. Failing to properly match the impedance can introduce signal loss into the readout circuitry (Wheeler et al. 1996).

9.2.8.2.12 Oscilloscope and Wave Analyzer

There are a variety of oscilloscopes available on the market today. An oscilloscope allows the direct measurement of signal properties such as period, pulse width, rise time, fall time, and bandwidth. Modern oscilloscope include additional features such as triggering, high sample rates, screen capture, multiple instrument synchronization, mixed signal capability, onboard memory, and resident analytical software. It is important to understand that while some of the typical properties can be measured by voltmeters, oscilloscopes are used because they are much more effective when the output of a sensor is changing very quickly. Wave analyzers are voltmeters with a narrow spectral bandwidth that can be tuned over a range of frequencies (up to around 50 kHz).

9.2.8.2.13 Coolers and Refrigerators

Essential elements in determining FOMs for detectors are coolers and refrigerators that control the temperature of the detector. Some detector noise terms are functions of temperature, and so a means for precisely controlling the detector's temperature environment, especially on its focal plane, is necessary. Some common temperatures that serve as reference points for detector performance are room temperature (298 K), the temperature of liquid nitrogen (77 K), and liquid helium (4.2 K). We will discuss coolers and refrigerators more in a later chapter.

9.2.9 Examples of Measuring Detector Attributes, Characteristics, and Figures of Merit

In this section, we discuss how the test set is applied for measuring some of the key detector attributes, characteristics, and detector FOM. For the purposes of illustrating the calibration of a basic detector test set, a lead selenide detector will be the subject of discussion for any following examples. The detector will be cooled to an operating temperature of 77 K.

9.2.9.1 Measuring the Active Area of the Detector

One of the essential detector parameters that need to be characterized is the active area of the detector (e.g., the *detector area*). In imaging cameras, the "fill factor" is also an important quantity. The fill factor is a measure of how much of the active surface is used for readout electronics. If the readout electronics are mounted on the back of the detector active area, the fill factor can be 100%. If they are located on the surface, they need to be accounted for in determining the detector area used in detector response calculations. The fill factor in context of imaging detectors is the ratio of light-sensitive detector area to total detector area. For instance, if a detector has a physical area of 1 cm^2 with a fill factor of 80%, then the available area used in responsivity calculations would be 0.8 cm^2. As discussed in earlier chapters, an optical comparator or a toolmakers microscope may be used for detector area measurements.

9.2.9.2 Characterizing a Detector's Operating Point

When talking about the operating point of a detector, the detectors can be divided into two separate classes: one that requires an external bias supply and one that self-generates the bias. Certain detectors that require the use of an external bias supply

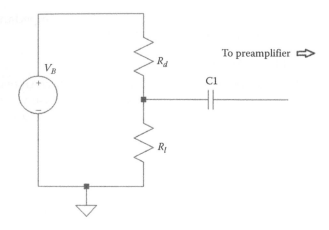

FIGURE 9.21
Method of applying external bias to a photoconductive detector. (Courtesy of Christopher Cox. With permission.)

include the following: photoemitters, bolometers, and photoconductors. A bias voltage is placed in series with the load resistor in these detectors. Figure 9.21 shows the placement of these components. In Figure 9.21, R_d is the variable detector resistance and R_l is the load resistance. Any changes in the incident flux changes the detector's conductivity, which results in changes in the current flowing through the circuit. The presence of the coupling capacitor C1 is to prevent disturbing the signal across the load resistor.

The following provides the voltage across the load resistor:

$$V_{Rl} = V_B \left(\frac{R_l}{R_d + R_l} \right), \tag{9.24}$$

where
  V_B is the bias voltage
  R_l is load resistor resistance
  R_d is detector resistance

Incident flux on the biased detector changes the detector's conductivity. The resultant magnitude of the change in the signal voltage that is present over the load resistor can be obtained by differentiating Equation 9.24 with respect to the detector resistance. The result is shown (Hudson 1969) in the following:

$$V_s = \frac{V_b R_d R_l}{\left(R_d + R_l \right)^2} \frac{\Delta R_d}{R_d}. \tag{9.25}$$

This equation is useful in understanding the effects of the load resistor and bias voltage on the output signal. When the load resistor and the detector match, the maximum signal voltage is generated across the load resistor. Table 9.5 shows the performance characteristics of a lead sulfide (PbS) photoconductor. For this particular photoconductor, detection of IR radiation will be optimum at approximately 2200 nm. When compared to common photovoltaic detectors that generate a current when light strikes their surface, the photoconductive material has a reduced electrical resistance when exposed to light.

TABLE 9.5

Lead Sulfide Photoconductor Specifications

Electrical Specifications		
Detector		PbS
Active area		3.0×3.0 mm^2 (9 mm^2)
Wavelength range	λ	1000–2900 nm
Peak wavelength	λ_p	2200 nm (typ)
Peak sensitivity[1]	$\Re(\lambda)$	2×10^4 V/W (min)
		5×10^4 V/W (typ)
Rise time[2] (0–63%)	τ_r	200 µs
Detectivity[3] (λ_p, 600, 1)	D^*	$1 \times 10^{11} \frac{cm * \sqrt{Hz}}{W}$ (typ)
Dark resistance	R_D	0.25–2.5 MΩ
Bias voltage	V_B	100 V
General		
Package		TO-5
Operating temperature		−30°C to 65°C
Storage temperature		−55°C to 65°C

Source: Reprinted from Thor Labs, http://www.thorlabs.us/thorcat/24700/
FDPS3X3-Manual.pdf, p. 8, 2014c, accessed November 29, 2014. With
permission.

[1] Measured at chopping frequency of 600 Hz, Bias Voltage of 15 V, RD = RL.
[2] Rise Time is measured from 0 to 63% of final value.
[3] Measured at chopping frequency of 600 Hz, Peak Wavelength.

The introduction of light produces an increase in electron mobility, which lowers the detector resistance and causes a measurable voltage change. This is why the responsivity (peak sensitivity) is expressed in units of V/W. The responsivity is graphically shown in Figure 9.22.

9.2.9.3 Characterizing a Detector's Operating Point That Requires Bias

For a detector that needs bias, the operating point can be determined using a test set that contains many high-precision load resistors of low noise. The DC bias supply will be capable of providing a DC voltage that can be varied from 1 to 500 V. It is important to ensure that the test set does not have any added signals that would be produced by a DC load supply. The resistance of the detector can be determined by mounting the detector to the test set and measuring the voltage across the load resistor. The measured load voltage and the known load resistance permit the calculation of the detector load resistance. After determining the detector resistance, a matched load resistor is typically used to determine the optimum detector bias. The purpose of this is to develop the relationship between noise and signal voltages as they relate to bias. Successive signal measurements are made at various detector bias levels. Turning off the signal and measuring the noise at each bias level make corresponding noise measurements. Optimum bias is that value of bias that maximizes the SNR measured at the output of the detector. By identifying the bias that produces the highest SNR at the detector output, the detectivity is maximized.

Figure 9.23 shows representative SNR, noise voltage, and signal voltage as the bias current changes. Note that at low bias current, the noise values increase slower than the signal values. As bias current increases, the noise values increase much faster than the signal values.

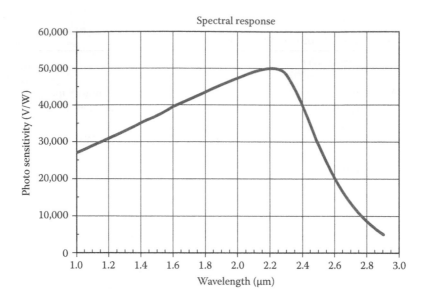

FIGURE 9.22

Peak sensitivity of a lead sulfide photoconductor. (Obtained from ThorLabs with permission, 2014a: http://www.thorlabs.us/thorcat/24700/FDPS3X3-Manual.pdf, p. 9, 2014, accessed November 29, 2014.)

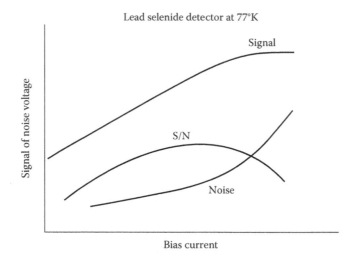

FIGURE 9.23

Signal and noise voltages of a lead selenide detector.

9.2.9.4 Introducing a Calibrated Voltage into the Test Configuration

Figure 9.24 illustrates how a calibrating voltage can be introduced into the circuit containing a photoconductive detector and the bias supply. The calibrated attenuator (shown at the bottom of Figure 9.24) develops calibrated voltages across the 100 Ω resistor. The larger kilo-ohm resistor provides a proper load for the calibrated attenuator. Compared to the resistance values R_d and R_l, the voltage that is calibrated is tiny so as to not perturb the condition of the bias that has been set. In this fashion, calibrated voltages can be introduced into the detector circuitry.

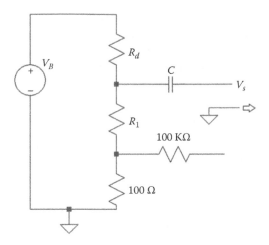

FIGURE 9.24
Method of introducing a calibrating voltage into a detector circuit. (Courtesy of Christopher Cox. With permission.)

9.2.9.5 Determining the Effects of the Chopping Frequency on the Detector Output Signal

In this test the chopping frequency is varied, and the effect this has on the detector output signal voltage is determined. Initially, a low chopping frequency is set (say 30 Hz), and a wave analyzer is tuned into this frequency. The chopping frequency is increased, and the tuning process with the wave analyzer is repeated. The detector output voltage is plotted relative to the initial low-frequency detector output voltage response. The response looks relatively flat out to some cutoff frequency after which the detector output voltage begins to decline.

9.2.9.6 Measuring Spectral Noise Characteristics

This section describes how the spectral characteristics of the detector noise can be determined. The wave analyzer is once again an excellent device for this application since it has an exceedingly small noise bandwidth. In this test, the sources are not turned on or used, and any potential unwanted signal from the source is blocked. The wave analyzer tunes through the identical frequencies that were used in determining the spectral signal response. The result is then divided by the square root of the wave analyzer noise bandwidth, and the root power spectrum is obtained.

9.2.9.7 Determining the Detector Time Constant

The response time of a detector is the time it takes the detector to respond to changes in the input power. When a variation of incident flux is induced, the detector response time is not instantaneous. Recall that the responsivity is the metric that describes the detector response as a function of input power. The equation that follows shows how the responsivity at low frequencies (R_0) varies as a function of the chopping frequency f:

$$R_f = \frac{R_0}{\left(1 + 4\pi^2 f^2 \tau^2\right)^{1/2}}. \tag{9.26}$$

The chopping frequency can be incrementally increased, and a wave analyzer is used to determine at what chopping frequency the responsivity falls to 0.707 of its initial value. When the responsivity drops to 0.707 of its initial value, the associated frequency f_1 is used to determine the detector response time (τ). Namely, τ is just the inverse of this radial frequency ($2\pi f_1$).

9.2.10 High-End Photon Detectors

In this section, we consider high-end photon detectors in terms of their D^* and D^{**} FOM. By high end, we mean that sufficient development has occurred to eliminate production process issues and that the associated excess noise term is eliminated and that the photon detector performance matches relatively well with theoretical predictions. The noise terms that remain after excess noise is eliminated are usually $1/f$ noise and GR noise. By employing a chopper with frequencies higher than the 500–1000 Hz range, the associated $1/f$ noise can be minimized. GR noise results from fluctuations in the charge carriers. Cooling the detector can minimize the GR noise by reducing the number of charge carriers produced by lattice vibrations, and consequently only photon noise remains. If there is no signal present on the detector, the photons come from the surrounding background environment or clutter. We assume that there are no clutter signals and that proper shielding is present to prevent stray photons from reaching the detector. In this case, only background photons cause a photon detection event, and we have a BLIP detector. In a background-limited photoconductive detector, the fluctuations in the rate at which the background photons arrive give rise to fluctuations in the rate at which the charge carriers are generated. During recombination, photovoltaic detectors do not show these fluctuations, thereby reducing the limiting noise to that of 40% of the photoconductive detectors (Hudson 1969).

The theoretical value of D^* for a background-limited photoconductive detector at a particular wavelength is given by

$$D_\lambda^* = \frac{\lambda}{2hc}\left(\frac{\eta}{Q_b}\right)^{1/2},\tag{9.27}$$

where
η is the quantum efficiency
Q_b is the photon flux from the background

If we substitute for h (Planck's constant) and c (velocity of light) and express λ in microns, we get

$$D_\lambda^* = 2.52 \times 10^{18} \lambda \left(\frac{\eta}{Q_b}\right)^{1/2}.\tag{9.28}$$

For a photovoltaic detector, recombination noise is absent. So, this equation is increased by the square root of 2 giving

$$D_\lambda^* = 3.56 \times 10^{18} \lambda \left(\frac{\eta}{Q_b}\right)^{1/2}.\tag{9.29}$$

These two equations are valid for the calculation of D^* at the spectral peak. The spectral peak is the same as the cutoff wavelength for photon detectors. The units for Q_b are photons cm^{-2} s^{-1} and Q_b is a broadband quantity.

For a detector that is considered ideal, the quantum efficiency would be equivalent to unity. For photon detectors that are commercially available today, the quantum efficiency η can vary from 0.1 to 0.4. Modern, high-end, single-photon-counting detectors, however, can have quantum efficiencies in the vicinity of 90% or experimental versions even higher. For instance, a laser-interferometric gravitational-wave detector that can measure distance changes on the order of 10^{-18} m RMS from separation distances of 1 km reported a quantum efficiency of 93% (Goda et al. 2008).

9.2.10.1 Shielding the Detector against Radiation

Radiation shielding is significant in that, if applied, it can raise the D^* value of a background-limited detector by dramatically reducing the background photon flux. For radiation shielding, one can apply a methodology of using either a cooled shield or cooled filter.

The purpose of using a cooled shield is that it is done with a big enough limiting aperture to allow a conical flux onto the optical device. The goal is to position the detector in this guard such that signal photons are the only things received. By doing so, the background photon flux is significantly reduced due to the size of the aperture. To the detector, D^* is now a function of the angular field of view, and the quantity D^{**} accounts for this dependence.

As opposed to using cooled shields, using cooled filters is a method that, as the name suggests, uses a cooled optical filter directly in front of the detector. The spectral bandpass of this filter is chosen so as to reject as much of the background photon flux as possible, while rejecting minimum target flux. Assuming the system is to be within the atmosphere of the earth, the atmospheric transmission needs to be considered in determining the filter bandpass. In Figure 9.25 shown, the relationship between temperature and performance

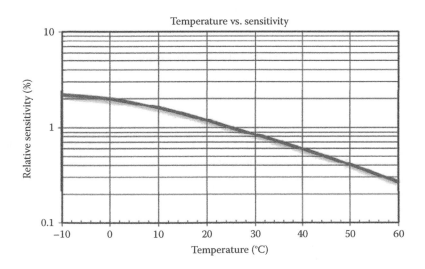

FIGURE 9.25

Temperature and sensitivity relationship of a PbS photoconductor. (Obtained from ThorLabs, http://www.thorlabs.us/thorcat/24700/FDPS3X3-Manual.pdf, p. 11, 2014b, accessed November 29, 2014. With permission.)

is clearly evident. As temperature increases, the sensitivity of a photoconductor decreases. The effort of cooling and shielding is critical to optical engineering as are the mitigation of noise and the management of the radiation flux.

9.2.11 Thermal Detectors and Haven's Limit

The performance of a thermal detector is calculated in two stages. The purpose of the first stage is to derive the incident radiation as related to a change in temperature by considering a system's thermal characteristics. In the second stage, the thermal change is related to the change in the output signal of the detector. The corresponding change in the output relative to the change in input irradiance determines the responsivity of the detector. While the calculations associated with the first stage are common for any thermal detector type, this is not true for the second-stage calculations because of variations in detectors.

By studying these various detectors, Haven's limit was derived, not as an intrinsic limit, but rather an estimation created by engineers. This value is used to quantify the performance of a thermal detector. Haven's limit is quite conservative and has not been exceeded by any room temperature bolometers or thermocouples. Haven originally concluded that the lowest possible energy that can be detected is identical for any type of thermal detector. Haven's limit is given by

$$(\Delta P)(\tau) = 3 \times 10^{-12}, \tag{9.30}$$

where ΔP is the smallest optical power that can be detected in watts, τ is the time (s) of the illuminating pulse, and the active detector area is considered 1 mm^2. The units of Haven's limit are joules. The quantity ΔP can be converted to the NEP by normalizing the detector area to 1 cm^2 and equating τ to the response time of the detector:

$$\text{NEP} = 3 \times \frac{10^{-11}}{\tau}. \tag{9.31}$$

Recalling the definition of D^* from this equation and repeated here for convenience,

$$D^* = \frac{(A_d \Delta f)^{1/2}}{NEP}, \tag{9.32}$$

the active area of the detector is assumed to be 1 cm^2, and so does not affect the value of D^* in this case. For the thermal detector, the relationship between the bandwidth and the time constant is often (Hudson 1969)

$$\Delta f = \frac{1}{4\tau}. \tag{9.33}$$

Substituting these values into the equation for D^*, the value of D^* for Haven's limit is calculated to be

$$D^* = \left(1.67 \times 10^{10} \tau^{1/2}\right). \tag{9.34}$$

9.2.12 Important Considerations in Selecting a Detector

For optical systems, an important systems design consideration is the selection of an appropriate detector. Apart from considering the different FOMs discussed previously, the following are some points to consider when selecting a detector (Hudson 1969):

1. Mission or application
2. Atmospheric transmittance (if applicable)
3. Detector type: thermal (e.g., thermistor, thermopile, and Golay cell) and photon detectors
4. Detector material such as lead sulfide (PbS), mercury cadmium telluride (HgCdTe), gallium arsenide (GaAs), silicon, and their spectral response curves
5. Operating mode: thermoelectric, bolometer, and gas expansion; photovoltaic, photoconductive, and photoelectromagnetic
6. Useful wavelength range ($\Delta\lambda$)
7. Wavelength of peak response (λ_p)
8. Noise levels and noise spectrum
9. Linear response range of detector
10. Detector cooling (size, weight, and power permitting)
11. Expected detector output signal levels (dynamic range)
12. Detector resistance (Ω) and readout circuitry
13. Packaging
14. Maintenance
15. Reliability
16. Cost

This list is intended as a collection of macroscopic considerations that should be undertaken by the development team. The biggest driver is the mission or intended application. Ultimately, the detector needs to provide reliable information that satisfies the intent of the stakeholders. The mission and the atmospheric transmittance will determine the type of detector that needs to be used. The noise characteristics in relation to the expected signal levels along with size, weight, and power considerations will determine if cooling is necessary and/or possible. Comparing datasheets of various manufacturers can greatly simplify the selection process for the systems engineer. However, in some cases, the optical systems engineer may need to order samples from the manufacturer and make measurements under conditions that simulate those expected in the final system. The detector resistance and readout circuitry are important in getting the signals from the material to the signal processing circuitry. The readout electronics also contribute to the overall optical system noise and are often separately categorized and quantified (e.g., the number of readout noise electrons). Other electrical components such as amplifiers, A/D converters, buffers, and digital-to-analog converters are used in the signal processing function that manipulates the detector output so that it is compatible with the display, storage, or transmission mechanisms. Signal processing will be discussed in a later chapter. The reliability and maintainability of the detector along with its cost are also import drivers in detector selection. In the next section, we apply the concepts presented in this chapter to an integrated case study involving the design of an optical system on an unmanned aerial vehicle (UAV) platform for a border patrol mission scenario.

9.3 UAV Case Study Application

This section will demonstrate typical dialog among the development team and stakeholders. Note, during the conversations that occur in this case study, facts are thrown around as being common knowledge of the speakers. However, in order to show factual integrity, they are explained or cited within the case study for accuracy. Also, note that although this is a hypothetical depiction of various types of engineer–stakeholder interaction, the scenarios are quite often encountered and can be learned from. Also, attempts have been made to be as realistic and accurate where possible, but some design parameters, which are not readily available, were manufactured. The intent is to demonstrate the process and concepts of this chapter with realistic examples, but, as always, facts should be checked and numbers verified.

As seen throughout the case study already expanded on in other chapters, Fantastic Imaging Technology is tasked by the Department of Homeland Security to design a human bipedal traffic monitoring system that is required to be operational in daytime and nighttime environments (Department of Homeland Security 2010; Jeffry 2011). In this section of the case study, the engineers at FIT are faced with developing TPMs and associated metrics that can be used in predicting the performance of the detector and the optical system and that are useful for analyzing, modeling, evaluating, and ultimately selecting the detector itself.

The following characters are involved in this scenario: Tom Phelps, FIT chief technical officer (CTO); Karl Ben, FIT senior systems engineer; Jennifer O. (Jen), FIT systems engineer; Ron S., FIT systems engineer (new hire); Amanda R., FIT optical engineer; Christina R. (Tina), FIT optical technician, and Phil K., FIT software engineering.

(We find Ron holding on to a stack of papers and fumbling for the door to the optics lab while his coffee mug teeters dangerously.)

Jen: "Give me that before you drop it. Amanda will skin us if we mess up her lab."

Ron: "Thanks Jen. I should probably get myself a man purse or something. Hey Amanda! Nice digs!"

Amanda: (Looks up) "Hey Ron, Jen, come on in! The detector characterization results are just in!"

Jen: "Great! How do things look?"

Amanda: "Actually, they look pretty good. Come over here and I'll walk you through it. Ron, you need to set down your mug of coffee over there, no food or drinks around the equipment."

Ron: "Great…."

Jen: "Did you run the tests using the nominal mission parameters yet? Karl and Tom were interested in those to establish a sensor performance baseline. Every time I see Karl, he gives me his famous 'Is it done yet?' look."

Amanda: (Laughs) "Yeah, we just ran those. Christina is putting the parameters into our optical systems model. Oh, and Phil said that our new detector module now integrates with the other model-based systems engineering (MBSE) tools so you should now have the latest detector parameters available to you."

Jen: "That's great! Phil is still working on our modeling piece in systems to identify and link the results from your integrated optical systems model to the requirements. He is also using the integrated MBSE tools to dynamically update and provide status on our measures of effectiveness (MOEs), measures of performance (MOPs), key performance parameters (KPPs), and technical performance measures and metrics (TPMs). Ron and I are supposed to understand your detector characterization process and make sure the appropriate effectiveness and performance metrics get into our MBSE tools and enterprise architecture (EA) artifacts correctly."

Amanda: "Nice! OK, we will walk you through the infrared detector characterization. Tina, can you give them a rundown of what you did?"

Tina: "Sure! We reconfigured the geometry to simulate the nominal conditions that Karl and Phil wanted. The numbers we were told to simulate were 5-mile separation from sensor to target at an angle of 45 degrees. That puts the height of the UAV at 5689.9 m or about 18667.7 ft. Since we are looking at the nighttime system, we looked at the spec sheets and found that HgCdTe detectors had the best D^* values in the spectral region that had the peak radiance for a human. We had a few HgCdTe detectors left over from another project, and so we borrowed one for this preliminary characterization study."

Ron: "Can we use these detectors for our project?"

Amanda: "No, they don't have the pixel density that we need, but the material properties are close enough for us to give us a good performance estimate for the radiometric aspects of the detection process. Based on the modeling and characterization results, we can then conduct a trade study to see if we want to use a commercial-off-the-shelf system (COTS) or build our own. We will need your help in systems to do that trade study. But anyways, Tina, keep going."

Tina: "Right. Anyways, for the human, bipedal detection and classification mission, an infrared photovoltaic imager is what's needed. We need to be able to spatially resolve the target in the long-wave infrared (LWIR) part of the electromagnetic spectrum. Of course, we are not testing for that right now, but the human detection mission requirement coupled with the need to search the scene drove us to look at photon detectors (specifically imaging detectors) instead of a thermal detector. The detector material, HgCdTe, has a spectral pass band from 6 to 15 μm and peaks at 10.6 μm. We looked at the atmospheric transmission window in a narrow band around the human body peak wavelength of 9.5 μm. Running our models for an effective human body temperature of 305°K, we are assuming nighttime detection with a reduced effective body temperature because of cool environmental conditions and clothing. We used a narrow-band filter from 8.75 μm to 9.5 μm. One sec, let me put this on the screen."

(Tina's fingers fly over her iPad Mini and the results pop up on the flat screen on the far wall.)

Tina: "Our models gave us the following relevant radiometric and basic detector results assuming a diffuse object with a 2 m² emitting surface area:

- Radiance at peak: 10.59 Watts/(m² sr μm)
- Radiance (blackbody): 153.07 Watts/(m² sr)
- Radiant emittance: 480.89 Watts/m²
- Narrow-band filter range (in band range): 8.75 μm–9.5 μm
- In band radiance: 7.90 Watts/(m² sr)
- In band radiant emittance: 24.82 Watts/m² (in band)
- In band optical power (radiant flux): 49.64 Watts (in band)
- In band radiant intensity: 15.8 Watts/sr (in band)
- In band atmospheric transmittance: 0.77

- Sensor look angle: 45 degrees
- Sensor-to-target separation: 8046.7 m
- Entrance pupil diameter: 20 inches (0.508 m)
- Entrance pupil area: 0.203 m^2
- Entrance pupil irradiance (free space; without [with] look angle effect): 244.02 [172.55] nWatts/m^2
- Optical power on detector (free space; without [with] look angle effect): 49.54 [35.03] nWatts
- Entrance pupil irradiance (in atmosphere; without [with] look angle effect): 187.90 [132.86] nWatts/m^2
- Optical power in entrance pupil (in atmosphere; without [with] look angle effect): 38.14 [26.97] nWatts
- Transmittance of imaging system optics: 0.975
- Optical power on detector (in atmosphere; without [with] look angle effect): 37.19 [26.3] nWatts
- IR detector responsivity: 4×10^4 Volts/Watt
- IR detector signal output (without [with] look angle effect): 1.49 [1.05] mV
- IR detector NEP: 0.089 nWatts
- Active detector area: 1 cm^2
- Electrical bandwidth: 16 Hz
- Detectivity (D): 1.124×10^{10} Watts^{-1}
- $D^*(\lambda_{9.5}$ μm, 77 K, 100): 4.5×10^{10} cm^2 Hz$^{1/2}$ Watt^{-1} (measured value)
- Detector electrical resistance: 35 ohms (measured)

We are only showing the radiometric and basic detector functions. We have separate modules for visualization, scanning, spatial resolution, and other optical analysis, but these give a good overview of our results."

Ron: "How did you get all these numbers from modeling or testing?"

Tina: "Both. Some we got from our basic test setup here. Those are shown as measured. The rest, we determined analytically using our modeling software. I'll walk you through the modeling results."

Jen: "Great!"

Tina: "We first calculated the amount of power radiated by a human body by using

$$P = \varepsilon \sigma A T^4, \tag{9.35}$$

where σ is the Stefan–Boltzmann constant, A is the surface area of the radiator, T is the temperature of the radiator, and ε is the emissivity (0.98 for a human body). For a human body with an effective temperature of 305 K and an area of 2 m^2, we get a power of 961.78 Watts and a radiant emittance of 480 W/m^2. To help reduce noise, we used a narrow-band filter, between 8.75 μm and 9.5 μm, and determined the amount of optical power in this band of wavelengths to be 49.64 Watts. The radiant emittance in this region is 24.82 W/m^2, and the radiant intensity is

15.8 Watts/sr in this band of wavelengths (in band). For a diffuse target, the irradiance falls off as the cosine of the look angle, and so we get a radiant intensity in the direction of the simulated UAV optical sensor of 11.17 Watts/sr (in band). The following equation gives the irradiance at the entrance pupil of the imaging system:

$$H = \frac{J}{r^2}. \tag{9.36}$$

Following the calculation for this equation, $H = 11.17 \text{ W}/(8046.7)^2 \text{ m}^2 = 172.6 \text{ nW/m}^2$.

Looking at Figure 9.26, we see the effect of the atmosphere on different wavelengths of the spectrum. From that, we find that the actual irradiance in the infrared part of the electromagnetic spectrum is approximately $0.77 \times 172.6 \text{ nW/m}^2 = 132.9 \text{ nW/m}^2$.

The flux at the entrance pupil is calculated as

$$\phi = H \times A_{ep}, \tag{9.37}$$

where H is the irradiance in the entrance pupil and A_{ep} is the area of the entrance pupil. To find the area of the entrance pupil, we know that the entrance pupil aperture is circular with a radius of 20 inches (0.508 m), and so the entrance pupil area is $A_{ep} = 0.203 \text{ m}^2$. Consequently, the optical power in the entrance pupil

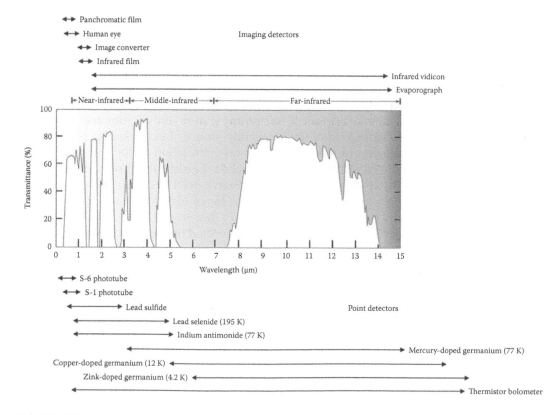

FIGURE 9.26
Atmospheric transmittance as a function of wavelength.

is 132.9 nW/m² × 0.203 m² = 26.98 nW. The transmittance of the optical system optics is 0.975, and so the optical power that falls on the detector is 26.31 nW. The chosen detector has a measured $D*$ value of 4.5×10^{10}. From that, we can find the NEP using

$$D* = \frac{\left(A_d \Delta f\right)^{1/2}}{NEP}, \tag{9.38}$$

where A_d is the active area of the detector = 1 cm² and Δf is the measured electrical bandwidth of 16 Hz,

$$NEP = \frac{\left(A_d \Delta f\right)^{1/2}}{D*} = \frac{\left(1\,cm^2 \times 16\,Hz\right)^{1/2}}{4.5 \times 10^{10}} = 0.089 \times 10^9\,W = 0.089\,nW. \tag{9.39}$$

The NEP value found in this equation is very low compared to the optical power on the detector that corresponds to the target. Therefore, we have ample signal to detect the human target for this nominal detection scenario. We are still working on including the individual detector noise terms like photon noise, shot noise, $1/f$ noise, and popcorn noise, but the NEP gives us what we need for now.

Next, we simulated the wide-field-of-view tip and queue system by adding a square wave chopper to the test set. We used a calibrated, tunable blackbody source, narrow-band filters, neutral density filters, and apertures to simulate the signal level in the entrance pupil plane and checked entrance pupil and detector plane optical power levels using a NIST traceable detector. The resulting optical power on the detector P_d is

$$P_d = \alpha 0.45 \frac{\sigma T^4 A_s}{\pi d^2} = 26.31 \text{ nW}, \tag{9.40}$$

where σ is Boltzmann's constant, T is the temperature of the calibrated source, A_s is the exit slit aperture area of the blackbody source, d is the source to entrance pupil plane separation in meters, and α is an attenuation factor that reflects the combination of filters and apertures to simulate the expected power on the detector. The size of the spot was made smaller than a pixel to simulate the large field of view for the tip and queue system, and the factor of 0.45 accounts for the chopper. The calibrated detector was used to verify that we did, indeed, have the right power level placed on the HgCdTe pixel. We verified that we would have a detectable modulated signal present. The chopper frequency we used was 100 Hz.

Next, we determined the time constant of the detector. We first measured the low chopper frequency responsivity and obtained 4.0×10^4 V/Watt. Next, we increased the chopper frequency until the responsivity dropped to 0.707 of its original value. This occurred at a frequency f of 318 Hz. We then used

$$\mathfrak{R}_f = \frac{\mathfrak{R}_0}{\left(1 + 4\pi^2 f^2 \tau^2\right)^{1/2}} \tag{9.41}$$

to solve for the time constant τ and obtained 0.5 ms. In this equation, R_f is the responsivity at the chopping frequency f (318 Hz), R_0 is the responsivity at zero frequency (4×10^4), and τ is the time constant of the detector (0.5 ms)."

Jen: "Wow, Tina. Nice summary! It looks like we are OK on the radiometric side. We now need to do the trade study to see if we can find a HgCdTe imaging sensor with the right combination of radiometric and imaging qualities available on the commercial side or if we will have to build our own."

Ron: "Imaging qualities?"

Amanda: "Yes. Things like pixel uniformity, fill factor, pixel dimensions, pixel area, and dynamic range."

Ron: "Oh, I see. Don't forget to look at the detector's reliability, maintainability, and availability metrics!"

(Jen smiles.)

Amanda: "That's why we have you systems folks to help us!"

Ron: "Looks like there are plenty of TPMs that we can use. All the figures of merit (FOMs) such as detectivity; D^*; responsivity; NEP, along with the detector technical parameters such as the pixel size, the effective focal length, detector area and dimensions; entrance pupil diameter and so forth all work together to determine the overall performance of the optical system. I'll start to organize these and link over to them from the model we are developing in systems. Some of these TPMs will feed into the KPPs. As you know, KPPs are system performance attributes that are critical to mission success. To be successful, the optical system must be able to 'see/detect' the human target at all expected ranges, and it must be able to provide information of sufficient quality to the operator to classify the target as a human. In order to do this, there are two considerations: (1) the minimum detectable signal that corresponds to a human target at maximum range, and worst detection conditions and assumptions, and (2) the spatial resolution capability of the optical system to correctly classify the human. Each one of these considerations could be turned into a KPP with a minimum threshold value and a target value (e.g., nominal conditions). I will start working on getting the TPMs into our MBSE tools and EA artifacts.

Amanda: "What EA artifacts?"

Ron: "Well, for instance, a consolidated, organized, dynamic, and linked set of our MOEs, MOPs, FOMs, KPPs, and TPMs for this program along with links to requirements, design, and test artifacts were appropriate. The KPPs all link to MOEs and both the MOEs and KPPs generate the MOPs. The TPMs are typically more detailed, technical measures of performance that stem from the MOPS and often directly relate to requirements."

Jen: "All right, let's go ahead and wrap things up. I have to let Karl know how the test went. Thanks for running us through all this!"

(Ron glares at his cold cup of coffee.)

Acronyms

CCD	Charge-coupled device
FIT	Fantastic Imaging Technology
HOQ	House of quality
HST	Hubble Space Telescope
IR	Infrared
KPP	Key performance parameters

MOE	Measures of effectiveness
MOP	Measures of performance
MTTF	Mean time to failure
NEDT	Noise-equivalent differential temperature
NEI	Noise-equivalent irradiance
NEP	Noise-equivalent power
PDF	Probability density function
QFD	Quality function deployment
RNC	Reflective null corrector
SiTF	Signal transfer function
SNR	Signal-to-noise ratio
TPM	Technical performance measure
TPMs&Ms	Technical performance measures and metrics

9.A Appendix: MATLAB® Code 1

```
%% Gaussian Distribution Example
% Author: Carlos J. Rivera-Ortiz
% Date: June 12, 2007
% Course:  SYS5380, IR Systems Engineering
% Instructor: Dr. William Arrasmith
%
%% Generate a 1000-elements of random number added to a sine wave
n=1000;
%% Generate the x-axis of n-elements in radians
a = linspace(1,10*pi,n);
%% Generate random numbers with a Gaussian distribution
b = randn(n,1);
%% Generate a pure sine wave of amplitude +/-0.5
c = 0.5*sin(a)';
d=b+c;
%% plot the combined random numbers
figure(1)
plot(a,d)
hold on
plot(a,c,'k','Linewidth',3)
grid
hold off
xlabel('time')
ylabel('amplitude')
legend('Combined signal','Pure sine wave')
%% Display a histogram of the random numbers with a bin-size of 25
figure(2)
hist([d b],15)
legend('combined signal','pure random signal')
s = std(b)
```

9.B Appendix: MATLAB® Code 2

```
%% Video Noise Example
% Author: Carlos J. Rivera-Ortiz
% Date: June 17, 2007
% Course:  SYS5380, IR Systems Engineering
% Instructor: Dr. William Arrasmith
%
%% load image into Matlab workspace
%% Image source: Wikipedia
photo=imread('Lenna.png');
%% Convert to gray scale
photo=rgb2gray(photo);
photo=im2double(photo);
[x y]=size(photo); %% get image size
%% Display images
subplot(2,2,1)
imshow(photo)
title ('Ideal Image')
%%
subplot(2,2,2)
photo_vh=1+0.2*randn(x,y);
photo_vh=photo.*photo_vh;
imshow(photo_vh)
title('Example of fix-pattern noise')
%%
subplot(2,2,3)
noise=1+0.15*randn(1,y);
line_noise=repmat(noise,x,1);
photo_v=line_noise.*photo;
imshow(photo_v)
title('Example of column noise')
%%
subplot(2,2,4)
noise=1+0.15*randn(x,1);
line_noise=repmat(noise,1,y);
photo_h=line_noise.*photo;
imshow(photo_h)
title('Example of row noise')
```

References

Allen, L., J.R.P. Angel, J.D. Mongus, G.A. Rodney, R.R. Shannon, and C.P. Spoelhof. 1990. The Hubble Space Telescope optical systems failure report. NASA report. Washington, DC: NASA.

Bahill, T. 2004–2009. *Technical Performance Measures*. Tucson, AZ: University of Arizona.

Behn, R.D. 2003. Why measure performance? Different purposes require different measures. *Public Administration Review*, 63(5): 586–606.

Blanchard, B.S. and W.J. Fabrycky. 1990. *Systems Engineering and Analysis*, Vol. 4. Englewood Cliffs, NJ: Prentice Hall.

Bradshaw, P.R. 1963. Improved checkout for IR detectors (Detector test measurements essential for stability and uniformity in design, manufacture and checkout procedure of airborne infrared detectors). *Electronic Industries*, 22: 82–86.

Cask05. 2006. *House of Quality*. http://en.wikipedia.org/wiki/File:A1_House_of_Quality.png (accessed November 29, 2014).

Cox, C. 2014. *Basic test configuration of EO/IR detectors*. Adapted from Biezl [2006]. http://en.wikipedia. org/wiki/File:Lock-in_amplifier_experimental_setup.svg (accessed November 29, 2014).

Defense Acquisition University. 2001. *Systems Engineering Process*. Ft. Belvoir, Washington DC: Defense Acquisition University Press.

Department of Homeland Security. 2010. Fact sheet: Southwest border next steps. Washington, DC: Department of Homeland Security. http://www.dhs.gov/news/2010/06/23/fact-sheet-southwest-border-next-steps (accessed March 6, 2014).

Goda, K., O. Miyakawa, E.E. Mikhailov, S. Saraf, R. Adhikari, K. McKenzie, R. Ward, S. Vass, A. J. Weinstein, and N. Mavalvala. 2008. A quantum-enhanced prototype gravitational-wave detector. *Nature Physics*, 4: 472–476.

Guerra, L. 2008. *Technical Performance Measures Module*. Austin, TX: University of Texas at Austin.

Hagan, G. 2009. *Glossary of Defense Acquisition Acronyms and Terms*. Ft Belvoir, VA: Defense Acquisition University Press. http://www.dau.mil/pubscats/pubscats/13th_edition_glossary.pdf (accessed November 29, 2014).

Holst, G.C. 1998. Testing and evaluation of infrared imaging systems. In *Testing and Evaluation of Infrared Imaging Systems*, Vol. 1, G.C. Holst, Ed. Winter Park, FL/Bellingham, WA: JCD Pub./ SPIE Optical Engineering Press.

Hudson, R.D. 1969. *Infrared System Engineering*. New York: John Wiley & Sons.

Jeffrey, T.P. 2011. Federal auditor: Border patrol can stop illegal entries along only 129 miles of 1,954-mile Mexican border. http://www.cnsnews.com/news/article/federal-auditor-border-patrol-can-stop-illegal-entries-along-only-129-miles-1954-mile (accessed March 6, 2014).

Kasunic, K. 2011. *Optical Systems Engineering*. New York: McGraw-Hill.

Krishnavedala. 2012. *Collimator*. http://en.wikipedia.org/wiki/File:Collimator.svg (accessed November 29, 2014).

Lichiello, P. and B.J. Turnock. 1999. *Guidebook for Performance Measurement*. Seattle, WA: Turning Point.

Markov, M.N. and E.P. Kruglyakov. 1960. Zonal sensitivity of PbS photoconductors. *Optics and Spectroscopy*, 9: 284.

Mitchell, J.S. 2010. Physical Asset Management Handbook. Key Performance Indicators. London, England: Springer-Verlag, pp. 319–322.

Oakes, J. 2004–2005. *Technical Performance Measures*. San Diego, CA: BAE Systems.

Oppenheim, A.V. and R.W. Schafer. 1975. *Digital Signal Processing*. Englewood Cliffs, NJ: Prentice-Hall, Inc.

Pedrotti, F.L. and L.S. Pedrotti. 1993. *Introduction to Optics*, 2nd edn. Upper Saddle River, NJ: Prentice Hall.

Polidore, M.J. 2002. F-22 raptor—A transformational weapon that continues to meet and exceed all key performance parameters. Farnborough, U.K.: Lockheed Martin Aeronautics Company.

Pruett, G.R. and R.L. Petritz. 1959. Detectivity and preamplifier considerations for indium antimonide photovoltaic detectors. *Proceedings of the IRE*, 47(9): 1524–1529.

Roedler, G.J. and C. Jones. 2005. Technical measurement. A Collaborative Project of PSM, INCOSE, and Industry, INCOSE -TP -2003 -020 -01. Practical Software and Systems Measurement (PSM) and International Council on Systems Engineering (INCOSE) Measurement Working Group. https://www.incose.org/ProductsPubs/pdf/TechMeasurementGuide_2005-1227.pdf (accessed November 29, 2014).

Tangen, S. 2003. An overview of frequently used performance measures. *Work Study*, 52(7): 347–354.

ThorLabs. 2014a. *Peak Sensitivity of a Lead Sulfide Photoconductor*. Rev. B, September 9, 2014, p. 8. http://www.thorlabs.us/thorcat/24700/FDPS3X3-Manual.pdf (accessed November 29, 2014).

ThorLabs. 2014b. *Temperature and sensitivity relationship of a PbS photoconductor*. Rev. B, September 8, 2014, p.11. http://www.thorlabs.us/thorcat/24700/FDPS3X3-Manual.pdf (accessed November 29, 2014).

ThorLabs. 2014c. *Lead Sulfide Photoconductor Specifications*. FDPS3X3 *Lead Sulfide Photoconductor Users Guide*. Rev. B, September 8, 2014, p. 8. http://www.thorlabs.us/thorcat/24700/FDPS3X3-Manual.pdf (accessed November 29, 2014).

Wheeler, A.J., A.R. Ganji, V.V. Krishnan, and B.S. Thurow. 1996. *Introduction to Engineering Experimentation*. Englewood Cliffs, NJ: Prentice Hall, p. 159.

10

Functional Analysis and Detector Cooling

Form and function should be one, joined in a spiritual union.

—Frank Lloyd Wright

Systems engineering (SE) uses "an iterative process of translating systems requirements into detailed design criteria" through the use of functional analysis (Blanchard and Fabrycky 2011). SE is "an interdisciplinary approach and means to enable the realization of successful systems" (INCOSE 2004). Through the SE process "the functional, performance, interface, and other requirements identified through requirements analysis are transformed into a coherent description of system functions that can be used to guide the design synthesis activity that follows. In order to make this happen, the designer will need to know what the system must do, how well it should be done, and what constraints will limit design flexibility" (INCOSE 2004). Functional analysis helps to define these three needs by breaking down all higher-level functions into lower-level functions, arranging the functions in a logical sequence, and identifying the resources required by the functions. Functional analysis creates the functional architecture, the map that serves as the baseline for all of the subsequent design activities. Functional analysis is used to develop functional block diagrams (FBDs). FBDs are used to iteratively break complex systems functions into successive levels of detail. FBDs can also be used to show relationships between system elements and identify the resources needed at each part. Many of the SE processes such as reliability analysis, maintainability analysis, human factor analysis, maintenance, support, producibility, and economic analysis are directly dependent on inputs from functional analysis.

This chapter explores the concept and process of functional analysis, with references to associated tools and techniques that demonstrate the value of using the functional analysis process within a variety of engineering disciplines. A demonstration of functional analysis on an existing optical system has also been included to aid the discussion. Subsequently, this chapter applies the processes of functional analysis in a case study of an unmanned aerial vehicle (UAV) optical system for use in aerial surveillance and detection of humans traveling by foot. As a representative example, this chapter will cover the functional analysis of the thermal control system for the optical sensors within the UAV optical system. We will cover the UAV optical systems design in this chapter with the continuation of the case study from previous chapters involving the fictitious company called Fantastic Imaging Technologies (FIT) whose members find themselves in the need of accomplishing a functional analysis. Overall, this chapter is organized into the following major sections:

- Section 10.1 provides an introduction to functional analysis discussing its value, process, and some of its tools. Included in this section is a discussion on the use of the three main tools of the functional analysis process: concept diagrams, functional flow block diagrams (FFBDs), and FBDs. The section closes with an illustrative example describing the use of the functional analysis tools discussed in previous subsections for conducting functional analysis of a lenticular imaging system.

- Section 10.2 contains the technical background necessary to understand detector cooling concepts and methods in relation to detector noise. This section provides the theory behind detector cooling, detectors, and coolant. It then explores different cooling systems to include open-cycle refrigerators, closed-cycle refrigerators, and solid-state refrigerators.

- Section 10.3 ties the previously discussed concepts together with a case study concerning a UAV optical system that is intended for a U.S. border patrol application.

10.1 Functional Analyses and Their Requirements

The SE process starts with the problem definition step. Problem definition is the most important part of the development process. Poorly defining the problem will result in rework, which historically has an impact on cost and schedule. Norman Augustine in his famous "Augustine's laws" stated that "Any task can be completed in only one-third more time than is currently estimated" (Augustine 1983). The conceptual design phase of a project converts stakeholder needs into requirements, which will be the basis for the systems development (Blanchard and Fabrycky 2011). When identifying "what the system needs to do," the first requirements that come to mind are almost always functional requirements. However, there are many other design considerations when developing requirements. These include affordability, producibility, quality, reliability, maintainability, disposability, technological feasibility, social/political feasibility, environmental sustainability, expandability, and interoperability. Human factors and safety also need to be considered. The resulting requirements that are based on these considerations are called nonfunctional requirements.

A function can be defined as a discrete action that is necessary to satisfy the problems and needs of the stakeholders. Because functions are essential in meeting the needs of the stakeholders, they directly relate to the requirements. Functional requirements at any requirements level are synthesized from higher-level requirements. The functional analysis process translates the higher-level functions into expanded functions at the lower level that will support the generation of more detailed requirements. It is an iterative process used to define hierarchical levels of functions and functional requirements, increasing the level of detail to the extent required in order to meet the needs of the stakeholder. The goal of functional analysis is to be as concise as possible, yet complete.

It is important to understand the desired end result of the functional analysis process. The end product of the functional analysis becomes the input into the subsequent stages of the SE process. This means that the functional analysis plays a major role in analysis of the system modularity or "packaging," reliability, maintainability, producibility, disposability, life-cycle cost, and so forth. For this reason, the functional analysis and functional decomposition should be performed with the entirety of the system in mind.

Blanchard stresses the importance of describing all aspects of a system within the system requirements such that the functional requirements capture all of the functionality required during the systems life cycle (Blanchard and Fabrycky 2011). Additionally,

International Council on Systems Engineering (INCOSE) emphasizes that functional analysis involves the external interface and system; subsystems; and components, assemblies, and parts (INCOSE 2004). In the components section, INCOSE extends the reach of functional analysis by referring to personnel requirements in terms of functional skill sets and the definition of nonpersonnel resource requirements (INCOSE 2004). During the process of functional analysis, the functions that fully support the operational needs are identified.

As the SE process continues, the system is "architected" by assigning functions to architectural components and subsystems. This process is known as functional allocation, and it straddles the line between functional analysis and requirements allocation (Kamrani and Azimi 2010). In other words, it is a top-down approach where the functions and requirements are decomposed into greater detail to establish subsystems (Project Management Institute 2013). Functional analysis begins as a top-down process in the conceptual design phase and ultimately, as the team reaches the preliminary design phase, continues from the bottom up (Blanchard and Fabrycky 2011).

This approach to engineering a system, dealing initially with a top-down process followed by a bottom-up process, may be a departure from established practices, so it is worthwhile to consider the purpose and the value of the processes. The goal of functional analysis is to depict the overall system from a functional viewpoint and to expose the functional requirements of each component. This is accomplished by systematically decomposing the system into smaller functional components. Each function along the hierarchy of functions has its own requirements and functional allocations. Thus, each component has technical performance measures (TPMs) that were allocated from the system's TPMs. A comprehensive functional analysis considers the entire systems life cycle. The functional analysis process presents great advantages in terms of identifying gaps in the initial systems concept, use case scenarios, and initial requirements. This information and detail will enable subsequent design activities to be executed efficiently and with lower risk.

The Project Management Body of Knowledge (PMBOK) also stresses the need for tools that enable cross discipline and cross-cultural communication (Project Management Institute 2013). The functional analysis process provides additional value here by enabling improved communication between systems and across disciplines (Blanchard and Fabrycky 2011). It is an effective tool that yields readily understandable work products for stakeholders in a variety of disciplines: development engineering, test engineering, as well as other nontechnical stakeholders.

Using easily understandable work products, such as the FFBDs, will better enable the execution of a system's successive decompositions. Studies by Arlitt discuss the benefits of using functional analysis with a focus on establishing a functional baseline as introduced by Hirtz (Hirtz et al. 2002; Arlitt et al. 2011). They concluded that using a functional analysis provides improved support in the development of sound architectures in subsequent design steps. Work by Eriksson describes the functional architecture by using a use case realization technique, which blends traditional use case analysis with functional analysis (Eriksson et al. 2006). Eriksson also emphasizes that the process of functional analysis provides a useful basis for communication and also discusses in his paper its effectiveness in a case dealing with the development of software-intensive defense systems.

10.1.1 Model-Based Systems Engineering Tools for Creating Functional Analyses

The importance and value of functional analysis become very apparent in an enterprise architecture, where each subdivision or department integrates to form the enterprise. As discussed earlier, functional analysis is a progressive elaboration of a system's functions and, thus, is used in all the design phases of the system, that is, the conceptual design phase, the preliminary design phase, and the detailed design and development phase. Eventually, functional analysis through all the design phases helps manifest documentation for the entire system that is readily understood by all the stakeholders. Blanchard refers to the ideal state where the SE process is executed with tools that fully integrate the entire workflow for the process (Blanchard and Fabrycky 2011). These tools exist for all engineering disciplines and are at various levels of maturity. One such example is in the field of software engineering, where there are integrated tool suites that are intended to integrate SE processes, from defining system requirements to software testing processes.

10.1.1.1 Rational Tool Suite

As an example, one such tool suite that is readily available for the fields of systems and software engineering is IBM's Rational tool suite. This tool suite is a standard in model-based SE tools and has a variety of useful integrated systems architecting, systems/software development, requirements management, configuration control and management, and testing components. A few of the more well-known components are System Architect for enterprise architecture applications, the Dynamic Object-Oriented Requirements System (DOORS), Synergy and Clear Case for software configuration management and change control, Rhapsody that provides a UML modeling environment, and Functional Tester that is an automated software testing tool. The tool suite can be used for the functional decomposition of a system and the generation of functionally based models during the preliminary design phase (Hoffmann 2009). As the SE process continues, the same models can be refined and used to automatically generate code to prototype as well as to generate usable code for the final design. As a result, the boundary between requirements analysis, functional analysis, and stages of detailed design and development becomes a continuum. The IBM Rational Suite delivers a comprehensive set of integrated tools that use SE and software engineering best practices and span the entire systems and software development life cycle (Rational Software Corporation 2002). The Rational Suite by IBM also improves communication within teams and departments and thus reduces development time and improves the quality of the final system/software. Systems and software engineering tool suites that provide similar functionalities also exist from other vendors as well. For instance, No Magic provides MagicDraw, which is a UML/SysML-based architecting, collaboration, business process modeling, and software and systems modeling tool with requirements management, Department of Defense Architectural Framework (DODAF), and Ministry of Defense Architectural Framework (MODAF) plug-ins, among others.

10.1.1.2 SysML

SysML is a widely distributed modeling language used by systems engineers for a broad range of applications. In the 2010 INCOSE Conference Tutorial, Paredis described SysML as an integrated set of languages that could be used to represent system components (Paredis 2011). SysML helps aid the development of functional descriptions of a system by providing constructs for behavior diagrams, activity diagrams, sequence diagrams, use case diagrams, and block diagrams. It has multiple constructs or graphical tools that will help create diagrams to enable teams to decompose high-level functionality into lower-level functional descriptions. Ultimately, this decomposition can be converted into a model that will help in selecting design alternatives. Paredis also comments on the use of spreadsheets in analyzing systems claiming that their use is error prone and can impact one's business (Paredis 2011).

10.1.1.3 Theory of Inventive Problem Solving

Bonnema discusses the use of the theory of inventive problem solving (TRIZ), an invention by a soviet inventor and science fiction author Genrich Altschuller, as means of capturing the higher-level functions in a system (Bonnema 2006, 2010). Here, Bonnema primarily focuses on the process of developing architectural design; however, he also stresses and discusses the importance and need to perform functional analysis before beginning the architectural design. He further explains through his papers that functional analysis creates a common understanding and helps provide specific information for all the system requirements and assures the availability of this information. He describes the use of TRIZ in creating a link between functional elements of the system and performance measures by using coupling matrices. Similar kinds of correlation matrices are also employed in the use of the quality function deployment (QFD) method to describe the relationships between the functions and means of delivering the functionality and hence marking a start to functional allocation in a system (Kinni 1993).

10.1.1.4 Simulink® and MATLAB®

Simulink®, a simulation and model-based design tool in MathWork's MATLAB®, "is a block diagram environment for multi-domain simulation and design" (MathWorks 2013). Simulink consists of a range of toolkits that are designed to cater to modeling of linear and nonlinear systems. Simulink has a set of libraries that can be customized, solvers that provide the ability to modeling and simulating static and dynamic environments (both continuous and discrete), and a very useful drag-and-drop graphical editor. It has a graphical front end making the process of modeling and development easy and intuitive. In addition to its utility in design, modeling, and simulating the technical systems aspects, it has the ability to automatically generate code, and it has the capability to employ continuous testing and verification of embedded systems. There have been continued efforts toward creating more integrated MATLAB-based tools to help reduce cost of development and produce reusable components (Brisolara et al. 2008; Shi-Xiang 2010). Brisolara, who has investigated the ability to integrate the Unified Modeling Language (UML) sequence diagrams into MATLAB Simulink models (Brisolara et al. 2008), conducted one such study. This helps ease the process of moving from the functional analysis of a system to the system design. Tools like these allow engineers to work with a domain of matrix mathematics and then move to creating executable code (Banerjee et al. 2000). Kennedy states, "There is a need to create standard libraries for basic functions that are required to transition these system models to executable and fast software (Kennedy 2005)."

10.1.2 Functional Analysis Tools

With the help of suitable tool sets, teams within an organization will be able to directly connect functional analysis to feasibility studies where design concepts and alternatives can be assessed using modeling tools. There are also capabilities to connect the functional analysis output directly to the development process; however, there can be constraints based on the technology involved.

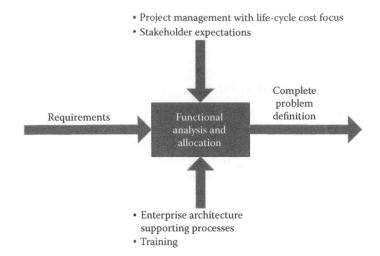

FIGURE 10.1
Concept diagram for functional analysis wording. (Courtesy of Andrea S. Rivers, with permission.)

Simple tools such as spreadsheets, word documents, and graphical drawing tools are not recommended for executing the functional analysis and allocation process, since studies conducted in 2005 showed that 94% of them contained errors (Panko 1998). However, these can be employed for use when more advanced and integrated tools are not available. Irrespective of the level of automation of the tools that are used for functional analysis, these tools provide a means of converting the system requirements into a complete problem definition at a high level (INCOSE 2004). Figure 10.1 depicts example inputs, outputs, controls/constraints, and mechanisms of the functional analysis and allocation process.

Functional analysis activities can generate many different types of work products. As suggested by the INCOSE Handbook, it is a good practice to have more than one form of documentation as a work product. This chapter further focuses on the use of concept diagrams, FFBDs, and block diagrams along with the functional analysis and allocation process as a means of describing an optical system in a fictional case study scenario. In the following section, we will describe the process of functional analysis and allocation using FFBDs, before discussing the example case study. The various potential forms of diagram types and functions that are useful in functional analysis and allocation are listed in Table 10.1.

10.1.2.1 Concept Diagrams

Concept diagrams are typically used to capture and describe the highest-level functions during the conceptual design phase; however, a concept diagram can be constructed for any block of an FBD (Blanchard and Fabrycky 2011). Concept diagrams clearly depict the inputs, outputs, controls/constraints, and enabling mechanisms of a function that are evaluated during functional analysis. An example concept diagram is depicted in Figure 10.2.

Any of the systems design considerations previously discussed could be constraints and controls if they impose limitations on the function. These could include technical, environmental, and interoperability considerations, for example, does it violate the laws of physics, does it have to operate in extreme environmental conditions, and does it have to be backward compatible with legacy equipment, respectively? Any resources required by

TABLE 10.1

Diagram Types and Functions

Diagram Type	Function
Behavior diagrams	Describe stimulus and response exchanges. The focus is on the relationship between events and functions in terms of time and conditions (Bernard 2005).
Concept diagrams	Show the inputs, outputs, mechanisms, and constraints for each of the functions.
Control flow diagrams	This refers to an entire class of diagrams including box diagrams, functional flow diagrams, IDEF charts, PERT charts, state transition diagrams, and more.
Data dictionaries	Used to define the data elements, structure of databases, and the flow of data through the system (Bernard 2005).
Data flow diagrams	Help in depicting the interfaces between components and the behaviors of the components. These diagrams are hierarchical in nature and can be used in conjunction with concept diagrams (Bernard 2005).
Entity relation diagrams	Depict the logical relationship between the functions and architectural components of the system (Bernard 2005).
Functional and concurrent thread analysis	Provides a graphical indication of the response path a system would assume based on the supplied initial conditions. This type of an analysis is used to a great extent especially in software-based systems (Kharboutly 2008).
Functional flow block diagrams (FFBDs)	Associate a hierarchical set of system components and their behaviors while providing graphical detail to depict the flow between components (Blanchard and Fabrycky 2011).
Integrated definition diagrams (IDEFs)	Process control signals are shown at the top of each function block, while inputs that are related to supporting mechanisms enter from the bottom. The inputs that are acted upon enter from the right and outputs exit from the left of each functional block.
Model-based systems engineering (MBSE) diagrams	Act as abstractions and are used to simulate system components. The output of models can be used to assess the capability of a proposed design.
N-squared charts	Describe the complexity of a systems functions and components by showing the interrelation between each of them (Simpson 2009).
Timing diagrams	Provide graphical detail of the time duration involved with functional operations and in between functional operations as well.

the function are referred to as enabling mechanisms of the function. These could include human and computer resources, materials, facilities, utilities, maintenance, and support (Blanchard and Fabrycky 2011).

10.1.2.2 Functional Flow Block Diagrams

FFBDs depict the flow between the functional components of the system and flow based on logical conditions. In an FFBD, blocks with solid lines represent all the functions. The flow of the diagram is generally from the top to bottom and left to right. Connecting each of these functions with horizontal lines depicts sequential functions. Parallel and alternate paths are typically depicted using circles containing logical operations such as "and" or "or" and typically diagonal lines connecting the blocks. Some FFBD, such as those depicting maintenance-related processes, use "Go" and "no go" labels to indicate a true state or false state path following a logical function. Figure 10.3 shows the described FFBD constructs. If the outcome of the function is true, then the "go" path is chosen and executed, and if the outcome is false, then the "no go" labeled path is chosen and executed.

Numbers are used to provide a unique identifier to each functional block. This numbering is not an indication of the order of operation; it is only to aid an observer in following

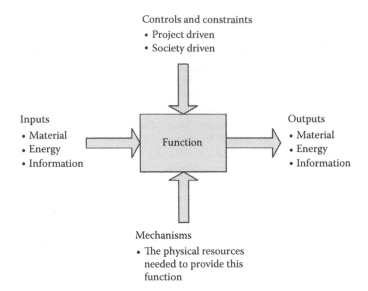

FIGURE 10.2
Classic concept diagram. (Courtesy of Andrea S. Rivers, with permission.)

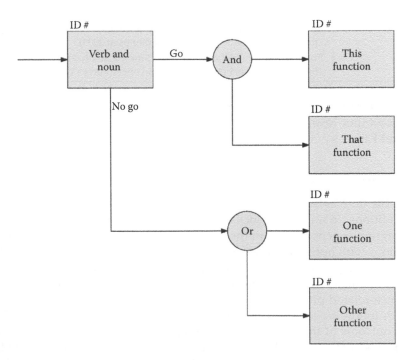

FIGURE 10.3
Flow block diagram constructs. (Courtesy of Andrea S. Rivers, with permission.)

the decomposition of the functional analysis. Each functional block of an FFBD can be decomposed into lower layers of greater detail, until the intended function of the system is adequately described. While decomposing each function, a systematic technique must be used to number the decomposed functions to maintain clarity. If the primary function is numbered as 2.0, the decomposed functions directly under it will be numbered as 2.1, 2.2, 2.3, and so on. This technique supports and aids in better understanding of the hierarchical nature of the functional analysis process.

As this hierarchical process continues, the functional blocks are aggregated based on the discretion of the development team. This partitioning of functional blocks into various groupings (e.g., subsystems) is based on many conditions such as location, minimization of interaction between parties, minimization of interfaces to reduce complexity, or emphasis on system reuse. The purpose of aggregation of functions is to start building the systems architecture.

10.1.2.3 Integrated Definition Diagrams

The acronym IDEF stands for Integrated DEFinition and it refers to a family of modeling languages. IDEF0 diagrams are used for functional modeling and they have certain diagram construction rules and a methodology for defining the way they have to be generated (Department of Commerce 1993). The National Institute of Standards and Technology (NIST) set the diagram construction rules and methodology for generating IDEF0 diagrams.

10.1.3 Functional Analysis Application to a Lenticular Optical Imaging System

In this section, the process of functional analysis using concept diagrams and FFBDs is illustrated using a lenticular imaging system as an example. First, the concepts behind the lenticular imaging system are explained. Afterward, the functional analysis process described in the previous section is applied to the imaging system.

10.1.3.1 Lenticular Optical Imaging System

The term "lenticular" refers to any material with a series of rounded ridges patterned on one side. The use of images delivered on lenticular media allows a viewer to see a series of images using a single-image card. A simple top and side view drawing of a representative piece of lenticular media is shown in Figure 10.4. Each of these rounded ridges is termed "lenticules" and acts as the lenses for the imaging system.

Lenticular images appear frequently in everyday objects, such as the seemingly "animated" images on prizes in breakfast cereal boxes or early advertising signage. A lenticular image is actually an interleaving of multiple images that are placed on the flat side of the media and viewed through the lenticules. The lenticular images that appear in cereal prizes and billboards are of relatively low quality. Higher-quality lenticular images require precise alignment of the lenticular material and the images that are viewed through them. Morton explains the fundamental principles of image delivery using lenticular media (Morton 2000). Referring to Figure 10.5, one can see that this is a simple process, whereby the lines of each image are sequentially placed in the composite image file.

For an accurate composite image (where the dimension and alignment are effectively perfect), the shape of the lenticule must be well controlled and the lenticule must be accurately registered with the image. Afterward, if the card is rotated from a suitable viewing

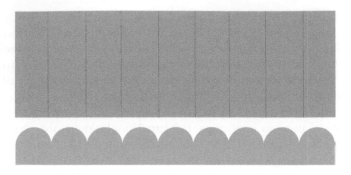

FIGURE 10.4
Lenticular media side and top view. (Courtesy of Andrea S. Rivers, with permission.)

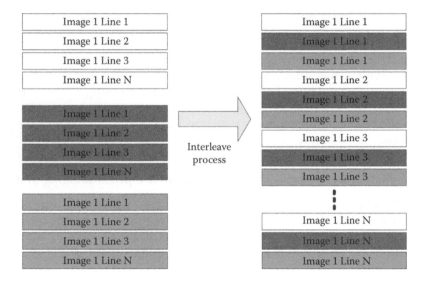

FIGURE 10.5
Image interleaving. (Courtesy of Andrea S. Rivers, with permission.)

distance, it will result in a selection of lines from one of the images such that the viewer can see one of the original images.

As indicated by the graphic of the eye in Figure 10.6, two viewing angles exist. The first viewing angle only allows viewing of the selections of image lines corresponding to the image represented by the dark shading. Similarly, the second viewing angle only allows the selection of lines corresponding to the lighter-shaded image to be seen. As an alternative to viewing the image from different angles, the process of viewing the image can be made easier by rotating the image media to achieve a similar effect. This allows the viewer to see the sequence of images that were previously interleaved under the lenticules. It is this effect that animates cereal boxes and other marketing materials. Cobb's study states "30 image lines could be interleaved and printed with each lenticule provided imaging lines are sufficiently small relative to the lenticular pitch of 0.015 inches" (Cobb et al. 2004).

The optical system in these lenticular imaging systems is designed to print images directly on to sensitized plastic lenticular media (Cobb et al. 2004). An illustration of the sensitized lenticular material is shown in Figure 10.7. The figure represents structured

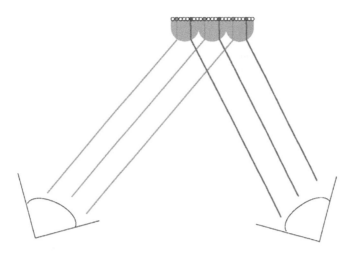

FIGURE 10.6
Viewing lenticular images. (Courtesy of Andrea S. Rivers, with permission.)

FIGURE 10.7
Sensitized lenticular imaging material. (Courtesy of Andrea S. Rivers, with permission.)

media consisting of sensitized layers, which are graphically depicted by a thin colored stripe coated onto the flat side of the media, such that a color could be imparted to the media using a tricolor laser patterning system.

Figure 10.8 is a higher-level graphical depiction of the lenticular imaging system as described by Cobb (Cobb et al. 2004). The figure describes the path of the imaging tricolor laser beam. The tricolored laser beam first passes through the beam-shaping optics and then through a polygon-shaped scanning element followed by folding mirrors that are used to direct the beam to the lenticular platen that holds a sheet of sensitized lenticular media. The polygon scanner rotates, resulting in a sweep of the tricolor beams (for our purposes, this motion would move the beam in and out of the page) along the long axis of the lenticules. If the lenticular platen is moved through the imaging region while the beam is scanning and if the beam is being modulated based upon image content, the sensitized media will be altered such that a photographic image can be developed on the lenticular media. If one rotates the developed lenticular media along an axis parallel to the long axis of the lenticules, one will view a specific portion of the image that has been applied to the media. If the images applied are structured in an interleaved manner where multiple images are interleaved into a single image, the sequence of images will be available for viewing to the viewer as he or she rotates the media.

A tight tolerance has to be maintained with respect to the registration of the imaging lines with the lenticules. This ensures the images during rotation are in a coherent sequence. As described in the patent by Cobb, a calibration laser and detector are used to scan and monitor the dimensions of the lenticular material before the printing process actually begins (Cobb et al. 2004). This prescan facilitates thorough control of the registration of the imaging tricolor beam throughout the process.

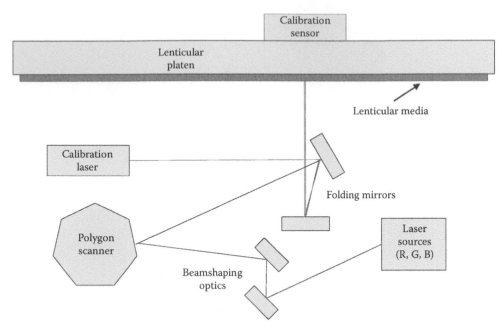

FIGURE 10.8
Lenticular imaging system. (Courtesy of Andrea S. Rivers, with permission.)

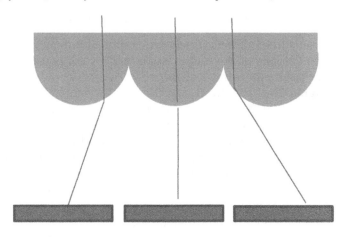

FIGURE 10.9
Calibration detector beam deflection. (Courtesy of Andrea S. Rivers, with permission.)

During the prescan, an infrared (IR) calibration laser beam is generated and directed toward the lenticular media. A sketch of the calibration beam is shown in Figure 10.9. As previously described, the lenticule media as a function of its shape deflect the beam. The sensor beneath the media helps identify the precise position of the lenticular media. The position-sensing detector is placed at a fixed point, while the media is moved through the path of the fixed calibration beam. These fixed-position sensors indicate, via electrical signals, when the beam is detected. The signal is processed with simple signal processing to derive a rate at which the lenticules moved past the optical beam. Humidity and temperature will play a role in changing the shape and dimension of the plastic-based media. Thus, a level offset and a start of page reference are derived from the electrical

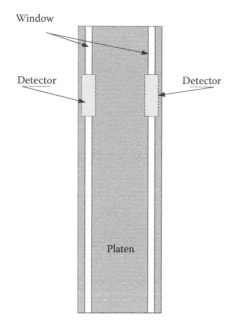

Window

Detector   Detector

Platen

FIGURE 10.10
Rotational alignment detection system. (Courtesy of Andrea S. Rivers, with permission.)

signals to be able to synchronize the imaging beam scan with the lenticules and in order to ensure the transport process operates with a low velocity flutter. The level offset is used to calibrate the servo system. The servo system controls the movements of the lenticular media past the imaging tricolor beam in the imaging scan. The imaging scan is executed after the prescan.

A similar sensing concept is used to rotationally align the lenticular media that is held by the platen. One must ensure that the starting image line is placed correctly, that is, having the correct vertical registration to the long axis of the lenticule. It is also necessary to ensure that the angle at which the tricolor laser beam writes on to the media is parallel to the long axis of the lenticule. If either of these is not correct, a fractured image will appear on the media when rotated. Figure 10.10 provides a graphical interpretation of the top view of the imaging platen. The figure shows two windows that extend for the length of the platen. Two detectors are shown as dashed lines to indicate that they are below the platen. These detectors are fixed. When a scanning beam is directed to the platen surface, the signals from the detectors can be compared in order to derive an angular rotation of the lenticules with respect to the scanning axis of the scanning beam. Flexures are used to mount the platen and are connected to a servo that rotates the platen relative to the calibration beam. The data from both sensors are used to determine the correct rotation of the platen. The combination of placement error of the media and the tolerance of the media from the manufacturer must be less than one lenticule.

10.1.3.2 Functional Analysis of the System

Now that we have a clear understanding of the concepts behind the lenticular imaging system, we can focus on the process of functional analysis and its use on the system. We begin with the system's concept of operations (CONOPS), shown in Figure 10.11. The lenticular imaging system is a subsystem within another system. A consumer uses

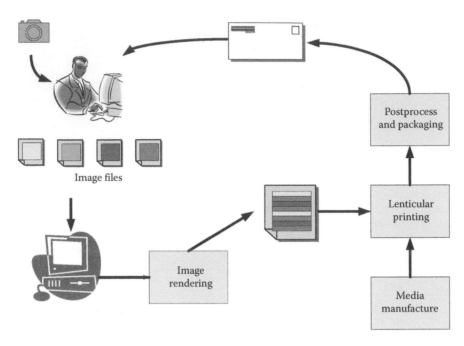

FIGURE 10.11
Lenticular fulfillment system. (Courtesy of Andrea S. Rivers, with permission.)

the lenticular imaging system to generate a sequence of images. These images are then submitted through the Internet to the lenticular website. On receiving the images from the consumer, the company then finishes the images and returns them to the consumer.

A high-level FFBD depicting the life cycle of the lenticular printing system is shown in Figure 10.12. Notice the horizontal lines connecting each function indicating that the functions are all sequential in nature. A high-level functional flow diagram helps precisely identify and list the requirements associated with each system during each phase of its existence. Besides the functional requirements, FFBD will draw attention to the nonfunctional requirements, also known as the "ilities," such as testability, manufacturability, and maintainability. During the design and development phase of the system, each of these

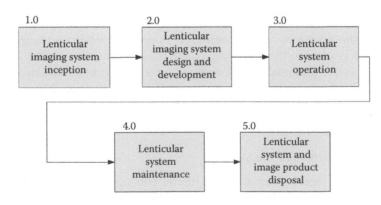

FIGURE 10.12
Top-level FFBD. (Courtesy of Andrea S. Rivers, with permission.)

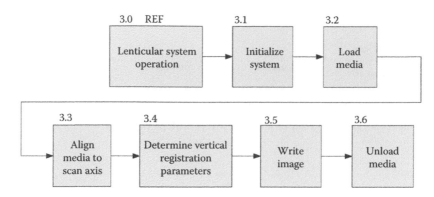

FIGURE 10.13
FFBD lenticular system: operation. (Courtesy of Andrea S. Rivers, with permission.)

subsystems will be prototyped and tested to ensure that the subsystems will be able to meet the lower-level requirements. The requirements associated with building, aligning, calibrating, and testing of the subsystems will be developed as all the functions needed at each point in the systems life cycle are considered.

However, for the purpose of this chapter, in order to better understand the goals and values of functional analysis and its tools, the scope will be limited to Block 3.0, "lenticular system operation," which involves operation of the system discussed earlier. As such, this chapter only focuses on lenticular printing and in particular on the operational functions of the optical system.

Figure 10.13 further breaks down the "lenticular system operation" block into subfunctions. We can see the primary functions of the lenticular systems listed sequentially. As mentioned previously, these functions can be further decomposed to the level of detail required.

Let us now focus on the functional breakdown of the "load media" block (3.2 in Figure 10.13). For the media to be loaded on the platen, the transport is commanded to the load position, and a separate system picks up media and places it onto the platen. This process is quite complex and can be decomposed into many steps dealing with the sensors, registration features, compliant members, a vacuum system that is used to retain the media after transport, and removal of the loading material. Figure 10.14 shows the load media function process broken down into three subfunctions each performed sequentially.

The functionality shown earlier in Figure 10.13, Function 3.3: "align media to scan axis," involves directing a nonimaging beam through the scanning path such that the location of the actual scanning beam can be detected by the position-sensing detectors that are located on both the left and right sides of the lenticular media. Figure 10.15 further refines this functionality into three component tasks: 3.31 "move media," 3.32 "detect rotation,"

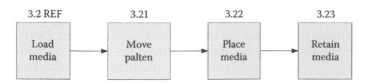

FIGURE 10.14
FFBD load media. (Courtesy of Andrea S. Rivers, with permission.)

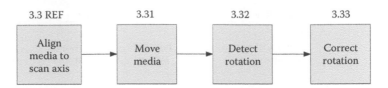

FIGURE 10.15
FFBD align media to scan axis. (Courtesy of Andrea S. Rivers, with permission.)

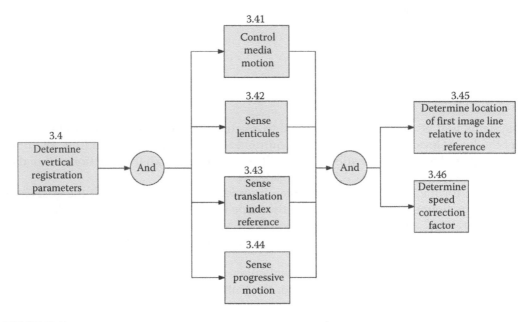

FIGURE 10.16
FFBD determine vertical registration. (Courtesy of Andrea S. Rivers, with permission.)

and 3.33 "correct rotation." Also, notice the numbering system in each of the FFBDs. It complies with the numbering pattern explained in Section 10.1.2.2.

Figure 10.16 is the FFBD that describes the process of aligning media relative to the imaging beams. This process involves detection of the lenticules as they pass through a fixed IR beam during the prescan and an imaging beam during the printing process. It involves monitoring the progressive motion of the platen during the process and detecting a reference index from which the start of page will be referenced with the help of the position sensors. Once complete, signal processing of the sensor data can be used to define the start of the first image line and check for a correction factor for the transport system. This will ensure the synchronization of image lines with the lenticules.

Figure 10.17 is a decomposition of the writing process. Notice the use of the AND labels and the concurrent functions running parallel to one another to signify functions that are executed in parallel. The transport (media) is moved, while the system monitors for the signal from the sensors that trigger the start of the writing process. Once the writing process starts, the system then monitors the signal from the sensors to maintain and ensure the proper printing onto the lenticular media by modulating the laser beams with respect to the position.

It becomes apparent that there are a number of common functions during the process under each subfunction. Multiple functions in the decomposition discussed earlier deal

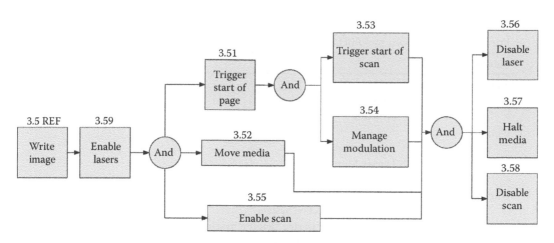

FIGURE 10.17
FFBD write image. (Courtesy of Andrea S. Rivers, with permission.)

with the motion of the transport. We have motion of the transport to scan the position, to load the media, to run the prescan on the media, and, while actually imaging the media, to use the tricolor beams. These common functions provide the opportunity to use common modules when developing the systems architecture.

As the process progresses toward establishing the systems architecture, the FFBDs and an understanding of the manner in which the team decides to partition the functionality allow the generation of a functional breakdown diagram, as shown in Figure 10.18.

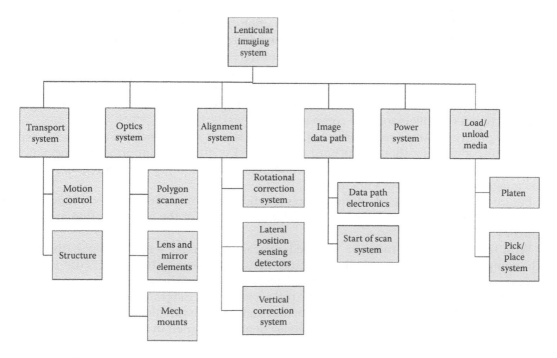

FIGURE 10.18
Functional breakdown. (Courtesy of Andrea S. Rivers, with permission.)

TABLE 10.2

Functional Breakdown by FFBD Function ID

Subsystem	Component	Functional ID
Transport		
	Motion Control	3.21, 3.31, 3.41, 3.52, 3.57
	Structure	
Optics		
	Polygon Scanner	
	Lenses, Mirrors	
	Mechanical Mounts	
Alignment system		
	Rotational Correction System	3.33
	Lateral Position Sensing Detector System	3.32, 3.42
	Vertical Correction System	3.43, 3.44, 3.45, 3.46
Image Data Path		
	Data Path Electronics	3.59, 3.56, 3.51, 3.55, 3.58, 3.54
	Start of Scan system	3.53
Power System		
Load/Unload Media		
	Platen	3.23
	Pick and Place system	3.22

Source: Courtesy of Andrea S. Rivers, with permission.

The columns shown in Table 10.2 relate the functional components to their FFBD by using the unique identifiers (numbers assigned to each of the functions and subfunctions). Notice that Table 10.2 is incomplete. If the functions related to the generation of suitable laser power and spot size and shape were considered, the optics would have additional functions allocated. Also, if vibration were considered with respect to the alignment system, then we would have to include a compensating measure such as vibrational isolation components. As additional information becomes available from the evolution of the design, the details of Table 10.2 can be expanded and filled in. This would result in assigning functional identifiers to these components.

As we progress toward requirement allocation of the system, Blanchard describes a step called functional allocation (Blanchard and Fabrycky 2011). An important performance consideration in the case of the lenticular imaging system is the quality of registration of the imaging beam with the lenticules.

Figure 10.19 graphically illustrates the registration TPM allocations for each functional component. In this case, we express this TPM in the parts of the lenticules that are indicated by the unit, L.

With the performance measure included, we can now create a system block diagram. The complete block diagram for the lenticular imaging system is shown in Figure 10.20. If necessary, additional documentation associated with each of these functions can also be created. This documentation supports the full system block diagram by providing greater detail during the design process. In the documentation, each function should be identified with its reference designator for the functional element assigned in the FFBDs. Apart from this, the documentation should include a list of inputs, outputs, constraints, and resource requirements for each functional element. This description can be presented in the form of a table or in the form of a concept diagram as shown in Figure 10.20.

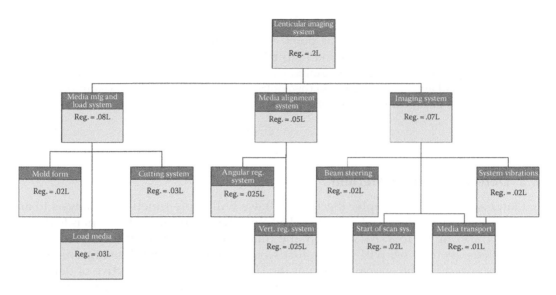

FIGURE 10.19
Lenticular registration TPM allocation. (Courtesy of Andrea S. Rivers, with permission.)

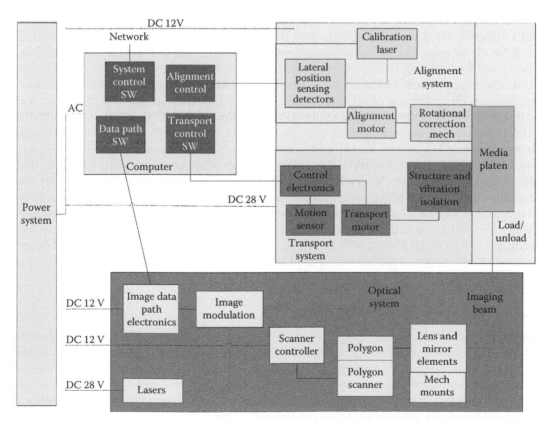

FIGURE 10.20
Lenticular imaging system block diagram. (Courtesy of Andrea S. Rivers, with permission.)

In this section, we presented an example of a functional analysis application using a lenticular imaging system as a focus point. We presented some fundamental concepts with regard to functional analysis and highlighted the use of FBDs and functional flow diagrams. In anticipation of the integrated case study application in Section 10.3, which applies functional analysis methods to the cooling system of an optical system on a UAV, we present some background information on optical detector cooling.

10.2 Detector Cooling Methods

In a later section of this chapter (Section 10.3), we will apply functional analysis to an optical system that is designed for a UAV. The objective of this optical system is to detect humans traveling by foot. As we apply the functional analysis concepts discussed in the previous section, we will mainly focus on the environmental conditioning that will be required for the optical system of the UAV to achieve its objective of detecting humans. The environmental conditioning unit will regulate factors such as pressure, radiation levels, light conditions, temperature, humidity, and dust. In the upcoming case study, we will focus on one aspect of temperature regulation—the cooling of the IR imaging detector.

Petrie's paper on Airborne Digital Imaging Technologies gives an overview of the numerous existing imaging sensors that are suitable for the UAV application (Petrie and Walker 2007). Research has indicated that complementary metal-oxide semiconductor (CMOS) sensors have an advantage over the charge-coupled device (CCD) sensors when used for imaging technologies applied in UAVs (Sumi 2006; Axis Communications 2010). The cheaper cost of the CMOS sensors, their ability to readily support windowing (i.e., sending portions of an image), and the fact that they generate less heat makes them a very viable option for UAV applications. Cooling of CMOS detectors is required in order to control image noise sources (e.g., heat sources affecting the quality of the image). Thermoelectric (TE) coolers are typically employed to aid in the cooling of the CMOS sensors (Allen 1997). TE coolers have the capability to provide cooling down to −70°C. The advantage they hold over the cryocoolers is that they do not have parts that wear out and generate vibration. Modeling tools can be used to analyze the impact of noise sources on image quality in CMOS sensors. Such tools would be useful in evaluating potential sensor choices and cooling solutions of the imaging sensor. Gow described a comprehensive tool produced in MATLAB (Gow 2007). In this section, we present a general overview of detector cooling concepts and methods and our UAV imaging scenario is used as an application example in Section 10.3.

10.2.1 Introduction to Detector Cooling

Detectors, like all electronic equipment, generate heat or radiation. With highly sensitive optics, additional heat can render images overexposed or noisy and make the data useless for detection purposes. To prevent this issue, the detector is cooled using one of many available refrigeration technologies. The FFBD is developed to correctly identify the best relationship between sensor function, refrigeration, and cost. Using the previously defined analytical hierarchy process (AHP), a predefined rating system can be applied to the different cooling options. These values are then normalized and compared to select the best optical sensor and cooling system a given set of criteria.

Johnson noise, or equivalently thermal noise, or Nyquist noise, is considered the noise floor of photon detectors. This noise type results from the motion of charged particles excited by heat in material that has electrical resistance (Dereniak and Devon 1984). Since the received photons must induce a voltage greater than the noise floor, lowering the noise floor will cause the photon detector to be more sensitive, increasing its detectivity. We provide a short review of the relevant noise terms discussed in previous chapters.

The noise voltage due to Johnson noise is given by Equation 10.1. The only parameters in an IR detector that influence Johnson noise, and are within the designer's control, are the detector resistance, its electrical bandwidth, and its operating temperature. This noise is directly proportional to absolute temperature; the lower the temperature, the lower the noise:

$$V = \sqrt{4kTR\Delta f},\qquad(10.1)$$

where
  k is the Boltzmann's constant = 1.38×10^{-23} J/K
  T is the temperature in K
  R is the resistance in Ω
  Δf is the effective noise bandwidth in Hz

Another factor affecting the noise floor of IR detectors is generation–recombination (G&R) noise. G&R noise is caused by the fluctuation in the rate at which charge carriers are generated and recombined (Hudson 1969). This type of noise also exhibits dependence on temperature. As these examples indicate, the detector noise and consequently the detectivity of an IR detector strongly depend on the operating temperature of the device (Wiecek 2005). Therefore, noise reduction methods are an important consideration in designing an IR detection system. Increasing detector performance by reducing its noise level has long been a common practice in high-end detection platforms.

This section details the various techniques used for cooling electro-optic and especially IR detectors. Without cooling, IR detectors would be flooded by their own radiation. When dealing with low-temperature regimes, the term cryogenic is often used. It is derived from the Greek term "cryo" meaning extreme cold and the suffix gen meaning "producer of." The cryogenic region starts approximately at 125 K and continues to absolute zero. Detectors that are cooled to low temperatures are contained in a vacuum-sealed case or Dewar, which is then cryogenically cooled. The two main reasons why the detector needs to be inside a chamber, or Dewar, are to (1) prevent water in the air from condensing on the image plane ruining the image and (2) allow the detector to cool more efficiently by limiting thermal conductivity with the surroundings.

10.2.2 Detector Cooling Vessels: The Dewar

Cooled detectors often come in an insulated package that is known as a Dewar flask. The Dewar, named after its inventor, James Dewar, is a container, very much like a thermos bottle, which has two walls. The outer wall typically interfaces with the ambient environment and the inner wall with the detector. The space in between the two walls is vacuum.

In building a Dewar, some important considerations include the selection of materials for the inside and outside of the Dewar, the choice of window material, the manner in which the detected signals are extracted from the Dewar, and how the detector is mounted. In removing the detected signal, very thin wires are used to connect the detector elements to contacts. These wires are kept as short as possible to reduce unwanted signals arising

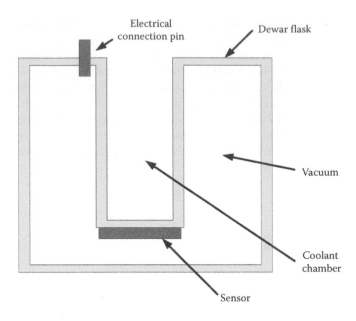

FIGURE 10.21
Typical cooled package detectors. (Courtesy of Chris Schuchmann, 2013.)

from the vibrations of the wires (microphonics). The external outputs of the Dewar are connected to the detector contacts with specially conducting paint. The optical window to the Dewar is coated for high transmission. A typical cooled detector package as shown in Figure 10.21 can readily achieve cooling temperatures down to 50 K. Achieving lower temperatures requires careful design.

In assembling the final product, care must be taken in the parts fusion and evacuation processes since the detector element is very sensitive. Excessive heat or mechanical pressure can also lead to damage near the pin locations and result in leaks.

10.2.3 Typical Coolants Properties

Two of the most commonly available coolants are liquid nitrogen and liquid helium. Because of their availability, these types of coolant are used more often than other types. Another reason why some other materials may not be so prevalent is because of their safety concerns. Most everyone has seen the experiments where liquid nitrogen is used to freeze a banana or some other substance. The banana is so hard that it can be used to drive a nail! Liquid nitrogen and other liquid gases can be extremely hazardous and great care must be used in handling them. Table 10.3 has some interesting properties of coolants. The first column in Table 10.3 is the boiling temperature, which marks the transition temperature between the liquid and gas state. Liquid nitrogen has a boiling temperature of 77.3 K and so can cool the detector in the vicinity of this temperature in its liquid state. Notice that liquid hydrogen can cool to the neighborhood of 20.4 K and liquid helium to temperatures of 4.2 K. The second column in Table 10.3 is the vaporization capacity. This column gives the volume of material (in cm^3 units) that is vaporized by applying 1 W h of thermal energy. This column is quite interesting since it points out that there is a big difference in the vaporization capacity for the higher temperature liquids than the lowest ones. The higher the vaporization capacity, the more of a thermal load it takes to dissipate the 1 cm^3 volume of

TABLE 10.3

Common Physical Parameters of Low-Temperature Coolants

Coolant	Boiling Temperature (K)	Vaporization Capacity (1 W h (cm³))	Weight–Density (kg/m³)	Triple Point (K)
Ice	273.2	—	—	—
Solid carbon dioxide	194.6	—	—	—
Liquid krypton	120.0	14	423	115.77
Liquid oxygen	90.2	15	1141	54.40
Liquid argon	87.3	16	1393	83.85
Liquid fluorine	85.0	14	1502	53.54
Liquid nitrogen	77.3	22	809	63.15
Liquid neon	27.1	35	1205	24.57
Liquid hydrogen	20.4	114	71	13.96
Liquid helium	4.2	1410	125	—

Source: Data from Hands, B.A., ed., *Cryogenic Engineering*, Academic Press, London, U.K., 1986.

material. If a large thermal load were placed on the liquid helium cooler for instance, much more liquid helium would be required to dissipate it than, say, liquid nitrogen. The weight density is useful in determining how much the coolants weigh, and the triple point is the temperature in which there are equal amounts of solid, liquid, and gas.

10.2.4 Cooling with Open-Cycle Refrigerators

Open-cycle refrigerators cool the detector by transferring its heat to a liquefied gas that has a cryogenic boiling point. The heat of vaporization of the liquefied gas absorbs the heat, and its boiling point determines the detector temperature. These refrigerators are often used in a laboratory and no attempt is made to collect and reuse the refrigerant after it has absorbed heat from the load. Since the spent refrigerant is not collected and reused, these types of refrigerators are labeled as refrigerators with an open cycle.

10.2.4.1 Cooling with Liquefied Gas

A simple liquefied gas refrigerator uses the Dewar that was discussed previously to hold the coolant and a line connected to the Dewar, which delivers the coolant to the detector. In this configuration, no attempt is made to capture and reuse the coolant (hence the open cycle). Figure 10.22 shows the operation of a representative refrigerator that uses the vaporization of liquid helium to remove heat. The system can also operate using liquid nitrogen.

The system is designed to cool samples all the way down to 2.0 K. However, by carefully controlling the amount of liquid helium that is transferred, temperature ranges up to 300 K can be selected. The sample (detector element) is exposed to liquid helium that is regulated to flow at a constant rate over the sample. The overall process is highly efficient. A needle valve is used to precisely regulate the flow rate to within a temperature stability of ±0.01 K.

This cooler has two major parts: a Dewar with an associated flexible helium transfer line and a cold-end assembly. Two sets of flows (one for the tip and one for the shield against the external environment) are generated. A small tube applies the liquid helium to the sample. After cooling the sample, the liquid helium also cools a radiation shield. If the operations require cooling down to 4.2 K, the spent sample is released into the atmosphere. However, in applications where the sample needs to be cooled below this value, vacuum pump recaptures the helium and the temperature is lowered through controlled pressurization.

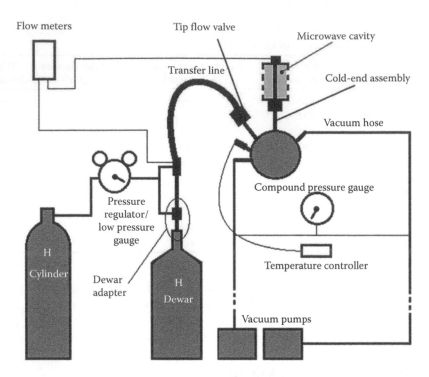

FIGURE 10.22
Operation of liquefied gas refrigerator. (Courtesy of Ty Phillips, 2013; adapted from publications of Advanced Research Systems, Inc.)

10.2.4.2 Cooling with the Joule–Thomson Effect

These refrigerators have a special device that is responsible for the cooling action—the Joule–Thomson cryostat. The Joule–Thomson cryostat is a small device that is capable of liquefying gas. It is small enough to be placed inside the coolant chamber alongside the detector. The Joule–Thomson cryostat provides cooling by direct expansion of gas from one or more pressurized bottles. The cooling process works by expanding the gas under high pressure using a throttle valve. A counter current heat exchanger is used to precool the expanded gas until it liquefies as shown in Figure 10.23. The cryostat can be readily installed in such a way that the liquefied coolant from the cryostat strikes the back surface that houses the detector element.

To change a gas to its liquid state, its temperature must be sufficiently dropped so that it falls below its boiling temperature. The Joule–Thomson coefficient μ is useful if the drop in temperature is achieved by the expansion or throttling of the gas. The Joule–Thomson coefficient is given by

$$\mu = \left(\frac{\delta T}{\delta P} \right)_h ,$$ (10.2)

where δ represents a change in the temperature (T) with respect to a change in the pressure (P) and constant enthalpy value (h). The enthalpy is a thermodynamic property

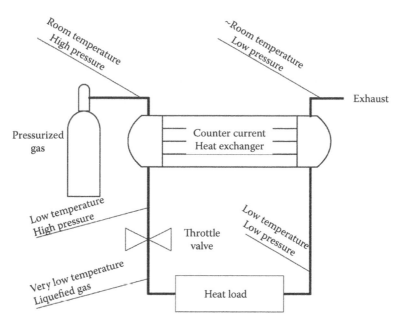

FIGURE 10.23
Ideal Joule–Thomson open-cycle refrigerator. (Courtesy of Chris Schuchmann, 2013.)

related to the total heat content of the system. Given Van der Waals model of the gas, the Joule–Thomson coefficient is

$$\mu = \left(\frac{\delta T}{\delta P}\right)_h \approx \frac{2}{5R}\left(\frac{2a}{RT} - b\right),\tag{10.3}$$

where
  R is the universal gas constant, 8.314 J/(mol K)
  T is the temperature in K
  a and b are constants for every gas

If μ is positive, then the gas cools. Conversely, if μ is negative the gas heats up. From Equation 10.3, we can conclude that the Joule–Thompson coefficient is negative given $2a/RT < b$, and this is strictly temperature dependent. The transition temperature is called the inversion temperature T_i. Above the inversion temperature, the gas heats up during expansion. Below the inversion temperature, the expanded gas cools. Table 10.4 shows the Joule–Thomson constants, coefficient and inversion temperature for typical gases.

The Joule–Thomson refrigerators have two major advantages. First, there are no moving parts in the system in contact with the detector, which reduces the possibility of microphonics. Second, a compressor (if used) can be located quite a distance away from the cryostat and consequently ease in packaging as compared to other systems. Even though relatively simple, the Joule–Thomson cycle has certain disadvantages. To begin with, it is an irreversible process that renders it relatively inefficient. The second disadvantage is contamination, which can occur, in the small orifice, which is about 0.002–0.007 in. in diameter. Any condensable impurity tends to freeze and block the orifice. This in turn calls for the use of a nonlubricated compressor or elaborate cleaning techniques.

TABLE 10.4

Joule–Thomson Constants and Coefficient

	a (Jm^3/mol^2)	b (m^3/mol)	μ (K/Pa) $T = 300$ K	T_i (K)
He	0.003457	2.37×10^{-5}	-1×10^{-6}	35.1
Ne	0.021349	1.71×10^{-5}	1.40×10^{-9}	300.3
H$_2$	0.024764	2.66×10^{-5}	-3.25×10^{-7}	223.9
Ar	0.136282	3.22×10^{-5}	3.71×10^{-6}	1018.1
O$_2$	0.137802	3.8×10^{-5}	3.49×10^{-6}	872.4
N$_2$	0.140842	3.91×10^{-5}	3.56×10^{-6}	866.5
CO	0.150468	3.98×10^{-5}	3.89×10^{-6}	909.5
CH$_4$	0.228285	4.28×10^{-5}	6.75×10^{-6}	1283.1
CO$_2$	0.363959	4.27×10^{-5}	1.20×10^{-5}	2050.4
NH$_3$	0.422525	3.71×10^{-5}	1.45×10^{-5}	2739.7
Air	0.1358	3.64×10^{-5}	3.48×10^{-6}	897.5

Sources: Data from Min, G. et al., *J. Phys. D: Appl. Phys.*, 37, 1301, 2004; Crowe, T.J. et al., *Int. J. Quality Reliab. Manag.*, 15(2), 205, 1998.

10.2.4.3 Cooling with Solids

Methane, argon, carbon monoxide, nitrogen, neon, and hydrogen are some of the most commonly used solid-refrigerant coolers used primarily to cool IR detectors on space vehicles. These refrigerants can operate under zero-gravity conditions, consume minimum power, are highly reliable, and weigh less than other cooling solutions. These properties make them ideal candidates for use in space environment. These refrigerants are stored in an insulated container and a thermally conducting rod is used to transfer heat from the detector to the space. The pressure maintained in the storage container defines the storage temperature of the refrigerant. Tables 10.5 and 10.6 list some interesting physical properties of some of the common refrigerants.

10.2.4.4 Cooling by Radiating Heat

Radiative-transfer coolers operate on the principal of radiating the heat away from the detector. This is often practical in space where the environment surrounding the spacecraft can plummet to cryogenic temperatures. The heat that is generated from the detector is thermally conducted to the exterior of the spacecraft where it is then radiated into outer space. This cooler is capable of cooling a spacecraft passively by radiating the heat

TABLE 10.5

Boiling Point and Freezing Point for Common Refrigerants

Refrigerant	Boiling Point (K)	Freezing Point (K)
Methane	111.5	90.6
Argon	87.0	84.0
Carbon dioxide	194.6	216.5
Nitrogen	77.4	63.15
Neon	27.1	24.5
Hydrogen	20.4	13.9

TABLE 10.6

Molecular Mass and Specific Heat of Common Refrigerants

Refrigerant	Molecular Mass (g/mol)	Specific Heat (J/kg/K)
Argon	39.948	520.33
Carbon dioxide	44.01	~720
Nitrogen	28.0134	1,040
Neon	20.179	1,030
Hydrogen	2.016	14,300
Helium	4.0026	5,193.1

into space. The thermal-balance equation for a cone shield and a patch system can be used to determine radiator size. The patch thermal-balance equation is shown in the following:

$$\sigma A_p T_p^4 = \sigma A_p F_a \varepsilon_{cp} T_p^4 + K_{cp}(T_c - T_p) + Q_d, \tag{10.4}$$

where

A_p is the external area of patch, cm^2
T_p is the absolute temperature of patch, K
T_c is the absolute temperature of the cone, K
F_a is the cone-to-patch configuration factor (dependent on the geometrical configuration of the patch and the cone)
ε_{cp} is the effective cone-to-patch emissivity
K_{cp} is the thermal conductance from cone to patch, W/K
Q_d is the thermal inputs from detector, W
σ is the Stefan–Boltzmann radiation constant, W/cm^2/K^4

The thermal-balance equation for the cone is

$$\sigma \varepsilon_c A_c T_c^4 = A_c a_s (E_s + E_a) + A_c E_{ER} \varepsilon_c + Q_{ac}, \tag{10.5}$$

where

ε_c is the emissivity of the cone
A_c is the effective area of the cone, cm^2
E_s, E_a, E_{ER} is the irradiance due to direct solar radiation, albedo (earth solar radiation reflection), and earth IR emissions, respectively, W/cm^2
α_s is the solar absorptivity of the cone
K_{cp} is the thermal conductance from cone to patch, W/K
Q_d is the thermal inputs from spacecraft to the cone—including radiative and conductive mechanisms, W

Radiators can be used, theoretically, down to about 60 K under ideal conditions. However, below about 100 K, their rejection capability falls dramatically because of the T^4 radiation law of heat transfer. In this case, their overall feasibility is dependent on the aircraft orbit, orientation, and attitude-control limitations. A typical application involves cooling detectors to temperatures sufficient for effectively conducting operations. For example, in the case of quantum detectors, the detector must be sufficiently cooled so that the photon energy $h\nu$ is greater than the thermal energy noise of the detector.

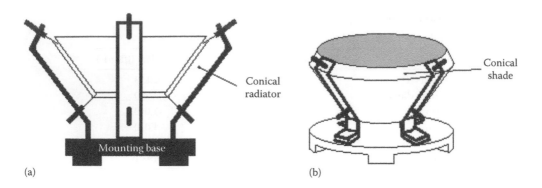

(a)　　　　　　　　　　　　　　　　　　　　　　(b)

FIGURE 10.24
Radiative-transfer cooler. (a) Depicts a side view with shade removed, whereas (b) depicts a slant view with shade. (Courtesy of Ty Phillips, 2013; adapted from NASA; Rickman, S.L. et al., 2010, Passive radiative cooler for use in outer space, http://www.techbriefs.com/component/content/article/9-ntb/tech-briefs/physical-sciences/6996, accessed June 25, 2014.)

Figure 10.24 shows a radiative IR cooler at Johnson Space Center, Houston, Texas. This radiative cooler has three critical components—a conical shade, a conical radiator, and a base for mounting the radiator. A high-IR emittance finish is used on the conical radiator to increase the radiative transfer to space. Tensioned, nonmetallic cords are used to suspend the radiator within the device.

The detector is placed in a sample holder and connected to a radiative cooler. The cooler itself is mounted on the spacecraft in such a fashion that it does not face toward any object that could heat it up (e.g., sun, planet, and objects in the near vicinity). In this way, the radiative element experiences a cold environment and can effectively radiate away heat from the detector. The angle of the cone is chosen so that it optimally shields the view of the detector from surrounding heat sources and is capable of effectively rejecting heat and is affixed to the sample tray, and the passive radiative cooler is mounted on a spacecraft structure that faces away from the sun or a planet. As the spacecraft orbits in a specified attitude, the sample tray and sample are cooled radiatively. The cone angle of the radiator is chosen to afford adequate radiative heat rejection while enabling the cone to shield the sample from viewing other bodies (the earth, the sun, or nearby objects) that could adversely affect heat balance of the sample. As an example of performance, the passive radiator was able to thermally isolate a sample under test at 116 K wherein the mounting surface temperature was 240 K (NASA 2010).

10.2.5 Cooling with Closed-Cycle Refrigerators

10.2.5.1 Cooling with the Joule–Thomson Closed Cycle

These types of refrigerators are similar to the open-cycle Joule–Thomson refrigerators in operation. An open-cycled J–T cooler can be converted into a closed-cycle cooler by adding a compressor, which repressurizes the gas leaving the cryostat as shown in Figure 10.25. The compressor is the most important component of this system. The compressor recycles the spent refrigerant by pressurizing the spent refrigerant. This recycled refrigerant needs to have practically no contamination. The high-pressure gas leaves the compressor and flows through a recuperative heat exchanger, which precools the gas. Additional cooling of the gas is produced by expansion to a lower pressure through the expansion valve. The fluid flows to an isothermal heat exchanger, where the load is applied. Then, after again

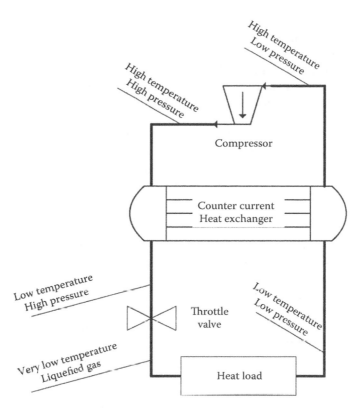

FIGURE 10.25
Joule–Thomson closed cycle. (Courtesy of Chris Schuchmann, 2013.)

passing through the recuperative heat exchanger where it absorbs the heat from the pressurized gas, the fluid returns to the compressor, where it is pumped back to high pressure, and the cycle repeats.

Depending on the operating time, different compressors can be used. The life of an oil-lubricated compressor is limited by the lifetime of the filters that prevent oil contamination. On the other hand, dry (unlubricated) compressors have a relatively short lifetime. An appealing alternative is a diaphragm pump, but this requires determining an acceptable diaphragm material.

Like its open-cycle counterpart, closed-cycle coolers are simple to construct. Also, the effects of microphonics are not observed in these refrigerators, since they do not contain any moving parts in the low-temperature side. The line connecting the compressor to the liquefier carries only high-pressure gas and thereby needs no thermal insulation. Thus, in an aircraft or ship, the compressor can be located wherever space permits. This makes the refrigerators portable.

10.2.5.2 Cooling with the Claude Cycle

Refrigerators based on the Claude cycle are more appropriate for medium scale refrigeration. They are similar in principle to the Joule–Thomson refrigerators, in that both use a counter current heat exchanger to cool. The initial Claude system invented by French engineer Georges Claude in 1902 used a reciprocating expansion engine. The operating principle was very similar to the conventional steam engine. Force can be generated using the heat energy produced by the high-pressure helium gas that is expanded. The resulting force moves a flywheel that actives a generator or motor. In this application, however, the objective is not motion but rather a drop in temperature caused by the spent energy.

Because the Claude cycle is reversible, it is more efficient than the Joule–Thomson cycle. The gas also does not need to be cooled below its inversion point. Similar to Joule–Thomson cycle, contamination can become a problem with the Claude cycle, even though it is not as serious as the Joule–Thomson.

Claude refrigerators do not offer a very good reliability when used in airborne IR detector sets due to the fast-acting parts. There is also a likelihood that the moving parts will introduce microphonics. When miniaturized, long life in a cryogenic unlubricated environment is difficult.

10.2.5.3 Cooling with the Stirling Cycle

The Stirling cycle owes its origins to a reverend from Scotland, Robert Stirling, who obtained a patent for a hot air engine in 1816. This engine was capable of converting heat to work through a reversible process involving fluids at different temperatures that were repeatedly compressed and expanded. The key element of the patent was the regenerator, a highly efficient, thermally isolating heat exchanger for oscillating gas flow.

The Stirling cycle is shown in Figure 10.26. First, work is done to push the retracted compressor piston, as shown in step (a), halfway into its chamber to compress the gas in the system. Meanwhile, the fully extended expander piston is held in position and does not move. The heat of compression in the piston chamber is removed through an after cooler heat exchanger. The system gas volume decreases and the pressure increases. Since the Stirling cycle is an isothermal process, the chamber is at a constant temperature. Second, the compressor and the expander move together to displace the compressed gas through the regenerator and into the expansion chamber, as shown in steps (b) and (c). The gas is transferred at constant volume, an isochoric process, from the compression to expansion chamber. Through this part of the process, the temperature and pressure decrease. Finally, the expander piston moves to its fully extended position as shown in step (d). This movement causes a lowering of system pressure, a cooling of the working fluid, and absorption of heat. Refrigeration is only produced in this part of the cycle. Last, the compressor and the expander move together to displace the compressed gas out of the expansion chamber, as shown in steps (d) and (a).

In 1834, John Hershel developed a Stirling engine based on the closed-cycle Stirling process. He used this Stirling engine to make ice. The process involved using the Stirling engine to reject thermal energy Q_c generated from a supplied work input W_c that was generated during isothermal compression:

$$Q_c = mRT_c ln\left(\frac{V_{max}}{V_{min}}\right) = W_c, \tag{10.6}$$

where
  m is the mass
  R is the gas constant
  T_c is the compression gas temperature
  V_{max} and V_{min} are the maximum and minimum volume

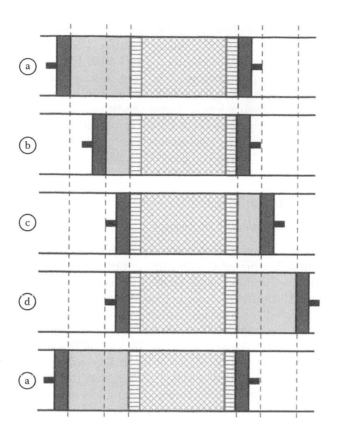

FIGURE 10.26
Workings of a Stirling refrigerator. (Reprinted with permission from Adwaele, The Stirling cycle in four steps, Wikipedia, https://en.wikipedia.org/wiki/File:Stirling_Cycle_Cryocooler.jpg, accessed April 10, 2014.)

Heat energy Q_e is absorbed during the isothermal expansion process. The following equation provides the expression for the heat absorbed:

$$Q_e = mRT_e ln\left(\frac{V_{max}}{V_{min}}\right) = W_e,$$ (10.7)

where
  T_e is the expansion gas temperature
  W_e is the expansion energy

The coefficient of performance (COP) is an important figure of merit concerning refrigeration. The COP is defined as the amount of heating absorbed to the total work consumed or

$$COP = \left(\frac{Q_e}{W}\right) = \left(\frac{Q_e}{W_c - W_e}\right) = \left(\frac{T_e}{T_c - T_e}\right).$$ (10.8)

The basic idea behind the Stirling engine has been unchanged since its inception. However, improvements made in seals, bearings, and materials used in the regenerator have resulted in a useful and reliable cooler. Stirling refrigerators used in cooling IR detectors primarily consist of pistons and cylinders and a thermal regenerator. The thermal regenerator passes the gas back and forth between the cylinders. The lowest temperature achievable with the Stirling refrigerators is approximately 15 K. Two other benefits are that the Stirling refrigerators are (1) not as susceptible to contamination as other refrigerators and (2) they work under relatively low pressure levels. These conditions allow for long operating life. A disadvantage of Stirling refrigerators is that they can be noisy since the detector mounts on the expander. Even so, the Stirling refrigerators can produce cooling in one stage down to 20 K and can even cool to 4.2 K with additional stages.

10.2.6 Cooling with Electric or Magnetic Effects

10.2.6.1 Cooling with Thermoelectric Properties

The working principle of a TE cooling system is based on the Peltier effect (Keyes 1977). If a current is passed through a material with different metals, some interesting effects can be observed. Based on the direction of the current, one junction in the material is heated, while the other is cooled. Simply by changing the direction of the current, this effect can be reversed with the "heated" and "cooled" junctions switched. TE devices have very low efficiency and a relatively low-temperature delta. However, they do not generate any vibrations and that is well suited for thermal detectors (Keyes 1977).

The Peltier effect is based on the difference of the charge carriers' energy in p-type and n-type semiconductors and the metal. At the cold junction in a TE refrigerator, the electrons flow from a p-type semiconductor element, having low energy, to an n-type element, having a higher energy. At the hot junction, we observe the exact opposite phenomenon, where the flow is from an n-type to a p-type element. This process results in energy generation and absorption. The required energy for the electrons to move through the circuit is provided by the power supply.

Figure 10.27 shows the cross section of a TE cooler. This cooler is constructed from doped semiconductor materials (most often bismuth telluride). TE cooler can be made using either n-type or p-type materials and operates by pumping heat from the cold junction material to the hot junction. The amount of heat that is dissipated is related to the current that flows through the electrical circuit and is also dependent on how many junctions are used.

Figure 10.28 shows TE coolers connected to form a cooling module. A multistage TE refrigerator can be constructed using several coolers cascaded such that the hot and cold junctions of the adjacent coolers are in good thermal contact with each other. The colder the operating temperature desired, the more stages are required. This increases the size of the device and power consumption also rises. From a practical point of view, three or four stages are the maximum limit to which the coolers can be cascaded.

The large size limits this refrigerator from fitting in a conventional Dewar. Alternatively, various foamed insulations are used to form the package containing the detector and TE cooler. Since no more than three or four stages of coupling can be achieved, sufficiently low temperatures cannot be attained using these refrigerators.

10.2.6.2 Cooling with Thermomagnetic Properties

These refrigerators are based on the Ettingshausen and the Nernst–Ettingshausen effects. These two effects are linked dynamically and a resulting cooling effect is achieved. El Saden performed an analysis concerning the characteristic and response of a cooling

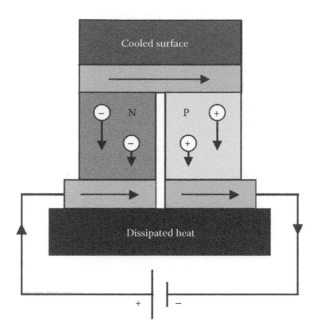

FIGURE 10.27
Cross section of a typical TE cooler. (Reprinted with permission from Brazier, K., A diagram of a thermoelectric cooler, Wikipedia, http://en.wikipedia.org/wiki/File:Thermoelectric_Cooler_Diagram.svg, 2008. Derived from Cullen, C., Thermoelectric Cooler, http://en.wikipedia.org/wiki/File:ThermoelectricCooler.jpg, 2007.)

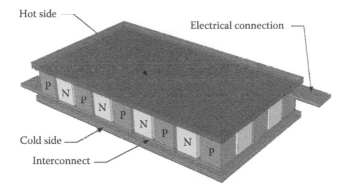

FIGURE 10.28
Typical TE module assembly. (Reprinted with permission from Michbich, Schematic of a Peltier device, 2010, Wikipedia, http://en.wikipedia.org/wiki/File:Peltierelement.png.)

system based on the Ettingshausen effect. Primarily, there are four transverse magnetic effects that take place, the Ettingshausen, the Nernst–Ettingshausen, the Hall, and the Righi–Leduc effects. Only the first two will be taken into consideration for the analysis. The latter two have to be maintained under static conditions.

Figure 10.29 demonstrates the Ettingshausen effect for a material with a positive Ettingshausen coefficient. The block is thermally insulated at the top and bottom sides and electrically insulated at the left and right sides. If an electric current density J is flowing in the y direction, and the magnetic field H in the z direction, the Ettingshausen temperature gradient $(dT/dx)_E$ appears in the x direction.

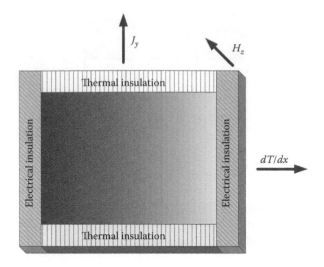

FIGURE 10.29
Configuration of a thermomagnetic cooling system. (Courtesy of Chris Schuchmann, 2013.)

In the absence of any thermal currents, the Ettingshausen temperature gradient is given as

$$\left(\frac{dT}{dx}\right)_E = PJH,$$ (10.9)

or, approximately,

$$\Delta T_E = \left(\frac{PIH}{c}\right),$$ (10.10)

where
   P and ΔT_E are the Ettingshausen coefficient and temperature difference, respectively
   I is the total electric current
   c is the dimension of the block in the z direction

Figure 10.30 shows the equivalent circuit for this effect. We can observe that if a thermal current of average density q is allowed to flow, the temperature gradient drops. A constant

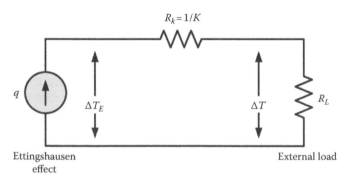

FIGURE 10.30
Circuit diagram for the Ettingshausen effect. (Courtesy of Chris Schuchmann, 2013.)

temperature gradient is represented by the current source. The quantity $R_K = 1/K$ is the internal thermal resistance of the material in the direction of heat flow, K is the thermal conductivity, and R_L is the external thermal load.

Therefore,

$$\left(\frac{dT}{dx}\right)_E = \left(\frac{dT}{dx}\right) + \left(\frac{q}{K}\right),$$ (10.11)

and

$$\Delta T_E = \Delta T + \left(\frac{aq}{K}\right),$$ (10.12)

where
  dT/dx and ΔT are the temperature gradient and temperature difference, respectively, while a thermal current is flowing
  a is the dimension of the block in the x direction

This section has provided some background on cooling properties and discussed various types of coolers and cooling mechanisms for electro-optical (EO) and especially IR detectors. The effect is to reduce some of the prevalent noise terms and thereby increase the detector detectivity. In the next section, we apply the functional analysis principles from Section 10.1 and the detector cooling principles of Section 10.2 to a UAV border patrol optical systems detection scenario.

10.3 UAV Integrated Case Study: Integrating the Detector and the Cooler

As in previous chapters, the following is a case study based again on the fictional company, Fantastic Imaging Technologies (FIT), a middle-sized company in developing EO/IR systems for a variety of applications. In this scenario, we find FIT in the midst of a functional analysis of the cooling aspects of the detector. Also, in order to reduce the noise on the detector element, FIT has to choose the type of cooler that needs to be used.

The characters that are part of this scenario are Karl Ben, FIT senior systems engineer; Ginny R., FIT UAV program manager; Jennifer O. (Jen), FIT systems engineer; Ron S., FIT systems engineer (new hire); and Amanda R., FIT optical engineer.

(It is 10 AM, Tuesday morning, and a technical interchange meeting [TIM] between the design team and the systems team is in full swing.)

Amanda: "No, no, we have to cool the detector. Room temperature won't work."
Ron: "But why? The radiometric analysis shows we have plenty of signal for the nominal imaging scenario. Remember, we are getting mV signal levels out of the detector for a sensor to target standoff range of 8046.7 m and at a look angle of 45 degrees."

Amanda: "Yes, but the performance specs that we are using for the HgCdTe detector assume that it is cooled to 77 K. We wouldn't be getting those signal levels if we operated the sensor at room temperature (295 K). The detectivity would drop, and so would the responsivity. Also, the noise terms would increase, especially Johnson noise and the generation and recombination noise."

Ron: "Ah, OK."

Jen: "Also, we shouldn't be basing decisions on the nominal condition case. We should be looking at the worst-case scenario with regards to the noise."

Amanda: "Good idea! Ron, flip open your iPad and log into OPSI."

Ron: "OPSI?"

Amanda: "Yeah, the optical propagation systems integration (OPSI) module. We just named it. Oh, and pop it on the screen so we can all see it, please."

Ron: "OPSI…right. We have to work on that."

(Ron hits a few macros and the OPSI graphical user interface [GUI] is displayed on the Samsung large screen on the wall of the conference room.)

Amanda: "Ok, now select the HgCdTe sensor type and load the default sensor parameters. These are the same as those that we got from the lab test the last time (short pause while the screen loads). Now, change the standoff range to 18616 m and the look angle to 15 degrees—that's our worst case. Go ahead and run the radiometric propagation model."

Ron: "By just changing the look angle and the standoff range and keeping all the other parameters the same as before, I get an optical power on the detector of 6.711 nWatts. That is still way above the detector noise equivalent power of 0.089 nWatts. There is quite a bit of margin there, even for the worst-case scenario."

Amanda: "Yes, but that NEP is for the detector operating at 77 K, not room temperature. The dark current density can increase several orders of magnitude as a function of temperature. Look at this plot of noise effects in a CMOS image detector that we plotted from an IEEE paper in 2010. It shows what I mean. Look at the plotted signal levels of the noise as a function of temperature range (Lin 2010) (Figure 10.31)."

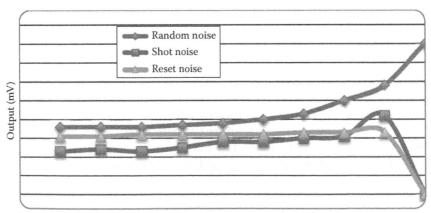

FIGURE 10.31

Noise versus temperature for representative CMOS imaging sensor. (Based on Lin, D.-L., 2010, *IEEE Trans. Electr. Dev.*, 57(2), 422, 2010.)

Amanda: "Gow mentions that for every 6 to 8 degrees rise in temperature the dark current doubles (Gow 2007). This dark current is often a major noise contributing to poor image quality. We will need to cool our sensor to 77 K to manage temperature related detector noise effects and provide acceptable detector performance."

Ron: "Ok, got it. So, how do we handle the cooling?"

Amanda: "That's what we need to hammer out. Let's start with looking at a comparison of the different types of cooling mechanisms. Ron, can you click on the cooling block in OPSI and then click on the documentation link and then select cooler comparison, please?"

Ron: "Sure. (Table 10.7)"

Amanda: "Here is a summary of some representative candidates along with some rough performance envelopes, benefits, and disadvantages. We didn't list the coolers that we didn't think would work for this application, like the thermomagnetic coolers or some of the cutting edge methods like optical cooling. We conducted an AHP based on stakeholder preferences (performance, reliability, maintainability, and size, weight, and power) and we chose the Stirling refrigerator. You can click on the AHP analysis under the cooler trade study link."

Ron: "How does it work?"

Amanda: "The Stirling cycle is reversible process and is highly efficient. The same piston is used for both compression and expansion; and so the operating efficiency approaches the theoretical max efficiency value. This process is performed in a simple machine, wherein no valves are needed between the compressor and expander. The design is lightweight due to the low operation pressure and low compression ratio used."

Ginny: "Ok, let's go with the Stirling cooler then."

Karl: "Now that we settled on the cooler, let's look at a functional breakdown of our system. We need to understand the functional requirements and their effect on the operations. Jen, can you bring up the functional flow diagrams and functional block diagrams we are developing and give us a rundown please?"

Jen: "Right. Let's start with Figure 10.32; this concept diagram identifies the inputs, outputs, controls/constraints, and mechanisms of a function."

Jen: "The function that needs to be achieved is shown using a box at the center, while the inputs and outputs of the system are shown on the left and right side, respectively. Also, the inputs at the top are essentially constraints on the function. The resources that are required for the function to be achieved are shown at the bottom of the figure. This concept diagram can be applied to any level in a functional decomposition and is helpful in finding missing constraints, mechanisms, inputs, outputs, and ultimately missing requirements."

Jen: "Now moving on to the FFBDs. FFBDs show the various functions of the system and the process flow between these functions. It consists of many blocks with text at the center describing their functionality. Then, arrows connecting the blocks indicate how the functions in the system are ordered. Logical relationships like AND or OR nodes can be used. This process follows a hierarchical approach, that is, each of the blocks is numbered. Each block is further broken down into smaller units that are used to depict what is required for the higher-level function. Also, when the blocks are numbered, the number of the higher-level block is taken as a prefix for the numbers of its lower-level blocks. Here, let me draw it for you."

(Jennifer drew Figure 10.33 depicting the optical subsystems.)

TABLE 10.7

Summary of Common Coolers

Cooler		Description	Temperature (Typical)	Heat Lift (Typical)	Advantages	Disadvantages
Passive coolers	Radiative	Radiates thermal energy from a sample into a cold reservoir like space.	116 K	0.1 W	Reliable Low vibration Long lifetime	Limitations on the heat load and temperature.
	Cryogen	Can achieve temperatures lower than radiators. Uses Dewars and cryogenic material like liquid helium.	4.2 K	0.25 W	Excellent temperature stability Low vibration	Short lifetime (limited to the amount of cryogen stored).
Active coolers—thermodynamic cycles are used to transport heat up a temperature gradient to achieve lower cold-end temperatures at the cost of electrical input power.	Joule–Thomson	Uses high-pressure gas expansion through a throttle valve and a counter current heat exchanger to cool gas to its liquid state.	77 K	0.25 W	Low vibration No moving parts in the system in contact with the detector Low noise levels	Irreversible process that renders it relatively inefficient. Contamination that can occur in the small orifice (any condensable impurity tends to freeze out and block orifice).
	Claude	Has similar principle to Joule–Thomson but the Claude uses a reversible engine expansion mechanism to replace the irreversible Joule–Thomson expansion mechanism.	Single stage: 35 K Multiple stages: 4.2 K		No liquid Relatively tolerant of contaminants	Medium pressures. Parts must operate at low temperatures. Complex Reliability issues.
	Stirling	Uses two isothermal processes known as the Stirling cycle wherein repetitive compression and expansion are used in the cooling process.	1-stage cycle: variable (down to 20 K) 2–3-stage cycle: 4.2 K	1-stage cycle: 0.8 W 2–3-stage cycle: 0.2 W	Tolerant of contaminants Light weight Low pressure Efficient 2-stage cycle: Low temperature	Noise. Vibrations. Effects of microphonics.
	Thermoelectric	The principle of these coolers is based on the Peltier phenomenon that occurs when current flows through the junction of two materials.	170 K	1 W	Lightweight Low vibration	High temperature. Low efficiency. High currents.

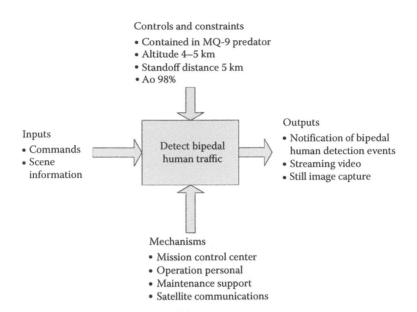

Controls and constraints
- Contained in MQ-9 predator
- Altitude 4–5 km
- Standoff distance 5 km
- Ao 98%

Inputs
- Commands
- Scene information

Detect bipedal human traffic

Outputs
- Notification of bipedal human detection events
- Streaming video
- Still image capture

Mechanisms
- Mission control center
- Operation personal
- Maintenance support
- Satellite communications

FIGURE 10.32
Concept diagram of UAV optical system. (Adapted from Blanchard, B.S. and Fabrycky, J.W., *Systems Engineering and Analysis*, Prentice Hall, Upper Saddle River, NJ, 2011.)

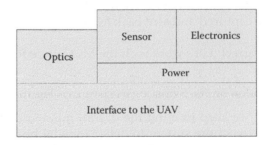

Sensor

Electronics

Optics

Power

Interface to the UAV

FIGURE 10.33
High-level Functional Blocks for UAV Optical System. (Courtesy of Andrea S. Rivers, with permission.)

Jen: "This functional block diagram gives a high-level functional description of the optical system and how it relates to the rest of the UAV. This is a good starting point for our functional decomposition."

Ron: "Well, we know this already. Do we really have to write this down? It seems common sense to me."

Jen: "Patience, young Jedi, and all will become clear. Functional analysis is a powerful approach to understanding stakeholder requirements and allocating them to lower-level subsystems and subsystem requirements. You start out by establishing a very complete understanding of all of the functionality that the system needs to provide. Defining and analyzing the system over its entire life cycle from concept through development, production, deployment, operation, and even disposal does this. This allows us to avoid surprises like not having a way to test a subsystem or missing a key interface that will be used in calibrating or running diagnostics. This process also helps everyone to develop a shared understanding

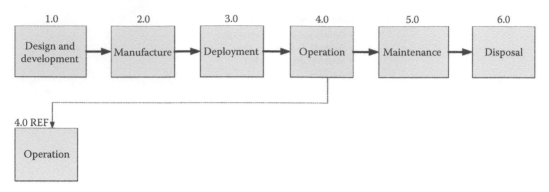

FIGURE 10.34
FFBD UAV program life-cycle phases. (Courtesy of Andrea S. Rivers, with permission.)

of the deliverable, to find alternate approaches to delivering that functionality. FBDs and FFBDs are the tools to accomplish the functional decomposition."

Ron: "Sounds good, but I think I need to see it."

Jen: "All right. So let's begin building the FFBD. Let me start by showing the different phases in the life cycle of the system and their connectivity. We will begin by working through the operational FFBD and then come back to the other phases. The functional flow diagram shown in Figure 10.34 shows the UAV systems life cycle."

Jen: "So, what is the desired operational function of the system?"

Ron: "Images need to be captured and sent back to the ground station or mission control center (MCC)."

Jen: "That's essentially correct. Let's work with the MCC piece for now and develop that further."

(Jen captures the information on the Samsung as shown in Figure 10.35.)

Karl: "It is essential that we keep in mind the overall mission that the system needs to accomplish. We need the system to detect bipedal humans. Once an image is captured, it is sent back to the MCC. But we need to realize that the image sent might be only a part of the image. Also, we need to ensure that the image transfer occurs in 5 seconds as required. As long we keep these issues in mind, we won't be constrained by a particular solution right away."

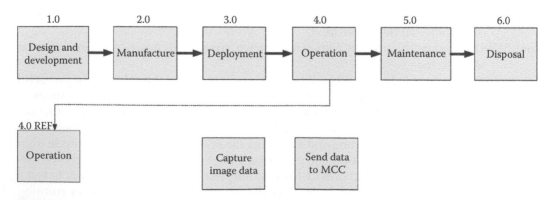

FIGURE 10.35
FFBD UAV optical system operational flow progression 1. (Courtesy of Andrea S. Rivers, with permission.)

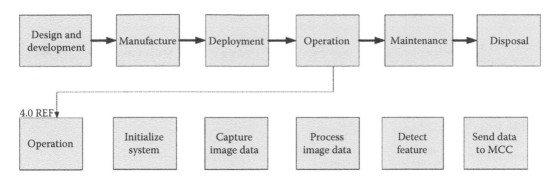

FIGURE 10.36
FFBD optical system operational flow progression 2. (Courtesy of Andrea S. Rivers, with permission.)

Ron: "What about the system operation when the system is between the point when it is on the ground and the point where the image data will be captured? I think we might have more functions here. Do we have functions related to powering up the system? Does it need to perform preflight operational checks?"

Jen: "Now you are getting the idea!"

(Jen adds several blocks to the MBSE model on the Samsung as shown in Figure 10.36.)

Karl: "Yes that is a valid point. Often the initialization of the system is forgotten. We also need to monitor the system when it is up in the air. Also, might I suggest the addition of redundancy to the system as the requirements specify that it needs to be operational 24 hours a day. Wait, the monitoring of the system should go under maintenance."

Ron: "Maybe the operational modes, manual or automatic, need to be a part of the FFBD as well?"

Karl: "Yes that is true. Jen, please go ahead and update it. Also go ahead and number the blocks we have listed. Let's flesh out this thing!"

(They continue working on updating the optical system FFBD. Jen displays the finished product on the Samsung [Figure 10.37].)

Ginny: "Let's not forget what Karl said about monitoring the health status of the optical system. Even though the monitoring occur during operations, I think we can put these tasks under maintenance, since they keep the system operational."

Karl: "Jen, can you draw a high-level FFBD for maintenance? (Figure 10.38)"

Amanda: "We also need to account for the condensation on the on-board optics as the system will travel from the ground to its operational altitude."

Karl: "Can you expand the compensate temperature block?"

Jen: "Certainly."

(Jen updated the temperature compensation FFBD as shown in Figure 10.39.)

Karl: "Some other ways the system could be used in operations are not taken into account by this FFBD. If we stop to consider the possibility that the system fails, we will have to ensure that the system can tolerate the failure, because a 98% operational availability is another one of our requirements. If we are unable to meet our availability target for the system, we may have to include backup components that would automatically engage if the primary component fails. A diagnostic check to see if the switch up is needed can also be performed."

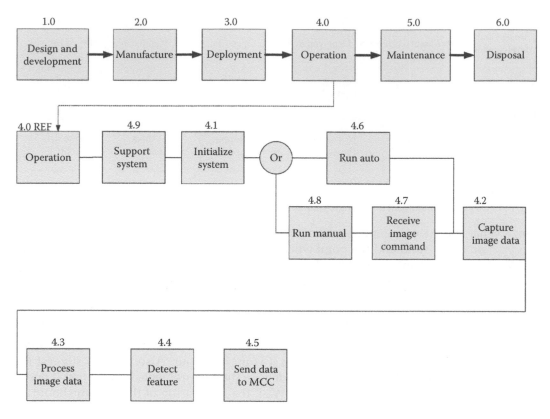

FIGURE 10.37

FFBD operational flow progression 3. (Courtesy of Andrea S. Rivers, with permission.)

(Jen incorporated the backup system into Figure 10.40.)

Karl: "This is good enough for now, let's move on to functional allocation. We have generated a lot of TPMs and we need to allocate them through the various levels of the functional decomposition. For instance, our TPM for the optical system's operational availability needs to be allocated down to the subsystems level and eventually down to the components/assemblies/parts. The chart showing the allocation of TPMs down to the next level will ultimately get expanded and fully filled in, as the design evolves (Blanchard and Fabrycky 2011). The optical system needs to be designed so that the allocated TPMs in each block are satisfied."

(Jen shows the allocated TPMs [Figure 10.41].)

Jen: "The concept diagram view of the cooling system in tabular form, based on our discussion and FFBD's, is as follows."

(Jen brings up Figure 10.42.)

Jen: "As you can see, the unique identifier from the FFBD we did previously has been referenced. As we move further along the process, we start to see the design taking shape. Later, we will begin working bottom-up unlike the top-down approach we have taken so far. For example, for one of the hardware subsystems, we might decide to use the RS-232 connector and hence the interface for that input is defined. However, now we just want to recognize that functionally that needs to exist."

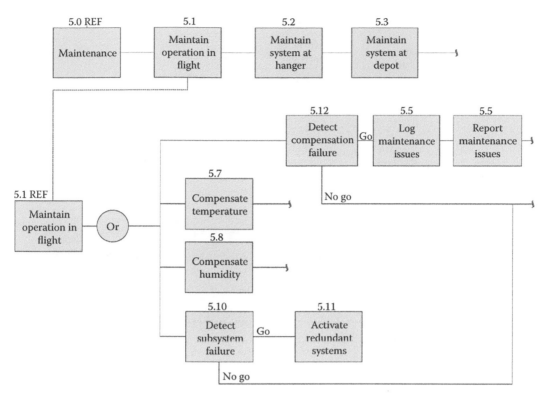

FIGURE 10.38
FFBD maintenance. (Courtesy of Andrea S. Rivers, with permission.)

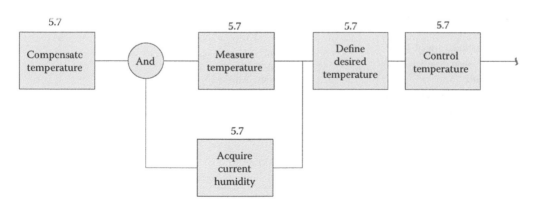

FIGURE 10.39
Compensate temperature FFBD. (Courtesy of Andrea S. Rivers, with permission.)

Jen: "To see the high-level elements of the system in one place, we can capture a high-level block diagram."

Jen brings up a block diagram as shown in Figure 10.43 where all the major subsystems in the UAV optical detection system are depicted. The communications interface between the main control system, the local control for each component of the system, and the power system are also included in this diagram. As the process continues, each of the blocks in

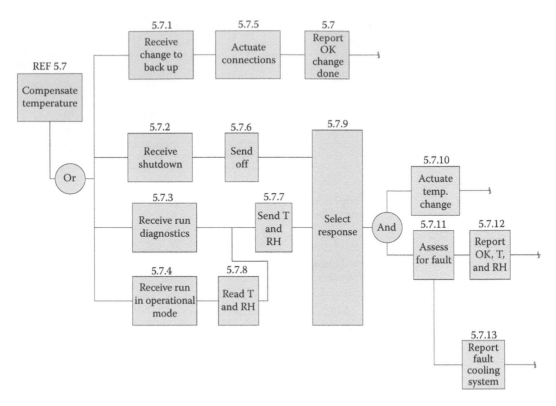

FIGURE 10.40
Compensate temperature FFBD progression 2. (Created by Andrea S. Rivers, with permission.)

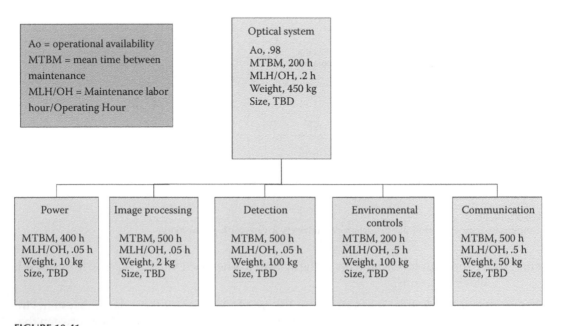

FIGURE 10.41
Functional allocation of TPMs. (Reprinted with permission from Andrea S. Rivers, 2013; adapted from Blanchard, B.S. and Fabrycky, J.W., *Systems Engineering and Analysis*, Prentice Hall, Upper Saddle River, NJ, 2011.)

Cooling 5.7	Commands from the system controller	MTBF, 200 h MLH/OH, 0.5 h Weight, TBD Size, TBD	Mechanical structure	Status messages
	Humidity detector signal (TBDV DC)		Thermal interface to sensor	Temperature change
	Temperature sensor signal (TBD V DC)		Power (TBD V, TBD A)	
	Current sense from cooling driver circuit		Communications cabling	

FIGURE 10.42
Concept diagram cooling system in tabular form. (Created by Andrea S. Rivers, with permission.)

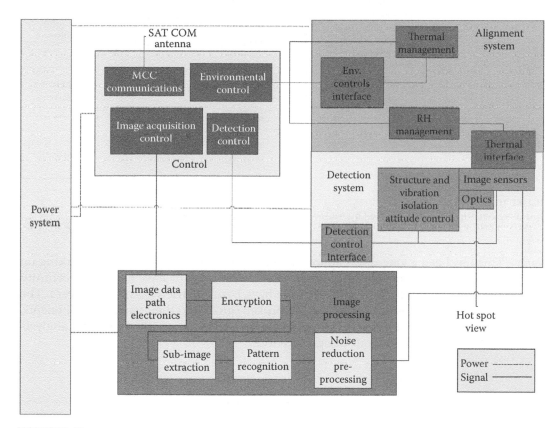

FIGURE 10.43
UAV optical detection system block diagram. (Created by Andrea S. Rivers, with permission.)

the diagram will be broken down into lower levels like with the FFBD's depicting the components, assemblies, parts, and interfaces of each subsystem.

Jen: "We can also show which functions fall under which section of the system by creating a functional breakdown diagram." Jen shows Figure 10.44. This is yet another way to describe the system."

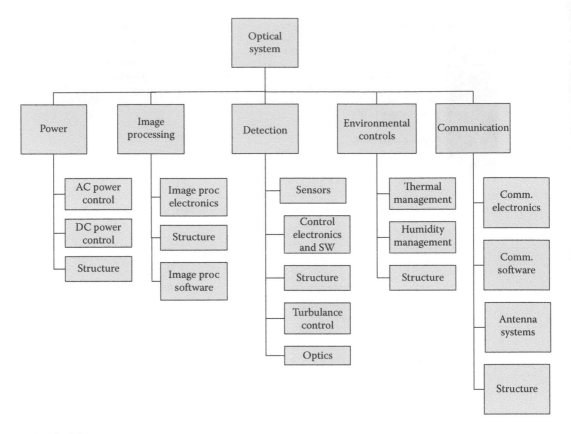

FIGURE 10.44
Optical system top-level FBD. (Created by Andrea S. Rivers, with permission.)

Karl: "This has been a long day. I appreciate your patience. We have started working on some FFBD's, concept diagrams, and FBDs and all of us have had an opportunity to practice with them. However, we have barely started with this process. The high-level FBDs need to be further broken down into their various components, assemblies, and parts. As you can see, we will be using the functional analysis process throughout our design activities."

References

Adwaele. 2010. The Stirling cycle in four steps. Wikipedia. https://en.wikipedia.org/wiki/File:Stirling_Cycle_Cryocooler.jpg (accessed April 7, 2014).

Allen, W. A. 1997. Thermoelectrically cooled IR detectors beat the heat. Laser Focus World. Vol. 33. http://www.laserfocusworld.com/articles/print/volume-33/issue-3/world-news/thermoelectrically-cooled-ir-detectors-beat-the-heat.html (accessed December 1, 2014).

Arlitt, R.M., K.D. Balinski, C.H. Dagli, and K. Grantham. 2011. Functional analysis of systems using a functional basis. Paper presented at *2011 IEEE International Systems Conference (SysCon)*, Montreal, Quebec, Canada, April 4–7, 2011, pp. 563–568.

Augustine, N.R. 1983. Norman Ralph Augustine. http://www.qotd.org/search/search.html? aid=5218&page=3 (accessed April 12, 2014).

Axis Communications. 2010. CCD and CMOS sensor technology. http://www.axis.com/files/ whitepaper/wp_ccd_cmos_40722_en_1010_lo.pdf (accessed April 1, 2014).

Banerjee, P., N. Shenoy, A. Choudhary, and S. Hauck. 2000. A MATLAB compiler for distributed, heterogeneous, reconfigurable computing systems. Paper presented at *IEEE Symposium on Field Programmable Custom Computing Machines*, Napa Valley, CA, April 17–19, 2000, pp. 39–48.

Bernard, S. 2005. *An Introduction to Enterprise Architecture*, 2nd edn. Bloomington, IN: Author House.

Blanchard, B.S. and J.W. Fabrycky. 2011. *Systems Engineering and Analysis*. Upper Saddle River, NJ: Prentice Hall.

Bonnema, G.M. 2006. *Function and Budget Based System Architecting*. Delft, the Netherlands: Delft University of Technology.

Bonnema, G.M. 2010. Insight, innovation, and the big picture in system design. *Systems Engineering*, 14(3): 223–238.

Brazier, K. 2008. A diagram of a thermoelectric cooler, January 11, 2008. Wikipedia. http:// en.wikipedia.org/wiki/File:Thermoelectric_Cooler_Diagram.svg. Derivative work of Cullen C. Thermoelectric Cooler. http://en.wikipedia.org/wiki/File:ThermoelectricCooler.jpg, 2007 (accessed April 7, 2014).

Brisolara, L.B., M.F.S. Oliveira, and R. Redin. 2008. Using UML as front-end for heterogeneous software code generations strategies. Paper presented at *Design Automation and Test in Europe*, Munich, Germany, March 10–14, 2008, pp. 504–509.

Cobb, J., J. Hawver, A. Rivers, and R. Morton. 2004. Detection of pitch variation in lenticular material. US Patent 6727972, filled April 5, 2002 and issued April 27, 2004.

Crowe, T.J., J.S. Noble, and J.S. Machimada. 1998. Multi-attribute analysis of ISO 9000 registration using AHP. *International Journal of Quality and Reliability Management*, 15(2): 205–222.

Department of Commerce. 1993. *Draft Federal Information Processing Standards Publication 183: Integration Definition for Function Modeling*. Gaithersburg, MD: Department of Commerce, National Institute of Standards and Technology.

Dereniak, E. and G.C. Devon. 1984. *Optical Radiation Detectors*, 1st edn. New York: Wiley.

Eriksson, M., J. Börstler, and K. Borg. 2006. Performing functional analysis/allocation and requirements flow down using use case realizations—An empirical evaluation. Paper presented at *16th International Symposium of the INCOSE 2006*, Orlando, FL, July 2006.

Gow, R. 2007. A comprehensive tool for modeling CMOS image-sensor noise performance. *IEEE Transactions on Electron Devices*, 54(6): 1321–1329.

Hands, B.A., ed. 1986. *Cryogenic Engineering*. London, U.K.: Academic Press.

Hirtz, J., B.S. Robert, A.M. Daniel, S. Simon, and L.W. Kristin. 2002. A functional basis for engineering design. NIST Technical Note 1447, National Institute of Standards and Technology, US Government Printing Office, Washington DC., pp. 65–82.

Hoffmann, H.-P. 2009. Rational statemate from code to concept. Armonk, NY: IBM Corporation. https://publib.boulder.ibm.com/infocenter/rsdp/v1r0m0/topic/com.ibm.help.download. statemate.doc/pdf46/concept_to_code.pdf (accessed April 1, 2014).

Hudson, R.D. 1969. *Infrared System Engineering*. New York, NY: John Wiley and Sons.

INCOSE. 2004. INCOSE home page. www.incose.org (accessed July 30, 2014).

Kamrani, A. and M. Azimi, eds. 2010. *Systems Engineering Tools and Methods*. Boca Raton, FL: CRC Press.

Kennedy, K. 2005. Telescoping languages: A system for automatic generation of domain languages. *Proceedings of the IEEE*, 93: 387–408.

Keyes, R.J. 1977. *Topics in Applied Physics—Optical and Infrared Detectors*. Berlin, Germany: Springer-Verlag.

Kharboutly, R.A. 2008. *Architectural-Based Performance and Reliability Analysis of Concurrent Software Applications*. Ann Arbor, MI: ProQuest LLC.

Kinni, T.B. 1993. What's QFD? Quality function deployment quietly celebrates its first decade in the US. *Industry Week*, 242: 31–32.

Lin, D.-L. 2010. Quantified temperature effect in a CMOS image sensor. *IEEE Transactions on Electron Devices*, 57(2): 422–428.

MathWorks. 2013. Simulink simulation and model based design. MathWorks. http://www.mathworks.com/products/simulink/ (accessed April 1, 2014).

Michbich. 2010. Schematic of a Peltier device. Wikipedia. http://en.wikipedia.org/wiki/File:Peltierelement.png (accessed April 7, 2014).

Min, G., D.M. Rowe, and K. Kontostavlakis. 2004. Thermoelectric figure-of-merit under large temperature differences. *Journal of Physics D: Applied Physics*, 37: 1301–1304.

Morton, R. Apparatus for image display utilizing lenticular or barrier screens. US Patent 6078424, filed July 13, 1998 and issued June 20, 2000.

NASA; S.L. Rickman, R.G. Iacomini et al. 2010. Passive radiative cooler for use in outer space. http://www.techbriefs.com/component/content/article/9-ntb/tech-briefs/physical-sciences/6996 (accessed June 25, 2014).

Panko, R.R. 1998. What we know about spreadsheet errors. *Journal of End User Computing's*, 10(2): 15–21. (Revised 2008).

Paredis, C. 2011. System analysis using SysML parametrics: Current tools and best practices. Atlanta, GA: Georgia Institute of Technology Model-based Systems Engineering Center, Georgia Tech. http://www.modprod.liu.se/MODPROD2011/1.252922/modprod2011-tutorial4-Chris-Paredis-SysML-Parametrics.pdf (accessed April 1, 2014).

Petrie, G. and A.S. Walker. 2007. Airborne digital imaging technology: A new overview. *The Photogrammetric Record*, 22(119): 203–225.

Project Management Institute. 2013. *Project Management Body of Knowledge*. Newtown Square, PA: PMI, Inc.

Rational Software Corporation. 2002. Rational suite, Version. 2002.05.00. ftp://ftp.software.ibm.com/software/rational/docs/v2002/rs_intro.pdf (accessed April 1, 2014).

Shi-Xiang, T. 2010. The conceptual design and simulation of mechatronic systems based on UML. Paper presented at *Second International Conference on Computer Engineering and Technology (ICCET)*, Chengdu, China, April 16–18, 2010, pp. V6 188–V6 192.

Simpson, J.J. 2009. System of systems complexity identification and control. *IEEE International Conference on System of Systems Engineering, 2009 (SoSE 2009)*, Albuquerque, NM, May 30–June 3, 2009, pp. 1–6.

Sumi, H. 2006. Low-noise imaging system with CMOS sensor for high quality imaging. *International Electronic Devices Meeting 2006 (IEDM)*, San Francisco, CA, December 11–13, 2006, pp. 1–4.

Wiecek, B. 2005. Cooling and shielding systems for infrared detectors: Requirements and limits. *27th Annual Conference of the IEEE Engineering in Medicine and Biology*, Shanghai, China, September 1–4, 2005, pp. 619–622.

11

Requirements Allocation

It's not enough that we do our best; sometimes we have to do what's required.

—**Winston Churchill**

11.1 Requirements Allocation Process

According to systems engineering and analysis, systems engineering is "A top-down approach that views the system as a whole" (Blanchard and Fabrycky 2010). Systems engineering integrates all phases of the life-cycle development effort and unites diverse disciplines, organizations, and groups into a cohesive team. It also provides a structured design and development process that consists of the conceptual design phase, preliminary design phase, and the detailed design and development phase (Ascendant Concepts LLC-Home 2013). There are three main elements of a system during these three design phases: the parts of a system, attributes or properties of the components, and relationships or how the components are connected. The production/manufacturing/construction phase, operations and support phase, and the system retirement and disposal phase make up the remainder of the standard systems engineering life-cycle phases. Requirements allocation activities are fundamentally connected to the design activities during the three systems engineering design phases mentioned earlier, and so we focus our attention on these three design phases.

Requirements lie at the heart of systems engineering methods, principles, and techniques. In fact, systems engineering developments are requirements-driven activities, and the systems engineering methods and processes are the basis for effectively developing a system that meets stakeholder needs. Stakeholder needs evolve into stakeholder requirements during the *problem definition* process. The conceptual design, preliminary design, and detailed design and development phases all have requirements generation processes in ever-increasing detail. The initial requirements/requirements baseline is the driving agent for the design activity in each phase. For example, the stakeholder requirements are the initiating drivers in the conceptual design phase. The purpose of the conceptual design phase is to satisfy the stakeholder requirements by generating system-level design concepts, down-selecting to an optimal conceptual design, and develop, approve, and baseline corresponding system-level requirements (e.g., the A spec) for the down-selected concept. The functional baseline is established with the approval of the A spec at the Systems Requirements Review (SRR). This marks the end of the conceptual design phase.

The system-level requirements contained in the A spec then serve as the drivers for the design activities during the preliminary design phase wherein the subsystems are designed. A set of subsystem requirements is generated for each of the hardware and software subsystems along with interface control documents (ICDs). These requirements are documented in development specifications (B specs), and the allocated baseline is established at the Preliminary Design Review (PDR) marking the end of the preliminary design phase.

In the final design phase, the detailed design and development phase, the B spec requirements documents are used to drive the detailed design and development phase activities. The requirements outputs of this phase are product specifications (C specs), process specifications (D specs), and material specifications (E specs) during the product baseline milestone event, the Critical Design Review (CDR).

Within each phase, a set of conceptually repetitive processes occurs. At the beginning of any given phase, a requirements document or set of requirements documents under configuration control are used to drive the phase's design activities. First, a functional analysis is accomplished to decompose the functions specified in the requirements documents to the next lower level. Then the requirements allocation process is used to generate the requirements for the functions at the decomposed level.

For example, at the system level, requirements allocation is the process of distributing higher-level system requirements into multiple lower-level subsystem requirements.

Requirements allocation is a repetitive and essential part of both the requirements generation and the design processes. Requirements allocation first begins during the conceptual design phase, and the requirements allocation process repeats in every design phase. Figure 11.1 shows where requirements allocation fits with respect to other fundamental systems engineering processes.

In Figure 11.1, the pathway to requirements allocation starts with the definition of the requirements themselves. As previously stated, the starting point in the conceptual design phase is the stakeholder requirements. In the preliminary design phase, the A specification

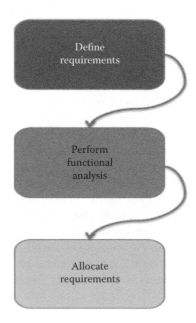

FIGURE 11.1
Pathway to requirements allocation. (Courtesy of Katharine King.)

is the starting point, and in the detailed design phase, the B specifications are the starting points. The second block in Figure 11.1 shows that a functional analysis is accomplished on the defined requirements. Among other things, the functional analysis decomposes higher-level functions into lower-level functions. Functional block diagrams (FBDs) and functional flow block diagrams (FFBDs) are useful in this regard. Functional analysis was discussed in a previous chapter. Once the lower-level functional decompositions are accomplished, the higher-level requirements from the top block of Figure 11.1 can be allocated to the lower-level functional decompositions in the "Allocate Requirement" step of Figure 11.1. These three steps are interconnected and repeated at each design phase in the development of a system, ensuring that requirements at each level of the system are distributed among the lower-level system decompositions in an efficient and complete manner.

11.1.1 Derivation, Allocation, and Apportionment

In the requirements allocation process, the functional blocks that have been identified by the functional analysis are grouped into logical partitions. Where possible, similar functions identified through separate requirements are grouped together. Once the functionally decomposed blocks are properly partitioned, the higher-level requirements are connected to the lower-level partitions in three ways: (1) derivation, (2) allocation, and (3) apportionment. Derivation is where a new, lower-level requirement is derived to define what is needed in the lower-level functional decomposition. As an example, consider a functional decomposition from the system-level operational requirement, "The system shall be capable of detecting and distinguishing human subjects in both day and nighttime environments." Let us assume that feasibility studies and trade studies resulted in a passive daytime visible telescopic imaging system and a passive telescopic infrared (IR) nighttime system being chosen as technical solutions to this requirement. The functional analysis could define two blocks as (1) daytime imaging sensor subsystem and (2) nighttime imaging sensor subsystem. We know that both systems are telescopic, so we can further decompose each of these blocks into (1) telescopic subsystem, (2) relay optical subsystem, (3) detection subsystem, (4) signal processing subsystem, (5) control subsystem, (6) communications subsystem, (7) power subsystem, and (8) the environmental conditioning subsystem. Each of these blocks could be further decomposed; however, let us look at the telescopic subsystem. For this block, additional trade studies are required. What kind of telescopic subsystem should be designed? Gregorian? Herschelian? Cassegrainian? Newtonian? The answer depends on a trade study that must be conducted that evaluates design considerations such as size, weight, power, interface mechanisms and locations for the relay optics, mounting, command and control, integration with other needed components, reliability, maintainability, supportability, and packaging issues, to name a few. Once the trade study is completed, a set of requirements must be derived through the requirements derivation process that fully describes the necessary attributes of the telescope subsystem so that the design and development team has the information needed to accomplish their tasks. A series of subsystem-level requirements, such as "The telescopic system for the daytime visible sensor shall have a primary mirror radius of curvature of 3 meters," will need to be derived for both the daytime visible telescopic system and the nighttime IR telescopic system.

An apportioned requirement is one in which a particular metric is shared/distributed through the lower-level functional decompositions. An example would be a weight budget. If the system-level requirement stated that, "The optical subsystem weight shall not exceed 120 lbs.," and then the technical performance measure (TPM)—weight—needs to be apportioned throughout all the subsystems of the optical system. For example, what is the maximum weight of the telescopic system? What is the maximum weight of the daytime sensor, the nighttime sensor, the daytime sensor relay optics, the nighttime sensor relay optics, the signal processing subsystem, and so forth? An engineering analysis would determine these values, and the sum would be ≤120 lb (typically less, since there is usually a reserve margin set to offset unexpected overages). Other examples of apportioned requirements include power budgets, reliability budgets, jitter budgets (the total acceptable amount of jitter that the optical system will experience), and cost.

Allocated requirements are similar to apportioned requirements in that multiple lower-level requirements can be generated from one higher-level requirement. For instance, in the previous example, the system-level operational requirement that the system must be able to detect and distinguish human subjects, in daytime and nighttime conditions, resulted in two representative, functional, decomposed blocks (the daytime visible sensor and the nighttime IR sensor), wherein both blocks need to comply with this requirement. The requirement to detect and distinguish human subjects is allocated to both the daytime visible sensor and the nighttime IR sensor.

Whether through allocation, apportionment, or derivation, enough requirements need to be generated at the functionally decomposed level (e.g., subsystem level in this case) so that the design team knows exactly how to proceed and that all the required information for the design at the functionally decomposed level has been specified. If this is not the case, then additional feasibility studies, trade studies, and/or requirements need to be generated at the functionally decomposed level. All higher-level requirements and functional and nonfunctional, quantitative, and qualitative requirements undergo the requirements allocation process. All performance factors, human factors, physical factors, and factors associated with the "ilities" such as producibility, supportability, and disposability must be included. Traceability of the requirements must also be established. Requirements at the lower-level functional decomposition must be traceable to the higher-level driving requirement and vice versa. Requirements must also be traceable to the qualification program and its artifacts including the test plans, test procedures, and (eventually) test reports.

11.1.2 Levels of Requirements (Leveling) and Requirements Allocation

As previously pointed out, the requirements allocation process occurs in every design phase of the system. Figure 11.1 shows the connection of the requirements allocation process with the requirements generation and functional analysis processes in the first two blocks of the figure. These three processes repeat and occur in every systems design phase, and they can be grouped into two complementary activities: (1) generate requirements and (2) design based on the requirements. These two processes are often referred to as the "whats" and the "hows." In the first block of Figure 11.1, the requirements are defined. These requirements are written in terms of what needs to be done—the "whats." The requirements generation process is used to define the requirements at any particular requirements level (stakeholder, system, subsystem, or component/assembly/part). Given a baseline set of requirements, the emphasis shifts to design. The newly generated requirements become the basis for driving the design. The design activity consists of the functional analysis and requirements allocation processes along with feasibility studies, trade studies, modeling and simulation, prototyping, and analytical activities needed to fully specify the lower-level functional decomposition requirements (Robertson and Robertson 2012). In effect, the design process generates the lower-level requirements. These lower-level requirements specify "how" the higher-level driving requirements will be accomplished, but they are still written as requirements. They specify "what" needs to be done at the lower level. An example is in order to illustrate this point. In our previous example for the visible sensor "telescopic subsystem" block, one of the lower-level-derived requirements was "The telescopic system for the daytime visible sensor shall have a primary mirror radius of curvature of 3 meters." This requirement was derived from the system-level requirement that the optical system needed to detect and distinguish human subjects. The higher-level system requirement defined "what" needed to be done, and the lower-level sub-system requirement showed one aspect of "how" to accomplish it (e.g., provide a mirror with a 3 m radius of curvature). This subsystem requirement was generated as part of the design process responding to the system-level requirements. On the other hand, at the subsystem level, this same requirement in combination with other derived, apportioned, and allocated subsystem requirements specifies "what" needs to be done. What needs to happen? A mir-ror needs to be procured, or developed, that has, among other things, a radius of curvature of 3 m. There will be numerous other requirements that need to be defined at the subsystem level before such procurement/development can take place. Requirements that address the size and weight of the primary mirror, the material of the mirror, the flatness and uniformity of the surface, its coatings, how it will be mounted, its relative positioning to other optical components, transport, storage, testing, installation, cleaning, repair, spares, and more must all be generated. Lower-level requirements must be generated for each of these attributes, and a make-or-buy trade study will need to be accomplished before the mirror is acquired.

Note that the design process is linked to, and driven by, a particular requirements level. There are four levels of requirements in the systems engineering process: (1) stakeholder-level requirements, (2) system-level requirements, (3) subsystem-level requirements, and (4) component/assembly/part-level requirements. Each of these levels is more detailed than the previous level, and each has functional analysis and requirements allocation processes that are associated with them. The stakeholder requirements are the input to the concep-tual design phase, and the corresponding functional analysis and requirements allocation processes produce the A specification (system specification) as previously stated. The pur-pose of the stakeholder-level requirements is to define what needs to be accomplished in the system development effort (e.g., what the system must accomplish and the qualities that the system must possess) (Robertson and Robertson 2012). Once stakeholders are identi-fied and their needs established, a list of stakeholder requirements could be compiled for further evaluation and transformation into system requirements (Halaweh 2012). As an example, some of the stakeholder requirements for the unmanned aerial vehicle (UAV) optical system case study discussed in Section 11.3 are shown in Table 11.1.

TABLE 11.1

Stakeholder Requirements

Reference	Stakeholder Requirement
SR-1	Shall provide image products to the MCC within 5 s of image acquisition
SR-2	Shall provide on-demand pictures and/or video
SR-3	Shall provide bipedal human detection and classification capability
SR-4	Shall provide capability to support an operational tempo of 24 h per day, 7 days a week
SR-5	Shall provide capability to integrate with existing systems (e.g., communications systems, control systems, power systems, and pointing and tracking systems) of the UAV platform using standard connectors and interface protocols
SR-6	Shall provide a capability that is easily maintainable by the stakeholder's maintenance division
SR-7	Shall provide a capability that supports future upgrades from the vendor
SR-8	Shall provide a capability that provides for both an automatic and a manual control mode
SR-9	Shall provide a means to store locally (on-board) 2 h of mission critical data

These requirements express what the system is expected to do. The system-level design process starts with the stakeholder requirements, CONOPS, and supportive information and involves several key steps that transform the stakeholder requirements into a feasible and optimal conceptual design of the system. Once the stakeholder requirements have been identified, system planning and architectural design are initiated, and a program management plan (PMP) is developed that provides guidance for managerial and technical activities to follow (Leonard 1999). The PMP provides required guidance for the management of the systems development effort as a whole and provides guidance for the development of the systems engineering management plan (SEMP). Some of the fundamental activities in the conceptual design phase are the development of the system-level architecture, functional architecture, operational requirements (A spec), conducting feasibility analyses, evaluating alternative technical concepts, developing the maintenance and support concepts, TPMs, and functional analysis and requirements allocation from

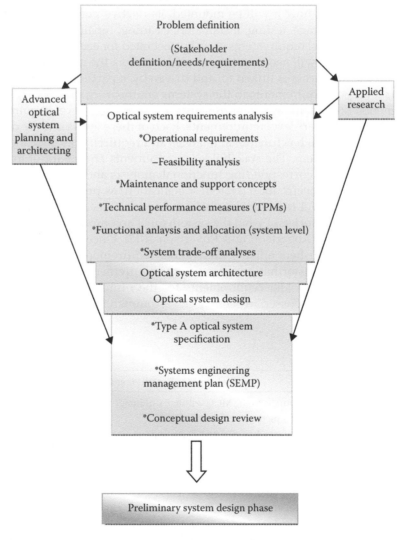

FIGURE 11.2
System-level requirements definition and allocation. (Courtesy of Katharine King.)

the stakeholder level down to the system level. This process of conceptual design leads to the creation and *baselining* of the system specification, also known as the type A or A spec (Blanchard and Fabrycky 2010). These resulting system-level requirements provide a framework for the initiation of the preliminary design phase. Figure 11.2 summarizes the major conceptual design phase activities for a general optical system.

The upper blocks in Figure 11.2 indicate the problem definition step wherein stakeholders are identified along with their needs and approved wants resulting in a set of stakeholder requirements for the optical system. In the next block, these requirements are analyzed for feasibility (e.g., technical, cost, operational, motivational, schedule, and organizational) and related to risk. The functional analysis and requirements allocation processes are used to develop the system-level operational requirements, and the maintenance and support concept planning activities are initiated. As the system-level requirements are generated through the conceptual design process, high-level technical performance parameters and key performance parameters (KPPs) emerge and complement established measures of effectiveness (MOEs) and measures of performance (MOPs).

Trade studies and feasibility studies are essential activities in the conceptual design process to eliminate unacceptable conceptual design options and determine the optimal concept. Once the optimal concept is selected, the architectural components are selected (enterprise architecture, systems architecture, and physical architecture), and the functional analysis and requirements allocation processes are used to generate the system requirements (A spec) for the optical system. The SEMP is produced, and the A spec is placed under configuration control at the conceptual design review milestone event, typically the System Requirements Review (SRR). Approval of the A spec at this review traditionally signifies the end of the conceptual design phase and the start of the preliminary design phase.

11.1.3 Commercial Off-the-Shelf Considerations and TPM Allocation

In the requirements allocation process, new requirements are generated at the functionally decomposed level by apportionment, derivation, or allocation. Typically, the process progresses through each design phase of the systems development effort. However, it is possible to skip a requirements level under certain circumstances. The preliminary design phase is intended to capture high-level design information. So what happens if no design is required for a particular component such as for a commercial off-the-shelf (COTS) item? In this case, no preliminary design is necessary for the COTS item, and the requirements flow directly from the system spec (A spec) to the product spec (C spec). Figure 11.3 illustrates this concept and shows the results of the TPM allocation process for three COTS products that are used in the optical system. The TPMs at the system level are identified at the top optical system block and allocated downward to the optical system's three components (component 1, component 2, and component 3). The components can be further decomposed, if necessary, to the assembly or piece level if needed. Note that this additional decomposition is still at the detailed design and development level and no additional design phases are generated. In Figure 11.3, we show one piece associated with each component for illustrative purposes. Note how the TPMs are apportioned downward through the functional decomposition. For instance, notice how the sum of the costs of the three components totals to the total cost in the optical system block. Notice also how the other TPMs seem to satisfy the higher-level optical system TPM allocations. But wait! There is a problem with the size! The component sizes will not fit in the $3 \times 3 \times 3$ ft optical system–level size allocation. Perhaps COTS components were not such a good idea! The engineering team must fix this issue! Functionally decomposing the TPMs helps identify problems like this.

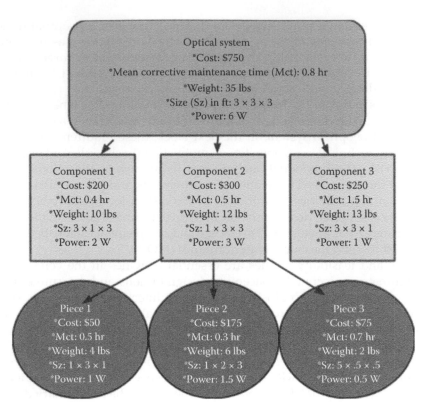

FIGURE 11.3
Requirements allocation at the subsystem level. (Courtesy of Katharine King.)

Note that there is only one A spec, but there likely will be multiple B specs, if design is involved. If COTS items are present, there are functional decompositions directly from the A spec to the C specs. If the engineering team needs to do design work, then the preliminary design phase cannot be skipped. If there is a mixture of design work and the integration of COTS products, then the functional decomposition will involve B specifications for all the design work and direct links to the respective product specifications (C specs) for the COTS items. The allocated requirements at any level can be improved or changed during any regular requirements review or through calling meetings of the configuration control board (CCB). Additional requirements allocation may also take place during the production/construction phase, the operational use, and system support phase of the systems development life cycle in order to incorporate new requirements, modifications, and any requirements changes deemed necessary by the CCB (Wasson 2006).

11.1.4 Connection with Functional Analysis

Functional analysis represents the intermediate step prior to the requirements allocation process, defining transformation of the system's requirements at each level of the system development into a detailed set of design criteria and identifying the resources required for system operation and support (Grady 2006; Blanchard and Fabrycky 2010). This functional analysis plays a critical role in establishing feasible system architecture and identifying an acceptable technical solution to the established set of driving requirements.

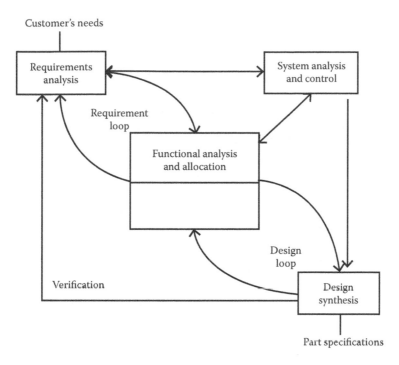

FIGURE 11.4
Functional analysis in systems engineering. (Courtesy of Dean Smith; adapted from DAU, Systems Engineering Fundamentals, Supplementary Text, Defense Acquisition Press, Fort Belvoir, VA, 2001.)

Figure 11.4 shows the role of functional analysis/requirements allocation in relationship to the fundamentally important requirements analysis and design synthesis processes.

Functional analysis and requirements allocation can be seen in the central box of Figure 11.4 and are shown as part of both the requirements analysis loop and the design synthesis loop. For example, after the identification of the customer or stakeholders' needs, the requirements analysis begins, as shown by the requirements analysis box at the top left of Figure 11.4. An iterative loop is formed between the requirements analysis block and the functional analysis/allocation block. As requirements are functionally analyzed, needful changes are looped back into the requirements analysis block, and resulting revised requirements are sent back to the functional analysis/allocation block. A similar loop occurs between the functional analysis allocation and the design synthesis blocks. The functional decompositions or allocated requirements may impact the design and vice versa. The systems analysis and control block monitor all the processes and artifacts. At completion, a new specification results appropriate for the particular design level. Overall, Figure 11.4 highlights the importance of the functional analysis and requirements allocation processes through its central location and its interconnections to the other critical systems engineering process blocks.

Two important tools in the functional analysis process are the FBD and the FFBD. FBDs represent a hierarchical, visual decomposition of functions from one requirements level to the next. The generation of the FBD can occur at any level starting with the stakeholder requirement but are most prevalently seen at the system requirements level and below. As an example, in Figure 11.5, we show an FBD for an optical subsystem on a UAV that will be showcased in Section 11.3. Notice that, as shown, the functional decomposition is from the perspective that the *optical system* is a subsystem of the UAV itself. The optical *subsystem*, therefore, has its

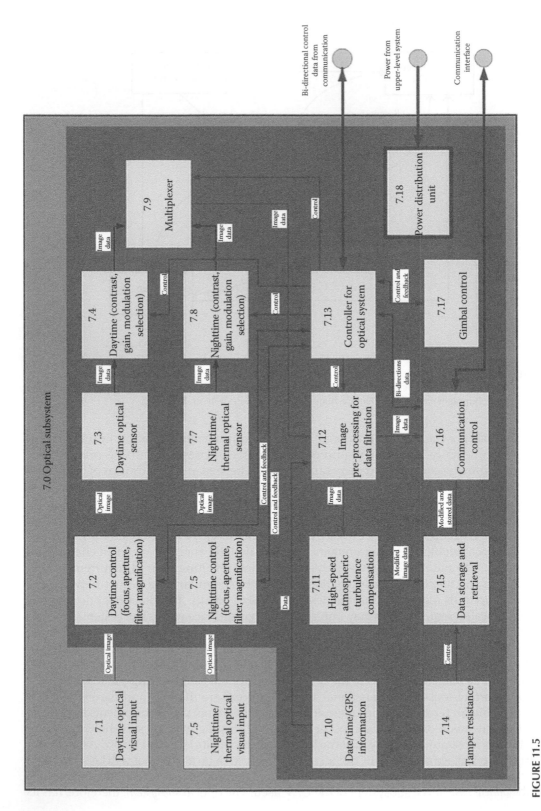

FIGURE 11.5
(See color insert.) Optical system FBD. (Courtesy of Dean Smith.)

own block in a larger FBD of the UAV. Other UAV system-level functional blocks might be the UAV's communication subsystem, flight control subsystem, power subsystem, and so forth. There may, however, be stakeholders that are only interested in the optical system itself. These stakeholders may consider the UAV as a host platform in which the optical system resides. For these stakeholders, the optical system is the primary system of interest and so would have its own set of stakeholder requirements. A representative set of stakeholder requirements for the optical system is shown in Table 11.1. Let us consider the functional decomposition from the perspective of the stakeholders interested in only the optical system itself.

In Figure 11.5, blocks labeled 7.1 through 7.18 represent the required functions of the optical system. These capture the *basic functions* that the system must accomplish. As an example, Block 7.18 within the figure represents the power grid supplied by the power distribution unit.

In addition to the decomposed functions, the control lines, power lines, and data lines between the decomposed functional blocks are also shown, as are some of the external interfaces. In addition to FBDs, some of the most useful tools in functional analysis are FFBDs. These block diagrams show sequential relationships and order of precedence. A conceptual difference between FBDs and FFBDs is that the blocks in the FBD can all be active at the same time, whereas the blocks in the FFBD are sequentially ordered. For instance, in Figure 11.5, the controller for the optical system (Block 7.13), the communications control (Block 7.16), and the data storage and retrieval (Block 7.15) could all be operating simultaneously and independent from each other. FFBDs are useful in breaking down higher-level functions into many lower-level functions when required to complete the system flow description in functional terms (Blanchard and Fabrycky 2010). FFBDs show the logical relations among functions derived from functional analysis in a graphical flowchart. The FFBDs have "AND" gates that signify parallel functionality that must be accomplished before proceeding through the flowchart and also "OR" gates that signify a single functionality path that can be used to traverse the flow (Pineda and Smith 2010). Figure 11.6 shows the operational steps needed to produce an image from a given initiating command sequence. The gimbal system, as part of the pointing and tracking system, aligns the boresight of the optical system with its intended target.

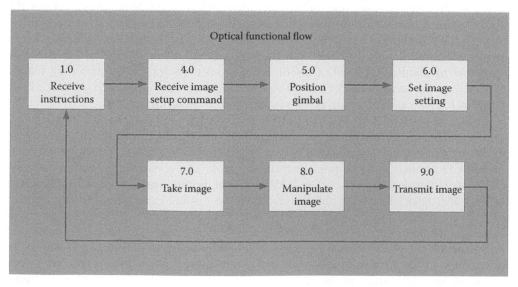

FIGURE 11.6
Optical system FFBD. (Courtesy of Dean Smith.)

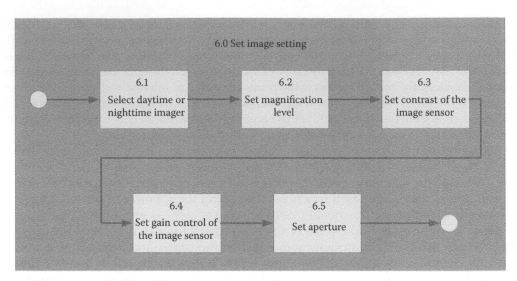

FIGURE 11.7
Set image setting FFBD. (Courtesy of Dean Smith.)

FFBDs can be further refined to provide successive levels of detail. As an example, Figure 11.6 has a block labeled "6.0 Set Image Setting" that can be further broken down into lower-level functions as shown in Figure 11.7. In Figure 11.7, the "Set Image Setting" is broken into five separate sequential steps that define the required functions and functional flow of the image-setting function of the optical system. The image-setting function selects the type of image sensor to use (daytime vs. nighttime) and accordingly sets the magnification level, contrast, gain control, and aperture settings of the imaging system in that particular order (hence a functional flow).

Similarly, Block 6.2 of Figure 11.7, "Set Magnification Level," can be further broken down into functional blocks as shown in Figure 11.8. As seen in Figure 11.8, the logic flow exists to make sure whether the received command for the magnification level needs to be adjusted or not. If no change is required, then the drive motors that adjust the system magnification are not engaged.

Through the formation of FFBDs, the functional analysis provides information on the required functional flow at ever-increasing levels of detail. Once the FBD/FFBD has been decomposed to a sufficient level of detail, what remains is for the engineering team to determine how the functional blocks will be accomplished. A useful mechanism for determining this for the optical system is the system concept diagram (SCD).

For the SCD, each block of an FBD/FFBD is analyzed for inputs and outputs. Additionally, external controls and constraints are determined, as are the mechanisms or physical resources required for accomplishing the intended function. Figure 11.9 shows an example of an SCD applied to the inputs, outputs, controls and constraints, and mechanisms for the top-level optical system on the UAV platform. Some inputs needed by the optical system are the associated information related to the gimbal controller and the required magnification settings for the optical system. Similarly, outputs are the images and videos generated by the optical system. Controls and constraints could include any political, technical, or security issues that would limit the development choices of the optical system. For instance, the UAV itself cannot violate borders and so must remain on U.S. soil. This places a limit on the standoff range for the optical sensor that is using the UAV as a host platform. Some other constraints are the physical limitations of the optical system, such as the physical size, weight, and power (SWAP) requirements of the optical system.

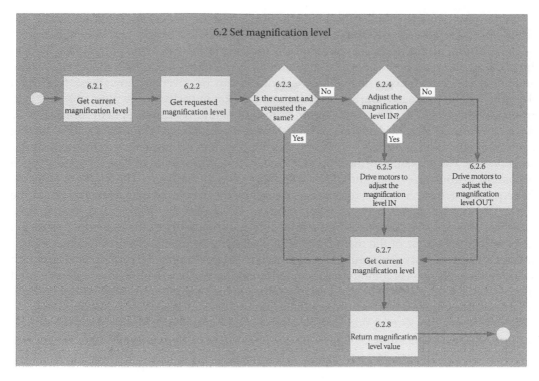

FIGURE 11.8
Set magnification level FFBD. (Courtesy of Dean Smith.)

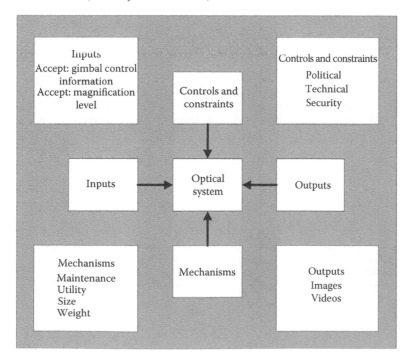

FIGURE 11.9
Optical SCD. (Courtesy of Dean Smith.)

Mechanisms are the resources required to implement the function. These resources can include facilities, special test equipment, skill sets, access to test ranges, and operational sites. By determining all the applicable mechanisms for each of the functions of the FBDs/FFBDs, the physical planning of the optical system and its elements begins to take shape. Once these functional elements become defined, consideration can be given to how these functional elements will be packaged. During this functional packaging activity, the main objective is to group closely related functions into common resources and packages. System elements may be grouped based on considerations such as sharing a common geographical location, having a common operating environment, and by sharing common equipment, or providing similar functionality. The packaging of the individual system functions should be kept separate as much as possible to minimize the required interactions between the packages. The benefit of reducing the required interactions between packages is to reduce the overall complexity of the system, the required communications between components, and simplify the testing and maintenance activities. This can be accomplished even if the internal complexity of the individual packaged element increases. One of the main benefits of packaging common elements together in this fashion is to pursue an open architecture approach when developing the system. This approach includes the application of common and standard modules that have well-defined interfaces and allow the system elements to be grouped in an efficient manner. Figure 11.10 shows an example of an optical system that has been functionally decomposed into logically useful partitions.

Some of these blocks may end up being the actual packaged component for the optical system, whereas others may be useful as logical constructs. In the next section, we present

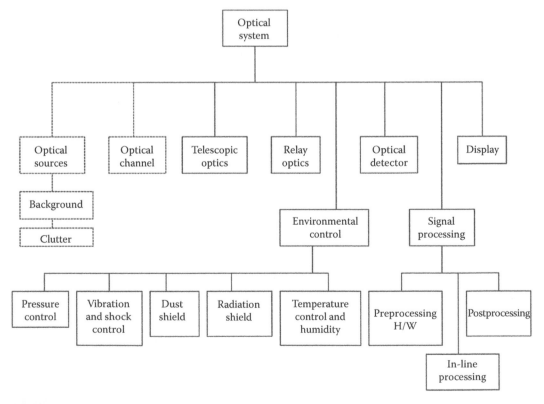

FIGURE 11.10
The functional breakdown of the system into components.

some optical system TPMs and discuss their potential connection to KPPs for the optical systems. The intent is to provide a representative set of TPMs as an example. A complete set of TPMs is application specific, and the reader can expand on this list as needed. Where practical, we also provide the associated fundamental analytical expressions.

11.2 Optical Systems Building Block: Representative TPMs and KPPs

The analysis of optical systems involves the identification and allocation of TPMs, along with special metrics, such as KPPs. The system functional analysis and requirements allocation processes package the system into functionally distinct and compatible subsystems, so that the lower-level requirements for each subsystem can be determined through a lower-level requirements definition process. TPMs are quantitative values that describe the system performance, measuring the attributes and/or characteristics that are inherent within the design of the system (Blanchard and Fabrycky 2010). TPMs include details of design life, weight, maintainability, cost, availability, and supportability. These measures emerge after definition of system operational requirements and the maintenance and support concept. They are given a relative importance value depending on the stakeholder desires or needs. This relative importance ranking is necessary for allocating requirements as some of the specified TPM values may be at odds with each other when determining the characteristics that will be incorporated into the design of the system. For example, the need for reducing detector noise may prompt the desire to add a cooler to the optical system, but this may adversely affect SWAP TPMs. The design solution may instead be to use a different detector, with lower noise characteristics if possible. The optimum system design configuration must be able to optimize the TPMs relative to the given and/or intended applications. To aid in the tracking of the TPMs and to help ensure that the TPMs are appropriately adopted in the design process, the TPMs are allocated to lower-level requirements. Configuration control and requirements management methods are used to ensure that the requirements and TPMs are traceable to the lowest requirements levels and linked to the design. Another critical system MOP is the KPP. The acquisition encyclopedia (ACQuipedia) from the Defense Acquisition University (DAU) gives the definition of a KPP as

> Performance attributes of the system considered critical to the development of an effective capability. The KPP normally has a threshold representing the minimum acceptable value achievable at low-to-moderate risk, and an objective, representing the desired operational goal but at higher risk in cost, schedule, and performance. KPPs are contained in the Capability Development Document (CDD) and the Capability Production Document (CPD) and are included verbatim in the Acquisition Program Baseline (APB). KPPs are considered Measures of Performance (MOPs) by the operational test community (ACQuipedia 2005).

In the following, we present some TPMs that are common in many optical systems (especially optical imaging systems) along with some analytical expressions for determining their value. These TPMs are commonly used in performance prediction analysis and can often be linked to KPPs in some applications, as needs dictate, either singly or in combination. For example, angular or spatial resolution TPMs may be linked to a KPP concerning the nominal and maximum standoff range of the optical sensor from the target of interest (TOI).

These TPMs are not intended to be all inclusive, since the choice of TPMs is application specific, but are to be interpreted as representative values that are useful in a variety of applications. These TPMs should be selectively applied, adapted, and/or augmented with necessary TPMs as needed for a given application. We include a table at the end to summarize and gather the results in one convenient place. We also illustrate how to adapt and augment these TPMs in our integrated case study in Section 11.3.

Source radiant emittance: Source radiant emittance is the optical power (or flux) that is radiating from the surface of a source per unit area (W/m^2). The relationship between the radiant emittance, the radiance (W/(sr m^2)), and the intensity (W/sr) is given as

$$W = \pi N = \frac{\pi J}{A}, \tag{11.1}$$

where
   W is the radiant emittance
   N is the radiance
   J is the intensity of the source (Hudson 1969)

It should be noted that the units on intensity apply strictly to a point source, and so this should be thought of as an equivalent intensity when working with objects of finite spatial extent.

Source radiance: The radiance measures the radiated power that is emitted from the surface of a finite extent source, per unit area, and that falls within a given solid angle in a specified direction. The SI unit for source radiance is watts per meter squared per steradian (W/(m^2 sr)). The radiance, in terms of the radiant emittance and the intensity, for diffusely scattering objects is given as

$$N = \frac{W}{\pi} = \frac{J}{A}, \tag{11.2}$$

where
   N is the source radiance
   W is the radiant emittance
   J is the intensity

The source area A is the surface area of the radiating finite extent object (Hudson 1969).

Optical invariant: The optical invariant states that the product of the source area and the angle from the source to the optical system's entrance pupil (assumed circular) is equal to the product of the image area on the detector and the angle from the optical axis at the image plane (detector location) to the edge of the exit pupil of the optical system. The optical invariant is expressed as

$$A_s = \frac{A_i \Omega_i}{\Omega_s}, \tag{11.3}$$

where
   A_s is the area of the source
   A_i is the area of the image in the image plane
   Ω_s is the source solid angle viewing the entrance pupil aperture of the optical system
   Ω_i is the detector solid angle from the perspective of the detector viewing the exit pupil
     (Guenther 1990)

Peak wavelength of source (for thermal radiators): The peak wavelength is the wavelength at which the highest amount of radiation is emitted from a thermally radiating source. The peak wavelength is given by Wien's law (Joos and Freeman 1987) as

$$\lambda_m = \frac{2898}{T}, \tag{11.4}$$

where

λ_m is the wavelength in micrometers at the point of maximum radiation
T is the source temperature in kelvins (Rudramoorthy and Mayilsamy 2013)

Note that when working with photons directly such as in photon counting cameras, the numerator in Equation 11.4 becomes 3669.73 and the calculated wavelength corresponds to the maximum of the photon emittance (Hudson 1969).

Radiant emittance as a function of temperature (for thermal radiators): The Stefan–Boltzmann law provides the total radiant emittance from a source as a function of temperature (Joos and Freeman 1987):

$$W = \sigma T^4, \tag{11.5}$$

where

σ is the Stefan–Boltzmann constant ($5.670373(21) \times 10^{-8}$ W/m^2/K^4)
T is the source temperature in kelvins (NIST 2014)

Spectral features of source: Source spectral features are simply radiation properties that are functions of frequency or wavelength. For instance, the spectral radiant emittance provides the optical power per unit area in a particular spectral band. As an example, Wien's law gives the wavelength at which a blackbody source emits its peak radiation. By measuring a source radiant emittance as a function of optical wavelength (or optical frequency), we describe the spectral features of the source. The term spectral radiant emittance would be used in this case. Similarly, the spectral radiance and the spectral intensity would provide corresponding radiometric quantities that are functions of optical wavelength or optical frequency (Hudson 1969). For a blackbody radiator, Planck's law provides the spectral distribution of the radiant emittance W_λ:

$$W_\lambda = \left(\frac{c_1}{\lambda^5} \right) \left(\frac{1}{e^{c_2/\lambda T} - 1} \right), \tag{11.6}$$

where

c_1 is the first radiation constant given as $c_1 = 2\pi hc^2 = 3.7415 \times 10^4$ W cm^{-2} μm^4
c_2 is the second radiation constant given as $c_2 = ch/k = 1.4388 \times 10^4$ μm K

In the previously shown expressions, h is the Planck's constant (6.6256×10^{-34} W s^2), c is the speed of light (2.9979×10^{10} cm/s), and k is the Boltzmann constant (1.38054×10^{-23} W s/K).

Optical system operational altitude: The optical system altitude is the height of the optical axis measured at the entrance pupil of the optical system with respect to a ground reference. The altitude can be expressed with regard to local ground height or a fixed reference such as sea level. For example, in a UAV application, the optical axis runs through the center of the primary mirror housed in the UAV gimbal pod. The optical system altitude is the distance from the center of the primary mirror to the ground reference.

Sensor-to-source/target distance: This metric is the distance from the center of the optical system's (e.g., sensor's) entrance pupil to the point where the optical axis intersects the source or TOI. Among other things, this distance is useful in projecting radiometric quantities from the source or target to the optical system's entrance pupil plane and for calculating the spatial resolution performance of the optical system. The assumption is that the optical system is boresighted on the TOI. If the optical system is located on the ground, and the TOI is also on the ground, then the sensor-to-source distance is just the range from the center of the entrance pupil of the optical system to the TOI.

Angular classical diffraction limit: The angular classical diffraction limit is a measure of an optical system's angular or spatial resolution. This limit is the theoretical "best" resolution that a perfect imaging system would produce (e.g., no optical aberrations, no system noises, and no atmosphere). The angular classical diffraction limit for a circular entrance pupil is given as

$$\theta = 1.22 \times \frac{\lambda}{D}, \tag{11.7}$$

where
   θ is the angular diffraction limit
   λ is the wavelength in centimeters or meters
   D is the diameter of the optical system's entrance pupil in centimeters or meters (Jackson 1975)

The interpretation is that angular features that are smaller than the classical diffraction limit could not be resolved by the imaging system. If the small angle approximation holds (e.g., if the target spatial extent is much smaller than the sensor-to-target distance), then the smallest spatial feature that is detectable by the imaging system is given by the product of the angular diffraction limit in Equation 11.7 and the sensor-to-source/target distance. If the entrance pupil was not circular and was square instead, the factor of 1.22 would drop out of Equation 11.7. Note these arguments hold only if the imaging system is operating in vacuum. If the imaging system is operating in the atmosphere, then the entrance pupil diameter is replaced by the Fried parameter r_0 (see Equation 11.8), and the angular resolution of the optical system becomes

$$\theta_\alpha = 1.22 \frac{\lambda}{r_0}, \tag{11.8}$$

where λ is the mean wavelength of the illuminating light. The classical diffraction-limited spatial resolution can be determined by multiplying Equation 11.8 by the sensor-to-target standoff distance. For large entrance pupil apertures, the atmosphere, and not the classical diffraction limit, most often limits the angular and spatial resolutions of the optical system.

Angular pixel size (transverse and longitudinal): This is the angle that a pixel makes in the transverse or lateral direction as seen from the exit pupil of an imaging system. By multiplying this angle by the imaging system's magnification (in the transverse or longitudinal direction), the angle of the pixel as projected from the entrance pupil to the target plane is determined (projected angular pixel size). In well-designed optical systems, the projected angular pixel size is smaller than the spatial resolution of the optical system (diffraction-limited spatial resolution for imaging systems in vacuum and the Fried parameter–modified spatial resolution for imaging systems operating in the earth's atmosphere). In poorly designed optical systems, the pixel size may be larger than the spatial resolution of the optical system, and so the projected pixel size itself limits the spatial resolution of the optical system.

Type of telescope: The major purpose of a telescope is to collect light from a large primary mirror and focus it either into an objective for viewing or onto relay optics that shape the collected light and place it on the detector surface. There are different types of telescopes that depend on how the light is brought to a focus. The simplest telescope is a Herschelian telescope, where a slightly curved primary collecting mirror is placed at a slight angle to the incoming light and so focuses the collected light at some distant center of curvature for the spherical primary mirror surface. A Newtonian telescope has a spherically curved reflective surface that focuses the incoming light onto a flat mirror, which, in turn, reflects the collected light through a small opening on the side of the telescope. A Cassegrainian telescope has a reflective primary surface with a hole in the center. Light is reflected from the primary mirror onto the surface of a convex mirror that intercepts the light before it gets to the primary mirror focus. The light from the convex mirror is focused through the hole in the primary mirror behind the telescope. A similar principle is used with the Gregorian telescope wherein a concave mirror is placed after the light from the primary mirror has passed through its focus. The concave reflective secondary mirror focuses the light through the opening in the center of the primary mirror. There are also specialty telescopes that are built for applications such as satellites and UAVs. For instance, the compact dual field-of-view (CDFOV) telescope has two distinct fields of view (FOVs; narrow and wide) that can be located in different parts of the optical spectrum (from UV to long-wave IR [LWIR]). In combination with a dual wavelength sensor, simultaneous narrow and wide FOV images can be captured on the focal plane of the imaging system. The CDFOV telescope also has advantages in size and weight, and the assembly techniques allow for fast fabrication (Peterson and Newswander 2009).

Generalized imaging model parameters: This analytical model allows an optical system with numerous components to be reduced to a "black box" with a few well-defined parameters in order to determine useful performance characteristics of the imaging system (Pedrotti and Pedrotti 1992; Born et al. 1999; Barrett and Myers 2003). This model is valid under geometrical optics approximation conditions and is an analog to the Norton's and Thevenin's equivalent circuit methodology that is so well known and practical in the electrical engineering discipline. In the generalized imaging model, the optical designer needs to know the size and location of the entrance pupil, exit pupil, and aperture stop. The location of the primary and secondary principle planes and the effective focal length (EFL) are the parameters needed to generate this model. The chief ray and marginal ray are useful ray tracing concepts that help establish important imaging properties such as the image plane location and the size of the image on the detector. The chief ray is an imagined ray that runs from the edge of the object through the center of the aperture stop and emerges from the exit pupil at an angle. If the detector is placed in the focal plane of the imaging system, the chief ray gives the size of the image on the detector. The marginal ray is an imagined ray that originates from the optical axis at the object and passes through the edge of the aperture stop and intersects the optical axis in the image plane. The entrance pupil is the image of the aperture stop as viewed from the object side of the optical system, and the exit pupil is the image of the aperture stop as viewed from the detector side of the optical system.

Entrance pupil diameter: The entrance pupil of the optical system is typically where the primary collecting aperture (usually the primary mirror in telescopic imaging systems) is placed. The entrance pupil diameter is then just the diameter of the primary mirror. This metric is important since it is used in many of the critical optical calculations such as the diffraction-limited performance of the optical imaging system. Knowing the location and diameter of the entrance pupil is essential in developing a generalized imaging model of the optical system. The first and second principal planes are planes perpendicular to the optical axis wherein rays originating from the focus emerge as parallel rays from the corresponding principal plane and rays that are parallel when crossing the principal plane end up converging on the focus. The first principal plane is associated with the optical system's front focal length, and the second principal plane is associated with the back focal length. The points at which the principal planes intersect the optical axis are the principal points. In air and vacuum, the distances from the principal planes to where parallel rays intersect the optical axis at the front and rear of the optical system are the front focal length and the back focal length, respectively. If the imaging system has symmetric lens elements, then the pupil planes and principal planes will coincide. The term symmetric lens elements means that there are the same lens elements before and after the aperture stop of the optical system. If this is not the case, then the principal planes and the entrance and exit pupils will be in distinct locations.

Aperture stop diameter: The aperture stop is the aperture that limits the light throughput in an optical system. The entrance pupil is the image of the aperture stop as viewed from the object side of the optical system. In well-designed imaging systems, the size of the entrance pupil matches the size of the imaged aperture stop. Larger entrance pupils would serve no purpose since the aperture stop would not let this light through the optical system and onto the detector. Smaller entrance pupil sizes would effectively not use light that could have made it through the optical system. The primary purpose of the telescope is to collect as much light as possible, and so having a mismatched entrance pupil diameter with the imaged aperture stop diameter could unnecessarily limit the throughput of the optical system.

Entrance pupil sample spacing: To model an imaging scenario on a computer, the entrance pupil must be adequately sampled (Oppenheim and Schafer 1975; Goodman 1988). One way to determine the sample spacing in the entrance pupil is to use information about the desired object detail as follows. Assuming we have an object that is in the far field of the optical imaging system, then the largest spatial feature on the object would determine the smallest sample spacing needed in the entrance pupil to resolve this large feature. From Fourier analysis, the expression for the maximum sample spacing in the entrance pupil is given as

$$\Delta x_p = 1.22 \frac{\lambda}{D_o} z, \tag{11.9}$$

where
  λ is the mean optical wavelength (m)
  D_o is the maximum diameter of the object (m)
  z is the imaging system to object separation (m)

Equation 11.9 provides the size scale in the entrance pupil that could be used to recover spatial frequency information from the object for object sizes up to D_o. Features that are larger than D_o would not be sufficiently sampled. A good rule of thumb is to have two samples across Δx_p in the entrance pupil for adequate spatial sampling in the entrance pupil.

EFL: The EFLs are the distances from the first principal plane (front focal length) and the second principal plane (back focal length) to the respective front and back focal planes of the optical imaging system. In symmetric lens component imaging systems, the EFL is also the distance from the exit pupil to the back focal plane. Sometimes these focal lengths are hard to measure because the principal planes are located somewhere within the optical housing. In this case, a flange focal length is sometimes defined as the distance from a measureable flange reference point to the focal plane. The EFL is sometimes also called the equivalent focal length of the optical system.

Instantaneous field of view (IFOV): The angle that a ray from the edge of the detector in the image plane of the optical system makes with the principal point can be used to provide the IFOV angle by the following expression:

$$\alpha_{IFOV} = 2m_a \tan^{-1}\frac{D_o}{2f_{eff}}, \tag{11.10}$$

where
   m_a is the angular magnification of the optical system
   D_o is the diameter of the detector
   f_{eff} is the EFL

The IFOV provides the largest angle in object space that the detector can see in one dimension at any given instant. If the detector is square, then Equation 11.10 holds in both linear dimensions (e.g., longitudinal and transverse directions). If the detector has different dimensions in the longitudinal and transverse directions, then two IFOV angles need to be calculated corresponding to the longitudinal and transverse directions. In this case, the D_o in Equation 11.10 is, respectively, replaced with D_l and D_t for the longitudinal and transverse dimensions of the detector.

FOV: The FOV is identical to the IFOV for staring imaging systems that do not move or scan. For gimbaled optical systems that can be rotated in the longitudinal and transverse direction, the transverse angular FOV is some factor c_t (≥1) of the angular IFOV in the transverse direction ($IFOV_t$). Equation 10.11 provides the FOV in the transverse direction as

$$FOV_t = c_t IFOV_t. \tag{11.11}$$

Similarly, the longitudinal FOV is some factor c_l (≥1) of the angular IFOV in the longitudinal direction ($IFOV_l$):

$$FOV_l = c_l IFOV_l. \tag{11.12}$$

For staring optical systems that are on moving platforms, the IFOV is often used and projected onto the target scene. As the platform moves, so does the projected IFOV in the target scene. This mode of operation is often called the "push-broom" mode analogous to pushing a broom in a given direction. The typically square or rectangular pattern of the collective pixels projected on the ground is the IFOV, and the motion of the platform moves the IFOV in a linear direction.

f-number (*f/#*): The *f/#* is a measure of an optical system's ability to collect optical power. The term is analogous to the term speed of the lens in optical photography. The *f/#* is given by Saleh and Teich (2007) as

$$f/\# = \frac{f_{eff}}{D}, \tag{11.13}$$

where
   f_{eff} is the EFL
   D is the diameter of the aperture stop

For a given EFL, an increasing diameter allows more light through the optical system and produces a smaller *f/#*.

Numerical aperture (NA): The NA of an optical system is a useful metric that relates to its light gathering capability. Unlike the $f/\#$, this metric has the advantage of increasing as the optical throughput increases. The NA can be related to the $f/\#$ as follows:

$$NA = \frac{1}{2f/\#} = n'\sin\theta, \tag{11.14}$$

where
     n' is the index of refraction between the last optic and the back focal plane
     θ is the half-angle associated with the marginal ray (Hudson 1969; Saleh and Teich 2007; Naidu 2009)

Isoplanatic angle: The isoplanatic angle is an angular measure over which the atmospheric turbulence statistics is assumed to be the same. For angles that are smaller than the isoplanatic angle, atmospheric turbulence compensation methods that are based on statistical properties of the atmosphere could be used to correct for the effects of atmospheric turbulence. Atmospheric turbulence is often the limiting factor in a well-designed optical imaging system with large apertures. When the angular IFOV is larger than the isoplanatic angle and atmospheric turbulence compensation methods are required to provide acceptable spatial resolution, the object scene must be processed in patches not larger than those given by the isoplanatic angle and *stitched* together. The following equation describes the isoplanatic angle as a function of the Fried parameter r_0 as follows:

$$\theta_{iso} \approx 0.31\frac{r_0}{h}, \tag{11.15}$$

where
     θ_{iso} is the isoplanatic angle
     r_0 is the Fried parameter
     h is the height of the turbulence layer that most strongly contributes to r_o (Roggemann et al. 1996)

Atmospheric coherence length (Fried parameter): The Fried parameter is a measure of the turbulence scale over which the atmospheric statistics are correlated. The Fried parameter is sometimes called the atmospheric coherence length and basically gives the diameter of a patch of atmosphere, which has similar statistical properties. The Fried parameter typically lies in the range of 1–20 cm and, as seen in Equation 11.8, plays an essential role in determining the spatial resolution of an imaging system that is looking through the earth's atmosphere.

Transmittance of the atmosphere: Molecules and aerosols absorb and scatter radiation as it propagates through the atmosphere. As a result, energy at certain wavelengths is strongly attenuated, whereas energy at other wavelengths is hardly affected. Strong absorption due to CO_2 molecules results in parts of the spectrum that are opaque and not good for imaging through the atmosphere. The atmospheric transmittance at other parts of the spectrum can be quite high. If the atmospheric transmittance is high over a range of wavelengths, we use the term "atmospheric window" to indicate that this wavelength band would be good for imaging through the atmosphere, since relatively little optical power would be lost. The atmospheric transmittance spectrum provides a plot as a function of wavelength of what percentage of optical power would be transmitted at a given wavelength passing from sea level to space. The atmospheric transmittance τ_a at a certain wavelength can be used to determine the atmospheric effects at a given wavelength to radiation that propagates significant distances in the atmosphere. For instance, if a point source in space has a given intensity, the power on a detector at sea level in vacuum could be determined through straightforward analytical means. The effect of the atmosphere could be included and modeled by multiplying the calculated power at the detector surface by the atmospheric transmittance at the wavelength of interest.

Transmittance of optics: Similar to atmospheric transmittance, the elements in the optical system can be represented by a coefficient τ_o that describes the transmittance of the optics. This transmittance is also a function of wavelength and is expressed as a percentage over a wavelength range. The effect of the optical elements on the optical power at the input of the optical system (e.g., entrance pupil of a telescopic imaging system) can be modeled by multiplying the input power by the transmittance of the optics.

Noise-equivalent temperature difference (NETD): For thermal sensors, this is the minimum temperature difference between two neighboring elements in an object plane scene that gives rise to an output signal on the detector with a signal-to-noise ratio of one.

Noise-equivalent irradiance (NEI): NEI is the amount of power per unit area incident on the detector that gives rise to a detector output signal-to-noise ratio of one (Bass et al. 2009). The *NEI* is given as

$$NEI = \frac{NEP}{A_d},\qquad (11.16)$$

where
   NEP is the noise-equivalent power
   A_d is the active area of the detector and the units are W/cm^2

NEP: NEP is the optical power in watts that is required to produce an output signal that is equivalent to the detector RMS output noise. NEP values should be stated at specific wavelength, modulation frequency, detector area, temperature, and detector bandwidth. NEP is the optical power that generates a detector output signal-to-noise ratio of one. The NEP applies to a given detector at a given data-signaling rate and effective noise bandwidth. It is the minimum detectable power. Consequently, the smaller the NEP value, the more sensitive the detector is for the specified operating conditions. Equation 11.17 describes how to find the NEP by using the detector area, its wavelength, and its specific detectivity (defined in the following):

$$NEP = \left(\frac{\sqrt{A_d \times \Delta f}}{D^*} \right).\qquad (11.17)$$

Here
   D^* is the specific detectivity
   A_d is the active area of the detector
   Δf is the detector's electrical bandwidth

The unit of NEP is watts.

Detector type: Optical detectors are generally divided into two types: photodetectors and thermal detectors. In photodetectors, photon energy interacts directly with the material to produce charge carriers. The photons must have enough energy to produce free electrons from the detector material (Jackson 1975). The wavelength response of the photodetectors cuts off sharply at the cutoff wavelength. In contrast, thermal detectors react to the thermal energy deposited by the incident radiation on the detector surface. Thermal detectors work by responding to material effects that depend on temperature such as the material's resistance. Consequently, the response of thermal detectors depends only on the total heat energy deposited on their surface and is independent of wavelength. There are several types of photodetectors (Ready 1991):

- *Photoconductive*: This type of detector generates free electrons from the interactions of the optical light with the detector material. The optically generated current changes the electrical conductivity of the material resulting in an output signal that is proportional to the optical power on the detector's active area. These types of detectors are often made from semiconductors like silicon.
- *Photovoltaic*: This type of detector generates a voltage by creating charge separation in a p–n junction of a semiconductor material. The generated voltage is proportional to the optical power that strikes the detector surface.
- *Photoemissive*: This type of detector works by the illuminating light directly releasing electrons from the material surface. The photoelectric effect is used to release the electrons, which are subsequently gathered using external circuitry.

Detector material: The choice of detector material affects the spectral region where the detector responds to the incoming optical radiation and also affects the strength of the response. Different types of material are used to fine-tune the detector response region and provide detector performance characteristics that are optimum for a given application. For instance, HgCdTe is a three-element semiconductor material that permits selecting the cutoff wavelength by the relative proportions of the constituent elements. By changing the alloy composition, the peak wavelength and range can be adjusted within the IR spectrum. If the energy of the photons is greater than the bandgap energy of the material, then the electron is freed into the conduction band. Consequently, the material's conductivity is increased. In order to ensure that the bias current is uniformly distributed throughout the active detector material, square or rectangular shapes are typically used. The HgCdTe material, in various alloy concentrations, is used in a variety of applications such as laser detection, missile guidance, and also high-performance IR optical detectors.

Filters (narrowband, low pass, high pass, notch, matched, polarization): Filters are important in discriminating signals from noise and have great application to signal processing. Optical filters can be hardware based and placed prior to the detector. They can also be software based and part of the analytical postprocessing activities. Some of the more common filter types are as follows:

- *Narrowband*: Narrowband filter is a filter that passes signal over a narrow spectral band. The filter rejects information outside of the narrow spectral band.
- *Low pass*: Low-pass filter is a filter that allows only low-frequency signals from 0 Hz to its cutoff frequency to pass. It drastically reduces signals that are higher than the cutoff frequency.
- *High pass*: High-pass filters, as the name implies, permit only signals with higher frequencies than a cutoff frequency to pass. Frequencies that are lower than the cutoff frequency are blocked.
- *Notch*: Notch filters remove frequencies within a small spectral band and pass all other frequencies. A notch filter can be made from a combination of a low-pass filter and high-pass filters where the cutoff frequency from the low-pass filter is lower than (but close to) the cutoff frequency from the high-pass filter.

- *Matched*: Matched filters are specially designed filters that match the specific noise conditions for a given application. For additive noise, the matched filter optimally maximizes the signal-to-noise ratio. In image processing, 2D matched filter implementations are frequently used for this purpose (Keller 2013).

- *Polarization*: A polarization filter removes one polarization state from the signal, for example, removing the horizontal component of incoming radiation, and leaves it with just its vertical component and vice versa.

Detectivity (D): Detectivity (D) is defined as the reciprocal of the NEP (Hudson 1969). The detectivity is a metric that increases as the NEP decreases and so gives some measure of the quality of the detector (a higher detectivity means a lower NEP and consequently a more sensitive detector for the specified operating conditions). The detectivity is given as

$$D = \frac{1}{NEP}.$$ (11.18)

The unit of detectivity is W^{-1}. This metric will vary depending on the detector area used and the electrical bandwidth of the detector itself. A modified detectivity metric was developed, called D^*, that normalizes out these varying effects (Trishenkvo 2010).

Specific detectivity (D*): The specific detectivity D^* (dee-star) is a metric that normalizes out detector-to-detector variations such as differing detector active areas and different electrical bandwidths. This allows for an "apples-to-apples" comparison for different detectors assuming the same input conditions. The NEP metric depends on both the detector area and the electrical bandwidth of the detector. Since the noise power is proportional to the electrical bandwidth Δf, the noise signal is proportional to $\Delta f^{1/2}$ (Hudson 1969). The D^* metric is expressed as

$$D^* = \frac{\sqrt{A_d \Delta f}}{NEP},$$ (11.19)

where
A_d is the detector area
Δf is the electrical bandwidth
the unit of D^* is cm $Hz^{1/2}$/W (Joshi 1990)

D-double star (D**): The D^{**} metric provides an additional normalization with regard to the input conditions. Since detectors may be affected by background noise, normalization with regard to the view angle of the input radiation is necessary. D^{**} includes an angular correction factor that accounts for an assumed effective viewing solid angle of a hemispheric surround. The following equation relates D^{**} to D^* as

$$D^{**} = D^* \sin\theta,$$ (11.20)

where
D^* is the specific detectivity
θ is the half angle of the solid angle that the detector makes with its limiting aperture

Responsivity: The responsivity is a measure of the detector's ability to respond to input power on its surface. The responsivity is the ratio of the output voltage (or current) to the input power. It has units of either amperes/watts or volts/watts, depending on whether the detector output signal is an electrical current or a voltage. The responsivity for both current and voltage outputs is expressed as

$$R = \frac{v_o}{HA_d} = \frac{i_o}{HA_d},$$ (11.21)

where
v_o is the detector output voltage
i_o is the detector output current
H is the irradiance on the detector
A_d is the detector's active area (Razeghi 2010)

Electrical bandwidth: The electrical bandwidth is the range of frequency over which the electrical circuit of the detector responds. It ranges from 0 Hz to a cutoff frequency. Electrical signals above the cutoff frequency are sharply attenuated. The electrical bandwidth when used in noise calculations is called the noise-equivalent bandwidth.

Sensor environmental conditions: The detector performance is often highly dependent on its environmental conditions. One such condition is temperature. The sensor's temperature can critically affect the noise characteristics of the sensor. For instance, shot noise is a function of temperature and can be reduced by cooling the sensor. Other environmental conditions include pressure, humidity, shock, vibration, dust, electrical interference, and radiation.

Wavelength range of detector: The detector's wavelength range gives the range of wavelengths over which the detector will provide a measurable response. Energy inside this wavelength band is considered detectable, whereas energy at wavelengths outside this wavelength range is strongly attenuated and considered not detectable.

Center wavelength (CWL): Typically used in narrowband applications, the CWL is typically the power-weighted mean wavelength. The CWL is often at or near the peak of the narrowband under consideration and is used as the representative wavelength in wavelength-dependent calculations.

Mean time between failures (MTBF): The MTBF is a statistical measure related to the reliability of a device. It is the average time that the device is expected to work before experiencing a failure. The MTBF is an important metric relative to maintenance activities.

Mean time to repair (MTTR): The MTTR is a metric relating to how long it takes to repair an item under consideration. Based on statistics, the MTTR provides the average repair time in hours. This metric is useful in reliability-centered maintenance activities.

Mean time between maintenance (MTBM): The MTBM is a statistical quantity related to the amount of time expected between maintenance activities. This metric is useful reliability analysis and reliability-centered maintenance analyses.

Operational availability (Aₒ): Operational availability is a measure for the operational "uptime" of a system. It measures the mean time that a system is in an operationally capable state

TABLE 11.2

Summary of Important Optical System TPMs

Name	Equation	Units	Description
Source radiant emittance	$W = \pi N = \dfrac{\pi J}{A}$	W/m²	Optical brightness emitted from source.
Source radiance	$N = \dfrac{W}{\pi} = \dfrac{J}{A}$	W/m²/sr	The radiance measures the radiated power that is emitted from the surface of a finite extent source, per unit area, and that falls within a given solid angle in a specified direction.
Optical invariant	$A_s = \dfrac{A_i \Omega_i}{\Omega_s}$	m²	Source area.
Peak wavelength of source	$\lambda_m = \dfrac{2898}{T}$	μm	The wavelength at the point of maximum radiation.
Radiant emittance	$W = \sigma T^4$	W/m	The total radiant emittance from a source as a function of temperature.
Spectral radiant emittance	$W_\lambda = \left(\dfrac{c_1}{\lambda^5}\right)\left(\dfrac{1}{e^{c_2/\lambda T} - 1}\right)$	W/m²/μm	Source spectral features are simply radiation properties that are functions of frequency or wavelength.
Angular diffraction limit	$\theta = 1.22 \times \dfrac{\lambda}{D}$	°/rad	A measure of an optical system's angular or spatial resolution.
Angular resolution of the optical system in atmosphere	$\theta_a = 1.22 \dfrac{\lambda}{r_0}$	°/rad	The angular resolutions are most often limited by the atmosphere and not the classical diffraction limit.
Entrance pupil sample spacing	$\Delta x_p = 1.22 \dfrac{\lambda}{D_0} z$	m	The maximum sample spacing in the entrance pupil.
IFOV angle (one linear dimension)	$\alpha_{IFOV} = 2m_a \tan^{-1} \dfrac{D_0}{2f_{eff}}$	°/rad	The IFOV provides the largest angle in object space that the detector can see in one dimension at any given instant.
FOV, transverse direction	$FOV_t = c_t IFOV_t$	°/rad	The FOV in the transverse direction.
FOV, longitudinal direction	$FOV_l = c_l IFOV_l$	°/rad	The FOV in the longitudinal direction.
f-number	$f/\# = \dfrac{f_{eff}}{D}$	Dimensionless	The $f/\#$ is a measure of an optical system's ability to collect optical power.
NA	$NA = \dfrac{1}{2f/\#} = n' \sin\theta$	Dimensionless	A measure of an optical system's ability to gather light.
Isoplanatic angle	$\theta_{iso} \approx 0.31 \dfrac{r_0}{h}$	°/rad	An angular measure over which the atmospheric turbulence statistics are assumed to be the same.
NEI	$NEI = \dfrac{NEP}{A_d}$	W/cm²	The amount of power per unit area incident on the detector that gives rise to a detector output signal-to-noise ratio of 1.

(Continued)

TABLE 11.2 (*Continued*)

Summary of Important Optical System TPMs

Name	Equation	Units	Description
NEP	$NEP = \left(\dfrac{\sqrt{A_d \times \Delta f}}{D*} \right)$	W	The optical power required to give a detector output signal equivalent its noise.
Detectivity	$D = \dfrac{1}{NEP}$	W^{-1}	Detectivity is defined as the inverse of the NEP.
Specific detectivity	$D* = \dfrac{\sqrt{A_d \Delta f}}{NEP}$	cm Hz$^{1/2}$/W	A metric that normalizes out detector-to-detector variations such as differing detector active areas and different electrical bandwidths.
D-double star	$D** = D* \sin\theta$	cm Hz$^{1/2}$/W	The $D**$ metric provides an additional normalization with regard to the angle of light on the detector.
Responsivity	$R = \dfrac{v_o}{HA_d} = \dfrac{i_o}{HA_d}$	Amperes/W or V/W	A measure of the detector's ability to respond to input power on its surface.
Operational availability	$A_o = \dfrac{MTBM}{MTBM + MMT + MLDT}$	%	A measure for the operational "uptime" of a system.

of performing an assigned task under stated condition. A_o considers the effect of system reliability, system maintainability, and delay times such as mean logistics delay time (MLDT):

$$A_o = \frac{MTBD}{MTBM + MMT + MLDT},$$

(11.22)

where

$MTBM$ is the mean time between maintenance

MMT is the mean maintenance time (Defense Department 1997)

We provide Table 11.2 that summarizes the main attributes of these metrics. We provide the name of the metric, the associated equation, units, and a short description.

In the next section, we show how these TPMs can be integrated with the higher-level systems engineering principles and methods discussed in Section 11.1.

11.3 Integrated Case Study

In this section, we find FIT in the process of reviewing some TPMs as part of a requirements allocation exercise. For this meeting, FIT has gathered an integrated product team (IPT) to review and discuss the TPMs. This is so that there is adequate organizational representation for all involved disciplines and over all life-cycle phases in the TPM discussions. For this scenario, the following characters are participating: Dr. Bill Smith, FIT chief executive officer; Dr. Tom Phelps, FIT chief technical officer (CTO); Garry Blair, FIT chief information officer (CIO); Karl Ben, FIT senior systems engineer; Ginny R., FIT UAV program manager; Jennifer O. (Jen), FIT systems engineer; Ron S., FIT systems engineer (new hire); Amanda R., FIT optical engineer; Marie C., FIT requirements management and configuration control; Kyle N., FIT maintenance and support engineer; Andy N., FIT maintenance support; Steven E., FIT technical support and IT; Rodney B., FIT field service; Malcolm P., FIT production; Warren M., agency liaison; and Doris H., FIT admin support.

Bill: "Welcome everyone! Let me just say a few words before we get started. I think everyone is doing a great job, and I appreciate everything you are doing. I have seen some of the model-based systems engineering (MBSE) models, the enterprise architecture (EA) artifacts, recent trade study results, and more, and this project is coming along great. We have been hard at work generating requirements for the subsystems, conducting functional analysis, and the functional block diagrams, and functional flow block diagrams are coming along nicely. As a result of all this hard work, we are generating many technical performance measures that need to be tracked. We will not be able to track all the technical parameters that arise as part of our design efforts, but we need to track those that matter. I have brought you together today to consider everything we have done to this point and identify the technical performance parameters that we need to track. While you are doing this, please see if any of these TPMs links to the KPPs that we have been given by our sponsor. Once identified, we will need to allocate these TPMs, along with requirements, down to the next functional level. I have asked each of you here because you all have a unique perspective on this program. I want to be sure we have as complete a set of TPMs as possible at this point. Once we have them, Garry can enter them into the EA, for central online storage, and your MBSE tools can access them. Any questions?"

Marie: "Will we be placing these under configuration control?"

Bill: "Good question. Let's keep an official configuration-controlled list of the TPMs with view privileges for all UAV project team members. But, since the parameters are likely to change for a while, let's have a working copy of the TPMs on the system, with write access for functional leads. For instance, we know that Amanda will be working with the integrated optics model...what's it called again?"

Amanda: "OPSI."

Bill: "Uh, OK. OPSI. Well, Amanda will likely need to tweak the optical parameters based on the evolution of the design, and so she is a logical lead for that MBSE tool. Karl, to start things off, can you give us the definition for TPMs and KPPs again please?"

Karl: "Sure, Bill. Let me bring it up on the system. We are using the Department of Defense Architectural Framework (DODAF) for this program, and it has an integrated dictionary that defines terms. Here you go."

(The following definitions appear on the Samsung large screen in the conference room.)

Technical Performance Measures (TPMs): TPMs are "quantitative values that describe the system performance, measuring the attributes and/or characteristics that are inherent within the design of the system" (Blanchard and Fabrycky 2010).

Key Performance Parameters (KPPs): "Performance attributes of the system considered critical to the development of an effective capability. The KPP normally has a threshold representing the minimum acceptable value achievable at low-to-moderate risk, and an objective, representing the desired operational goal but at higher risk in cost, schedule, and performance. KPPs are contained in the Capability Development Document (CDD) and the Capability Production Document (CPD) and are included verbatim in the Acquisition Program Baseline (APB). KPPs are considered Measures of Performance (MOPs) by the operational test community" (ACQuipedia 2005).

Bill: "Thanks Karl. Warren, what were the KPPs that we were given by our sponsor?"

Warren: "There were five KPPs that were provided to us:

1. Standoff range (nominal [low–med risk], maximum [med–high risk])
   a. Nominal: 8,046.7
   b. Maximum: 18,616.0
2. Data latency (nominal [low–med risk], target [med–high risk])
   a. Nominal: 5 seconds
   b. Target: 3 seconds
3. Probability of false alarm (nominal [low–med risk], minimum [med–high risk])
   a. Nominal: <5%
   b. Minimum: <2%
4. Probability of failure to detect (nominal [low–med risk], maximum [med–high risk])
   a. Nominal: <3%
   b. Minimum: <1%
5. Operational availability (nominal [low–med risk], maximum [med–high risk])
   a. Nominal: 98%
   b. Maximum: 99.9%

As you can see, these KPPs touch on practically every aspect of our system. The first KPP deals with the operational standoff range of the system. The intent is for nominal or maximum standoff range; our system has to be able detect and classify walking humans. The second KPP relates to response time of our information. We must be able to get our images and/or video stream back to the mission control center (MCC) or mobile command center (MoCC) in a timely manner so that they can send response units. The third KPP relates to how often we incorrectly classify a target as human. The fourth KPP is related in that we just don't detect the human, when, in fact, the human is in the scene. The last KPP is based on the expected operational tempo."

Bill: "I want you to all consider these KPPs as we go into this exercise. I want to link appropriate TPMs that we develop to these KPPs. OK, let's get started!"

Jen: "As we decide on the TPMs, let's organize them into the corresponding functional decompositions where possible; that way, the lead for the functional decomposition can take ownership of the TPM."

Bill: "Good idea Jen. Let's break into our functional groups and generate a list of TPMs. You can use the white boards here if you want. We'll meet back in an hour, and then we will all review the consolidated list. Doris, can you please consolidate the list in a table?"

Doris: "Sure, Dr. Smith."

Bill: "All right folks, let's get busy. Ginny, can I talk to you for a minute?"

    (At the completion of the exercise, the following TPM table was generated.)

Bill: "Welcome back! Karl, can you please summarize the results before we move on to the requirements allocation process?"

Karl: "Yes. As you can see, Table 11.3 displays the consolidated list of TPMs that we collectively came up with. These aren't complete yet because we are still waiting for some trade studies to complete before adding them. Anyways, the first column gives the category of the TPM that can be used to identify the functional lead, along with the TPM name. The second column indicates the symbol for the TPM and provides the requirements level that the TPM is connected with. Level 0 is the stakeholder, level 1 is the system level (A spec), level 2 is the subsystem level (B specs), and 3 is the components/assemblies/parts level (C specs, D specs, and E specs). When we enter these into our requirements management system, we can link them directly to the appropriate requirements as well as to design artifacts and test documentation. The third column shows the units of the TPM. The fourth column provides the nominal value (e.g., metric) for the TPM. The fifth column, where appropriate, provides the range of the TPM. The sixth column provides an identifier if the TPM is connected with a KPP along with a numeric identifier in which KPP is affected. Once we get these TPMs into our MBSE tools, additional attributes and connections can be made as needed. Notice, we put 'to be determined (TBD)' in places where we didn't yet know the value of the TPM. This may be because we haven't completed the trade study yet or haven't sufficiently evolved the design at this point. If we sort by requirements level, the amount of TBDs provides a quick insight as to the maturity of the design."

Ginny: "I can also use these in periodic status meetings to see how the design and development effort is progressing and, combined with the MBSE models, project the performance of our system."

Bill: "Very true. This list seems to have a lot of TBDs associated with the visible sensor. Why is that?"

Amanda: "We haven't really nailed down the visible sensor yet. We have been focusing on the IR imaging system since that has the worst resolution. We will start focusing our attention on the visible system soon. Also, the imaging models haven't been finalized yet, so we can't provide some useful TPMs like the f/#, numeric aperture, and such. We haven't wrestled much with the actual atmospheric turbulence since that predominantly affects the visible sensor. So TPMs like the isoplanatic angle aren't there yet either. We will add them once we get to that point. Also, you should know that there are a few estimated values in the TPM chart at the moment that still have to be sorted out. Some of the ranges, for instance, are estimates since they ultimately will depend on our detector selection."

Marie: "Well, you need to color code the estimated values or flag them somehow so that the rest of the development team knows that they may change. How about using green, for 'accepted/baselined' values, yellow for 'working numbers/current best estimate,' and red for placeholder, ballpark, or unverified numbers? We could also remove everything that isn't an accepted value and replace that with a TBD."

Bill: "I agree with Marie. Having a shared TPM list where we are uncertain of the entries will most certainly lead to problems. We need to either clearly flag the uncertain entries or remove them and replace with TBDs like Marie suggested. It seems like the color coding gives us some extra information such as how confident we are in the actual TPM values. I will leave it up to Ginny and the systems group on whether or not to use the green, yellow, red color scheme or some other flagging method."

Ginny: "I think the color-coding method has promise. I will let you know what we come up with."

TABLE 11.3

UAV Program TPMs

Functional Category/Name	TPM (Symbol: Levels [0,1,2,3])	Units	Nominal Value	Range	KPP Link (KPP#)
Operational					
Standoff range (daytime, nominal)	z: 0,1	m	8,046.72	±1	1,3,4
Standoff range (nighttime, nominal)	z: 0,1	m	8,046.72	±1	1,3,4
Entrance pupil (nominal height)	$h_{ep, nom}$: 2	m	5,689	±1	1,3,4
Entrance pupil (min height)	$h_{ep,min}$: 0,1	m	4,000	±1	1,3,4
Entrance pupil (max height)	$h_{ep,max}$: 0,1	m	15,240	±1	1,3,4
Target scene area	A_{ts}: 0,1	m^2	100×10^6	±1	1,3,4
Video data rate	0,1	Hz	30	±1	2
Onboard storage (color video)	sto_{cv}: 2	h/day	1	±1/60	
Onboard storage (color images)	sto_{ci}: 2	Images/ day	144	±1	
Onboard storage: on-demand video	sto_{odv}: 2	min/day	10	±1	
Onboard archival period (max)	t_{arch}: 2	Days	1	1–1.000694	
UAV/MCC/MoCC data latency	t_{lat}: 2	Seconds	5	Target: 2 Nominal: 5	Partial KPP 2
Max number of hot spots	N_{hs}: 0,1	#	10	9–10	
Max weight	w_{os}: 1	kg	54.431	±0.01	
MTBF	MTBF: 1	Years	10	±0.001%	5
Instantaneous reliability	R_{inst}: 1	%	70	±1	5
MTTR	MTTR: 1	Hour	8	±1	5
Operational availability	A_o: 0,1	%	98	±0.1	Is KPP 5
Operational life (+option)	t_{ol}: 1	Years	10 (+15)	±0.1	5
Number of systems	N_{os}: 1	#	22	±0	
Number of spares	N_{sp}: 1	#	8	±0	
Data encryption	N_{sec}: 2	bits	256	±0	
Optical system					
EFL (visible/IR)	f_{eff}: 2	m	1	0.2–2/0.2–2	1
Transverse magnification (visible/IR)	m_t: 2	#	1	1–25/1–25	1
Longitudinal magnification (visible/IR)	m_l: 2	#	1/1	1–25/1–25	1
Entrance pupil diameter (visible/IR)	D_{ep}: 2	m	0.508/0.508	±0.001	1
Exit pupil diameter (visible/IR)	D_{ex}: 2	m	TBD/TBD	TBD	1
Peak wavelength (visible/IR)	λ_p: 2	μm	0.5/9.5	±0.1/0.1	1
Atmospheric transmittance (at λ: visible/IR)	T_{λ_p}: 2	%	65/80	±2%	1

(Continued)

TABLE 11.3 (*Continued*)

UAV Program TPMs

Functional Category/Name	TPM (Symbol: Levels [0,1,2,3])	Units	Nominal Value	Range	KPP Link (KPP#)
Average in-band atmospheric transmittance (visible/IR)	T_{IB}: 2	%	TBD/77	±2%	1
Average in-band transmittance of optics (visible/IR)	T_{opt}: 2	%	TBD/97.5	±0.1	1
Source radiant emittance (IR)	H_s: 2	W/m²	480.89	±TBD	1,3,4
Source radiance (IR)	N_s: 2	W/m²/sr	153.07	±TBD	1,3,4
Source intensity (IR; in band)	J_s: 2	W/sr	15.8	TBD	1,3,4
Source equivalent area	A_s: 2	m²	2	±10%	1,3,4
Narrowband filter range (visible/IR)	$F_{NM,IR}$: 2	μm	—	(8.75–9.5/TBD)	1,3,4
Atmospheric coherence length (visible/IR)	r_o: 2	cm	10/342	(8–12/273–411)	1,3,4 (Vis Only)
IFOV (longitudinal; visible/IR)	$IFOV_l$: 2	Rad	Various	(0.0025 0.625/0.0025–0.625)	1,3,4
IFOV (transverse; visible/IR)	$IFOV_t$: 2	Rad	Various	(0.0025–0.625/ 0.0025–0.625)	1,3,4
Detector					
Dimensions: IR (lateral/transverse)	$D_{det,l}$: 2	cm	1	±0.05	1,3,4
Dimensions: visible (lateral/transverse)	$D_{det,t}$: 2	cm	1	±0.05	1,3,4
Wavelength range (visible/IR)	$\Delta\lambda_{det}$: 2	μm	—	0.5–1.1/6–15	1,3,4
Detector array temperature (visible/IR)	T_{det}: 2	K	295/77	±1	1,3,4
Chopping frequency (IR tip and queue)	$F_{c,tq}$: 2	Hz	100	±0.1	1,3,4
Resistance (visible/IR)	Ω_{det}: 2	Ohm	TBD/45	±1%	1,3,4
Electrical bandwidth (visible/IR)	Δf_e: 2	Hz	16	±1	1,3,4
Responsivity (visible/IR)	R_{det}: 2	A/W or V/W	(0.7 A/W, 4×10^4 V/W)	±5%	1,3,4
Uniformity (visible/IR)	PIU_{det}: 2	%	90/90	±1	1,3,4
Detectivity (visible/IR)	D: 2	W⁻¹	TBD/1.124×10^{10}	±2%	1,3,4
Specific detectivity (visible/IR)	D^*(9.5 μm, 77, 100): 2	cm/ Hz/W	TBD/4.5×10^{10}	±3%	1,3,4
Noise-equivalent power (visible/IR)	NEP: 2	W	TBD/0.089×10^{-9}	±8%	1,3,4
Number of pixels (visible/IR)	Npx, Npy: 2, 2	#	2,046 × 2,046/ 2,046 × 2,046	—	1,3,4

Bill: "Thanks Ginny. This is a nice start for collecting and organizing our TPMs. As we identify more of them, we will add to the list. I'd like to see updates to this list presented during our monthly program review. Also, I'd like to see this list sorted according to the TPM's requirements level to get an idea how the design is evolving and how well we are progressing towards our next milestone event. Speaking of that, Karl, can you provide a status of the requirements allocations activities before we break?"

Karl: "Absolutely. As you know, we started with a set of stakeholder requirements that are listed in Table 11.1. Notice how these are framed in terms of capabilities that our stakeholder wants. These are considered the driving requirements in the conceptual design phase. Next, the development team conducted a functional analysis, and we generated the required optical system functions to meet the stakeholder requirements. We used FBDs and FFBDs to conduct the functional analysis and hierarchically decompose the optical system to the system level. Figure 11.5 shows our top-level FBD for our optical system."

Marie: "Wait a minute. This block diagram shows our optical system as a subsystem. Isn't a subsystem a B spec level document?"

Karl: "That is a good point. In fact, you are right. As part of our project documentation, we were given a higher-level functional block diagram for the UAV. In that functional decomposition, the optical system is just one subsystem of the larger UAV (e.g., UAV power subsystem, UAV communications subsystem, UAV flight control subsystem and so forth). We were asked to develop our numbering scheme to be consistent with their higher-level decomposition. So, instead of starting our top-level optical systems block as 1.0, we are starting it with 7.0, which is the UAV optical subsystem. In essence, the UAV optical *subsystem* is our *system* level block. This is just a requested numbering mechanism from our stakeholder to maintain consistency in their documentation. We will use our internal systems engineering processes and the traditional systems engineering design phases, along with our MBSE tools and methods, to develop this optical sensor system. Jen, can you briefly discuss the results of the functional analysis?"

Jen: "Sure. If you look at Figure 11.5 and mentally adjust the word subsystem to system to reflect the FIT perspective, you can see that the blocks illustrate the functional characteristics that our system needs to satisfy. For instance, our optical system must provide a day and night imaging capability. Our optical system must also be able to compensate for the effects of the atmosphere and interact with the power and communications systems from the UAV. Notice that the grayed-out section defines the boundary for our optical system. In support of this FBD, Figures 11.6 through 11.8 show sequential FFBDs as representative examples essential process flows. Notice that we are free to use a separate numbering sequence for the FFBDs since they are different functional analysis mechanisms and are used for separate purposes (e.g., FBDs provide a functional decomposition, whereas FFBDs provide a description of the functional flow). Regardless of the chosen numbering scheme, the FBDs and FFBDs must each be hierarchically decomposed and numbered so that lower-level decompositions trace back to higher levels. After the functions are decomposed into lower-level functions, the requirements allocation process 'bundles' those functions into logical lower-level partitions. Once the functional analysis is accomplished, then the requirements at the higher level (e.g., stakeholder-level requirements in our case) can be apportioned, or allocated, to the lower-level functional blocks (in our case the system level), and new system-level requirements can be generated as needed. These system-level requirements are captured and eventually baselined in our system specification (A spec). The apportioning, allocation, or derivation of requirements at the lower level can be one to one, many to one, or one to many, highlighting the importance of good requirements traceability methods. For example, the stakeholder requirement,

> *SR 8: Shall provide a capability that provides for both an automatic and manual control mode,*

can give rise to at least two system-level requirements:

> *AR 001: The optical system shall have an automatic control mode.*
> *AR 002: The optical system shall have a manual control mode.*

As an example of derived requirements, a series of requirements need to be generated to describe the attributes of the FBDs and FFBDs. For example, to describe Block 7.10 in Figure 11.5, requirements like the following are needed:

> *AR 003: The optical system shall provide header information for captured video that includes date, video start time, and GPS coordinates for all recorded videos.*
> *AR 004: The optical system shall provide header information for captured images that include date, image capture time, image frame number, and GPS coordinates for all recorded images.*
> *AR 005: The optical system shall provide header information for captured image segments that include date, image capture time, image frame number, image segment number, and GPS coordinates for all recorded image segments.*
> *AR 006: The format for date information shall be dd-mmm-year (e.g., 07-JUN-2014).*
> *AR 007: The format for time information shall use a 24 h clock as follows: hr:mn:sc (e.g., 22:54:38).*

Note the change from a capability perspective in the stakeholder requirements to a systems perspective in the system-level requirements. Of course, more requirements are needed to fully describe the attributes of Block 7.10 in Figure 11.5 as well as the other blocks."

Karl: "Thanks Jen. Not only do we need to generate the functional requirements at the system level as Jen just stated, but we also have to generate nonfunctional requirements like safety requirements, reliability requirements, maintainability requirements, and availability requirements, to name a few. We also need to ensure that our TPMs are correctly allocated and apportioned to the system level and new TPMs are generated if needed. The system-level requirements will all be organized along with other descriptive information in the A spec. The finished A spec will be internally reviewed and presented to our stakeholder as part of our Systems Segment Review (SRR) process."

Bill: "Thanks Karl and Jen. Ginny, please work with the functional heads to resolve open TBDs as quickly as possible and let me know if you need anything. Malcolm, Rodney, Julian, and Andy, please work with the systems folks to make sure we have a good set of system-level requirements and TPMs for your respective areas. Good job everyone!"

11.4 Summary

Requirements allocation is an integral part of the systems engineering discipline and is the third activity in a three-step process (requirements definition, functional analysis, and requirements allocation). At each level of requirements, the requirements definition process is followed by functional analysis that determines the functional decomposition one level below the current requirements level (e.g., from the stakeholder requirements level, functional analysis determines the functional partitions at the systems requirements level). Functional analysis also provides supporting technical details needed by the design team. Requirements allocation bundles the partitions into logical groupings and then expands the higher-level requirements into the bundled functional partitions through the processes of apportionment, allocation, and/or derivation. The three-step (requirements generation, functional analysis, and requirements allocation) process occurs in every design phase in the systems engineering process (conceptual design for stakeholder requirements to system requirements level, preliminary design for systems requirements level to subsystem-level requirements, and detailed design and development for subsystem-level requirements to the bottom-level developmental requirements). At each requirements level, specifications are developed that capture essential requirements. (System-level requirements are captured in the A specification, subsystem requirements in B specifications, and detailed design and developmental requirements in C specifications.)

References

ACQuipedia. 2005. Key Performance Parameters (KPPs). *ACQuipedia.* https://dap.dau.mil/acquipedia/Pages/ArticleDetails.aspx?aid=7de557a6-2408-4092-8171-23a82d2c16d6 (accessed February 25, 2014).

Ascendant Concepts LLC-Home. 2013. Systems engineering support for industry and government. http://ascendantconcepts.com/index.html (accessed March 11, 2014).

Barrett, H.H. and K.J. Myers. 2003. *Foundation of Image Science,* 1st edn. Hoboken, NJ: Wiley-Interscience Press.

Bass, M., C. DeCusatis, J. Enoch, V. Lakshminarayanan, G. Li, C. MacDonald, V. Mahajan, and E.V. Stryland. 2009. *Handbook of Optics,* 3rd edn. Vol. II: *Design, Fabrication and Testing, Sources and Detectors, Radiometry and Photometry.* New York: McGraw-Hill Professional Press.

Blanchard, B.S. and W.J. Fabrycky. 2010. *Systems Engineering and Analysis.* Upper Saddle River, NJ: Prentice Hall Press.

Born, M., E. Wolf, A.B. Bhatia, P.C. Clemmow, D. Gabor, A.R. Stokes, A.M. Taylor, P.A. Wayman, and W.L. Wilcock. 1999. *Principles of Optics: Electromagnetic Theory of Propagation, Interference and Diffraction of Light,* 7th edn. Cambridge, U.K.: Cambridge University Press.

DAU. 2001. Systems Engineering Fundamentals, Supplementary Text. Fort Belvoir, VA: Defense Acquisition Press.

Defense Department, Defense Systems Management Co. 1997. *Acquisition Logistics Guide.* Washington, DC: United States Government Printing Office Press.

Goodman, J.W. 1988. *Fourier Optics.* New York: McGraw-Hill Press.

Grady, J.O. 2006. *System Requirements Analysis.* San Diego, CA: Academic Press.

Guenther, R.D. 1990. *Modern Optics.* New York: Wiley Press.

Halaweh, M. 2012. Using grounded theory as a method for system requirements analysis. *JISTEM—Journal of Information Systems and Technology Management,* 9(1): 23–38. http://www.jistem.fea.usp.br/index.php/jistem/article/view/10.4301%252FS1807-17752012000100002 (accessed April 16, 2014).

Hudson, R.D. 1969. *Infrared System Engineering.* New York: John Wiley & Sons.

Jackson, J.D. 1975. *Classical Electrodynamics,* 2nd edn. New York: Wiley Press.

Joos, G. and I.M. Freeman. 1987. *Theoretical Physics.* New York: Dover Publications.

Joshi, N.V. 1990. *Photoconductivity: Art, Science, and Technology.* New York: CRC Press.

Keller, J. 2013. Persistent surveillance with UAV-mounted infrared sensors is goal of DARPA ARGUS-IR program. *Military and Aerospace,* 21(2), p:1.

Leonard, J. 1999. *Systems Engineering Fundamental: Supplementary Text.* Fort Belvoir, VA: Diane Press.

Naidu, S.M. 2009. *A Textbook of Applied Physics.* Delhi, India: Pearson Education India Press.

NIST. 2014. The NIST reference on constants, units, and uncertainty. http://physics.nist.gov/cgi-bin/cuu/Value?sigma (accessed June 28, 2014).

Oppenheim, A.V. and R.W. Schafer. 1975. *Digital Signal Processing.* Englewood Cliffs, NJ: Prentice Hall.

Pedrotti, F.J. and L.S. Pedrotti. 1992. *Introduction to Optics,* 2nd edn. Englewood Cliffs, NJ: Prentice Hall.

Peterson, J. and T. Newswander. 2009. *Compact Dual Field-of-View Telescope for Small Satellite Payloads.* Logan, UT: Utah State University Press.

Pineda, R.L. and E.D. Smith. 2010. Functional analysis and architecture. In: *Systems Engineering Tools and Methods,* 5044: 35–79. Engineering and Management Innovation. New York: CRC Press.

Razeghi, M. 2010. *The MOCVD Challenge: A Survey of GaInAsP-InP and GaInAsP-GaAs for Photonic and Electronic Device Applications,* 2nd edn. New York: CRC Press.

Ready, J. 1991. Optical detectors and human vision. Fundamentals of photonics. https://spie.org/Documents/Publications/00%20STEP%20Module%2006.pdf (accessed April 25, 2014).

Robertson, S. and J. Robertson. 2012. *Mastering the Requirements Process: Getting Requirements Right.* Upper Saddle River, NJ: Addison-Wesley Professional Press.

Roggemann, M.C., B.M. Welsh, and B.R. Hunt. 1996. *Imaging Through Turbulence.* Boca Raton, FL: CRC Press.

Rudramoorthy, R. and K. Mayilsamy. 2013. *Heat Transfer: Theory and Problems.* Coimbatore, India: Pearson Education India Press.

Saleh, B.E.A. and M.C. Teich. 2007. *Fundamentals of Photonics,* 2nd edn. Hoboken, NJ: John Wiley & Sons, Inc.

Trishenkvo, M.A. 2010. *Detection of Low-Level Optical Signals: Photodetectors, Focal Plane Arrays and Systems,* Vol. 4. Solid-State Science and Technology Library. Norwell, MA: Kluwer Academic Publishers.

Wasson, C.S. 2006. *System Analysis, Design, and Development: Concepts, Principles, and Practices.* Hoboken, NJ: John Wiley & Sons, Inc.

12

Introduction to Systems Design

> When you want to know how things really work, study them when they're coming apart.
>
> **—William Gibson**

Systems engineering is a discipline for realizing systems/services/products/processes, maintaining and supporting them over their entire life cycle, and seeing them safely and effectively disposed/recycled/converted for other use at the end of their life cycle. Systems engineering covers the entire system, over its entire life cycle, and is requirements driven. Systems engineering has a set of principles, methodologies, techniques, and tools for effectively bringing these systems into being. These systems can be large and complex such as a satellite development program and involve many stakeholders. The systems can also be much smaller, such as a local, individual business. In the latter case, the systems engineering principles, methods, tools, and techniques are tailored and adapted for the smaller system. Systems engineering applies to systems, such as complex space or defense systems; services, such as service providers like high-end graphics shops; products, such as commercial and industrial items; or even processes such as petroleum or chemical production facilities. Throughout this chapter, we use the word "system" in its most general sense to mean a system, service, product, or process, since any one of these things can be thought of as a system. Where necessary and appropriate, we will distinguish between systems, services, products, and processes but for brevity's sake, we use the broad interpretation of system where possible.

This chapter describes the systems design process that is one of the more fundamental processes in the systems engineering methodology. The systems design process is not only concerned with the immediate design itself but also considers the production, manufacturing, and construction, operations, and support, and the retirement and disposal aspects of the system, as integral design drivers.

Systems design consists of conceptual, preliminary, and detailed design phases. These three phases progressively add detail while simultaneously narrowing down design choices. This results in a final design that will be used to develop the system and support and maintain it throughout its life cycle. The key point is that the systems engineering methodology is holistic and encompasses every aspect of the system throughout its life cycle. As such, the design, development, retirement, and disposal of the system is a highly multidisciplinary activity that requires a thorough understanding of the systems engineering principles, methods, and techniques, a strong sense of teamwork, and excellent communications skills.

This chapter will cover systems design in three parts. Section 12.1 begins with a brief overview of the systems development life-cycle (SDLC) process in the context of how the SDLC relates to the core topic of systems design. Next, the three systems design phases are discussed in terms of inputs, outputs, activities, and milestones to illustrate how the design phases relate to each other and the overall SDLC. In Section 12.2, we present complementary material for analyzing optical systems. This analytical formalism is essential for modeling optical systems performance and provides useful information to the optical systems designer in designing the optical system. In this section, systems engineering–oriented expressions for understanding imaging systems and nonimaging systems are presented, and a basic range equation is developed that expresses the optical systems standoff range as a function of source radiometric properties, channel effects, and high-level optical systems technical parameters. The resulting range equation can be used in the design process to conduct trade studies between key systems performance parameters.

The final part of this chapter, which is described in Section 12.3, is used as a case study in a narrative form to illustrate selected topics from Sections 12.1 and 12.2. Section 12.3 illustrates aspects of the design process using a fictional unmanned aerial vehicle (UAV) optical system as a representative example. We will use our fictional company, Fantastic Imaging Technologies (FIT), as the backdrop for our applications-oriented scenario. The main purpose of this section is to understand how to apply the conceptual topics and optical methods learned in this chapter to a simulated real-life optical systems development activity.

12.1 Systems Engineering Design Process

The systems engineering design process (SEDP) is an essential aspect of the overall SDLC. Each phase within the SEDP has its unique role and is discussed in detail in this section. However, before discussing the SEDP, we need to define an elementary set of perspectives from which to view the SDLC and the SEDP and its inherent activities. Table 12.1 provides a basic set of roles to illustrate some of the essential dynamics.

In Table 12.1, the stakeholder includes external entities that are responsible for initiating the systems development effort and have the authority to approve the requirements for the system. Each of these roles has their own perspective. The systems engineering role represents the perspectives from the design and development team as well as the technical interface perspective with the stakeholder and user. This is just a representative set of roles for illustrating different perspectives and motivations. Other roles such as the program manager, designer, manufacturer, logistics, administration, legal, and testing are part of the design, development, and support activities but still share a "systems design and development and support" perspective, as does the representative systems engineer role, and so are not separately listed. As a result, the roles shown have fundamentally different perspectives.

TABLE 12.1

Systems Engineering Roles

Role	Description
Stakeholder	Primary entity defining the need to initiate the design process
System	Entity being designed
Systems engineer	Member of the systems engineering team representative of various roles in analysis and design activities
User	End user of the system

Source: Data from L. Szatkowski, 2013.

12.1.1 Systems Engineering and the Systems Life-Cycle Process

The main purpose of this section is to understand the overall function of systems design. The systems design process is described in detail in the international standard, life-cycle management systems life-cycle processes, International Organization for Standardization/ International Electrotechnical Commission (ISO/IEC) 15288 (ISO/IEC 2008). This systems engineering standard covers life-cycle processes and life-cycle phases and categorizes systems life-cycle processes into four broad categories: enterprise processes, agreement processes, project processes, and technical processes.

The first set of processes described in the systems engineering life-cycle processes standard is the enterprise processes. These processes are concerned with management of fundamental enterprise processes. There are five processes that make up the enterprise processes category: (1) enterprise environment management process, (2) investment management process, (3) systems life-cycle processes management process, (4) resource management process, and (5) quality management process (ISO/IEC 2008). Notice that these processes are management processes. For instance, the systems life-cycle processes management process describes how the systems life-cycle processes themselves are managed. The agreement category has two processes, (1) acquisition process and (2) supply process, and covers how items are obtained and delivered. The project processes category covers project-related planning, assessment, control, decision making, and essential management activity processes. There are seven processes that make up the project processes category: (1) project planning process, (2) project assessment planning, (3) project control process, (4) decision-making process, (5) risk management process, (6) configuration management (CM) process, and (7) information management process (ISO/IEC 2008).

Perhaps the most recognizable processes in typical systems engineering activities are the technical processes. The ISO 15288 standard lists 11 technical processes for conducting a wide range of fundamental systems engineering activities. These processes include (1) stakeholder requirements definition process, (2) requirements analysis process, (3) architectural design process, (4) implementation process, (5) integration process, (6) verification process, (7) transition process, (8) validation process, (9) operation process, (10) maintenance process, and (11) disposal process (ISO/IEC 2008).

Table 12.2 summarizes the stages of the systems life cycle according to the ISO 15288 standard. Notice that the concept and development stages are an abbreviation of the traditional systems engineering phases of conceptual design phase (CDP) (concept stage), preliminary design, and detailed design and development (development stage). The conceptual and development stages contain the highest content of developmental engineering activities during the life cycle; however, the stages from production to retirement also contain engineering activities.

All of these stages are complementary and are essential in successfully developing a system and supporting it throughout its life cycle.

Life-cycle stages play an important role in the direction, activities, and interactions of the development team. The development team is organized around the activities that must occur in the individual life-cycle stages. In fact, the name of the stage provides an indication of the types of activities that the development team will engage in. For instance, in the concept stage, the development team will understand the stakeholder's needs and approved wants, develop corresponding requirements that capture the needs/approved wants, and develop various concepts that satisfy the stakeholder requirements. The set of concepts are then further analyzed and evaluated with regard to feasibility, risk, and optimality and down-selected into a particular chosen concept. The chosen concept is then

TABLE 12.2

Systems Life-Cycle Phases

Life-Cycle Stage	Purpose
Concept	Identify stakeholders' needs
	Concept exploration
	Propose and evaluate potential options
	Select and document best option
Development	Generate collective requirements
	Design system
	Build and evaluate prototypes
Production	Construct/produce/manufacture system
	Test, deliver, and transfer system
Operations and support	Conduct systems operations and sustain system over remaining life cycle
Retirement	System disposal, recycling, and/or retirement

Source: Data from L. Szatkowski, 2013.

fully described in a set of system-level requirements captured in the A spec. Figure 12.1 shows a breakdown and clarification of high-level activities and roles for the concept and development life-cycle stages.

Systems design focuses on the concept and development stages and the processes executed in each. All the process categories (i.e., enterprise, agreement, project, and technical) are critical in the design activities, although systems design predominantly focuses on the technical process areas. Enterprise processes must be in place to engage in teaming agreements and conducting basic operations. The agreement process is needed by the development team to provide and distribute the products and services consumed during the design. The project processes are needed for basic project management and control, and the technical processes provide guidance for essential technical activities during systems design. Figure 12.2 summarizes the technical, project, agreement, and enterprise processes.

12.1.2 Systems Documentation and Baseline

One of the main functions of systems engineering is to document and preserve the design activity. The collection of design documentation is referred to as the technical data package. The technical data package contains the artifacts of the design process and is helpful for sustaining production, operations, and support of the system. It is also useful for process improvement, quality engineering, and establishing lessons learned for future efforts.

In the process of creating a design, multiple requirements specifications are sequentially generated. The top-level specification is the systems specification, which is also called the "A spec." Derived from this specification, there are many other requirements specifications that are needed to fully describe the system's requirements, such as subsystems specifications, component/assembly/parts specifications, process specifications, and material specifications. Table 12.3 describes these specifications.

Systems specifications are supported by other elements of the technical data packages, which includes diagrams, drawings, models and simulations, design documentation,

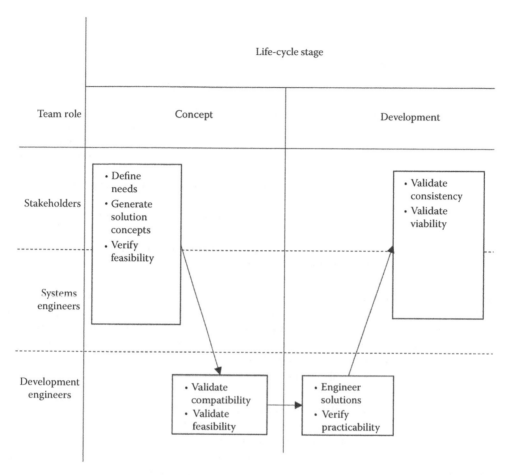

FIGURE 12.1
Roles versus life-cycle stages. (Created by and courtesy of Jay J. Pierce.)

Enterprise processes			
	Agreement processes	Project processes	Technical processes
• Enterprise management • Investment management • System life-cycle management • Resource management	• Acquisition • Supply	• Planning • Assessment • Control • Decision making • Risk management • Configuration management	• Stakeholder needs definition • Requirements analysis • Architectural design • Implementation • Integration • Verification • Transition • Validation • Operation and maintenance • Disposal

FIGURE 12.2
Systems life-cycle processes. (Based on information in Systems engineering—System life cycle processes, ISO/IEC 15288:2002(E), 1st edn., November 1, 2002, Figures D.7 and D.8, pp. 60 and 61; may be used without restriction from L. Szatkowski, 2013.)

TABLE 12.3

Specification Types

Specification Type	Specification Name	Specification Contents
A spec	Systems specification	Technical and mission requirements and constraints at the systems level
B spec	Development specification	States requirements for engineering development, sufficiently detailed to describe effectively the performance characteristics
C spec	Product specification	Technical requirements for configuration items below the systems level; may be functional or fabrication oriented as appropriate
D spec	Process specification	Requirements applicable to a service that is performed on a product or material
E spec	Material specification	Requirements applicable to raw (chemical compound), mixtures (paints), or semifabricated (electrical cable) materials

Sources: Based on U.S. Department of Defense, Standard Specification Practices, 1985-06-04, MIL-STD-490A, pp. 6–13; Created by and courtesy of Jay J. Pierce.

training, operating manuals, and maintenance procedures (ISO/IEC 2008). The type and number of specifications and the contents of the technical data package depend upon the scope of the systems design. The systems documentation itself will not answer all the questions relating to how documentation is maintained. CM is a project process that describes the process to record changes and also identifies the version of a document. CM is mentioned in this chapter to show the role it plays in organizing and maintaining documentation and also its relationship to baselines.

Baselines are very important in a systems development effort. Transitions from one SDLC phase to the next is marked by a milestone event where a particular requirements baseline is established. By baseline, we mean that an artifact undergoes formal configuration control and is considered to be the definitive artifact. For instance, if the A spec is "baselined," that version of the A spec is considered to be the definitive systems requirements document, and any further updates or changes must be formally approved and accepted, and the document is strictly controlled. This milestone event occurs after completion of the CDP and is called the functional baseline. The functional baseline is established to provide a documented starting point for the development phase. The ISO/IEC in ISO/IEC 15288 defines a baseline as "a specification or product that has been formally reviewed and agreed upon, that thereafter serves as the basis for further development, and that can be changed only through formal change control procedures" (ISO/IEC 2008). Before a document has reached baseline status, it can have multiple review cycles. The review process can happen throughout or at the end of the phase. After a document has reached baseline status, it cannot be changed without a formal approval. The main baselines in systems design are the functional baseline, allocated baseline, and product baseline. The allocated baselines are created at the end of the preliminary design portion of the development phase. The product baseline is generated at the end of the development phase when the systems design has been completed and production may begin.

The systems design life-cycle phases delineate two design phases: concept and development. In the ensuing discussion, the development phase will be further broken down into preliminary design and detailed design and development, to provide more granularity to the design process and to better illustrate how the allocated baseline fits into the systems design phases. Figure 12.3 shows the systems design life-cycle phases.

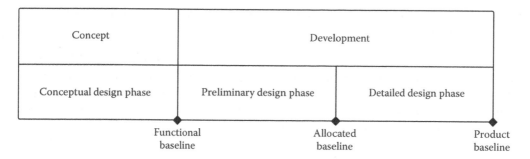

FIGURE 12.3

Design phases and associated baselines. (Created by and courtesy of L. Szatkowski, 2013.)

12.1.3 Three-Stage Design Process: Conceptual Design, Preliminary Design, and Detailed Design and Development

Systems design consists of three phases: the CDP, the preliminary design phase, and the detailed design and development phase. Each of these phases is described below in Sections 12.1.3.1 through 12.1.3.3 the perspective of its inputs, outputs, and major activities. When information is required/presented at the start of the phase, it is considered an input, items produced within the phase are considered outputs of that phase and inputs to the next phase, and the actions that occur in the phase are described as activities.

12.1.3.1 Conceptual Design Phase

The CDP is where the overall systems concept is developed, matured, and shown to be feasible. Conceptual design is the first and arguably the most important phase in the systems design process. This phase defines and bounds the problem, establishes the goals and objectives for the design, provides necessary inputs for the following phases, and dictates how the system will perform and interact within its environment, throughout its life cycle.

The major input to the CDP is the expression of a need for the system. During the conceptual phase, various alternatives will be developed and analyzed, and one concept will ultimately emerge. The systems specification (A spec) is the primary output from this phase. Table 12.4 shows the systems specifications as well as other important documents generated during the CDP.

As we have seen thus far, the systems design process has a number of process-oriented activities. Some of the technical activities, such as the Systems Engineering Management

TABLE 12.4

Outputs, Conceptual Design Phase

Documentation Type	Document Name	Description
A spec	Systems specification	Technical, performance, operational, and support requirements at the systems level
SEMP	Systems engineering management plan	Supports A spec requirements from a management perspective
TEMP	Test and evaluation master plan	Supports A spec requirements from a validation perspective

Source: Created by and courtesy of L. Szatkowski, 2013.

Plan (SEMP) and the Test and Evaluation Master Plan (TEMP), support the A spec requirements and the design activities.

12.1.3.1.1 Systems Engineering Management Plan and the Test and Evaluation Master Plan

The SEMP focuses on the planning aspects of the design effort and includes information regarding systems engineering functions and tasks, technical program planning, CM, risk management, and other management concerns (Blanchard 2008). The TEMP includes information regarding requirements for test and evaluation, test categories, procedures for accomplishing test, and other test planning concerns (Fox 2011). Most organizations will provide a template SEMP and TEMP as part of their enterprise process descriptions to assist in the development of these documents. Templates are also freely available from multiple U.S. government agencies via their websites.

A high-level view of the activities occurring in the CDP is shown in Figure 12.4 and described in the following sections. As shown in Figure 12.4, conceptual design is an iterative process that provides successive feedback to the preceding activities. Feedback takes place when a conceptual design is refined or adapted and information is needed by the preceding activity blocks to make proper changes.

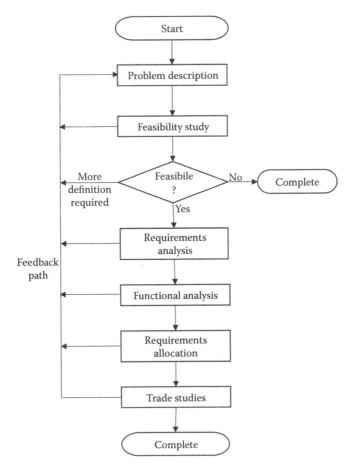

FIGURE 12.4
Activities in the CDP. (Created by and courtesy of Jay J. Pierce.)

12.1.3.1.2 Problem Description and Needs Identification

The first activity in the CDP is problem description. The problem description activity may use many mechanisms such as creating a concept of operations (CONOPS), identifying stakeholder needs/requirements, specifying goals, and developing a corresponding sequential set of measurable objectives that achieve the goals.

In the process of identifying and describing the problem to be solved, one of the primary requirements is to identify the stakeholders. There are two convenient stakeholder classifications: primary and secondary stakeholders. Primary stakeholders can directly have positive and negative impacts on the current systems developmental effort. Secondary stakeholders do not have a direct impact on the current program, but may have subsequent interest once the program is successfully demonstrated. An example would be "spin-offs" from a current program or future customers that may want to adapt the system for their own purposes. Note that sometimes the primary stakeholder is not revealed for security reasons. To have accurate requirements for the systems design, it is essential to have good communications with the stakeholders and approved written stakeholder requirements to ensure that the problem description is accurate. Even when the primary stakeholder is not revealed, a communications channel must exist to complete the problem statement and obtain the necessary approvals. If this step is not accomplished, the developer risks solving the wrong problem or developing a system that is not wanted or needed.

Robertson and Robertson (2006) further identify four stakeholder classification categories: intended system, operational work area, containing business, and wider environment. These different categories help in identifying and classifying specific stakeholders for a particular system. The "intended system" category is an end-user category that applies to stakeholders that will have a direct impact or benefit from the system. Stakeholders who are involved in maintaining and managing a system fall under the "operational work area" category. Stakeholders that will benefit from the system even though they are not actively involved fall into the "containing business" category. And finally, included in the "wider environment" category are all stakeholders who are outside the scope of the system, but who may exert influence on or have an interest in the systems design. The core development team should also be listed as stakeholders, because these individuals are involved in the design and developmental activities at all levels. Table 12.5 shows some representative examples using Robertson and Robertson's classification scheme.

Blanchard and Fabrycky suggest that stakeholders should be matrixed to classes of knowledge to reveal knowledge gaps in stakeholder representation as shown in Table 12.6 (Blanchard and Fabrycky 2011). The matrix also helps provide a road map of where questions for stakeholders should be routed.

When time is spent in considering the stakeholder, needs and wants, as Blanchard and Fabrycky have stated, there can be a tendency on the part of the design team to believe that they know what a stakeholder wants without consulting the stakeholder (Blanchard and Fabrycky 2011). Not only the design team but also the core team can make this error. Table 12.7 provides a representative list of core team stakeholders.

Without an active involvement from the stakeholders during the design process, the development team risks technical problems, and budget and schedule overruns resulting from misunderstanding the stakeholders wants, needs, and intentions (Freeman 2010). If a design process reaches its final phase and has not recognized and correctly incorporated stakeholder needs and approved wants into the design solution, drastic consequences, to include potential program termination, can occur.

TABLE 12.5

Potential System Stakeholders

Wider environment	Negative stakeholders	Competitor Hacker Public opinion
	External consultants	Auditors Security specialist COTS supplier
	Customer	Department manager Another organization Member of the public
Containing business	Internal consultant	System architect Marketing specialist Technology expert
	Functional beneficiary	Manager of operations Business decision maker User of reports
	Client	Investment manager Strategic program manager Chief executive
Operational work area	Interfacing technology	Existing software systems Existing hardware Existing machines
	Maintenance operator	Hardware maintainer Software maintainer Mechanical part maintainer
	Operational support	Help desk Trainer Installer
	Normal operator	Operation technical users Operation business users Members of the public

Source: Original table data from Robertson, S. and Robertson, J., *Mastering the Requirements Process*, 2nd edn., Addison-Wesley, Upper Saddle River, NJ, 2006, pp. 525–528; Courtesy of L. Szatkowski.

After the stakeholders have been defined, the problem and needs identification process can begin to gather the stakeholder requirements that will drive the systems design process. The problem definition and needs identification process is a bridge between the stakeholders "needs/wants" and the resulting stakeholder requirements. A complete problem definition will cover all necessary information needed to begin the systems design process. These aspects include mission definition, stakeholder level performance and utilization parameters, environmental factors, operational deployment considerations, life cycle and effectiveness factors of the intended system in its environment, and interoperability factors. Table 12.8 summarizes these categories. The problem definition process in systems design is intended to be a succinct description of the intended system capabilities with sufficient detail to convey its functional and nonfunctional characteristics.

Needs identification uses the information gathered from the problem definition process to capture the stakeholders' perceptions of what the system must and should do.

TABLE 12.6

Stakeholder Knowledge Classes

Goals
Business constraints
Technical constraints
Functionality
Look and feel
Usability
Performance
Safety
Operational environment
Portability
Security
Cultural acceptance legal
Maintenance
Estimates
Risk
Design ideas

Sources: Original table data from Robertson, S. and Robertson, J., *Mastering the Requirements Process*, 2nd edn., Addison-Wesley, Upper Saddle River, NJ, 2006, pp. 528–529; Table provided by L. Szatkowski.

TABLE 12.7

Core Team Stakeholders

Core team	Project manager
	Business analyst
	Requirements analyst
	System analyst tester
	Technical writer
	Systems architect
	Systems designer

Source: Original table data from Robertson, S. and Robertson, J., *Mastering the Requirements Process*, 2nd edn., Addison-Wesley, Upper Saddle River, NJ, 2006, pp. 525–528; Courtesy of L. Szatkowski.

TABLE 12.8

Components of a Problem Definition

Component	Contents
Mission definition	Define mission, goals, and objectives
Performance and physical parameters	Systems functions, operating parameters, physical characteristics
Operational deployment	Quantity of resources and their expected geographical location
Operation life cycle	Operation life cycle anticipated period of time that system will be in operational use
Utilization requirements	The expected use of the system and associated components
Effectiveness factors	Measures of performance, measures of effectiveness, key performance parameters, figures of merit, technical performance measures
Environmental factors	Attributes of the environment wherein the system operates or interacts

Sources: Original table data from Blanchard, B.S. and Fabrycky, W.J., *Systems Engineering and Analysis*, 5th edn., Prentice Hall, New York, 2011, pp. 61–62; Table provided by L. Szatkowski.

It must be stated that, in addition to the stakeholder needs, stakeholder wants must also be considered to successfully manage expectations. As the problem is being defined, a needs analysis must be performed to translate needs/approved wants into more specific system-level requirements. The questions asked in the needs analysis are intended to discover what kind of systems requirements are needed. Needs identification also determines the primary and secondary functions needed in the systems design. Since the stakeholder requirements succinctly reflect the stakeholders' needs/approved wants, the identified stakeholder requirements are considered to be "the voice of the customer." Needs identification focuses on finding the systems design requirements needed to accomplish a particular solution (Laplante 2011).

12.1.3.1.3 Concept of Operations

In order to understand the scope of the systems design solution, a CONOPS diagram is used. This diagram helps the user in understanding the systems interface with other entities in its environment. The CONOPS may also take the form of an extensive written document and use cases or a series of interaction diagrams rather than just a simple context diagram. The form of the CONOPS must be selected to fully show the user interactions with the system. The consideration of the user interactions will often lead to a significant number of requirements at the A and/or B specification levels. As was discussed with other documents earlier, templates for CONOPS are available from multiple sources including IEEE and U.S. government agencies. Also, diagram styles that can be used as part of a CONOPS are described as part of the Unified Modeling Language.

12.1.3.1.4 Systems Planning

If a need for a new or major modification of an existing system is being initiated, then early stages of planning must also be initiated. To guide the execution and control of the development project, a program management plan (PMP) is developed as one of the initial program documents. The PMP is developed to provide a management plan for all the program activities. It should be noted that technical requirements are determined by the systems engineering process as defined in the SEMP as previously described. Although it is ideal to develop the PMP as the first document, and then flow all other documents from it, this is often not practical. Upon many occasions, the necessary project initiation/development activities are required in advance or parallel with the PMP in order to define the scope of the program and hence complete the PMP.

12.1.3.1.5 Feasibility Studies

To evaluate various design solutions and technology approaches, feasibility studies are carried out in the CDP. "It must be agreed that the 'need' should dictate and drive the 'technology'...." (Blanchard and Fabrycky 2011) is a good quote to keep in mind while considering the feasibility of any given requirement. There are three steps in conducting a feasibility study (Blanchard and Fabrycky 2011):

1. Identify system-level design approaches.
2. Find the most desirable approaches and evaluate them.
3. Recommend a preferred course of action.

These aforementioned steps appear simple, but a high-quality feasibility study looks beyond design. It also evaluates and studies the production, operations and support, and

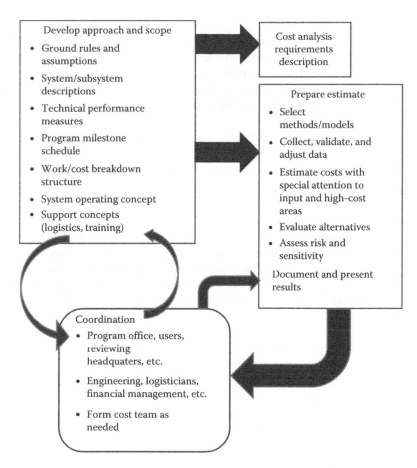

FIGURE 12.5
Overview of life-cycle cost analysis. (Redrawn from *Defense Acquisition Handbook*, Figure 3.7.F1; Additional data from Blanchard, B.S. and Fabrycky, W.J., *Systems Engineering and Analysis*, 5th edn., Prentice Hall, New York, 2011, Figure 17.5, p. 575.)

retirement life-cycle phases. For example, a life-cycle-costing model is a useful tool in describing and comparing alternatives. This model should identify high-cost elements and cost drivers and should evaluate cause-and-effect relationship in the design approach. Figure 12.5 shows an overview of the life-cycle cost analysis process.

The feasibility analysis should also provide evidence that all appropriate technology applications have been considered; existing technologies have been selected, where feasible; and selected technologies will be available when needed (Blanchard 2008). Feasibility studies provide results that impact the design and the technical approach taken. Feasibility studies also affect the resulting operational characteristics of the system and the support, maintenance, and disposal aspects of the system.

12.1.3.1.6 Systems Requirements Analysis

ISO/IEC 15288 describes requirements analysis as the process of transforming the stakeholder (needs-driven) perspective of a system into a technical representation (ISO/IEC 2008). After the design and needs identification approach has been finalized and agreed, the systems requirements analysis begins. Operational requirements are defined, and the functional analysis is implemented using the systems requirements. The A spec, which contains the operational requirements and other system-specific information, is the object of the systems requirements analysis at this stage. As discussed in Chapter 11, the functional analysis and requirements allocation processes generate the requirements at the next lower level and progressively defines the details of the evolving system under design and development.

Writing requirements that are complete, testable, and explicit can be quite difficult. However, requirements writing is often referred to as an art, but one in which it is possible to "learn the craft and get better with practice" (Gilb and Finzi 1988).

12.1.3.1.7 Systems Requirements

Systems requirements result from analyzing the stakeholder requirements and supportive documents such as the CONOPS and mission goals and objectives. These requirements describe the system in operational terms and are written in operator/user language. These requirements can be functional and/or nonfunctional requirements that fully describe what the system must do. In order to generate systems requirements, a four-step process is useful in systematically considering essential elements. This four-step process consists of the following parts:

Identification: In this step, the mission is analyzed along with any secondary or alternative missions. The intent is to determine what needs to be accomplished for mission success. The result is a set of operational profiles that identify the high-level goals and objectives of the mission.

Performance: This step defines the operational characteristics and functions needed by the system. Critical systems performance parameters and how these parameters are related to the mission scenarios are also addressed.

Distribution: This part of the process considers how to distribute resources such as equipment, personnel, facilities, and other system elements to the operational site(s). This stage of the process also includes the associated transportation and logistics requirements.

Constraints: Constraints are special types of requirements that affect the scope of the systems design in terms of budget and schedule (Robertson and Robertson 2006). Constraint requirements limit the technical solutions and should be identified early on in the design process.

Considering these four steps helps ensure that all the relevant operational details and mission essentials are addressed and captured by the systems requirements.

12.1.3.1.8 Maintenance and Support Concept

According to Blanchard and Lowery (1969), maintenance and support requirements are defined to develop and design systems and equipment that can be maintained with the least possible cost and minimum expenditure of resources, without affecting the performance and safety characteristics of the end system. The maintenance and support concept is addressed in the CDP and the A spec contains the systems maintenance and support requirements. Just like the system itself, the maintenance and support functions need to be designed and developed. The A spec contains all the system-level information required for planning, designing, and developing the maintenance and support infrastructure.

The maintenance and support infrastructure includes personnel, parts, equipment, facilities, spares, training, and logistics to support the maintenance and repair functions and activities throughout the systems life cycle.

12.1.3.1.9 *Technical Performance Measures*

Technical Performance Measures (TPMs) are, "measures for characteristics that are, or derive from, attributes inherent in the design itself" (Blanchard and Fabrycky 2011, pg. 37). As an example, maintenance and support TPMs can be found by analyzing maintenance and support concepts during the CDP. Analysis-generated design criteria are necessary for the development of operational maintenance and support requirements in the systems specification. The TPMs are used to guide the design effort and help ensure that the design meets the stakeholders' justified expectations.

Two types of design parameters that can be considered as candidate TPMs are the design-dependent parameter (DDP) and design-independent parameter (DIP). DDPs are used to develop the internal characteristics of a design and can be estimated in known quantities such as weight, size, and reliability. DIPs are used in designing the external characteristics of a design that can be estimated or forecasted such as labor rates or material costs. Both DDPs and DIPs are important characteristics of the design, but it is the DDPs that make the best TPMs (Fabrycky 2011).

In the process of developing TPMs, the systems engineer should consult the stakeholder to translate qualitative goals into quantitative measures that can be used to evaluate the systems design. For example, "easy to use" is a possible TPM when expressed in qualitative terms. A quantitative interpretation could be "no action requires more than three mouse clicks."

TPMs may include parameters such as size, weight, performance, interoperability, reliability, and many others. The objective when defining TPMs is to select specific measures that provide insight into the system's design or system's performance.

The design team has to rank the TPMs to get maximum benefit from them. It is not sufficient to just identify the TPMs. The stakeholders and design team members reevaluate rankings in all the phases of the SDLC.

An excellent tool in relating stakeholder requirements to a technical response is the "house of quality (HoQ)" in the quality function deployment (QFD) method. A high-level example of the HoQ is shown in Figure 12.6. One implementation of the HoQ is to list the stakeholder requirements along the rows of the HoQ (left block) and relate them to the corresponding technical response of the design and development team (columns). The cell entries at the intersection of rows and columns indicate the strength of the connection of the stakeholder requirement to the technical response. The TPMs are given at the bottom of each column and are relatable to the technical responses.

12.1.3.1.10 *Functional and Nonfunctional Requirements in the Design Process*

Functional and nonfunctional requirements are one of the more fundamental requirements categories. As the name implies, functional requirements deal with functional aspects of the system. For example, given an operational requirement such as "The detection system shall be capable of operating any time during the day or night" implies that there is a daytime sensing function, a nighttime sensing function, and perhaps a twilight sensing function that occurs during dawn and dusk. Nonfunctional requirements do not deal with functional aspects of the system. For example, the so-called ilities like reliability, maintainability, availability, supportability, and manufacturability do not affect the required functions of the system (e.g., the fundamental ability to detect and classify humans), but

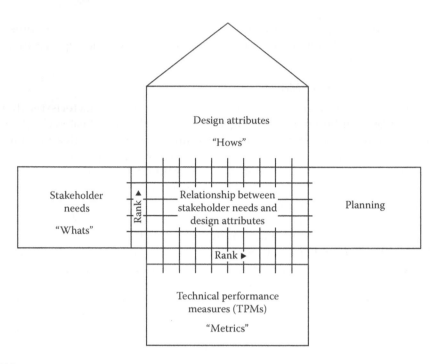

FIGURE 12.6
Modified HoQ. (Original drawing adapted from Blanchard, B.S. and Fabrycky, W.J., *Systems Engineering and Analysis*, 5th edn., Prentice Hall, New York, 2011; Created by and courtesy of L. Szatkowski, Figure 3.18, p. 84.)

instead relate to how well the system can accomplish its mission. An illustration is in the case of reliability requirements that relate to how long a system is capable of performing its functions and what redundancies are designed into the system. Various other useful requirements categories exist such as operational requirements, mission requirements, quality requirements, key requirements, and constraint requirements, and more but all of these can be classified into functional and nonfunctional requirements categories.

12.1.3.1.11 Fit Criteria

Fit criteria are closely related to requirements in that they indicate what needs to be accomplished in order for the requirements to be met. Typically, these are measures and metrics that are associated with the requirements such as weight, size, temperature, pressure, or range of values related to the requirements. For example, if a requirement states, "The system shall weigh no more than 120 US lbs.," then the measure associated with the requirement is weight in U.S. lb., the metric is 120 U.S. lb., and the fit criterion is "less than or equal to 120 US lbs." Requirements need fit criteria so that satisfaction arguments can be made, and establishing measures and metrics associated with the requirements are means to document the fit criteria.

12.1.3.1.12 Functional Analysis in the Conceptual Design Phase

The design team must effectively and efficiently design a solution in a complex, often international environment. The design process aims to provide lower-level, higher-detailed requirements for the functionally decomposed system. The process of developing the lower-level functional elements and analyzing the system with respect to the lower-level functional blocks is the functional analysis process. In the CDP, the starting level requirements are the stakeholder requirements. The functional analysis process and the subsequent requirements allocation process are used to transform the stakeholder requirements to the systems requirements (contained in the A spec). In developing the A spec and lower-level specifications, it is helpful to generate a specification tree to show the connection between higher-level and lower-level specifications.

The results of the functional analysis should provide the following minimum set of information captured in the A spec: (1) results of feasibility studies; (2) technical, operational, performance, and support characteristics of the system; (3) system-level TPMs and metrics; (4) a functional description of the system; and, with requirements allocation, (5) requirements allocated to the systems level. The purpose of functional analysis is to provide a design and analysis baseline and a foundation to flow physical resource requirements (Cogan 2012). The functional analysis also provides input to the functional packaging of the system elements. This can be important for technology insertion points, modularization, open architecture design, and equipment packaging. Functional analysis also provides information for reliability analysis activities such as building reliability models of the system and determining the system/subsystem/component mean time between failure (MTBF), fault tree analyses (FTAs), and failure modes effects analysis/failure mode effects and criticality analysis (FMEA/FMECA) and for conducting reliability predictions. Results of the functional analysis can also be used for maintainability analysis activities such as generating reliability-centered maintenance (RCM) models, level of repair analysis (LORA), maintenance task analyses (MTAs), and maintainability predictions. Human factors analysis also uses the results from functional analysis in generating operational sequence diagrams (OSDs), conducting operator task analyses (OTAs), safety/hazard analysis, and training requirements. Functional analysis is also used for planning and generating requirements for maintenance and logistics support activities such as determining the number of spares, parts, inventory requirements, special handling requirements, supportability analysis, transportation requirements, and maintenance requirements. Functional analysis also provides input to producibility analysis, disposability analysis, determining the quantity and type of maintenance personnel, test equipment, support equipment, and life-cycle cost analysis.

Functional analysis can be represented with a series of functional block diagrams (FBDs) and functional flow block diagrams (FFBDs). The difference between the two diagrams is that the FBD can show a simultaneous decomposition of the system (e.g., power subsystem, communications subsystem, control subsystem that can all be operational at the same time), whereas the FFBD applies a "flow" or sequence of events. Both FBDs and FFBDs are useful and have their place. Figure 12.7 decomposes a top-level function (e.g., Ref. Block 9.2, Provide Guidance) into several blocks that have a sequencing requirement.

For instance, Block 9.2.1 must occur before Blocks 9.2.2, 9.2.3, and 9.24. The "and" and "or" gates provide logic for combining the signal paths. As an example, inputs from both Blocks 3.5 and 1.1.2 are required as inputs to Block 9.2.1. The methodology for decomposition (e.g., generating lower-level blocks from the higher-level functions) is a critical aspect of the design process. Each block can be further decomposed to lower levels until enough granularity is present in the decomposition so that the developer can build the decomposed system element. Of course, to get to a sufficiently detailed level of functional decomposition requires several design phases (namely, the CDP, the preliminary design phase, and the detailed design phase). Notice in Figure 12.7 that tentative functions can be included in the FFBD by using dashed blocks. An FBD would look similar to Figure 12.7 except that there would not be sequence or flow required.

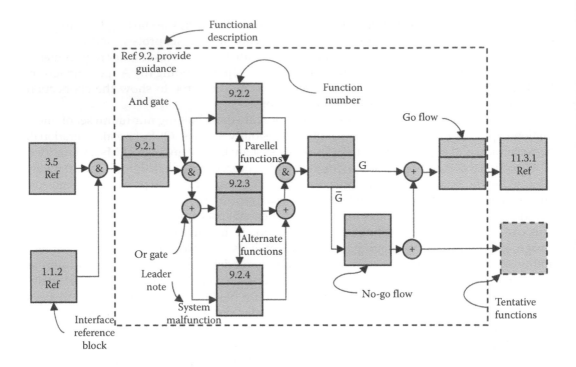

FIGURE 12.7
Generic FFBD. (From http://en.wikipedia.org/wiki/Functional_flow_block_diagram#mediaviewer/File:
Functional_Flow_Block_Diagram_Format.jpg; courtesy of Jay J. Pierce.)

After all functions have been identified in FBDs and FFBDs, the next step is to specify inputs, outputs, controls/constraints, and mechanisms necessary to realize each function. Figure 12.8 shows the relationships and representative considerations for each block.

Each block in an FBD or FFBD is analyzed with respect to its inputs, outputs, constraints, and mechanisms to ensure that proper resources are available to implement the functions within the block and that there are no missing functions or requirements. Once the functions have been identified, they are then analyzed to determine the functional packaging. Figure 12.9 shows a simplified and generic packaging example.

Notice that at the far right of Figure 12.9, logical groupings of functions are packaged together. There may different packaging required for different operations. For instance, a detector that provides range and bearing information may be needed whereas another application may only need bearing information resulting in two different functional packaging configurations. A sensor diagnostic function may be important to both functional packages and so may be bundled with each configuration.

12.1.3.1.13 Trade Studies in the Conceptual Design Phase

Trade studies are often used to compare design alternatives and consider potential design solutions. During the CDP, trade studies are performed to focus on the system-level design issues such as what is the best technology to use or which systems architecture to use. A high-level process description of a trade study is shown in Figure 12.10. There are multiple methodologies for executing a trade study, but many involve a tabular design option, versus a weighted evaluation criteria matrix. Quantitative methods are preferred and can usually be used. In many situations, relative scores are used rather than absolute values.

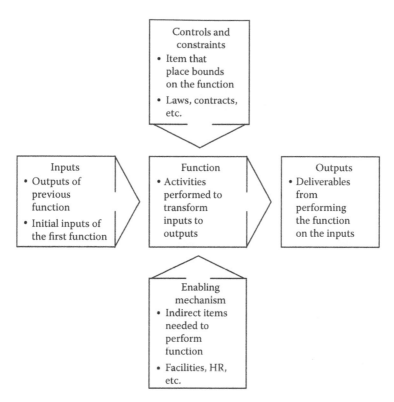

FIGURE 12.8
Function inputs, outputs, controls/constraints, and resources. (Created by and courtesy of Jay J. Pierce.)

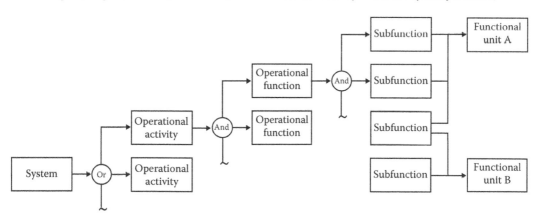

FIGURE 12.9
Functional packaging example. (Created by and courtesy of L. Szatkowski.)

This allows qualitative criteria to be assigned quantitative values and allows for various analytical methods such as weighted averages, statistical methods, and the analytical hierarchy method (AHP) to be used in evaluating the results.

12.1.3.1.14 Systems Requirements Review

The first formal systems engineering design review held during the CDP is the systems requirements review (SRR). However, there may be a series of informal technical,

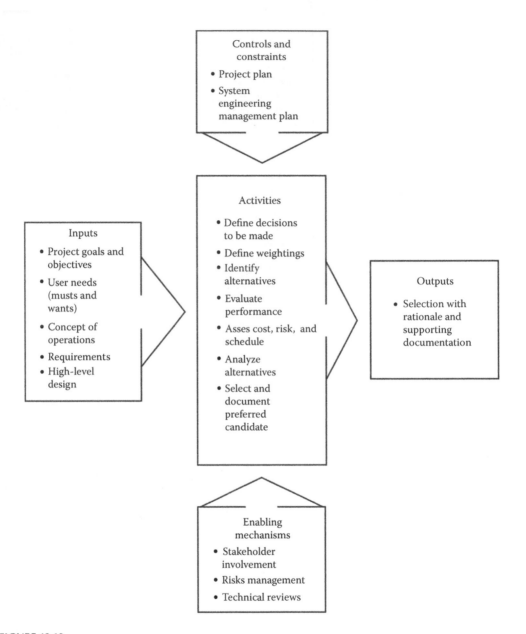

FIGURE 12.10
Trade study process. (From *Systems Engineering Guidebook for Intelligent Transportation Systems*, v3.0, p. 143. U.S. Department of Transportation, Federal Highway Administration—California Division, California Department of Transportation, Sacramento, CA; Adapted by J. Pierce. 2014.)

requirements, and design reviews throughout the CDP. The SRR is held at the conclusion of the CDP and constitutes the milestone event for transitioning from the CDP to the preliminary design phase of the systems development effort. This review emphasizes the system-level requirements and provides an assessment of the readiness of the system to move on to the preliminary design phase. At the successful completion of the SRR, the functional baseline of the system is established. The following list provides the expected deliverables that are reviewed during the SRR:

- Feasibility analyses
- Systems requirements
- Systems maintenance concept
- Functional analysis
- Applicable figures of merit (FOMs) and TPMs
- Systems specification (A spec)
- SEMP
- TEMP
- Design artifacts (trade studies, models, simulation results)
- Compilation of the systems-level design including interfaces
- Demonstrated compliance of the systems-level design with the systems-level requirements

This review also provides a formal evaluation of the entire system and the subsystems and components (e.g., commercial off-the-shelf items) included in the CDP with respect to the systems requirements. If there are many problems identified as part of the review, then the CDP may not be closed out until the problems are fixed and another formal review is completed. If few discrepancies are noted, the phase may be closed after the discrepancies have been closed out without need for another full review. The review serves as a validation process between various stakeholders. For example, the design team is given a chance to explain and justify the design approach.

The SRR review process also provides a configuration control mechanism that contains all the analyses, predictions, and trade-offs between different designs and also supports the finalized decision on the adopted design approach. The documentation package forms the system-level design baseline. One of the major objectives of the SRR is to baseline the systems requirements specification (A spec). A guideline for accepting the A spec is that at least 90% of the specification is complete with the additional caveat that no critical requirements are in the missing 10%!

12.1.3.2 Preliminary Design Phase

Once the system-level requirements have been defined in the CDP and approved at the SRR, the preliminary design phase starts. The focus in the preliminary design phase is on subsystems. During the initial parts of the preliminary design phase, the open items from the CDP are completed and the design sequence of functional analysis and requirements allocation is repeated and focused on the subsystems. The goal of this phase is to establish the high-level design of the system and its subsystems. During the preliminary design phase, all the needed subsystems are defined and their required functions are established. A set of more detailed requirements is written to fully describe what the subsystems must accomplish. The main criterion for this phase is to develop the design requirements for the subsystems from the system-level requirements. Completing the functional analysis and requirements allocation processes on the system-level requirements establishes the preliminary design requirements and associated design documentation such as interface

TABLE 12.9

Outputs, Preliminary Design Phase

Specification Type	Specification Name	Specification Contents
B spec	Development specification	States requirements for engineering development, sufficiently detailed to describe effectively the performance characteristics; one specification is written for each subsystem.
C spec	Product specification	Technical requirements for configuration items below the systems level; may be functional or fabrication oriented as appropriate; one specification is written for each item.
D spec	Process specification	Requirement applicable to a service that is performed on a product material; one specification is written for each process.
E spec	Material specification	Requirement applicable to raw (chemical compound), mixtures (paints), or semifabricated (electrical cable) materials; one specification is written for each material type.

Source: Data from J. Pierce, 2013.
Note: C, D, and E specs are drafts.

control documents (ICDs). This phase also identifies suitable tools required for engineering design and analysis (Blanchard and Fabrycky 2011).

Inputs to the preliminary design phase include the A specification and the technical package from the CDP. The main outputs of this phase are lower-level requirements specifications (B specs) that describe what each of the subsystems must accomplish. There typically are many B specs that are generated, usually one for each subsystem, and software deliverable. The B specs are collectively known as development specifications. Usually, draft versions of the lowest-tier (C, D, and E) specifications will be developed to uncover potential lower-level design issues as early as possible. A summary of the different specification types is provided in Table 12.9. The subsystems Type B specifications should be at least 90% complete to exit the preliminary design phase and enter the detailed design and development phase. Each of the specifications includes performance, effectiveness, and support characteristics that are mandatory in developing the design from system-level requirements (Blanchard and Fabrycky 2011).

In the preliminary design phase, subsystems are the main focus. The main design and trade activities that take place in this phase are shown in Figure 12.11. The figure illustrates that these activities are iterative and provide feedback to previous blocks.

12.1.3.2.1 Subsystems Functional Analysis

Starting from a completed system-level functional analysis, it should be possible to identify subsystems, lower-level packages, and possibly even some components. In the preliminary design phase, the functional analysis is extended from the system-level analysis to analyzing all the subsystems. The FBDs and FFBDs that were generated as part of the design activities in the CDP are now refined and extended to the subsystems level. In further breaking down the FBDs and FFBDs to the subsystems level, similar functions are identified and grouped together in the requirements allocation process. Breaking down the FFBDs from system-level specification into subsystems is used to find groups with identical and similar functions. The subsystems functional blocks are defined while maximizing independence of the subsystems, minimizing communications between subsystems, and maximizing use of standard interfaces and components (Blanchard and Fabrycky 2011).

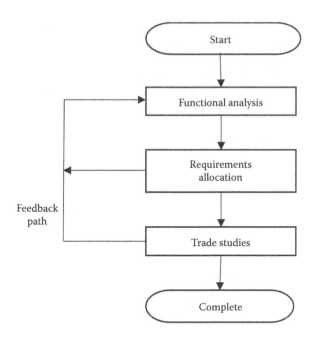

FIGURE 12.11
Activities, preliminary, and detailed design phases. (Created by and courtesy of Jay J. Pierce.)

12.1.3.2.2 Requirements Allocation in the Preliminary Design Phase

Requirements allocation is the process of allocating, apportioning, or deriving requirements and their associated design criteria down to the next lower level. In the preliminary design phase, the starting level is the systems level and the lower level is the subsystems level. It is a big part of the systems engineer's responsibility to allocate design criteria from level to level. Figure 12.12 shows an example of system-level design criteria allocated to functional units. This very simple example illustrates how some criteria are simply distributed among functional units, while other criteria exhibit a different relationship. Weight is additive, so the total weights apportioned to the functional units simply sums to the total

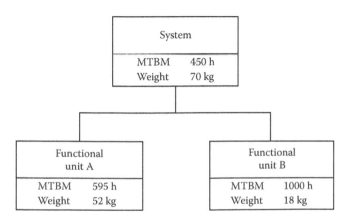

FIGURE 12.12
Requirements allocation. (Redrawn by L. Szatkowski, 2013.)

weight specified for the system. The mean time between maintenance (MTBM), on the other hand, must be allocated using the mathematical rules associated with maintainability analysis and are more complex than simple sums. The resultant allocations must meet or preferably exceed the total MTBM for the system to ensure that the system MTBM can be met. Note that some reserve may be left unallocated to allow assignment to subsystems that are unable to meet their initial allocation.

12.1.3.2.3 Trade Studies during Preliminary Design

As the design progresses, analysis and evaluation processes continue. The proposed design configurations and different elements of the system are generated, and trade studies are conducted to evaluate and determine if any alternative design approaches are preferred. Many analytical models can be used to evaluate alternative approaches to the design configurations of the subsystems. The analyses conducted here are at a more in-depth level than those performed during the CDP.

12.1.3.2.4 Preliminary Design Review

Primary reviews are scheduled throughout the preliminary design phase as needed. Although the preliminary design phase is predominantly concerned with subsystems, primary reviews continue to focus on system configurations and major system elements as well as the subsystems. This approach ensures that the A spec drives the preliminary design during this phase. There can be many reviews depending on the complexity of the system. Primary reviews may cover topics from functional analysis to allocation of requirements to the development of the product as well as processes involved and material specifications (Blanchard and Fabrycky 2011).

B specs generated during the preliminary design phase are placed under formal configuration control as part of the preliminary design phase milestone event—the preliminary design review (PDR). The baseline that is established when successfully passing the PDR is called the allocated baseline. From this point forward, changes made to the baseline A spec or B specs require a formal review and approval process. This practice ensures that any changes in design are formally identified, tracked, and controlled. Note that creation of draft C specs, D specs, and E specs may begin during the preliminary design phase, but these specifications are not normally part of the allocated baseline.

12.1.3.3 Detailed Design and Development Phase

Inputs to the detailed design and development phase include documentation reviewed in the preliminary design phase. Of primary importance are the B specs. In this phase, the major focus is in describing the subsystems, units, assemblies, lower-level components, and software modules and addressing their relationships (Blanchard and Fabrycky 2011). Primary outputs from the detailed design and development phase are specifications for associated components, processes, and materials as listed in Table 12.10.

The basic steps involved in the detailed design and development phase are shown in Figure 12.13. The detailed design block receives feedback from the activities in the detailed fabrication and verification block. Although the diagram is very basic, there is a considerable amount of activity during this phase. We provide further detail in the next section. For most of the detailed design activity, the design engineer has the lead, and the systems engineer is in a supporting role.

TABLE 12.10

Outputs, Detailed Design Phase

Documentation Type	Document Name	Description
C spec	Product specification	Technical requirements for configuration items below the systems level; may be functional or fabrication oriented as appropriate; one specification is written for each item.
D spec	Process specification	Requirements applicable to a service that is performed on a product or material; one specification is written for each process.
E spec	Material specification	Requirements applicable to raw (chemical compound), mixtures (paints), or semifabricated (electrical cable) materials; one specification is written for each material type.

Source: Data from J. Pierce, 2013.

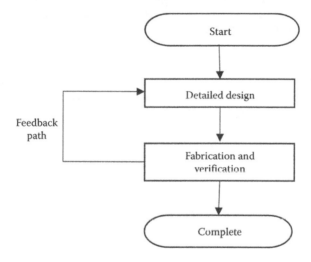

FIGURE 12.13
Activities, detailed design, and development phase. (Created by and courtesy of J. Pierce, 2013.)

12.1.3.3.1 Detailed Design and Development Activities

Detailed design introduces very important design and development processes within the systems engineering discipline. It develops the design requirements for all low-level components, assemblies, and parts of the system and verifies the design objectives are met. It provides a basis for integrating all the system elements. Requirements for this phase are derived from Type A specification and progress to lower-level specification such as Type B, Type C, Type D, and Type E (Blanchard and Fabrycky 2011). The design evolution advances as more details are specified. There are always checks and balances as the design progresses. There are inherent assessment and feedback mechanisms in each design phase to identify issues and allow for corrective measures to be taken.

12.1.3.3.2 Integration of System Elements

Integrating system elements is very important in the detailed design and development phase. The design engineer has to make decisions about how to best satisfy the requirements of each of the specifications. This includes performing trade studies for "make vs. buy" decisions in design solutions. The entire life-cycle cost is an important factor to keep in mind while making these decisions. What is expedient during design may result in increased costs over the entire life cycle.

12.1.3.3.3 Design Tools

To successfully complete the design process, appropriate tools are required to help the design team in accomplishing the design objectives in an efficient manner. Tools like enterprise architecture and model-based systems engineering (MBSE) tools, computer-aided engineering and computer-aided design (CAD), computer-aided manufacturing, and computer-aided logistics allow the development team, and the design engineers, to carefully evaluate their engineering designs with the help of computers. These tools drastically reduce the number of errors, speed up computations and analysis time, and increase accuracy. Validating the system becomes easier and more efficient. Simulation tools are used to develop problem instances just like real-life scenarios. In addition to these tools, a designer is provided with physical models to construct a realistic system configuration (Blanchard and Fabrycky 2011).

12.1.3.3.4 Data Collection

Methods of documenting designs are rapidly changing with the introduction of new technology. In today's world, even complex, 3D models no longer require a drafter to manually create them. It can be done using 3D software where detailed parts and elemental blocks are already created. With the incorporation of CAD tools, enterprise architecture tools, and MBSE tools, the collection, protection, storage, and dissemination of design data and critical program data have become much more reliable and efficient.

12.1.3.3.5 Critical Design Review

Technical reviews are scheduled throughout the detailed design and development phase as needed. Technical reviews focus on system elements such as equipment, hardware, software, components, assemblies, and parts, with each review typically concentrating on a single system element. The Critical Design Review (CDR) occurs after the other detailed design reviews at the end of the detailed design and development phase. The product baseline is established as part of the CDR.

Successful completion of the CDR completes the design process and allows transition into the production, construction, and manufacturing life-cycle phase. The product specifications (C spec), process specifications (D spec), and material specifications (E spec) are generated during the detailed design and development phase and are placed under formal configuration control at completion of CDR. From this point forward, any changes to the baselined A through E specifications must be made through a formal review and approval process. This practice ensures that any changes in design are formally identified, tracked, and controlled.

Note that the product baseline is sometimes referred to as the "design to" baseline. This moniker distinguishes it from the "as built" baseline, which may be different based on changes made to the design introduced during system production.

12.1.4 Alternative Design Processes

The three-stage design process earlier is often referred to as the "waterfall model." The name is derived from the way the process appears on a Gantt style schedule chart with the phase plotted as the ordinate starting with conceptual design at the top and time as the abscissa. It reflects the traditional process flow and has been, and still is, used on a significant number of development projects. However, it is not the only design process that may be employed.

The difficulties of successfully estimating the cost and schedule of software projects using traditional methods led to research into enhancements of, and alternatives to, the waterfall model. One particular branch of research that has garnered considerable momentum is the agile development method.

12.1.4.1 Agile Manifesto

In the 1990s, the software community was introducing alternatives to the waterfall model realizing that, by its very nature, software can be effectively built in smaller steps. In 2001, 17 software developers met to consider their individual approaches for achieving less cumbersome processes to the documentation-heavy waterfall model. They produced the Agile Manifesto that led to the adoption of the term "agile" for the new class of methodologies (Beck et al. 2001). The Agile Manifesto is short and is fully reproduced as follows (Beck et al. (2001):

> We are uncovering better ways of developing software by doing it and helping others do it. Through this work we have come to value:
>
> **Individuals and interactions** *over processes and tools*
> **Working software** *over comprehensive documentation*
> **Customer collaboration** *over contract negotiation*
> **Responding to change** *over following a plan*
>
> That is, while there is value in the items on the right, we value the items on the left more.

Kent Beck	*James Grenning*	*Robert C. Martin*
Mike Beedle	*Jim Highsmith*	*Steve Mellor*
Arie van Bennekum	*Andrew Hunt*	*Ken Schwaber*
Alistair Cockburn	*Ron Jeffries*	*Jeff Sutherland*
Ward Cunningham	*Jon Kern*	*Dave Thomas*
Martin Fowler	*Brian Marick*	

The agile methodologies have since been refined and practiced in the software industry on a widespread but far from exclusive basis. A decade later, the successes seen in the software arena prompted the methodologies to be investigated for use in full systems development.

12.1.4.2 Agile Development Cycle

The agile philosophy, as can be gleaned from the Agile Manifesto, is not one of planning cradle to grave before beginning development. It takes on the development in small chunks with the chunks further down the road having little to no definition of what they will contain at the outset of the project. The cycle is a closed loop with two inputs and one output that is passed through as many times as needed. The agile development cycle is an application of the "observe, orient, decide, and act (OODA)" loop developed by U.S. Air Force Colonel John Boyd (Stelzmann 2012). The four steps in the loop are roughly analogous to the waterfall model, just in much smaller scope and less rigorously documented. The analogies are noted in parenthesis in Figure 12.14. The connection to the Agile

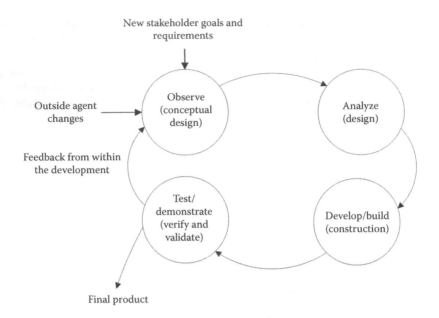

New stakeholder goals and requirements

Outside agent changes

Observe (conceptual design)

Analyze (design)

Feedback from within the development

Test/ demonstrate (verify and validate)

Develop/build (construction)

Final product

FIGURE 12.14
The agile development cycle. (Adapted from Stelzmann, E., Contextualizing agile systems engineering, *IEEE Aerospace Electron Syst. Mag.* May 2012, p. 17.)

Manifesto exemplifies itself through the way the loop is executed. The "individuals and interactions" is implemented through a less structured and highly collaborative working environment. The "working software" is implemented by an iterative development cycle wherein iterations are performed rapidly with minimal documentation so that the speed in development may be maintained. The "customer collaboration" is implemented by the frequent demonstrations and gathering of new or changed goals and requirements based upon the demonstrations. The "responding to change" is executed by passing through the loop a significant number of times through the development.

12.1.4.3 Specific Agile Development Methodologies

There are a number of agile methodologies that have been developed and are in use to varying degrees today. The more popular methods include Crystal, extreme programming, lean software development, feature-driven development, and Scrum (Stelzmann 2012). In order to gain more insight, a brief overview of the quite popular Scrum methodology follows.

The term "scrum" comes from the sport of rugby. Scrums are used in rugby to restart the game after minor infractions have occurred (Wikipedia 2014). The term is applicable to the agile process, as the development cycle shown in Figure 12.14 is repeated many times over the course of a project with a restart for every trip through the cycle.

The scrum team is made up of everyone on the project (or portion of the project within scope of the team) including the full time development staff and other stakeholders that are affected by the portion of the project being developed by this team (Hu et al. 2009). The "backlog" is a list of tasks that is held at various levels. There is a product/project backlog and a subset called a sprint backlog (Hu et al. 2009). The sprint is a short period, typically 30 days, where the sprint backlog is worked off by making a full pass through the development cycle focused only on satisfying the backlog items in scope for this sprint (Hu et al. 2009). At the end of the sprint, the results are demonstrated against the sprint backlog goals (Hu et al. 2009). As a part of the sprint, a daily, short meeting is held with all of the scrum team members. The team leader, called the scrum master, asks each team member three questions: (1) What progress have you made against your goals? (2) What problems did you encounter and what will help resolve them? (3) What will you be doing before the next meeting? (Hu et al. 2009) The purpose of the daily meeting is to expose and rapidly correct problems and errors in a cost- and schedule-effective manner (Hu et al. 2009).

The sprint structure is preceded by a planning and architecture phase and followed by a closure phase (Hu et al. 2009). The planning and architecture phase is not the full CDP from the waterfall model as one of the tenants of agile methodologies is "no BRUF" where BRUF means big requirements up front. However, each set of sprint backlogs is a set of smaller requirements that are developed just in time for the sprint (Murphy 2007).

The closure phase could occur at any loop completion. Each sprint produces a potentially shippable product (Johnson 2011). Therefore, management can determine when to release the product, given market and business pressures, and the product's current feature state.

12.1.4.4 Agile Methodologies Applied to Full Systems

Applying agile methodologies as instantiated by scrum is simply not practical for projects that involve hardware. It is quite cost prohibitive to produce a shippable grade hardware product every 30 days. In fact, it often takes many times longer than 30 days to procure the components required to build mechanical or electrical hardware. Due to the expenses and long lead times, the number of hardware iterations should be minimized. This does not mean that elements of agile cannot be applied to full systems. In fact, they can be used to reduce the number of hardware iterations (Huang et al. 2012). Two approaches, both used in appropriate amounts, can be used: (1) making the system itself agile (e.g., reprogrammable, modular) and (2) using agile design processes where possible, even in small subsets of the life cycle (Haberfellner and de Weck 2005).

The concept of small teams addressing particular subsets of a problem is applicable to, and could be injected into, the functional decompositions of the waterfall model. The decomposition would not lock in levels until lower levels could be explored sufficiently to reduce the risk at the higher levels rather than freezing level by level. Without the early lock in being performed, requirements discoveries from the lower-level investigations can be flowed to the affected areas and possibly change the higher-level models. Also, the architecture can be created to maximize the handling of uncertainty. The more functionality that can be placed into software and flexible hardware elements, such as field programmable gate arrays, the more change management can be economically executed. In the simulated environment, scrum-like design cycles could be effectively employed. Mechanically, the use of 3D models and 3D printing can also be used in scrum-like design cycles. These techniques in the hardware realm do not meet the goals of a potentially deliverable system at each iteration, but may reduce the schedule, expense, and number of actual hardware sets built when an overall agile philosophy is being employed.

The concept of agile systems engineering is still developing (Stelzmann 2012) and will continue to develop, as the fourth part of the Agile Manifesto, "responding to change," is implemented (Beck et al. 2001). Research is ongoing and will take a considerable time to obtain measured data on performance, due to the typically long cycles of full system projects.

12.1.5 Transition to Optical Systems Building Blocks

In the next section, we look at how the optical system's emerging design can be understood and influenced by integrating TPMs and other measures of performance (MOPs) into modeling activities. MOPs such as key performance parameters (KPPs) and TPMs are used to analyze, define, design, develop, and evaluate the performance aspects of our optical system. As an example, a basic range equation is developed that determines the optical systems standoff range as a function of high-level systems engineering parameters.

12.2 Optical Systems Building Block: Analyzing Optical Systems

12.2.1 Understanding the Analytical Problem

Building on all of the information presented in the previous 11 chapters, it is now time to look at the optical system as a whole. Our goal is to assemble essential material that we learned having to do with sources, propagation of radiation, atmospheric effects, optical system effects, and detector effects to come up with useful, integrated expressions related to the performance of the optical system. This is a challenging task since the approach taken is dependent on the particular application that needs to be addressed. Even the optical modality and basic concepts such as what part of the optical electromagnetic spectrum should be used depends on a given particular scenario. Conducting radiometric calibration tests in a laboratory is vastly different from remotely determining the composition of a gas cloud or modeling a tomographic imaging system for medical applications. In this section, we present an approach for defining the correct general optical model for the particular application and then illustrate how to implement the model, based on what we have previously learned. We cannot cover every possible imaging scenario or optical model because of their complexity and depth, but we will illustrate a useful approach in identifying and generating the correct macroscopic optical modeling approach. We will also provide a representative example of how to properly combine results from previous chapters into useful system-level expressions. In order to accomplish this, we must start at the beginning with the intended mission/application and the big picture.

12.2.2 Choosing the Modeling Framework

As we discussed in Chapter 5, the problem definition step is vital in correctly understanding the problem that needs to be solved. Correctly understanding the "big picture" and the intended application is a prerequisite to determining any particular approach to a solution. Systems engineering processes like stakeholder identification, needs analysis, and developing a viable CONOPS are vital steps in determining the correct technical approach to the given problem. For example, recalling our discussions from Chapter 2, if the application calls for accurately determining the electromagnetic fields scattered from the near vicinity of some relatively basic and simple objects, then modeling methods using classical electrodynamics and Maxwell's equations in either their integral or differential form may be appropriate. If the observables also happen to be very small (e.g., atomic level), then the framework of quantum electrodynamics is appropriate. Additionally, relativistic quantum electrodynamics may be necessary if small particles or objects are moving at speeds approaching the speed of light. On the other hand, if relatively large objects are considered (large with respect to the scale of the illuminating wavelength), classical electrodynamic modeling methods become too computationally burdensome. In the case of large, complex objects that are located in the near field or far field of the optical system, methods like Fourier optics and statistical optics are exceedingly useful. In a scenario that involves high-energy fields interacting with complex materials, nonlinear optics (NLO) methods are important. If the application involves the properties and transmission of photons, then photonic methods are appropriate. One can see that right from the start, a thorough understanding of the particular application is required to determine the correct analytical framework that needs to be applied. At this stage, a series of helpful questions are as follows:

1. What is the intended application (e.g., target tracking, surveillance, detection, categorization, classification, optical communications, medical imaging, imaging through the atmosphere)?

2. What type of information is requested (e.g., radiometric, photometric, photopic, scotopic, imaging, nonimaging, spectral, and thermal)?

3. What are the relevant operational parameters (optical system distance from intended observable, relative scale of the object with respect to the illuminating wavelength and separation distance from the optical system, daytime, nighttime, dusk, and dawn)?

4. What are the expected illumination source properties (e.g., high energy, low energy, high-light levels, low-light levels, thermal, directly illuminating target, indirectly illuminating target, passive illumination, active illumination, fully coherent, partially coherent, incoherent, directional, broadband, narrowband, monochromatic, quasi-monochromatic, black body, gray body, and selective radiator)?

5. What are the target properties (e.g., smooth or rough on the scale of an optical wavelength, reflected/absorbed/transmitted/radiated light, shape, area, volume, size with respect to the spatial resolution of the optical system)?

6. What are the environmental considerations (e.g., clutter objects in optical system instantaneous field of view [IFOV], background levels present, expected variations in clutter and background during observation period(s), atmospheric aberrations, systems aberrations, and noise, stray light, temperature, humidity, pressure, shock, vibrations)?

Answers to these questions will provide insight as to what modeling framework(s) should be used as well as provide additional useful information when implementing the models. Sometimes, more than one modeling framework is applicable and necessary. For instance, radiometric methods may be used to provide expected signal levels at points of interest such as the object plane, entrance pupil, or detector output. A simple geometric imaging model may be sufficient to determine important details like the expected size of an image on a detector in the image plane of the optical system. A linear, shift-invariant optical systems model may provide frequency space information such as the optical transfer function (OTF), coherent transfer function (CTF), and modulation transfer function (MTF), along with the point spread function (PSF), generalized pupil function (GPF), and impulse response of an optical system. Both Fourier optics models and statistical optics models are extremely useful in modeling long-range optical imaging scenarios. Once the modeling framework(s) have been decided, then the analytical process for determining appropriate model parameters begins. We next provide an example of selecting an application specific model and populating the model to obtain a useful analytical result. The process may be repeated for other modeling frameworks.

12.2.3 Developing the Model: An Application-Specific Example

Let us consider an application where our stakeholder asks us to develop a nighttime optical system for determining vehicle movement throughout a 5 km × 5 km grid from a helicopter that is flying at a nominal height of 6000 m above the local terrain. We know that the intended targets (vehicles) are much larger than the optical wavelength and much slower than relativistic speeds. We also know that the vehicles come in all sorts of shapes and sizes and are generally more complex than simple. Consequently, the direct application of Maxwell's equations is likely not the most practical approach. We can rule out classical electrodynamics, quantum electrodynamics, and relativistic quantum electrodynamics. Since our stakeholder does not want us to illuminate the targets with an active source like a laser (for safety reasons), we can rule out high-energy models such as the NLO models. Since our stakeholder wants to "see" the vehicle movement, we likely will use an imaging solution versus a nonimaging solution. For imaging applications, there are two major considerations that affect the choice of the modeling framework: (1) the detectable light levels and (2) the spatial resolution of the optical imaging system. Since our stakeholder wants to "see" the images/videos, we do not need to account for the response of the human eye (e.g., we are not interested in signals after the eye but rather in front of it on a display). Consequently, we do not need photometric, photopic, or scotopic models. We are left with radiometric models, Fourier optics models, statistical optics models, and geometrical optics models as viable candidates. We may also use some photonic modeling methods dependent on what particular approach is adopted. Likely, all of these will prove useful in the development effort. Since we know that our stakeholder likes SI units, an SI version of the radiometric model will be adopted and used for further illustration.

At this point, we know quite a bit about the intended application and its implication on the candidate optical system. We know that the targets are rough on the scale of an optical wavelength and so a diffuse scattering model may initially be appropriate. More complex models may be built as the development effort proceeds. We now want to use everything we have learned previously to understand how important systems parameters affect the radiometric aspects of the optical imaging system. We will conceptually break the imaging scenario into its constituent components (i.e., source effects, channel effects, collection aperture effects, imaging system optics effects, optical modulator effects, and detector effects). We will start with the source, pick a suitable radiometric quantity, and then propagate that quantity through the imaging channel, into the entrance pupil, through the optics and any devices in the imaging system, and onto the detector and determine the signal output at the back end of the detector in general terms. We will see that the detector output can be written in terms of a product of factors such as

$$S_o(\lambda) = F_S(\lambda) \, F_C(\lambda) \, F_{ep}(\lambda) \, F_{op}(\lambda) \, F_{det}(\lambda), \tag{12.1}$$

where
    S_o is the signal output from the detector
    F_S is the source factor
    F_C is the channel factor
    F_{ep} is the entrance pupil factor
    F_{op} is the optics factor
    F_{det} is the detector factor

We start with the source factor and sequentially proceed through all of the terms.

The first thing to establish is the particular radiometric quantity that best reflects the quantity of interest for the particular application. Since we are in an imaging mode, versus a nonimaging mode, we expect, in most cases, to have multiple pixels across our targets of interest (TOIs). Consequently, for nominal conditions, we expect to predominantly view extended objects, and not point sources, and so either the source radiance or source radiant emittance radiometric quantity is a suitable choice. However, at extreme distance, we may have the situation where a potentially interesting TOI is smaller than the resolution size of the optical system and so acts as a point source. We can write the source spectral radiometric quantity $RQ_S(\lambda)$ as

$$RQ_S(\lambda) = \begin{cases} N_S(\lambda), \\ H_S(\lambda), \\ J_S(\lambda), \end{cases} \tag{12.2}$$

where

$N_S(\lambda)$ is the extended source spectral radiance in units of W/m^2/sr/µm
$H_S(\lambda)$ is the extended source spectral radiant exitance, in W/m^2/µm
$J_S(\lambda)$ is the equivalent point source spectral intensity, in W/sr/µm

These spectral quantities are explicitly written to show their wavelength dependence. It is often a good idea to convert the spectral radiometric quantity of interest to the equivalent intensity form $J_S(\lambda)$, since the entrance pupil power can then readily be determined (e.g., by multiplying $J_S(\lambda)$ by the steradian subtended by the entrance pupil). This simple relationship holds for a vacuum. Channel effects and geometric factors will be treated shortly. Including the conversion in the terms of Equation 12.2 and assigning the target to entrance pupil distance to the upcoming channel factor, we get the following source factor in terms of starting spectral radiometric quantities:

$$F_S(\lambda) = \begin{cases} N_S(\lambda)A_S, \\ \dfrac{H_S(\lambda)A_S}{\pi}, \\ J_S(\lambda), \end{cases} \tag{12.3a}$$

where A_S is the source area in m^2. Notice that all three terms earlier give the equivalent source intensity for a diffuse source. Let us assume that we can identify a characteristic peak temperature, associated with the engine, or exhaust of the vehicles that gives rise to a peak wavelength in the IR part of the spectrum. Let us also assume that we will be using a narrowband filter in the optical system to reduce system noise. The spectral radiometric quantities shown in Equation 12.1 would then have to be integrated over the wavelength range in the spectral band (i.e., in-band quantities) after being combined with other factors. Since this is most often the case, the wavelength parameter λ indicates that the radiometric quantity has wavelength dependence.

But wait, why are we calling the target a source? Is it not possible to use another source, like the moon or starlight or street lamps perhaps? The answer is yes, and the model earlier would have to be adjusted with another factor F_{targ} after the source factor and also another channel factor between the source and the TOI. However, since the moonlight and stars disappear on a cloudy night, and the terrain could possibly be away from street lamps (in the countryside for instance), systems based on reflected light might not always work. The stakeholder wants a system that will work in total darkness (e.g., a system based on the thermal radiation properties of the target). In this case, the illumination source and target is the same thing. The idea is to pick a representative object in the scene and assign an appropriate radiometric quantity to the object. In this case, our TOIs are vehicles so selecting a representative vehicle and assigning a radiometric quantity is what is currently needed. We do not need a specific value yet for the radiometric quantity since we want to keep things general. We will determine this later.

The next things we need to consider are channel effects. This will include things like atmospheric transmittance and separation distance of the source from the imaging system. For channel noise due to channel scattering, or absorption of radiation (such as molecular, atomic, and aerosol absorption, and subsequent radiation at higher wavelengths), these effects are often treated as background source effects, just like clutter signals. For example, in a daytime imaging system that is looking for aircraft, the background signal might be measured by looking at a "blue patch" of air that has no target or clutter information in it. However, the sun is illuminating the atmosphere, and the scattering and absorption effects of the atmosphere become part of the background radiation. We are interested in the optical power collected by the entrance pupil of our well-designed optical system and so we convert the spectral radiometric quantities in Equation 12.2 to spectral power in the entrance pupil of the optical system. The channel factor for each of the respective terms is given by

$$F_C(\lambda) = \tau_a(\lambda)\frac{A_{ep}}{R_{S:EP}^2}, \qquad (12.3b)$$

where
   $\tau_a(\lambda)$ is the transmittance of the atmosphere as a function of wavelength
   A_{ep} is the area of the entrance pupil in m^2
   $R_{S:EP}$ is the range from the source to the entrance pupil of the optical system

Notice this does not include any geometrical factors due to view angle or scanning. We will include these as part of the entrance pupil factor. The fraction term shown on the right side of Equation 12.3b is an approximation for the steradian. In this example, for typical entrance pupil diameters for helicopter surveillance systems, the approximation is justified. Of course, A_{ep} could also represent the surface area from a segment of a sphere, with radius $R_{S:EP}$ and a cone angle of $\tan^{-1}(r_{ep}/R_{S:EP})$. Here, r_{ep} is the radius of the entrance pupil of the imaging system. In this case, the fraction term in Equation 12.3b is the actual solid angle in sr.

By multiplying Equation 12.2 by Equation 12.3, we get the optical power in the entrance pupil. If the source radiates uniformly in all directions, no geometric correction is required. However, if the source radiates diffusely into a hemisphere, then the angular dependent correction factor for the source spectral radiant intensity is obtained by multiplying the spectral radiant intensity by the cosine of the angle made between the normal to the radiating surface and the observer. For example, say that we model a vehicle as an ideal diffuse emitter where the surface of the radiating hemisphere is on the ground. The direction of maximum radiant intensity is in the direction normal to the ground or straight up. If the optical detection system on the helicopter is directly over the vehicle and is looking down at the vehicle, then the angle between the surface normal and the observer (e.g., helicopter optical system) is zero radians, and the correction factor is 1. On the other hand, if the helicopter is not directly overhead, then the correction factor is cos (θ), where θ is the angle between the surface normal and the observation platform, and the spectral radiant intensity is multiplied by this factor. We now can describe the entrance pupil factor by

$$F_{ep} = \cos(\theta). \tag{12.4}$$

The next factor is spectral transmittance of the optics given by

$$F_{op}(\lambda) = \tau_o(\lambda). \tag{12.5}$$

By multiplying the product of the previous factors shown in Equation 12.1 by Equation 12.5, we get the spectral power on the detector surface. The transmittance of the optics includes the spectral transmittance effects of any filters, apertures, or devices in between the entrance pupil and the detector surface.

The final factor needed at this point is the one for the detector. Recalling what we learned previously, the detector's spectral output signal is relatable to the input signal by the detector responsivity:

$$F_{det}(\lambda) = \begin{cases} \mathfrak{R}_v(\lambda) = \dfrac{v_{det}(\lambda)}{\Phi_{det}(\lambda)}, \\ \mathfrak{R}_i(\lambda) = \dfrac{i_{det}(\lambda)}{\Phi_{det}(\lambda)}, \end{cases} \tag{12.6}$$

where

$\mathfrak{R}_v(\lambda)$ is the responsivity of the detector at wavelength λ
$v_{det}(\lambda)$ is the output voltage of the detector at wavelength λ
$\Phi_{det}(\lambda)$ is the optical power on the detector surface at wavelength λ
$i_{det}(\lambda)$ is the detector output current at wavelength λ

For the detector factor $F_{det}(\lambda)$, the appropriate responsivity is chosen depending on whether the output signal of the detector is a voltage or a current. The output signal s_{det} from the detector can be obtained by integrating Equation 12.1 with the appropriate substituted terms in Equations 12.2 through 12.6 or

$$s_{det} = F_{ep} \int_{\lambda_1}^{\lambda_2} F_S(\lambda) F_C(\lambda) F_{op}(\lambda) F_{det}(\lambda) d\lambda, \tag{12.7}$$

where the factor F_{ep} was pulled out of the integral since it is not dependent on the wavelength. The limits of integration are from the lowest wavelength λ_1 to the highest wavelength λ_2 that makes it through the optical system. For instance, if a narrowband optical filter is limiting the spectral throughput, then the wavelengths associated with the cutoff frequencies of the narrowband filter can be used as the limits of integration in Equation 12.7.

This integral, unfortunately, is very difficult to solve since the atmospheric transmittance term is a function of both wavelength and range. Fortunately, replacing each of the wavelength-dependent terms in the previous equations with their average, in-band quantities introduces very little error (Hudson 1969). We continue forward with this assumption. If we replace the wavelength-dependent parameters by their in-band average quantities, we get an expression for the detector output signal in terms of important detection scenario parameters:

$$s_{det} = \cos(\theta) J_S \tau_{atm} \tau_o A_{ep} \frac{\Re}{R^2_{S:EP}}. \tag{12.8}$$

In Equation 12.8, the responsivity term R is the average, in-band responsivity that corresponds to the detector output signal (e.g., either a voltage responsivity or a current responsivity). Equation 12.8 is quite interesting; it is very general. Notice that it makes no difference on whether we are considering an imaging or a nonimaging scenario; Equation 12.8 is valid for both cases. Notice also that the term on the bottom of Equation 12.8 is the range from the collection aperture of the sensor platform to the TOI. If we divide both sides of Equation 12.8 by the root mean square (RMS) value of the system noise power (e.g., the average in-band noise signal) s_n, then we have an expression for the signal-to-noise (SNR) ratio at the detector output:

$$SNR_{det} = \frac{s_{det}}{s_n} = \cos(\theta) J_S \tau_a \tau_o \frac{A_{ep}}{R^2_{S:EP}} \frac{\Re}{s_n}. \tag{12.9}$$

Solving Equation 12.9 for the range, we get an expression for the separation of target and detection system in terms of SNR at the detector, and some useful detection scenario parameters:

$$R_{S:EP} = \left[\cos(\theta) J_S \tau_a \tau_o \frac{A_{ep}}{SNR_{det}} \frac{\Re}{s_n} \right]^{1/2}. \tag{12.10}$$

The general results for Equations 12.8 through 12.10 provide rich insight into the optical systems performance. These equations can also be further expanded in terms of optical systems parameters to understand the impact of important technical performance metrics. For instance,

1. The responsivity can be written in terms of the specific detectivity to provide insight into detector area and electrical bandwidth.
2. The numerical aperture can be related to the diameter of the entrance pupil (which is relatable to the entrance pupil area in the expression) and substituted to provide insight into the effective focal length parameter on the imaging scenario.
3. The IFOV can be introduced into the expression through the effective focal length and detector area.

These points just represent a few examples of how Equations 12.8 through 12.10 can be recast to provide interesting detection scenario insights. Some other possibilities are to express parts of the expression in terms of the noise-equivalent power/irradiance, D^{**}, or optical systems scanning parameters. In making these substitutions, care must be taken with the units since some of the parameters like D^* are usually expressed in centimeters rather than meters.

As an example, in Equation 12.10, we can first express the responsivity in terms of the detector output signal s_{det}, the detector irradiance H_{det}, and the detector area A_d. We can then relate the expression for D^* to the responsivity by replacing the NEP in D^*, with the ratio of the optical power at the input of the detector (in terms of the detector irradiance H_{det} and detector area A_d) to the SNR at the output of the detector (e.g., s_{det}/s_n). By also noting that the detector area can be written in terms of the IFOV of the optical system and its effective focal length f_{eff} (e.g., $A_d = IFOV\, f_{eff}^2$), expressing the entrance pupil area in terms of the entrance pupil area ($A_{ep} = \pi(D_o/2)^2$), and using the definition of the numerical aperture ($NA = D_{ep}/(2f_{eff})$), the following range equation can be obtained:

$$R_{S:EP} = \left[\cos(\theta) \frac{\pi D_{ep}\,(NA)D^*\,J_S\tau_a\tau_0}{2(IFOV\Delta f)^{1/2}SNR_{det}} \right]^{1/2}. \tag{12.11}$$

- D_{ep}: cm
- D^*: $(Hz)^{1/2}$ cm/W
- J_S: W/sr
- $IFOV$: sr
- Δf: Hz
- (NA): no dimensions
- τ_a: percent
- τ_0: percent

This expression provides the standoff range in terms of common source, channel, and optical systems parameters. An interesting application is to calculate the maximum standoff range for the optical system, by setting the SNR_{det} term in the denominator of Equation 12.11 to 1. Conversely, Equation 12.11 can be used to determine the minimum in-band source intensity for a given distance (by setting SNR_{det} to 1 and solving for J_S). Equations 12.8 through 12.11 are useful for trade studies and feasibility analyses and for evaluating decisions in the design process. These equations can be readily adapted for applications such as tracking systems, search systems, thermal-mapping systems, and for measuring radiometric properties (Hudson 1969).

How do all the various parameters come together in the design of an optical system? What sequence of steps need to be accomplished to evolve the optical systems design? The answer lies in how the requirements at higher levels are flowed down to lower levels of the design. Typically, the mission and operational parameters are defined first and are then used to decide top-level optical systems characteristics like wavelength regime, imaging, nonimaging, and other high-level design characteristics. For example, Figure 12.15 shows a representative requirements flow down that shows a process for evolving the optical systems design for a space-based infrared system (Lawrie and Lomheim 2001).

This figure illustrates a representative decision flow from the sensor-related requirements to the design of the infrared sensor. In the top block of Figure 12.15, sensor requirements are used to drive the sequence of design steps shown in the figure.

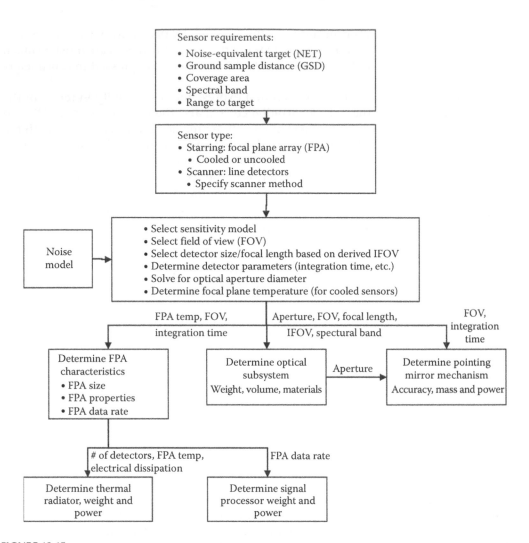

FIGURE 12.15
IR systems requirements flow down example. (Redrawn from Lawrie, D.G. and Lomheim, T.S., Advanced electro-optical space-based systems for missile surveillance, Sensor Systems Subdivision, Electronic Systems Division, Engineering and Technology Group, The Aerospace Corporation, El Segundo, CA, Report number TR-2001(8556)-1 for Air Force Contract Number F04701-00-C-0009; found at http://www.dtic.mil/dtic/tr/fulltext/u2/a400345.pdf, Figure 3, accessed April 19, 2014; drawn by Jay J. Pierce, 2014.)

Notice that the range to target is one of the requirements listed in the top block. The collection of these requirements is used to determine what type of sensor that will be used (e.g., a focal plane array staring sensor, single-detector element that is scanned through the scene, a push-broom line scanner, cooling considerations). Once these decisions are made, then optical systems parameters like the IFOV, effective focal length, noise models, and other detector parameters can be determined. The design continues by choosing the focal plane characteristics, size, weight, and volume of the optical system and then the characteristics of the pointing and tracking system. In the next section, we show how the FIT UAV-based optical system uses some of the operational considerations to drive the optical systems design.

12.3 Integrated Case Study: Application of Optical Analytical Model to FIT's System

This section continues with FIT entering the systems design stage of their UAV-based optical systems development effort. FIT needs to come to grips with some potential requirements fluctuations anticipated by Homeland Security and must provide rapid responses to DHS "what if" questions.

In this scenario, we see Dr. Bill Arrasmith, FIT chief executive officer (CEO); Dr. Tom Phelps, FIT chief technical officer (CTO); Garry Blair, FIT chief information officer (CIO); Karl Ben, Ginny R., FIT UAV program manager; Jennifer O. (Jen), FIT systems engineer; Ron S., FIT systems engineer (new hire); Phil K., FIT software engineer; Amanda R., FIT optical engineer; Christina R., FIT optical technician; and Warren M., FIT agency liaison.

Bill: "Welcome to the DHS UAV program weekly team meeting. This week, I want to talk about two things: (1) the projected performance of the IR optical system—especially how the expected signal level relates to range—and (2) possibly adopting some agile methods in anticipation of a possible mod to our existing contract with DHS. Let me start with the first point and tell you where this is coming from. This past week, I have been getting some queries from Homeland Security on some "what if" scenarios that they are running. Apparently, they have been very interested in the tip-and-queue approach that we have been considering for out IR optical system, and it has got them thinking on some new applications for our sensor. They have been talking with some of their silent partners and are hinting at a major modification. At this point, they said that because of the classification of the work, they can't tell us anything except that the delivery schedule can't change. We would have to be able to absorb the additional work, with increased funding but the original system and upgrades would have to be delivered at the same time. So, we need to look at possibly adapting our processes to make both things happen."

Ginny: "Do you want Warren or Arlene to take point on the discussions?"

Bill: "No, I'll do it. Wilford and I go way back, so I'll squeeze in the time to be the lead contact until we get this thing ironed out. At this point, my biggest concern is that they are really not settled on the requirements. The good news is they are relying on us to help them out. Wilford thinks that since this intended work falls under the existing scope of the current contract, he can use an Engineering Change Proposal (ECP). They want to do that because they have the delivery date for the UAVs and would like to have the optics ready to go for both applications as quickly as possible after they get the UAVs. He is checking with his legal folks."

Ginny: "Ugh, there goes my risk profile. The schedule is already tight and this is going to introduce requirements volatility."

Bill: "Yes we have to be careful there. I'd rather spin up a separate group here to handle it but the timeline won't allow it. Anyways, I have been getting questions like 'If the UAV was at 50,000 ft, could we still detect a human at 10,000 m from the fence line?' or 'If we had a human target that was partially obscured, so that we aren't getting the full intensity, what minimum intensity can we see with a target 5000 m in their territory, and our UAV flying at an altitude of 4000 m, and

300 m in from the fence on our side?' So, Tom, can we answer these questions, and more like it that are sure to come."

Tom: "The systems folks came up with a nice analytical model that ties into their integrated optical tool suite and the parametric database that we have working on. I'll let Jen tell you about it since she was working with the software and optics folks. Jen?"

Jen: "Thanks to the optical MBSE tool, our centralized parameter database, and the folks from the optics, and software groups, and of course Ron who did the legwork, we came up with an analytical construct that will quickly answer those types of questions. We derived the following equation that is a good starting point:

$$R_{S:EP} = \left[\cos(\theta) \frac{\pi D_{ep} \, (NA) D^* J_S \tau_a \tau_0}{2(IFOV \Delta f)^{1/2} SNR_{det}} \right]^{1/2}. \tag{12.12}$$

Notice that this equation has both the parameters that Bill mentioned in his earlier question (e.g., the intensity of the source, J_S, and the sensor to target stand-off range, $R_{S:EP}$. Given one, you can calculate the other. You can also do trades with the other parameters."

Amanda: "Yes, but you have to be careful."

Ginny: "What do you mean? The equation looks pretty straightforward."

Amanda: "Yes it is, but there are interrelated parameters. For instance, if you wanted to know the effect on the standoff range of increasing the entrance pupil diameter, D_{ep}, then you would also have to change the value of the numerical aperture, NA, and the value D^* since they are both functions of the entrance pupil diameter. Another consideration is one of interpretation. Look at the SNR_{det} expression in the denominator of the equation. If the detector signal-to-noise ratio increases, wouldn't you expect the standoff range to increase too? At first blush, an increase in the detector signal-to-noise ratio in the denominator appears to decrease the detection range. How could a better sensor, one that has a higher detector signal-to-noise ratio, produce a shorter detection range? This reflects a problem in interpretation. If the source intensity is kept fixed and all the operational parameters remain the same (e.g., look angle, optical systems parameters, atmospheric transmittance, IFOV, and electrical bandwidth of the detector), how would you increase the signal-to-noise ratio on the detector? The answer is fly closer! On the other hand, decreasing the system noise somehow increases the signal-to-noise ratio even for a constant detector output signal; however, D^* in the numerator is also increased. Finally, you have to be very careful with the units. We often consider some optical parameters in terms of meters (such as the irradiance in the entrance pupil is in W/m²). However, some of the terms in the expression are in units of centimeters such as D^*, which has units of cm Hz$^{1/2}$/W."

Ginny: "Ok, I see. Sounds like we need an optics expert looking over our shoulder."

Amanda: "That's why we are building the optics MBSE tool. The problem with the units goes away since the tool knows what parameters to adjust. If Bill, or any of us, wants to run a "what if" scenario, we plug in a particular parameter value, say entrance pupil diameter, and the tool automatically adjusts the related terms in the expression. It also guides you in what units to enter and ensures unit consistency throughout the calculation. It even has error checking to make sure we are not violating some of the basic assumptions of the model (e.g., it checks the size of the steradian used in the IFOV calculation and then uses the exact or approximate versions of the equation as appropriate and reports the error). Since the tool also connects to our current working set of parameters automatically, he will have the best estimate, and most accurate estimate that we can provide available to him."

Bill: "Perfect! Steven, can you load up the latest version on my MacBook?"

Steven: "Sure. We also are developing a mobile app for us with secure dial-in to our company cloud that you will be able to access with your mini-tablet."

Bill: "Nice! Let me know when that is ready. Running "what if" scenarios by the pool, what's this world coming to? All right, how do I determine the maximum standoff range for a given 'what if' analysis?"

Jen: "In that case, you set the SNR_{det} to 1 since smaller detector signal-to-noise values won't be detected. If the question is, what's the range for a given input intensity, then you load in our current parameters, put in the source intensity value, and solve for the range. Here, let me show you."

(Jen brings up the optical range tool on the network and ports it to the large screen.)

Jen: "The GUI is decent but we plan on adding more features. If you open the parameter database link, you can see our currently defined TPMs. Our current value for in-band source intensity is 15.8 W/sr. The IR entrance pupil diameter is 0.508 m and, watch this, it automatically converts it to 50.8 cm. Our effective focal length for the current optical design is nominally 1 m, now 100 cm, with a range of 20–200 cm. The atmospheric transmittance in-band is 77% and the transmittance of the optics is 0.975. For a 1 cm^2 active detector area, we have an IFOV of 7.853 × 10^{-5} steradians. The electrical bandwidth we are working with is 1000 Hz. Notice this isn't settled yet since it is color-coded yellow. We adopted the TPM color-coding scheme discussed in the last meeting (green for baselined TPM values, yellow for working numbers, and red for requiring validation). Our D^* value at 9.5 µm, and 77°K, with a chopping frequency of 100 Hz, is 4.5 × 10^{10} cm Hz$^{1/2}$/W. We are characterizing the D^* value at other chopping frequencies in the lab and will update the value as needed (hence, the yellow color rating). With the UAV flying at 50,000 ft (that's 15,240 m) at the fence line with the target 10,000 ground distance away, we get a look angle θ of 0.99 radians resulting in a maximum standoff range of 3.408747 × 10^6 cm, or about 34087.5 m. This result is much larger than the actual standoff range 18,227 m for this scenario and so the detector should have sufficient signal. At least on the surface."

Bill: "What do you mean on the surface?"

Jen: "Well, Amanda was telling me that this equation holds for both imaging and nonimaging scenarios."

Bill: "That's good news."

Jen: "Yes, but there is a difference in the optical equipment. The imager doesn't need the chopper that the IR tip-and-queue system needs and so the chopper term is not included in the equation. We are going to modify the tool to ask whether a chopper is used or not. If it is present, the modification to the equation is quite simple: we need to include a transmittance term for the optical modulator, τ_{om}" (Hudson 1969):

$$R_{S:EP} = \left[\cos(\theta) \frac{\pi D_{ep}\,(NA)D^*\,J_S \tau_a \tau_0 \tau_{om}}{2(IFOV\Delta f)^{1/2} SNR_{det}} \right]^{1/2}. \qquad (12.13)$$

Jen: "Tina characterized this parameter in the lab, and we are using a transmittance value of 0.48 for the optical modulator. Using this value, we get a maximum standoff distance of 43,046.6 m, and so we still have enough signal strength, even if the UAV is flying at 50,000 ft. To answer the question about the partially obscured target, we would have to decide whether we are using the imaging or nonimaging mode, determine the actual standoff range based on the imaging geometry, and then calculate the maximum range for the partially obscured target (e.g., use an estimated reduced source intensity). If the calculated maximum range estimate for the partially obscured target is larger than the actual range calculated from the imaging scenario, then there is sufficient signal strength to detect the partially obscured target."

Ginny: "This tool will be very useful. What about the effects of scanning? From my experience, some of the applications might involve tracking or searching. Can this tool be adapted for that?"

Amanda: "Good question Ginny. The answer is yes. Notice the IFOV term in the denominator of Equation 12.13. Recasting the IFOV term, at the bottom of Equation 12.13, in terms of scanning parameters, can include the effect of the scanning system on the range estimate. For instance, if τ_{dwell} is the time it takes in seconds for an image of a target to cross a pixel (assuming an unresolved target), Ω_{fr} is the extent of the field of regard in a given linear direction in steradians, and the number of detector elements in the given scan direction is n_{sd}, then the IFOV in steradians is given by (Hudson 1969)

$$IFOV = \frac{\tau_{dwell}\Omega_{fr}}{n_{sd}}. \tag{12.14}$$

Substituting this result into Equation 12.11 or 12.12 results in the range estimate now depending on scanning parameters. Notice that if the actual detector element only had one pixel, then n_{sd} would equal 1, and the dwell time τ_{dwell} would be the time it takes for the image of the target to traverse the entire detector length in a given linear direction. There are quite a number of applications, and the basic range equation can be readily adapted. In fact, we will be developing alternate expressions for different applications to our model as needed."

Bill: "That is good news. Make sure to validate the model in the laboratory. Great job everyone! Let's move on to the related issue of what to do if we get this work. I am somewhat concerned about meeting an already tight schedule, with the added commitments. Hiring new people may not cut it because of the spin-up time required. Is there something we can do with our internal processes to help?"

Ginny: "Well, we need to get some more information from our sponsor regarding the added requirements. Like I said earlier, the risk is likely to increase. Time to get the risk tool up and running."

Karl: "Oh Ginny, it may not be that bad. We can expand our modeling effort on this project."

Bill: "Remember, we are dealing with the federal government here. They don't have the best track record for timely decisions. We may have to deal with some volatile requirements in the short term."

Phil: "If this were a software job, I'd say it sounds like a good candidate for an agile approach. We tried scrum on a couple of projects where the customer didn't know all of the requirements up front and we helped them figure it out by seeing the iterations role out. It's not for every project, but it worked well on those two."

Karl: "Jennifer, didn't you work on those jobs with Phil?"

Jennifer: "Yes, Phil is right, it did work well. It was a bit different as the systems engineer. I can't figure out how we could do it on a hardware job like this though."

Karl: "Everybody in the house was gone last weekend except me. That was nice for a change; let me read some journals I had lying around. There is some activity out there on using agile techniques for projects that are more than just software. The amount of agile methods to use and what projects are good candidates are still not cut and dried. But we can look at it here. One of the main tenants of agile is to embrace change. I think we should do some study here and see how much if any of the agile methodologies works for us. Bill, we may have to waive some of our processes to try this. What do think? Will it affect our relationship with DHS?"

Ginny: "This is supposed to reduce risk? Why am I getting uncomfortable?"

Bill: "Ginny, you're getting uncomfortable because we are talking about going outside the Program Management Book of Knowledge here. That is your expertise and you are really good with that perspective. We all, well maybe except for the software staff, would be stepping outside our normal bounds. Learning is good for you, keeps you sharp. But I must give you credit on shedding some light on how DHS might react. As I said before, I've known Wilford for a long time. If we can prove to ourselves this route is the highest probability of success, then I am pretty sure he will convince his organization to work with us on this. If it blows up in our face because we missed something obvious or don't execute, yeah it will hurt us with DHS, and they have friends in other departments. But that is not our nature. We have a track record of incorporating new ideas into our flows. It is the continuous improvement cycle. You don't always have to make the improvement based upon a negative lesson learned. We don't have to decide this today. Ron, Jennifer, you two have been quiet. What say you?"

Jen: "I liked the way it worked on the software projects. I am still struggling with how to do the HW."

Ron: "I think it is an intriguing idea. I can see from the work I have done how if we freeze the baseline too early it will be painful. Maybe with some more extensive modeling...."

Karl: "Exactly!"

Bill: "Karl, I see the gears churning already. Can you find the time this week to look through more of those journals and come back to us with a proposal on how agile would work for us on this project?"

Karl: "Phil, you've lead agile projects. Do you have the time to help me with this?"

Phil: "Sure thing, bring in those articles for me to read."

Ginny: "Alright, I know what is going to take up our time next week."

(A week passes and the next weekly project meeting begins.)

Ginny: "OK let's get started. Oh wait, I have to dial in for Garry, he wanted to hear this too. (Ginny dials into the conference call.) I heard one other beep, is that you Garry?"

Garry: "Yes it is and I talked with Karl some yesterday before I got on the plane so I am up to speed."

Ginny: "Karl, the floor is all yours."

Karl: "Thanks Ginny. This was a very interesting little bit of research. The first thing we need to consider is if we have the right context to use agile methods. I found a great paper by Ernst Stelzmann out of Austria that really does a nice job with this question. In fact, it is the one that got me thinking about this in the first place. His summaries are so clean, I am just going to go ahead and quote him directly on the factors of agile feasibility:

'1. Nature of systems allows development in small steps (prototyping and testing can be done quickly and cheaply).
2. Customer is willing to support development in this way and is available for frequent feedback.
3. Product is not safety critical.
4. Implementing changes to the system is easy, can be done quickly and is cheap (Stelzmann 2012).'"

Bill: "Wilford called me so I floated a trial balloon past him on point two. He said that he would absolutely be available for feedback in the case here because of the dynamics, at least for the next couple of months. But I didn't go down the whole agile road with him."

Karl: "And we are not safety critical so we are OK on number three as well. That leaves us with one and four, which really come down to the same question for us, when do we need to freeze the baselines? This is in part where Garry's enterprise architecture tools come in. He has found some great modeling tools out there. Plus we already have our own MBSE models including the optical systems integration tool. The mechanical team has been doing all of their work with the 3D solid models for a long time and we have the high-grade 3D printer now. Yeah, we can't cool the model plastic like the real structures, but at least we can do real fit checks with the UAV. Plus, they bought a 3D scanner so we can get good models of the UAV if we can get to one. And the electrical and software teams can model the control algorithms in Matlab, even though we have never done that before."

Garry: "The trick is going to be gluing it all together into one model. I am not sure we can get there, but even with the individual pieces, we can play a lot of "what if" scenarios as the stakeholders throw potential changes at us."

Karl: "So, we think that we could use the modeling environment to execute agile methods."

Ginny: "I sense a "but …" in your voice."

Karl: "Oh, you are so perceptive! This is a two-way trade. The other question is do we really need to be agile here. Again, Stelzmann has some good questions to guide us through that decision, so I'll just directly quote him again:

'1. Rate of change is high.
2. Product is innovative.
3. Business environments are dynamic.
4. Corporate culture fits better to agile than heavyweight processes (Stelzmann, 2012).'"

Phil: "The reasons we used scrum on those two software projects before were reasons one through three on this list."

Bill: "And number four on corporate culture Phil?"

Phil: "Umm … well, we got it done after trading some of the people on the project with other projects."

Bill: "Thanks for the tactful but honest answer. If we go agile here, I can give it a little push and that should help out on the culture issue."

Karl: "So in this case, the product is not super innovative although there are some tricky areas with the atmospheric turbulence compensation approach, since this is modification to our existing contract, then the business environment is not going to be very dynamic. So that leaves us with requirements change rate."

Bill: "From what I am gathering, the change rate will calm down after a few months. If we can really get this simulation and modeling environment going, we can really show them system trades like how optical range affects the payload size and field of view. Maybe we could do a hybrid of agile methods through design and then let the traditional processes flow after critical design review."

Ginny: "We have to be careful about the long lead items; we will have to settle on them early."

Garry: "I am visiting some tool vendors as part of this trip. I'll keep working it from this angle."

Karl: "I want to see what Garry brings back before I make a final recommendation."

Ginny: "I am going to move forward with our known processes. And then, if we change, I'll make the profit be management reserve, OK Bill?

Bill: "Well, we'll talk about the management reserve. I was thinking of buying tools with the profit, which means your project still has to make it on the other aspects."

Karl: "Any other questions? No, then I'll put this out on the portal for your reference."

Bill: "We will see what Garry brings back, and then the executive team will make the call. See you next week."

12.4 Conclusion

Having the best available tools in the world is no substitute for a well-defined, well-executed process. The systems engineering life cycle provides a framework for describing, designing, and documenting a system from conception to retirement. The ISO/IEC life-cycle management systems life-cycle process provides the process guidance to effectively use that framework.

Conceptual design is the first and arguably the most important phase in the systems design process. This phase defines and bounds the problem, establishes objectives for the design, provides necessary inputs for the following phases, and dictates how the system will act and interact throughout its life cycle. The preliminary design phase includes all the systems design requirements and generates the requirements for the subsystems. Expanding the requirements to the lowest level and finishing the design process occurs in the detailed design and development phase. Successful design relies on a well-defined problem definition, careful consideration of design alternatives, and a life-cycle focus. It also requires a team approach and systems engineers with a broad knowledge of diverse engineering disciplines. For instance, Agile methodologies may provide effective, streamlined, alternative processes, particularly for software projects.

12.A Appendix: Acronyms

ABC	Activity-based costing
CAD/CAM	Computer-aided design/computer-aided manufacturing
CBS	Cost breakdown structure
CCC	Command control center
CD	Committee draft

CEO	Chief executive officer
CIO	Chief information officer
CM	Configuration management
CONOPS	Concept of operations
COTS	Commercial off-the-shelf
CTO	Chief technology officer
DDP	Design-dependent parameter(s)
DEA	Drug Enforcement Agency
DHS	Department of Homeland Security
DIP	Design-independent parameter(s)
DM	Data management
EA	Engineering architecture
EC	Electronic commerce
EDI	Electronic data interchange
FFBD	Functional flow block diagram
FIT	Fantastic Imaging Technologies, Inc., a fictitious company used to illustrate steps in the systems engineering design process.
FOM	Figure of merit
GOVT	Government
HOQ	House of Quality
IEC	International Electrotechnical Commission
ISBN	International Standard Book Number
ISO	International Organization for Standardization
IT	Information technology
KPP	Key performance parameter
MAINT	Maintenance
MCC	Mission control center
MEX	Mexico
MOP	Measure of performance
MTBM	Mean time between maintenance
MULTI	Multiple
OODA	Observe, orient, decide, and act
R&D	Research and development
REF	Reference
RUSS	Remote unmanned sensing system
SAT	Satellite
SEMP	System Engineering Management Plan
SYS	System
TEMP	Test and Evaluation Master Plan
TPM	Technical performance measure
UAV	Unmanned aerial vehicle
UAVs	Unmanned aerial vehicles
U.S.	United States
WBS	Work breakdown structure

References

Beck, K. et al. 2001. Manifesto for agile software development. http://agilemanifesto.org/ (accessed April 1, 2014).

Blanchard, B.S. 2008. *System Engineering Management*, 4th edn. Hoboken, NJ: John Wiley & Sons, Inc.

Blanchard, B.S. and W.J. Fabrycky. 2011. *Systems Engineering and Analysis*. Upper Saddle River, NJ: Prentice Hall Press.

Blanchard, B.S. and E.E. Lowery. 1969. *Maintainability: Principles and Practices*. New York: McGraw-Hill Book Company.

Cogan, B. 2012. *Systems Engineering Practice and Theory*. Rijeka, Croatia: InTech.

Fabrycky, W. 2011. System design evaluation: A design dependent parameter approach. Seminar *System Design Evaluation: A Design Dependent Parameter Approach*. Lecture, System Design Evaluation from Delft University of Technology, Delft, the Netherlands, September 9, 2011.

Fox, J.R. 2011. *Defense Acquisition Reform, 1960–2009: An Elusive Goal*. Washington, D.C.: Center of Military History.

Freeman, R.E. 2010. *Strategic Management: A Stakeholder Approach*. Cambridge, U.K.: Cambridge University Press.

Gilb, T. and S. Finzi. 1988. *Principles of Software Engineering Management*. Wokingham, U.K.: Addison-Wesley Pub. Co.

Haberfellner, R. and O. de Weck. 2005. Agile SYSTEMS ENGINEERING versus AGILE SYSTEMS engineering. *Fifteenth Annual International Symposium of the International Council on Systems Engineering (INCOSE)*, Rochester, NY, July 10–15, 2005.

Hu, Z.-G., Q. Yuan, and X. Zhang. 2009. Research on agile project management with scrum method. *2009 IITA International Conference on Services Science, Management and Engineering*, Zhangjiajie, China, pp. 26–29.

Huang, P.M., A.G. Darrin, and A.A. Knuth. 2012. Agile hardware and software system engineering for innovation. *2012 IEEE Aerospace Conference*, Big Sky, MT, March 3–10, 2012.

Hudson, R.D. 1969. *Infrared System Engineering*. New York: John Wiley & Sons.

ISO/IEC. 2008. Tools and support for ISO/IEC 15288 and related standards. Tools and Support for International Standards Organization. ISO/IEC 15288 and Related Standards. http://www.15288.com/index.php (accessed April 15, 2014).

ISO/IEC 15288:2002(E). 2002. Systems lifecycle processes, pp. 60–61. Geneva, Switzerland: International Standards Organization.

Johnson, S.S. 2011. *Agile Systems Engineering*. INCOSE Chesapeake Chapter presentation, Laurel, MD, September 2011.

Laplante, P.A. 2011. *Requirements Engineering for Software and Systems*. Boca Raton, FL: CRC Press.

Lawrie, D.G. and T.S. Lomheim. 2001. Advanced electro-optical space-based systems for missile surveillance. Aerospace Corporation Technical Report (DTIC: ADA400345), El Segundo, CA.

Murphy, M. 2007. Agile requirements—No BRUF just GRIT. Seilevel. http://requirements.seilevel.com/blog/2007/10/agile-requirements-%E2%80%93-no-bruf-just-grit.html (accessed April 15, 2014).

Pierce, J. 2014. Adapted from *Systems Engineering Guidebook for Intelligent Transportation Systems*, v3.0, p. 143. Sacramento, CA: U.S. Department of Transportation, Federal Highway Administration—California Division, California Department of Transportation.

Robertson, S. and J. Robertson. 2006. *Mastering the Requirements Process*, 2nd edn. Upper Saddle River, NJ: Addison-Wesley.

Stelzmann, E. 2012. Contextualizing agile systems engineering. *IEEE Aerospace and Electronic Systems Magazine*, May 2012.

United States Department of Defense. Standard Specification Practices (1985-06-04). MIL-STD-490A, pp. 6–13.

Scrum Alliance. 2014. Learn about Scrum. https://www.scrumalliance.org/why-scrum (accessed December 3, 2014).

13

Quality Production and Manufacturing

Quality is not an act, it is a habit.

—**Aristotle**

Quality engineering methods and techniques play an instrumental role in many of the notable achievements of a project, product, system, or service, such as an overall reduction in cost, an increase in quality, faster delivery, increased stakeholder satisfaction, and better brand recognition. We will use the word system in its general sense to mean either a project, product, system, or service for the sake of brevity. Accounting for the quality of the system and considering production and manufacturing processes during the early design stages in the systems development life cycle (SDLC) is crucial to producing both an effective and efficient product. Someone can come up with an intricate, complex, and elegant design, but this will be of no consequence unless they can bring the system into being. Even the best ideas fall short of their mark if there is no quality inherent in the final product.

This chapter is divided into two sections. The first section consists of two components: (1) a review of systems engineering aspects related to the production phase of the SDLC, with a particular emphasis on the production of optical elements such as lenses, and (2) a discussion on various concepts of quality throughout the SDLC, looking specifically for things such as loss of quality, robust design, statistical process control (SPC), and process capability. This is considered a high-level overview of the systems engineering life-cycle quality engineering processes. A focus on the production and manufacturing phase is presented in the first section of this document. The input and output products of this phase are highlighted with a high-level, discussion of basic lens design and fabrication. The lens design example is a representative example used to illustrate the concepts, methods, and techniques presented in this chapter. Ultimately, it needs to be understood that the purpose of these concepts, methods, and techniques is to reduce the variability of the manufacturing process, to improve product quality, and to reduce production costs throughout the life cycle of a system. This is often accomplished through concurrent engineering and total quality methods. In these methods, various techniques are shown that can reduce variability, increase quality, and reduce cost in the production and manufacturing phase as well as the utilization phase of the SDLC. Representative techniques include the identification, characterization, and remediation of the loss of quality, implementation of robust design methods and SPC, and the determination of process capability.

The following section, Section 13.2, is an integrated scenario wherein we integrate and demonstrate key concepts illustrated in Section 13.1 to the unmanned aerial vehicle (UAV) optical systems design integrated case study seen in previous chapters. More specifically, it shows how quality engineering tools and methods can be applied to a fictitious, but representative, UAV optical system in a simulated "real-life" scenario. This case study is important in helping the reader understand how the quality concepts, methods, and techniques in this chapter can be applied in a simple, practical, and realistic way.

13.1　Introduction to Manufacturing and Production

Systems engineering is truly an essential ingredient that is taking root in the design and development of successful systems, services, products, or processes. For years, the Department of Defense (DOD) has used systems engineering to increase efficiency of production and services; and it is finding increasing acceptance and usage in the commercial world. Until the middle of the 1970s, various companies took part in what is shown in Figure 13.1. This figure illustrates the traditional engineering life cycle, which is used by many companies involved in systems development, consisting of the following parts: design and development, production and manufacture, and support and maintenance (Wysk et al. 2000).

The traditional systems engineering life cycle consists of the design and development team recognizing stakeholder needs and then implementing the system accordingly. Once the totality of the development effort is finished, then all of the systems documentation is passed on to the team that actually manufactures the product, as well as the process engineers who evaluate the manufacturing processes. Once manufactured, appropriate systems documentation is, once again, passed to a different group, who provides systems maintenance and support to make certain everything continues to operate properly.

In traditional systems developments, some systems suffer from ineffective communications, limited interactions, and lack of continuity between the core processes (and sometimes leadership) in each of the traditional SDLC blocks shown in Figure 13.1. For example, in many DOD projects and programs, different organizations, funding lines, and leadership teams may be in charge of the acquisition phase, than those responsible for the utilization phase. This may exert pressure on the leadership team to optimize the system based on drivers and pressures in their own particular phase, as opposed to optimizing globally over the life cycle of the system. For instance, a program management team during the acquisitions phase may make decisions to alter the design of the system in such a way as to cut development costs in the acquisition phase, but the altered design introduces more required maintenance actions and, consequently, more costs during the utilization phase and/or during system retirement. Another example is the leadership team during the acquisition phase considering pushing some problems to the utilization phase to save budget and/or schedule. This locally optimized philosophy has often been shown to be a poor choice due to the fact that the system is very difficult and expensive to change, especially if the system is several stages into development or in manufacturing (Blanchard and Fabrycky 2011). It is important to change this way of thinking, especially when the major costs of supporting a system during its operational and support phase

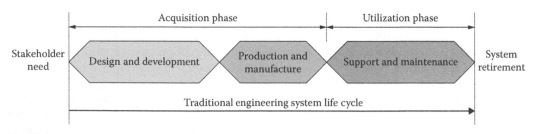

FIGURE 13.1
Traditional engineering system life cycle. (Adapted by Brian Zamito from Wysk, R.A. et al., *Manufacturing Processes: Integrated Product and Process Design*, McGraw-Hill, New York, 2000; Blanchard, B.S. and Fabrycky, W.J., *Systems Engineering and Analysis*, 5th edn., Prentice Hall, New York, 2011. With permission.)

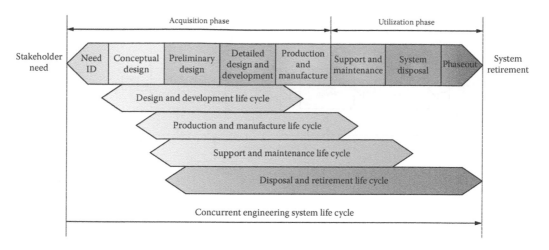

FIGURE 13.2
Concurrent engineering system life cycle. (Adapted by Brian Zamito from Wysk, R.A. et al., *Manufacturing Processes: Integrated Product and Process Design*, McGraw-Hill, New York, 2000; Blanchard, B.S. and Fabrycky, W.J., *Systems Engineering and Analysis*, 5th edn., Prentice Hall, New York, 2011. With permission.)

may be caused by poor design decisions. These support costs, over the long run, may be much more expensive than the actual systems development costs.

Figure 13.2 illustrates concurrent engineering interactions between phases during the SDLC and is somewhat of a deviation from the conventional development method. The development team focuses on the needs of the stakeholders by making sure they are documented in the form of requirements and that these requirements are implemented and supported throughout the entire life cycle of the system. A critical part of the progression of the developmental effort through the SDLC stages is that some of the production and manufacturing, maintenance and support, and even retirement and disposal activities are pushed forward into the design and development phases of the system. Contrary to the traditional sequential development method, there is a lot of feedback and communication between every area of development; this results in a higher-quality product, lower overall development and support costs, and an overall increase of efficiency during the life cycle of the system. With the concurrent engineering method, everyone involved in the project is seeing the big picture all throughout the entire process, not only at the end of production (Blanchard and Fabrycky 2011). The increased communications throughout the organization greatly benefits not only big companies but smaller ones as well.

The key difference that sets concurrent engineering apart from the traditional engineering method of development and production is that concurrent engineering requires that several steps of the product are worked on in parallel, as opposed to a linear progression seen in traditional engineering methodologies. Due to the fact that meeting stakeholder needs is the primary goal, the ultimate design of the product needs to minimize manufacturing issues and maximize quality, which is generally seen as a critical requirement in achieving customer satisfaction (Sage et al. 1999).

13.1.1 Manufacturing and Production Process

Figure 13.3 shows a more detailed diagram that explains some of the key products of concurrent engineering throughout the systems engineering development life cycle. This figure shows not only the feedback mechanisms but also the specification flow between

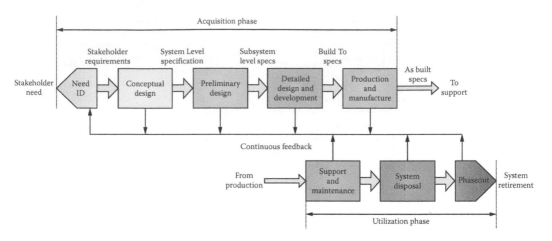

FIGURE 13.3
Systems engineering development life-cycle process. (Adapted by Brian Zamito from Blanchard, B.S. and Fabrycky, W.J., *Systems Engineering and Analysis*, 5th edn., Prentice Hall, New York, 2011; IEEE, Systems and software engineering—Systems lifecycle processes, IEEE STD 15288-2008, 2008. With permission.)

the systems development life-cycle phases. The needs identification step shown at the left side of Figure 13.3 produces a set of stakeholder requirements used in the conceptual design phase. These stakeholder requirements are then used to drive activities in the conceptual design phase to produce the system specification (A spec). The A spec is then used to initiate the high-level design in the preliminary design phase. Notice that each design phase has at its input and output a complete set of requirements specifications.

The systems engineering development life-cycle processes are continued until a set of "build to" design specifications are created. These specifications are then implemented in the production and manufacturing phase. Apart from producing the actual system itself, the major documentation resulting from this stage of the SDLC are the "as built" specifications. The "as built" specifications document the final system configuration and give a true depiction of the final postmanufacturing design of the system.

The IEEE STD 15288 document attempts to record, describe, and provide standardization for concurrent engineering processes. It also describes the production and manufacturing phase as a set of processes used in creating an element, which allows for verification of design requirements and research needs of the stakeholders (IEEE 2008).

Figure 13.4 describes the production and manufacture phase at a high level of detail and illustrates its complex, systems engineering processes. Earlier in the chapter, it was mentioned that requirements specifications are the key outputs of each of the design phases. In Figure 13.4, it is shown that the production and manufacturing phase has to begin with the "build to" specifications that are created from the detailed design and development phase. These specifications need to be used to develop production plans and processes and production schedules and oversee the labor, materials, and equipment needed to produce the system. They also are analyzed and provide important information to the total quality management (TQM) system. For instance, parts tolerances specified in the design documentation must be implemented and verified in the production phase. The TQM system ensures that the production processes are running under optimal conditions, parts are produced with minimal waste, the system is built in accordance with the design, and as efficiently and reliably as possible, and the produced system satisfies the stakeholder needs and approved wants.

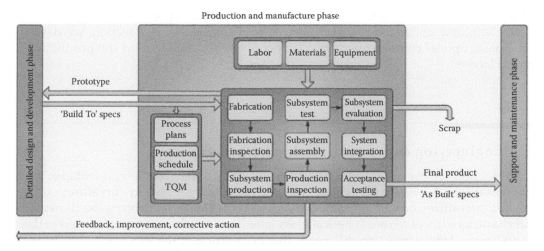

FIGURE 13.4
High-level view of the production and manufacture phase. (Adapted by Brian Zamito from Blanchard, B.S. and Fabrycky, W.J., *Systems Engineering and Analysis*, 5th edn., Prentice Hall, New York, 2011; Sage, A.P. et al., *Handbook of Systems Engineering and Management*, John Wiley & Sons, New York, 1999; IEEE, Systems and software engineering—Systems lifecycle processes, IEEE STD 15288-2008, 2008; Kasunic, K.J., *Optical Systems Engineering*, McGraw-Hill Professional, New York, 2011. With permission.)

When the system enters the production phase, there are many distinct activities. The production process starts with fabrication, which involves converting raw materials, such as glass or metal, into a usable item. After fabrication, the subsystems are produced, inspected, assembled, and tested. The subsystems are then integrated into the system configuration and the integrated system undergoes acceptance testing. Any changes made to the original specifications, due to required alterations during the production and manufacturing processes, are documented in the "as built" specifications. The final "as built" specifications are transferred, along with the system itself, to the user and the operations and support phase of the SDLC begins. It is essential to give the design and development team the necessary feedback concerning any deviations from the original design, due to manufacturing and production considerations. This feedback consists of improvements and corrective actions that have already been made, as well as overall feedback and recommendations to the design team. Finally, developmental elements such as prototypes are sent back to the detailed design and development team and each unusable piece is considered a loss. It should be noted that sometimes, the prototypes evolve into the actual fielded system. For instance, in the United States, the National Aeronautics and Space Administration (NASA) has a term, "protoflight," that is distinguished from prototype in that the prototypes are scrapped, whereas protoflight units are used in or may evolve into the actual flight unit.

In Figure 13.4, the high-level blocks indicate the major activities in the production and manufacture phase. These blocks can be further defined by breaking them down into subblocks using functional analysis. Subblocks allow more details to be seen at the lower levels of production and may be based upon the kind of unit that is being manufactured, as well as what is required by the final requirements/design specifications of the system. One representative example is, in lens fabrication, there are several activities that need to be accomplished to complete the fabrication process. Though they are not illustrated in the figure, the subprocesses involved for completion of the fabrication process must be specified and accounted for to achieve the production of a high-quality lens.

Prior to starting these production processes, all "build to" specifications must be complete, clear, and understood by the production team. In the next section, we describe some basic optical characteristics and factors needed to understand the production of optical lenses.

13.2 Engineering and Manufacturing Optical Devices

In this section, we consider the manufacturing of an optical lens as a representative application example. A lens may have sides that are flat, convex (outward curvature), concave (inward curvature), or various combinations thereof. Figure 13.5 shows the dimensions and characteristics of a basic spherical glass lens. The lens shown in the figure is called a plano-convex lens, where "plano" means that one surface of the lens is flat and "convex" means that the second surface has an outward curvature. This means, in simple terms, that the lens has a flat side and a curved side.

The basic function of a lens as shown earlier (Smith 2008; Kasunic 2011) is to capture the incoming light, which is shown entering on the left side of the figure, and direct the light in such a way that it travels to a primary focal point, usually on the optical axis seen as the horizontal dashed line bisecting the lens. The focal length is defined as the distance from the lens to its primary focal point. The focal length is found by using lens attributes such as the lens' material properties and physical properties such as size and shape, which can be seen in Figure 13.5. The dimensions of the lens can also be related its optical throughput through metrics like the numerical aperture and $f/\#$.

To quantify lens performance, the following fundamental equations can be used. Equation 13.1 is known as the lens grinder equation (Kasunic 2011). In this equation, the focal length (f) of a lens can be determined in terms of specific properties such as index of refraction (n), the lens center thickness (CT), and the curvature of the inner and outer lens surfaces (R_1, R_2). This process is also reversible in that desiring a specific focal length for a

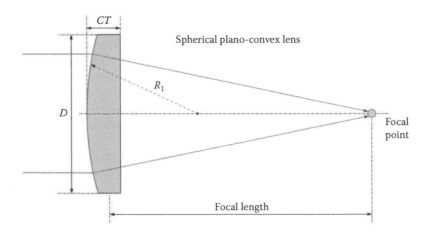

FIGURE 13.5
Basic spherical glass lens dimensions. (Adapted by Brian Zamito from Kasunic, K.J., *Optical Systems Engineering*, McGraw-Hill Professional, New York, 2011; Smith, W.J., *Modern Optical Engineering: The Design of Optical Systems*, 4th edn., McGraw-Hill, New York, 2008. With permission.)

lens of specific dimensions and made of a given material type (e.g., the index of refraction is known) and then paired radii of curvature can be calculated from the so-called lens grinder equation:

$$\varphi = \frac{1}{f} = (n-1) * \left[\frac{1}{R_1} - \frac{1}{R_2} + \frac{(n-1)(CT)}{(n)(R_1)(R_2)} \right]. \tag{13.1}$$

Equation 13.1 is an expression for the lens optical power, which is the inverse of the lens focal length. The optical power is given by φ and relates to the lens' light-bending ability (Kasunic 2011). The shorter the focal length, the better the lens can bend light, and the higher the optical power becomes. Optical power is difficult to measure directly; however, the focal length is relatively easy to determine. If we know the lens index of refraction, its thickness at the center, and respective radii of the lens' surface curvature, we are then able to find the lens focal length. In typical applications, a target focal length is desired and the optical designer must determine the physical attributes of the lens including its thickness, material properties, and radii of curvature. Table 13.1 provides a summary of the variables used in Equation 13.1.

In evaluating Equation 13.1, there are two important points to consider. The first point is for lens designs that have small center thicknesses when compared to their focal lengths (e.g., $CT \ll f$), wherein the center thickness may be considered to be zero resulting in the thin lens equation. Secondly, if any one of the surfaces is flat, then the corresponding radius can be considered as infinite. If either of these considerations is present, then Equation 13.1 is simplified.

The speed of the lens can be calculated using Equation 13.2, which shows the lens focal length f divided by the lens diameter D. This ratio is known as the f-number. Basically, it represents the amount of light-gathering power of the lens. For instance, if the speed of a given lens is 2.4, which can be written as $f/2.4$, then the focal length divided by the lens diameter is 2.4:

$$f - \text{number} = \frac{f}{\#} = \frac{f}{D}. \tag{13.2}$$

Now that the basic properties of a lens have been presented, its production process can be explored in further detail. Once the optical design of a given lens is complete, these properties must then be captured in the design documentation. This documentation needs to

TABLE 13.1

Basic Optical Equation Variable Descriptions

Description	Variables
Refractive index of lens material	n
Center thickness of lens	CT
Outside surface radius of lens	R_1
Inside surface radius of lens	R_2
Focal length of lens	f
Optical power of lens	φ
Aperture diameter of lens	D
F-stop number, relative aperture, speed of lens	$f/\#$

Source: Data from Brain Zamito, 2013.

clearly indicate the ratings, tolerances, material processes, and coatings of the product. These data are then transferred to the production team who determines if the lens should be prepared by conventional means, or by automated means, based on the current cost, time, and numbers expected. The lens fabrication process has been understood for over a hundred years (Fischer and Tadic-Galeb 2000). Depending upon what is desired, the process can be very low-tech and requires a lot of labor, or it can be a very high-tech and automated process. Usually, the need falls somewhere in between these two extremes. If done manually, this process allows for a less technical approach but is more labor intensive. If the work is done using an automated system known as a computer numerically controlled (CNC) machine, then the lens is produced by automated grinding and polishing methods, which can be much more accurate and can reduce the number of errors (Fischer and Tadic-Galeb 2000).

The upper part in Figure 13.6 shows a method for manufacturing glass lenses (Fischer and Tadic-Galeb 2000; Smith 2008; Kasunic 2011). The initial part of the process is molding, wherein the glass is separated into portions and placed in a mold after it is heated. The result is a newly formed black lens mold. The mold generation process is not a mandatory step and some manufacturers choose to skip that section and jump right into the process with a blank, plano-plano glass disk. At this point, the molds or blank disks are then ground into an extremely course shape that will be used to make the final lens, a process known as "generating." It should be noted that the lens has larger dimensions at this point than in its final form because some of the surface will be removed in the subsequent grinding and polishing processes.

Following the generating process, the "blocking" process is initiated. In this step, lens blanks are grouped or blocked together. After the blocks have been properly positioned, the lens-grinding process removes excess layers and leaves the lens in the general shape

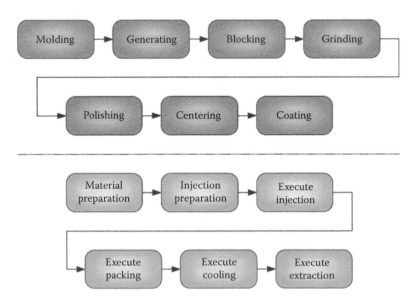

FIGURE 13.6
Glass and injection molding lens fabrication process. (Adapted by Brian Zamito from Kasunic, K.J., *Optical Systems Engineering,* McGraw-Hill Professional, New York, 2011; Smith, W.J., *Modern Optical Engineering: The Design of Optical Systems,* 4th edn., McGraw-Hill, New York, 2008; Fischer, R.E. and Tadic-Galeb, B., *Optical System Design,* McGraw-Hill, New York, 2000; Lo, W.C. et al., *Int. J. Adv. Manuf. Technol.,* 41, 885, 2009. With permission.)

needed by the customer. The grinding process is iterative using different granularity and both sides of the lens are ground until they are relatively smooth. After the completion of the grinding stage, the lens is polished. This is the stage wherein the surface imperfections are removed and the radii specified in the lens design documentation are attained.

Next, the optical and mechanical axes are positioned and the lens is centered. Afterward, the lens undergoes cleaning and inspection and required coatings are added. Then, the newly fabricated lens gets inspected, tested, assembled, and integrated into the rest of the optical system.

The steps in fabricating a lens using injection molding are shown at the bottom of Figure 13.6 (Lo et al. 2009). Even though it is a common thought that lenses made from plastic are low quality, this is not always the case. It is true that mass-producing plastic lenses are less expensive than their glass counterparts. However, plastic lenses are used in a variety of products such as mobile phones and low-cost video recorders to great effect. There are three high-level steps to the injection molding methodology—filling, packing, and cooling. A lens mold is made, which is then injected with hot, liquid plastic to form the lens shape. Afterward, the lens is then allowed to cool and is separated from the mold. Care must be taken during the entire process since injection molded plastic lenses are also susceptible to imperfections (Lo et al. 2009).

A different optical manufacturing method, though unrelated with the production of lenses, is the optical fiber extrusion process (Acquah et al. 2006). This process is used to generate the optical cables in fiber-optic communications. The optical fibers are pulled from a furnace and stretched until long enough. As it is being stretched, the optical fiber is repeatedly heated and cooled in a controlled fashion, and finally, a protective coating is added and the completed optical fiber is rolled up for storage, shipment, or for later implementation.

Some points to consider from these examples are that variations can be introduced at various steps in the fabrication process. Variations can be introduced through the material selection process, through the operational steps during fabrication, and through environmental conditions, material attributes, or the design itself (Fischer and Tadic-Galeb 2000). This variability can introduce a risk component to the fabrication process that must be properly managed. One way to accomplish this is to integrate quality engineering methods into the fabrication process.

13.2.1 Total Quality Management

The term quality was used several times throughout this chapter so far. What does the term "quality" mean? Unfortunately, there are different definitions and interpretations given by leading thinkers on the topic. Quality can be seen as one particular definition or even a collection of attributes such as compliance with the specifications, fitness for use, value for price paid, and support services (Reid and Sanders 2009). One view looks at quality in terms of conformance to requirements. In other words, if a company wants to develop a quality product, they have to make sure that it meets its design specifications and tolerances. Most often, the actual litmus test for what is meant by quality lies in the stakeholder expectations (e.g., their agreed needs and wants). If the stakeholder requirements properly reflect their needs and wants (both stated and hidden), and the system meets or exceeds the stated requirements, then the stakeholder will perceive that they have a quality product. For multiple stakeholders, there may be different needs and wants and some of these needs and wants may conflict with each other. That is why it is so important to have the needs and wants agreed up front. The stakeholder requirements capture the agreed needs and wants and the mutual understanding is if the stakeholder requirements are correctly implemented, then the collection of stakeholders must be satisfied. Regardless of the definition, quality is extremely important when it comes to successfully navigating, and constantly competing in, a global market (Gunasekaran et al. 1998).

Establishing an effective quality program for an institution is no simple matter. In order to implement a solid quality engineering program throughout, the SDLC requires resources and management commitment. Investing in quality not only improves an organizations bottom line, it also helps promote confidence in the organization's products and can boost the organization's reputation. These benefits can be vital to the long-term success of the organization. Stakeholders recognize quality when they see it—good or bad. Some stakeholders have long memories and sustained low-quality products may result in the mass loss of customers and damage to the organization's reputation (Reid and Sanders 2009).

Establishing a high-quality program within an organization has costs. However, ignoring quality or accepting low-quality results has its own costs and consequences. Some representative costs associated with high-quality products includes developing the quality plan, training personnel, quality accounting, process improvements, inspections, and tests. These are known as preventive and appraisal costs (Reid and Sanders 2009). On the other hand, some cost impacts associated with poor quality include rework, scrapping, adding materials, and potential delays to the schedule. In addition, there are the external costs such as returns, repairs, warranty, and liability (Reid and Sanders 2009).

It is well known that implementing quality in an organization may be costly. However, the main point to consider is whether or not the quality improvements are worth the cost. A beneficial result of establishing a quality program within an organization is that typically, costs are recovered over the life cycle of the product. Quality expert Philip Crosby summarized this idea, when he said "quality is free" (Reid and Sanders 2009). In essence, taking on the costs of preventive quality will lower the costs that could be realized in pursuing a reactive quality strategy. The overall effect should be that the preventive costs should be equal to or less than the reactive costs. One means to implement quality steps within an organization is through the concept of TQM. TQM uses stakeholder expectations as the litmus test for quality improvements of all levels of society and throughout the life cycle (Reid and Sanders 2009). Figure 13.7 shows the high-level categories, tools, and techniques that make up the characteristics of the TQM process. Every aspect of the organization, from customer, to employer, to suppliers, is addressed and various proven quality tools are implemented such as SPC, Taguchi robust design, and the quality function deployment (QFD). The intent of this chapter is not to dwell on the TQM paradigm itself, but to focus on representative quality tools that are useful in the production and manufacturing phase.

With regard to TQM, the implementation process has its own share of challenges (Reid and Sanders 2009). It requires management commitment and must be made a priority in order achieve success. Adopting a TQM methodology must be seen as a core philosophy and cultural change—a long-term commitment and not a fad.

We now provide some basic mathematical relationships that are useful in quality methods such as SPC. With a basic understanding of statistics and probability and the use of the equations starting with Equation 13.3 below, a number of quality tools become available that are useful in the production and manufacture phase. We start with some basic relationships. The mean of a set of observed data and the mean of a sample (subset) of the

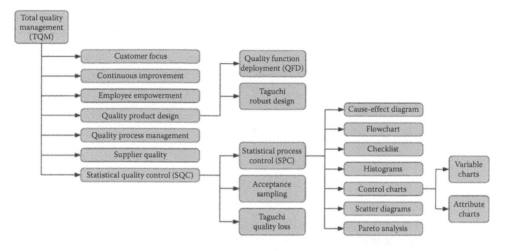

FIGURE 13.7
Characteristics of TQM. (Adapted by Brian Zamito from Reid, R.D. and Sanders, N.R., *Operations Management*, 4th edn., John Wiley & Sons, New York, 2009; Gunasekaran, A. et al., *Int. J. Qual. Reliab. Manage.*, 15(8/9), 947, 1998. With permission.)

observed data set can be determined by Equations 13.3 and 13.4, respectively. Ultimately, the mean measures the central tendency of a data set or sample:

$$\mu = \frac{\sum_{i=1}^{w}(X_i)}{w}, \tag{13.3}$$

$$\overline{X} \lim_{X \to \infty} = \frac{\sum_{i=1}^{n}(X_i)}{n}. \tag{13.4}$$

A performance parameter or the mean of a set of observed data is given by μ and $\overline{X}$ yields the sample mean ($n < w$). In Equation 13.3, the parameter w is over the entire observed data set and the parameter n is the number of samples in a sample set. Equations 13.5 and 13.6 show the variance of a given process and that of a sample set, respectively. These equations provide insight as to the degree of variation about the estimated mean of the data set:

$$\sigma^2 = \frac{\sum_{i=1}^{w}(X_i - \mu)^2}{w}, \tag{13.5}$$

$$s^2 = \frac{\sum_{i=1}^{n}(X_i - \overline{X})^2}{(n-1)}. \tag{13.6}$$

In Equation 13.5, the mean μ is known. In Equation 13.6, the data themselves are used to estimate the variance. Equations 13.7 and 13.8 describe how to calculate the process and sample standard deviation of a set of observed data. Equation 13.9 provides the standard deviation for a collection of n averaged samples in terms of the standard deviation from all the samples σ. The standard deviation measures the amount of variation of the data set from the central tendency:

$$\sigma = \sqrt{\sigma^2},$$ (13.7)

$$s = \sqrt{s^2},$$ (13.8)

$$\sigma_{\bar{X}} = \frac{\sigma}{\sqrt{n}}.$$ (13.9)

In the last term, Equation 13.9 shows the standard deviation associated with n samples of averaged data. Say, for instance, that there are 100 data points but we decide to work with 20 sets of 5 samples each. If we compute the average of each 5-point sample (sample mean), determine the overall mean of the 20 sample averages by adding all the sample means and dividing by 20, and then compute the standard deviation of the 20 sample means (shown on the left side of Equation 13.9), that would be equivalent to the true standard deviation σ of the full 100-point data set, divided by the square root of the number of sample means (square root of 20 in this case). Variables, along with their descriptions, for Equations 13.3 through 13.9 are listed in Table 13.2.

In applying some of these parameters in quality initiatives during the production and manufacture phase, Six Sigma methods come to mind. TQM and Six Sigma are related because they both aim to improve quality. TQM is a quality management goal that is generally discussed at an organizational level, whereas Six Sigma is one method toward implementing that goal. The term sigma defines the amount of standard deviation of a process from its central mean value. A sigma value of 6 means that all products produced within ±6 standard deviations of the mean (99.9999998%) are accepted. The defect rate of a true Six Sigma process is approximately two parts per billion. This means that during the manufacturing process, only two parts out of a billion would be rejected because of defects. Since Six Sigma is hard to attain, several companies use a three-sigma process.

TABLE 13.2

Basic Statistical Equation Variable Descriptions

Description	Variables
Performance parameter/observation data	X_i
Number of samples (population size)	w
Process mean	μ
Number of observations per sample (sample size)	n
Sample mean of observation data	$\bar{X}$
Process variance	σ^2
Sample variance of observation data	s^2
Process standard deviation	σ
Sample standard deviation of observation data	s
Standard deviation of the distribution of sample means	σ_s

Source: Data from Brian Zamito, 2013.

Similarly, a sigma value of 3 means that all products produced within ±3 standard deviations of the mean (99.74%) are accepted. The rejection rate of a 3-sigma process is on the order of 2600 parts per million.

Using a Six Sigma process may or may not initially seem to have much of an advantage over using a 3-sigma process. However, Motorola engineers discovered that a process tends to shift by approximately 1.5 sigma over time (Arnheiter and Maleyeff 2005). This affects a Six Sigma process minimally. It will still accept 99.99966% of parts and rejects only 3.4 parts per million. Note that, in this case, the Six Sigma process has been reduced to a 4.5-sigma process due to the long-term negative shift of 1.5 sigma. However, the effect can really be seen in a 3-sigma process. In this case, the acceptance rate will fall to about 93.32% and the rejection rate will increase to 66,800 parts per million. So even though a Six Sigma process may be hard to attain, depending on implementation cost, it may be worth it in the long run.

13.2.2 Taguchi Quality Engineering

To help with the process of creating a quality product, Taguchi, who is a renowned quality thinker and analyst, should be considered in this discussion. Genichi Taguchi was not only an engineer but also a statistician. He completed his doctorate in engineering at Kyushu University (Benton 1991). His methods are known for narrowing the steps, and testing that it takes, to achieve a solid quality product. Taguchi quality engineering, also known as the Taguchi method, is a quality process that consists of three focus areas as follows: systems design, parameter design, and tolerance design (Taguchi 1995). New technologies, available resources, and the top-level design are considered within the first step of systems design.

The last two steps of parameter design and tolerance design are often related to an idea called robust design (Wysk et al. 2000). Taguchi acknowledges that quality cannot be made into a product; it must be purposely designed in Benton (1991). Taguchi's method tries to reduce constant, incorrect changes and increase the overall quality of the product. This in turn reduces the cost that is associated with poor quality products.

13.2.2.1 Taguchi Quality Loss

The traditional interpretation of quality loss is relatable to a step function that represents accepted outcomes within the confines of a step function that is centered on the target mean (Wysk et al. 2000). The edges of the step function to each side of the mean represent the limits between accepted and rejected target values. If an item is produced with an attribute that falls within the limits of the step function, it is deemed acceptable, whereas items that fall outside of the step function limits are rejected. In this manner, items that are accepted are said to have zero quality loss. If an item falls outside of the tolerance limits, it is rejected and considered a quality loss. Figure 13.8 illustrates the traditional interpretation of the loss of quality.

Figure 13.8 shows the loss to society as the dark line. The target value is shown as T in the diagram and the dark line represents possible variations. Note that Unit 3 has been manufactured right on the target value and consequently has no loss to society and no cost impact. Notice also that Unit 2 has been produced near the specification limit and is far from the nominal target value. Because of the traditional acceptance criteria, wherein everything that falls within the specification limits is accepted, Unit 2 is also accepted with

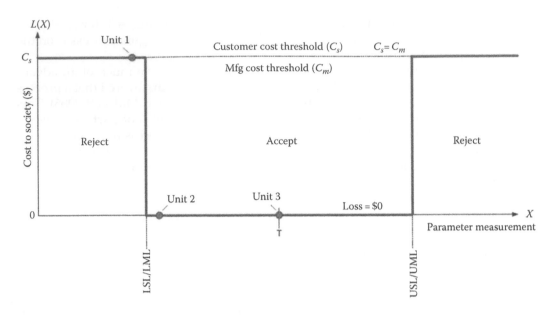

FIGURE 13.8
Traditional view of quality loss. (Adapted by Brian Zamito from Wysk, R.A. et al., *Manufacturing Processes: Integrated Product and Process Design*, McGraw-Hill, New York, 2000; Benton, W.C., *Int. J. Prod. Res.*, 29(9), 1761, 1991. With permission.)

zero quality loss and no assumed cost impact. As a final example, Unit 1 has crossed the specification limit by a small amount. In the traditional quality loss paradigm, units that exceed the upper specification limit (USL) or fall below the lower specification limit (LSL) are rejected regardless how close they fall to the USL or LSL.

With respect to quality, we might ask why would one unit be rejected and the other kept when they both fall so closely to the specification limit (Benton 1991)? In addition, why would not there be a loss assigned to Unit 2 since it was manufactured so far away from the target value (Benton 1991)? The Taguchi quality loss method addresses these concerns.

Taguchi devised a way of presenting this loss concept in simple mathematical terms. He developed expressions for the cases (a) smaller is best, (b) larger is best, (c) or nominal is best. Some physical examples would be in the case of number of defects, smaller is best; in the case of profits, larger is best; and in the case of human body temperature, nominal is best (98.6°F). We focus on the nominal is best approach since this is often used in a production environment.

For the nominal is best category, Taguchi developed a quadratic function to measure the "loss to society." A parameter related to the production process is identified and monitored. For instance, let's say a clamp was being manufactured that had to be 10 cm long plus or minus 0.01 cm. The identified parameter would be the length of the clamp in centimeters. Once a product is equal to the nominal target number (10 cm), it can then be deemed as satisfactory and effective without costing customers or manufacturers (i.e., society) more money. According to Taguchi, as the parameter value wanders further and further away from its nominal value, this directly increases the cost to society quadratically until it reaches the USL/LSL. With this method, products that measure beyond the specification (spec) limit will also be rejected causing some kind of cost loss associated with it. If the parameter value exceeds the specification limits even slightly, a different

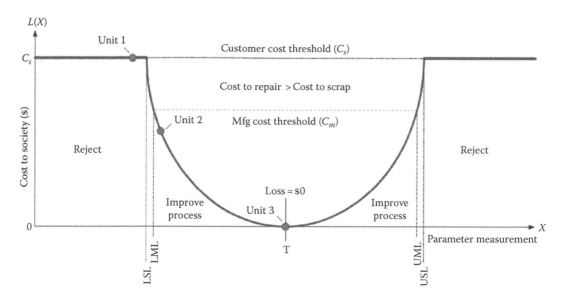

FIGURE 13.9

Taguchi view of quality loss. (Adapted by Brian Zamito from Wysk, R.A. et al., *Manufacturing Processes: Integrated Product and Process Design*, McGraw-Hill, New York, 2000; Benton, W.C., *Int. J. Prod. Res.*, 29(9), 1761, 1991. With permission.)

result occurs than if the parameter value lands on the specification limit. If the specification limit is exceeded by the parameter, a loss occurs that has a related cost. This cost can be quantified and used by management to understand the quality-related costs associated with their product, process, system, or service. Figure 13.9 shows the Taguchi quality loss function.

Looking at the same information from Figure 13.8 and comparing it to Taguchi's view, Unit 3 still has zero loss to society because it was produced right on target. On the other hand, Unit 2 has some loss associated with it, which differs from Figure 13.8 when it was treated the same as Unit 3. Unit 1 is still beyond the specification limits and is again rejected. In addition, though the range between the target and the specification limit is not rejected, it still shows that there is a cost consequence for producing units off of the target value. The concept of a manufacturing cost threshold is also introduced (Benton 1991). A manufacturing cost threshold determines the point of whether a manufacturer should either scrap or repair a unit. In this case, units with high variability from the center are rejected well before the customer would ever get a chance to see the final product or outcome. The associated lower manufacturing limit (LML) and upper manufacturing limit (UML) is shown in Figure 13.9.

The quality loss equations used to determine the "loss to society" and that are illustrated in the quality loss charts are now presented. Equation 13.10 illustrates this loss function for the situation where the target parameter is a nominal value, such as in the average temperature of the human body. The loss to society is shown in Figure 13.9 as the dark line with the quadratic shape. The nominal value given by specifications is shown as T on the figure. Deviations from the nominal value are shown by x_i. The factor k on the right side of Equation 13.10 has units of cost per parameter dimension squared and is used to map parameter variations to quality loss costs for a single unit. Equation 13.10 can be used to find single unit costs; however, Equation 13.11 permits calculating the cost loss for multiple units.

Being able to identify this information from the equations allows companies to better prepare and understand loss of quality costs and provides a better way of analyzing product quality:

$$L = k * (X_i - T)^2,$$ (13.10)

$$\overline{L} = k * [s^2 + (\overline{X} - T)^2].$$ (13.11)

Equations 13.12 and 13.13 are used to calculate the costs associated with their losses. The parameter C_s is the cost per unit if the unit is produced on the specification limit. Due to the fact that these costs are typically determined by returns, repairs, and servicing costs found by analyzing records and historical data, this coefficient may be hard to quantify:

$$k = \frac{C_s}{(\Delta_s)^2},$$ (13.12)

$$\Delta_m = \Delta_s * \sqrt{\frac{C_m}{C_s}}.$$ (13.13)

The parameter Δ_s is the distance from the target value T to the specification limit. C_m is the cost per unit when the unit is built on the manufacturing specification limit, and Δ_m is the distance from the target value to the manufacturing specification limit. Lastly, the manufacturing and specification limits, which are highlighted in Equations 13.14 and 13.15, represent the parameter values at the specification limits and the manufacturing limits, respectively. Items that fall beyond these limits are either reworked or scrapped well in advance of the product being shipped to the customer:

$$\text{Taguchi Spec Limit} = T \pm \Delta_S,$$ (13.14)

$$\text{Taguchi Mfg Limit} = T \pm \Delta_m.$$ (13.15)

The variables associated with the Taguchi quality loss are summarized in Table 13.3.

The quality loss function proposed by Taguchi provides a step in the right direction for companies looking to improve quality and cost by justifying how various levels of quality can affect their business.

Li et al. (2007) explains how to improve quality even further by adapting the quadratic loss function proposed by Taguchi. In the referenced paper, Li decides to look at several other functions that could approximate the loss due to voltage brownouts. He first looks at the inverse normal loss function (INLF) example, which is compared to the Taguchi loss function, seen in Figure 13.10. Li also investigates the inverse gamma loss function (IGLF) and the inverse beta loss function (IBLF). The INLF is more feasible to implement in a *nominal is best case.*

Ultimately, using the INLF further restricts the allowable deviations from the target product. This is an important thing to know because companies putting together a quality loss profile should attempt to use or develop a function that represents their particular

TABLE 13.3

Taguchi Quality Loss Equation Variable Descriptions

Description	Variables
Target specification value (nominal)	T
Performance parameter/observation data	X_i
Quality loss coefficient	k
Quality loss to society per unit (dollars/unit)	L
Average quality loss to society per unit (dollars/unit)	$\bar{L}$
Sample mean of observation data	$\bar{X}$
Sample variance of observation data	s^2
Process variance	σ^2
Cost loss threshold for manufacturing (dollars)	C_m
Cost loss threshold for society (dollars)	C_s
Manufacturing tolerance	Δ_m
Specification tolerance	Δ_s
Upper specification limit (max)	$USL = T + \Delta_s$
Lower specification limit (min)	$LSL = T - \Delta_s$
Upper manufacturing limit (max)	$UML = T + \Delta_m$
Lower manufacturing limit (min)	$LML = T - \Delta_m$

Source: Data from Brian Zamito, 2013.

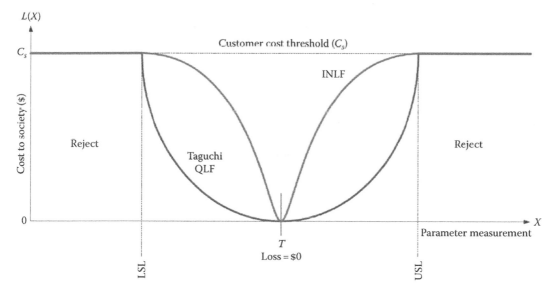

FIGURE 13.10
Modification of Taguchi quality loss function. (Adapted by Brian. Zamito from Li, G. et al., *IEEE Power Tech*, 617, 1509, 2007. With permission.)

product in order to obtain the most accurate results. The best way to do this would be to use a more defined and specific function that can be tailored for individual products. Finally, a novel idea was presented that proposes the use of the analytical hierarchy process (AHP) to calculate and quantify the cost loss coefficient *k* through a rigorous approach (Khorramshahgol and Djavanshir 2008).

13.2.2.2 Taguchi Robust Design

Earlier in this discussion, Taguchi suggested three methods for adding quality into the design process, repeated as follows: systems design, parameter design, and tolerance design. The systems design aspect is often a difficult process to undertake because it requires innovation on the part of the designer. A novel way to include innovation in the systems design process is to use the QFD method. This process is highlighted in Figure 13.11 (Bouchereau and Rowlands 2000).

The QFD method is useful in the systems design process because it takes the stakeholder requirements and generates lower-level technical responses that can be easily reviewed and managed. Generating the lower-level responses can be a very involved process, so Bouchereau and Rowlands proposed using Taguchi robust design methods to quantify the requirements before moving into the next phase of the QFD. This quantification can compare specification and tolerance levels to industry standards or industry leaders—a process known as benchmarking.

Robust design is a statistical design of experiments methodology used to determine how a set of parameters affects the mean and variation of a process. For example, the Taguchi method for robust design involves setting up process parameters in an orthogonal array and analyzing the results. The goal is to find which parameters instrumentally affect the process so that they can be adapted to reduce process variations and deviations from the process mean in a relatively reduced number of experiments. This would save cost. Figure 13.12 shows a flow chart of the Taguchi robust design process (Wu and Wu 2012; Ku and Wu 2013).

To perform a robust design experiment, an interested party would first define the objectives they are looking to achieve. Some examples to illustrate this idea include the optimization of polishing parameters for a glass substrate (Lien and Guu 2008), the optimization of fiber-optic sensor development (Chen et al. 2004), or the optimization of a miniature L-type optical lens (Sun et al. 2009). Once the objective is determined and written down and the output under consideration is determined, the input design parameters (sometimes called control factors) that determine the output are identified along with any noise factors (like temperature fluctuations and vibration) present. Generally, control factors are the input design parameters that are easiest to control in a process, hence the name control factors.

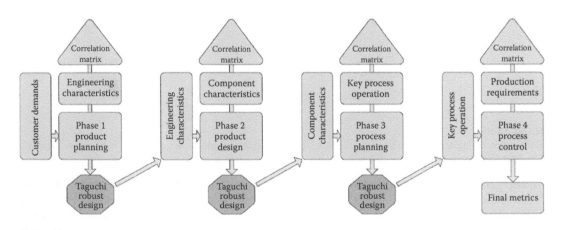

FIGURE 13.11
Taguchi influence on QFD. (Adapted by Brian Zamito from Bouchereau, V. and Rowlands, H., *Benchmarking Int. J.*, 7(1), 8, 2000. With permission.)

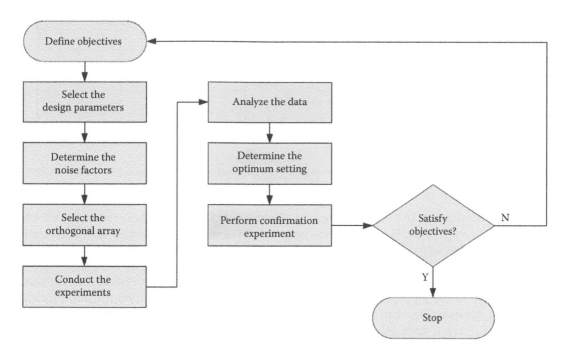

FIGURE 13.12
Taguchi robust design flow chart. (Adapted by Brian Zamito from Wu, H. and Wu, Z., *Energy*, 47(1), 411, 2012; Ku, H. and Wu, H., *J. Power Sources*, 232, 199, 2013. With permission.)

These are the factors that are varied to determine the response and quality of the output. The noise factors are the things that are either very difficult or costly to control. Once those factors are determined, as opposed to a full factorial design process, which contains all possible combinations of a set of factors, Taguchi uses the idea of an orthogonal array to determine the number of experiments to run, based on the level and number of input factors that have been given. This again varies from the full factorial design in which an experiment is run at every combination of the factor levels. For example, if an experiment was to be run with four input parameters that each had three setting levels, a full factorial experiment would have required us to perform a minimum of 81 experiments. How does this equate to 81? Well, with four input parameters and each having three setting levels, we would need to have 3^4 combinations ($3 * 3 * 3 * 3 = 81$). For example, if three trials where to be run, this would equal a total of 243 experiments. In essence if one complete trial entails 81 experiments, then three trials multiplied by the number of experiments would equal 243 experiments ($3 * 81 = 243$). Taguchi allows us to compute a similar analysis with only nine experiments. Why does one method take 243 experiments and another only 9? Well this is possible because of Taguchi's orthogonal array method and statistical analysis. This method finds the maximum defects and offers the most coverage with minimal test cases by using a systematic and statistical way of testing pair-wise interactions (Benton 1991). Figure 13.13 shows an orthogonal array, used for the optimization of a glass lens polishing process.

The required number of experiments, as well as the combination of factor levels (Taguchi's orthogonal array), can be looked up in a table provided by Taguchi in his robust design method. For example, according to a table generated by Taguchi, four factors that have three levels each require an L-9 orthogonal array, where the 9 indicates the number of

Control factor	Symbol	Level 1	Level 2	Level 3
Time (Min)	A	5.0	7.5	10.0
Pressure (kPa)	B	4.9	9.8	12.7
Platen speed (RPM)	C	40	50	60
Oscillation speed (RPM)	D	3	4	5

L-9 Array	Control factors			
Run	A	B	C	D
1	1	1	1	1
2	1	2	2	2
3	1	3	3	3
4	2	1	2	3
5	2	2	3	1
6	2	3	1	2
7	3	1	3	2
8	3	2	1	3
9	3	3	2	1

FIGURE 13.13
Taguchi L-9 array. (Adapted by Brian Zamito from Lien, C.H. and Guu, Y.H., *Mater. Manuf. Process.*, 23, 838, 2008. With permission.)

required experiments. In his work, Taguchi also provided the L-9 orthogonal array, which indicates, for each of the 9 experiments, what the level settings should be for the factors. Once the number of experiments and the factor levels are set, the experiment is run and the output response of each experiment is measured and analyzed as a signal-to-noise ratio (SNR), in order to have a better ability to minimize, nominalize, or maximize the output response to the input levels. Equation 13.16 shows the SNR equation for minimizing the output, for example, minimizing the surface roughness of a lens. Equation 13.17 is used to nominalize the output, which is essentially meeting a particular target value; Equation 13.18 is used to maximize the output (e.g., first pass test yield through a process):

$$SNR_{min} = -10 * \log_{10} \left(\frac{\sum_{i=1}^{n} X_i^2}{n} \right), \tag{13.16}$$

$$SNR_{tgt} = +10 * \log_{10} \left(\frac{\overline{X}^2}{s^2} \right), \tag{13.17}$$

$$SNR_{max} = -10 * \log_{10} \left(\frac{\sum_{i=1}^{n} \left[\frac{1}{X_i^2} \right]}{n} \right). \tag{13.18}$$

The variables used (and their meaning) for the SNR equations are summarized in Table 13.4.

Once the results are recorded and analyzed, the optimum factor levels are determined and confirmed through additional experimentation. If the defined objectives are met, then the design is complete; otherwise, the design should be evolved.

For example, an experiment was run to minimize the surface roughness of a glass lens during the polishing process (Lien and Guu 2008). The control factors and levels were identified as shown in Figure 13.13. The end result of the experiment (shown later) showed that Factor C had the most impact on the surface roughness, whereas Factor A had the least impact. The entire set of values was optimized to minimize surface roughness by setting Factor A at Level 3, Factor B at Level 1, Factor C at Level 1, and Factor D at Level 2. This example is expanded upon in the case study in Section 13.3.

TABLE 13.4

Taguchi Robust Design Equation Variable Descriptions

Description	Variables
Performance parameter observation data	X_i
Number of observations per sample (sample size)	n
Sample mean of observation data	$\overline{X}$
Sample variance of observation data	s^2
Taguchi minimized signal-to-noise ratio	SNR_{min}
Taguchi targeted signal-to-noise ratio	SNR_{tgt}
Taguchi maximized signal-to-noise ratio	SNR_{max}

Source: Data from Brian Zamito, 2013.

13.2.3 Statistical Process Control

Walter Shewhart is widely regarded as being the "grandfather of quality control (Reid and Sanders 2009)." His work recognized that variation is inherent in any manufacturing process. He is also widely known for creating charts to track this variation. These "control charts" are used to determine whether a particular process had controlled, random variation (in control) or uncontrolled assignable causes of variation (out of control). SPC can therefore be thought of as a methodology that uses control charts and other quality tools to analyze the variability inherent in a manufacturing process.

The ability to control this variability and target it on predetermined product specification and tolerance levels leads to a higher confidence that a product that is produced under SPC will be of greater quality (Blanchard and Fabrycky 2011). The advantages of implementing SPC include conformance to design parameters; an increase in product quality; the reduction of scrap, rework, and returns; and the ability to quantify quality in order to meet potential stakeholder requirements (Benton 1991).

The main disadvantage with SPC is that it is often an offline quality tool implemented after the product has been designed, and potentially leading to a higher cost product. Unlike the concepts of Taguchi that were discussed earlier, SPC typically has little impact on the product design stage. A novel idea was put together to make SPC more of an online quality tool by combining SPC with engineering process control (EPC) and the idea of Taguchi quality loss (Duffuaa et al. 2004). This concept of integrated process control (IPC) is summarized in the flow chart of Figure 13.14.

The SPC methodology provides the capability to manage this variability, based only on predetermined product specifications and an acceptable level of confidence (Blanchard and Fabrycky 2011). The advantages of the SPC method are numerous and include establishing the compliance of design parameters with specifications, improving product quality by the reduction of process deviation from the mean, and process variation, waste reduction, and the ability to quantify aspects of quality in the production phase (Benton 1991).

To briefly summarize, an SPC and EPC function receive data from manufacturing. An integrated processor is used by the EPC function to determine the cost impact for process variations. For the SPC function, $\overline{X}$ charts can be used to decide if the process in or out of control. $\overline{X}$ charts as well as other useful SPC charts are discussed later in this text.

If the process is determined to be out of control, it is then necessary to find the root cause for the out-of-control process and adjust the process accordingly. The Taguchi quality loss equation discussed earlier can be used to determine quality loss costs for processes that are in control.

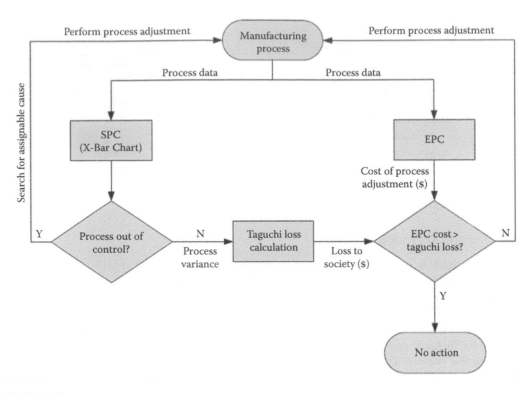

FIGURE 13.14
IPC flow chart. (Adapted by Brian Zamito from Duffuaa, S.O. et al., *Int. J. Prod. Res.*, 42(19), 4109, 2004. With permission.)

A comparison can then be made between the costs of fixing the process variations and the loss of quality costs. If the quality loss costs are greater than that for fixing the process variation costs, then process adjustments are supportable and should be made if practical.

Figure 13.7 shows SPC can integrate several very useful tools: Pareto charts, control charts, scatter diagrams, flow charts, cause-and-effect diagrams, histograms, and check-lists. These tools are very useful in identifying and analyzing quality-related issues prior to becoming too expensive and costly to the company and possibly the stakeholder (Reid and Sanders 2009).

Cause-and-effect diagrams, also known as fish bone diagrams, are useful for tracking program defects to its source. These diagrams are problem-solving tools used by quality teams who brainstorm the reasons why poor quality is occurring (Reid and Sanders 2009). Cause-and-effect diagrams can be used either before or after an issue has been found, depending on how proactive the team is. Figure 13.15 gives an example of how quality teams use cause-and-effect diagrams. In this diagram, the causes for a potential defect in plastic lenses are explored. For this example, the issue is identified on the far right as a "lens defect (Lo et al. 2009)." Working from the right to the left side of Figure 13.15, potential categories that gave rise to the lens defect are identified (e.g., machine, mold, fabrication, and operator). Each category has its individual "causes" (illustrated by the twigs connecting to the categories). As an example, a potential cause of the lens defect could be an issue with several fabrication items. This diagram can be adjusted to account for as many categories and causes as required.

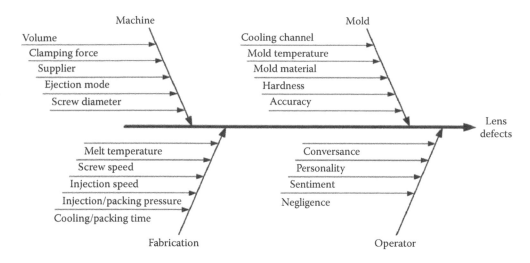

FIGURE 13.15
Cause-and-effect diagram. (Adapted by Brian Zamito from Lo, W.C. et al., *Int. J. Adv. Manuf. Technol.*, 41, 885, 2009; Reid, R.D. and Sanders, N.R., *Operations Management*, 4th edn., John Wiley & Sons, New York, 2009. With permission.)

An excellent method for determining, quantifying, and prioritizing quality-related issues is to conduct a Pareto analysis. The Pareto analysis is based on the fundamental assumption based on observations that the majority of quality problems in a product can be attributed to relatively few root causes. The Pareto analysis rule of thumb can be expressed as the so-called 80–20 rule where 80% of the defects or problems result from 20% of the quality-related root causes (Reid and Sanders 2009).

In conducting a Pareto analysis, first problems are rank ordered in terms of largest to smallest. Often classifying and tallying the number of defects in a product can do this. Assuming each type of defect has equal importance, the problem associated with the largest number of defects is addressed first. This approach has the benefit of providing a logical way to proceed in prioritizing and resolving quality issues. On the other hand, if the problems are not of equal importance, then a weighting method may be used to determine problem relative priority. Figure 13.16 provides insight to the Pareto analysis process and reflects a representative application. In Figure 13.16, 2400 lenses are inspected for defects after being manufactured (Henderson 2011). Upon inspection, 110 lenses were rejected because of noncompliance with requirements. In Figure 13.16, the columns indicate the type of observed problem along with their frequency of occurrence. As an example, approximately 45 lenses had problems with their coatings representing 40% of the total defects. Upon investigating the root cause for this failure, the lens manufacture isolated the problem to a coating supplier. After resolving the root cause, a subsequent Pareto analysis showed that coating problems dropped from the highest problem category to just in the sixth category thereby demonstrating a significant reduction in coating related errors.

Control charts are very useful in a production environment. Typically, two types of control charts are extensively used and these will be discussed in more detail in the next section. Variable charts are used when quality characteristic can be expressed by a variable (e.g., temperature, length). An attribute chart is used when only summary or classification type of information is available (e.g., pass or fail). Typically, attribute charts are used to handle a broad set of issues, and a variable control chart is used to further investigate

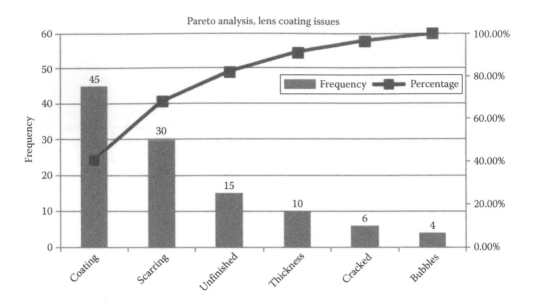

FIGURE 13.16
Pareto analysis chart. (Adapted by Brian Zamito from Henderson, G.R., *Six Sigma Quality Improvement with Minitab*, 2nd edn., John Wiley & Sons, Chichester, U.K., 2011. With permission.)

specific parameters. For example, a common use is to identify if a given manufacturing or industrial process variability is or is not in control. Process variation is a naturally occurring event and some variation is to be expected. However, processes that are out of control cause measurable excessive process variations for a variety of root causes such as component variations, machine assembly or tolerance issues, process errors, environmental conditions (e.g., temperature, pressure, humidity), and sequencing errors. All of these natural occurrences are classified as being common variations (Reid and Sanders 2009). It is considered normal to have process variation. In fact, it would be unnatural if there were no variation at all. The purpose of SPC methods is to find and remove special causes of variation in a process. If these special causes are present and severe enough, at this point, the process is said to be out of control and the root cause of the error must be found. Once identified, these are considered to be the assignable causes of variation. Removing the special causes results in the statistical process having an acceptable amount of variation and being in control.

13.2.3.1 Variable Control Charts

One important method in SPC is the variable control chart. This chart is used when a process variable can be identified, for example, weight, length, or size. Variable control charts typically come in pairs. A common pair of variable control charts is the $\bar{X}$ chart and the R-chart. The $\bar{X}$ chart determines the variations of the means of a set of sample sets. Take, for instance, our previous example of 100 samples that were broken into 20 sets of 5 samples each. Say that each of the samples consists of samples from a different production line. We have 20 production lines each producing 5 samples. The $\bar{X}$ chart looks at the variations of the means of each sample about the average of all the means, and the R-chart measures the variation of the ranges about the average range over all sample sets. We will look at the mathematical expression starting with Equation 13.19 below. The $\bar{X}$ and associated R-chart are commonly used in SPC methods (Reid and Sanders 2009).

Control charts, as stated before, provide an excellent method to understand process variability and are one of the fundamental SPC quality tool. Their primary purpose is to measure parameter variations in a process and indicate whether or not the process is in control. The control charts have four essential features consisting of control limits (upper and lower), the centerline, and parameter data that are plotted on the control chart. To determine the centerline, the mean of the plotted parameter data is calculated. The variation of the plotted parameter data about the centerline provides insight into parameter variability. By having carefully chosen the observable parameter, insight into process variability is obtained.

As an example, the $\overline{X}$ chart shown in Figure 13.17 indicates variations in the lens diameter from an optical lens manufacturer. The graph on the left of Figure 13.17 shows all of the data points within the upper and lower control limits (UCL and LCL). The indicative process parameter, the lens diameter, in this case is under control and the variations can

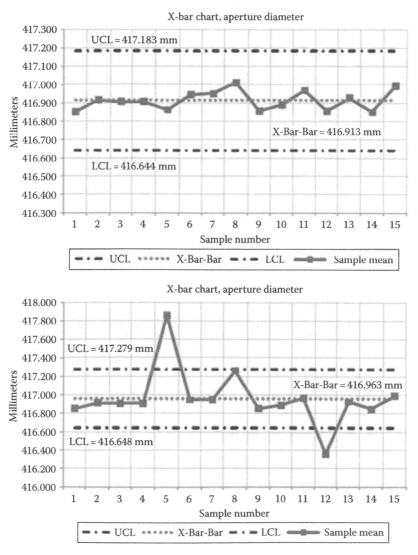

FIGURE 13.17
$\overline{X}$ control chart comparison. (Created by and courtesy of Brian Zamito.)

be attributable to natural statistical process variations. On the other hand, the graph on the right side of Figure 13.17 shows points that fall outside of the control limits. The conditions that lead to the point falling outside of the UCL need to be investigated to see if there is any root "special causes."

Equations 13.19 through 13.23 provide the essential details for constructing an $\overline{X}$ chart. To illustrate, Figure 13.17 shows a particular manufacturing run that produced 15 batch runs of lenses. Each batch (set) had four lenses for a total of 60 lenses (15 batches times 4 lenses each). In terms of the $\overline{X}$ chart, each batch can be considered a sample, and the four lenses in each sample can be considered individual observations. The average diameter for each batch is represented by $\overline{X}$ and can be obtained by averaging the diameters of the four lenses in each batch/sample. Equation 13.19 provides the mean of the sample averages or $\overline{\overline{X}}$:

$$\overline{\overline{X}} = \frac{\sum_{i=1}^{w}\left(\overline{X}_i\right)}{w}.$$

(13.19)

This equation takes the sample mean—the average of each sample—and computes an overall average of the sample means. $\overline{\overline{X}}$, which in this case is 4.169 m, becomes the center-line or the mean for the entire set of sample means. The fifteen sample means are plotted on the chart to illustrate how far they deviate from the centerline. In this case, w is 15, and $\overline{X}_i$ is the mean of the 4 lens aperture diameters at each of the 15 samples. In this case, the UCL and LCL cannot directly be specification limits since we are working with averaged data and not the parameter data themselves (e.g., the averages of the lens diameters, not the actual lens diameters themselves). There are two ways to calculate the control limits for the $\overline{X}$ chart. Given knowledge of the standard deviation of the sample means ($\sigma_{\overline{X}}$), Equations 13.20 and 13.21 provide the UCL and LCL. The parameter z represents how many standard deviations away from the centerline are tolerated and found acceptable for process variation. As an example, if the value z is chosen to be 3 (a common number in the United States), then the UCL and LCL are separated from the centerline by 3σ and the probability of making a type I error is 0.27%. A type I error represents a false alarm/positive in the sense that a sample appears to be out of control but in reality, it resulted from a naturally occurring variation. If the USL is at the $+3\sigma_{\overline{X}}$ point, then 99.73% of the naturally occurring observations will fall within the UCL and LCL. This means, however, that there is a smaller percentage chance (0.27%) that the observations will naturally fall outside of the control limits. This would indicate that there may be one or more special causes present when, in fact, there is not (false alarm):

$$UCL_{\overline{X}} = \overline{\overline{X}} + (z * \sigma_{\overline{X}}),$$

(13.20)

$$LCL_{\overline{X}} = \overline{\overline{X}} - (z * \sigma_{\overline{X}}).$$

(13.21)

However, since samples are being taken from an overall process, the standard deviation for the process is not often known. Consequently, an alternate method for determining the UCL and LCL for the $\overline{X}$ chart is given by the following equations:

$$UCL_{\overline{x}} = \overline{\overline{X}} + (A_2 * \overline{R}),$$

(13.22)

$$LCL_{\overline{x}} = \overline{\overline{X}} - (A_2 * \overline{R}).$$

(13.23)

TABLE 13.5

$\bar{X}$ Chart Control Equation Variable Descriptions

Description	Variables
Sample mean of observation data	$\bar{X}$
Number of samples (population size)	w
Average of the sample means of observation data	$\bar{\bar{X}}$
Standard deviation of the distribution of sample means	$\sigma_{\bar{X}}$
Standard normal variable (number of standard deviations)	z
Average range of sample observation data	$\bar{R}$
Statistic value based on standard normal variable	A_2
$\bar{X}$ chart upper control limit	$\bar{\bar{X}} + (A_2 * \bar{R})$
$\bar{X}$ chart lower control limit	$\bar{\bar{X}} - (A_2 * \bar{R})$

Source: Data from Brian Zamito, 2013.

In these two equations, the average sample range $(\bar{R})$ is used instead of the sample means $\sigma_{\bar{X}}$. The multiplicative factor A2 is a statistical parameter related to the number of sample observations. These values, and other useful statistical constants, are readily available from any good SPC book or from the Internet. For instance, a table of statistical constants that contains A_2 for various sample sizes can be found from the University of Delaware website as well as many other sites (Udel 2014). Some useful parameters related to the $\bar{X}$ chart are described in Table 13.5.

Just like other control charts, the R-chart is constructed in a similar fashion and follows the same rules. The R-chart is typically paired with the $\bar{X}$ chart since the same data set is used. An example of the R-chart is provided in Section 13.2. Just like the $\bar{X}$ chart, the samples consist of four observations. The range for one of the 15 samples is found by subtracting the lowest lens diameter value in the sample from the highest lens diameter value. The average of these $\bar{R}$ values is then determined by Equation 13.25, which is very similar to the calculation needed for the $\bar{X}$ chart. The average $\bar{R}$ value becomes the centerline for the R-chart. The collective 15 sample ranges are plotted in a similar fashion to the $\bar{X}$ chart and visually illustrate the variation of the range from its centerline:

$$R_i = \text{Max}(X_i) - \text{Min}(X_i), \tag{13.24}$$

$$\bar{R} = \frac{\sum_{i=1}^{w}(R_i)}{w}. \tag{13.25}$$

Here, the subscript i denotes the ith set of four lens aperture diameter observations. The UCL and LCL are determined by Equations 13.26 and 13.27. Similar to the parameter A_2 earlier, D_3 and D_4 are statistical parameters related to the number of sample observations. These values can be found in any good book on statistical quality control or found on the Internet (Udel 2014). We will provide these values where needed for the examples in this text. Parameters commonly used in developing R-charts are presented in Table 13.6:

$$UCL_R = (D_4 * \bar{R}), \tag{13.26}$$

TABLE 13.6

R-Chart Control Equation Variable Descriptions

Description	Variables
Performance parameter observation data	X_i
Range of sample observations (largest–smallest)	R
Number of samples (population size)	w
Average range of sample observation data	$\bar{R}$
Statistic value based on standard normal variable	D_3
Statistic value based on standard normal variable	D_4
R-chart upper control limit	$UCL_R = (D_4 * \bar{R})$
R-chart lower control limit	$UCL_R = (D_4 * \bar{R})$

Source: Data from Brian Zamito, 2013.

$$LCL_R = (D_3 * \bar{R}). \tag{13.27}$$

If the R-chart LCL is negative, then it is set to zero. Though the $\bar{X}$ chart and the R-chart are used to measure different components of the data, both sets of charts use the same set of data to make their measurements and findings. Often, they are used together since neither of them is mutually exclusive (Reid and Sanders 2009). Based upon the different components that these charts measure, the $\bar{X}$ chart could indicate that the observable process variations are not in control. For the same data, the R-chart could show that the variations in the standard deviation are in control. Since these charts measure different aspects of the data, both are needed to determine whether or not a particular variable is in control.

There are several other useful variable control charts such as the X-chart, the moving range chart (R_M), the exponential weighted moving average and deviation charts (EWMA and EWMD), and the cumulative sum (CuSum) chart. The X-chart is useful in looking at individual data and not averages of the data. An interesting point is that the UCL and LCL can be directly related to specification tolerance limits in the X-chart but not for the $\bar{X}$ chart since the variation in the $\bar{X}$ chart is the variation of the means and not the actual data themselves. R_M charts are useful when evaluating ranges of moving averages such as in streaming data. EWMA and EWMD can be used when data values in the moving average need to be weighted. The EWMA method is useful in detecting shifts in the mean of a data set (such as slowly rising temperature), whereas the EWMD chart finds shifts in the variability (same mean temperature but variations about the mean temperature). The CuSum chart can find slight shifts in the data. We refer the reader to one of the many excellent books on SPC for further information.

Ultimately, production issues need to be addressed if either of the two charts shows data that are out of control, driving the quality team to identify and fix the root causes for the process variations and get the process backed up and running as soon as possible. Section 13.3 provides and applications example for variable control charts.

13.2.3.2 Attribute Control Charts

Another set of important quality control charts that require consideration is attribute control charts. Similar to the way variable control charts are practical for plotting variables, attribute control charts can be used for determining aggregate quality characteristics that may be broadly classified and are countable. For instance, attributes like how many lenses failed (or passed) inspection and the number of pits or scratches on the optical lenses are representative examples. The attribute control charts are constructed in a similar fashion as variable control charts and obey similar rules. The concepts of the centerline, UCLs, LCLs, and data variations about the centerline are similar to those used in the variable control charts. The difference between attribute control charts and variable control charts has to do with the type of measured data and that variable control charts are used for single-variable data, whereas attribute charts can deal with multiple quality characteristics at once.

The P-chart is one of the most used attribute control charts. It can be used in tracking how many defects there are in a rejected (or passed) sample (Reid and Sanders 2009). One particularly useful applications example is in determining the first pass yield (FPY) of a product undergoing a manufacturing process. Useful parameters needed for generating P-charts are shown in the following equations:

$$\bar{p} = \frac{\sum_{i=1}^{m}(p_i)}{m}, \tag{13.28}$$

$$\sigma_p = \sqrt{\frac{\bar{p}*(1-\bar{p})}{n}}, \tag{13.29}$$

$$UCL_p = \bar{p}+(b*\sigma_p), \tag{13.30}$$

$$LCL_p = \bar{p}-(b*\sigma_p). \tag{13.31}$$

The quantity p_i is the proportion of defects in sample set i, and m is the number of sample sets, and n is the number of samples within each sample set (assumed the same for each sample set). The value $\bar{p}$ is then the average number of defects across all samples, and σ_p is the standard deviation of the proportion of defects. The parameter b is a multiplier that determines how many standard deviations of σ_p there are between the UCL and LCL of the attribute chart. The following is a basic example for how to construct a P-chart. Let's say that we have a total number of 300 units that are being produced in a 15-day period. The quality engineer wants to investigate a particular manufacturing process that produces 20 test articles every day. Of particular interest is how many units failed or passed inspections on a daily basis. Analyzing this information provides the quality engineer with a good indication on whether the production processes were working as expected or not.

Equation 13.28 determines the ratio of the number of failed units per day to the total number produced that day and averages this result over the 20 days. This equation gives you $\bar{p}$, the proportion defective, and also the central line of the P-chart. The average number of units that passed over the 20-day period is given by $(1-\bar{p})$. Given the standard deviation for the proportion of defects is known, the process for determining the UCL and LCL is similar to that of the $\bar{X}$ chart. In this case, Equation 13.29 gives the standard deviation of the proportion of defects. The variables used to construct the P-chart are described in Table 13.7. An application example of the P-chart is given in Section 13.3 of this text.

The C-chart is another commonly used quality tool that falls into the attribute chart category. The C-chart is practical when it comes to counting defects on a per-unit basis. This value basically measures defects on a per-unit or per-sample basis. The underlying

TABLE 13.7

P-Chart Control Equation Variable Descriptions

Description	Variables
Proportion of defects per ith sample	p_i
Total number of sample sets	m
Average proportion of defective items per population	$\bar{p}$
Number of observations per sample (sample size)	n
Standard deviation of the proportion of defects	σ_p
Number of standard deviations	B
P-chart upper control limit	$UCL_p = \bar{p} + (b * \sigma_p)$
P-chart lower control limit	$LCL_p = \bar{p} - (b * \sigma_p)$

Source: Data from Brian Zamito, 2013.

TABLE 13.8

C-Chart Control Equation Variable Descriptions

Description	Variables
Number of defects per unit	c_i
Sample time	M
Average number of defects over sample time	$\bar{c}$
Standard normal variable (number of standard deviations)	z
C-chart upper control limit	$UCL_c = \bar{c} + (b * \sqrt{\bar{c}})$
C-chart lower control limit	$LCL_p = \bar{p} - (b * \sigma_p)$

Source: Data from Brian Zamito, 2013.

statistical distribution is the Poisson distribution wherein the mean and variance are the same. Consequently, the standard deviation for the C-chart is just the square root of the average number of defects for the unit. Note also that the sample size is the actual unit itself (e.g., equals 1). The equations for the C-chart are very similar to those of the P-chart and are shown in Equations 13.32 through 13.34. The parameters used in generating C-charts are shown in Table 13.8:

$$\bar{c} = \frac{\sum_{i=1}^{m}(c_i)}{m}, \tag{13.32}$$

$$UCL_c = \bar{c} + (b * \sqrt{\bar{c}}), \tag{13.33}$$

$$LCL_c = \bar{c} - (b * \sqrt{\bar{c}}). \tag{13.34}$$

The decision to use either the P-chart or the C-chart depends on if the proportion of defects is known or not (Reid and Sanders 2009). For C-charts, the size of the sample unit must be kept the same from sample to sample. Other useful kinds of attribute charts are the NP-chart and the U-chart. The NP-chart measures the number of defective items versus the proportion of the defective items in the P-chart. Both the P-chart and the NP-chart use the binomial distribution and are useful in measuring multiple defects or characteristics. The U-chart measures average defects per sample instead of the number of defects per sample unit in the C-chart. The C-chart and U-chart use the Poisson distribution.

13.2.3.3 Process Capability

Control charts, as discussed previously, are useful tools for the quality team and production team for assessing whether or not a given process is in or out of control. If the process is in control, then it is also stable and if it the process is out of control, it is unstable. The quality team or production team must find the root cause and fix the associated issues as quickly as possible so that the process, once again, is under control. It must be pointed out that just because a data point falls outside of the control limit, it does not mean that the process is out of control. Since there are statistical variations associated with the deviation of data points from their centerline, there is a finite probability that some of the data points will naturally fall outside of the established control limits. Data points that naturally fall outside of the control limits are called falls alarms, and their probability of occurrence can be estimated by integrating the "wings" of the probability density function that falls outside the control limits. Conversely, even though the data points fall within the control limits, that does not mean that the process is capable of producing acceptable results for the quality and production teams. It is possible to have an in-control process but the process itself is not capable. Consequently, we need a separate means for determining process capability and then strive to ensure that a given process is both in control and capable.

A good way to estimate process capability is to directly compare the specification limits to the process variation as shown by Equations 13.35 and 13.36 (Reid and Sanders 2009). The units of the numerator and denominator are both in standard deviations so the resulting capability is unitless. Equation 13.35 gives the potential capability that is inherent in the process and Equation 13.36 gives the actual capability realized from the production process. The control limits in both expressions can be related to specification limits in a requirements document. The engineers, during the design process, implement these specification-based tolerances during their design activity:

$$C_p = \frac{\text{Spec Width}}{\text{Process Width}} = \frac{(USL - LSL)}{(6 * \sigma)}, \tag{13.35}$$

$$C_{pk} = \min\left(\frac{(USL - \mu)}{(3 * \sigma)}, \frac{(\mu - LSL)}{(3 * \sigma)}\right). \tag{13.36}$$

Notice, we are assuming a common three-sigma process as seen from the denominators in the aforementioned two equations. A representative illustration is a target voltage requirement of 5.0 V with a ±10% peak-to-peak tolerance level. The USL is determined by adding 5% of the target value to the specified voltage requirement (USL = 5.5 V) and similarly subtracting 5% of the target voltage (LSL = 4.5 V). For the denominator, the design team determines the required process standard deviation and associated multiplier (6 and 3 in the aforementioned two equations) to ensure that the resulting process is capable. The parameters used in the Equations 13.35 and 13.36 are summarized in Table 13.9.

In determining process capability, typically three separate limits are used (Reid and Sanders 2009). A process is considered to be minimally capable if the process capability C_p is exactly equal to 1.00. When a process is only minimally capable, then the width of the

TABLE 13.9

Process Capability Equation Variable Descriptions

Description	Variables
Upper specification limit (max)	$USL = T + \Delta_s$
Lower specification limit (min)	$LSL = T - \Delta_s$
Process standard deviation	σ
Process mean	μ
Process capability index (potential)	C_p
Process capability index (actual)	C_{pk}

Source: Data from Brian Zamito, 2013.

control limits is equal to the width of the established process variations. The term minimal is used since there is no margin between the specified specification limits and the process variation. Exactly 6 standard deviations (3 on each side of the mean) span the specification limits resulting in an expectation of 99.73% acceptable process results. The expected unacceptable results are then 0.27%. This percentage may be too high, especially when there are large numbers of units involved. Recall also the shift in the process standard deviation that was previously mentioned that could result in an increase of unacceptable results over time to 6.7%! Notice that if the statistical variations of the process are larger than the bounds established by the numerators in Equations 13.35 and 13.36, then the C_p and C_{pk} values fall below 1.0 and we have a "not capable" process. The variations in the process are higher than what is required resulting in an unacceptably high defect rate. If the process is in control but not capable, changes must be made to the process or the driving tolerance requirements need to be relaxed with stakeholder consent.

Figure 13.18 shows examples of processes that are "not capable" and "minimally capable." One interpretation is that the minimally capable processes are barely acceptable and may still be improved upon, whereas not capable processes require action to address their root cause(s).

Equation 13.36 provides the actual capability of the process C_{pk} and accounts for a shift in the mean of the process. The difference between the potential capability C_p and the actual capability C_{pk} is that the potential capability assumes that the process mean is adjustable

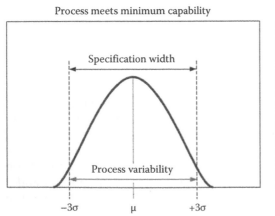
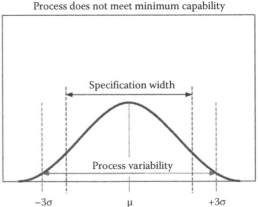

FIGURE 13.18

Comparison of minimum capability versus failing capability. (Adapted by Brian Zamito from Reid, R.D. and Sanders, N.R., *Operations Management*, 4th edn., John Wiley & Sons, New York, 2009. With permission.)

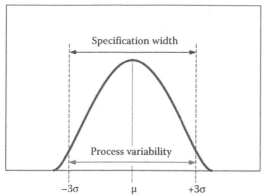

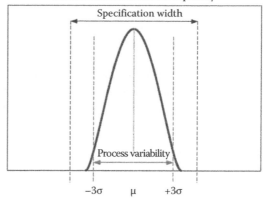

FIGURE 13.19

Comparison of minimum capability versus exceeding capability. (Adapted by Brian Zamito from Reid, R.D. and Sanders, N.R., *Operations Management*, 4th edn., John Wiley & Sons, New York, 2009. With permission.)

to target value. In this case, the process variations are symmetric about the target value. C_{pk} accounts for deviations in the process mean and provides an estimate of the actual capability. For example, if a machine in a production line has acceptable variations but has drifted from the target value (perhaps due to lack of calibration), then C_p would be greater than or equal to one reflecting the fact that the process is potentially capable by calibrating the machine and moving the mean of the statistical variations back to the target value. In this case, the value C_{pk} would likely indicate a not capable system since the minimum term in Equation 13.36 would be less than one. An example comparison between a minimally capable process and a capable process is shown in Figure 13.19. Notice the difference in the height between both processes. The more capable the process, the higher the height and smaller the width showing that more units are built at the target value while variations about the target value are reduced.

Some authors have tried to further define the concept of a capable process and apply process capability to specific manufacturing applications. For instance, Chen attempted establishing a connection between the time it takes to manufacture a product and process capability. In so doing, these authors defined four process capability ranges: 1.00–1.33 as "capable," between 1.33 and 1.50 as "satisfactory," between 1.50 and 2.00 as "excellent," and greater than 2.00 as "super" (Chen et al. 2006). Subsequently, Chen's results indicated that the focus of the manufacturer should be time to manufacture once acceptable process capability is achieved. Quality and lead time can sometimes be of equal importance to the stakeholder (Chen et al. 2006).

13.3 Integrated Case Study and Application: UAV Optical System Project

This chapter has introduced the reader to broad systems engineering concepts that high-light the manufacturing and production phase of the SDLC. In the first section of this chapter, a lens design fabrication example was used to showcase quality methods and techniques and introduce the reader to the typical inputs and outputs of this stage of the SDLC. Concepts such as concurrent engineering and total quality were discussed with an emphasis on quality methods that reduce defects and process variations leading to a higher-quality product and an overall reduction in cost. Representative topics appropriate to this phase in the SDLC were discussed, such as Taguchi loss functions, robust design methods, SPC concepts, and techniques for determining whether or not a process is capable.

We now introduce an integrated case study that uses a fictitious but realistic company to provide application examples of the concepts and methods presented earlier in this chapter. This case study is a continuation of the running application examples from previous chapters. In this section, we find our company, Fantastic Imaging Technologies (FIT) facing manufacturing issues for their high-end optical system that needs to be integrated with a fleet of Department of Homeland Security (DHS) UAVs. Just as in previous sections, the approach will be to provide a fictitious scenario that presents essential chapter in a manner as might be expected in a working environment. The characters involved in this scenario are Dr. Bill Smith, FIT chief executive officer (CEO); Tom Phelps, FIT chief technical officer (CTO) and optics expert; Karl Ben, FIT senior systems engineer; Amanda *R.*, FIT optical engineer; Carlos *R.*, FIT quality manager; Malcolm P., FIT production manager; Wilford Erasmus, chief of operations and acquisitions, United States Customs and Border Patrol; and Simon Sandeman, Chief, Acquisitions and Procurements, Central Intelligence Agency (CIA).

13.3.1 Case Study Background

FIT's work for the DHS's United States Border Patrol to develop and field high-end optical systems was a complete success. In the end, roughly 50 units, including some prototypes, were manufactured, tested, and delivered. All of the stakeholders were happy with the delivered systems, but not so satisfied with the 6-month delay it took to deliver the systems.

Similarly, the leadership at FIT was satisfied that they had made much progress in capturing the interest of government stakeholders and were hopeful to parlay their success into additional government contracts. On the other hand, they were not so satisfied having lost some of their incentive money on the cost plus incentive fee contract due to their late delivery. A cost plus incentive fee contract is a cost-reimbursement contract that provides for an initially negotiated fee to be adjusted later by a formula based on the relationship of total allowable costs to total target costs (Ku and Wu 2013). The biggest worry that they had, even greater than their reduced incentive fee, was the potential business impact the late delivery could have with any future work with USBP or other government groups interested in high-performance optical systems. FIT had expertise in optical systems and solid systems engineering practices ingrained as part of their core concepts. However, at the time of the DHS contract, the quality program at FIT needed improvement.

FIT's experience was in mass-produced, lower-quality, and low-cost optical systems for commercial applications or low volumes of high-quality optical systems for scientific applications. With the DHS contract, FIT found themselves for the first time having to produce relatively large numbers of optical systems in a short period of time and their quality and production practices were not up to the task. FIT encountered a number of unfortunate problems throughout the planning and production stages of the DHS effort, which ultimately ended with various units having to be altered or deleted; this had led to extended production times, higher production costs, and much frustration. The issues associated with the quality variations in the production processes of the UAV optical systems came to a head shortly after the contract was fulfilled. The CEO of FIT, Dr. Bill Smith, ruled that the Six Sigma quality program would be adopted and implemented throughout FIT to increase performance, improve quality, and enhance their reputation and image with the company's stakeholders.

It has been almost 2 years since the completion of the DHS effort. FIT wants to reengage DHS for a new contract by telling them about some recently patented face recognition technology that they have advanced since previously working with one another. Since FIT expected the evolution of this technical breakthrough during the original design of their optical system, they prepared for the developments to come. In the original design of the hardware and software, FIT added the necessary hooks to the needed equipment so that it could be easily upgraded when a new technology upgrade was released. Ultimately, FIT wants to regain the trust and attention of their primary stakeholder, USBP, by efficiently presenting a short-term contract to upgrade USBP's existing units in addition, to building bridges with potential new stakeholders in the future.

With this new face identification technology and the successful implementation of Six Sigma, Dr. Smith calls Mr. Erasmus personally to go over this power-packed opportunity. Mr. Erasmus found himself very excited about the new opportunity with the identification technology and the fact that he could upgrade the quality of his current product for better results. Mr. Erasmus asked Dr. Smith to come visit him in our nation's capital to present the face recognition approach to his team. Unknown to Dr. Smith, Mr. Erasmus had invited a special customer to sit in on the meeting.

13.3.2 Initial Customer Meeting

Bill: "Hello Wilford! I'm glad that I got a chance to run into you again today. How's your wife and children doing?"

Wilford: "Hey Bill. I'm glad that we got this opportunity too. All are doing well. My wife is wonderful and the kids are just growing too fast; thanks for asking. And what about you? How has the business been going? I hope you have had a nice and relaxing trip?"

Bill: "Well Wilford, I have to say that the trip up here was pretty bland, but I am glad to be up here in this beautiful warm weather. Living in Melbourne, it gets too hot too quickly. How has business been treating you so far?"

Wilford: "Honestly, we have seen better times. The sequestration has truly made it more difficult for us to be able to get more funding approved from the government. It's getting to the point where I can't even get a decent cup of coffee here!"

Bill: "I'm sorry to hear that. It seems that everyone is running into hard times these days. Most of my government contacts are fighting to keep what they have going, while some are actually having to eliminate programs altogether. I apologize for being so blunt about this, but why would you ask me to come here if USBP is so tight?"

Wilford: "I know that may seem strange, but don't worry because I asked you here for a good reason. Despite all the cuts that are going on around here, I'm still extremely interested in going forward with this new identification methodology we discussed earlier on the phone. I know it would be tough to get this approved on my own, that's where you come in. Another party I invited became more interested after I told them you would be joining us. My hope is that together, we will be able to push this request through and gain approval. As a matter of fact, let me introduce you to Mr. Simon Sandeman who is with the CIA."

Simon: "Hello Bill, I have been looking forward to finally meeting you."

Bill: "Hi Simon. It's a pleasure to meet you. I hope you don't mind me asking this, but what do you do in the CIA?"

Simon: "Well, Bill I am afraid that is classified information...just kidding! I get people all the time with that one. Believe it or not my job is very similar to what Wilford does with USBP, though the CIA is just on a much larger scale. Ultimately, I want to put together a huge fleet of UAVs for needed surveillance reasons outside of the country. For example, North Korea is really picking up fire, but that's all the details I can share right now, the rest of that information actually is classified."

Bill: "Interesting! Well, all I know is that FIT is always willing to assist the CIA in any way possible. Did Wilford get a chance to bring you up to speed on why I am here?"

Simon: "Yes we are all on the same page. We both reviewed your technical brief together earlier this morning and decided that this is clearly an idea that deserves to be pursued. The CIA is constantly looking for better ways of identifying people in different ways."

Wilford: "Bill, I have to say that I am so happy you called. The current detection UAV that you installed has been working very well for us. That system has helped in catching several bad guys before they entered the country, greatly reducing the amount of unwanted visitors."

Bill: "I'm truly glad that we were able to help you all with that issue and make our country that much safer. Did you want to share the specifics about what you are looking for and the time frame we have to operate in?"

Wilford: "Bill, before we step into that territory, I do have to say that our senior advisers were not too happy with the little delivery debacle that occurred a few years back. On the upside, we did hear about your implementation of Six Sigma; that will assist us in convincing them to do business with you all again. That's a great thing because I know this reconnection will be vital to getting this contract to go through. Simon?"

Simon: "Yeah Bill. I know we will need a total of 2000 units to be delivered over the next 2 years. These units account for the needs of CIA and USBP. Basically, there will be a delivery of about 83 units every month. I know you are just completing Six Sigma, but we thought that with your track record combined with the heavy number of units we are requesting, any contract would be contingent on FIT proving that they have implemented statistical process controls to improve quality, for reassurance and peace of mind, on our part."

Wilford: "Bill we want to present all the facts so I have to say that these requirements are not negotiable due to the current economic climate. I know that the quantity we are requesting is pretty high, but honestly, this is about the only way I see that we could get this done. Do you think your people are up for the task?"

Bill: "I know that we are up for the task! This is an amazing opportunity and don't want us to miss out on it. I want to discuss this with my team in Melbourne and will contact you all within the week about our next step to be taken. How does that sound?"

Wilford: "That sounds good to me. Just keep us updated and I hope that we will be signing some paperwork in the near future. Take care and I hope you have safe travels home."

Simon: "Bill, again, it was great to meet you. Here is my card with all of my information. I look forward to hearing from you soon."

13.3.3 FIT Executive Team Meeting

Bill: "Gather around team. I appreciate you all coming out on such short notice. As most of you know, I just got back from Washington were I met with USBP and the CIA who may be potential customers! I needed to meet with everyone because we have some very important decisions to make concerning this awesome opportunity. The big question we have today is can we realistically produce 2000 units periodically in a 2-year period? What kind of systems do we have placed to prove that we are fully accommodating Six Sigma?"

Tom: "At this point, I believe that the design we have put together is pretty solid. I think what we all have to do is drop the new identification components into the new design of what USBP and the CIA want and go from there. What are your thoughts about that Karl?"

Karl: "Though the majority of our design is stable, some recent testing done on one of the prototypes showed the fact that the main lens needs to be redesigned to accommodate the new identification system. I don't foresee this being too much of a problem since my best designer is on the job. This should be a low-risk item since our company does well with lens fabrication."

Carlos: "Sorry Karl, but I have to disagree with you on that point. Looking from a quality perspective, we had many defects the last time we ran the UAV lens through the fabrication process just a few years ago. These defects played a major part in why our program was delayed in the past."

Bill: "Though this is true Carlos, we have to remember that we are Six Sigma compliant now. Due to that fact we should no longer have any of those issues at all."

Carlos: "Bill, do you realize it will take our company several years to truly have fully compliant Six Sigma program? This is no small task to accomplish overnight. Yes, in part, we have introduced many Six Sigma concepts, but at the end of the day, we are mainly following a system of total quality management. Our main goals are to improve quality, reduce cost, and increase stakeholder satisfaction. Since we have made quality a key priority of our company, there have been major strides taken toward improvements. Despite these facts, we need to know that we still have a ways before being completely Six Sigma compliant. We are currently operating effectively at 3-sigma level, and over the next 2 years, we want to move up to the four-sigma level."

Bill: "Wow! I didn't know we were not Six Sigma compliant! How am I supposed to explain to the government that I basically lied to them? This is not going to be a very pleasant conversation with them."

Carlos: "From my understanding I thought the CIA and USBP only wanted proof that SPC methods were being used to improve the quality of our systems. In that respect, we are most definitely in compliance. SPC was the first thing that we put in place with our emphasis on high-quality standards."

Bill: "I can breathe a little better now, but I still need to know what would be the smartest way to share with them that our current production processes are only working at the 3-sigma level? What do we mean by that?"

Carlos: "Actually, I created and sent you some charts from a recent report that would make this information easier to understand. Figure 13.20 is a drawing of the current sigma model. Looking closely, the dark line in this drawing represents the normal distribution; most of the items we produce will be in the white area of this chart."

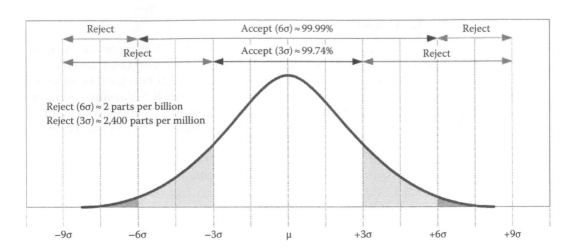

FIGURE 13.20
Sigma model. (Adapted by Brian Zamito from Arnheiter, E.D. and Maleyeff, J., *TQM Magazine*, 17(1), 5, 2005. With permission.)

Carlos: "Consider this, if the requirement for one of our lenses is to have an effective focal length of 1 m, then most of the lenses we build will come out of our production processes close to this value. Others, however, won't meet the specified tolerances.
　　For example, if we specify a lens to have a focal length of 1 m, then a majority of the lenses produced will be close to 1 m and the rest drops off from there. According to our 3-sigma process, approximately 99.74% of the lenses will fall within 3-sigma of the 1 m target value. Anything outside of the 3-sigma level is considered to be noncompliant and is rejected. We expect about 0.26% of our lenses to be out of tolerance and rejected in this fashion. In other words, about 2400 part rejected out of 1 million produced. These facts are contrasted with Six Sigma standards, which require even higher performance and reduced variability. In this case, 99.99% of the produced items are free of defects! This means that out of 1 billion parts, only 2 parts will be rejected. These are the reasons why it is very difficult and time consuming to truly implement Six Sigma in to a manufacturing process."

Bill: "Well that doesn't add up to me. Why are we going through all of this extra trouble to achieve Six Sigma? What is wrong with staying at 3 sigma? Why should we invest that amount of extra money when there isn't a major difference in the amount of return?"

Carlos: "I understand your questions, but there is actually a very good reason to do so. Take a look at Figure 13.21, showing the shifted sigma model. The first model shown presumes that the mean never shifts. Motorola, who is the company that originally invented Six Sigma, observed that over time, the mean drifts up or down 1.5 sigma."

Carlos: "Basically, this means that operating at a 3-sigma process largely increases the potential for more errors, and with more errors means more rejected products, which leads to a more expensive cost in the long run correcting them, a very nasty cycle. For example, if our manufacturing processes are operating so our initial 3-sigma processes are providing a 99.74% product acceptance percentage (0.26% product rejection), and if the previously discussed shift in sigma were realized, then our acceptance percentage would drop to 93.33% and our product rejection"

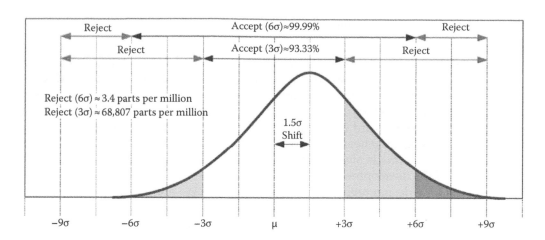

FIGURE 13.21
Shifted sigma model. (Adapted by Brian Zamito from Arnheiter, E.D. and Maleyeff, J., *TQM Magazine*, 17(1), 5, 2005. With permission.)

percentage would increase to 6.67%. The sigma shift also affects Six Sigma processes; however, the sigma shift is much less noticeable. For the Six Sigma case, the shift reduces the acceptance percentage to 99.99% (0.01% products rejected). As a comparison, the parts rejected jumps to 68,807 out of 1 million produced for the 3-sigma process, whereas the parts rejected for the Six Sigma process are 3.4 parts per million. That is the reason behind our push toward implementing Six Sigma in our manufacturing processes. As far as the government goes, we just need to tell them that we currently have statistical process control methods in place and that rolling in the Six Sigma program."

Bill: "Ok, I can see your rationale. I understand the overall concept, so my question would be do you have any recommendations?"

Carlos: "I do! It is clear that most of our issues arise in the fabrication process, so I would like to try running a Taguchi analysis that focuses on the polishing process. This would help us determine the optimal machining parameters to put in place that would reduce the variability of the surface roughness. This was the main issue that we had 2 years ago with our products."

Bill: "That sounds like a good plan to me. I think we have a good start and can move toward wrapping this meeting up. Malcolm, in your opinion, do you think that we can produce the high number of quantities for the customer? To meet these quotas, we would need to ship out 83 units every month, which breaks down to at least 20 units being produced per week."

Malcolm: "Bill, though these are high numbers, I think we will be able to produce the units in the necessary time frames. The factory has already implemented a lot of the quality principles that Carlos has shared with us. Running the prototypes for engineering has given us a jump in starting on these improvements so I think we can get there."

Bill: "Great to hear. Okay everyone I am getting the answer that we can take on and succeed with this project. I am going to go ahead and contact USBP and the CIA so that we can start writing up a rough contract outline. Thanks everyone, I really appreciate your time."

13.3.4 Second Customer Meeting

Bill: "Hello Wilford, this is Bill calling from FIT. I spoke with my team and we had a great conversation about your proposal. We've currently implemented statistical process control as well as some other quality measures in our manufacturing processes and we are progressing well toward our Six Sigma targets. My team has reassured me that we can take on this project and meet all of the requirements. We can move forward with drawing up the official contract when you are ready."

Wilford: "Bill I'm so glad to hear that and I know that Simon will be too. I know that you have already made significant strides in the design process by having funded most of the development with your own finances. Would you be able to be prepared for a design review in the next 2 months? We are pushing to get this project approved before any more money is taken away from our budget."

Bill: "Wilford, I know it will not be an issue for my team to be ready by that date for you. I will be sending you and Simon a personal invitation down here to Melbourne for this presentation."

Wilford: "That sounds great Bill! We look forward to seeing you soon and take care!"

13.3.5 Design Review

Bill: "Hello team! I want to take the time to thank everyone for all of the long hours and hard work you all have put in over the last 2 months to complete our design package for our special guest. Everyone, I would like you all to please welcome Mr. Wilford Erasmus from USBP and Mr. Simon Sandeman from CIA. Gentleman, I hope you all enjoy the presentation; feel free to interject or ask questions at any time. For Simon's benefit, I would like to start the presentation by giving a brief overview of our systems engineering (SE) process here at FIT. The V-diagram in Figure 13.22 should be very familiar. This figure describes the FIT systems engineering technical process model."

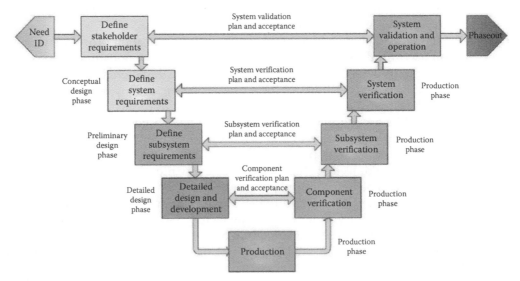

FIGURE 13.22
(See color insert.) FIT systems engineering technical process model (V-Model). (Adapted by Brian Zamito from Blanchard, B.S. and Fabrycky, W.J., *Systems Engineering and Analysis*, 5th edn., Prentice Hall, New York, 2011; IEEE, Systems and software engineering—Systems lifecycle processes, IEEE STD 15288-2008, 2008. With permission.)

Bill: "With customer needs being identified, this allows us to define a clear set of stakeholder requirements. The purpose of these stakeholder requirements are to push the development to the system level, subsystem level, and eventually the component/assemblies/parts level so that a complete set of requirements (e.g., 'build to' specs) and associated documents are available for system production. Requirements at every level must be verified and approved. For example, all requirements must be baselined at each level by passing a formal review. Are there any questions so far?"

Wilford: "Well Bill I have one question…is this actually going to be a briefing?"

Bill: "I got the hint Mr. Wilford. Let's continue, allow me to introduce Malcolm, our production manager. Malcolm will tell you a little about our production operations and processes Malcolm?"

Malcolm: "Thank you Bill. If everyone could please refer to Figure 13.23; this illustration shows a clear flow chart of the manufacturing process we take for this product."

Malcolm: "For this optical system, we have three major components that are being manufactured: an electrical assembly, the optics assembly, and the mechanical subsystems. The manufacturing processes begin with raw materials for the lens being received by the OP100. We have an in-process inspection team at OP110. Afterward, the lens is fabricated at OP200. The mechanical assembly is integrated with the optical components at OP500. After this, the integrated assembly is aligned and tested at OP530. The tested assembly is then

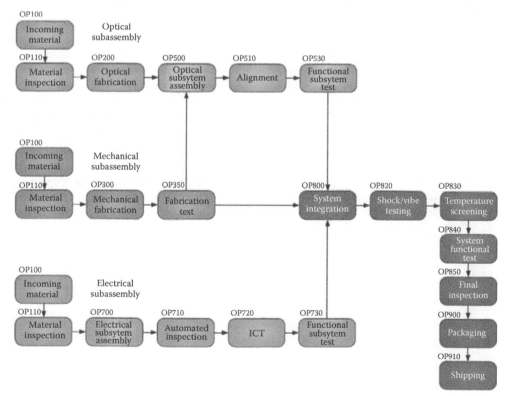

FIGURE 13.23

FIT UAV optical system manufacturing flow. (Adapted by Brian Zamito from Kasunic, K.J., *Optical Systems Engineering*, McGraw-Hill Professional, New York, 2011. With permission.)

ready for system integration along with other mechanical components. The process for the electrical assembly is similar to this. However, one distinction is that we are not fabricating anything for the electrical system. Instead, we are assembling circuit cards with commercial-of-the-shelf electrical components, inspect the completed circuit boards, and then send them on for in-circuit test (ICT) prior to sending them on to be functionally tested and integrated with the system. The system-level integration occurs at OP800 wherein the electrical, optical, and mechanical systems are integrated and undergo system-level testing. At this stage, the integrated system is then subjected to several tests such as shock, vibration, and temperature testing. Following these battery of tests, a final functional test is administered before shipping the unit off to our customers."

Wilford: "I love this figure! It clearly lays out your production flow and lays out your testing and inspection points."

Malcolm: "Thank you sir, your opinion really means a lot to us. We truly are striving to produce a superior and high-quality product. With that said, allow me to introduce you all to our quality manager, Carlos R."

Carlos: "Hello gentleman. As you may recall, the fabrication of our original UAV optical system lens had some quality issues, which resulted in our delivery date being delayed. After this instance, we took the time to identify the source of this problem and determined that our polishing process needed to be updated in order to produce a lens within its specifications on a consistent basis. Figure 13.24 provides my rationale for the Taguchi robust design of experiments analysis. This analysis allowed me to identify three key levels of four varying control factors that are vital and easily controllable within the lens polishing process. Based on the number of control factors, and the number of levels of each of the control factors, Taguchi provided a cross-reference matrix to look up what type of an orthogonal array must be used. This results in the required number of experiments and the variations of the control factors that are needed. In our case, these factors have been put into an L-9 Taguchi array system that performed nine experiments to determine the optimum settings for the four factors."

Carlos: "We used Taguchi's methods to determine the actual number of experiments that we would need to run versus using the full factorial approach that runs an experiment for every single possible combination. The results of the analysis are shown in Figure 13.25."

Control factor	Symbol	Level 1	Level 2	Level 3
Time (Min)	A	5.0	7.5	10.0
Pressure (kPa)	B	4.9	9.8	12.7
Platen speed (RPM)	C	40	50	60
Oscillation speed (RPM)	D	3	4	5

L-9 Array	Control factors			
Run	A	B	C	D
1	1	1	1	1
2	1	2	2	2
3	1	3	3	3
4	2	1	2	3
5	2	2	3	1
6	2	3	1	2
7	3	1	3	2
8	3	2	1	3
9	3	3	2	1

FIGURE 13.24
Taguchi robust design control factors for lens polishing. (Adapted by Brian Zamito from Lien, C.H. and Guu, Y.H., *Mater. Manuf. Process.*, 23, 838, 2008. With permission.)

L-9 Array	Control factors				Surface roughness (nm)			Mean	SNR Min
Run	A	B	C	D	Trail 1	Trail 2	Trail 3		
1	1	1	1	1	1.41	1.80	2.32	1.84	−5.4861
2	1	2	2	2	2.05	2.76	1.91	2.24	−7.1232
3	1	3	3	3	3.32	5.52	4.17	4.34	−12.9286
4	2	1	2	3	1.55	2.18	2.02	1.92	−5.7346
5	2	2	3	1	4.46	4.97	5.08	4.84	−13.7044
6	2	3	1	2	1.52	2.10	1.09	1.57	−4.2097
7	3	1	3	2	2.08	2.42	2.70	2.40	−7.6524
8	3	2	1	3	2.48	2.04	1.94	2.15	−6.7135
9	3	3	2	1	1.90	2.49	2.30	2.23	−7.0186

Factor A

L1	2.8067	−8.5126
L2	2.7744	−7.8829
L3	2.2611	−7.1282
Range	0.5456	1.3845

Rank: 4

Factor C

L1	1.8556	−5.4698
L2	2.1289	−6.6255
L3	3.8578	−11.4285
Range	2.0022	5.9587

Rank: 1

Factor B

L1	2.0533	−6.2911
L2	3.0767	−9.1804
L3	2.7122	−8.0523
Range	1.0233	2.8893

Rank: 2

Factor D

L1	2.9700	−8.7364
L2	2.0700	−6.3284
L3	2.8022	−8.4589
Range	0.9000	2.4080

Rank: 3

FIGURE 13.25
Taguchi robust design results for lens polishing. (Adapted by Brian Zamito from Lien, C.H. and Guu, Y.H., *Mater. Manuf. Process.*, 23, 838, 2008. With permission.)

Carlos: "Ultimately, the reason we chose the Taguchi method was so that we could cut cost and save time, while still getting effective results. Applying the results to our lens polishing problem, the optimized values determined through the Taguchi robust design analysis were used to optimize our processes and increase quality. Factor A was set to operate at Level 3, Factor B was set to run at Level 1, Factor C ran at Level 1, and Factor D ran at Level 2."

Simon: "Carlos that was an excellent overview! I can see that you have made great strides in your quality program and I am quite confident that you have things well in hand."

Carlos: "Thanks! Now I'd like to introduce you all to our optical designer Amanda R. She will be sharing how the new lens will be designed so that our optical system can be used for recognition and identification purposes."

Amanda: "Thank you Carlos. After running our prototype testing, we were able to determine that some adjustments needed to be made in order for the lens to be used in the identification application. The lens was determined to be the critical component in the optical train so for brevity, I'll only summarize that. Figure 13.26 shares the 'build to' specifications for this particular lens."

Karl: "The lens we designed has an effective focal length of 1 m with a center wavelength of 486.1 nm in the visible band. The target $f/\#$ for this lens is 2.4. We are using BK7 glass from Schott for the lens material and it has a specified index of refraction of 1.52238 ± 0.0005 (measured at 486.1 nm). This lens will provide us with the diffraction-limited spatial resolution needed for identification and recognition applications using our high-speed adaptive optics/atmospheric turbulence

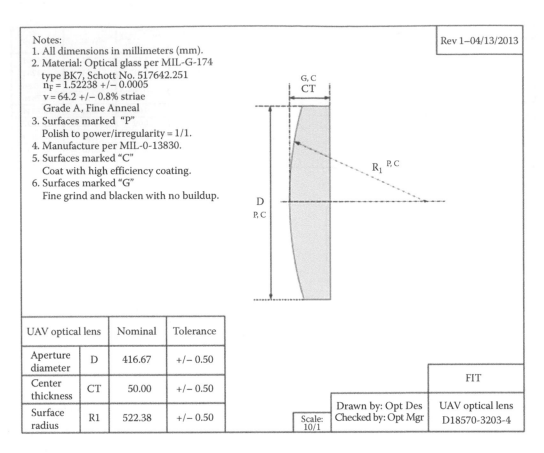

FIGURE 13.26
FIT UAV optical system lens specification. (Adapted by Brian Zamito from Kasunic, K.J., *Optical Systems Engineering*, McGraw-Hill Professional, New York, 2011; Smith, W.J., *Modern Optical Engineering: The Design of Optical Systems*, 4th edn., McGraw-Hill, New York, 2008. With permission.)

compensation methods. We used Zemax to develop the optical systems design for the relay optics to the focal plane. Now, I am handing things back over to Carlos and he will go over our statistical process control methods."

Carlos: "Karl, thank you very much. We have worked very hard here at FIT to implement a robust statistical process control capability for our manufacturing processes. We are applying the SPC methods to control the critical system parameters in our production processes. We are not only implementing SPC methods; however, FIT has made a commitment to implement total quality management (TQM) methods to improve the overall quality of our products. Besides making improvements to our product line, the TQM methods also focus on methods to improve quality for our stakeholders and our employees. Our quality assurance methods also extend to our suppliers and we hold them to the same high standards. For instance, if our suppliers aren't providing us with high-quality parts and services, then either work with them to fix the problem or find new suppliers. Some of our fundamental quality processes that we have implemented and regularly employ, and that we think that you might be interested in, are the seven Japanese quality tools: control charts, checklists, Pareto analysis, cause-and-effect diagrams, flow charts, histograms, and scatter diagrams. We would like to tell you about our control chart processes but does anyone have any questions before we move to this?"

Wilford: "Yes Carlos I do. I have gotten pretty comfortable with the details of flow charting that Malcolm talked about earlier, and also control charts, but I am still somewhat foggy concerning the other two methods. Could you possibly explain those a little better?"

Carlos: "Wilford, I'd be happy to provide some additional information. Though there is not an example in this briefing, I can explain this information in some detail for you. These two quality tools are classified as 'offline' quality methods. To use these tools, data are obtained from previous manufacturing runs and analyzed offline. If a quality issue is discovered during the data analysis, we use the cause-and-effect diagram to identify the root causes for the quality issue. The Pareto analysis is used to determine the relative priority of the discovered quality issues. If the quality problem is serious enough, we may shut down the production line until the relevant issues are fixed. The cause-and-effect quality tool is great for brainstorming wherein a comprehensive look at possible causes for an observed quality problem are identified by the quality team. This can be either a proactive or reactive quality method in that it can be applied before or after the quality problem is identified. The cause-and-effect diagram (or fishbone chart) is generated as part of the analytical process. It is called a fishbone chart since the completed chart looks somewhat like the skeleton of a fish. In a typical application, a quality issue is identified and becomes the focus of the fishbone chart. The quality team then brainstorms on the reason(s) that led up to the quality issue, and these are documented on the chart in a series of lines and intersecting line segments that identify the possible causes for the observed quality issue. In the case of the Pareto analysis, the discovered quality issues are categorized and rank ordered in terms of their importance. The fundamental concept/assumption behind Pareto analysis is that the majority of quality-related problems result from a few root causes. Consequently, focusing on the issue that results in the largest number of significant observed defects can result in biggest improvements in product quality and can also provide many downstream quality improvements. In this way, the Pareto analysis serves as an excellent tool for rank-ordering quality problems. Since all quality issues may not be of equal importance, an engineering review of the relative priority of the defect categories is necessary to determine the proper relative importance of the defects. For instance, a single crack in our lens is likely of more concern than four superficial scratches on the edge of the lens underneath the mounting hardware. Does this help?"

Wilford: "Thank you Carlos, yes it does. I now have a better understanding of that information. Please feel free to move forward to the control charts. I'm free for as long as it takes!"

Carlos: "With that said let's keep moving forward. There are two main parameters that we measure using three different types of charts. Initially, we start with the variable control charts so that we can control the aperture diameter of the new lens that Amanda discussed. Due to our past experiences, we concluded that this issue had the highest risk. We set up an $\bar{X}$ chart to monitor the average lens diameter that we produced. We also used an R-chart to plot the associated range. Additionally, we are monitoring the first pass yield at OP840 with an attribute chart (P-chart). This P-chart is applied at the final system functional test on the integrated optical system. I have to mention that...."

Simon: "I'm sorry to interrupt you Carlos, but could you please explain what in the world an $\bar{X}$ chart is before you go any further?"

Carlos: "Simon please forgive me for assuming that everyone was familiar with the term. Let me back up a bit and share the basics concerning control charts that would make them easier to understand. Basic control charts are very similar to the charts that I have mentioned. They are ideally suited for finding process variations such as a shift in the process mean or a change in the process range or variance. One major reason for using these charts is to make an analytical assessment of whether a given process is or is not in statistical control. Variable control charts are particularly useful if a process variable can be identified that is relatable to the observed quality issue. Variable parameters such as volume, length, temperature, height, or weight make good representative examples. For our application, the variable of interest is the diameter of our manufactured primary lens. We use an $\bar{X}$ chart to plot sample means and we use an R-chart to illustrate variations in the lens diameter through plotting the sample range or standard deviation. The $\bar{X}$ and R-chart are very commonly used in practice. Another type of control chart is the attribute chart. Attribute charts overcome the single-variable restriction of variable control charts and can be used to monitor many quality-related parameters at once. Attribute control charts can be used to characterize quality issues that can be placed into two categories such as pass or fail. Another example would be the monitoring the proportional number of defective units. For our application, the P-chart is being used to determine the number of lenses that pass inspection the first time compared with the number of units that fail inspection. The rules for generating attribute charts are very similar to those for generating variable control charts. Both methods determine a centerline and plot the variations of the data with respect to the centerline. Both variable control charts attribute control charts have upper and lower control limits and can be used to determine whether or not a particular process is in statistical control. The major difference between the two types of charts is in the nature of the data that are being analyzed: variable data versus categorized quality characteristics. I know this is a lot to take in, so let me provide an example. Figure 13.27 shows a representative $\bar{X}$ chart (left) and a corresponding R-chart (right)."

Carlos: "Recall that the $\bar{X}$ chart plots the variations of the sample mean for a series of samples. In our case, this is the variation in the mean of a batch of lens diameters plotted over a series of batch runs. Similarly, the R-chart plots the corresponding variations in the range of the lens diameters within a particular batch, over a series of batch runs. Despite their differences, both of these control charts have the following in common: a centerline, UCL, LCL, and the same starting variable data set that is being analyzed and plotted. The data mean is represented by the centerline and the actual plot of the sample mean lens diameter over all the samples/batches illustrates the variation in the sample/batch mean. If the plotted $\bar{X}$ and R charts have points that cross the UCL or LCL, then it is possible that the underlying manufacturing process may be out of statistical control. Our goal is to have the plotted points on the stay within the UCL and LCL limits and only allow natural statistical variations to exceed these limits. This means we need to make a determination whether or not a point that falls outside of the UCL/LCL is there because of a natural statistical variation or if it is due to some underlying cause that indicates a quality problem."

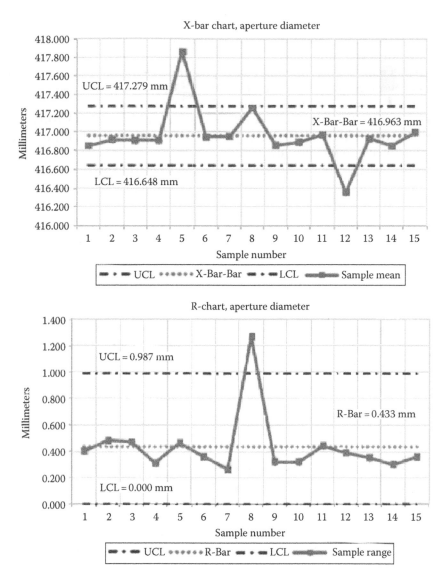

FIGURE 13.27
UAV optical system lens fabrication: out of control. (Created by and courtesy of Brian Zamito.)

Simon: "I think I get it. I am going to take a shot at this. If our goal is to have a process that is in control, then we would want to have most of our points fall within the UCL and LCL. I am still a bit fuzzy on how to interpret a point that falls outside of these limits. You are saying, even though the point is outside of the UCL/LCL, it might still be in statistical control?"

Carlos: "Simon that is correct. If we see that a monitored data point exceeds the UCL/ LCL, we shut down the production line if the potential impact is severe enough. We then analyze the situation surrounding the issue and determine whether or not the occurrence was due to natural statistical variations or if there is some root cause related to a quality issue. In our example, we determined that our test inspection personnel used the wrong alignment procedure. This caused wrong data to be reported. However, since we had the statistical process control methods in place, we found these issues straight away and so our quality loss and downtime was minimized."

Simon: "Ah that makes sense! So these charts give you a clear and concise means to detect quality problems before they become a costly outrageous mess! That is good news! Continue with your briefing please."

Carlos: "Thank you. Figure 13.28 shows the control charts taken at OP840 showing the first pass yields (FPYs) for our optical lens system. We had just made some adjustments to our production line and this was our starting point before we made any corrections so don't let these results alarm you."

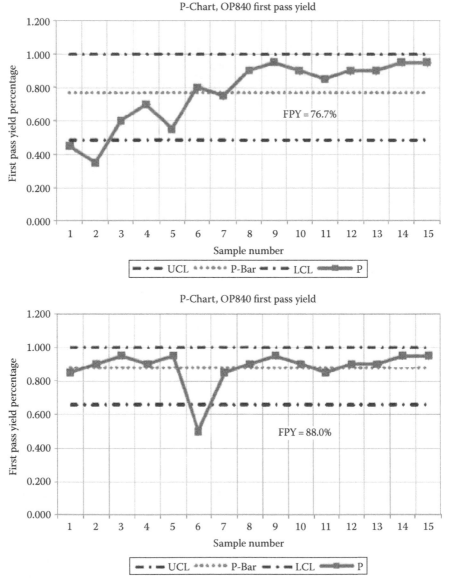

FIGURE 13.28
UAV optical system FPY through OP840. (Created by and courtesy of Brian Zamito.)

Carlos: "In Figure 13.28, the chart on the left has two points (first and second) that have exceeded the LCL indicating that there may be a quality issue. Our FPY for these data was 76.7%. Notice also that the entire data set appears to have a positive slope. We looked into the reason behind these results and found that the production line had a calibration error. After fixing this, we ran the test again a week later. The chart on the right shows these results. Even though our FPY increased to 88.0%, there was still one point that exceeded the LCL. We found that operator error was behind this result and we modified our training program to prevent this problem from occurring again."

Wilford: "Carlos I am extremely pleased with all of this wonderful information you are sharing with us. It is clear to me that FIT has turned a sharp new corner when it comes to quality. I appreciate your time and fully understand that the process will need time to vamp up so all the kinks can be worked out. Honestly, I am just happy to hear that the issues that did occur where quickly identified and corrected. Bill I am very impressed by the progress you all have made. Simon, is there anything you wanted to add?"

Simon: "Wilford I have to agree with you. I too am very impressed by the work that FIT has done and presented here today. I am no expert when it comes to quality. Probability and statistics never were my strong suit. I appreciate you all breaking the information down so that I could understand it better. All I can say is keep up the great work and it seems that everything is on track."

Bill: "Wilford and Simon, I'm glad that you all are pleased with our efforts so far. We strive to satisfy every need that our customers have. As for the FIT team, thank you for all of your hard work. Does anyone want to say anything before we wrap up this meeting?"

Wilford: "Actually, I have a request. I am very interested in tracking your production results. I know we're in good hands but I would just like to be kept in the loop and follow your progress. As Ronald Reagan said a while ago, 'Trust but verify!' It is great to hear about the awesome work you all are doing."

Bill: "No problem. As a matter of fact, we can schedule a follow-up a month from now. Thanks for everything and have a safe trip home."

13.3.6 First Quality Meeting

Bill: "Hello Wilford. It's Bill. As promised, we will give you an update on our production processes. I have Carlos and Malcolm on the line with me. If you recall, Carlos is our head of quality, and Malcolm runs our production department. They have been working great together. Is Simon with you?"

Wilford: "Ah, no. Simon got tied up with some other work at the moment and unfortunately won't be here today. I told him I would fill him in later. So how is everything coming along in the sunshine state of Florida?"

Bill: "Wilford I must say that things are developing nicely here. Despite the few expected bumps in our testing and production processes, our quality system is now working great. We have noticed an increase in our FPY, a reduction in variability in our production lines, and an overall decrease in the costs of production. On top of that, some of the process improvements that Carlos implemented around here has led to some schedule improvements and shaved off time on our projected ship date. As it stands, we will be able to deliver two months early!"

Wilford: "That is the best news I have heard all week. The political environment is becoming more and more unpredictable so it's good to know that you all are really on top of things. Have you encountered any issues so far?"

Carlos: "Hello Wilford, this is Carlos. I can update you on that question. Please look at Figure 13.29 in your briefing packet. We conducted a Pareto analysis last month at the integrated systems test station (OP840 if you recall). At that time, we ran 50 units through our production line early in the production phase and about 20 units had defects. Clearly, a yield of 60% isn't good enough."

Carlos: "As we looked into what was going on, we found that 9 failures had blurry images. We knew that something was going on in the optical assembly process.

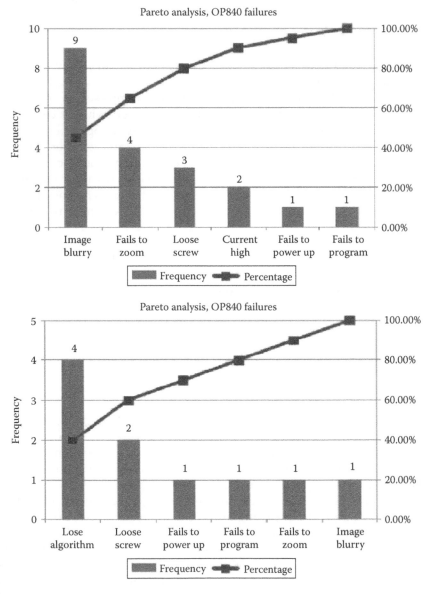

FIGURE 13.29
Pareto analysis of OP840 failures. (Created by and courtesy of Brian Zamito.)

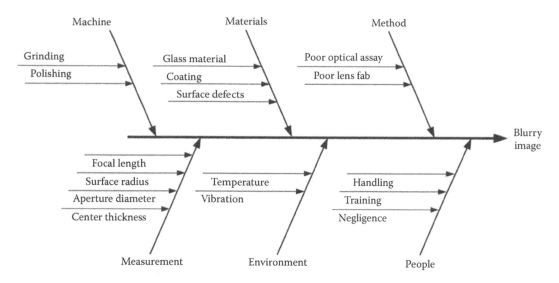

FIGURE 13.30

Cause-and-effect analysis of OP840 failures. (Adapted by Brian Zamito from Lo, W.C. et al., *Int. J. Adv. Manuf. Technol.*, 41, 885, 2009; Reid, R.D. and Sanders, N.R., *Operations Management*, 4th edn., John Wiley & Sons, New York, 2009. With permission.)

Compared to the other problems, we decided that this issue was the most critical and so we focused our attention to solving that one first. We quickly found the root cause for the blurry images and fixed the problem. We repeated the test after the fixes were implemented and had a production run of 100 units with a 10% defect rate. Our follow-up Pareto analysis, shown on the right side of Figure 13.29, showed that we fixed the problem. Interestingly, by fixing the blurry image problem, we also fixed the 'fails to zoom' problem. In fact, these two issues were related. Our yield then jumped to roughly 90%, which is a drastic change from the previous 60%. The quality team, members of the production team, and I got together and we generated a cause-and-effect diagram to help us understand the underlying causes for the blurry image issue. Figure 13.30 shows our results. The 'effect' was the blurry image, and the team assessed all the possible causes that could result in this effect. The team looked at possible environmental problems, how we fabricated the item, the assembly processes, and handling of the system and we identified that not enough testing was occurring until the integrated system testing at OP840. The ultimate problem was attributed to issues with aligning the optics that were introduced during vibration testing at OP820. I will let Malcolm give you more information on how we fixed this problem."

Wilford: "Thank you Carlos. That was great information. It's good hearing that the quality tools you implemented are working and helping you find problems and improve your production processes. This just reassures me even more that FIT is taking quality seriously. I also appreciate seeing one of those fishbone charts too!"

Carlos: "Not a problem at all! I am always happy to help. Here is Malcolm."

Malcolm: "Hello Wilford. I will not downplay the fact that the problems with the blurry images caused quite a stir around here when we found them. We were concerned that we might have a major defect. A 60% yield was not sustainable over the long term and so several of us were quite concerned at the time. Our quality team really came through for us though. The fact that we were able to identify

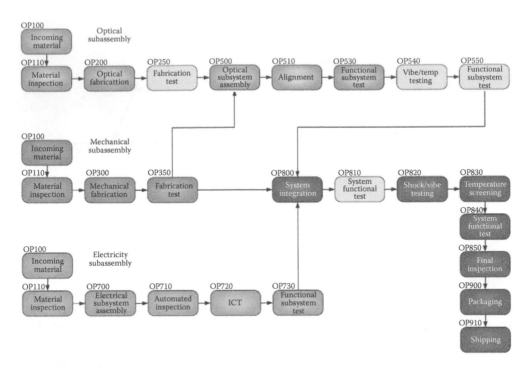

FIGURE 13.31

FIT updated UAV optical system manufacturing flow. (Adapted by Brian Zamito from Kasunic, K.J., *Optical Systems Engineering*, McGraw-Hill Professional, New York, 2011. With permission.)

the problem so early in the process enabled us to make minor adjustments to the production flow. We introduced an early test station that let us verify the proper alignment of the optics. Figure 13.31 shows our updated process flow. This diagram is similar to what I showed you in the last meeting with the exception that some additional operational steps were added. These operational steps are shown as the light-gray boxes in the diagram and illustrate the additional tests we added. I will give you a short overview and explain these blocks."

Malcolm: "When we analyzed the original production flow, we noticed that we actually didn't test the lens fabrication until OP530 where we functionally test the optical subsystem. However, this test occurred after the optical system is assembled at OP500. As a consequence, we didn't know whether problems found in the functional test at OP530 resulted from assembly problems introduced at OP500, or lens, or metalwork fabrication problems at OP200. We solved this by introducing a fabrication test (OP250) after the optical fabrication step (OP200) to determine whether or not we had problems with the fabrication process. Carlos pointed out that the root cause of the blurry image problem was due to alignment issues that were aggravated by the vibrational testing. Previously, we only did vibrational testing after the optical system was completely assembled at OP820. It took us a bit of time to understand the connection between the vibrational test and the blurry image. Were the problems associated with optical system itself, the mechanical system, or the electrical system? After discovering the alignment issue, we wanted to isolate the optical system and so introduced a vibration test for the optical subsystem at OP540 and functional testing of the optical subsystem at OP550. We also added a system functional test at OP810 to make sure that the integrated system was functional prior to final environmental testing at OP820 and OP830. This procedure gives us added confidence and lets us isolate problems better when they occur."

Wilford: "Awesome Malcolm. It looks like you all handled this well and got on top of the issue fast enough before it developed into a serious problem. Is there any new information on the control aspects of the production line that you discussed before?"

Carlos: "Yes Wilford, as a matter of fact, we do. If you would please go down to Figure 13.32, these charts show our latest results coming from the production floor. Notice that our lens fabrication processes for the lens diameter are now looking good. We are producing on the centerline with acceptable variations. Our first pass yields are now on the order of 90% as we mentioned earlier today."

Wilford: "That's wonderful! Does this mean that the factory is operating as efficiently as possible?"

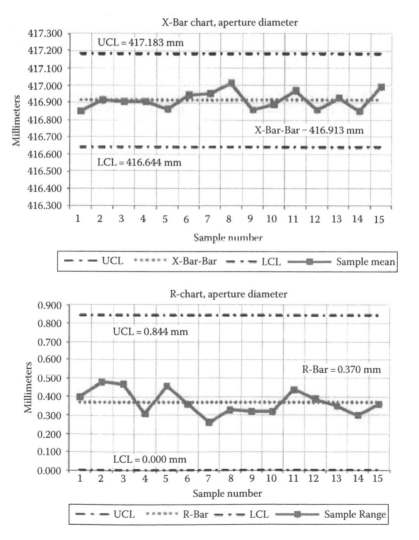

FIGURE 13.32
UAV optical system lens fabrication: in control, not capable. (Created by and courtesy of Brian Zamito.)

Carlos: "I wouldn't say that just now. It is great that our production processes are now in control; however, there may be more issues that could arise and require us to tweak the process again. Currently, the lens diameter variations fall within the 3-sigma levels of the centerline and we are happy with this start. However, we would like to reduce these variations some more by introducing some equipment changes. We also have some process changes in mind that we will implement over the near term."

Wilford: "That sounds great to me. Thanks again good news and I look forward to having the same next time. Continue to push forward. You all are doing a great a job!"

13.3.7 Second Quality Meeting

Bill: "Hello Wilford. It's Bill and I am on the line with Carlos. Are things going well with you in Washington these days?"

Wilford: "How would Tony the tiger say it? 'I'm ggrreeaattt!' We got your shipment of the UAV optical identification system and we are very happy with what we got. It's quite the buzz around here! You all have done a fantastic job in getting us these systems. Simon first wanted me to apologize for him being absent again, but he is anxious to hear back from me about your overall progress. He fears that there is a possibility that you all could run into a large amount of variation over time."

Carlos: "We have some good news for you! Please look at Figure 13.33 in your briefing packet. As you can see, we have optimized our production processes and are now achieving a FPY exceeding 90%. The results show our FPY through OP840."

Wilford: "That is great news Carlos. It is nice to see that you all are indeed providing excellent products, especially since these systems are going on the front lines where lives are at stake. Do you have any more updates for me? Is the lens fabrication coming along well?"

Carlos: "We have been working hard on the lens fabrication process. After much research, we made some adjustments to the process enough and now can regularly make low-variation lenses. By that, I mean that we have a capable process in which the

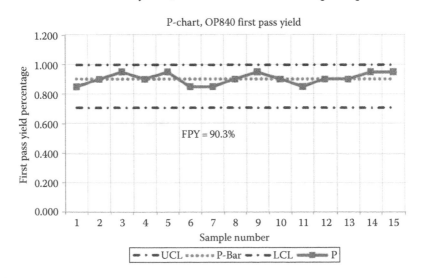

FIGURE 13.33
UAV optical system FPY through OP840 with flow update. (Created by and courtesy of Brian Zamito.)

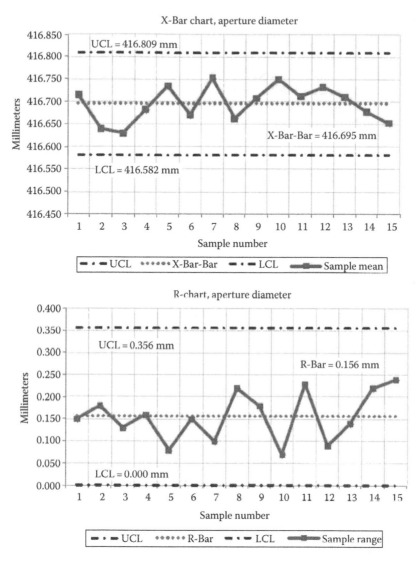

FIGURE 13.34
UAV Optical system lens fabrication: in control, capable. (Created by and courtesy of Brian Zamito.)

3-sigma variations in our lens diameters are less than the control limits in our variable control charts. I now have no doubt in my mind that FIT will be able to produce high-volume and high-quality systems for Simon. Figure 13.34 show our latest results."

Wilford: "Awesome! Again Bill, I have to say that you all are doing a wonderful job and you really came through. Continue with this superior work and I am sure folks around here will be lining up for your systems after we get through these budget crises!"

Bill: "Wilford, thanks for your time and we are glad that you all have found the presentation to be satisfactory. Take care Wilford, and please send our best wishes to Mr. Simon. It would be great to see him the next time we get together."

Wilford: "Thank you! We look forward to working with you again!"

13.A Appendix: Acronyms

AHP	Analytical hierarchy process
CBP	U.S. Customs and Border Protection
CEO	Chief executive officer
CIA	Central Intelligence Agency
CNC	Computer numerical control
CTO	Chief technical officer
DHS	Department of Homeland Security
EPC	Engineering process control
FIT	Fantastic Imaging Technologies
FPY	First pass yield
IBLF	Inverse beta loss function
ICT	In-circuit test
IGLF	Inverse gamma loss function
INLF	Inverse loss function
IPC	Integrated process control
OP	Operation
QE	Quality engineering
QFD	Quality function deployment
QLF	Quality loss function
SE	Systems engineering
SNR	Signal-to-noise ratio
SPC	Statistical process control
SQC	Statistical quality control
TQM	Total quality management
UAV	Unmanned aerial vehicle
USBP	U.S. Border Patrol

13.B Appendix: Variable Descriptions

F/#	F-stop number, relative aperture, brightness or speed of lens
α	Type I error
β	Type II error
Δ_m	Manufacturing tolerance
Δ_s	Specification tolerance
Σ	Process standard deviation
σ_p	Standard deviation of the average proportion of items
σ_s	Standard deviation of the distribution of sample means
σ^2	Process variance
μ	Process mean
φ	Optical power of lens
A_2	Statistic value based of standard normal variable

C_i	Number of defects per unit
$\bar{c}$	Average number of defect over sample time
C_m	Cost loss threshold for manufacturing (dollars)
C_p	Process capability index (centered)
C_{pk}	Process capability index (uncentered)
C_S	Cost loss threshold for society
CT	Center thickness of lens
D	Aperture diameter of lens
D_3	Statistic value based on standard normal variable
D_4	Statistic value based on standard normal variable
f	Focal length of lens
k	Quality loss coefficient
L	Quality loss to society per unit
$\bar{L}$	Average quality loss to society per unit (dollars/unit)
LCL_C	C-chart lower control limit
LCL_P	P-chart lower control limit
LCL_R	R-chart lower control limit
$LCL_{\bar{R}}$	$\bar{X}$ chart lower control limit
LML	Lower manufacturing limit (min); $LML = T - \Delta_m$
LSL	Lower specification limit (min); $LSL = T - \Delta_s$
m	Total number of observations
n	Number of observations per sample (sample size)
η	Refractive index of lens material
P_i	Number of proportioned units per sample
$\bar{p}$	Average proportion of items per population
R	Range of sample observation (largest–smallest)
$\bar{R}$	Average range of sample observation data
R_1	Outside surface radius of lens
R_2	Inside surface radius of lens
s	Sample standard deviation of observation data
s^2	Sample variance of observation data
$SNR_{\min}$	Taguchi minimized signal-to-noise ratio
SNR_{tgt}	Taguchi targeted signal-to-noise ratio
$SNR_{\max}$	Taguchi maximized signal-to-noise ratio
T	Sample time
T	Target specification value (nominal)
UCL_C	C-chart upper control limit
UCL_P	P-chart upper control limit
UCL_R	R-chart upper control limit
UCL_X	$\bar{X}$ chart upper control limit
UML	Upper manufacturing limit (max); $UML = T + \Delta_m$
USL	Upper specification limit (max); $USL = T + \Delta_s$
W	Number of samples (population size)
x_i	Performance parameter observation data
$\bar{X}$	Sample mean of observation data
$\bar{\bar{X}}$	Average of the sample means of observation data
z	Standard normal variable (number of standard deviations)

13.C Appendix: Control Charts and Taguchi Robust Design Data

This appendix shows the Microsoft Excel data used to generate the control charts and Pareto charts in this document. The complete Taguchi calculation was shown in the main body of the document in Figure 13.24 and Figure 13.25.

X-Bar chart/R-chart, In control, not capable

Sample Number	Observations				Average X-Bar	Range R
	1	2	3	4		
1	416.77	416.82	416.71	417.11	413.85	0.40
2	417.17	416.69	416.84	416.96	416.92	0.48
3	416.87	417.14	416.95	416.67	416.91	0.47
4	417.00	417.04	416.73	416.86	416.91	0.31
5	416.70	417.16	416.70	416.89	416.86	0.46
6	416.92	416.96	417.13	416.77	416.95	0.36
7	417.06	417.05	416.80	416.90	416.95	0.26
8	417.02	417.14	416.81	417.08	417.01	0.33
9	416.69	416.82	416.91	417.01	416.86	0.32
10	417.06	416.74	416.83	416.93	416.89	0.32
11	417.15	416.89	416.71	417.13	416.97	0.44
12	416.69	417.08	416.94	416.72	416.86	0.39
13	416.82	416.79	416.96	417.14	416.93	0.35
14	416.98	416.68	416.78	416.96	416.85	0.30
15	417.02	417.05	417.13	416.77	416.99	0.36

All units are in millimeters

416.913	0.370
X-Bar-Bar	R-Bar

Upper control limit (UCL):	417.183	0.844
Lower control limit (LCL):	416.644	0.000

Number of samples (w):	15	15
Number of observations per sample (n):	4	4

A2 value for number of observations:	0.729	
D3 value for number of observations:		0
D4 value for number of observations:		2.282

Nominal value (T):	416.670
Upper tolerance:	0.500
Lower tolerance:	0.500

Upper spec limit (USL):	417.170
Lower spec limit (LSL):	416.170

Process mean (μ):	416.913
Process std deviation (σ):	0.1554

Process capability (C_{pk}):	0.5507

X-Bar chart/R-chart, out of control, Not capable

Sample Number	Observations				Average X-Bar	Range R
	1	2	3	4		
1	416.77	416.82	416.71	417.11	416.85	0.40
2	416.17	416.69	416.84	416.96	416.92	0.48
3	416.87	417.14	416.95	416.67	416.91	0.47
4	417.00	417.04	416.73	416.86	416.91	0.31
5	417.70	418.16	417.70	417.89	417.86	0.46
6	416.92	416.96	417.13	416.77	416.95	0.36
7	417.06	417.05	416.80	416.90	416.95	0.26
8	417.02	417.14	416.81	418.08	417.26	1.27
9	416.69	416.82	416.91	417.01	416.86	0.32
10	417.06	416.74	416.83	416.93	416.89	0.32
11	417.15	416.89	416.71	417.13	416.97	0.44
12	416.19	116.58	416.44	416.22	416.36	0.39
13	416.82	416.79	416.96	417.14	416.93	0.35
14	416.98	416.68	416.78	416.96	416.85	0.30
15	417.02	417.05	417.13	416.77	416.99	0.36

All units are in millimeters

416.963	0.433
X-Bar-Bar	R-Bar

Upper control limit (UCL):	417.279	0.987
Lower control limit (LCL):	416.648	0.000

Number of samples (w):	15	15
Number of observations per sample (n):	4	4

$A2$ value for number of observations:	0.729	
D3 value for number of observations:		0
D4 value for number of observations:		2.282

Nominal value (T):	416.670
Upper tolerance:	0.500
Lower tolerance:	0.500

Upper spec limit (USL):	417.170
Lower spec limit (LSL):	416.170

Process mean (μ):	416.963
Process std deviation (σ):	0.3547

Process capability (C_{pk}):	0.1942

X-Bar chart/R-chart, In control, capable

Sample Number	Observations				Average	Range
	1	2	3	4	X-Bar	R
1	416.77	416.62	416.71	416.76	416.72	0.15
2	416.57	416.69	416.74	416.56	416.64	0.18
3	416.67	416.54	416.64	416.67	416.63	0.13
4	416.60	416.64	416.73	416.76	416.68	0.16
5	416.70	416.76	416.70	416.78	416.74	0.08
6	416.62	416.66	416.63	416.77	416.67	0.15
7	416.76	416.75	416.80	416.70	416.75	0.10
8	416.52	416.74	416.71	416.68	416.66	0.22
9	416.79	416.72	416.61	416.71	416.71	0.18
10	416.80	416.74	416.73	416.73	416.75	0.07
11	416.65	416.86	416.71	416.63	416.71	0.23
12	416.69	416.78	416.74	416.72	416.73	0.09
13	416.65	416.79	416.66	416.74	416.71	0.14
14	416.69	416.68	416.78	416.56	416.68	0.22
15	416.76	416.53	416.55	416.77	416.65	0.24

All units are in millimeters

416.695	0.156
X-Bar-Bar	R-Bar

Upper control limit (UCL):	416.809	0.356
Lower control limit (LCL):	416.582	0.000

Number of sample (w):	15	15
Number of observations per sample (n):	4	4

A2 value for number of observations:	0.729	
D3 value for number of observations:		0
D4 value for number of observations:		2.282

Nominal value (T):	416.670
Upper tolerance:	0.500
Lower tolerance:	0.500

Upper spec limit (USL):	417.170
Lower spec limit (LSL):	416.170

Process mean (μ):	416.695
Process std deviation (σ):	0.0772

Process capability (C_{pk}):	2.0505

P-chart, Ramp UP

Sample Number	Pass Observations (p)	Defective Observations ($/p$)	Observations per Sample (n)	Defective Proportion (d)	First Pass Yield
1	9	11	20	0.55	0.45
2	7	13	20	0.65	0.35
3	12	8	20	0.40	0.60
4	14	6	20	0.30	0.70
5	11	9	20	0.45	0.55
6	16	4	20	0.20	0.80
7	15	5	20	0.25	0.75
8	18	2	20	0.10	0.90
9	19	1	20	0.05	0.95
10	18	2	20	0.10	0.90
11	17	3	20	0.15	0.85
12	18	2	20	0.10	0.90
13	18	2	20	0.10	0.90
14	19	1	20	0.05	0.95
15	19	1	20	0.05	0.95

230	70	300
Total Passing Observations	Total Defective Observations	Total Observations (m)

P-Bar:	0.767
σ_p:	0.095

Upper control limit (UCL):	1.000
Lower control limit (LCL):	0.483

Number of samples (w):	15
Number of observations per sample (n):	20

P-chart, Out of Control

Sample Number	Pass Observations (p)	Defective Observations ($/p$)	Observations per Sample (n)	Defective Proportion (d)	First Pass Yield
1	17	3	20	0.15	0.85
2	18	2	20	0.10	0.90
3	19	1	20	0.05	0.95
4	18	2	20	0.10	0.90
5	19	1	20	0.05	0.95
6	10	10	20	0.50	0.50
7	17	3	20	0.15	0.85
8	18	2	20	0.10	0.90
9	19	1	20	0.05	0.95
10	18	2	20	0.10	0.90
11	17	3	20	0.15	0.85
12	18	2	20	0.10	0.90
13	18	2	20	0.10	0.90
14	19	1	20	0.05	0.95
15	19	1	20	0.05	0.95

264	36	300
Total Passing Observations	Total Defective Observations	Total Observations (m)

P-Bar:	0.880
σ_p:	0.073

Upper control limit (UCL):	1.000
Lower control limit (LCL):	0.662

Number of samples (w):	15
Number of observations per sample (n):	20

P-chart, In Control

Sample Number	Pass Observations (p)	Defective Observations ($/p$)	Observations per Sample (n)	Defective Proportion (d)	First Pass Yield
1	17	3	20	0.15	0.85
2	18	2	20	0.10	0.90
3	19	1	20	0.05	0.95
4	18	2	20	0.10	0.90
5	19	1	20	0.05	0.95
6	17	3	20	0.15	0.85
7	17	3	20	0.15	0.85
8	18	2	20	0.10	0.90
9	19	1	20	0.05	0.95
10	18	2	20	0.10	0.90
11	17	3	20	0.15	0.85
12	18	2	20	0.10	0.90
13	18	2	20	0.10	0.90
14	19	1	20	0.05	0.95
15	19	1	20	0.05	0.95

271	29	300
Total Passing Observations	Total Defective Observations	Total observations (m)

P-Bar:	0.903
σ_p:	0.066

Upper control limit (UCL):	1.000
Lower control limit (LCL):	0.705

Number of samples (w):	15
Number of observations per sample (n):	20

UAV Optical System, OP840 Failures

Item	Failure	Frequency	Percentage	Cumulative Frequency	Cumulative Percentage
1	Image blurry	9	45.00%	9	45.00%
2	Fails to zoom	4	20.00%	13	65.00%
3	Loose screw	3	15.00%	16	80.00%
4	Current high	2	10.00%	18	90.00%
5	Fails to power up	1	5.00%	19	95.00%
6	Fails to program	1	5.00%	20	100.00%
	Total	20			

UAV Optical System, OP840 Failures

Item	Failure	Frequency	Percentage	Cumulative Frequency	Cumulative Percentage
1	Lose algorithm	4	40.00%	4	40.00%
2	Loose screw	2	20.00%	6	60.00%
3	Fails to power up	1	10.00%	7	70.00%
4	Fails to program	1	10.00%	8	80.00%
5	Fails to zoom	1	10.00%	9	90.00%
6	Fails to Blurry	1	10.00%	10	100.00%
	Total	10			

References

Acquah, C. et al. 2006. Optimization of an optical fiber drawing process under uncertainty. *Industrial & Engineering Chemistry Research*, 45(25): 8475–8483.

Arnheiter, E.D. and J. Maleyeff. 2005. The integration of lean management and Six Sigma. *The TQM Magazine*, 17(1): 5–18.

Benton, W.C. 1991. Statistical process control and the Taguchi method: A comparative evaluation. *International Journal of Production Research*, 29(9): 1761–1770.

Blanchard, B.S. and W.J. Fabrycky. 2011. *Systems Engineering and Analysis,* 5th edn. New York: Prentice Hall.

Bouchereau, V. and H. Rowlands. 2000. Methods and techniques to help quality function deployment (QFD). *Benchmarking: An International Journal*, 7(1): 8–19.

Chen, K.S., M.L. Huang, and P.L. Chang. 2006. Performance evaluation on manufacturing times. *International Journal of Advanced Manufacturing Technology*, 31: 335–341.

Chen, X., Y. Zhang, G. Pickrell, and J. Antony. 2004. Experimental design in fiber optic sensor development. *International Journal of Productivity and Performance Management*, 53(8): 713–725.

Duffuaa, S.O., S.N. Khursheed, and S.M. Noman. 2004. Integrating statistical process control, engineering process control and Taguchi's quality engineering. *International Journal of Production Research*, 42(19): 4109–4118.

Fischer, R.E. and B. Tadic-Galeb. 2000. *Optical System Design*. New York: McGraw-Hill.

Gunasekaran, A., S.K. Goyal, T. Martikainen, and P. Yli-Olli. 1998. Total quality management: A new perspective for improving quality and productivity. *International Journal of Quality & Reliability Management*, 15(8/9): 947–968.

Henderson, G.R. 2011. *Six Sigma Quality Improvement with Minitab*, 2nd edn. Chichester, U.K.: John Wiley & Sons.

IEEE. 2008. *Systems and Software Engineering—Systems Lifecycle Processes*, IEEE STD 15288–2008.

Kasunic, K.J. 2011. *Optical Systems Engineering*. New York: McGraw-Hill Professional.

Khorramshahgol, R. and G.R. Djavanshir. 2008. The application of analytic hierarchy process to determine proportionality constant of the Taguchi quality loss function. *IEEE Transactions on Engineering Management*, 55(2): 340–348.

Ku, H. and H. Wu. 2013. Influences of operational factors on proton exchange membrane fuel cell performance with modified interdigitated flow field design. *Journal of Power Sources*, 232: 199–208.

Li, G., M. Zhou, B. Zhang, and J. Yang. 2007. Economic assessment of voltage sags based on quality engineering theory. *IEEE Power Tech 2007*, 617: 1509–1514.

Lien, C.H. and Y.H. Guu. 2008. Optimization of the polishing parameters for the glass substrate of STN-LCD. *Materials and Manufacturing Processes*, 23: 838–843.

Lo, W.C., K.M. Tsai, and C.Y. Hsieh. 2009. Six Sigma approach to improve surface precision of optical lenses in the injection-molding process. *International Journal of Advanced Manufacturing Technology*, 41: 885–896.

Reid, R.D. and N.R. Sanders. 2009. *Operations Management*, 4th edn. New York: John Wiley & Sons.

Sage, A.P., W.B. Rouse, and K.P. White. 1999. *Handbook of Systems Engineering and Management*. New York: John Wiley & Sons.

Smith, W.J. 2008. *Modern Optical Engineering: The Design of Optical Systems*, 4th edn. New York: McGraw-Hill.

Sun, J.H., B.R. Hsueh, Y.C. Fang, and J. MacDonald. 2009. Optical design and extended multi-objective optimization of miniature L type optics. *Journal of Optics A: Pure and Applied Optics*, 11: 1–11.

Taguchi, G. 1995. Quality engineering (Taguchi methods) for the development of electronic circuit technology. *IEEE Transactions on Reliability*, 44(2): 225–229.

Udel. 2014. Table of statistical process control constants. University of Delaware, Newark, DE. http://www.buec.udel.edu/kherh/table_of_control_chart_constants.pdf. (accessed July 7, 2014).

Wu, H. and Z. Wu. 2012. Combustion characteristics and optimal factors determination with Taguchi method for diesel engines port-injecting hydrogen. *Energy*, 47(1): 411–420.

Wysk, R.A., B.W. Niebel, P.H. Cohen, and T.W. Simpson. 2000. *Manufacturing Processes: Integrated Product and Process Design*. New York: McGraw-Hill.

14

Optical Systems Testing and Evaluation

> Manufacturing is more than just putting parts together. It's coming up with ideas, testing principles and perfecting the engineering, as well as final assembly.
>
> **—James Dyson**

According to the International Council on Systems Engineering (INCOSE), a test is "an action by which the operability, supportability, or performance capability of an item is verified when subjected to controlled conditions that are real or simulated" (INCOSE 2006). This usually requires the product's parametric data from the verification process to be used directly in the evaluation process (Goldberg et al. 1994).

The qualification process is an essential aspect in the overall systems development effort. This process runs throughout the systems development life cycle and includes activities such as the generation of test plans, test procedures, reviews, inspections, analysis, modeling, and simulation, in addition to the actual testing of the components, subsystems, and the system itself (Hull et al. 2005). By implementing a thorough qualification program throughout the systems development life cycle, we can ensure the stakeholders that their needs and approved wants are being met at logical progression points at every developmental stage.

There is an intrinsic connection between requirements and the testing process. How can we determine if a requirement will be met? What test and evaluation (T&E) methods should be used for different scenarios? These questions lie at the heart of the qualification process and are something the systems engineer must understand to ensure that the system will function properly in its intended application. Throughout this chapter, we will use optical systems as the backdrop for our discussions.

In developing a general optical system, the *requirements definition* process is considered the starting point at any given systems engineering requirements level (e.g., stakeholder level, systems level, subsystems level, and component/assembly/part level). For instance, the problem definition step in the conceptual design phase that was discussed earlier first identifies the stakeholders and their needs and wants and then defines the stakeholder requirements. The methodology is then to produce lower-level requirements through repetitive cycles of requirements definition, functional analysis, and requirements allocation. The systems requirements evaluation cycle shown in Figure 14.1 is complementary to the systems development process in that it ensures that the system is built in accordance with the requirements generated during the development of the system. Once the requirements at any particular level have been defined, level-appropriate modeling activities commence as illustrated by the systems modeling block in Figure 14.1. These models are descriptive and, where possible, predictive, qualitative, and quantitative and cover the breadth of the systems development effort.

Some examples include enterprise architecture models that define the system in terms of its larger environment and relationships to other systems and entities, model-based systems engineering models, analytical and simulation-based models, graphical models, reliability models, logistics and support models, physical mock-ups, and more.

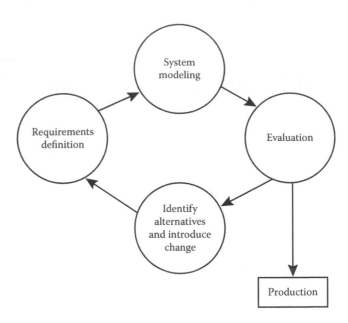

FIGURE 14.1
Systems requirements modeling cycle. (Created by and courtesy of Mirko Ermini.)

With regard to evaluation, these models provide descriptive information and often predicted performance that is useful in comparative analysis of the test results. For instance, if a test result or evaluation does not provide the expected result, the modeling activity can often be used in explaining the results and determining the cause. During the evaluation process, if the results are not as expected, then alternatives are identified to resolve the issue.

At times, problems may be found with the actual test itself and test plans and procedures may need to be changed. At other times, the design may not produce the intended results and the design team needs to reevaluate and modify the design to accomplish the requirements. Problems may also occur with the requirement itself, in which stakeholder involvement is needed for modifying, eliminating, or waiving the requirement. Establishing modeling methods and integrating modeling activities throughout the systems development life cycle are very helpful in solving problems as well as useful in preventing problems from occurring in the first place.

This chapter focuses specifically on optical systems and discusses some of the tools used to test and evaluate these systems. The chapter starts by introducing the process of testing and evaluating an optical system, followed by a detailed look at some of the tools used, and lastly, a case study is presented to apply the concepts and demonstrate what we have discussed.

14.1 General Concepts in Testing and Characterizing Optical Systems

According to *The Infrared and Electro Optical Systems Handbook*, "Electro-optical imaging systems convert electromagnetic radiation at optical wavelengths to electrical signals for source detection and/or analog visual display" (Dudzik 1993). Electro-optical (EO) systems can be used to detect, amplify, and manipulate light received from a target in either broad or narrow frequency bands. These optical systems are sensitive to electromagnetic radiation and, combined with infrared (IR) systems, cover the ultraviolet, visible, IR parts of the electromagnetic spectrum. Unfortunately, because of the small wavelengths involved, these systems tend to be very sensitive to any sort of interference, noise, or defect. Therefore, EO/IR systems require robust, specific, and well-thought-out T&E methods to ensure proper operation.

14.1.1 Systems Life Cycle and Test

The life cycle of a system is divided into multiple phases, where each phase plays a unique role in the development, operational use and support, and eventual retirement and disposal of the system. Steps in the systems development effort can be executed in either series or parallel, depending on the nature of the system, the current stage of the systems development, and whether or not concurrent engineering practices are being used in the development effort. A thorough qualification program is essential in successfully developing the system and verifying and validating that the system meets the requirements-driven design and satisfies the stakeholders' needs and approved wants. Consequently, the qualification program must be carefully integrated throughout the systems development life cycle and must be one of the core components in terms of importance.

As shown in Figure 14.2, the optical system's life cycle can be broken into two major phases: the acquisition phase and the utilization phase, in order to differentiate the

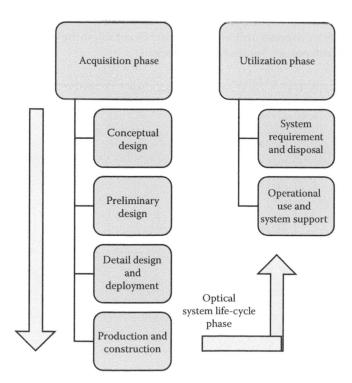

FIGURE 14.2
Systems life-cycle phases. (Created by and courtesy of Patrick Zinter.)

activities of the system designer and the system user (Blanchard and Fabrycky 2011). The system is then further broken down into what is conventionally known as the traditional systems engineering life-cycle phases: conceptual design phase, preliminary design phase, detailed design and development phase, production and construction phase, operational use and systems support phase, and the system retirement and disposal phase.

Misconceptions and misunderstandings about testing can lead to a costly and poorly integrated adoption of test practices in a system's development. From *Cost Management: Accounting and Control*, "Cost reduction strategies should explicitly recognize that actions taken in the early stages of the lifecycle could lower costs for later production and consumption stages" (Hansen and Mowen 2006). This idea is directly applicable to the qualification program in the sense that the earlier the qualification program is implemented, the less risk there is of an associated flaw emerging in a later developmental stage.

As you can see in Figure 14.3, the greatest opportunities to reduce life-cycle costing (LCC) usually occur in the early phases of a developmental activity. At the initiation of a developmental activity, the committed LCCs increase rapidly whereas the overall LCCs increase relatively slowly. Consequently, positive changes in the early-stage development can have significant impacts in the operational and support phase. For instance, a problem found in the requirements generation phase of a systems development effort might be solved with a modification or elimination of the requirement, whereas a problem found in the operational phase may result in extensive redesign, rework, and retesting at great expense.

With regard to the qualification program, its early adoption and careful integration into the systems development effort provides the ability to make early and ongoing assessments of the system's progress and is at the heart of establishing system compliance with program goals, objectives, and stated requirements.

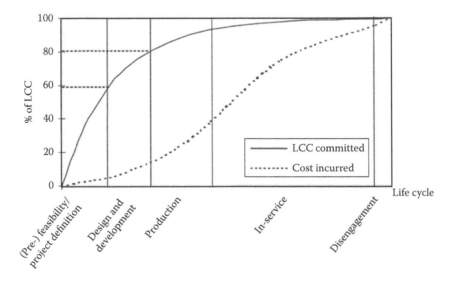

FIGURE 14.3
Traditional LCC committed versus incurred cost curve. (From Task Group SAS-054, *Methods and Models for Life Cycle Costing*, Defense Technical Information Center, Fort Belvoir, VA, http://www.dtic.mil/cgi-bin/GetTRDoc? AD=ADA515584, 2007, accessed February 5, 2014.)

14.1.2 Test Verification versus Validation

Test verification and validation are complementary terms that are fundamental to the testing process. Each of these terms correlates to a very specific and separate role in the system evaluation process. According to the *NASA Systems Engineering Handbook*, "The type of verification completed will be a function of the lifecycle phase and the position of the end product within the system structure" (NASA 2007). Verification aims to demonstrate that all requirements at a given requirements level have been successfully implemented in the developed system or system element. Verification typically occurs through the testing process when an observer, such as a systems engineer, quality engineer, test engineer, or technician, uses inspection, analysis, demonstration, test, or some other methods to ensure that the system, service, product, or process under consideration meets the stated requirements in a given requirements document.

The order of precedence is that verification testing occurs first followed by validation testing. After a system is built, verification testing occurs at the lowest level of the decomposed system—the component, assembly, and/or parts level. At this level of decomposition, verification tests the systems component/assemblies/parts against the detail-level design requirements (C specs), and, where necessary, the process requirements (D specs), and material requirements (E specs). Once the components/assemblies/parts are themselves verified, the verification testing process continues with the integration of these components (integration test), wherein the integrated components/assemblies/parts are verified against the corresponding B spec requirements. Finally, the system itself is verified against the A spec requirements. These verification activities are commonly known as developmental T&E (DT&E).

Validation is concerned with checking whether the system will meet the stakeholder's actual needs, and it involves making assessments of how well the system addresses these needs in a real-world environment. During the systems design and development activities, portions of a system may be validated by activities such as modeling, simulation, prototyping, and user evaluation. An example of a validation-related qualification activity would be to create a prototype system as part of the detailed design and development phase activities and test it under simulated real-world or real-world conditions. After the system has been produced and has successfully passed verification testing as part of the DT&E activities, the system must be tested in its operational environment under real-world conditions. This is commonly known as validation testing and occurs during the operational T&E (OT&E) activity that marks the end of the developmental activity and the system turnover to the customer/user stakeholders. The purpose of OT&E is to demonstrate to the stakeholder(s) that the system satisfies all requirements and is capable of meeting their needs and approved wants.

Most well-designed products can complete validation testing after completing the system verification process. However, as described by the NASA handbook, there are many systems that complete verification and then fail in a critical portion of the field validation (NASA 2007). The verification process ensures that all requirements have been correctly implemented, but that does not ensure that the final system will meet stakeholder needs. Since the cost of fixing a problem typically increases the later in the systems development cycle the problem is found, it is beneficial to validate the system at intermediate stages, rather than at the end of the development. One way of doing this is by using validated models that sometimes can be used in lieu of actual testing. For instance, given some antenna parameters, sophisticated validated electromagnetic models may be used to verify and validate certain performance aspects of an antenna. The stakeholder may deem these results acceptable without actually conducting a physical test on the actual antenna itself.

14.1.3 Test Often, Test Early: Integrated Testing

The lack of cohesion between developmental and operational testing has long been a point of concern in the systems engineering community (Wilson 2009, pp. 375–380). Many successful projects have used integrated testing to ensure that all aspects of the system being developed will function properly. Integrated testing seeks to "improve the complementary relationship between development and operational testing throughout the system life cycle" (Wilson 2009, pp. 375–380). Department of Defense policy states "Developmental and operational test activities need to be integrated whenever possible to improve overall T&E efficiency with increased emphasis on operational relevance" (Wilson 2009, pp. 375–380). Test integration is useful in identifying operational issues early and helps reduce the impact of required corrective actions.

Successfully integrated test programs all share the following three methodologies: (a) collaborative planning, (b) collaborative execution, and (c) shared data. According to the National Defense Industrial Association (NDIA), collaborative planning involves good coordination between the system developers and the various testing/certification agencies (Wilson 2009, pp. 375–380). The key focus is that teamwork must start as early as possible; the longer this takes, the more difficult it will be to successfully collaborate at a later time. Early test involvement is accomplished by "conducting integrated exercises during the initial requirements development program phase" (Wilson 2009, pp. 375–380). The development of an integrated test events matrix can provide a solid foundation for test execution by providing a single structured document to drive test execution.

Next is "collaborative execution," which focuses on "...chartering and implementing integrated test teams" where "these teams provide the coordination and cooperation for executing an integrated test strategy" (Wilson 2009, pp. 375–380). In order to do this, the project management must create well-thought-out teams that can implement an integrated test plan. It requires budget allocation, sharing of resources, and dedicated time at every step in the process. Even though this is an upfront investment, "teamwork of this kind provides the opportunity for collaborative testing which can greatly expand the effectiveness [of the testing effort]" (Wilson 2009, pp. 375–380).

The final step is "shared data," which is described as "maximizing the data that [is] available and usable for all participants" (Wilson 2009, pp. 375–380). Participants include all of those involved in the testing process. In this step, the objective is to get clear, concise, and accurate data from test programs in a well-thought-out, standardized reporting structure. Sharing data increases productivity by eliminating unnecessary communications and also increases the ability to detect existing problems early. It is essential to implement integrated testing methods early in the systems development life cycle. Doing so has many advantages, such as decreasing downtime, avoiding unnecessary costs, avoiding duplication of efforts, and detecting problems earlier on in the development life cycle.

14.1.4 Categories of Systems Test and Evaluation

INCOSE lists four main categories for testing a system or system element: development, qualification, acceptance, and operational testing (INCOSE 2006). Figure 14.4 shows the four basic categories and provides a brief description. Development testing is typically used to explain feasibility for concepts, while qualification testing focuses on prototypes or first articles. Note that qualification testing here is distinct from the earlier notion of establishing a qualification program. The qualification testing here is, in fact, one small aspect of the larger qualification program (Hull et al. 2005). The first of the performance

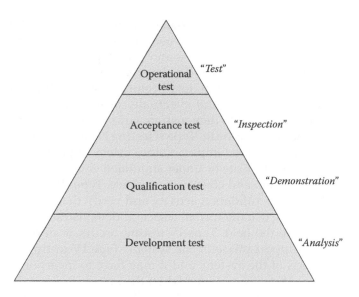

FIGURE 14.4
Four basic test categories. (Created by and courtesy of Patrick Zinter.)

test categories is acceptance testing, which assesses whether or not contract or specification requirements have been successfully completed. Depending on the results of acceptance testing, the customer can decide if the system meets the defined specifications and whether or not to approve the acquisition. The last category is operational testing, which verifies if the system meets its specifications under real-world conditions (INCOSE 2006).

These four basic test categories can be further broken down and organized by the traditional systems engineering life-cycle phases. Figure 14.5 shows the systems engineering life cycle phases through the operational and support phase and the applicable tests for each phase. Each phase is assigned a "type" depending on its location in the development life cycle. Analytical testing and development of analytical models start early in the conceptual design phase. The purpose of modeling in the conceptual design phase is to develop high-level models for conceptual development and understanding. Computer-aided design (CAD),

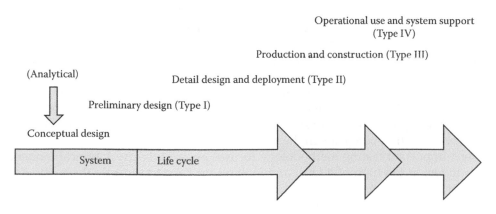

FIGURE 14.5
Testing types in the systems life cycle. (Created by and courtesy of Patrick Zinter.)

computer-aided engineering (CAE), computer-aided manufacturing (CAM), and computer-aided logistics support (CALS) models are also used along with the analytical models. Type I testing is performed on breadboard models and prototypes usually in the preliminary design phase and early detailed design and development phase.

Type II testing occurs on production quality prototype or protoflight units. A protoflight unit is a NASA term where test unit eventually evolves into an actual flight unit. Prototypes are typically developed to demonstrate key system capabilities and are traditionally scrapped once the system is built. Type II testing is more detailed than Type I and should expose problems with limited impact on the system. Type III testing involves production units at a designated test site or under simulated environmental conditions and occurs as part of the production and construction phase. Type III tests focus on a design review or other verification or validation activity and verify that the system is ready for production. Type III tests are usually formal tests and demonstrations closely integrated with validation activities in the field. Type IV testing occurs at an operational site during the operations and support phase of the system. Type IV testing is for operational use and sustaining efforts of the system, and it may have to incorporate new technologies or capabilities. Type IV testing is used when the system needs to be improved and upgraded after it has been successfully deployed. The scope of testing depends on the system complexity and the reliability of the system. Also, when a new part or component is introduced in the system, testing methods are applied to reevaluate the system and demonstrate that the system is still compliant with requirements and able to perform. However, "there may be system elements that are tried and true (commercial) off the shelf (COTS) parts which may not require all types of system tests previously discussed" (Blanchard and Fabrycky 2011).

The key point to effective systems testing is to determine the optimum amount of testing needed to verify proper operation. Too much testing wastes time and money, but too little testing may result in downstream mistakes that can degrade systems performance and jeopardize the intended operations.

14.1.5 Systems Test Methodologies and Problem-Solving Tools

In many cases, the main reason that T&E efforts fail is due to the culture of the organization. Some organizations tend to be internally divided and lack adequate communication between organizational elements. This lack of communication can lead to a weak T&E program and hinder progress toward an effective integrated testing paradigm. Methodologies that promote a robust T&E capability and an integrated test environment should be encouraged. According to Hopkins, "the methodologies should continually enforce that system test is meant to help with system development" (Gearhart and Vogel 1997). The feedback from collected test data can provide improvement opportunities for the design team. This should be a quick feedback, which means that an elegant test setup may not be practical at the start. If feedback does not come quick enough, the design team may not benefit from the results, leading to potential design flaws and mistakes being carried throughout the development and discovered late in the process.

The next important testing methodology focuses on modularity. A typical tendency is to begin to test at too high of a level and looking for "one test to rule them all." "There is no silver bullet when it comes to a complex system; tests need to be modularized" (Gearhart and Vogel 1997). The culture surrounding simulations is also a hot topic in T&E efforts. Simulations are very useful when it comes to testing. It is possible to simulate various conditions, which may stress the system and aid in detecting potential weaknesses.

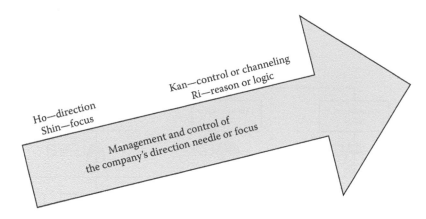

FIGURE 14.6
Hoshin kanri. (Created by and courtesy of Patrick Zinter.)

Also, validated models and simulators may produce results that count as actual test results. This can simplify the testing process and reduce overall cost.

The final methodology touches on isolating systems issues. As a test organization, it is necessary not only to verify and validate the system but also to "probe the system for vulnerabilities" (Gearhart and Vogel 1997). Testing has the aim of demonstrating the ability of the system to meet acceptance criteria and stress the system under conditions that are not necessarily nominal to look for reliability and performance shortcomings.

Other ways to institutionalize a good testing philosophy in an organization are to understand and adopt quality engineering principles and philosophies. One of the more effective philosophies used in industry today is known as "Hoshin Kanri" (Dhillon 2004). This is a Japanese term that focuses on a systems approach to change the organization using logical and organized planning. As Figure 14.6 shows, hoshin kanri involves organizational alignment and direction. This philosophy states that goals (or in this case, requirements) must be measurable, and key tests must be identified that will meet these requirements. "Team members need to reach a consensus at the end of each lifecycle phase, and processes should be challenged to improve capabilities" (Dhillon 2004).

There are many problem-solving tools that fall into the realm of system T&E. Some may be related to analytical design tools, statistical tools, or even system safety and reliability tools (Goldberg et al. 1994). Some examples include design-related tools like sensitivity analyses, statistical-related tools like analysis of variance (ANOVA), and reliability tools like fault trees just to name a few. For determining system robustness, a sensitivity analysis is a great tool. A sensitivity analysis varies many parameters in the system and then looks for potential stack up issues (Goldberg et al. 1994). ANOVA investigates variations between sample sets in order to identify system shifts in test. Another useful tool is the fault tree analysis, as seen in Figure 14.7. With this method, possible causes of failure are determined in a top-down analytical approach. It can also be applied to quality aspects of a system in the same manner. Fault tree analysis is a highly effective logical problem-solving technique to find a failure's root cause (Goldberg et al. 1994). To identify the root cause of a failure, multiple causal factors have to be evaluated, which may affect the failure conditions.

Fault tree analysis is very useful to highlight potential failure modes. The fault tree starts with the failure, such as the test symptom or the inability to take the appropriate measurement. The potential causes of the failure are listed at the top of the tree and are expanded into further detail at lower levels of the tree. This is a widely used tool in

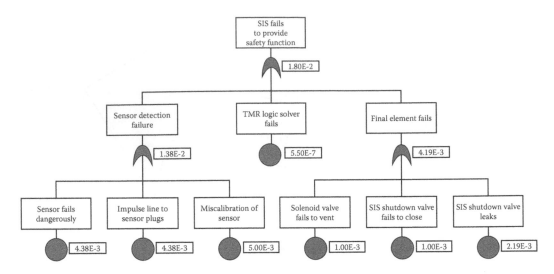

FIGURE 14.7
Fault tree example. (From Suttinger, L.T. and Sossman, C.L., *Operator Action within a Safety Instrumented Function*, National Technical Information Center, Washington, DC, 2002, http://sti.srs.gov/fulltext/ms2002091/ms2002091.html, accessed February 5, 2014.)

failure analysis that can be used to identify the root cause of the failure. The fault tree can also be combined with statistical methods to estimate the probability that the failure will occur or that one of the conditions leading up to the failure will occur.

14.1.6 Planning and Preparation for Systems Test

"Verification plans are critical to the realized system meeting customer requirements, and should be one of the first steps in a test plan" (NASA 2007). The verification plan should describe all T&E efforts from beginning to end in the systems life cycle and can take many different forms (NASA 2007). The verification plan should start with, and be driven by, the requirements, which are tied to the systems engineering test plan and evaluation process (Kasser 2007). As seen in Figure 14.8, there should be four main points used to test each requirement. The first point, acceptance criteria, will confirm whether the system has met the specified requirement or not. The next listed point is the planned verification methodologies, which include descriptions of the demonstration methods and analysis methods that are to be used. Following that is the testing parameters point, which defines the relevant test parameters and helps define the boundaries of test plans and procedures. The last step is to identify resources needed for testing. This can include test equipment, time to perform the testing, personnel needs, and even lab space (Kasser 2007).

Before any requirement testing can take place, the test team has to perform validation planning, which determines the proper approach and methods to validate the system. As shown earlier in Figure 14.5, there are four main categories to choose from: test, inspection, demonstration, and analysis. Each requirement must be tested by using at least one of these testing categories, depending on what method is the most effective in relationship with the requirement under test.

A list of requirements, or collection of requirements, that must be validated must be identified early in the planning stage. This list should have associated validation criteria, methodologies, parameters, and resources of detailed information (e.g., in the performance specifications portion, the planning team should put summaries of

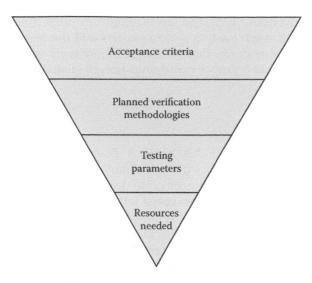

FIGURE 14.8
Four test plan points. (Created by and courtesy of Patrick Zinter.)

required accuracy, precision, sensitivity, specificity, range of results, normal values, and limitations on the list).

There are other important aspects of validation that may relate to the test environment/conditions or even specific tests that need to be performed. As seen in Table 14.1, there are some examples of specific tests that may be needed to satisfy systems requirements (NASA 2007). The acronyms HALT, HAST, and HASA stand for highly accelerated life test, highly accelerated stress test, and highly accelerated stress analysis, respectively, and are tests for simulating long-term effects and reliability testing. It is important that the validation plan is reviewed with appropriate stakeholders before it is applied. This is a critical step to confirm that the plan will take into account their expectations. The greatest benefit of this is to ensure that the stakeholders are part of the decision process and that they are supportive of the validation test results.

As an aside, the team should also plan to "verify the manual, as well as the product" (Carrico 2009). "The technical writer should continually edit and update the user's manual to ensure its validity during the product development cycle (International Validation Forum)." After this process is finished, managers can use the user's manual to determine whether or not the system, service, product, or process can be released. This process is

TABLE 14.1

Examples of Test

Testing Examples		
Leak Test	**Vibration**	**Operational Test**
Environmental	HALT, HASS, HASA	Voltage limits
Aerodynamic	RF emissions	Security tamper
Systems test	Salt spray	Acceptance tests
Burn-in	Fungus growth	Drop testing

Source: Created by and courtesy of Patrick Zinter.

important but often overlooked. System users should evaluate the user's manual with access to the system and reach back to system experts and the technical team. The users should use the literature to guide them through the systems intended functions. This evaluation process can ensure proper communication of customer centered publications (Carrico 2009).

14.1.7 Reporting and Feedback

As Dr. Joseph E. Kasser states in his book, *A Framework for Understanding Systems Engineering*, immediate reports "reduce the risk of non-delivery and non-compliance to the requirements" (Kasser 2007). He even goes on to dedicate an entire chapter on reporting feedback entitled "What do you mean you can't tell me how much of my project has been completed?" (Kasser 2007). This chapter explains that one of the fundamental reasons for reporting feedback is to show project status. Depending on the number of variables and the project complexity, this could be an extremely complicated process. Kasser promotes the transition from the traditional role of T&E to a new methodology where differences at each step along the systems development life cycle are monitored and reported, leading to a more accurate estimation of project completion percentage.

One approach of feedback reporting, which is coined as the Categorized Requirements in Process (CRIP) approach, was created to track and report on the status of meeting customer requirements (Kasser 2007). The CRIP approach can be broken into four steps:

1. Categorize the requirements.
2. Quantify each category into ranges.
3. Place each requirement into a range.
4. Monitor the differences in the state of each of the requirements.

Based on the cost of implementation, CRIP first categorizes requirements and then customer needs. Implementation of test solutions and their effectiveness are then reported back as an estimated project completion percentage. This style of reporting has many advantages including easier tracking due to requirement buckets, increased knowledge of the current status of a project, and an ordering of each requirement in terms of importance (Kasser 2007).

There is another important aspect of feedback reporting, which is the verification report itself (NASA 2007). A verification report should be created for each major test activity. In particular, this report should include functional performance in the lab, compatibility within the overall system, and environmental tests. Verification reports should contain similar information, including the overall objectives of the test, the description of the testing, configuration setups noting differences versus field testing, and descriptions of deviations from expected results. Moreover, specific results should be gained from each test run. This information can be communicated through graphical analysis, pictures, or even data tables. In the end, each report should have conclusions and recommendations for a course of action based on the results of the testing. The test report can also contain final recommendations for follow-up actions to be carried out in order to improve systems performances.

In addition to reporting test feedback, test data should be analyzed for consistency, validity, and quality. "Any anomalous readings should be reviewed and explained" (NASA 2007).

14.1.8 Clean Test Environment

As the common idiom goes, "Cleanliness is next to Godliness," and it still holds true, especially in the world of high-precision optics. An unclean test environment can skew the results of testing of an optical system. Worse than that is the fact that unclean testing can permanently damage the system being tested. Even touching certain lenses can destroy the lens coating, rendering it inoperable.

Manufacturers have implemented the instantiation of "5S" programs in order to improve the efficiency and cleanliness of the work environment. It is a program that begins with cleaning but ends with an improved and efficient work place (Harkins 2009). Lots of innovative organizations have implemented 5S for the goal to improve performance and the overall morale of the workplace. The 5S's refer to five steps in the program to continuously improve the work environment: sort, straighten, shine, standardize, and sustain (Harkins 2009) (Figure 14.9). Sorting of tools, equipment, and the subjects of tests themselves improves efficiency and prevents loss of equipment. The next step is straighten, which involves establishing a logical organizational structure for equipment and resources. These may include a floor plan and an optimized process flow. Moreover, labels, signs, and bulletin boards can help to make sure everything is in its place. The next step is systematic cleaning, which emphasized the importance of establishing "an ongoing pattern of cleanliness" (Harkins 2009). This cleanliness mentality will reflect on every product through precision and consistency.

Standardization efforts can include standardized test practices, station calibration methods, production practices, and more. Doing all of the aforementioned, in a standardized fashion, can help increase interchangeability, which will increase efficiency and reduce costs. The last step is sustain, which is the action that must be taken to assure proper adherence to the previous four steps.

As a representative example, NASA's space telescope provides an excellent illustration of the 5S methodology. "Particulate and molecular contamination can cause significant degradation in a space telescope's reliability, overall performance, and clarity of images" (National Aeronautics and Space Administration Marshall Space Flight Center 2007).

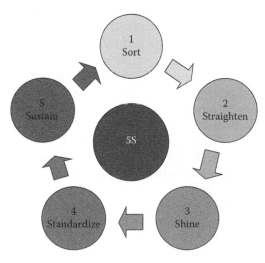

FIGURE 14.9
5S. (Created by and courtesy of Patrick Zinter.)

TABLE 14.2

Major Sources of Contamination

Particulate Contamination	Molecular Contamination
Airborne particles settle out from the air.	Residue from manufacturing processes
Paint spray, insulation fibers, fabric fibers.	Outgassing from materials
Trapped particulate on inner surfaces that is redispersed from vibration.	Oxidation (from O_2 molecules in low orbit)
Propulsion system exhaust plumes and water release may cause residual particulate clouds.	Space vehicle propulsion plumes depositing material on optical surfaces
	Exposure to volatile materials in the environment during assembly

Source: Created by and courtesy of Patrick Zinter.

Particulates can cause clouds, which can hamper the image quality of the optical mirror or optical sensors.

Table 14.2 illustrates some typical sources of contamination in the testing process. NASA advocates numerous 5S-like practices to control the contamination of optical systems (National Aeronautics and Space Administration Marshall Space Flight Center 2007), such as proper material selection, precleaning of systems components, and cleaning upkeep throughout the manufacturing/transport processes. Contamination control procedures must be developed and used from conceptual design all the way to the launch of the orbital system (National Aeronautics and Space Administration Marshall Space Flight Center 2007).

14.1.9 Static-Free Test Environment

Besides the need for a clean environment, the optical system must be free from electrostatic discharge (ESD) events, while being assembled, tested, and shipped. ESD, which is the phenomenon of a rapid discharge of electrons, can permanently damage or destroy sensitive components within the system. ESD may cause an immediate failure in the factory or, even worse, a failure in the field at an inopportune time. Establishment of an ESD control plan is important to ensure that the quality and reliability of the system are maintained.

The American National Standards Institute (ANSI) has developed an industry standard for protecting against ESD—ANSI-ESD-S20-2. This document is focused on "the development of an electrostatic discharge control program for the protection of electrical and electronic parts, assemblies and equipment" (ANSI/ESD 2007). There are three main principles used to control ESD and implement and maintain control in the standard test environment. The first principle is that the system must be electrically grounded or be at the same potential as conductors in the protected area. The second principle states that the nonconductors that are permitted in the protected areas cannot lose their charge when connected to ground. Materials like Styrofoam, paper, and wood, which can build a charge, then rapidly discharge it, are not allowed in protected areas. Lastly, it must be enclosed in static-protective material when being transported outside of a protected area (ANSI/ESD 2007).

The ANSI standards are used to describe administrative and technical requirements for the implementation of an ESD control program. This includes personal ground systems, work surface requirements, flooring, seating, equipment grounds, and ESD packaging. ESD control applies to all manufacturing processes but especially applies to testing. ESD control should always be considered for the development of these sensitive EO and IR systems.

14.1.10 Optical Table

The optical bench is one of the staples of the optical testing world. Typically, the optical bench consists of a special table that holds optical components, optical sources, detectors, and specialized equipment. The optical components can be as simple as a basic source and detector or can have numerous optical components including spectral and polarization filters, calibrated sources and detectors, beam-forming and beam-shaping optics, aberration control and correction, environmental conditioning equipment, and a host of other specialized diagnostic and test equipment.

A very simple example is a source, collimator, test article, imaging optics, and a microscope. The collimator is used to generate a beam, or column, of light from a source. The column of light can be used to uniformly illuminate the test article. The imaging optics collects the scattered light from the test article and forms an image of the test article, while the microscope is used to evaluate the image. In other configurations, the microscope could be used to precisely specify length, inspect an object for physical damage, or verify the curvature or other physical properties of a lens.

The optical table is not just an ordinary table. It has certain characteristics that reduce the amount of noise that is introduced in the test environment. The optical table is designed to isolate the test from vibrations and shock. At optical wavelengths, even slight vibrations and shocks can throw off a test or calibration measurement. Optical tables are typically very heavy and use an isolation mechanism like pneumatic springs to eliminate most vibrations. Sometimes, the optical table itself must be on an isolated pier to prevent vibrational and shock coupling through the ground. For example, an elevator can cause minute vibrations in the frame of a building that can couple through to an optical table. To mitigate these effects, a giant concrete slab that is physically decoupled from the building and is anchored on a solid foundation is sometimes necessary. Essentially, the building can move and shake on a microscopic scale but the isolated pier and the optical table itself are sheltered from these vibrations and shocks. The optical table surface must be perfectly flat and level and is often made of highly polished metal. The surface contains a myriad of holes that are used to mount fixtures that hold the optics and optical components. When the optical table has its isolation mechanism engaged, it is said to be "floating."

The optical table is key to performing many optical tests or calibration measurements. Performing a test or calibration on a surface other than an optical table could cause tolerance issues in the final product and may be sufficient reason for negative results or failure of a particular test.

14.1.11 Optical Calibration Standards

The complexity and certification processes of the calibration of optical systems have changed dramatically over the past few decades. In the age of the film camera, the camera was calibrated only by its "imaging geometry" (Sandau 2010). In calibrating the camera, only a few critical components, such as the lens, a marker, and pressure plate were required (Sandau 2010). The complex inner working of newer digital technology requires more stringent overall system calibration. Since the advent of digital cameras, calibration procedures have undergone a complete overhaul, as they now not only involve the optical system but the digital image recorder as well. Laboratory precision is required to calibrate the effective resolution and the spectral responsivity of these optical systems. This has given birth to a number of accredited calibration laboratories that are used to qualify digital imaging systems (Sandau 2010).

14.1.12 General Concepts in Testing and Characterizing Optical Systems Summary

So far, we have covered the importance of establishing a qualification program and how the qualification program relates to the systems development cycle, different types of testing, the test environment, and the optical bench. The next section will cover test methods specifically for optical system in a much more detailed manner. Section 14.2 will cover the four basic methodologies to optical testing: interferometry, spectrometry, polarimetry, and radiometry. Additionally, some of the tools and required facilities used in these testing methods will be described.

14.2 Optical Systems Testing Methods

In order to successfully develop optical systems, it is essential to have (or have access to) optical testing facilities and ensure that the optical system functions as intended. In this section, we present an overview of the optical systems testing, techniques, tools, and facilities needed to measure essential fundamental optical system properties. There are four basic optical testing techniques: interferometry, spectrometry, polarimetry, and radiometry. Each one of these techniques provides unique information to the tester about the relative performance of the system. Not all of these techniques need to be applied in every case. For instance, if the optical system does not have any sensitivity to polarization, then polarimetric methods are not necessary. Interferometric methods are useful in determining mirror and lens characteristics and diagnosing coherence effects such as in laser beam quality. Spectrometry techniques are useful in characterizing the spectral response of sources, detectors, or substances. Polarimetric testing methods are useful in evaluating the interaction of different polarization states of electromagnetic radiation with materials. Radiometric testing methods are useful in understanding the signal levels along the optical path from source, through the optical channel, through the optical system, and onto the detector itself. In the next section, we take a closer look at each of these fundamental optical testing methods along with the associated required equipment and facilities.

14.2.1 Interferometry

Interferometry is a useful tool used to find, among other things, the topographic measurements of lenses and mirrors. This testing methodology is important in measuring with high precision the narrow dimensional tolerances needed to produce high-quality lenses, mirrors, and optical components. One example that shows the relative importance of interferometric measurements is the near-catastrophic issue with the Hubble Space Telescope's mirror. When manufactured, the mirror was out of tolerance by a mere 2.3 µm; however, this was enough to cause severe spherical aberrations, causing light from the edges of the lens to focus on a point other than the center of a mirror. Figure 14.10 shows the before and after shots of the Hubble Telescope when a fix was finally applied. In the figure, COSTAR stands for Corrective Optics Space Telescope Axial Replacement and refers to the fix for the discovered problem.

"Interferometry is a family of techniques in which waves, usually electromagnetic, are superimposed in order to extract information about the waves" (Bunch and

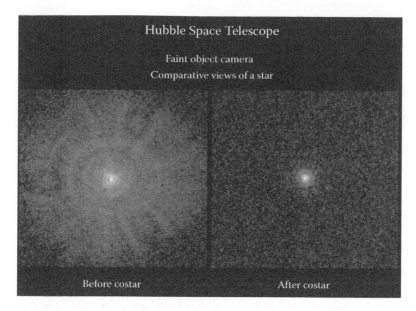

FIGURE 14.10
COSTAR Team. 1994. (Comparative View of a Star Before and After the Installation of the Corrective Optics Space Telescope Axial Replacement (COSTAR). http://hubblesite.org/newscenter/archive/releases/1994/08/image/a/, accessed December 3, 2014.)

Hellemans 2004, p. 695). Both spatial and temporal coherence properties of a source or material can be measured. For instance, spatial variations in the surface of a mirror would lead to changes in the interference pattern of a spatially sensitive interferometer like the Mach–Zehnder interferometer. Some interferometers use the physical phenomenon known as Newton's rings to calculate the topographic measurements of a material's surface. This phenomenon occurs when monochromatic light from a flat surface coherently interferes with light from a spherical surface that rests on the flat surface. A pattern of concentric light and dark circles is created on the flat surface. Since light possesses a wavelike motion, when optical waves are coherently superimposed on each other, this creates an interference pattern seen in interferometry (Goldberg et al. 1994). Figure 14.11 shows an example of Newton's rings.

The Newton rings result from the interference of light from the spherical surface of the lens with the optically flat surface upon which it rests. The radius of the dark rings for light at wavelength lambda that is normally incident on the spherical lens is given by

$$\rho_m^2 = \frac{m\lambda R_l}{n_g} \tag{14.1}$$

where
ρ_m is the radius of the mth dark fringe
R_l is the radius of curvature of the spherical lens surface adjacent to the flat bottom plate
n_g is the index of refraction of the material in the gap between the curved surface and the flat bottom plate surface

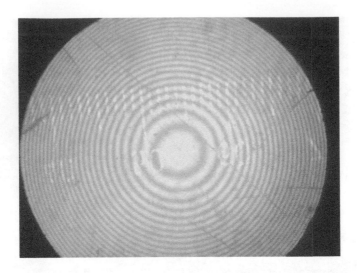

FIGURE 14.11
Example of Newton's rings phenomenon. (From Warrencarpani, Newton's rings as observed through a microscope, Wikimedia commons, 2011, http://commons.wikimedia.org/wiki/File:20cm_Air_1.jpg, accessed December 5, 2014.)

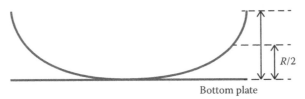

Bottom plate

FIGURE 14.12
Simple interferometer. (Created by and courtesy of Patrick Zinter.)

Figure 14.12 shows an example of the simple interferometer discussed earlier.

In Figure 14.12, the height of the gap can be given by

$$R = \frac{\rho_m^2}{2R_l} \tag{14.2}$$

where R is the gap height measured from the flat surface to the curved surface on the spherical surface. This simple interferometer is very useful in measuring the properties of spherical lenses. Aberrations of the lens show up as distortions in the concentric circles. We now discuss different types of conventional interferometers that are commonly used in practice today.

Two-beam interferometry is one of the simplest of interferometry techniques and is consequently the most common method used in interferometric testing. In a two-beam interferometer, a coherent beam of light is split into two beams by a beam splitter. The split beams traverse separate paths and then are recombined on a flat surface. If the difference between the optical path lengths between the different split beams is zero, then the beams interfere constructively with each other and have concentric circles like the Newtonian rings earlier or fringes of light and dark regions (depending on the type

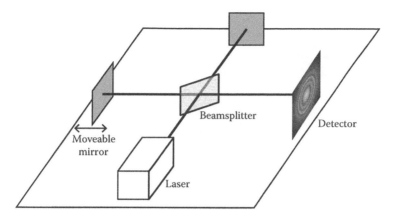

FIGURE 14.13

Michelson interferometer. (From ColinEberhardt, The Michelson interferometer experimental apparatus, 2004, Wikipedia, http://en.wikipedia.org/wiki/File:Michelson-interferometer.png, accessed December 5, 2014.)

of source and alignment of the source with the optical axis of the interferometer). An example of a two-beam interferometer, known as a Michelson interferometer, is shown in Figure 14.13.

If the movable mirror is changed, the optical path lengths begin to differ and a bright band moves outward from the center aperture with the following brightness:

$$I = I_1 + I_2 + 2\sqrt{I_1 I_2} \cos\left(\frac{2\omega d}{c}\right) \tag{14.3}$$

where
I_1 is the brightness of the first split beam of light
I_2 is the brightness of the second beam of light
ω is the optical radial frequency $2\pi c/\lambda$
d is the separation difference between the two optical paths
c is the speed of light

Sometimes, a corrective plate is inserted in one of the split optical paths to account for phase changes made due to surface reflections. The Michelson interferometer is sometimes aligned for equal optical path lengths and then a test object is inserted into one of the optical paths to analyze its temporal coherence properties. The relative movement of the concentric circles or fringes indicates the optical path difference introduced by the test object. Equation 14.3 applies until the temporal coherence properties between the two optical paths are no longer coherent, which results in the elimination of the concentric circles or fringes. As an example, for white light sources, the optical path difference between the two split optical paths is very small—on the order of the size of the wavelength—and the optical path difference must be smaller than this for fringes to form. Laser sources, because of their high-power potential for high coherence and directionality, can tolerate substantially larger optical path differences than white light sources and are consequently often used in testing applications. Michelson interferometers can therefore be used to determine temporal coherence properties of sources and/or test objects and are a necessary component in an optical testing facility.

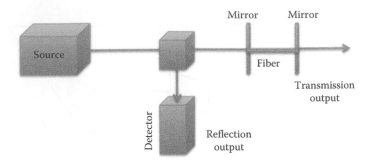

FIGURE 14.14
Multiple-beam interferometer. (Created by and courtesy of Maryam Abdirad.)

14.2.1.1 Multiple-Beam Interferometers

"The technique of multiple-beam interferometry is based upon placing two highly reflective, parallel surfaces in close proximity with each other, injecting a directional, coherent beam, and using a lens to converge the emerging beams which have undergone multiple-reflection between the surfaces" (Komatsu et al. 2000). An advantage of this method is that the interference pattern fringes become extremely narrow thereby increasing the achievable resolution and precision of the test station. An example of a multiple-beam interferometer is shown in Figure 14.14 (Optique Online Courses 2009). This approach uses a classic Fabry–Pérot type of interferometer with highly polished opposing mirrors and uses optical fibers to make the multiple reflections.

Multiple-beam interferometer method can produce fringes that are roughly 50 times thinner than those produced in the regular two-beam method. With a multibeam setup, the limiting resolutions are on the order of 0.5 nm. Achieving these resolutions requires special considerations such as ensuring that the reflecting plates have high-reflection and low-absorption coatings. The distance between the two surfaces should be kept as small as possible. The input light should be collimated to a parallel beam with a divergence less than 3°, and the source should be as close to the reference plate as possible (Komatsu et al. 2000).

14.2.2 Spectrometry

Spectrometry is the study of optical phenomena as a function of wavelength. It has many uses from discovering the composition of a material in chemistry to understanding the construction of stars in astronomy. This section presents some basic concept of spectrometry such as finding the spectral power distribution of a source, spectral absorption/transmission characteristics of the atmosphere or a given substance, or the spectral response of an optical device. An optical spectrum analyzer is an optical device that is capable of separating spectral components of the optical signal. For example, a prism separates light into its constituent color components. Likewise, a monochromator can select light over a narrow band of wavelengths and tune this narrow band over a broader spectral range. As such, by combining a radiometer with a tunable monochromator, the result is the capability of measuring the optical power in narrow bands of light over a particular wavelength range. A typical block diagram for an optical spectrum analyzer is shown in Figure 14.15.

The diagram shows that a broadband input signal is processed by a tunable band-pass filter then detected by a photodetector. The monochromator mentioned earlier is an example of the tunable band-pass filter and the radiometer is an example of the photodetector.

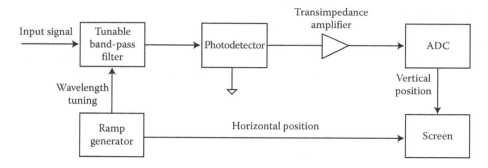

FIGURE 14.15
Spectrum analyzer block diagram. (Adapted from Agilent Technologies, Optical spectrum analysis basics, http://cp.literature.agilent.com/litweb/pdf/5963-7145E.pdf [accessed April 09, 2014], 1996.)

The ramp generator is used to tune the tunable band-pass filter and the amplifier and analog-to-digital converter processes the signal for display. Most optical spectrum analyzers can be grouped into one of two categories: diffraction-grating-based spectrum analyzers and interferometer-based spectrum analyzers (Agilent Technologies 1996).

14.2.2.1 Interferometer-Based Optical Spectrum Analyzer

Interferometer-based optical spectrum analyzers are created with one of two architectures: Michelson and Fabry–Pérot. As discussed earlier, a Michelson-based spectrum analyzer creates an interference pattern of the input light with a delayed version of itself. In essence, by taking a sequence of measurements over different path length separations, this is equivalent to taking the temporal autocorrelation of the input signal. By taking the Fourier transform of the sequence of delayed interference patterns, the power spectra of the input signal are determined. This type of interferometer is used for coherence time and coherence length measurements (Agilent Technologies 1996). The Fabry–Pérot-based spectrum analyzers use highly polished mirrors as a resonant cavity that is similar to a laser. A diagram of a Fabry–Pérot-based spectrum analyzer is shown in Figure 14.16.

As can be seen in Figure 14.16, the Fabry–Pérot cavity is placed in between two lenses. The first lens focuses the parallel components of a diffuse input source into the cavity where the gap spacing is such that the reflected light interferes constructively at the cavity surface on

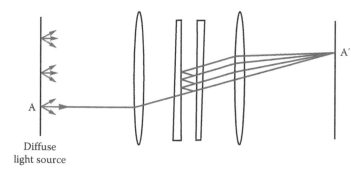

FIGURE 14.16
Fabry–Pérot-based spectrum analyzer. (From Stigmatella aurantiaca, Fabry perot interferometer diagram, 2012, Wikimedia commons, http://commons.wikimedia.org/wiki/File:Fabry_Perot_Interferometer_-_diagram.png, accessed December 5, 2014.)

the right. The transmitted light from the cavity is in phase and is focused onto the observation screen where very thin fringe widths are possible. The benefit of the Fabry–Pérot architecture is that it provides high spectral resolution. This is important for measuring phenomenon that has a high degree of signal variations over a small range of frequencies such as in a laser chirp. In a chirp signal, the signal increases or decreases rapidly over short spectral intervals.

14.2.2.2 Diffraction-Grating-Based Optical Spectrum Analyzer

The majority of optical spectrum analyzers implement a diffraction-grating-based architecture. In this type of optical spectrum analyzer, a set of fine lines (reflecting grooves) is placed on a material like aluminum or gold, which diffracts light in different directions like a prism. Just like a prism, different wavelengths are diffracted in different directions and only the wavelengths that escape from the small exit orifice of the spectrum analyzer make it to the detector. The optical spectrum analyzer selects the particular center wavelength based on the scattering angle from the diffraction grating, and the size of the opening determines the wavelength range (Agilent Technologies 1996). Diffraction-grating-based optical spectrum analyzers have a greater separation between wavelengths than prism-based optical spectrum analyzers, and so diffraction-grating-based spectrum analyzers are capable of higher spectral resolution.

14.2.3 Polarimetry

Polarimetry is the measurement and interpretation of the polarization of transverse waves (Mishchenko et al. 2011). Transverse waves are electromagnetic waves that oscillate in the normal plane to the direction of propagation of the electromagnetic wave. The polarization of the transverse waves can be measured one of two ways: mechanically or electronically. One method that uses linear polarization involves two mechanical polarizers. In this method, a polarization filter is used to pass one polarization state (e.g., horizontally polarized or vertically polarized state). Another polarizer is placed crosswise at some distance from the initial polarizer preventing light from passing through. Test articles that may change the polarization state of the initially polarized electromagnetic wave are then placed in between these polarizers and the changed polarization state can be measured by the amount of optical power that makes it past the second crossed polarizer.

To test the polarization effects of a sample, equipment such as the polarization measurement system in Figure 14.17 can be used. A linear polarizer is placed at the beginning of the polarization measurement system to allow only one linearly polarized state (horizontal or vertical) to enter the measurement system. The Faraday modulator is used to change the orientation of the incoming polarization state with respect to the sample.

Consequently, the angle of the applied linearly polarized light to the sample under test can be controlled. The analyzer shown at the right side of Figure 14.17 is typically aligned at 45° to the incoming polarized light, without the sample present. In this case, no light is transmitted through the analyzer and no signal falls on the detector. Now, by introducing the sample, the linearly polarized light injected into the sample is altered, and the detector then sees the polarization component that is aligned with the analyzer. By rotating the analyzer, the effect of the sample material on the polarization states of the referenced light can be determined.

To measure the polarization properties of a remote source, the vertical and horizontal polarization components can be directly measured by introducing a linear polarizer in

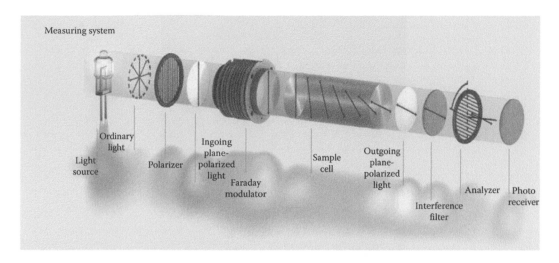

FIGURE 14.17
Agilent light wave polarization analyzer. (From Antonpaar, Functionality of a polarimeter, 2013, Wikimedia commons. http://commons.wikimedia.org/wiki/File:Polarimeter_measuring_system.jpg, accessed December 5, 2014.)

front of the detector element, measuring the result, and then rotating the linear polarizer 45° to get the orthogonal polarization state measurement. This method works if there are no consequential changes in the source polarization over the time period it takes to rotate the linear polarizer. Alternatively, a polarizing beam splitter can be used to separate out the polarization states of the incoming light, and two detectors can be used to simultaneously measure the vertical and horizontal polarization components of the source.

14.2.4 Radiometric Testing

As discussed in earlier chapters, a photodetector is a device used as a transducer between incoming light and output electricity; it converts light energy into a proportional electrical current or voltage. A radiometer can be used to test the radiometric characteristics of an optical system. Crookes radiometer is an example of a simple radiometer. Crookes radiometer can be made with an airtight glass bulb with light-sensitive vanes in a partial vacuum, mounted on a rotating axis with very low friction. The greater the irradiance on the vanes, the faster the vanes will spin. An image of a Crookes radiometer is shown in Figure 14.18.

Since the rpm of the radiometer is proportional to the optical power on the vanes, an accurate measurement can be derived through this linear relationship. A variety of electronic radiometers are available on the market today. Like the Crookes radiometer, they respond linearly to optical power. When considering a radiometer, its noise-equivalent power (NEP), responsivity, and dynamic range need to be adequate to measure the signal of interest with adequate resolution over its expected range.

In many radiometric applications, it is also necessary to measure the optical power in a narrow spectral band or over a defined range of wavelengths. In this case, instruments such as the spectroradiometer are used. As discussed previously, the combination of a monochromator and a broadband radiometer provides the capability of measuring optical power in a narrow wavelength band, which can be tuned over a broad range of wavelengths.

FIGURE 14.18
Crookes radiometer. (From Timeline, Crookes radiometer, 2005, Wikipedia, http://en.wikipedia.org/wiki/File:Crookes_radiometer.jpg, accessed December 5, 2014.)

14.2.4.1 Application of Optical Systems Testing

With any kind of optical sensor, production inconsistencies or lens misalignment can cause variations from unit to unit (Smith 1997). In order to ensure the quality and consistency of these parts, rigorous testing is needed. Typically, the more complex the detector, the more complex the testing will need to be. As systems become more complex, it is essential to have fundamental hardware, software, and integrated systems testing capabilities. In this section, we present some common application examples of the optical testing methods and tools just discussed to gain some familiarity with how these techniques and tools are applied.

One important application area is in using interferometers to measure the flatness of optical components like mirrors and lenses. This becomes especially challenging for the large mirrors that exceed 3 m in diameter wherein the surfaces are grinded down, polished, and coated so that the surface deviations are much less than an optical wavelength! Surface deformations and nonuniformities can be accurately determined from the interference pattern from the interferometer. Similarly, imperfections and aberrations in lenses can be observed using the interferometer.

Radiometric measurements are prevalent in many disciplines. In its simplest form, a radiometric just measures the power in an optical signal. The radiometer can be calibrated against a National Institute of Standards and Technology (NIST) traceable source and afterward can be used to determine absolute power measurements at strategic points in the test configuration. For instance, the calibrated detector can measure the optical power in the entrance pupil to verify theoretical or simulated calculations. It can be used to determine the absolute power at any location that the detector can be placed. In this fashion, the calibrated radiometer can be used to determine the actual optical power that falls at a particular location. A test object that can be placed in between the source and the calibrated radiometer would then measure transmitted power through the test object. By knowing the before and after power on the detector, the transmittance properties of the device can be determined as well as how much of the light was both absorbed and reflected from the test object. By combining a spectral filtering device such as a monochromator, or a spectrally narrow source like a tunable laser with the radiometer, the transmission, absorption, and reflection characteristics of a test object can be determined as a function of wavelength or optical frequency. This is one of the fundamental principles underlying optical methods used in forensic analysis methods that determine the absence or presence of certain substances in a material sample. Substances respond differently to optical radiation, and spectroradiometric methods can be used to detect, track, classify, and/or identify the material (or absence of the material). A radiometer can basically be thought of as a "light bucket" that responds to the total optical power that strikes its surface.

An imaging camera can be thought of as a collection of light buckets called pixels that have a known spatial relationship with the other pixels. In addition to the radiometric properties of the imaging camera such as its NEP, detectivity, responsivity, uniformity, electrical bandwidth, and dynamic range, the resolving power of the optical system is needed. The pixel dimensions themselves introduce a resolution limit since when the pixel is projected into the object space, features smaller than the projected pixel size will not be resolved. On the other hand, the best theoretical spatial resolution of the optical system (assumes perfect optics, no aberrations, no atmosphere, and no system or detector noise) is given by the so-called classical diffraction limit. The practical spatial resolution limit may be much larger than the classical diffraction limit. For instance, the turbulent atmosphere may be the limiting factor for an optical system that does not employ an adaptive optics system or an atmospheric turbulence compensation system. Very often, increased spatial resolution can be obtained by removing the effects of the atmosphere.

In an imaging system, the scale of the pixels projected into the object space should be less than the smallest scale that is being examined (Austin 2010). To test the spatial resolution of an optical imaging system, standard test items have been developed such as the U.S. Air Force resolution bar chart that was developed in 1951 and adopted in MIL-STD-150A. This chart has a calibrated set of bars of various sizes, spacing, and orientations. When imaging this chart, the spacing that can just barely be visually observed determines the spatial resolution of the imaging system. Figure 14.19 illustrates that the quality of an image under test depends not only on radiometric considerations but also on the geometric calibration of the optical system.

When imaging systems collect images at different wavelengths, they are known as spectral imaging systems. These tend to be very high-end optical sensing devices. If images are captured at only a few wavelengths (on the order of 6 or less), then the imaging system is called a multispectral imaging system. If the images are collected at 100s of different wavelengths, the imaging system is called a hyperspectral imaging system. If the images are collected at 1000s of different wavelengths, the imaging system is called an ultraspectral imaging system. These systems fuse the spatial imaging properties of optical imaging systems with the material discrimination capabilities of spectral systems. In essence, each pixel provides a spectral response along with its ability to provide an image. The downside to these high-end imaging systems is the vast amounts of data that are generated and need to be processed! Take for instance a 1024 × 1024 pixel ultraspectral imaging system that collects data at 1000 different wavelengths. If each pixel has 24 bit resolution, then the amount of data collected over 5 min at video rates (30 Hz) would be on the

Determination of
image quality

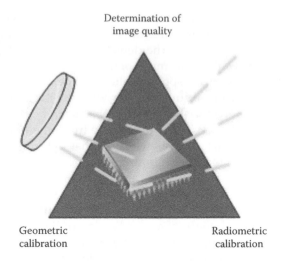

Geometric
calibration

Radiometric
calibration

FIGURE 14.19
Determination of image quality. (Created by and courtesy of Patrick Zinter.)

order of 1024×1024 (pixels^2/cycle/wavelength) $\times$ 30 Hz (cycles/s) $\times$ 24 (bits/pixel) $\times$ 1000 (wavelengths/pixel) $\times$ 5 min $\times$ 60 s/min or 2.265×10^{14} bits! Let us go process that! Polarization states can also be used in a similar fashion as spectral systems to gain insight into material properties.

In setting up a test station, the correct alignment of the optical components is critical. When setting up the test bench, visual alignment of optical elements is a good first step. However, more precise measurements are needed, and therefore, it is recommended to use a laser as an alignment tool (Smith 1997). The lenses must be aligned to an optical axis to make sure that the image they are testing will be rightly focused (Smith 1997). Moreover, the use of a test sled made from steel bars can be a cheap and effective aid to secure lenses on axis for test.

Lasers are also used when very thin line widths are needed in making spectral measurements. They are also used in providing reference beams with a constant phase as diagnostic tools. They can also be used in optical testing to send pulses of optical light over a range and/or through a test station such as in laser detection and ranging (LADAR) applications. Some more esoteric uses are to scatter a high-power laser beam from the sodium layer of the atmosphere to simulate a star for adaptive optics purposes (the so-called laser guide star). The lasers utility and applications are too numerous to list. They are essential components in an optical lab and/or optical testing facility.

In a no frills approach, author Warren J. Smith describes how to measure and categorize EO systems (Smith 1997). In his book, *Practical Optical System Layout*, Smith says that "The preliminary design of a complex optical system is often carried out, not by a specialist in lens design but by someone who is a systems engineer or a generalist (Smith 1997)." He illustrates several useful tools that can be applied to many different optical devices from telescopes to microscopes. One section of his book deals specifically with incorporating stock lenses into the design rather than fabricating them in house. Like any other system element, the use of COTS optical elements needs to be verified before adding them to the system. There can be many different imperfections that can severely degrade systems performance in a stock lens. Table 14.3 lists some of the common reasons for imperfections in stock lenses. In some applications, however, adopting optical COTS components may be warranted and cost effective, and the design robust enough to deal with these imperfections.

TABLE 14.3

Reasons for Lens Imperfections

Common Optical Defects in Lens Systems
Axial Chromatic Aberration
Transverse Chromatic Aberration
Spherical Aberration
Coma
Astigmatism
Field Curvature
Geometrical Distortion

Source: Created by and courtesy of Patrick Zinter.

Some of the most critical components in optical systems may be the lenses. There are many different methods for verifying the functionality and quality of a lens. One simple and effective method is using a ring spherometer, which is a linear dial with a bell-shaped indicator at the tip (Smith 2000). The radius of curvature of the lens can be measured by zeroing out the dial on a flat surface, then by measuring a "sag" distance when affixed to the lens under evaluation. This is a quick and useful method for many small lenses (Hilton and Kemp 2010). The second method relates to the measurement of the lens's modulation transfer function (MTF) with a test plate (made from optical glass). In this method, the lens under evaluation is placed in contact with an optical test plate above a light source. The MTF instrument then measures the number of interference fringes, confirming the lens's surface flatness, focal length, and image quality (in line pairs per millimeter) on/off the optical axis (Hilton and Kemp 2010). Another method is usually reserved for high-quality, high-precision optics and involves the use of a laser-based interferometer. A representative example is a laser interferometer that uses a precise HeNe laser (emitting at 0.6328 μm) to measure lens characteristics on a properly calibrated optical table (Hilton and Kemp 2010). A variety of other interferometers such as Twyman–Green, Fizeau, Fabry–Pérot, holographic, phase shifting, and many other types of interferometers are currently available.

Scanning electron microscopy (SEM) can also be a great lens evaluation tool (O'Shea et al. 2004). SEM uses a high-energy electron beam to generate a gamut of signals from the surface of a solid object (Science Education Research Center at Carleton College 2013). SEM measurements are ultraprecise and can produce high-spatial-resolution images. A magnification of 60,000× is readily achievable for a system like this! It is possible to 3D map the surface of a lens to micrometer tolerances with this kind of precision (O'Shea et al. 2004). Care must be taken with lens samples that are subjected to the SEM process since the high-energy electrons can damage the lens. Usually, a conductive overcoat layer is deposited on the sample to prevent damage to the optical surface (O'Shea et al. 2004). A downside to using SEM is that this method is very costly. A SEM instrument is usually in the hundreds of thousands of dollars, making it impractical for some testing facilities. Sending samples out for third-party SEM evaluation may be the most cost-effective method in this case. Figure 14.20 shows a modern Scanning Electron Microscope.

14.2.4.2 Optical Systems Test Equipment

Besides the equipment for implementing some of the optical testing techniques discussed earlier, there are specialized testing components that may be required for a given application. Some of these special tests include environmental testing, vibration and shock

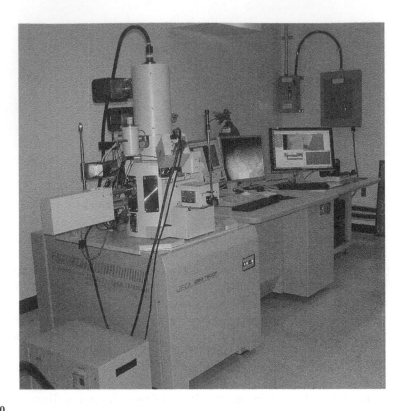

FIGURE 14.20
Phase-shifting interferometer. (From NASA ARES SEML, JEOL JSM-7600F scanning electron microscope, NASA Johnson Space Center ARES Scanning Electron Microscopes Laboratory, http://ares.jsc.nasa.gov/images/new_ares_images/SEM-JEOL-JSM7600F.jpg, 2014.)

testing, thermal cycle testing, vacuum testing, and radiation testing. Optical systems may be placed in a variety of environments, and the optical system must demonstrate that it can operate in the expected environment. Optical systems that undergo vibrations and/or shocks will typically have vibration- and shock-related requirements that need to be tested. Vibration tables and shock towers are sometimes used for this purpose. Putting the optical device in an environmental chamber that cycles the temperatures (thermal cycle tests) is often used for reliability-related testing. If an optical device needs to operate in an environment that exposes the optical device to large amounts of radiation, then radiation testing will be required to demonstrate that the optical element can function in that environment. Vacuum chambers are used to test optical elements that need to operate in a vacuum. An example would be optical systems that are launched into space. Additionally, a variety of environmental conditions may need to be simulated to demonstrate that the optical system can function in its intended environment. Things like moisture, humidity, condensation, leaks (e.g., systems put into water), dust, stray light, temperature variations, background noise, and clutter may all need to be tested to demonstrate that the optical device under test is compliant with requirements and can perform as expected.

14.2.4.3 Optical Systems Testing Methods Summary

The material discussed in Section 14.2 is intended to provide an overview of some of the well-known optical testing methods and provide examples of some of the essential test equipment needed for conducting optical systems test. The intent is not to provide any detailed optical calculations but rather to introduce the optical testing methods and present necessary tools. Any standard text on optics such as Pedrotti and Pedrotti's (1993) book, *Introduction to Optics*, will provide the basic principles of mathematical foundation. Some excellent advanced texts are the classic texts from Born and Wolf (1999), *Principles of Optics*, and Barrett and Myers (2004), *Foundations of Image Science*. So far, we have provided an overview of four common optical testing techniques: interferometry, spectrometry, radiometry, and polarimetry. We have presented some typical instruments for these methods and have seen that each instrument has a specific purpose in optical testing. We also presented a short discussion on how to apply these techniques as well as discussed some special test equipment that may be needed in the optical testing facility.

The final section of this chapter will incorporate the information we have learned in the previous two sections into a case study. This case study will integrate the systems engineering and optical systems test concepts discussed so far in a practical UAV optical systems application. The intent is to demonstrate how the previously described techniques, methods, and tools might be applied in the real world.

14.3 Integrated Case Study: Testing the FIT Optical Systems

In this section, we continue with our integrated case study and show how the systems engineering principles learned earlier are used in a simulated realistic technical setting. In this chapter, we see FIT coming to grips with the need of expanding their testing capabilities to handle some possible new work resulting from their success with the Department of Homeland Security, U.S. Customs and Border Patrol. The characters used in this scenario are Tom Phelps, FIT chief technical officer (CTO) and optics expert; Karl Ben, FIT lead systems engineer; Jennifer O. (Jen), FIT systems engineer; Ron S., FIT systems engineer (new hire); Amanda R., FIT optical engineer; Christina R., FIT optical technician; Arlene G., FIT business development; Carlos R., FIT quality manager; Phil K., FIT software engineering; Kari A., FIT test manager; Malcolm P., FIT production manager; and Julian F., FIT product service.

(It is bright and early Monday morning and Karl and James arrive at Tom's office.)

Karl: "Hey Tom, I brought Kari with me to talk to you about something if you have a few moments."

Tom: "Sure, what's up?"

Karl: "Bill just told me he is getting calls from different organizations wanting us to help them out. Apparently, they like our high-speed, high-resolution systems, and the work we did for Homeland Security, Border Patrol, and they want us to adapt our systems for their applications."

Tom: "Well, that's good news isn't it?"

Karl: "Yeah, but some of these applications Bill has been talking about have me a bit worried."

To: "How so?"

Karl: "Some of these potential customers are talking about putting our systems on the back of ATVs, trucks, balloons, underwater, in space, in helicopters, in man packs, and other platforms and I don't think we have the facilities to properly test them."

Kari: "You are right on that point. Between the development labs and our testing facilities, we have a good start. However, we are not equipped to tackle some of the tougher applications you mentioned."

Tom: "Let's take a walk over to the main optics lab and talk with Amanda. She's our lead designer and she runs the developmental labs. She'll be able to give us an idea of what we have."

Karl: "Good idea."

(After a few winding turns through the building, and down a few steps, Tom swipes into the lab.)

Tom: "Hey folks! Keep on doing what you're doing. Karl, Kari, and I just stopped by to check on some equipment. Amanda and Christina, can you take a break for a little bit, or are you in the middle of something?"

Amanda: "I have time. I was just entering some of our updated test parameters into our MBSE tool."

Christina: "I'll join you in a few minutes, I am just about finished aligning this laser for some spatial resolution experiments we are doing."

Tom: "Ok. Join us when you get to a good stopping point."

Christina: "Will do."

Karl: "Wow, this lab is spotless! What did you guys do around here?"

Amanda: "Carlos has been whipping us into shape with his 5S program."

Karl: "Ah, that reminds me, we will need to talk about updating our processes and best practices to handle the new work."

Amanda: "New work?"

Karl: "That's why we came over. Because of some prospective new customers, we are considering some potential upgrades to our test equipment and we wanted to see what we had first."

Amanda: "Like preparing the ship before the storm."

Karl: "That's a good analogy! I like it. Amanda, can you give us a quick overview and help us get some insight into what equipment we still need please?"

Amanda: "Sure, Tom. I need an Aston Martin DBS."

Tom: "Nice. I will pick one up for you the next time I am at the toy store."

Amanda: "Hah! Well, we are well stocked for conducting normal business around here. Besides a basic set of mirrors, mounts, lenses, filters, apertures, polarizers, and other basic components, we have what we need to test our detectors and optical systems that we design and develop. We have plenty of optical benches (about 5 in this lab and 4 in the one across the hall) and that doesn't include Kari's facilities. We have monochromators, several radiometers, electro-optical and acousto-optical modulators, blackbody sources, reticles, and a variety of lasers. We have a mini shake table. The one we have is nothing like the one that is over in Kari's group, but it is good enough to introduce some vibrations into the optical test environment and see how things hold together. See that's it over there."

(Amanda points to a little table with holes in the top and pneumatic vibrators attached to the legs.)

Amanda: "Kari also has a drop tower that we can use to check out shocks when we need to. Over in that corner is our ZYGO interferometer for checking out mirror flatness and lens aberrations (Zygo 2014). We have a couple of other Fizeau interferometers that we use for high spectral resolution measurements. In the cabinets on the wall over there, we have a variety of light sources, both calibrated and uncalibrated types. We have some NIST-traceable blackbody sources and also a few NIST-traceable spectroradiometers. Over on that optical bench in the corner where Christina is, we do our high-spatial-resolution experiments and tests. We have various spatial resolution masks that work with both reflected and transmitted light. That one you see on the table is the Air Force spatial resolution bar chart. We also have several targets, from simple to complex, which can be illuminated by ambient light or coherent light. We can simulate a variety of optical scenarios on that spatial resolution table. On the large bench on the opposite wall, the one that has the enclosure and light shielding, is where we characterize our detectors and determine their performance characteristics. We have coolers for the detectors and a little environmental chamber to control the temperature. It isn't like the one that we have in production that does the thermal cycles for our reliability tests but it works for our applications. In the secured lab across the hall, we have all of our atmospheric turbulence compensating test equipment. We have a basic adaptive optics system complete with a wave-front sensor and deformable mirror. We can compare some of our high-speed software-based atmospheric turbulence compensation methods to the hardware-based traditional adaptive optics system. We can simulate far-field atmospheric turbulence, near-field atmospheric turbulence, and distributed turbulence in that room. All of our signal and image processing occurs in there too, including our proprietary algorithms and our parallel processing, distributed processing, and cloud computing work. That's the majority of what we have on the hardware side. On the software side, we have our optical design programs, as well as our own analytical tools, like our integrated optics tool. We have a lot of MATLAB™, Mathematica™, ENVI™, and LabVIEW™ algorithms and GUIs as well as the stuff that the folks are coming up with in the software group. We have software to run and integrate the equipment, conduct rapid prototyping, and analyze our results. There is also our MBSE software, and our EA software. That's it on the big stuff. Of course, Kari has more in her testing facilities. She has the big stuff that's needed for production and operational test and evaluation (OT&E). We basically design and develop the system, build the prototypes, work out the kinks, and get it ready for first article production. We then transition to the Malcolm's group in production and they handle low rate of initial production (LRIP) and beyond. Kari has her own facilities that handles the big stuff."

Kari: "Yeah, like the drop tower, environmental testing, reliability tests, stress tests, on the final article and things like that."

Amanda: "What kind of customers are we talking about?"

Karl: "Well, we are in the preliminary stages at the moment but we are getting interest in adapting our systems for all kinds of applications—everything from imaging under water, to ground vehicles, to helicopters, to space systems, science stations, and even the media. We want to start planning for this new business and start to adapt smartly for the new demands."

Kari: "Wow. In that case, we will definitely need to make some new arrangements. For underwater experiments, we will need to have access to a pool, one that we can instrument. Depending on the depth, we may need a pressure chamber to simulate pressures under water."

Amanda: "Yes, and we will need blue-green lasers and build or buy watertight housing that we can fit our systems into and instrument, batteries, and communications equipment to get our data out."

Kari: "We will also need access to in situ facilities along with equipment to operate there. For space, we will need access to a radiation chamber to qualify our parts for the space environment. We also need to understand the space qualification and certification processes and make sure that we have the equipment, and training, to conduct the appropriate tests. A vacuum chamber will also be needed."

Amanda: "Tom, don't forget that the further away we get from a target, the larger the diameter of the entrance pupil needs to be. If we are talking about imaging deep space objects from the earth, or even extremely distant objects within the Earth's atmosphere, then the mirror sizes that we will deal with can increase dramatically. Our instruments work for the smaller lenses and mirrors that we have been working with so far, but we may run into a problem if the required optical elements become too large. Christina, can you think of anything else?"

Christina: "Well, the polarization equipment that we have is just useful for some very simple measurements. We may need a polarization measurement system for both remote objects and materials that can be put into a sample cell. Also, we don't have much of a multispectral, hyperspectral, or ultraspectral imaging testing capability yet. If we work with these kinds of systems, then we will need some of these special detectors and also set up the ability to collect, process, and analyze the data. The ultraspectral imaging system can generate massive amounts of data that someone will have to go through. It's quite an intensive effort and it needs to be properly planned, otherwise the data will just sit there because no one has the time, tools, or energy to deal with it. A scanning electron microscope would also be nice to have if we are going all in. If we had one of those, we could exactly see what's going on with our lenses. That's all that I can come up with off the top of my head right now."

Karl: "That's quite a nice list coming off the top of your head! Thanks Amanda and Christina. I am going to get Arlene from business development to give us an idea of the likely prospects and what we know of their needs. If you all can take that information and generate a list of what test equipment we need to support them, that would be helpful. We will plan on how to get it, or at least get access to it, to support our new customers. It's getting time for the update to our strategic plan. This will be a good opportunity to phase in some of the equipment that we need, especially if we tie it into our expanding customer base and our EA future state."

Karl: "This is a good start. Let's all meet in the conference room after lunch and discuss how we need to evolve our processes. I will invite the quality team and folks from business development, production, and support to join us. Thanks everyone."

Tom: "All right. Good job everyone!"

(Later on that afternoon, Karl's meeting is in full swing.)

Karl: "Ok, let's hear some of the ideas for improving our processes around here so that we can efficiently deal with the expected workload that we are about to get. Kari?"

Kari: "One thing we should do is move toward integrated testing. Right now, we are following the traditional testing paradigm of a developmental test and evaluation

(DT&E) followed by a separate operational test and evaluation (OT&E) activity. Integrated testing improves the complementary relationship between development and operational testing. This will typically improve overall test and evaluation efficiency. Think about it, if you wait too long to test, you could have missed a few big hiccups that you now have to go back and correct. With integrated testing, you are testing periodically, using benchmarks as stopping points to commit to test and evaluation."

Amanda: "I have seen that that simulation is a good, inexpensive form of testing. I suggest that we integrate simulation and modeling as much as possible. This would be particularly applicable for our new atmospheric turbulence compensation software. We could quickly simulate a multitude of atmospheric turbulence scenarios without having to invest in extra equipment or having to conduct separate tests."

Kari: "That's a great idea. We will have to verify and validate the models, and we will have to work with the stakeholders to make sure that the modeling and simulation results are acceptable to them as test results."

Karl: "Great. The modeling and simulation approach fits in well with our MBSE approach and should be heavily supported from DOD customers."

Christina: "I think we could use flexible test stations, which implies modular testing with simplified operating parameters. These test stations would be able to accommodate any of our UAV optical test needs, while still being able to be applied to new products in the future. Furthermore, the modularization means that tests can be recombined and altered quickly to meet our testing needs."

Phil: "The modularity idea is analogous to our modular design methodology in software. We intentionally design and package our software modules to be reusable in other programs and as self-contained as possible. We pass in parameters to the subroutines and they return results. For instance, we have a centroid module that calculates the centroid of a group of pixels. This module is very modular and has a general set of parameters that are passed into and out of the module. It allows us to calculate the centroid for a variety of different circumstances (e.g., entire image block, in a region of the image, for a particular wavelength, for a set of wavelengths, over separate image frames). There is also extensive error checking and health and status information in the routine. Other routines and other programs that need centroid information can call this routine as needed."

Kari: "We will have to look at our testing processes to see how we can streamline and modularize the processes. We may need to bring in some human factors and operational research folks to help."

Carlos: "What about the product manuals? Do we have a test strategy for that? I didn't hear anything about testing the product manuals so far."

Kari: "Well, no. We figured that once we start production, we could just leverage a lot from past product manuals and have Ron fill in the blanks. Why would we have to test the manual?"

Carlos: "This is something that is quite often overlooked. The manual needs testing just as much as the product does. Ron shouldn't be filling in the blanks after the fact. The subject matter experts, like Amanda for the optical system, should be filling in the blanks right now and making recommendations for corrections if needed. We then need to have a typical user to evaluate the manual in tandem with the UAV optical system. New guy Ron would be perfect for that role. This will make sure there isn't disconnect between the manuals intention and the perceived interpretation."

Karl: "Good point Carlos. The early review and testing of the product manuals and simulated user testing are both good recommendations. By the way, I noticed that the lab is spotless. Amanda was saying it was because of your emphasis on the 5S program. What is that?"

Carlos: "It's a program that starts with cleaning but ends with an improved work place. There are 5 steps here: sort, straighten, shine, standardize, and sustain. Each of these steps has a specific task. For example, standardization seeks to include standardized test practices, station calibration methods, production practices, and more. Doing all of the above, in a standardized fashion, can help increase interchangeability, which will increase efficiency and reduce costs."

Karl: "So what's the benefit for us to use the 5S program?"

Carlos: "Well, lots of innovative organizations have implemented 5S for the goal to improve performance and overall morale of the workplace. We think that this should apply to FIT's efforts as well. In fact, that was one of the initiatives we started this year. We have seen defects drop, as well as a huge morale boost to boot. I was wondering whether your new test efforts could incorporate the 'standardize' part of the 5S program?"

Kari: "You mean like standardization of test practices, station calibration methods, stuff like that? That shouldn't be a problem."

Malcolm: "What about clean manufacturing techniques? A clean work environment is one of the pillars of lean production. I read in some place that the workplace should be comparable to an operating room."

Karl: "Well, I tell you that I sure was impressed walking into the optics lab earlier today. I think the 5S approach will pay off. It will also make quite an impact when we parade customers through here. What about our information flow? Is there anything that we can do to improve that?"

Carlos: "Kari and I were just talking about that yesterday. We have always been pushing the idea of quick feedback, which would give the designers a chance to see test data in real time, without having to wait for a weekly or even longer feedback report."

Malcolm: "We have started a product scorecard on the production server to accomplish this instant feedback. This scorecard is automatically updated with test data from any automated stations and can be filled in by technicians for manual tests. This allows us to see exactly where we are in the process of production."

Jen: "I think it's great. But I would like to see more of a tie-in to the requirements. I read a book by Kasser, 'A Framework for Understanding Systems Engineering (Kasser 2007).' There's a section in there that might help you with this new scorecard you came up with. The section describes an approach to feedback reporting called CRIP or categorized requirements in process. There are some good methods in CRIP for requirement status reporting that I'd like to see. Specifically, the CRIP chart uses a good color scheme to show the progress of validating each requirement. Ron, I want you to read that section tonight and then create a CRIP template for the team."

Ron: "Can you elaborate more on CRIP reporting?"

Jen: "Sure. The four steps in the process are in this order: (1) categorize the requirements, (2) quantify each category into ranges, (3) place each requirement into a range, and (4) monitor the differences in the state of each of the requirements (Kasser 2007). Implementation of test solutions and their effectiveness are then reported back as an estimated project completion percentage."

Ron: "Ok, but what are the advantages of this process?"

Jen: "Well there many advantages, including easier tracking due to requirement buckets, increased knowledge of the current status of a project, and an ordering of each requirement in terms of importance. Most importantly it allows the tracking and reporting of the status of meeting customer requirements."

Ron: "Cool! Thank you."

Karl: "Well, I think we should wrap this up. There have been some great ideas presented today. Jen, can you take the lead on pulling these ideas together? I want an internal review to hammer things out and then Tom and I will bring the results to the executive board for approval and implementation. Great job everyone!"

14.A Appendix: Acronyms

QA	Quality assurance
PMD	Polarization mode dispersion
LCC	Life-cycle costing
5S	Five "S": sort, straighten, shine, standardize, and sustain
ANOVA	Analysis of variance
ANSI	American National Standards Institute
CCD	Charge-coupled device
CMOS	Complementary metal oxide semiconductor
CONOPS	Concept of operations
COTS	Commercial off-the-shelf
CRIP	Categorized requirements in process
DOD	Department of Defense
ESD	Electrostatic discharge
FIT	Case study company: Fantastic Imaging Technologies
HALT	Highly accelerated life test
HASA	Highly accelerated stress audit
HASS	Highly accelerated stress screen
INCOSE	International Council on Systems Engineering
IR	Infrared
ITEA	International Test and Evaluation Association
MTF	Modulation transfer function
NASA	National Aeronautics and Space Administration
OCR	Optical character recognition
OT&E	Operational test and evaluation
QE	Quality engineering
SDLC	Systems development life cycle
SE	Systems engineering
SOS	System of systems
T&E	Test and evaluation
TOD	Triangle orientation discrimination
UAV	Unmanned aerial vehicle

References

ANSI/ESD. 2007. *ESD Association Standard for the Development of an Electrostatic Discharge Control Program for Protection of Electrical and Electronic Parts, Assemblies and Equipment (Excluding Electrically Initiated Explosive Devices)*. ANSI/ESD S20.20-2007. Rome, Italy: American National Standards Institute/Electrostatic Discharge Association.

Antonpaar. 2013. Functionality of a Polarimeter. Wikimedia Commons. http://commons.wikimedia.org/wiki/File:Polarimeter_measuring_system.jpg (accessed December 5, 2014).

Austin, R. 2010. *Unmanned Aircraft Systems: UAV Design, Development, and Deployment*, 1st edn. John Wiley & Sons, Hoboken, NJ.

Agilent Technologies. 1996. Optical spectrum analysis basics. http://cp.literature.agilent.com/litweb/pdf/5963-7145E.pdf (accessed April 09, 2014).

Barrett, H. H. and K. J. Myers. 2004. *Foundations of Image Science*. Hoboken, NJ: John Wiley & Sons.

Blanchard, B. S. and W. J. Fabrycky. 2011. *Systems Engineering and Analysis*, 5th edn. Boston, MA: Prentice-Hall.

Born, M. and E. Wolf. 1999. *Principles of Optics*, 7th edn. Cambridge, England: Cambridge University Press.

Bunch, B. H. and A. Hellemans. April 2004. *The History of Science and Technology*. New York: Houghton Mifflin Harcourt, 695pp.

Carrico, R. J. 2009. *Back to basics–Testing, testing, 1, 2, 3*. http://asq.org/quality-progress/2009/09/back-to-basics/testing-testing-1-2-3.html. Quality Progress Magazine (online), September 2009: American Society of Quality (accessed December 5, 2014).

ColinEberhardt. 2004. The Michelson Interferometer Experimental Apparatus. Wikipedia. http://en.wikipedia.org/wiki/File:Michelson-interferometer.png (accessed December 5, 2014).

Dhillon, B. S. 2004. *Reliability, Quality, and Safety for Engineers*, CRC Press, Boca Raton, FL.

Dudzik, M. C. (ed.). 1993. *The Infrared & Electro Optical Systems Handbook*, Vol. 4: Electro-Optical Systems Design, Analysis, and Testing. Ann Arbor, MI: Infrared Information Analysis Center, Environmental Research Institute of Michigan (ERIM)/Bellingham, WA: SPIE Optical Engineering Press.

INCOSE-TP-2003-002-03. June, 2006. *Systems Engineering Handbook*, Version 3. Cecilia Haskins (ed.). San Diego, CA: INCOSE—International Council on Systems Engineering. http://www.incose.org (accessed April 09, 2014).

Gearhart, S. A. and K. K. Vogel. 1997. Infrared system test and evaluation at APL. *Johns Hopkins APL Technical Digest*, 18(3): 448–459.

Goldberg, B. E., K. Everhart, R. Stevens, N. Babbitt III, P. Clemens, and L. Stout. December 1994. *System Engineering "Toolbox" for Design-Oriented Engineers*. NASA Reference Publication 1358. Huntsville, AL: National Aeronautics and Space Administration Marshall Space Flight Center MSFC.

Hansen, D. R. and M. M. Mowen. 2006. *Cost Management: Accounting and Control*, 5th edn. Mason, OH: Thomson/South-Western.

Harkins, R. 2009. One good idea–Gimme five. http://asq.org/quality-progress/2009/08/one-good-idea/gimme-five.html, *Quality Progress Magazine* (online), August 2009, American Society of Quality (accessed December 5, 2014).

Hilton, A. R. and S. Kemp. 2010. *Chalcogenide Glasses for Infrared Optics*.

Hubble Site. 1994. Comparative view of a star before and after the installation of the Corrective Optics Space Telescope Axial Replacement (COSTAR). HubbleSite. Space Telescope Science Institute. http://hubblesite.org/newscenter/archive/releases/1994/08/image/a/ (accessed April 09, 2014).

Hull, E., K. Jackson, and J. Dick. 2005. *Requirements Engineering*, 2nd edn. London, U.K.: Springer.

Kasser, J. E. 2007. *A Framework for Understanding Systems Engineering*. Cranfield, U.K.: Right Requirement Ltd.

Komatsu, H., T. J. Felleres, and M. W. Davidson. 2000. Principles and applications of multi-beam interferometry. Nikon MicroscopyU. http://www.microscopyu.com/articles/interferometry/multibeam.html (accessed April 09, 2014).

Mishchenko, M. I., Y. S. Yatskiv, V. K. Rosenbush, and G. Videen (eds.). 2011. *Polarimetric Detection, Characterization and Remote Sensing, Proceedings of the NATO Advanced Study Institute on Special Detection Technique (Polarimetry) and Remote Sensing*, Yalta, Ukraine. September 20–October 1, 2010, Series: NATO Science for Peace and Security Series C: Environmental Security, 1st edn.

NASA ARES SEML. 2014. JEOL JSM-7600F Scanning Electron Microscope. Obtained from NASA Johnson Space Center ARES Scanning Electron Microscopes Laboratory. http://ares.jsc.nasa.gov/images/new_ares_images/SEM-JEOL-JSM7600F.jpg (accessed December 5, 2014).

National Aeronautics and Space Administration, NASA Headquarters. December 2007. NASA/SP-2007-6105: *NASA Systems Engineering Handbook*.

National Aeronautics and Space Administration MARSHALL SPACE FLIGHT CENTER. 2007. Preferred reliability practices—Contamination control of space optical systems, NASA PRACTICE NO. PD-ED-1263.

O'Shea, D. C., T. Suleski, A. Kathman, and D. Prather. 2004. *Diffractive Optics: Design, Fabrication, and Test*. Bellingham, WA: SPIE Press.

Optique Online Courses. 2009. Multibeam interferometers. http://www.optique-ingenieur.org/en/courses/OPI_ang_M06_C04/co/Contenu_08.html (accessed April 09, 2014).

Pedrotti, F. L. and L. S. Pedrotti. 1993. *Introduction to Optics*, 2nd edn. Upper Saddle River, NJ: Prentice-Hall.

Sandau, R. (ed.) 2010. *Digital Airborne Camera Introduction and Technology*.

Science Education Research Center at Carleton College. 2013. Scanning Electron Microscopy (SEM). Geochemical Instrumentation and Analysis. http://serc.carleton.edu/research_education/geochemsheets/techniques/SEM.html (accessed April 09, 2014).

Smith, W. J. 1997. *Practical Optical System Layout: And Use of Stock Lenses*. New York: McGraw-Hill.

Smith, W. J. 2000. *Modern Optical Engineering: The Design of Optical Systems*, 3rd edn. New York: McGraw-Hill.

Stigmatella aurantiaca. 2012. Fabry Perot Interferometer Diagram. Wikimedia Commons. http://commons.wikimedia.org/wiki/File:Fabry_Perot_Interferometer_-_diagram.png (accessed December 5, 2014).

Suttinger, L. T. and C. L. Sossman. 2002. Operator Action within a Safety Instrumented Function. Obtained from National Technical Information Center. http://sti.srs.gov/fulltext/ms2002091/ms2002091.html (accessed February 5, 2014).

Task Group SAS-054. 2007. Methods and Models for Life Cycle Costing. Obtained from Defense Technical Information Center. http://www.dtic.mil/cgi-bin/GetTRDoc? *AD=ADA515584* (accessed February 5, 2014).

Timeline. 2005. Crookes Radiometer. Wikipedia. http://en.wikipedia.org/wiki/File:Crookes_radiometer.jpg (accessed December 5, 2014).

Warrencarpani. 2011. Newton's rings as observed through a microscope. Wikimedia Commons. http://commons.wikimedia.org/wiki/File:20cm_Air_1.jpg (accessed December 5, 2014).

Wilson, B. 2009. Integrated testing: A necessity, not just an option. *International Test and Evaluation Association (ITEA) Journal*, (30): 375–380.

Zygo. 2014. *Laser Interferometers*, Zygo Metrology Solutions Division, Middlefield, Connecticut. http://www.zygo.com/?/met/interferometers/ (accessed April 09, 2014).

15

Optical System Use and Support

> Logistics is the most important thing in the world. It is what creates and sustains civilization. Without logistics, the world as we know it would cease to exist.
>
> **—James V. Jones (2006)**

A great deal of information has been presented in the previous 14 chapters of this text surveying the major points along the waterfall model of the systems development life cycle (SDLC). After the system has been designed, built, and tested—culminating with the previous chapter—we now enter into the operational use and systems support phase. This is a lengthy stage, which lasts the life of the product. It is the final life-cycle phase before the disposal and retirement phase, which will be the topic for the next and final chapter. In this chapter, the topics of system use, integrated logistics, service and support, and systems modification/sustaining maintenance are highlighted. During the activities performed in this phase, usage, performance, and reliability data are collected and analyzed and used as part of the next system design to allow for continual improvements of performance and reliability.

This chapter is broken into three parts. The first part covers the systems engineering (SE) management aspects—"Big M." The second part focuses on the optical signal processing subject matter expert engineering topic—"Big E." The third part attempts to illustrate the blending and application of the material covered in the previous two sections through an integrated case study of a fictitious company, Fantastic Imaging Technologies (FIT), which needs to upgrade some signal processing capabilities for one of their main clients, the Department of Homeland Security (DHS), U.S. Customs and Border Patrol. FIT developed a high-resolution optical system for use in day and night border patrol operations on a fleet of unmanned aerial vehicles (UAVs) intended for use on the U.S.–Mexican border. To begin this chapter, let us set the stage by considering what it was like for some of the first border patrol agents tasked with physically monitoring the U.S. borders and explore why a modern UAV approach makes sense.

Imagine yourself a border patrol person, in a hot, secluded area, near the border. You are alone. The only sounds heard in the night are an occasional bird or airplane flying in the distance. Suddenly, you hear a noise. Your heart starts pounding. Your mind begins to race. What was that noise? Is it an animal roaming in this isolated field? Is it someone approaching and attempting to cross the border illegally?

This was the life of Texas Ranger, Jeff Milton, as the first border patrolman from the 1880s. His job description back then is similar to the job description of the U.S. Border Patrol today. The goal or objective back then is the same as today: to protect our borders from the enemy or anyone that does not rightfully belong on our country's soil. Today, the U.S. Border Patrol serves to protect our borders from the drug cartel, illegal immigrants, terrorists, or any other potential threats to the United States and their citizens (Department of Homeland Security 2013). This is a dangerous job, and the U.S. DHS, like the defense, intelligence, and law enforcement agencies, faces these deadly kinds of threats as a normal part of their service.

With the increasing danger to those protecting our borders came the U.S. DHS increasing interest in using UAVs to protect our borders. UAVs have proven to be useful in observing ground activities without putting soldiers or pilots in danger. UAVs can carry optical sensor devices, such as cameras and forward-looking infrared (FLIR) sensors to supply images of what is happening on the ground in an area of interest. These optical devices are harder to detect than radar, which adds to the safety features and benefits of using UAVs (Schwartz et al. 1990).

UAVs combined with optical sensor devices and their ability to safely and efficiently monitor the border day and night for the border patrol department is the premise of the case study discussed in Section 15.3.

15.1 Introduction to System Use and Support

The goal of SE is to bring a product or system to life in order to meet the needs of the stakeholder(s). Once a system has been developed, it is ready to be used by the stakeholder and may require modifications and/or support. Support is typically required to help the stakeholder understand how to use the system and apply it appropriately to meet their needs. This often comes in the form of training and technical support. Support may also be in the form of preventive maintenance activities or servicing random system hardware failures. Modifications may be required because of latent design errors that do not emerge until the system is fielded. Other modifications may occur due to new insights of how to employ the technology or emerging technologies or process changes. To properly support a system, service, product, or process (system for short), SE principles and processes should be followed throughout the SDLC (cradle to grave).

15.1.1 Background and Definitions

To effectively develop a new product or system in a cost-effective manner, the complete SDLC needs to be considered. The development and final system costs are not the only costs that need to be considered. If support costs such as maintenance, warranty repair, and consumer training are not considered and factored into the cost analysis in the beginning, they can prove to be quite costly and turn an overall company profit into a loss. The SDLC process considers all phases of a product or system from initial concept to disposal (Asiedu and Gu 2010). The SDLC is divided into three phases: acquisition, utilization, and retirement/disposal/recycle, as shown in Figure 15.1.

The acquisition phase is where the stakeholder's needs are identified and concept design, preliminary design, detailed design and development, and production/construction/manufacturing are completed. It encompasses the design and development and production

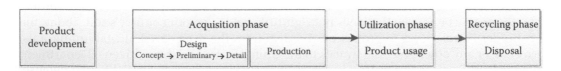

FIGURE 15.1
Systems development product life cycles. (Adapted from Gu, P. and Asiedu, Y., *Int. J. Prod. Res.*, 36(4), 883, 1998.)

phases of a system and results in a final manufactured product. Once the design has culminated into a produced system, the utilization phase is initiated.

The utilization phase is where the consumer/user takes possession of the product or system and begins to use it for its intended purposes. During this stage, the consumer/user will operate, and perhaps stress, the system and assess its performance and reliability. In addition, the user may also determine alternate uses of the system and identify new or enhanced performance needs. This information will drive system improvements, enhancements, and inspire block upgrades. A block upgrade is where a set of planned features and capabilities (blocks) are implemented in the operational system as part of preplanned product improvement (PPPI) activities.

The recycling phase is the phase where the system is retired, disposed, and, where possible, salvaged or recycled. Once a system has been deployed and is actively being used by the end user, its effectiveness is continuously evaluated to determine if system performance still matches operational requirements and needs. During this time, the possibility of system retirement or maintenance is considered, and the cost of repairs is weighed against the cost of replacement (Koopman 1999). As an example, "all embedded systems are eventually retired, discarded, or replaced. Designing the system to be retired gracefully can significantly reduce costs to the manufacturer, the user, or society as a whole" (Asiedu and Gu 2010, p. 29). Once it is determined that the system is no longer meeting its performance requirements and maintenance is no longer cost effective, the decision is made to retire the system. At this time, the system will be disposed of as appropriate based on local laws.

When developing a system, it is important to consider both utilization and retirement to ensure the proper trade-offs are made during the development process. Properly focusing on, and planning for, support and operational activities ensures that the system continues to meet stakeholder needs and has proper and cost-effective systems reliability, maintainability, and availability (SRMA) as well as logistical support throughout its useful life. With respect to the recycling phase, some countries, for which the system may be used, may have special disposal requirements and laws around material usage, known as Restriction of Hazardous Substances (ROHS) of the system. If these requirements are not met, the producer of the system may be responsible for disposing of the system. If this is not considered and properly addressed at the beginning of the SDLC, the result can prove to be an added and unexpected cost that can have dramatic consequences on a company's bottom line. It is very important to develop the system processes and logistics in parallel with the systems development as shown in Figure 15.2. Developing the manufacturing processes and configurations, support and maintenance processes, and the retirement processes all in parallel throughout the SDLC ensures they are considered early on driving systems requirements and factoring into the overall development of the product to ensure that proper system trade-offs are made. This approach can result in significant cost savings. System designs should be continuously improved using historical data collected during previous SDLCs. "Experience over many decades indicates that a properly functioning system that is effective and economically competitive cannot be achieved through efforts applied largely after it comes into being. Accordingly, it is essential that anticipated outcomes during, as well as after, system utilization is considered during the early stages of design and development" (Asiedu and Gu 2010, p. 29).

The next section will focus on the utilization phase. Figure 15.2 depicts the activities that occur during this phase. These activities include the use, support, modification, and maintenance of the system.

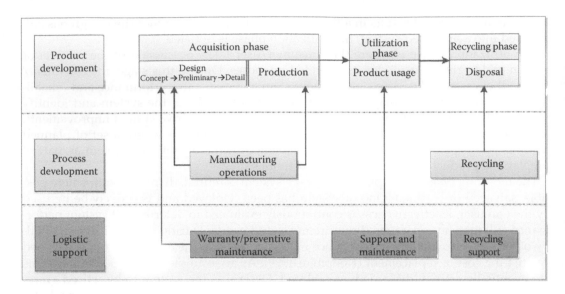

FIGURE 15.2
Parallel life cycles in parallel development. (Adapted from Gu, P. and Asiedu, Y., *Int. J. Prod. Res.*, 36(4), 883, 1998.)

15.1.2 Using, Modifying, Supporting, and Maintaining Systems

The utilization phase of a system is initiated once the design has culminated into a constructed product or system that is manufactured, accepted by the stakeholder, and distributed for use. In order to transition into the utilization stage, a formal design review is held, establishing a system baseline and verifying the system meets all of its requirements and is ready for use. Even after a system is moved into the utilization phase, changes can still be introduced to the system. Changes to the system design can be required for many reasons throughout the SDLC. For example, functional issues that are not part of the current requirements baseline or scope of work may be discovered during a design review, such as the critical design review (CDR), and design improvements might be planned to be phased in during the utilization phase (e.g., planned block upgrade[s]). Other examples for system updates are the availability of new technology or a result of compatibility issues that may arise once the system is in use. Changes may also occur as a result of misunderstanding stakeholder requirements or the applications of the system or from new requirements imposed or requested by a stakeholder. Logistics support is also required throughout the introduction and implementation of these proposed changes, to ensure that an efficient, operational system is available as required by the user.

15.1.3 Modifying Systems with the Engineering Change Proposal

Modifications or improvements occur during the utilization phase when a new need or performance deficiency is identified. When this occurs, the established baseline of the product or system is revised to meet the required changes. The baseline is defined as a "set of approved and released parameters and documentation that represents the definition of a product or system as it is designed" (Mottier 1999). Regardless of the size of the change, proper configuration change management must be adhered to. Small changes to the baseline of a system can have a large impact. Even though some changes to hardware, software, data, or a process might be initially viewed as low impact on system performance, they often have hidden impacts since the change affects the entire system. "For instance, a change in the design configuration of prime equipment (e.g., a change in size, weight, repackaging, and added performance capability) will likely affect related software, design of test and support equipment, type and quantity of spares/repair parts, technical data, transportation and handling requirements, and so on" (Blanchard and Fabrycky 2011).

A change in any single item can have an impact on another item, or it can have an impact on the overall system. At times, multiple changes are required in conjunction with one another. If more than one change is required at a given time, tracking and maintaining requirements become increasingly difficult. All changes must be carefully managed, and the required documentation must be updated, and retesting of the requirements must occur. Anytime requirements and/or the design is changed, regression analysis must be conducted to determine how the system has been impacted and what tests need to be rerun to ensure proper performance. Once the retesting indicates the new requirements have been met, the updates can be rolled out to the consumer. Some of the changes may be voluntary, while others may be required for proper, safe operation or dictated by the system contract. These added activities come as a direct cost to the developer unless it can be determined that the developer is not contractually responsible for the update. In the majority of cases, the later the change is introduced in the SDLC, the higher the costs. In general, changes are considered late if they are introduced after a particular baseline has been established. For instance, if the systems requirements specification (A spec) has been placed under configuration control, after having been approved at the systems requirements review (SRR) at the milestone 1 event, any subsequent change to the A spec is considered late.

That being said, changes are inevitable in the SDLC and so must be properly planned. These changes to the system baseline may affect various system elements, or the system as a whole. All of these changes result in life-cycle cost impacts. The key to minimizing these cost impacts is to implement changes in their proper life-cycle phase. In order to avoid introduction of changes at the wrong phase, or the incomplete implementation of a change, a formalized change process is needed. Having a formalized change process ensures traceability and reproducibility from one baseline to another and throughout the SDLC.

The engineering change proposal (ECP) is the formal process that provides governance to proposed changes and is depicted in Figure 15.3. According to the military handbook, MIL-HDBK-61A (Department of Defense 1999, MIL-HDBK-61), "an ECP is the management tool used to propose a configuration change to a Configuration Item (CI) and its Government baseline performance requirements and configuration documentation during acquisition (and during post-acquisition if the Government is the current document change authority (CDCA) for the configuration documentation)" (Department of Defense 1999, MIL-HDBK-61).

Incorporation of proposed changes must align with the requirements of the system. Any change that is made must either adhere to the requirements as written or drive a change to the requirements. It is important that the system performance meets the requirements as defined in the current revision of the requirements. To ensure this happens, the configuration control board (CCB) reviews all changes proposed through the ECP process, prior to implementation. The evaluation of each change must consider the impact(s) on the overall system. Approved changes by the board will lead to the implementation of the changes and updating of all the supporting documentation, such as requirements documents, detailed design documentation, test procedures/reports, installation instructions,

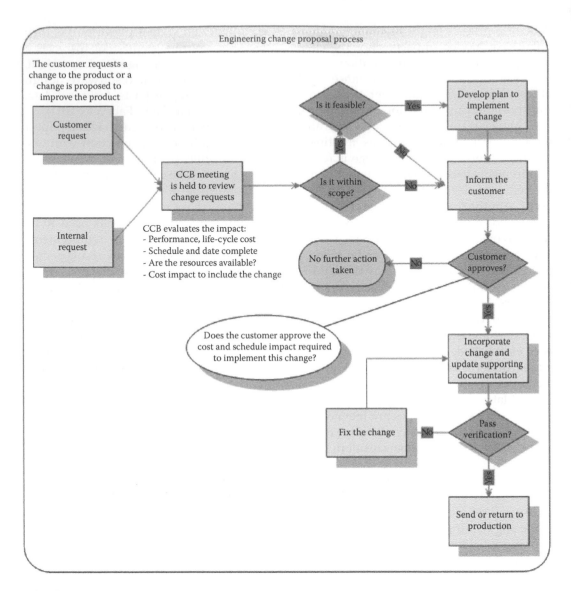

FIGURE 15.3
(See color insert.) ECP process. (Redrawn from Fabrycky, W.J. and Blanchard, B.S., *Systems Engineering and Analysis*, 5th edn., Pearson, Upper Saddle River, NJ, 2011, p. 147, ISBN 13: 978-0-13-221735-4.)

and operation manuals. "Accordingly, there needs to be a highly disciplined configuration management (CM) process from the beginning and throughout the entire system life cycle. This is particularly important in the successful implementation of the systems engineering process" (Blanchard and Fabrycky 2011, p. 147).

Configuration management (CM) facilitates the CCB and manages any changes to the documentation, hardware drawings, or software baselines, associated with the system, as approved through the ECP process. ECPs specify the details and impact of the proposed changes, with varying levels of priority and severity. They address all possible reasons for change: defects, planned product enhancements, and system improvements.

Throughout the SDLC, strict adherence to the ECP process is a must to ensure accurate documentation, properly performing systems, and consistency between requirements, design, and end product.

15.1.4 Planned Improvements

During the SDLC, there are many reasons why a system will require modification or change. To predict these changes at the beginning of systems development is difficult, if not impossible. Products and systems evolve as a stakeholder's needs/wants change, technology improves, and knowledge about the product or systems increases. Consequently, SE processes have evolved to mitigate risk associated with both planned and unexpected changes.

A combination of thoughtfully planned releases and improvement strategies is needed to overcome challenging engineering problems related to

- Requirements that are not fully defined or completely understood at program start
- Implementation of technologies that are not fully developed or implementation of technologies that are emerging
- Changing stakeholder needs, needs that have increased and been upgraded, are different, or are in flux
- Requirements changes due to modified government policies, standards, or regulations
- Requirements changes due to operational philosophy, logistics support philosophy, or other planning or practices
- Technology advancements that allow the system to perform better and/or less expensively
- Potential reliability and maintainability upgrades that make it less expensive to use, maintain, or support, including development of new supply support sources
- Service life extension programs that refurbish and upgrade systems to increase their service life (Defense Acquisition University 2001)

If safety-related issues occur, these are often unplanned and must be corrected immediately. Products and systems must be developed with the consideration of future requirements, upgrades, or anticipated changes. These necessary changes are recognized in performance and supportability upgrades throughout the SDLC and result in rebuilds or configuration changes. Correcting performance issues or improving system performance is the basis of utilization phase activities, where product use, support, and maintenance occur.

Product improvement strategies occur in different phases of the SDLC. These strategies include evolutionary acquisition, PPPI, and open systems.

15.1.4.1 Evolutionary Acquisition

Aldridge defines evolutionary acquisition as "An acquisition strategy that defines, develops, produces or acquires, and fields an initial hardware or software increment (or block) of operational capability. It is based on technologies demonstrated in the relevant environments, time-phased requirements and demonstrated manufacturing or software deployment capabilities. These capabilities can be provided in a shorter period of time, followed by subsequent increments of capability over time that accommodate improved technology and allow full and adaptable systems over time" (Aldridge 2002).

Evolutionary acquisition is an approach with a goal to deliver useful capability to the stakeholder as rapidly as possible. The focus is to deliver 80% of the required functionality. There are two main approaches to evolutionary acquisition: incremental and spiral development. With incremental development, the end product is known, but the performance or the features are phased into the system after the utilization phase has begun. Incremental development is also referred to as PPPI and is discussed in the following section. Spiral development works on the premise that the full performance requirements, or needed features, are unknown and user input is needed to provide insight into the final system configuration. Once the user starts to interface with and uses the system, this will drive additional performance requirements and system updates. When using this approach, the initial release of requirements focuses on the foundational system, with an understanding that this central piece will be built upon and evolved until the entire system has met the needs of the user. This method focuses on defining the core performance, and an upgrade path, to implement the additional features. Typically, this approach uses an open system or modular design approach to allow for facilitating the upgrades. This also allows updates in technology to be integrated into the product or system and helps control costs. Evolutionary acquisition sets the stage for planning product improvements throughout the life cycle of a product or system. This method is preferred by the Department of Defense (DOD), which resulted in the publication of DOD Directive 5000.1 (Department of Defense 2000, Directive 5000.1, 2000) and DOD Instruction 5000.2 (Department of Defense 2000, Directive 5000.2). Evolutionary acquisition ensures that the DOD remains technically advanced in their military weapons and defense mechanisms by allowing their products or systems to be upgraded or synchronized with the fast-evolving technology of today.

15.1.4.2 Preplanned Product Improvement

PPPI permits introduction of changes throughout the SDLC. This approach is also known as P3I. This strategy intentionally defers development of requirements or improvements that are difficult to achieve within the defined schedule (Pinkston 2000, p. 18). It is also used to defer the implementation of those requirements that are not fully understood or those that provide advanced features. As stated earlier, the key with this approach is to deliver 80% of the system performance to the user as fast as possible. The Army used PPPI as a "method to reach the Army goals of increased capability, increased reliability, and increased equipment life span to upgrade existing (legacy) Army systems" (Pinkston 2000, p. 18). P3I can be used when interim solutions to features and capabilities can be provided or implemented, while other portions of the system remain in the development phase. This is also sometimes referred to a staged feature release. The key to the success of this approach is well-defined interface requirements that are considered an integral part of the design. Integration of the interfaces should be completed when they become available. PPPI should be considered when the system is complex and has varying levels of performance or features. The PPPI approach is also used to implement technology advancements or simple performance enhancements. To implement these system enhancements, the key is a modular design with well-defined interfaces. The better the interfaces are defined, the easier it is to develop the replacement modules, which integrate easily and successfully into the existing system. This approach should also be considered for long-term, parallel efforts that incorporate modular equipment or open systems as part of the design.

15.1.4.3 Open Systems Approach

According to Institute of Electrical and Electronics Engineers (IEEE) POSIX 1003.0/D15 as modified by the Tri-Service OSA Working Group, Nov. 1995, the definition of an open system is

> a system that implements sufficient open specifications for interfaces, services, and supporting formats to enable properly engineered components to be utilized across a wide range of systems with minimal changes. An open system is characterized by the following (James and McFadden 2010, p. 115):

- Well-defined, widely used, nonproprietary interfaces/protocols
- Use of principles that are developed/adopted by industrially recognized standards
- Definition of all aspects of system interfaces
- Explicit provision for expansion or upgrading

The goal of the open systems design approach is to allow new technologies to easily integrate modifications into an existing system using interface management. To successfully use this approach, the design must account for future changes and anticipate technological advancements that are not currently available. The system must be designed in such a way that modifications are easily introduced and implemented, as well as, facilitate efficient absorption of technology.

> As a preferred business strategy, the open systems approach is becoming widely applied by commercial manufacturers of large complex systems. It has the attention of DOD management who has mandated its use by DOD systems developers to maintain continued superior combat capability affordably. System designs incorporating open systems concepts and principles more readily accommodate changing technology to achieve cost, schedule, and performance benefits by promoting multiple sources of supply and technology insertion (Larson et al. 2002, p. 2).

It is important to identify supportability benefits and challenges that arise from an open systems methodology. Open systems design uses an increased amount of commercial-off-the-shelf (COTS) products, provided by vendors that do not require permission to update their products. Some of the updates may improve performance, while others can change required performance. It is important to ensure awareness of any changes that occur with the COTS, by requesting updates by the manufacturer. With the success of an open systems approach being highly dependent upon interface management, this requires more rigorous interface industry standards and a need for businesses to keep abreast of the latest technology. An added benefit of the open systems use of standardized protocols and interfaces allows flexibility and increases the number of off-the-shelf solutions that can be used. This can result in improved availability and/or cost savings. Use of standardized interfaces often provides a more robust and flexible solution than an internal homegrown solution. More planning, research, testing, and data management may be required to ensure a quality product or system that is reliable and operationally available when using COTS systems.

Leveraging commercial products in an open systems approach presents new challenges to the logistics community, while offering the benefit of lower development costs. Quality assurance, CM, and data management need to anticipate how new technologies will affect supply support and properly adjust resources. Similarly, system integration and test will need to adjust resources to accommodate the introduction of new components to an existing or legacy system. If this is not properly managed, the use of COTS products can increase the overall life-cycle cost. Updating documentation, drawings, specifications, and training materials is a challenge the systems development and support teams face in an open-source approach. The documentation needs to be created with the open system design approach in mind, to enable efficient updating throughout the SDLC. Good CM is critical in the success of this approach. Properly estimating or predicting the amount of time and the cost of resources required to support these activities is challenging and many times underestimated.

Another benefit to the planned improvements approach is that it allows the more challenging stakeholder requirements to mature, prior to implementing them. Once the system is in the utilization phase, questions about usage and performance needs will become better understood, resulting in an understanding of the stakeholder require-ments and how best to fulfill them. This leads to better implementation of requirements and results in a system that meets the real needs of the stakeholders, and is not over- or underdesigned.

Planned improvements involve three different strategies categorized as evolutionary acquisition, P3I, and the open systems approach. Each of these categories uses *open systems*, which uses COTS or nondeveloped products. As a result, each of these strategies offers the benefit of lower life-cycle costs, along with presenting new challenges for the logistics support community.

15.1.5 Integrated Logistics Support

Before analyzing the challenges presented to the logistics community by using open sys-tems, logistics support must first be defined. According to the MITRE Corporation, a not-for-profit organization that applies their expertise in SE concepts and technology, logistics support is "the management and technical process through which supportability and logistic support considerations are integrated into the design of a system or equipment and taken into account throughout its life cycle. It is the process by which all elements of logistic support are planned, acquired, tested and provided in a timely and cost-effective manner" (MITRE Organization 2014). In other words, integration of changes required throughout the life cycle of a product or system are managed, planned, and supported by integrated logistics support (ILS). A strong technical background and skillset is required to effectively support all aspects of ILS throughout the SDLC. The areas supported by ILS, as depicted in Figure 15.4, are as follows:

- Training support
  - Training materials development and course management
  - Developing trainers
  - Help line or support group
- Supply or inventory management
  - Spares, repairs, or obsolete parts
- Technical documentation development
  - Creation and delivery of operator manuals, quick start guides, etc.
  - Development of maintenance or service bulletins

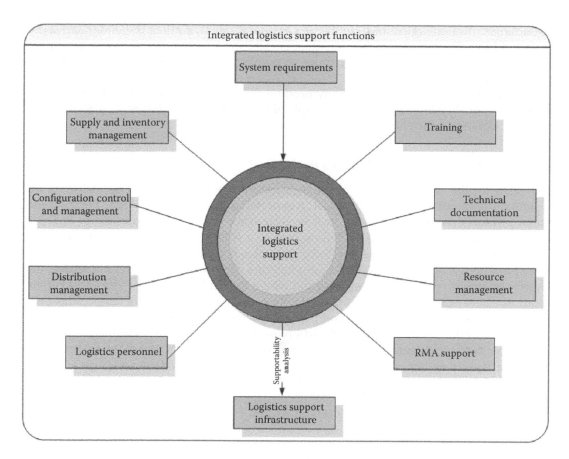

FIGURE 15.4
Integrated logistic support functions. (Jeri Feltner redrew this figure from Fabrycky, W.J. and Blanchard, B.S., *Systems Engineering and Analysis*, 5th edn., Pearson, Upper Saddle River, NJ, 2011, p. 504, ISBN 13: 978-0-13-221735-4.)

- Configuration control and management
  - Support of hardware and software baselines and changes in the form of specifications, drawings, code baselines, etc.
- Resource management
  - Management of test equipment and resources
- Distribution management
  - Packaging, handling, and transportation of the product or system
- Logistics personnel management
  - Personnel required to provide logistics and maintenance support
- SRMA support
  - Determines the probability of system/subsystem/component/assembly/part failure and associated failure rates
  - Determines required preventive and corrective maintenance
  - Determines warranty periods and what is covered

Included as input in Figure 15.4 are the SE considerations that affect the integrated logistics paradigm. To ensure proper ILS, a supportability analysis, as shown as an output at the bottom of Figure 15.4, needs to be accomplished (Blanchard and Fabrycky 2011, pp. 503–531).

Logistics are the key to success of any product or system, especially long-term complex projects. Optical systems are a good example of complex systems that rely on SE methods to fulfill logistics requirements.

15.1.5.1 Systems Support and Servicing Background

Providing for the correct amount of systems support depends largely on the nature of the system, its environment, and the needs and expectations of the stakeholders. For example, many products on the market have moved away from long-term supportability or traditional SE service levels and started falling back to a *cheaper, faster* paradigm versus either of the other two paths in the *cheaper, faster, better* development strategy. The *cheaper, faster, better* paradigm is often discussed in the engineering community, and other industries, and focuses on how to properly do business and where to find balance in product and system creation. There are numerous books and references showing that in recent history, there has been a movement away from *better* and that best practices dictate we should reincorporate better into our development strategies to decrease issues in the future. The basic concept of the *cheaper, faster, better* paradigm assumes that you can implement two of the three aspects but not all three together: cheaper, faster, and better. Lately, as stated by numerous individuals in the industry including author Michael Hammer, cheaper and faster are winning out over better, and engineers and professionals are discussing ways to change that and shift our industries back to best practices (Hammer and Hershman 2010). It begs the question on whether or not some products and systems are better served with meeting the cheaper and faster methodology. For example, with the ever-increasing technology we see today, markets such as smartphones and MP3 players may be better served with the cheaper and faster methodology. These markets deal with continuously improved technology and with a consumer market that largely wants the latest and greatest in product features. Most of the products have a 2-year life with consumers, before they trade them in on the next newer model. This is often referred to as throwaway products. These consumers would not want to pay the additional cost for longer life or logistic support past their required 2-year life. This places emphasis on time to market and removes emphasis on durability or serviceability. On the other hand, many systems such as military systems, and systems where people's lives are on the line, may be planned to be in service for long periods of time and would place emphasis on the need for durability, upgradeability, and serviceability. Part of the logistic support is to determine what level of support is required for the product or service being provided. Not all products require or even desire the same level of logistic support. Special attention must be placed on the development of the logistic support system/product requirements to ensure they meet what the consumer or stakeholder actually requires.

15.1.5.2 Integrated Logistics Support from Past to Present

Support for SE and planned logistics is not anything new. In fact, these concepts have been used throughout the ages and date back thousands of years. When people in Egypt built the pyramids, they did not build them with the thought of needing the Pharaoh to just approve the final product, they built them with the idea that they must last over the centuries for the Pharaoh to live on in the next life. In fact, quite often, it was an honor for the support staff themselves to be entombed with the pharaohs to help them in the afterlife. The Egyptian's concept was for all eternity. The logistics behind the pyramid system was incredibly choreographed and fine-tuned and included teams of foreman, thousands of workers, and a constant flow of supplies from long-distance quarries over decades, all using stones that weighed nearly 1000 lb to build pyramids that would withstand the tests of time. In fact, the pyramid builders are considered system engineers of their time, and much research has been done concerning their engineering and construction, including in-depth research on the supply chain and the remarkable continuity of the system for the production of the pyramids themselves. For example, a study done by Bill Jacobs of the University of Maryland cited the commonalities between current SE practices and axioms to that of the ancient builders (Jacobs 2002).

The idea of supportability was naturally engrained in their systems development processes. When the Romans built the aqueducts, their concern was not how cheaply and quickly they could build them. Rather, they were concerned with how to best get freshwater to their cities. These aqueducts had to be built to withstand wars and disasters and supply water from distances that spanned thousands of miles. So, once again, systems support was naturally built into the systems design of the aqueducts. The aqueducts were designed as a series of subsystems that even relate to our SE principles of today. The Romans used the classic roman arches to brace and support the aqueducts. This was such a sturdy structure that it continues to persist today and is found in many of today's building techniques and styles and is still considered one of the "super structures in existence," as seen in Mark Denny's book *Super Structures: The Science of Bridges, Buildings, Dams, and other Feats of Engineering* (Denny 2010).

Over time, and more noticeably in the last few hundred years, systems have become more and more complex. Between this and the need for faster and cheaper systems, system supportability has sometimes taken a backseat, has been left out, or is underutilized in many modern designs. Before, when systems were smaller, less complex, and did not impact so many other systems, it was easier to determine and therefore implement the necessary factors to incorporate supportability into the design process. However, our systems and technologies continue to evolve at a rapid pace, and so does our need to develop and improve our thought processes. Even Moore's law has broken out of its original technical ties to transistors on integrated circuits (ICs) and has expanded to include technological advances not only in semiconductors but also in hardware, software, and systems. Furthermore, it has the potential to redefine whole industries and economies as described in a recent report written by D.E. Liddle on *The Wider Impact of Moore's Law* (Liddle 2006).

In advancing as quickly as we do and being inventors and creators of new systems, we can no longer afford to look at simply the deliverable; we must look at the whole picture and effectively design the logistics and support structure behind it. The ability not only to support systems after being implemented but also to service these systems in the most efficient and cost-effective way possible is becoming more and more important in today's and future complex, integrated technologies.

15.1.5.3 Integrated Logistics Support Definitions

ILS as taught by the Defense Systems Management College (DSMC) is defined as "a disciplined, unified, and iterative approach to the management and technical activities necessary to (1) integrate support considerations into system and equipment design; (2) develop support requirements that are related consistently to readiness objectives, to design, and to each other; (3) acquire the required support; and (4) provide the required support during the operational phase at minimum cost" (Defense Systems Management College 1994). Furthermore, according to the DOD, the military definition is "those aspects of military operations that deal with: a. design and development, acquisition, storage, movement, distribution, maintenance, evacuation, and disposition of materiel; b. movement, evacuation, and hospitalization of personnel; c. acquisition or construction, maintenance, operation, and disposition of facilities; and d. acquisition or furnishing of services" (Department of Defense 2006, Joint Publication 1-02).

Why look at these definitions? In the United States, the defense community along with many other governmental agencies has long been strong SE proponents and has pioneered and developed much of the field's key processes, principles, and methods. Military and government products/systems are much like those of the pyramids and the aqueducts, where durability and maintenance are highly valued. Almost all documents describing systems and systems acquisitions within the government or military make reference to some form of SE and hence supportability models. For example, the U.S. Coast Guard, in their *System Integrated Logistics Support Policy Manual*, defines four iterative phases: (1) acquisition phase, defined by influencing the material, supplies, and specifications of the system, defining related support requirements, and then developing, acquiring, and delivering those resources with the initial operation of the system; (2) sustainment phase, defined as continued life-cycle cost improvements and replenishing the materials and components of the system; (3) contingency phase, defined as ensuring that the system is easily and readily duplicated and/or rebuilt; and (4) disposal phase, which ensures minimal impact to environmental and health concerns as well as ensuring proper disposal that the items do not fall into unauthorized hands (United States Coast Guard 2002). The Air Force further states that due to the improvements in capabilities for managing the Air Force supply chain in implementing ILS, they have received hundreds of documented requests based on interest for further improving the process (Tosh et al. 2009). Even the Army contains documentation on ILS stating that it "is the process that facilitates development and integration of all the logistics support elements to acquire, test, field, and support Army systems" (Department of the Army 2012).

15.1.6 Elements of Support

There are many variables that must be considered when determining how to support a system and when creating a comprehensive maintenance plan for the long-term use of the system. The current industry standards show nine elements to systems support that must be factored into the systems development and support plan. Figure 15.5 shows the collection of these elements along with some associated considerations.

These elements are not all-inclusive but provide a good starting point when working with a system. Together, they cover the basis for determining a well-rounded support plan for the system. Many elements need to be considered to properly provide ILS. From maintenance and support planning to support facilities to processes, equipment, facilities, resources, and personnel to maintain, measure, supply, train, test, handle, and transport equipment and other resources, ILS brings together these key areas and provides the necessary capabilities to ensure that systems continue to effectively meet their mission/operational requirements throughout the utilization phase.

When considering ILS for a particular system, the complexity and scale of the system drive the type of support, maintenance activity, and logistics required. For example, a simple optical system like the microscope shown in Figure 15.6 would have relatively simple support requirements.

Microscopes are common and used daily in a variety of environments, from elementary schools to chemistry labs, to medical fields across the world. They are simple optical systems

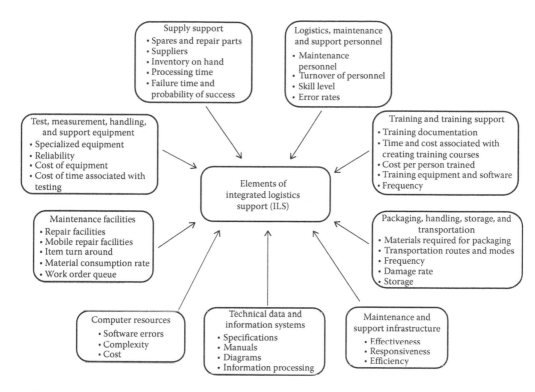

FIGURE 15.5
Elements of ILS. (Created by and courtesy of Kristine Traver.)

per the definition of optics and system, but their commonality and simplicity result in a relatively simple ILS structure. Consider a basic, simple microscope shown in Figure 15.6 from the perspective of the ILS elements in Figure 15.5. Many of the tasks, shown in Figure 15.5, simply do not need to be implemented as part of the procurement and support strategy for this small-scale optical device. For instance, from the consumer's perspective (e.g., someone who buys the microscope for a particular application), we likely would not have to build maintenance facilities for the microscope. If something happened, such as a cracked lens, we would simply replace the lens under warranty or obtain a replacement lens from some other supplier. If we had many microscopes, such as in a lab, and if one microscope broke, another might be readily available while the other one was repaired or replaced. There would not be an extensive logistics operation to support the microscope from the consumer perspective.

On the other hand, if we considered a more complex systems example, such as the Hubble Space Telescope (HST) as shown in Figure 15.7, then a more involved and detailed ILS process would be required (Loftin 1995, Mattice 2005).

The HST is one of the most recognized telescopes in space to date. Due to its size and numerous parts, plus the difficulty in physical workings of these systems, the required ILS for the HST is much more complex and involved than that for the small-scale system example of the microscope. For instance, the HST has specialized test equipment, must have redundant and reliable systems, must be space qualified, and has very specialized parts. The personnel working on it need special training, and the ability to perform maintenance is exceedingly limited! It should be helpful to the reader to keep these two examples in mind in the following sections and consider how the topics discussed would differ in application between the two systems.

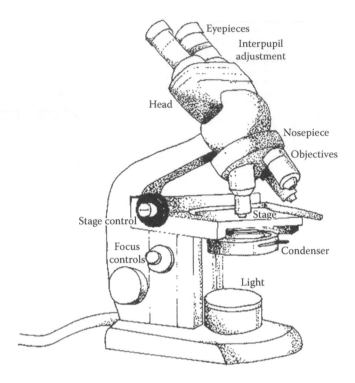

FIGURE 15.6
A labeled diagram of a compound microscope. (Obtained from P. Coleman, 1994. Wikimedia. http://commons.wikimedia.org/wiki/File%3ALabelledmicroscope.gif, accessed December 6, 2014.)

15.1.6.1 *Maintenance and Support Planning*

Maintenance and support planning is the umbrella category that addresses the planning for all of the maintenance and support activities required to sustain the system in the utilization phase. This lays the groundwork for all of the other elements of ILS. Maintenance and support planning is an iterative process that results in a detailed maintenance and support plan that defines and deals with a multitude of different scenarios. There are three useful categories to consider when conducting support planning: effectiveness, responsiveness, and efficiency.

The system support and maintenance activity has to be effective and reliable in maintaining the system at the desired level of performance. Effectiveness involves having the right resources, know-how, abilities, and motivations (e.g., tools, personnel, equipment, facilities, processes, training, and time) to accomplish the needed maintenance and support functions. If the system needs to be operable 24/7, then the effectiveness of the support must match that performance criteria, and backup or redundant systems are required. For example, a fiber-optic network must generally remain operational even when there are breaks or issues with the system equipment, since they support large groups of people, consumers, and industries. To support this, most systems have redundancies built into them deploying not only redundant fibers but also redundant optical generators (Bischoff et al. 1996). This ensures that a system remains operational when faults are detected in the main lines allowing the maintenance personnel the time necessary to fix those faults.

Responsiveness pertains to the time it takes to fix the system when an issue arises, for example, how long it takes for the redundant system to kick in and the time it takes the maintenance or repair personnel to get to the system and resolve the issue.

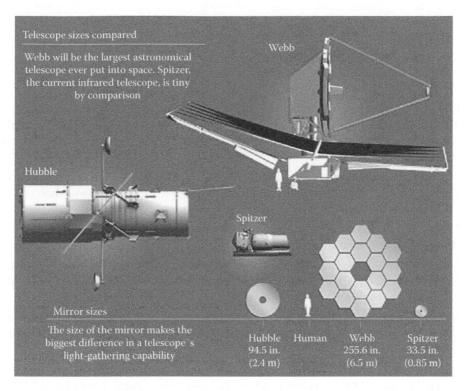

FIGURE 15.7
Space Telescope Sizes Compared. (Obtained from NASA James Webb Telescope, 2014. Flickr. http://www.flickr.com/photos/nasawebbtelescope/6802406019/sizes/o/in/photostream/, accessed December 7, 2014.)

Efficiency relates to useful work performed relative to the total work. For highly efficient maintenance activities, a high percentage of the work expended is useful (e.g., low errors, low rework, and low breakage). Efficiency can also be thought of in terms of the system itself, in that if the useful work output of the system were low compared to the repair work that is necessary to sustain the system, then it would be deemed an inefficient system and considered for possible replacement.

There are a variety of factors that determine the responsiveness, effectiveness, and efficiency of the maintenance function: factors like how long does it take to (a) discover the problem, (b) actually prepare the system/subsystem/component for maintenance, (c) get the item(s) to the maintenance facility, (d) process the item for maintenance, (e) perform the maintenance, (f) verify the maintenance action, (g) return the maintained item to the operational site, and (h) install the item (if necessary) or place into inventory for future use.

A series of useful metrics are used to quantify and address the effectiveness, responsiveness, and efficiency of maintenance and support activities. Many of these metrics have been previously discussed and are summarized here (Blanchard and Fabrycky 2011):

- *Maintenance downtime (MDT)*: This is the total time that a system is not available for operations (includes administrative delay time [ADT] and logistics delay time [LDT] as well as actual repair time).
- *ADT*: The amount of time related to administrative reasons that maintenance is not occurring (e.g., logging the system in, scheduling the maintenance, and wait time based on priority).

- *LDT*: Time it takes to get the system to the maintenance facility, waiting on parts or equipment, or resources needed to conduct the maintenance.
- *Mean active maintenance time (MAMT)*: The average time it takes to perform maintenance activities (both preventive and corrective). This is also called mean time to repair (MTTR).
- *Mean preventive maintenance time (MPMT)*: Average time it takes to conduct preventative maintenance on a system. This is a function of the frequency of preventive maintenance and the time it takes to conduct each of the preventive maintenance actions.
- *Mean corrective maintenance time (MCMT)*: Average time it takes to conduct corrective maintenance on a system. This is a function of the failure rate of each correctable item and the time it takes to repair that item.
- *Mean time between maintenance (MTBM)*: Average time between all maintenance activities (scheduled and unscheduled).
- *Mean time between replacement*: Average time between when an item is replaced. This metric affects spares and logistics support and deals with both preventive and corrective maintenance activities.
- *Mean time between failure (MTBF)*: The average time it takes for the system to fail. This is the inverse of the failure rate and is a central parameter in reliability analyses.
- *Mean time to failure (MTTF)*: Average time from the current time (time since the system was placed into service) to when the system is expected to fail (i.e., $MTTF = MTBF - t$, where t is the current time).
- *Operational availability (A$_o$)*: The probability that the system works in an operational environment. This metric is given by the MTBM divided by the sum of MTBM with MDT.
- *Inherent availability (A$_i$)*: The probability that the system works in an ideal environment. This metric is given by the MTBF divided by the sum of MTBF with MCMT.
- *Achieved availability (A$_a$)*: This is a similar definition to inherent availability but includes preventive maintenance. This metric is given by the MTBM divided by the sum of MTBM and MAMT.
- *Maintenance labor hour (MLH)*: Labor time unit applied to maintenance activities.
- *Cost-effectiveness (CE)*: Reflects how effective the system is with regard to fulfilling its mission in terms of life-cycle cost. There are a variety of figures of merit (FOMs) that are related to CE such as system benefits versus life-cycle cost, supportability versus life-cycle cost, and availability versus life cycle.

Statistical analysis can be applied to estimate these metrics, and different probability density functions (PDFs) are useful for different maintenance and support actions. For instance, the normal distribution applies for standard types of routine maintenance actions; the exponential distribution is used in methods that involve substituting parts; and the log-normal distribution is used when maintenance activities have several tasks associated with them.

It is important to remember that maintenance and support planning lays the groundwork for all others logistic and support activities. In addition, as with all other aspects of the SE life cycle, it is important to plan for maintenance during the conceptual design to ensure that it is viewed through all phases of the engineering process. During the first phase of the SE life cycle, the conceptual phase, support requirements are developed. As the process is reiterated through subsequent phases, the final outcome is a detailed maintenance plan that guides the supportability of the system.

15.1.6.2 Logistics, Maintenance, and Support Personnel

The second element to support includes logistics, maintenance, and support personnel. This includes all the personnel required for supportability of the system over the SDLC. It includes information such as the number of personnel required and their skill levels, the turnover rate for personnel, how many hours it takes to complete a given action or piece of maintenance, the error rates of those personnel, and the cost per person per organization. The maintenance strategy also needs to be determined to fully understand the number and types of personnel required. All these factors come into play when deciding and planning maintenance and personnel for system support.

15.1.6.3 Supply Support

Supply support is the third and probably one of the most important and complex factors in system support. This includes a wide range of items and possible issues. The most common issue considered is the availability of the part or the lead time. There are numerous other factors involved though, such as whether those parts are always available, how many should be kept in inventory, and how much it will cost to store those in inventory. Other factors that should be accounted for are the number of vendors that can supply the part, the processing or procurement time for parts, whether the parts will fix the issues with the system, and the turnover rates for the inventory. If we take a closer look at supply support, we can break down some of the more important factors. Supply parts are required, not only for scheduled maintenance, such as replacing rubber gaskets on telescopes, but also for unscheduled maintenance such as replacing cracked lenses.

One function of supply support is to determine initial quantities of spares for supporting the system. When determining those initial levels of inventory for a given system, the following must be considered: the quantity of spares required for corrective and preventive measures, additional stock levels to compensate for repairing items or further maintenance, additional quantities to cover for long lead times in procuring items, and additional items or parts used when an existing part is completely scrapped or determined nonrepairable. It is often challenging in determining the amount of spares you need. If you have too few spares, this will cause the MTTR and system downtime to increase, causing a less efficient system. On the other hand, if you order too many spares, you are incurring additional cost for the company such as the cost associated with purchasing and storage for the parts. If parts obsolescence is thrown into the equation, you can see why spares determination can be a difficult endeavor to pursue.

The first step to determining supply parts and inventory is to determine the probability that a system works given that there is only one failure. The assumption here is that the failure is the catalyst for the maintenance action. Recalling that the Poisson PDF is used in determining the system reliability as a function of the system failure rate, number of failures, and time, an expression for the probability that the system works given one failure is given by (Blanchard and Fabrycky 2011)

$$P = e^{-\lambda t} + (\lambda t)e^{-\lambda t}, \tag{15.1}$$

where
   λ is the failure rate
   t is the time since the item (e.g., system, subsystem, component, assembly, and part) has been placed in service

P becomes the probability of success within the system given that there is only one spare to fix one failure. For example, if the reliability of a part over the period t was determined to be 0.8 and the value of λt was 0.223, then the probability that the system works with only one failure would then be 0.9784 or 97.84%. Another way to phrase this is, given that at a certain point in time t (measured from when the system was placed in service and operated), the probability that the system works is 0.8 (80%). If it is operationally acceptable to have one or less failures in time t, then the probability of success jumps from 80% to 97.84%.

The aforementioned equation is representative of a system with one spare. Each additional spare adds a term in the Poisson expansion. If there are n items in the system that have an equal failure rate, then a generalized expression for the probability of success (e.g., the system works) for n items in the system and x spares (assuming that all x spares can be expeditiously repaired without impacting the operations) is given by (Blanchard and Fabrycky 2011)

$$P(n,x,\lambda,t) = e^{-n\lambda t} + (n\lambda t)e^{-n\lambda t} + \frac{(n\lambda t)^2 e^{-n\lambda t}}{2!} + \cdots + \frac{(n\lambda t)^x e^{-n\lambda nt}}{x!}. \tag{15.2}$$

This assumes that all spares are interchangeable. Based on this equation, it must then be determined how many parts should be stored on hand to maintain a certain probability of system operational success. Also, adjustments need to be made for lead times for obtaining parts. In other words, the number of spares kept on hand depends on how many spares are required to keep a system operational for a given probability of success, including the time it takes to get and stock the spares. Assuming that all spares are on hand and that all repairs can be made without impacting operations, the probability of success is given in compact form by

$$P(n,\lambda,t) = \sum_{x=0}^{s} \frac{(n\lambda t)^x e^{-n\lambda t}}{x!}, \tag{15.3}$$

where
   $P(n, \lambda, t)$ is the probability of success of the system
   s is the number of spares in stock
   n is the number of parts to the system (e.g., parts in the system that can fail and require the spares in stock)

All of this leads into the supply required for a given system from initialization through primary maintenance and throughout the life cycle of the system.

Finally, when looking at how much inventory to have, we need to understand the criticality of that part in the system overall and the cost of these parts. One part may render a system inoperable, while another may simply be an inconvenience. One part may cost $20, whereas another may cost $20,000. All this leads to the fact that there is a fine balance on how many spares are required for each part in order to keep the system operational for the specified time.

The complexity and size of the system will dictate the extent of parts or repairs and must be anticipated and designed into the support process.

15.1.6.4 Training and Training Support

Training and support drive the need for specialized personnel. There needs to be technicians and service personnel that are trained in all aspects of the system from general everyday maintenance to troubleshooting miscellaneous aspects. There is significant cost associated in the training of these individuals. Several questions must be considered with regard to support personnel such as how long it takes to train individuals; the number that can be trained at a given time; how often they need to be trained, or retrained, or updated on new and emerging technology regarding the system; the cost of creating a training program and all its associated documentation; what equipment needs to be used for training; what software needs to be used to train the personnel; and ultimately how much will it cost to train the required personnel. For example, a system that must conduct 24/7 operations requires more than one support technician. Even if it is a small system, a single person cannot be expected to be available at all times. Support personnel need to be available and capable of repairing the system in an efficient and effective manner to ensure the availability requirements are met.

15.1.6.5 Test, Measurement, Handling, and Support Equipment

This area includes all equipment that is used in the system for the express purposes of testing, measurement, handling, diagnostics, calibrating, support, and things of this nature. After the system is built, part of the maintenance activity is to ensure that all parts of the system operate within the specified tolerances and norms. To ensure this, special equipment is normally needed to verify these tolerances are met and that the equipment is maintained properly. To determine the quantity of test equipment required, the following elements must be taken into consideration: the availability of the specialized equipment on hand, the reliability of the equipment, the cost per test, the cost per hour of usage, and the reliability of the equipment being tested or repaired. Maintenance of test equipment needs to be considered as well, since in some cases technical calibration equipment is costly and must be properly cared for and handled. The test equipment may also require calibration and maintenance to keep it in good working order.

15.1.6.6 Maintenance Facilities

Maintenance facilities are all physical locations where maintenance and logistics processing is performed. The location, size, and layout of these facilities must be considered early on in the SDLC to ensure they are ready prior to the system entering the utilization phase. Depending on the system, it may not be possible to send the system to a maintenance facility to be repaired. These systems, due to size or where they are installed, may require maintenance to be performed on-site. Some systems may require maintenance personnel to go on-site to determine the cause of the failure and replace the faulty components. The maintenance individual may send the faulty component to a maintenance facility to be repaired. Maintenance facilities can range from vans, to shops, to laboratories and must all be accounted for in the overall supportability plan. Common items to include when designing a plan regarding such facilities would be the number of items processed in each facility, the item processing time, turnaround time, length of work queue, material consumption and consumption per period, and the cost per maintenance action.

15.1.6.7 Packaging, Handling, Storage, and Transportation

Packaging and handling along with storage and transportation can create a lot of the cost associated with a system. Many optical systems have very delicate pieces and must be treated differently from most other storage requirements. A plush toy, for instance, does not have to worry about how it is shipped as long as it is covered, but a mirrored optical system will be very sensitive to scratches and cracks, jars, and bumps, and therefore, a lot of time and effort must be put into creating packaging that will maintain it in pristine condition throughout shipping. How the items are transported is also important as costs can rise when transporting outside normal means or under expedited conditions. Planning for this element requires consideration of transportation routes and sources, the frequency of transportation, cost, packaging containers and safety or specialty containers, cost of containers and reusability, reliability of transportation, and time for transport, environmental conditions during shipping (shock and vibration, temperature and humidity), and package damage rate.

15.1.6.8 Computer Resources

Planning for supportability must also involve computer resources or more specifically computer hardware and software. It is important to consider and define the necessary hardware/software requirements, including upgradability, networks, special facilities, environmental conditions, and interfaces between the software and hardware. Computer resources also have a need for parts inventory, tracking repair history, tracking field issues, performing maintenance and repair, and updating and maintaining documentation. Hardware and software reliability needs to be considered as does software/hardware complexity and cost. Since software can often function, even with some errors, reliability is often measured by the fact that the software operates in a given environment, for a certain amount of time, without any flaws or errors.

15.1.6.9 Technical Data and Information Systems

Finally, technical data and information systems must be factored into the overall supportability model. This includes any documentation and procedures for the system. It includes technical spreadsheets or system data, reports, instructions, supplier lists, support database, parts inventory, tracking repair history, tracking field issues, updating and maintaining the unit history file, and anything else that can be considered important documentation to support the overall system. Things to reference when discussing technical data and information for supportability include the number of items per system, the format and capacity, the access time, database size, processing time, and change implementation. As an example, when you purchase technically complex items, you typically receive a manual (e.g., user's manual) for the system that provides information on how to set up, configure, and operate the system. It also contains information on how to troubleshoot the system in the event of operational issues and provides a number to call for technical support.

15.1.7 Logistics Support in the Overall System Life Cycle

In the previous section, we discussed the details of systems support and logistics. In this section, we will address the overall big picture of logistic support in the overall SDLC. The following sections will tie all specific elements together and provide an overview of the process behind system supportability and logistics. Supportability analysis is viewed as one integrated, iterative, and defining process that will lead to the overall documentation showing how the system will be supported and maintained throughout the SDLC. The two main functions of supportability analysis include (1) influencing the design of the system beyond the initial deliverable and guiding it to be flexible and supportable well after initial delivery and (2) helping in identifying logistics and maintenance resources and criteria for the system. Figure 15.8 details the supportability model in conjunction with the SDLC.

When explaining the process for performance-based logistics (PBL) through requirements, analysis, review, test and verification, and finally the support and maintenance plan, refer back to Figure 15.8 to reference where these processes are in the overall design process. As seen previously, after the supportability review, shown at the bottom right of the figure, if systems support is found to either be not feasible or incomplete, then the supportability requirements or supportability analysis process can be repeated to refine and fix any issues.

15.1.7.1 System Support Requirements

The beginning of the supportability analysis process starts, as seen in Figure 15.8, when operational and support requirements are defined for the system. At the beginning of the

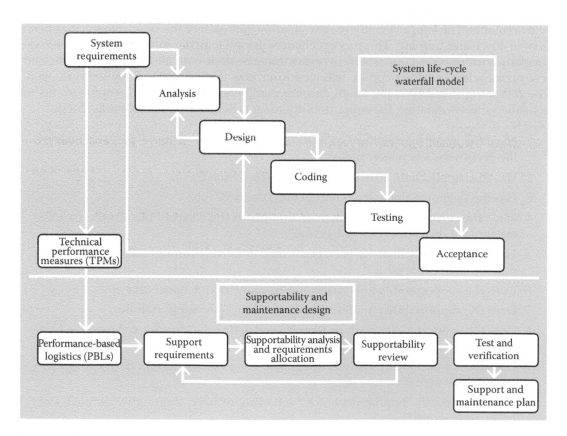

FIGURE 15.8
Supportability and maintenance design. (Created by and courtesy of Kristine Traver.)

systems development effort, the systems requirements and technical performance metrics are generated such that logistics considerations are included as part of the design considerations. This concept is known as PBL. Consequently, there are technical performance measures (TPMs) that relate to maintenance, logistics, and supportability. In turn, the PBLs lead to the creation of requirements for the various elements in the supportability and maintenance design process.

15.1.7.2 Supportability Analysis and Requirements Allocation

Supportability analysis and requirements allocation deal with taking the higher-level support requirements and transforming them into lower-level design criteria. Analysis is performed to determine how the aforementioned requirement will be met. For instance, based on the system-level logistics requirements, the number of support personnel, spares, and various time factors such as MTBR, MTTR, and skill levels must be determined. Once these factors are specified, they can be used as "design to" criteria for the lower-level design, and the lower-level requirements can be written for the various logistical decompositions.

15.1.7.3 Supportability Review

Once allocated, each requirement must be analyzed and flowed into the lower-level requirements. The supportability review ensures that each of the requirements has been flowed down and fulfilled. This process ensures that all logistics considerations have been properly addressed. A supportability review must be done over all the requirements, and a supportability checklist can be followed to ensure that nothing major is overlooked. The following checklist is not an all-encompassing checklist. It provides some points to consider when evaluating the supportability model (Blanchard and Fabrycky 2011):

1. Have the major logistics and support functions for the entire system and over its life cycle been adequately defined?
2. Has the supply chain structure been properly defined?
3. Has the system maintenance concept been defined?
4. Have the correct TPMs been defined and hence the correct PBLs for the various elements of the support infrastructure?
5. Has supportability analysis been performed iteratively throughout the design and development process?
6. Does the supportability analysis evolve from the maintenance concept?
7. Does the supportability analysis justify the design for supportability?
8. Does the supportability analysis identify and define the logistics and maintenance resource requirements for the system?
9. Does the supportability analysis integrate the various models for design and various analysis areas?
10. Have the specific requirements for each element been thoroughly defined and adequately scoped?
11. Has the correct design to requirements been specified?

These guidelines are further detailed in Blanchard and Fabrycky's (2011, pp. 530–531) book on *Systems Engineering and Analysis*. When considering supply chain management, one area to key in on, which can seriously impact systems support, is part availability. The individual part life cycles must meet or exceed the life cycle of the system that they must support. At times, this may not be possible, but it should always be a conscious decision and not left to chance.

15.1.7.4 Test and Evaluation

Once the system support analysis and design is considered feasible, then system validation is implemented. In this step, the resulting logistics, maintenance, and support structure is evaluated to make sure it meets the original supportability requirements. Supportability-related validation testing is often part of the system testing activities and can involve various types of testing (usually from type 2 [detailed design and development phase] through type 4 [utilization phase]). Some examples of the types of tests that occur include (Blanchard and Fabrycky 2011) the following:

- *Reliability qualification test*: The system elements are tested for reliability in the same environment conditions as would be expected under operational conditions.
- *Maintainability demonstration*: They are tested for logistics and maintenance tasks expected in the supported operational environment.
- *Personnel test and evaluation*: This test determines if the personnel needed for operations and maintenance tasks are adequate. Items such as number of personnel and skill levels are evaluated along with the time taken to repair given items.
- *Test and support equipment compatibility*: This test shows that the equipment used to test and calibrate the system elements will function as intended.
- *Logistics validation*: This test verifies the process for logistics-related activities like purchasing, handling, transportation, material flow, warehousing, and packaging system elements.

The further along in the SDLC, the more insight obtained and details examined. Some of the integrated testing, which requires developer/customer/supplier cooperation and interaction, can only be tested once the system is in its actual operational environment (type 4 testing).

15.1.8 Support and Maintenance Summary

This completes the first section of this chapter describing the utilization phase, where system support and maintenance occurs. This section focused on the major elements of system support and highlighted the importance of creating and maintaining a comprehensive plan for logistics support and maintenance throughout the SDLC. ILS was discussed along with PBL. The PBL-related TPMs and supportability-related systems requirements are used in the supportability analysis and allocation process to generate design criteria that evolve into the specification of lower-level logistics, maintenance, and support requirements. The section ended with a short discussion on testing and presented several different useful tests that occur later on in the systems development and continues into the utilization phase.

15.1.9 Transition to Optical Systems Building Blocks

On a daily basis, we use very complex optical systems, as a part of our everyday life. Probably the most common optical system that most everyone shares is the human eye and is illustrated in Figure 15.9.

As Figure 15.9 shows, the human eye is a system of optical building blocks consisting of a lens, retina, optic disc, optic nerve, and more. These building blocks work together to provide an optical signal to the brain, where the signal is processed and interpreted. When light hits a common object, an electromagnetic wave is reflected. If the electromagnetic wave enters the instantaneous field of view (IFOV) of the eye, it is captured and focused onto the retina. The retina converts the detected optical information into an electrical signal. This electrical signal is then sent to the brain for processing (McBride 2010). The human brain is a key building block in the eye's optical system. It receives the signal, filters it, and processes it into an image. The brain is a high-end, signal processor that processes many different types of input signals, including auditory, taste, and feelings in parallel.

Other examples of common optical systems are microscopes, binoculars, telescopes, projectors, magnifying glasses, and cameras. These optical devices use similar optical building blocks, such as lenses, detectors, and signal processing to collect and display the information in a usable form.

In this section, we will focus on the optical systems building block that is highly appropriate and important during the utilization phase—signal processing. During operations, the optical system is fulfilling its purpose in collecting data/images and transforming it into an actionable form to be used by the user/customer stakeholders.

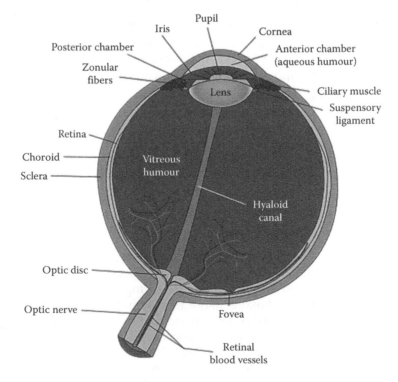

FIGURE 15.9
The eye as an optical system. (Obtained from Rhcastilhos, 2007. Wikimedia Commons. http://commons.wikimedia.org/wiki/File:Schematic_diagram_of_the_human_eye_en.svg, accessed December 7, 2014.)

15.2 Optical Systems Building Block: Using the Detected Signal–Signal Processing

Signal processing is one of the most essential building blocks in the optical system. Without it, we have no usable signal or information. If it is poorly done, we can miss essential data or introduce artifacts into the data that can obscure, clutter, or change the signal that we are trying to detect. Analog Devices made the following comment: "Signal processing building blocks are at the core of optical networks. Dense wavelength division multiplexing (DWDM) schemes and optical components that rely on high performance signal processing are enabling today's high capacity optical networks. Both active optical components (which emit, receive, and modify light) and passive components (which reroute, split, and combine light) rely on high performance signal processing semiconductors to provide the speed, precision, reliability, and cost efficiency" (Analog Devices, Inc. 2010).

Signal processing accomplishes a host of useful and necessary functions. It takes a relatively weak signal and amplifies it so that it can be detected or converted into a usable form. Signal processing is involved in converting analog signals to digital signals and vice versa, filtering unwanted components from the detected information, reducing system noise, and converting the signal for display, storage, and/or transmission. Signal processing is at the heart of routing information, formatting, modulating, encoding, and encrypting activities. Signal processing basically manipulates and transforms the currents or voltages at the back end of a detector into something we need. Image processing is signal processing performed on typically an array of sensors that has some useful spatial connection between the individual sensors in the array. Signal processing is a key building block of any optical systems.

15.2.1 Definition of Signal Processing

What is a signal? Simply put, a signal is something that conveys data, information, or instructions. Signals are everywhere and are essential parts of our everyday lives. They are used to transmit music from the car radio and sound to our ears and display images to computer monitors for professors reviewing midterm papers (Karu 1995). Signals are either discrete or continuous in relation to time. Discrete-time signals are a series of signal values that are only present at discrete points in time, as shown in Figure 15.10.

An example of discrete-time signals is a computer image composed of picture elements (pixels) (Karu 1995). Notice that the horizontal axis (time) is discontinuous and individual samples occur at discrete times. Also, notice that with respect to the vertical axis, there is no restriction of where the sample points could fall. In essence, the projection of the sampled points onto the vertical axis could intercept the vertical axis at any point. Continuous-time signals, as illustrated in Figure 15.11, are defined along a continuum of time and are often called analog signals.

Notice that in the continuous-time signal, both the vertical and horizontal axes can have points along a continuum. Discrete-time signals can be derived from continuous-time signals by sampling the continuous-time signal along the time axis. A digital signal is a discrete-time signal where time and the amplitude are both discretized, as shown in Figure 15.12.

Digital signal processing is when a signal is manipulated after it has been converted from an analog signal (analog-to-digital [A/D] conversion) into a digital form, like a number (Smith 1998). Digital signal processors (DSPs) take signals that have been digitized and manipulate them mathematically. In other words, DSPs are created to add, subtract, multiply, or divide a signal at a fast rate.

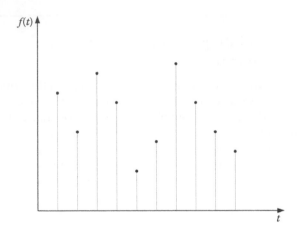

FIGURE 15.10
Discrete-time signal. (Redrawn from MIT Open Courseware.)

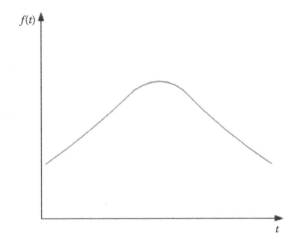

FIGURE 15.11
Continuous-time signal. (Created by and courtesy of Jeri Feltner.)

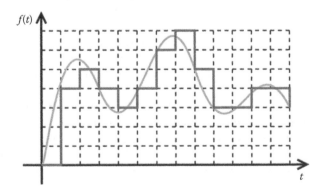

FIGURE 15.12
Digital signal. (Obtained from Wdwd. 2010a. Wikimedia Commons. http://commons.wikimedia.org/wiki/
File:Digital.signal.svg, accessed December 7, 2014.)

The IEEE Signal Processing Society provides a long description of signal processing in their organizing documents. According to the IEEE Signal Processing Society Constitution, Article II,

> Signal processing is the enabling technology for the generation, transformation, and interpretation of information. It comprises the theory, algorithms, architecture, implementation, and applications related to processing information contained in many different formats broadly designated as signals. Signal refers to any abstract, symbolic, or physical manifestation of information with examples that include: audio, music, speech, language, text, image, graphics, video, multimedia, sensor, communication, geophysical, sonar, radar, biological, chemical, molecular, genomic, and medical, data, or sequences of symbols, attributes, or numerical quantities. Signal processing uses mathematical, statistical, computational, heuristic, and/or linguistic representations, formalisms, modeling techniques and algorithms for generating, transforming, transmitting, and learning from analog or digital signals, which may be performed in hardware or software. Signal generation includes sensing, acquisition, extraction, synthesis, rendering, reproduction and display. Signal transformations may involve filtering, recovery, enhancement, translation, detection, and decomposition. The transmission or transfer of information includes coding, compression, securing, detection, and authentication. Learning can involve analysis, estimation, recognition, inference, discovery and/or interpretation. Signal processing is essential to integrating the contributions of other engineering and scientific disciplines in the design of complex systems that interact with humans and the environment, both as a fundamental tool due to the signals involved and as a driver of new design methodologies. As such, signal processing is a core technology for addressing critical societal challenges that include healthcare, energy systems, sustainability, transportation, entertainment, education, communication, collaboration, defense, and security (IEEE Signal Processing Society Constitution 2012).

A simpler explanation of signal processing is taking a signal and changing it to a different form, so data can be extracted or interpreted from the original signal. Since the beginning of time, humans have been the model signal processor. All of mankind performs signal processing when they use their brain to process data or information. The human brain is a 25 W processor that uses 10 W to process data (Taylor and Williams 2006). Figure 15.13 depicts a sound signal being transmitted and processed from one human's voice to another human's ear.

"In the real-world, analog products detect signals such as sound, light, temperature, or pressure and manipulate them" (Skolnick et al. 1995). An A/D converter uses an analog signal at its input to create an output in the digital format of 1s and 0s. Then, the DSP uses the digitized information as input, processes it, and outputs the digitized information so that it can be used for practical applications. All of these take place at very high speeds.

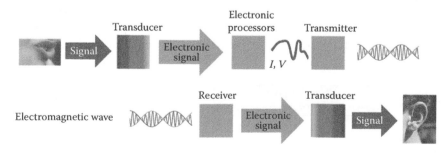

FIGURE 15.13
Signal processing. (Obtained from Brews Ohare, 2012. Wikimedia Commons. http://commons.wikimedia.org/wiki/File:Signal_processing_system.png, accessed December 7, 2014.)

Computers can use information from a DSP to directly control useful electronic and mechanical devices. In order for signals to be transmitted quickly, they need to be compressed, as done in teleconferencing. In teleconferencing, both the video and audio signals are sent over the telephone line. Signals can also be improved or enhanced by manipulating them. This allows the signal to provide information that humans are unable to detect, such as the noise cancelling ability of a good set of headphones or computer-enhanced images. Digital signal processing allows signals to be interpreted at a high rate of speed with highly accurate results.

DSPs can be used for various applications, such as filtering (Abilove 1999, Douglas et al. 1999, Skolnick et al. 1995, Wang 1982). With filtering being one of many applications of signal processing, it is easy to understand its importance as a building block for optical systems. Digital signal processing can be seen in the entertainment business, transportation, broadband communications, control systems, medical devices, video imagery, and military technology.

The military often uses analog signal processing devices to operate at optical frequencies, since analog devices can perform at high frequencies, are faster than digital systems, and can be potentially more accurate since analog signals have no discretization of the vertical and horizontal axes. These devices are used in military operations or scenarios for navigation, imaging, and radar, as seen in Section 15.3.

15.2.2 Signal Processing for Optical Systems

Detectors are used to convert optical signals into electrical analog signals. The signal that is produced by typical detectors is so minimal that it must be amplified before it can be sent to a central processing unit and onto the graphical user interface. Though this may seem as easy as putting an amplifier to subwoofers in the back of your father's car, it is really a fight against electronics themselves to ensure that the signal displayed is the signal received. This chapter will introduce some common terminology associated with signal processing, compare traditional signal processing equipment versus contemporary, provide methods for noise reduction in a signal, and discuss some current display technologies.

Quite often, it is the role of the signal processing function to take a low-level signal received at the detector, optimize the detector output signal-to-noise ratio (SNR), amplify the signal, limit its bandwidth, convert an analog signal to a digital signal, process the data, and put it into a usable format for an end device (e.g., display, graphical user interface, transmission device, storage, and/or analytical tool) as shown in Figure 15.14. Dynamic use of the system may require some system operational flexibility. Consequently, the signal processing function often must be adaptable and compatible with a given application.

We assume a linear, time-invariant system in our discussion of signal processing. For a linear system, its output is relatable to its input by application of scaling and superposition principles.

With regard to time invariance, a time delay in the inputs causes a corresponding time delay in the output of the linear system. Consequently, if we apply an input now, or t seconds later, the output will be the same in a time-invariant system. Hence, the input signal

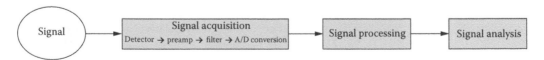

FIGURE 15.14
Signal acquisition and processing. (From Karagiannis, A. et al., Biomedical time series processing and analysis methods: The case of empirical mode decomposition additional information, In *Advanced Biomedical Engineering*, G.D. Gargiulo and McEwan, A. Eds., InTech, 2011.)

is treated as a continuum of point sources where the resulting signal is a superposition of weighted impulse responses. Assuming superposition here means a linear combination of solutions to the system is also a solution to the same linear system.

15.2.2.1 Gain

In electronics, gain is usually taken as the ratio of the output signal over the input signal and usually involves the mean of the signal values. A gain of 5 would imply that the voltage, current, or power is increased by a factor of 5. Unfortunately, the amount of gain necessary for computers and displays to cope with the output signal from the detector is many times larger than a ratio of 5 to 1; thus, it is common to use a logarithmic scale. Originally, the "bel" unit was used:

$$G_{bel} = \log_{10}\left(\frac{P_{out}}{P_{in}}\right) \qquad (15.4)$$

where
  G_{bel} is the gain in bels
  P_{out} is the power out of the system
  P_{in} is the power into the system

With the technology that exists today, it is possible to achieve much larger gain values, so the decibel is used instead

$$G_{dB} = 10\log_{10}\left(\frac{P_{out}}{P_{in}}\right) \qquad (15.5)$$

or expressed as a signal ratio

$$G_{dB} = 20\log_{10}\left(\frac{S_{out}}{S_{in}}\right), \qquad (15.6)$$

where S_{out} and S_{in} are the system output signal and input signal in terms of voltages or currents (whatever is appropriate), respectively. A similar unit using natural logarithms is called the neper.

15.2.2.2 Amplifiers

Amplifiers have gone through considerable change since the early days of electronic amplification. Among the first amplifiers were vacuum tube amplifiers, which modify signals by controlling the electron motion in an evacuated tube. When heated, a filament (cathode) undergoes thermionic emission releasing electrons into the evacuated space. As a result, an electron cloud forms in the vacuum (space charge) that has a negative charge. A positively charged metallic plate in the vacuum tube attracts the electrons, and consequently, a current is generated. This process is one directional since the positively charged plate cannot be heated.

When transistors came along, vacuum tubes were quickly replaced by this new technology. Figure 15.15 shows some examples of each.

This transition happened for similar reasons to why they were adopted in other devices; namely, the transistors can be made smaller, for less money, consume less power, run at a cooler temperature, have a faster response time, and last longer. Also, transistors are

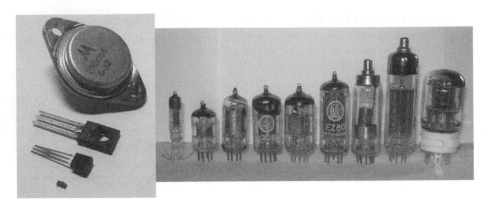

FIGURE 15.15
Vacuum tube diode and transistors. (Adapted from Stefan Riepl. 2008. *Eine Zusammenstellung von Elektronenröhren.* Wikipedia. http://en.wikipedia.org/wiki/File:Elektronenroehren-auswahl.jpg, and, Transisto. 2008. Assorted Discrete Transistors. Wikipedia. http://en.wikipedia.org/wiki/File:Transistorer_(croped).jpg, accessed December 7, 2014.)

versatile and can be used in many applications other than amplification, such as signal control, voltage stabilization, and signal modulation.

ICs brought transistors down to the nanometer scale. For example, Intel says it has manufactured a 153 Mbit static random access memory (SRAM) chip using transistor line widths of 45 nm. ICs became popular for the same reasons transistors quickly became popular.

15.2.2.2.1 Operational Amplifiers

With the invention of the IC came operational amplifiers. Operational amplifiers, or op-amps, are a great collection of electronic circuits that can perform many functions such as addition and filtering, all based on the properties of ideal amplifiers. Initially, differential amplifiers where often used as preamplifiers. An example can be seen in Figure 15.16.

The differential amplifier measures the difference between two input voltages. Operational amplifiers characteristically have an enormous input resistance and a very small output resistance. Similarly, instrumentation amplifiers are differential amplifiers that have high input impedance, a low bias current, and a programmable gain.

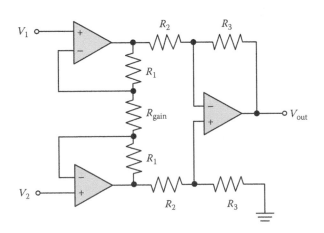

FIGURE 15.16
Op-amp instrumentation amplifier. (Obtained from Inductiveload. 2009. Wikipedia. http://en.wikipedia.org/wiki/File:Op-Amp_Instrumentation_Amplifier.svg, accessed December 7, 2014.)

15.2.2.2.2 *Instrumentation Amplifiers*

Instrumentation amplifiers are perfect for the preamplifier position in signal processing. However, one drawback is that the resistors and source impedance need to be precisely matched to get the best possible common-mode signal cancellation. The common-mode signal is the signal from the environment, such as noise, which is common to both input terminals. This type of preamplifier uses only the differential-mode signal found across the detector. Unfortunately, some common-mode noise will still make it through the instrumentation amplifier since there are natural variations in the noise signal at both terminals. Therefore, the common-mode rejection ratio (CMRR) is used to gauge how much common-mode noise has come through the instrumentation amplifier in the signal. If the following first equation is true (differential-mode voltage equation), then the second equation (CMRR) can be used:

$$v_{out} = A_{dm}(v_b - v_a) + A_{cm}\left(\frac{v_b + v_a}{2}\right), \tag{15.7}$$

$$CMRR = \left|\frac{A_{dm}}{A_{cm}}\right| = 20\log_{10}\left(\left|\frac{A_{dm}}{A_{cm}}\right|\right). \tag{15.8}$$

where
   A_{dm} is the differential-mode gain
   A_{cm} is the common-mode gain
   $CMMR$ is the common-mode rejection ratio

The first equation is obtained by summing the output voltages for the common-mode term and the differential-mode term. The second equation is just the definition of the CMRR (Holt 1978). Ideally, the CMRR would be infinite, thus making the differential-mode gain much larger than the common-mode gain. The CMRR can often be written in terms of the resistance values of the instrumentation amplifier, but this is specific to the particular instrumentation amplifier circuitry. Typically, values run from 70 dB at low frequencies to 120 dB at high frequencies (Analog Devices, Inc. 2008).

As mentioned earlier, there is a need for adjustable signal processing. For example, there may be a need for adjustable gain to compensate for the effect distance has on the target's signal. If the detector is flying toward the target source, the gain will need to be decreased for the increasingly higher levels of electromagnetic radiation received at the detector. Using an amplifier with a digitally configurable gain can set programmable gain.

Today, it is no longer required to build discrete, instrumentation amplifiers. Instrumentation amplifier ICs can be purchased off the shelf. Table 15.1 provides a representative sample list of instrumentation amplifiers, which are commercially available.

15.2.2.2.3 *Differential Amplifier Example*

One of the common signal processing functions is signal amplification. Differential amplifiers are of great utility in detector technology in that they simultaneously amplify the signal of interest and reduce the common noise in the signal. The following provides a simple example. Figure 15.17 shows a common, basic, differential amplifier that can be used to amplify the signal and reject common noise.

This differential amplifier amplifies the difference between the input terminals of an incoming weak signal while rejecting the common-mode voltage. The common-mode

TABLE 15.1

Instrumentation Amplifier ICs

Analog Devices	Texas Instrument
AD8422	INA121
AD8479	INA827
AD8232	INA333
AD8219	INA129
Linear Technology	
LT1167	
LTC1100	
LT1168	
LTC2053	

Source: Created by and courtesy of Joelle Rudnick.

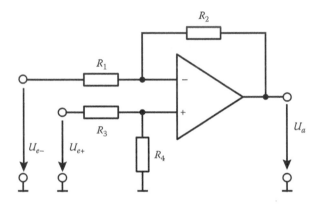

FIGURE 15.17
Differential amplifier. (Obtained from Daniel Brown. 2007. Wikipedia. http://commons.wikimedia.org/wiki/File:Differential_Amplifier.svg, accessed December 7, 2014.)

voltage consists of the bias voltage and common-mode noise. The difference between the two inputs is amplified by the gain and output by the differential amplifier. The relationship between the output signal U_a and the differential input voltages, U_{e+} and U_{e-}, is given by (Braun 2007)

$$U_a = \frac{(R_1 + R_2)R_4}{(R_3 + R_4)R_1}U_{e+} - \frac{R_2}{R_1}U_{e-}. \tag{15.9}$$

If $R_1 = R_3$ and $R_2 = R_4$, we get

$$U_a = (U_{e+} - U_{e-}) \times \frac{R_2}{R_1}. \tag{15.10}$$

As we can see in this equation, only the difference in the input terminal voltages gets amplified, and the noise, being a common factor to both input terminals, gets eliminated as a difference. One example of how this amplifier is useful is to put nonsignal-dependent noise voltages on the negative terminal and the detected signal with approximately the same noise contributions across the positive terminal. By looking at Equation 15.10, the noise is cancelled and the signal is amplified by the gain, R_2/R_1.

15.2.3 Bandwidth

Determination of the optimum bandwidth is an important decision in signal processing. An operational amplifier's bandwidth is the highest frequency it can process. When selecting the bandwidth needed, the spectral characteristics of both the signal and the noise must be considered. Figure 15.18 depicts how the bandwidth is defined for a band-pass filter with 3 dB frequencies at F_1 and F_2. For a low-pass filter, or an operational amplifier, the bandwidth is the frequency range $(0–F_2)$ at which the output drops by 3 dB. For an amplifier, the output amplitude is ideally constant until the 3 dB cutoff frequency at F_2. An amplifier output, r, which is uniform in strength and sharply cuts off within the specified window is preferred; however, the op-amp performance is generally not ideal, so we select an op-amp with the smallest feasible bandwidth.

15.2.4 Signal Conditioning

The signal leaving the detector is preamplified because the signal is small and the processing unit may be located some distance from the detector. As the signal travels to the processor, it will naturally lose some of its signal strength in the wiring to the data processor as well as encounter additional noise effects. Therefore, the typical connections are made using wires with shielding, such as a coaxial configuration, with impedance that matches the preamplifier. Depending on the number of detector elements, heavy shielded wiring may not be suitable. Instead, multiplexers can take many signals and pass them through one wire and split them up again closer to the data processor. In essence, the signal is amplified close to the detector to maximize signal to noise. For many electro-optical and infrared (IR) systems, such as on airplanes, there is a need to put the data processor somewhere else such as inside the fuselage, in order to preserve the balance and aerodynamic performance of the airplane. In cases like this, shielding the signal and amplifying the signal are essential.

Once the signal arrives at the data processor, the signal is then converted to a digital signal using an A/D converter. Figure 15.19 shows a continuous analog signal, which is sampled to obtain discrete numbers at discrete times (e.g., a digital signal). The resolution of this process is based on the number of bits representing the signal. For 12 bits, there are $2^{12} = 4096$ quantization levels. The resolution per level, or bit, for a 10 V signal is 2.44 mV. In sampling a signal, the Nyquist–Shannon sampling conditions must be met. The Nyquist–Shannon sampling theorem states that in order to avoid aliasing, reproduction

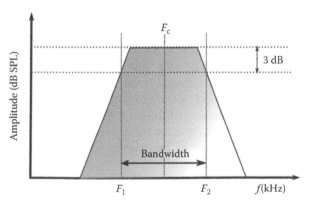

FIGURE 15.18
Band-pass filter. (Obtained from Mike Lifeguard. 2009. Wikimedia Commons. http://en.wikipedia.org/wiki/File:Band-pass_filter.svg, accessed December 7, 2014.)

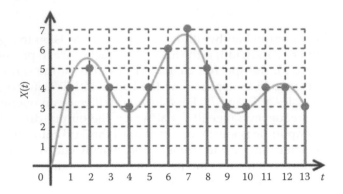

FIGURE 15.19
(See color insert.) A/D Conversion. (Obtained from Wdwd. 2010b. Wikimedia Commons. http://commons. wikimedia.org/wiki/File:Digital.signal.svg, accessed December 7, 2014.)

of an analog signal requires a sampling rate faster than twice the highest frequency of interest in the signal.

To this point, noise has been minimized to permit the maximum possible signal level. In effect, this will determine the contrast of the display, ultimately governing whether or not the target can be distinguished from its background. The display of the system is generally the final destination of the signal if it is not processed by any other data processors, stored, or transmitted to another location.

15.2.5 Displays

The laptop computer industry has driven the development of a variety of vividly contrasting displays that far exceed the performance of the common cathode-ray tube. Currently, the active-matrix liquid-crystal display (AMLCD) is a common display technology in the computer industry for laptops. This flat-panel technology has a good range in colors, responds quickly, provides good quality images, and is lightweight. The active matrix part of the technology refers to each pixel in the display having a dedicated transistor. The resulting matrix of transistors is implemented on a thin film and, when combined with polarizing and color filters, is capable of directly addressing and supplying power to each pixel of the liquid-crystal display (LCD). The transistors, along with capacitors, can also store the current state of the pixel, and state transitions can be localized to the changing pixels themselves.

Another flat-panel display technology is the plasma display panel (PDP), which uses fluorescent lamps to provide a high-quality picture. This technology uses noble gases between glass plates where a gas discharge from the plasma interacts with a phosphor to produce the image. The gases used in the PDP are an inert mixture of the noble gases xenon and neon and contain no mercury (unlike the AMLCD). Plasma displays have better viewing angles than LCDs, a higher contrast ratio, and less motion blur. Some disadvantages include higher power used than LCDs, higher weight than LCDs, potential issues with altitude, and possible RF interference.

An interesting display technology is digital light processing (DLP), which uses small controllable mirrors on a chip to form the image. Originally developed in 1987 at Texas Instruments, this technology uses a digital micromirror device (DMD) that has individually controllable, microscopically small mirrors that are in an array configuration on a semiconductor and act as pixels to produce the image. Advantages include replaceable light source technology, exceedingly long life using light-emitting diode (LED) or laser technology, and excellent contrast, and it has more colors available, does not use fluids, and is lighter than LCDs and plasmas. Some downsides include potentially using more power than plasma or LCD technologies, not as thin as LCDs and plasma technologies in rear-projection implementations, the available view angle less than LCDs and plasmas, and the response time when converting lower resolutions to high definition is longer than in their LCD and plasma counterparts.

15.2.6 Systems Engineering Tools and Techniques

Typically, the system designer for a display should have certain questions answered up front by the user:

- What is the resolution requirement?
- How big do you want your screen?
- What should the aspect ratio be?
- What is the input voltage requirement?
- What is the refresh rate requirement?
- What technology should be used in this application?
- What polarization model would be optimal?

To help with the task of simulating these requirements, Optical Research Associates have developed a software package called Lighttools. Lighttools allows a systems engineer to model a display system in 3D space through the use of computer-aided design (CAD). One example of its use could be modeling a LCD backlight for a personal digital assistant (PDA) device. This software will allow the simulation of multiple polarizations to test for contrast of the backlight. Another feature is its ability to simulate screen coatings to show what happens to the LCD's color with different coating materials. This is just one of many optical simulation tools available to a system engineer.

15.2.7 Example of Signal Processing for an Optical System

The following is an example for a generic optical detector. Radiometric calculations can be used to determine the amount of optical power that is incident on the detector. The responsivity of the detector can provide the output signal level(s) of the detector (assumed analog here in this example), and this is the starting point for the signal processing system.

15.2.7.1 Gain Calculations for the Detector

As a representative example, assume we are using a MCT-1000 preamplifier. This preamplifier provides typical gain values in the range of 50–1000. The gain is typically expressed as the ratio of the output voltage to the input voltage. Knowing the upper and lower bounds on the gain, we can determine what effect this preamplifier has on our signal in terms of power gain in decibels. Recalling Equation 15.6,

$$G_{dB} = 20 \log_{10} \left(\frac{V_{out}}{V_{in}} \right). \tag{15.11}$$

The expected lower gain provided by this preamplifier is

$$G_{dB\_LB} = 20\log_{10}(50) \tag{15.12}$$

$$G_{dB\_LB} = 33.978 \text{ dB}. \tag{15.13}$$

Similarly, the upper bound for the gain for this detector is

$$G_{dB\_UB} = 20\log_{10}(1000) \tag{15.14}$$

$$G_{dB\_UB} = 60 \text{ dB}. \tag{15.15}$$

From the these calculations, it can be concluded that a single preamplifier can amplify a signal up to 1000 times, and thus, the lower and upper bounds for the gain values are 33.978 and 60 dB, respectively. Using detector information from previous chapters,

$$D^*(9.5 \text{ μm}, 77 \text{ K}, 100) = 4.5 \times 10^{10} \text{ cm Hz}^{1/2}/\text{W} \tag{15.16}$$

$$A_d = 1 \text{ cm}^2 \tag{15.17}$$

$$\Delta f = 16 \text{ Hz} \tag{15.18}$$

Equations 15.16 through 15.18 can be used to determine the noise-equivalent power (NEP):

$$NEP = 8.9 \times 10^{-11} \text{ W} \tag{15.19}$$

Given the responsivity, the NEP can be converted into noise-equivalent voltage at the output of the detector. A representative responsivity R was given in previous chapters:

$$R = 4 \times 10^4 \text{ V/W}. \tag{15.20}$$

If we assume that we have optical power on the active area of the detector that is a factor of 10 larger than the *NEP*, we would have 8.9×10^{-10} W on the detector. Given the responsivity shown in Equation 15.20, the detector output voltage would be 35.6 μV. What signal level would this produce on the other side of the MCT-1000 preamplifier using the highest gain setting? If we use Equations 15.11 and 15.15, we get

$$60 \text{ dB} = 20\log_{10}\left(\frac{V_{out}}{35.6\,\mu\text{V}}\right) \tag{15.21}$$

$$V_{out} = 0.0356 \text{ V} \tag{15.22}$$

An output voltage of 0.0356 V is insufficient to drive a typical display, which for LCDs is on the order of volts. Consequently, additional amplification will be required to drive the display for this very low signal. In designing the amplification circuitry, the dynamic range of the detector output signal, due to variations in the detection scenario (e.g., different standoff ranges, target signal strength levels, clutter signal levels, and differences in optical system settings), must be considered and incorporated in the design. In the end, the range of expected detector output signals must be mapped into the acceptable voltage ranges to drive the display device.

15.2.8 Optical Systems Block Summary

Throughout this chapter, different concepts on signal processing and displays were introduced. The first concept dealt with the role of signal processing in an optical system. We described some of the essential electronic components needed in the signal processing circuitry. From there, gain and amplifiers were presented as a means for increasing the signal strength sampled from the working environment. Then signal conditioning concepts were presented that are useful in signal transport, as well as illustrating how to transform signals into useful information. We presented some useful expressions for understanding the performance of a commonly used differential amplifier. A short overview discussion on display technologies was presented, and we ended this section with a gain calculation example that used some of the parameters presented in previous chapters. In the next section, we apply representative material from both Sections 15.1 and 15.2 to the UAV-based optical system in our integrated case study.

15.3 Integrated Case Study: Signal Processing on the FIT Optical System

The previous sections laid the foundation for understanding SE methods for modifying, using, and supporting a system and also developed the concepts surrounding optical signal processing. This foundation will be applied to the UAV-based optical system that is the basis for our integrated case study. In this part of the integrated case study, we find our fictitious company, Fantastic Imaging Technologies (FIT), having to deal with some issues related to the signal processing subsystem as well as some related logistics and maintenance issues. The characters used in this scenario are Bill Smith, FIT chief executive officer (CEO); Karl Ben, FIT senior systems engineer; Jennifer O. (Jen)., FIT systems engineer; Lena A., FIT logistics analyst; Amanda R., FIT optical engineer; Christina R., FIT optical technician; Kari A., FIT test manager; Rodney B., FIT field service; Julian F., FIT product service; Andy N., FIT maintenance and support; Wilford Erasmus, chief of operations and acquisitions, U.S. Customs and Border Patrol, DHS; Jean H., DHS operations manager; Kyle N., DHS optical systems user; Ben G., Depot Maintenance.

(It is Wednesday, 2 p.m., and Ginny has called a special meeting of the optical system support and service team, including some members from systems engineering and select member of the design team.)

Ginny: "Thanks for coming everyone on such short notice. We have a couple of problems that have developed. It appears that there has been a sudden increase in the failure rate of our optical system in the field, and we need to get to the bottom of it. The problems seem to be in the signal processing units. There's also a problem with one of the vetted suppliers moving some of their operations to a region in the world where labor rates are cheaper. They build the circuit card that handles the noise reduction, amplification, and signal conditioning between the detector output signal and the communications system. Consequently, the spare parts for the signal processing units are taking 3 weeks to get there when they used to be overnighted. We need to figure out what's going on with those failures and how this parts delay affects things. Andy, can you call your counterpart at the depot to see how this delay is affecting them? I want to know if this shipping delay will affect some of our logistics requirements."

Andy: "Sure, I'll call Ben at the depot and see what I can find out."

Ginny: "Thanks Andy. Kari, I've asked for some of the circuit cards manufactured overseas to be sent to us so that we can check them out. I can't believe it's coincidence that we are seeing problems with our signal processing unit, and at the same time, there is a significant change in one of our signal processing component suppliers. I've also asked them to send us the signal processing units from several of the systems that are experiencing problems. I expect them to arrive later on today."

Kari: "I'll get with Andy once they arrive, and we will see if it's the amplifier card or something else in the signal processing unit."

Ginny: "Great! Karl, can the systems engineering team get together with Andy's logistics person, Lena, to see how the delay impacts our support requirements? Also, we may want to have Bill call Wilford to give him a heads-up. I want Wilford to know we are on it and that we'll get to the bottom of things as quickly as possible."

Karl: "That's a good idea. You know how things got with that production issue we had. I'm sure we all don't want a repeat of that. Also, based on our success with DHS, we are getting ready to enter contract negotiations with several new customers. We need to show that we are capable of supporting problems when they arise. This could be a great opportunity to demonstrate that."

Ginny: "That's true. OK, Kari, please let me know when you and Andy are done looking at the returned signal processing units. Let's get together again after everyone has had a chance to dig into this further. Thanks!"

(The meeting resumes 2 days later after the testing has completed, and everyone tells what he or she has learned.)

Ginny: "Welcome back everyone. Oh! Hi Bill, I didn't expect you at this meeting."

Bill: "I just popped by for a few minutes on my way out to Washington. I am meeting with some special customers. I talked with Wilford, and he told me some things that might shed some light on things. Apparently, he is in a bad mood about sequestration, and also some funding cuts that came his way. He has had to cut his maintenance and support budget by 20 percent for the short term, and he's had to consequently make some changes to his operational procedures. He also says he appreciates the quick response. Anyways, good luck in this meeting. Skype me if you need to. I have to catch my plane."

Ginny: "Thanks Bill. I'll fill you in after the meeting. Enjoy your trip to DC!"

Bill: "Thanks!"

Andy: "That explains some things. I talked with Ben over at the depot, and they are saying that the DHS operations manager, Jean H, told him that there weren't as many optical systems spares, due to budget cuts, and so they've had to make more local repairs to support the 24/7 operational tempo of the systems out there. Coupled with the fact that it is taking longer to get the amplification boards from the support vendor, the on-site technicians are getting "creative" and conducting some repairs themselves instead of sending the boards back to the vendor for refurbishing and refreshment. They're even conducting preventative maintenance on some parts that normally call for vendor service. They simply don't have the modules on hand to hot swap them because of the budget cuts and because they can't get them fast enough when something goes wrong. Since they didn't have all the detailed documentation on the amplifier circuitry, they reverse engineered it by replacing parts with equivalent components."

Kari: "That explains what we saw in the failed signal processing units. It turns out that each of the problem units had in-field repairs made that replaced the differential preamplifier and its circuitry. Further, the new units made overseas are using inferior components. The tolerances on the resistors connected to the operational amplifier used in the preamplification circuit are out of spec. They are using 10% acceptable variations in the resistance when they should be 1% or better. Let me show you the problem.

(Kari puts Figure 15.17 on the conference room's Samsung large screen and writes Equations 15.9 and 15.10 on the whiteboard.)

Kari: "Christina, over in design, pointed out that the tolerance in the resistors have to be really tight; otherwise, you don't get the cancelations in Equation 15.9 to produce Equation 15.10. The resistance values R_1 need to be equal to R_3 within 1% as do the resistance values for R_2 and R_4. If not, you don't reliably approximate Equation 15.10, and you end up allowing noise to pass through the system as well as introducing variations in the gain. For instance, if R_2 has a value of 100 kΩ and is low 10% and R_1 is 100 Ω and is high by 10%, then the gain term R_2/R_1 would drop from a nominal value of 1000 to 818 that's an 18.2% drop in the gain! Compare that to a drop in gain of 980.2 if 1% tolerances were used on the resistors. Also, more of the system noise terms make it through the differential preamp because the signal and noise voltages are being scaled by the resistance values shown on the first term on the right of Equation 15.9. Consequently, the noise voltage terms are not being effectively cancelled and are amplified through the rest of the system along with the signal itself. The gain goes down; noise goes up, hence the observed problems."

Ginny: "Wow. Do the boards from overseas have this problem?"

Kari: "Yes they do. It seems the maintenance techs at the depot used the overseas cards when they reverse engineered the signal processing boards. Since they didn't have the detailed specs for the circuit card, they replaced the resistors with similarly out-of-spec resistance components. If they were lucky, the resistance values of the replaced components were close to where they needed to be and the system worked. If not, then the system exhibited problems for the weaker signal detection scenarios."

Ginny: "Great job, let's get that information back to the depot. They will have to swap out those 10% tolerance resistors with the 1% versions. They will also need to let their supplier know of the problem. I suspect they will need to tighten up their quality control procedures. How is the delay affecting our support requirements?"

Jen: "Lena, Kyle, and I looked at that. The first point is that the vendor who is causing the delay is one of the depot's support vendors not ours. This vendor shouldn't be using out-of-spec parts, and the depot shouldn't be conducting repairs that potentially can void their warranty with us. Under the circumstances, I can see their motivation. Before the budget cuts and reduction in spares, the depot had one spare for each essential component per UAV in the fleet (50 of them). Since the budget cuts, spares are down to critical items only for 60 percent of their fleet."

Lena: "That causes a problem since our reliability-centered maintenance support is planned around a policy that the user maintains at least one spare per component per UAV. If an operational UAV has a problem with the optical system, a backup UAV takes over the mission while the problem is fixed. We can swap out and repair any of our system components because of our modular-based design principles and get things repaired and tested within 5 hours of detecting the problem (assuming parts are on hand). Our maintenance downtime (MDT) is only 5 hours, and our mean time between maintenance is 30 days. This gives us an A_o of 99.31% exceeding our A_o requirement of 98% for the optical system. However, with the loss of a good portion of the spares and the increase in logistics delay time (LDT) relative to the amplification circuit card, the MDT has increased to 24 hours, and the A_o has dropped to 96.8%. This is below the required 98% operational availability for the optical system."

Ginny: "Lena, can you explain how we have such a high A_o and how the spares fit into it?"

Lena: "Sure! If the spare is assumed to be on hand and a failure occurred in a given UAV, the part could be used in the repair. The old part could either be repaired locally, if permitted by the warranty agreements, or a new part ordered that would arrive in overnight air. The next day, the replacement part would go into our spares inventory, and we are ready for the next event. However, since a good portion of the spare parts are not in inventory due to the budget cuts, if a failure occurs, then a spare UAV has to take over the mission until a repair is accomplished and the broken system is once again operational. The MDT is going to be affected by the logistics delay time (LDT) quite significantly. Our mean corrective maintenance time may stay the same; however, the addition of the LDT affects the operational availability. Also, the inherent availability A_i suffers also because the incorrectly repaired parts decrease the mean time between failure (MTBF). If you recall, the inherent availability is the MTBF divided by the sum of the MTBF and the mean corrective maintenance time (MCMT)."

Ginny: "How does this get fixed then?"

Karl: "DHS needs to ensure there are adequate spares as per our maintenance agreements. They also have to make sure their supplier fixes the problem with the circuit cards. This will restore our original MTBF values, and we will once again meet our inherent availability requirements. As far as the operational availability requirement, DHS will need to get the LDT down significantly. The supplier will either need to manufacture and ship stateside or fix the delay issue along with their technical issue. Alternatively, DHS will have to find a new supplier. To prevent this problem and similar ones from happening in the future, perhaps, we should talk to DHS about having us do the warranty work associated with anything related to the signal processing system related to the optical system."

Ginny: "Good idea. I will mention it to Bill."

15.A Appendix: Acronyms

AMLCD	Active-matrix liquid-crystal display
CCB	Configuration control board
CDCA	Current document change authority
CEO	Chief executive officer
CI	Configuration item
CM	Configuration management
CMRR	Common-mode rejection ratio
COTS	Commercial-off-the-shelf
DHS	Department of Homeland Security
DLP	Digital light processing
DMD	Digital micromirror device
DOD	Department of Defense
DSMC	Defense Systems Management College
DSP	Digital signal processor
DWDM	Dense wavelength division multiplexing
ECP	Engineering change proposal
FIT	Fantastic Imaging Technologies
FLIR	Forward-looking infrared sensors
IFOV	Instantaneous field of view
ILS	Integrated logistics support
IR	Infrared
PBL	Performance-based logistics
PDP	Plasma display panel
Pixels	Picture elements
RMA	Reliability, maintainability, and availability
TPM	Technical performance measurements
UAV	Unmanned aerial vehicle
U.S.	United States

References

Abilove, A., O. Tuzunalp, and Z. Telatar. 1999. Real-time adaptive filtering for non-stationary image restoration using gaussian input. *Proceedings of the 7th Mediterranean Conference on Control and Automation*, Haifa, Israel, pp. 2152–2160.

Aldridge, E.C. 2002. Evolutionary acquisition and spiral development. Memorandum for Secretaries of the Military Departments. The Under Secretary of Defense, Washington, DC.

Analog Devices, Inc. 2008. Tutorial MT-042: Op-Amp CMRR. http://www.analog.com/static/imported-files/tutorials/MT-042.pdf (accessed July 15, 2014).

Analog Devices, Inc. 2010. http://www.analog.com/static/importedfiles/overviews/172112347OpNet_Brochure_8-2-02.pdf (accessed April 15, 2010).

Asiedu, Y. and P. Gu. 2010. Product lifecycle cost analysis: State of the art review. *Internal Journal of Production Research*, 36(4): 883–908.

Bischoff, M., M.N. Huber, O. Jahreis, and F. Derr. 1996. Operation and maintenance for an all-optical transport network. *IEEE Communications Magazine* 34(11): 136–142.

Blanchard, B.S. and W.J. Fabrycky. 2011. *Systems Engineering and Analysis*, 5th edn. Boston, MA: Prentice Hall.

Braun, D. 2007. Operational amplifier. http://de.wikipedia.org/wiki/Operationsverst%C3%A4rker (accessed July 15, 2014).

Brown, D. 2007. Differential amplifier. Wikipedia. http://commons.wikimedia.org/wiki/File:Differential_Amplifier.svg (accessed December 7, 2014).

Coleman, P. 1994. A labeled diagram of a compound microscope. Wikimedia. http://commons.wikimedia.org/wiki/File%3ALabelledmicroscope.gif (accessed December 6, 2014).

Defense Acquisition University (DAU). 2001. *System Engineering Fundamentals.* Fort Belvoir, VA: Defense Acquisition University (DAU) Press.

Defense Systems Management College. 1994. *Integrated Logistics Support Guide.* Fort Belvoir, VA: Defense Systems Management College.

Denny, M. 2010. *Super Structures: The Science of Bridges, Buildings, Dams, and Other Feats of Engineering.* Baltimore, MD: The John Hopkins University Press.

Department of Defense (DoD). 1999. *Military Handbook—Configuration Management Guidance.* September 30, 1997. pp. 4–12. MIL-HDBK-61.

Department of Defense. 2000. *The Defense Acquisition System.* October 23, 2000. Directive 5000.1.

Department of Defense. 2003. *Operation of the Defense Acquisition System. 2003.* 12: s.n., May 2003. Directive 5000.2.

Department of Defense. 2006. *Dictionary of Military and Associated Terms. Joint Publication 1-02.*

Department of Homeland Security. 2013. Securing America's borders. http://www.cbp.gov/xp/cgov/about/history/legacy/bp_historcut.xml (accessed December 11, 2013).

Department of the Army. 2012. *Integrated Logistics Support, Army Regulation 700-127.* Washington, DC: Department of the Army.

Douglas, S.C. 1999. Introduction to adaptive filters. K.M. Vijay and D.B. Williams (eds.) *Digital Signal Processing Handbook*, p. 18. Boca Raton, FL: CRC Press LLC.

Fabrycky, W.J. and B.S. Blanchard. 2011. *Systems Engineering and Analysis*, 5th edn. Upper Saddle River, NJ: Pearson, pp. 147 and 504, ISBN 13: 978-0-13-221735-4.

Gu, P. and Y. Asiedu. 1998. Product life cycle cost analysis: State of the art review. *Internal Journal of Production Research*, 36(4): 883–908.

Hammer, M. and L.W. Hershman. 2010. Faster Cheaper Better, The 9 Levers for Transforming How Work Gets Done. New York, NY: Crown Business Publishing.

Holt, C.A. 1978. *Electronic Circuits.* New York: John Wiley & Sons.

IEEE. 2012. Signal Processing Society Constitution, Article II. http://www.signalprocessingsociety.org/about-sps/governance/constitution/ Piscataway, NJ: IEEE Signal Processing Society (accessed December 6, 2014).

Inductiveload. 2009. Op-Amp Instrumentation Amplifier. Wikipedia. http://en.wikipedia.org/wiki/File:Op-Amp_Instrumentation_Amplifier.svg (accessed December 7, 2014).

Jacobs, B. 2002. *Were the Ancient Egyptians System Engineers? How the building of Khufu's Great Pyramid Satisfies System Engineering Axioms.* Rockville, MD: B & I Computer Consultants, Inc.

James, A. and LTC W.J. McFadden, II. 2010. Open systems: Designing and defining our operational interoperability. http://www.dau.mil/pubscats/pubscats/AR%20Journal/ARJ53/Ash53.pdf (accessed February 14, 2014).

Jones, J.V. 2006. *Integrated Logistics Support Handbook*, 3rd edn. New York: Sole Logistics Press McGraw-Hill.

Karagiannis, A., P. Constantinou, and D. Vouyioukas. 2011. Biomedical time series processing and analysis methods: The case of empirical mode decomposition. In: *Advanced Biomedical Engineering*, G. Gargiulo (Ed.), InTech, DOI: 10.5772/20906. Available from: http://www.intechopen.com/books/advanced-biomedical-engineering/biomedical-time-series-processing-and-analysis-methods-the-case-of-empirical-mode-decomposition.

Karu, Z. 1995. *Signals and Systems Made Ridiculously Simple.* Cambridge, U.K.: ZiZi Press, ISBN: 0-9643752-1-4.

Koopman, P. 1999. Life cycle considerations. Carnegie Mellon University, Pittsburgh, PA. Spring, 18-849b Dependable Embedded Systems.

Larson, A.G., C.K. Banning, and J.F. Leonard. 2002. An open systems approach to supportability. p. 2. http://www.dtic.mil/docs/citations/ADA404574 (accessed February 14, 2014).

Liddle, D.E. 2006. The wider impact of Moore's law. *IEEE Solid-State Circuits Newsletter*, 11(5): 28–30.

Lifeguard, M. 2009. Band-pass filter. http://en.wikipedia.org/wiki/File:Band-pass_filter.svg (accessed December 7, 2014).

Loftin, R. 1995. Training the Hubble space telescope flight team. *IEEE Computer Graphics and Applications*, 15(5): 31–37.

Mattice, J.J. 2005. *Hubble Space Telescope Systems Engineering Case Study*. Wright-Patterson AFB, OH: Center For Systems Engineering at the Air Force Institute of Technology.

McBride, D. 2010. The human eye and vision. *Modern Miracle Medical Machines*. http://web.phys.ksu.edu/mmmm/student/vision.pdf (accessed February 12, 2014).

MITRE. 2014. *Systems Engineering Guide*. https://www.mitre.org/publications/systems-engineering-guide/systems-engineering-guide (accessed July 13, 2014).

Mottier, M. 1999. Configuration management—Change process and control. Revision 1.1, November 16, 1999, EST/ISS Document, p. 5. LHC-PM-QA-304.00.

NASA. 2014. Space Telescope Sizes Compared. Obtained from Flickr. http://www.flickr.com/photos/nasawebbtelescope/6802406019/sizes/o/in/photostream/ (accessed December 7, 2014).

Ohare, B. 2012. Signal Processing. http://commons.wikimedia.org/wiki/File:Signal_processing_system.png (accessed December 7, 2014).

Pinkston, D. 2000. P3I BAT preplanned product improvement: A simulation-based acquisition that meets the army's 2020 vision. Innovations in Acquisition. November–December 2000, p. 18.

Rhcastilhos. 2007. The eye as an optical system. http://commons.wikimedia.org/wiki/File:Schematic_diagram_of_the_human_eye_en.svg (accessed December 7, 2014).

Riepl, S. 2008. *Eine Zusammenstellung von Elektronenröhren*. Wikipedia. http://en.wikipedia.org/wiki/File:Elektronenroehren-auswahl.jpg (accessed December 7, 2014).

Schwartz, C.E., T.G. Bryant, J.H. Cosgrove, G.B. Morse, and J.K. Noonan. 1990. A radar for unmanned air vehicles. *The Lincoln Laboratory Journal*, 3(1): 119–143.

Skolnick, D., N. Levine, M. Byrne et al. 1995. *A Beginner's Guide to Digital Signal Processing*. http://www.analog.com/en/content/beginners_guide_to_dsp/fca.html (accessed February 12, 2014).

Smith, D. 1998. ARRL—The National Association for Amateur Radio. http://www.arrl.org/dsp-digital-signal-processing (accessed April 15, 2013).

Taylor, F. and A. Williams. 2006. *Electronic Filter Design Handbook*, 4th edn. New York: The McGraw-Hill Companies, Inc., ISBN: 978007141718.

Tosh et al. June 22, 2009. *The Free Library*. S.v. Improving air force enterprise logistics management tools. Retrieved from http://www.thefreelibrary.com/Improving+Air+Force+enterprise+logistics+management+tools.-a0212035133 (accessed December 6, 2014).

Tosh, G.R., A.Y. Briggs, R.K. Ohnemus et al. 2009. Improving Air Force enterprise logistics management tools. *Air Force Journal of Logistics*, 55–60.

Transisto. 2008. *Assorted Discrete Transistors*. Wikipedia. http://en.wikipedia.org/wiki/File:Transistorer_(croped).jpg (accessed December 14, 2014).

United States Coast Guard. 2002. *System Integrated Logistics Support (SILS) Policy Manual*. Washington, DC: U.S. Department of Transportation.

Wang, C.D. 1982. Adaptive spatial/temporal filters for background clutter suppression and target detection. *Optical Engineering*, 21(6): 216033.

Wdwd. 2010a. Digital Signal. http://commons.wikimedia.org/wiki/File:Digital.signal.svg (accessed December 7, 2014).

Wdwd. 2010b. File:Digital.Signal.svg. http://commons.wikimedia.org/wiki/File:Digital.signal.svg (accessed December 7, 2014).

16

Disposal and Retirement of Optical Systems

What goes up must come down.

—Attributed to Sir Isaac Newton

16.1 Introduction

The emphasis of most projects has been on getting them into the operational phases. Sometimes, however, not much consideration is given to what happens to projects once they have reached the end of their operational life. A program's current schedule and cost pressures may foster a near-term focus and impact proper consideration of end of life (EOL) issues. This chapter addresses retirement and disposal considerations for optical systems from a systems engineering point of view. We will use a space-based optical system as a complex, representative example and illustrate both conventional and unique aspects of system retirement and disposal. Space-based optical systems have unique challenges that require careful planning and so serve as a good example for a difficult retirement and disposal activity. Sometimes, simply disposing of the item is not sufficient and recovery actions must be undertaken. Some reasons for the need to recover the system (or parts of the system) are as follows: return of vitally important artifacts; evaluation and assessment of onboard data, specimens, or test results; protection and retrieval of classified information/trade secrets; and safety reasons. While the focus of this chapter concentrates on space-based optical systems, much of what is discussed herein can apply to complex earth-bound optical systems.

This chapter will first introduce some generic topics related to system retirement and disposal, including various methods of disposal and many factors to consider when planning for or performing EOL activities. Next, the chapter will discuss the engineering activities related to system disposal as applied to an optical satellite system example. Finally, the chapter will attempt to combine the generic system disposal topics and specific engineering activities using an integrated case study, which involves Fantastic Imaging Technologies (FIT), a hypothetical company involved in designing, developing, and supporting high-end optical systems for the commercial and government sectors.

16.1.1 System Retirement and Disposal

The disposal or retirement of a system is defined by the Department of Defense (DOD) as "the process of redistributing, transferring, donating, selling, abandoning, destroying, or other disposition" of property (U.S. Department of Defense 1997). This definition shows us that there are multiple options to consider for a system's EOL. For both earth-bound and space-based systems, there are challenges in planning for and executing a system's disposal process. Engineers must consider these challenges from the outset of the project.

Earth-bound systems, for example, are usually physically accessible through all phases of the system's life cycle. However, a satellite, for example, is practically inaccessible for two of its systems development life cycle (SDLC) phases: the operations and support phase and the phaseout and the retirement and disposal phase. Inaccessibility to the system provides for unique challenges for systems engineers, the design and development team, and the operational support, and disposal teams as well. The SE process model functional flow block diagram (FFBD) in Figure 16.1 lays out a recommended path for the entire SDLC in the form of an FFBD.

The terminology used for the naming convention of the blocks may be slightly different between authors but the general concepts identical. For instance, some authors aggregate Blocks 1.0 through 6.0 and call this the acquisition phase while calling Blocks 6.0 through 8.0 the utilization phase (Blanchard and Fabrycky 2011). We have put these blocks in separate gray ovals to emphasize this grouping. Others use the six traditional life-cycle phases: (1) conceptual design phase, (2) preliminary design phase, (3) detailed design and

System engineering process model
Level 1—functional flow diagram

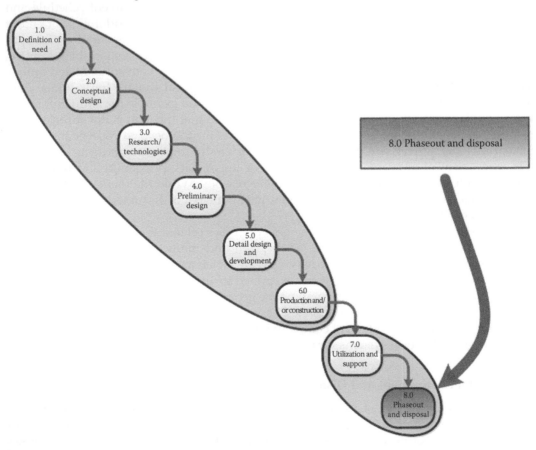

FIGURE 16.1
SE process model FFBD. (Created by and courtesy of Marty Grove.)

development phase, (4) production, manufacturing, and construction phase, (5) operations and support phase, and (6) retirement and disposal phase. Some authors use utilization and support instead of operations and support in Block 7.0. Other authors call Block 7.0 product use and support. We found the term utilization and support appropriate and to our liking since this term implies usage of an operational system, product, process, service, or combination of these things, while "support" is a complementary sustaining activity. Our use of the word utilization in Block 7.0 is not the same thing as other author's use of the term utilization phase for Blocks 7.0 and 8.0, and this will be clear in the context of our discussions. Since there are various terms, classifications, and groupings in practice today, we urge some flexibility in this regard.

As shown in Figure 16.1, there is a noticeable space between Blocks 6.0 and 7.0 in the SDLC model. This space is a representation of the difficulty of interacting with a satellite system, after it has been deployed in orbit. As such, Blocks 1.0 through 6.0 must include appropriate planning for the utilization and support phase (Block 7.0) through the phaseout and disposal phase (Block 8.0), which can be significantly more difficult and expensive depending on the system's accessibility. During these final phases, a system will undergo either a predefined disposal process or an assessment period to determine which disposal method to utilize.

In this chapter, our emphasis is on the last phase of the SDLC process, the system retirement, or phaseout and disposal phase shown by Block 8.0 in Figure 16.1. The next section will introduce some possible disposal methods for earth-bound systems and space-based systems alike.

16.1.2 Methods of System Disposal

Systems engineers have multiple options to consider when dealing with the EOL of their system. The different disposal methods can be divided into categories. These categories group methods based on the final state of the system after disposal activities are completed. The system might be left abandoned, placed in long-term storage, reused, recycled, or destroyed.

16.1.2.1 Abandonment

Abandoning a system is not a common practice planned for earth-bound systems, but it is a practice that is often considered among space satellite systems. Abandoning a system on earth is not as simple as turning it off and leaving it behind. Items must be properly disposed of. For example, aging aircrafts that are no longer airworthy are typically abandoned in areas often referred to as aircraft boneyards. These are similar to automotive junkyards where previous owners can relinquish control of the item to the owner of the junkyard or boneyard for them to do as they see fit. Abandoning a space-based satellite system is a bit more complicated. This usually involves maneuvering the satellite's orbit into a higher-altitude orbit referred to as a graveyard orbit and halting communication. Another option is to drive the satellite out of earth's orbit on a path that sends the system into deep space or into the sun.

16.1.2.2 Long-Term Storage

Placing a system or parts of a system in long-term storage is another common method of disposal. This is typically used when systems are too valuable to destroy, too complicated to reuse, too dangerous to abandon, or when no other means of disposal are practical. When placed in storage, careful consideration is made to avoid or lessen the degradation of the system as time passes. One example of long-term storage is placing the system in a museum. Decommissioned aircraft and spacecraft, such as the SR-71 Blackbird and the Space Shuttle Endeavour, are systems, which became museum exhibits. Even before the system enters its EOL, there are instances when parts of a system are placed in long-term storage. Such is the case with nuclear power plants. The spent nuclear fuel rods are processed to reduce the volume of waste and then stored in special packages and shielded to contain the radiation.

16.1.2.3 Reuse

Reusing an old system is often times a smart option for systems engineers to plan for during development. This method is typically used as a cost or time saving measure. It can also be a way to reduce risk when developing a new system. Also, reusing a system or components of a system will reduce the amount of material that has to be disposed of by some other means.

16.1.2.4 Recycle

In recent years, recycling has become an effective means commonly used to dispose systems. Recycling is a process that changes materials in such a way to prevent waste, reduce the consumption of new raw materials, reduce energy usage, and reduce pollution emissions. Recycling is also "an important means of reducing disposal costs and increasing total product value" (Blanchard and Fabrycky 2011). To effectively recycle a system, careful planning must be made to ease product recovery, ease disassembly, promote acceptance of recycled materials, and ensure that non-recoverable materials are disposed of in an environmentally friendly manner. There are many ways to recycle a system. Metals like aluminum or gold can be reclaimed. Plastics can be melted and reformed. The entire systems can be cleaned and sunk to the bottom of the ocean to become new marine habitats. Even subsystems can be used in new ways within a new system.

16.1.2.5 Destruction

Destroying a system is usually a last resort method of disposal. There are countless ways for a system to be demolished. Systems can be incinerated, pummeled, smashed, compacted, exploded, or even disintegrated. All of these options and more are available for the destruction of earth-bound systems. The U.S. Navy has been known to use all sorts of vessels for target practice. They have used derelict boats and even radio-controlled boats for testing their weapons systems and practicing their skills. Fewer options are available for space-based systems. Satellites can be destroyed in space by using a deorbit burn or ejected from orbit on a path toward the sun. Another, more costly, option is to attempt a recovery mission to bring the satellite back to earth. Once back on the surface, all the various disposal options can then be considered.

16.1.3 Considerations at the End of the Systems Life

At the EOL of a system, proper consideration must be given for the safe and environmentally friendly disposal and retirement of the system. As part of this consideration, a variety of risk factors must be addressed that can cause challenges for the phaseout and disposal team. If these risk factors are not properly addressed, complications can arise in the phaseout and disposal process that can lead to increased costs, safety issues, or even early termination of the mission. Some risks that require proper assessment are discussed in the following sections.

16.1.3.1 Corroded Metals

Corrosion is a gradual degradation of a material caused by reactions with the environment. There are different types of corrosion, such as oxidation, galvanic corrosion, and microbial corrosion. Corrosion can lead to the reduction of various physical properties, like structural integrity, electrical conductivity, or appearance. Corrosion is most common among metals and there are many prevention methods to consider in an effort to reduce risks during the utilization and support as well as the system retirement and disposal phases. Aluminum and magnesium alloys are lightweight and are used extensively in cases where weight is an important factor. However, they are highly reactive and therefore susceptible to corrosion. Plating or painting metals is often used to reduce corrosion vulnerabilities by shielding the reactive metal from the environment. Such is the case with aluminum and magnesium. Both metals are highly reactive, but their lightweight properties are very desirable. Therefore, they are extensively used in airborne and satellite fabrication applications with some form of coating applied to reduce the risk of corrosion. Sea vessels often use a sacrificial material to protect the steel hull from corrosion. Care must also be taken with the coating itself since some coatings can be bad for the environment or humans. For instance, cadmium in products and coatings is highly toxic and is a known carcinogen (OSHA 2014). Proper planning during early design will reduce the risk of corrosion and possible failures during system disposal.

16.1.3.2 Electronic Components

Electronic components are produced using various materials and are prevalent in an optical system's infrastructure. Properly addressing the disposal of hazardous materials associated with these electronic components is a critical consideration at the end of the optical systems life. It is also important, from a security standpoint, if some of these components (e.g., read-only memories [ROMs], random access memories [RAMs], and electrically erasable programmable ROMs [EEPROMs]) contain sensitive data. A modular design can make it possible to salvage electrical components and shift the goal of EOL activities from destruction to reclamation.

16.1.3.3 Liquid Metal Penetration

The penetration of liquid metals into the gain boundaries of metals can cause solid metals to become extremely brittle. The penetration typically requires some form of stress for the atoms of the liquid metal to enter the solid metal. That is not the case for certain aluminum alloys in contact with mercury. Liquid mercury is used as an integral substance in a satellite damping system. A damping system absorbs energy and helps stabilize mechanical aspects of the satellite. Besides being a hazardous material, one issue with liquid mercury is that it permeates the silver alloy joints of certain stainless steel tubing (type 304). With proper planning and design, the risk of liquid metal penetration can be reduced and complications during operations and disposal could be mitigated.

16.1.3.4 Structural Problems

Predicting what sort of structural problems might develop over the life cycle of an optical system is difficult. These problems could impose difficulty when deploying a disposal method. Two possible problems that might occur are fatigue and thermal expansion/contraction. Fatigue is a major concern because of the vibration and shock effects experienced by the subsystems, components, and parts while the optical system is being launched into space. Microscopic structural cracks can form in the material because of the forces and stresses experience during launch. Thermal expansion or contraction occurs when temperature changes cause the expansion or contraction of matter and consequently a change in the volume of the system. Differential thermal expansion or contraction occurs when the volume of bonded materials change at different rates. The rate at which different materials change is represented by their coefficient of thermal expansion. These defects might be relatively unimportant in systems on the ground but may become points of concern in systems in space.

16.1.3.5 Low-Pressure Environment

In a 1976, the U.S. Committee on Extension to the Standard Atmosphere published a book, which defined a mathematical model of the earth's atmosphere up to 1000 km (U.S. Committee on Extension to the Standard Atmosphere 1976). Using data derived from this mathematical model, Figure 16.2 shows how atmospheric pressure is affected by increasing altitude. Note that at a height of 35,000 ft, the pressure is less than ¼ of the pressure at sea level.

Operating in low-pressure environments has its challenges. There is a force that causes matter to move from high-pressure areas to lower-pressure areas. This force is most noticeable in liquids and gases and can be a point of concern while preparing for EOL activities.

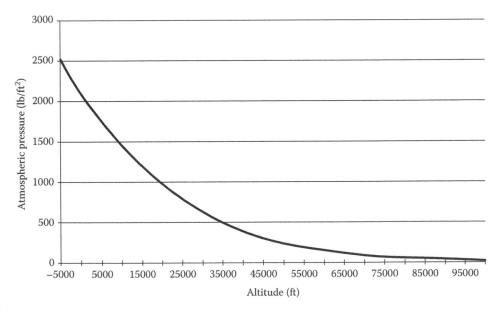

FIGURE 16.2
Altitude and atmospheric pressure. (Created by and courtesy of Blake Wilczek.)

Lubricants are one example of the possible liquids found in many systems. Various components in a satellite system require lubrication in order to perform properly. For example, deploying solar panels; changing optical properties such as the optical system magnification, the instantaneous field of view (IFOV), or the bore sight; and gimbal functions of the optical system all involve moving parts that require lubrication. Maintaining stability of the lubricating materials over long periods of time presents challenges, especially in a low-pressure environment like space. Losing lubricants due to pressure gradients can cause components to seize and hinder operations and disposal activities.

Gases can be stored in containers, frozen in solids, or suspended in liquids. Gases are most susceptible to the forces caused by pressure gradients, and reliable seals will keep the gases contained. When working in extremely low-pressure environments, outgassing is a different sort of concern than just sealing pressure vessels. Outgassing occurs when gases, which have been trapped in a material, are released. A common example is when carbon dioxide is released from a carbonated beverage when opened. It is often difficult to keep a low-pressure environment hermetically sealed. Consequently, materials such as solvents, polymer resins, plasticizers, and moisture can escape into the environment. These released materials can degrade performance or even cause failure in surrounding system components (Babecki and Frankel 2009). For example, materials released into the surrounding environment may attach to the surface of optical elements impacting their performance.

Power subsystems and fuel sources also require proper consideration when it comes to a satellite systems EOL. Proper containment for associated hazardous materials and dealing with escaping materials are challenging issues that must be planned for during the design and development phases. If not properly attended to, escaping materials during mission operations can threaten the mission itself, possibly degrade performance, and/or interfere with system disposal and retirement activities. Also, unrecovered material in space or resulting artifacts left in space because of poor or failed disposal mechanisms can pose a threat to future operations in space. Careful consideration and design must occur to ensure that subsystems remain operational throughout their expected life and that the containment and disposal subsystems work when needed.

16.1.3.6 Hazardous Materials and Special Handling

Most systems are composed of a variety of materials. Often, various satellite subsystem components have materials that are hazardous or need special handling throughout their life cycle and during disposal and system retirement activities. "Hazardous and solid waste management activities must comply with all applicable Federal, State, and local regulations" (NASA 2011). For example, the Guidelines and Assessment Procedures for Limiting Orbital Debris (NASA 1995) provide guidance in mitigating space debris when space systems reach their EOL. Some hazardous materials and their uses in space systems are as follows:

- Beryllium—a material that replaces aluminum
- Hydrazine, monomethylhydrazine (MMH), and nitrogen tetroxide (NTO)—needed for rocket fuel
- Nickel–hydrogen (NiH_2), lithium ion (Li-ion), lithium thionyl chloride (LiSOCl), hydrogen, nickel–cadmium (NiCd), and nickel–hydrogen (Ni-H_2)—used in storing power
- Radioisotopes—used in radioisotope power systems (RPSs), radioisotope thermoelectric generators (RTGs), and radioisotope heater units (RHUs)

The proper disposal of hazardous materials used by earth-bound systems is also a major concern. Many fluids used by complex systems pose a threat to the delicate balance of the ecosystem. Allowing oil, lubricants, fuel, or other synthetic fluid to be released into nature can damage all levels of the ecosystem and may take years, if not longer, to repair. Consequently, these fluids are typically drained and sent to facilities that specialize in their disposal. Motor oils can be refined and used as fuel for heaters, distilled into a diesel fuel, or reused as another type of lubricant.

Some of the driving factors behind the proper disposal of hazardous materials are various government regulations. Companies can face sizable fines if found in violation of the rules. Also, consumer opinion can be affected by the disposal methods a company uses to deal with hazardous materials. Therefore, proper planning can help to avoid damaging operations and the environment, avoid government fees, and avoid negatively affecting customer opinion.

16.1.3.7 Disposal Costs

The cost of disposal can be a limiting factor of the success of EOL activities. For many companies, paying for disposal might be considered a burden because it rarely returns a directly visible profit. An exception of course is companies that specialize in disposal processes and make this their business focus. However, spending the money to properly dispose of materials can be beneficial to a company's reputation. In recent years, companies have taken pride in earning the right to be called "green." A "green" company is environmentally friendly and pays the extra costs to ensure materials are properly disposed of or efficiently recycled. This "green" label is often used as an advertising tool or a means to increase prestige. With proper planning, a company can ensure that funds are available to successfully complete any disposal activities, while simultaneously protecting the environment, people, and leaving a "green" legacy.

Disposal cost varies based on the nature of the system and should be well thought out. For example, the disposal costs of space-based optical systems are typically higher than their terrestrial counterparts. The space-based systems' location prevents easy access and simple disposal of the components. Recently, satellite developers have given increasing attention to disposal considerations for their systems. In the past, satellite systems were often used until the end of their operational life and then abandoned in orbit. Afterward, these satellites would reenter the earth's atmosphere and hopefully burn up, or they would impact with another satellite or space junk and create more debris in low earth orbit (LEO).

Secondary systems have been developed to aid in the disposal of satellites that have reached their EOL. Some of the more exotic and expensive approaches require a separate launch vehicle to retrieve components and/or safely capture/deorbit/maneuver the satellite in a safe manner. A more practical solution is to include the disposal features in the satellite design and development.

A requirement already exists for satellite developers and manufacturers to provide for sufficient fuel for disposal activities and so the cost impact of the disposal system may not be so severe. As an example, an Italian vendor called D-Orbit produces a disposal system that attaches to the satellite system itself. At the satellite's EOL, this device safely deorbits the satellite so that it burns up in the earth's atmosphere. The CEO of the company Luca Rossettini states that because of the extra fuel requirement, "although the D-Orbit device represents additional mass to add at launch, it is not necessarily an additional cost" (Emanuelli 2013). However, it is interesting to note that at the geosynchronous earth orbit (GEO), the European Space Agency (ESA) estimates that 70% of the satellites in GEO orbit do not perform any EOL disposal maneuvers (Emanuelli 2013). Some reasons for this are that the fuel that was intended for EOL maneuvers ends up being used to extend operations or was lost because of system leaks. In any case, accounting for disposal of satellites in the design and development phases can mitigate potential disposal problems and corresponding financial impacts incurred at the satellite's EOL.

16.1.3.8 Security Considerations

Ensuring that an optical system is secure throughout its life cycle is often a driving requirement and a primary consideration for the design and development team. An optical system can collect a variety of sensitive or even classified data and therefore precautions need to be taken in safeguarding this information. Also, the construction of the optical system itself may employ trade secrets and require protection. There are four security areas that need consideration for satellite systems: "launch and control sites, communication links, electromagnetic spectrum (EMS), and the vulnerability of the satellite platforms themselves" (Military Technology—MILTECH 2008). These are areas of concern for any automated system. Launch and satellite control facilities are located at fixed sites and are therefore more susceptible to physical and online threats. Loss of control of any automated system can be catastrophic, especially if disposal procedures are preprogrammed and able to be activated remotely.

16.1.3.9 Historical Database

Another important issue is the historical database. In some cases, program data archives are maintained at a separate, remote facility. This site may contain sensitive data and often critical program information collected over an extended period of time. The amount of data collected depends on the available storage and the type, amount, and frequency of the program collection activity. The historical database often contains the entire mission data collected by the sensors and appropriate support information and is organized for easy retrieval with modern archival methods and security protocols. These archival methods provide comprehensive search tools, permit compression and encryption of data, and facilitate the organization of the data.

16.1.3.10 Transition to Optical Systems Building Blocks

Space-based systems pose severe and unique disposal challenges at their EOL such as the harsh space environment and the extreme distant location of the space asset. Space systems must be designed with redundancy in mind, provide their own serviceability, or be able to receive a vehicle launched with the purpose of providing maintenance or disposal assistance. With the large amount of planning and numerous details to consider, systems engineers need to include system disposal in the various design diagrams and documents during early development. The following section will show how system disposal is included in development activities using an optical system satellite example.

16.2 Optical Systems Building Block: Optical Systems in Space

Ground-based optical systems may have much in common with their space-based counterparts such as telescopes, lenses, mirrors, apertures, filters, and detectors. However, there are notable differences as well, such as ease of repair, required component reliability, and environmental conditions (radiation, extreme temperature fluctuations, and operating in a vacuum in the case of the space-based optical system). Space systems require considerable detailed trade-off analyses and are often constrained in the following areas:

- Weight
- Dimensions
- Serviceability
- Reliability
- Maintainability
- Availability
- Operability
- Disposability
- Cost

Space-based systems are confined to operate with available resources within the launch vehicle itself, and technical parameters such as weight, size, and power are often severely constrained as compared to their earth-bound counterparts. Solar panels must fold to be contained in the launch vehicle's payload bay and are deployed once the vehicle reaches its destination. Delicate equipment must be protected from the forces and stresses of launch and the harsh environment of space.

Servicing and maintaining the satellite also provide significant challenges. For ground-based systems, either a technician can directly service a failed optical systems component or the component can be sent to a service center. Once launched, space-based optical systems often deal with failure in terms of employing backup systems. Even if access panels are provided to perform maintenance in space, the remote location of the equipment, the requirement for space gear for the technician, and the lack of diagnostic equipment limit the quantity, type, and ease of repairs that can be made. Because of this, certain steps should be taken during the design and construction of space systems. The first step is to design the system for high reliability. If certain critical components fail, then backup systems must be engaged to ensure that the system is capable of fulfilling its mission. The system must be designed to provide high availability and extended operational service. Second, onboard diagnostic equipment must be part of the optical systems design to isolate faults and help to remotely diagnose problems. Third, the optical system must have redundant systems for critical subsystems and components.

This enables technicians, after determining component failure, to switch to a similar, error-free component or backup system to continue operations. Last but not least, the satellite's exterior must be accessible and serviceable. If an assembly, component, or part fails in the optical system, it might be possible to send a repair vehicle to rendezvous with the satellite to conduct repairs on the failed components if the proper accommodations have been made in the design.

Operability is the ability to keep a system in proper working condition. The operability of space systems is not significantly different from that of terrestrial systems, but some differences do exist. For earth-bound systems, access to the optical system is generally available for operators and repair/service technicians alike and so making functional changes to the earth-bound system is relatively easy. Space-based systems do not benefit from these conveniences and so stringent planning is required to address the scope of the mission and possible contingencies.

The disposability of a system should be a major concern because improper disposal can impact the environment or affect other systems. There are many options to choose from for the disposal of ground-based optical systems. Fewer options are available for the disposal of space systems, and the disposal activities are more difficult to plan and perform. Consequently, proper planning during early development phases is essential to ensuring the successful completion of disposal activities.

Lastly, the overall life-cycle cost is always of great interest when it comes to optical systems in space. As an example, the James Webb Space Telescope is currently scheduled to go operational in 2018 and has an estimated cost between $6.2 and $6.8 billion (NASA 2014). In contrast, the most expensive terrestrial optical systems project is the Atacama Large Millimeter/submillimeter Array (ALMA), which has an estimated production cost of $1.4 billion (Atacama Large Millimeter/Submillimeter Array 2013).

16.2.1 Background for Optical Space Systems

To the shock of the world, the first artificial satellite, Sputnik I, rocketed into space from the Tyuratam launch facility (Baikonur Cosmodrome) in the Soviet Union, on October 4, 1957. Sputnik I orbited the earth in about 92 min and performed roughly 1400 revolutions around the earth in a period of 2 months. Thus, the "space race" was born. Since the launch of the spherical, basketball-sized Sputnik I, there have been about 6910 satellite launches (Kyle 2014). Figure 16.3 shows the launch statistics associated with these satellites.

Figure 16.3 has statistics from 1957 to 2010 on the total satellite launches, satellites decayed from orbit, and the number of satellites remaining in orbit. Launch total accounts for both successful and unsuccessful launches originating from all countries. Decayed from orbit is the total number of satellites that have experienced orbital decay, a prolonged

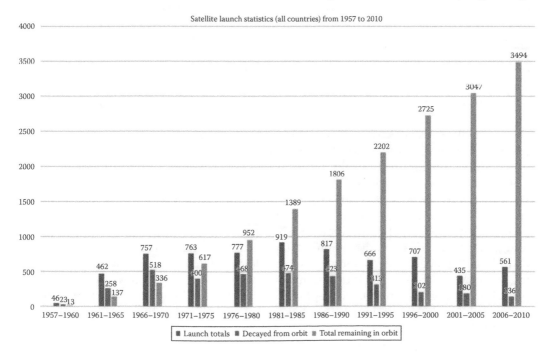

FIGURE 16.3
Satellite launch statistics. (Created by and courtesy of Marty Grove.)

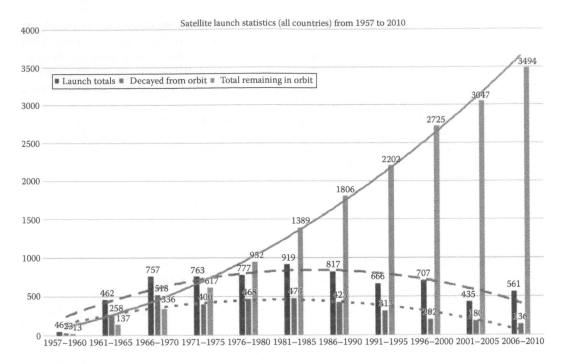

FIGURE 16.4
Satellite launch trending analysis. (Created by and courtesy of Marty Grove.)

period of diminishing altitude, typically concluding in the destruction of the satellite as it reenters the earth's atmosphere. The orbital decay of these satellites may have been intentional, unintentional (e.g., due to a catastrophic failure), or having reached its anticipated EOL. The "total remaining in orbit" category is the total number of satellites that have remained in orbit with their intended original orbital velocity and acceleration, regardless of their operational status. Another view of this graph that reveals interesting information on the status of these satellites is illustrated in Figure 16.4, which depicts historical trending analysis.

The solid black trend line in Figure 16.4 indicates a significant increase in remaining orbital satellites, while the dashed bell curve trend lines demonstrate a decrease in total satellite launches and orbital decay of existing satellites. As technologies in the various areas of satellite deployment developed, sustaining existing orbital systems has dramatically improved. With each new satellite deployment, expected EOL has been expanded. From 1957 to 2010, earth-orbiting satellite systems have increased by approximately 26.777%, of which the vast majority of these systems are rendered derelict. Having little to no consideration for an EOL process, they have been largely ignored and left in orbit. The need to consider a more complete systems life-cycle approach of managing existing and future satellite systems is clearly recognized. This complete systems approach must include an EOL concept of disposal (Janovsky et al. 2002). Figure 16.5 is the debris plot by NASA. It depicts the objects in the earth's orbit that are currently being tracked by NASA.

This rendition is quite illustrative of the problem of space debris and the threat it poses to operational satellites. The debris comes from the remnants of satellites that have reached their EOL but have not been properly disposed of or retired. "By January 2002,

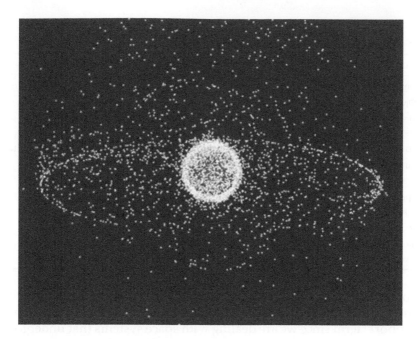

FIGURE 16.5
(See color insert.) Space debris plot by NASA. (Image courtesy of NASA Orbital Debris Program Office, Johnson Space Center. 2014. http://orbitaldebris.jsc.nasa.gov/photogallery/beehives.html, accessed December 8, 2014.)

a total of 4191 launches since 1957 had deployed 17,050 payloads, rocket bodies, and mission-related objects, which caused 27,044 detectable and tracked objects in Earth orbits. Of these 27,044 catalog objects, 18,051 had decayed into the atmosphere, leaving an on-orbit catalog population of 8993" (Wertz and Larson 2008).

16.2.2 Systems Engineering Principles in Space Projects

Systems engineering for space projects follows all the guidelines that systems engineering for on-the-ground projects includes. Space missions and goals vary from communication across the planet, to manufacturing items in space, to interplanetary exploration, and even burial in space. Even so, the planning for a space mission is much like planning the launch of any commercial, government, or educational venture on earth.

16.2.3 Space System Life Cycle

Typically, the life cycle of a space system progresses through the traditional systems engineering life-cycle phases:

1. Conceptual design phase
2. Preliminary design phase
3. Detailed design and development phase
4. Production, manufacturing, and construction
5. Operations and support
6. Phaseout and disposal

TABLE 16.1

Needs Analysis and Concept Development Table

Needs Analysis	Concept Development
Generate potential requirements based on:	Reassess potential requirements generated during needs analysis
Mission objectives	
Concept of operations	Develop and assess alternative mission operations concepts
Schedule	
Life-cycle cost and affordability	Develop and assess alternative space mission architectures
Changing marketplace	
Research needs	Estimate:
National space policy	Performance and supportability
Changing threat to national defense	Life-cycle cost and produce ability
Military doctrine	Schedule and funding profiles
New technology developments	Risk and return on investment
Commercial objectives	

We will provide a quick overview of these traditional phases and some of their main attributes as they pertain to space systems. Note that we are using the traditional "operations and support" term since we are dealing with space systems that undergo operations in space.

The *conceptual design phase* includes a "study" stage that results in a broad definition of the space mission and its conceptual attributes. The operators and end users define their needs and requirements and present them to the developer for concept development. Table 16.1 lists many of the needs analysis and concept development activities performed during this phase. The milestone event is the systems requirements review (SRR) where the systems-level requirements specification, the A-specification, is baselined.

The second phase is the *preliminary design phase* wherein the high-level subsystems design is developed along with subsystems requirements documentation. Although focused on the subsystems, the design at this level can be quite detailed including pseudo-code and mock-ups. The preliminary design review (PDR) is the milestone event wherein the B-level specifications are baselined. The *detailed design and development phase* is the last design stage wherein detailed designs are developed for the system based on the preliminary design requirements. Protoflight and prototypes are developed during this phase to mitigate risk. Detailed requirements documents are developed that are baselined during the critical design review (CDR)—the milestone event for this phase. Next, the *production, manufacturing, and construction phase* includes the actual construction of the ground and flight hardware and software and also includes the launch of the payloads. The next phase, *operations and support*, occurs once the payload is in orbit and includes the daily operations of the system, maintenance, and support. Finally, *phaseout and disposal* occurs when the system has entered its EOL stage. This phase encompasses the "retirement" of the system.

16.2.3.1 Space System Phaseout and Disposal

As mentioned, phaseout and disposal requires a great deal of planning. Using block diagrams is one way for systems engineers and the development team to organize their planning and engineering activities. Using the life cycle shown in Figure 16.1 as a starting point, the phaseout and disposal stage can be expanded in greater detail. The next level of the functional flow decomposition (level two of the phaseout and disposal FFBD) describes the sequential activity blocks that make up the phaseout and disposal stage and is illustrated in Figure 16.6.

In Figure 16.6, segment 8.0 (phaseout and disposal) is broken down into nine stages (8.1 through 8.9). Each of these nine stages is further broken down into subsystems (or child systems). These subsystems are designed to support the parent systems and describe them in more detail. Turning our attention to Figure 16.7, segment 8.1 (create plan for optical satellite disposal), we expand this block into Blocks 8.1.1–8.1.8.

Referring to Figure 16.7, there are three methods available for the disposal of optical system satellites: 8.1.4, deorbit burn; 8.1.5, graveyard orbit; and 8.1.6, deep space ejection. Which method the mission planning team chooses depends on considerations such as risk factors, available fuel to execute the required orbital maneuvers, safety, and overall cost. We will address these considerations in the form of an example in the case study section of this chapter (Section 16.3).

8.0 Phaseout and disposal
Level 2—functional flow diagram

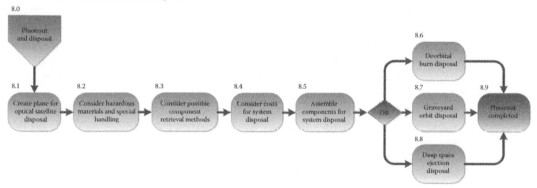

FIGURE 16.6
Phaseout and disposal FFBD. (Created by and courtesy of Marty Grove.)

8.1 Create plan for optical system disposal
Level 3—functional flow diagram

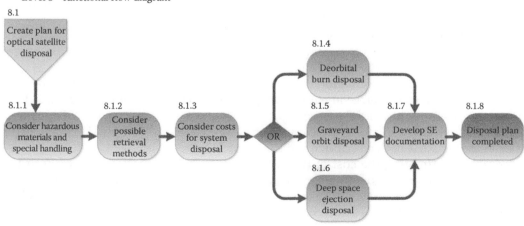

FIGURE 16.7
Create plan for optical system disposal FFBD. (Created by and courtesy of Marty Grove.)

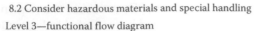

Level 3—functional flow diagram

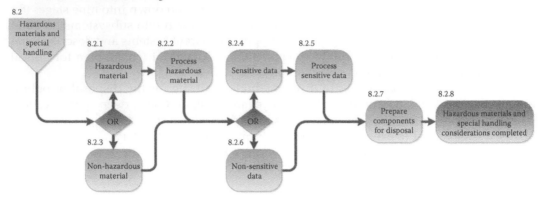

FIGURE 16.8
Consideration for hazardous material and special handling FFBD. (Created by and courtesy of Marty Grove.)

The deorbiting process involves slowing the speed of the satellites until the space vehicle reenters the earth's atmosphere and is incinerated. Graveyard orbit is an orbit that is located in high earth orbit (HEO). It is an orbit that is dedicated to satellite systems that have reached their EOL. Deep space ejection is a disposal method that sends the optical system on a path to either deep space or toward the sun where it will be incinerated. From an economic standpoint, the optimal choice in safely retiring the satellite depends on the required change in velocity (and consequently the amount of fuel needed) for the particular retirement/disposal maneuver.

Turning our attention to the second block of Figure 16.6, the expansion of Block 8.2 describes subsystems for the handling of hazardous material and other special components and is shown in Figure 16.8.

This figure indicates that sensitive data disposal would follow specific processes dependent upon data safeguarding and disposal requirements. Certain data may require retrieval, while other data may require a secure destruction method. Hazardous materials such as fuels, lubricants, and various chemicals need proper disposal methods to avoid possible threats to human life or other orbital objects.

Also an important function of the phaseout and disposal process, the FFBD illustrated in Figure 16.9 shows the points of considerations for the possible retrieval of specific optical satellite components, such as secure data.

Equipment and other information that require further terrestrial analysis must be considered and planned for retrieval early in the optical systems design phase. Figure 16.10 shows representative high-level cost considerations associated with the cost of disposing a system. The risk block (Block 8.4.4) includes technical, cost, and schedule risks. The specific order of the blocks is not prescriptive and can vary.

Accurately estimating the cost of the optical system on the satellite is critical in the successful, development, deployment, operations, support, and disposal of space systems. A variety of cost drivers are encountered over the life cycle of the optical satellite system that must be identified and characterized. These cost drivers can be used in holistic models, parametric models, and detailed bottom-up models to provide insight into expected costs at various stages of the optical satellite system's life cycle. As an example, parametric cost estimations models are used by NASA and the DOD and require

8.3 Consider possible component retrieval methods

Level 3—functional flow diagram

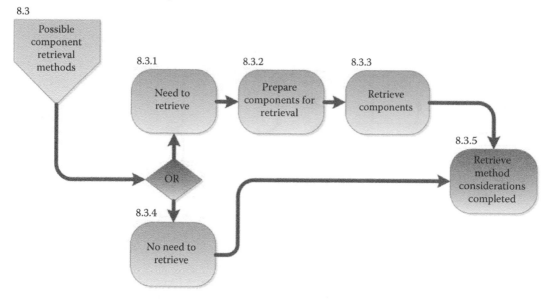

FIGURE 16.9
Consideration of possible component retrieval methods FFBD. (Created by and courtesy of Marty Grove.)

8.4 Consider costs for system disposal

Level 3—functional flow diagram

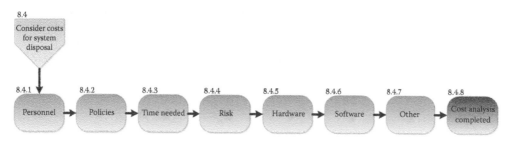

FIGURE 16.10
Consideration of costs for system disposal FFBD. (Created by and courtesy of Marty Grove.)

only high-level requirements/design information along with historical cost data, and regression analysis, to produce useful cost estimating results (Wertz and Larson 2008). Important considerations in the phaseout and disposal phase are the necessary resources such as personnel, equipment, facilities, and materials, schedule, contingencies, risks, and international treaties and laws. Figure 16.11 shows a representative functional flow diagram for the necessary components needed to dispose of the space system.

One consideration in the disposal and retirement process is whether or not a retrieval mission is required to retrieve critical mission data or items. Various methods such as ejecting from the satellite a protected reentry vehicle that contains the data/desired item or sending an autonomous or manned missed for retrieval may be considered. With the commercialization of space, some of these activities could possibly

8.5 Assemble components for system disposal
Level 3—functional flow diagram

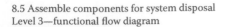

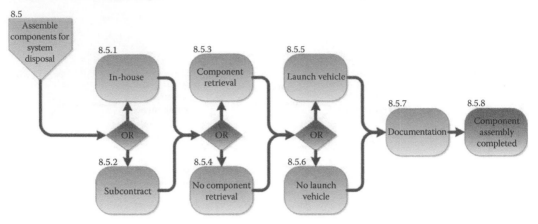

FIGURE 16.11
Assemble components for system disposal FFBD. (Created by and courtesy of Marty Grove.)

be performed by commercial entities that have the means to conduct these operations either in the form of satellite system add-ons or having the retrieval infrastructure in place. A representative process for deorbiting a satellite for system retirement is illustrated in Figure 16.12.

Since during the deorbiting procedure, the optical system will be reentering earth's atmosphere, one risk is that the satellite is not completely incinerated when it reenters the earth's atmosphere. Instead, the satellite would break into smaller pieces that either bounce off the atmosphere back into the LEO posing a threat to other spacecraft in this orbit or come careening down to the earth's surface and potentially creating a hazardous situation. Consideration of these types of contingencies is required during the design phase and the choice of the satellite's materials, construction, and the planned deorbit maneuver must be chosen to minimize reentry hazards.

8.6 Deorbital burn disposal
Level 3—functional flow diagram

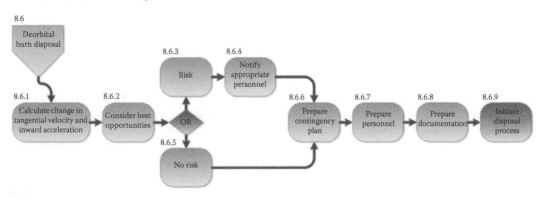

FIGURE 16.12
Deorbital disposal FFBD. (Created by and courtesy of Marty Grove.)

8.7 Graveyard orbit disposal
Level 3—functional flow diagram

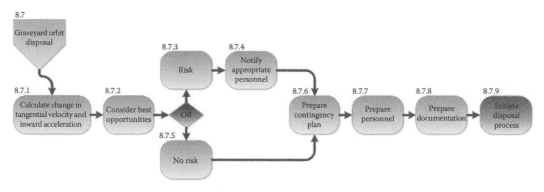

FIGURE 16.13
Graveyard orbit disposal FFBD. (Created by and courtesy of Marty Grove.)

Figure 16.13 illustrates the graveyard orbit disposal method for system retirement.

A relatively common approach in disposing of satellites is to place them in a graveyard orbit where they cannot possibly interfere with active satellites. This form of disposal poses less of a hazardous threat to life and systems on earth; however, this method still requires careful contingency planning to execute a proper system retirement maneuver in space. Issues such as potentially striking other active satellites, encountering other space debris, and consequently contributing to the space debris problem are all-important considerations.

Another possible consideration is to dispose of the satellite by ejecting it into deep space as shown in Figure 16.14.

Alternatively, the satellite could be sent on a path into our sun where it would be incinerated. The idea is to change the velocity of the satellite enough that it can break the earth's and sun's gravitational pull (in the case of the deep space ejection) or slow

8.8 Deep space ejection disposal
Level 3—functional flow diagram

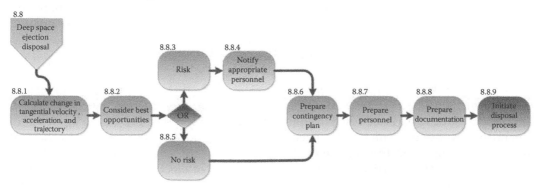

FIGURE 16.14
Deep space ejection disposal FFBD. (Created by and courtesy of Marty Grove.)

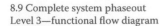

8.9 Complete system phaseout
Level 3—functional flow diagram

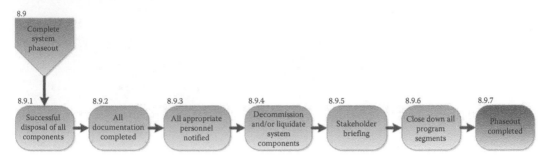

FIGURE 16.15
Complete the system phaseout FFBD. (Created by and courtesy of Marty Grove.)

the speed of the satellite so that it falls within the capture range of the sun's gravitational pull. In maneuvering the satellite for this purpose, similar risks as those for the graveyard orbit are encountered in that collision with other satellites or space debris needs to be avoided. We will see whether or not these options are viable in Section 16.3 of this chapter.

The FFBD in Figure 16.15 serves as a checklist to indicate the system phaseout and disposal completion. These considerations are important in ensuring all activities are complete and all stakeholders are aware of results.

16.2.4 Environmental Effects on Optical System Satellites

Once in orbit, the optical system will experience a number of environmental effects that are not commonly experienced by earth systems. These effects present challenges, which engineers must address during the design, operation, and disposal phases.

16.2.4.1 Plasmas and Spacecraft Charging

There are distinct regions in the space environment such as the ionosphere and magnetosphere wherein satellites must operate. Within the magnetosphere, there are separate and distinguishable energy levels. Each of these energy levels is created as solar winds come in contact with earth's magnetic field and get converted from kinetic to magnetic energy. Figure 16.16 illustrates the various layers of the earth's magnetosphere.

From time to time, the magnetic energy is released in the form of energized plasmas on the order of 5–20 keV. These plasmas then can potentially interact with satellites in their path. As these plasmas come in contact with a spacecraft's hull, charges can build, depending on the hull's material capacitance and grounding, to the point of electrical arcing occurring along the surface of the spacecraft. This effect then becomes harmful to a spacecraft's electrical equipment in the form of electromagnetic interference. Most vehicle systems do not have to worry about this effect until the energy differential reaches between 10 and 20 keV. When designing an EO/IR space system, a designer can take some precautious against this effect. First, all lens materials on the outside of the craft should be tested for their capacitive properties, and all that do not sufficiently

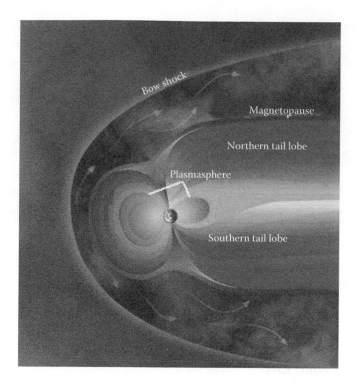

FIGURE 16.16
(See color insert.) Representation of the earth's magnetosphere. (Image courtesy of Dennis Gallagher. 1999. Wikimedia Commons. http://commons.wikimedia.org/wiki/File:Magnetosphere_Levels.jpg, accessed December 8, 2014.)

meet grounding requirements should be coated with a conductive coating if possible. Second, EMI shielding techniques should be undertaken when designing or selecting the electrical components of an EO/IR space system.

16.2.4.2 Trapped Radiation

When traveling to space, all systems must worry about radiation from the Van Allen radiation belts. Figure 16.17 gives a brief representation of the Van Allen belt layers.

The energies of particles trapped within the Van Allen belt are on the magnitude of 30 keV or greater. There are two models that describe the Van Allen belt energies: the AP8MIN and AP8MAX models. These models show the energies at the solar minimum and maximum. An electro-optical/infrared (EO/IR) systems designer will want to take into account the duration of the mission since the energy levels depend on the solar cycle. One needs to remember that radiation shielding is important; however, too much shielding could increase weight and provide little to no benefit. Some of the effects a spacecraft could experience with radiation dosage are enhanced IR background noise from low-energy particles leaving their energy on the spacecraft's surface and degradation in the effectiveness of paints and protective glass.

16.2.4.3 Solar Particle Event

Solar particle events are better known by their other name, solar flares. When a solar particle event occurs, a large rush of high-energy particles, between 1 MeV and 1 GeV, are sent toward the earth. If not protected against, effects from a solar particle flare

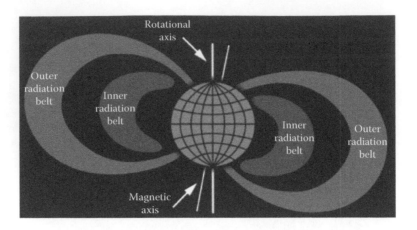

FIGURE 16.17
Van Allen Belt layers representation. (Image courtesy of Chris Martin (Booyabazooka). 2006. http://en.wikipedia.
org/wiki/File:Van_Allen_radiation_belt.svgl, accessed December 8, 2014.)

include "increased background noise for many types of electro-optical sensors and degradation in solar cell arrays" (Marietta 1993).

16.2.4.4 Galactic Cosmic Rays

Galactic cosmic rays are probably the most harmful and most frequent issue to space-based electrical systems. When a cosmic ray comes into contact with a memory or microprocessor, there is the possibility that the energy from the ray will be enough to cause a single-event phenomenon. Each single-event phenomenon can be classified as one of three types.

The first type is single-event upset. A single-event upset causes the state of a transistor to change. This is harmful because it essentially can corrupt any preloaded software or data stream from a device. Currently, technology has been developed that can detect these upsets and correct for them; however, most of it is slow in processing and very expensive. This type of event is very common and does not cause any physical damage to a part.

The second type is single-event latch-up. Single-event latch-up causes a part to hang up, drawing excessive amounts of electrical current and ceasing operation. The way to protect equipment from this type of event is to detect the excessive current and restart the device. Due to the excessive current, it is possible that a single-event latch-up can cause harm to a part; however, if the event is caught in time, damage is minimized.

The third type of effect is single-event burnout. This is the least common of events; however, it is the most harmful. A single-event burnout causes permanent damage to a part to the point of part failure.

Though single-event phenomena cannot be fully predicted, it is now possible to find probabilities of failure through one of three laboratory tests. The first type of test uses a computer program to find the probability based upon device factors, such as amount of shielding around the device, minimum charge required to cause an upset, and dimensions of the transistors. The second test uses a particle accelerator and bombards a working unit for some time. Then a computer counts the amount and types of failures. The final type of testing uses high-altitude environments and measures the amount of failures over an extended amount of time.

16.3 Integrated Case Study: FIT Special Customer

In this section, we present an integrated case study that demonstrates the principles and methods learned in the preceding sections of this chapter. We find our fictional optical systems development company, FIT, engaged in a prospective collaboration with their old-time customer from the Department of Homeland Security (DHS) and a new "special customer" with interests in putting the FIT high-spatial-resolution optical system into space. The topic at hand is how to dispose of the space-based optical system and how to potentially retrieve some important information from the optical system at its EOL. The characters involved in this scenario are Dr. Bill Smith, FIT CEO; Karl Ben, FIT's recently promoted director of systems engineering; Dr. Tom Phelps, FIT chief technical officer (CTO) and optics expert; Wilford Erasmus, chief of operations and acquisitions, U.S. Customs and Border Patrol, DHS; Glen H., DHS technology specialist; Rebecca H., "special customer," unnamed organization; and Muz S., special customer tech advisor.

Bill: "Hi, Wilford, Rebecca, and Muz! It's great seeing you all again. Thanks for making yourselves available for this videoconference. You all remember Karl, our director of systems engineering?"

Wilford: "Hey Karl. Congratulations on the promotion! How're things in Melbourne?"

Karl: "Hot and muggy."

Rebecca: "I asked Muz to dial in today. He's one of our tech advisors. He's currently on the west coast in a meeting with the Aerospace Corporation, so it's a little bit early for him. Thanks for joining us Muz."

Muz: "Glad to be here."

Bill: "We have the team in place to put our modified, high-spatial-resolution optical system on one of your special space platforms Rebecca. I think you will be pleased with the new capabilities it will provide. Wilford, this optical system will also provide DHS the ability to have dedicated space surveillance operations."

Wilford: "We can't wait for this system to go operational. It's going to give us a game changing capability. This system is getting attention all the way up, so we need to do things right."

Rebecca: "I just want to say we can't wait to get this new high-spatial-resolution imaging capability either. The systems you provided DHS and the DOD are amazing and we are quite impressed with your company. Just let us know what you need, and we will make it happen."

Bill: "Thanks Rebecca, we are thankful and grateful to be working with you all too. We will make sure that this optical systems development effort is carried out to the highest standards."

Wilford: "No doubt about that! Let's get things rolling."

Bill: "All right! The only thing we still need to do at this point is to discuss the phase-out and disposal plans for this new system, hence this videoconference. I know that there will be some sensitive information on the onboard memory, the recording system, as well as some highly classified algorithms and data that will need to be properly handled at the EOL of the system. To get things started, we'd like to understand some of the options that are available for the satellite EOL."

Rebecca: "Sure. Muz, can you give us a quick overview of some of the methods please?"

Muz: "OK. I will start with describing some of the EOL options for the larger satellite structure. Basically, we have three types of disposal methods to discuss. The options are to deorbit the satellite, send it into a graveyard orbit, or eject it into deep space (or the sun). Each method has its advantages and disadvantages. To deorbit, the satellite, with your optical system on it, will be slowly maneuvered into the earth's atmosphere. Upon entering the atmosphere, the atmosphere will heat things up, and depending on the materials of the system, it should be incinerated. This approach is very common in retiring and disposing of space objects in LEO. There are some inherent risks in the satellite not fully breaking up once it reenters the atmosphere. Also, there must be assurances that any hazardous materials or sensitive, secure, or protected information is destroyed in the reentry process. Economically, this is often the cheapest solution."

Muz: "A second option is to boost the satellite into a higher orbit that is out of the way of active satellites. This is the so-called graveyard orbit and it is located higher than the working orbit of our GEO. A graveyard orbit is further out than the typical orbit of the satellites. Whether or not this option is attractive or not depends on the amount of fuel required to reach the graveyard orbit. The metric we use is called Delta-v describing the change in velocity required to make the orbital maneuver."

Tom: "Isn't the change of velocity a function of the mass of the object that is moving?"

Muz: "Yes it is! Since there is so much variation in the mass of a spacecraft, the mass is not a good metric to use. The change in velocity, on the other hand, is a result that is a function of the mass of the spacecraft, the mass of the surrounding gravitational objects (like the earth, moon, sun, planets), the spacecraft starting velocity, its location and orbit, and its particular intended maneuver. For example, in a rocket engine, the fuel is continuously being spent and so the mass continuously changes. However, the end result, the change in spacecraft velocity is what is required to execute a particular maneuver. For instance, it takes a Delta-v of about 9,300–10,000 m/s for a spacecraft to go from the earth (Kennedy Space Center) to Low-earth-Orbit (LEO), regardless of the mass of the spacecraft (Delta-v 2014). Delta-v is quite useful in comparing spacecraft maneuvers."

Tom: "So, basically, the bigger the Delta-v, the more fuel is required for a particular maneuver?"

Muz: "That is generally correct, unless you can take advantage of some other physical phenomenon like the gravitational forces from surrounding gravitational bodies to increase your velocity in the desired direction. Sort of like a slingshot!"

Tom: "Ok, I get it."

Muz: "Anyways, whether or not this option is attractive depends on the Delta-v involved for the spacecraft to get it to its final graveyard orbit as compared to the Delta-v required to deorbit the satellite. For example, a satellite in GEO requires a Delta-v of about 1500 m/s to deorbit the satellite versus a Delta-v of 11 m/s to push it into the super-synchronous graveyard orbit (Graveyard Orbit 2014). Higher Delta-v values typically mean higher fuel required and so the lowest Delta-v required for the orbital maneuvers is usually the way to go. However, some other issues involved with the graveyard option include safeguarding classified or sensitive information, risks associated with onboard hazardous materials, and the need for continued or periodic monitoring of the orbit and/or station keeping requirements. Not that there is a lot of risk of collisions with active assets, accidentally deorbiting, having the earth's gravitational attraction deorbit the satellite from its graveyard orbit, or having our data compromised; however, from a security and safety perspective, guaranteed total disintegration or permanent removal from possible access is the preferred option. Speaking of permanent removal, that brings me to the next method."

Muz: "The last disposal option is to consider sending the satellite into deep space. The idea is to increase the velocity of the satellite in a trajectory that would break both the earth and sun's gravitational fields and send the satellite into deep space far beyond our solar system. From a security perspective, this method would provide the best option apart from sending the satellite directly into the sun. Additionally, we wouldn't have a need for monitoring the satellite once it left our solar system."

Wilford: "Won't there be additional costs to launch the satellite with an amount of fuel reserved for the trajectory change?"

Muz: "Likely, yes. However, if we include it in our initial planning, there may be some different types of propulsion technologies that may be useful, and we might be able to use gravitational assist to increase the velocity of the spacecraft and change its direction, that is, assuming we can generate enough Delta-v to maneuver our spacecraft to take advantage of the gravity assist in the first place."

Tom: "Gravity assist?"

Muz: "The 'gravity assist' method basically uses the gravitational pull of a planet, moon, or large gravitational object to slingshot the spacecraft into a new direction with an increased velocity and without using fuel. The velocity of the spacecraft can be increased or decreased by the relative velocity of the planet through space, using the planet's gravitational attraction on the spacecraft. There is also the powered slingshot, where the spacecraft motors are fired just at the right time to enhance the slingshot effect. The gravity assist method has been used in the Voyager program (Delta-v Budget 2014) and also in the Apollo moon trips. In the end, it is still a question of overall Delta-v required for the particular maneuver in question."

Muz: "To get an idea of the Delta-v's involved, let me first discuss the orbital zones that must be considered. There are three well-known and distinct orbital zones: LEO, medium earth orbit (MEO), and HEO. There are other important orbital points like Lagrange points but those aren't that relevant for this disposal discussion. The LEO zone ranges from about 160 km (or 99.4 miles) to roughly 2000 km (or 1242.7 miles). The MEO orbital zone occurs between 2,000 km and 35,900 km (or 22,307.2 miles). The HEO orbital zone goes from 35,900 km and beyond (NASA 1995). The three orbital zones discussed are illustrated in Figure 16.18."

Muz: "Most of the satellites today are located in the LEO zone and this is the intended operational location for your satellite. Within this zone, there is a particular orbit that provides a constant solar illumination angle known as the sun-synchronous orbit (Marietta 1993). This type of orbit requires a polar orbit and fixes the angle of the sun with respect to the satellite in order to provide relatively constant viewing and illumination conditions."

Muz: "As illustrated in Figure 16.19, in a sun-synchronous orbit, the orbital plane of the satellite makes an angle of about 37.5° to the plane through the center of the earth and sun. The sun-synchronous orbit provides excellent illumination conditions for optical imaging applications in the visible part of the electromagnetic spectrum."

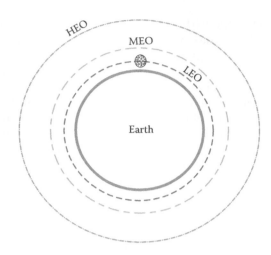

FIGURE 16.18
(See color insert.) Orbital zones. (Created by and courtesy of Marty Grove.)

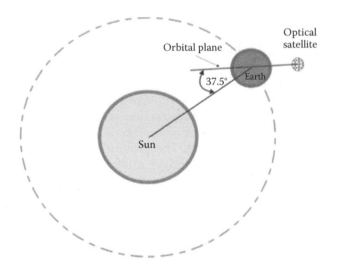

FIGURE 16.19
Sun-synchronous orbit. (Created by and courtesy of Marty Grove.)

Karl: "Ok, so our spacecraft starts in LEO in a polar orbit that fixes the relative position of the sun for optimal imagery. That is cool and I get it. I'm still somewhat unsure about the Delta-v's involved for the various options that you were talking about. Can you go over that please?"

Muz: "Sure, I was just getting to that. Let me start by describing the deorbit process. In Figure 16.20, we can observe the forces that are acting on the satellite as it revolves in a standard orbit about the earth. The parameter r is the distance from the earth's center to the circular satellite orbit, v is the velocity of the satellite, the gravitational force exerted by the earth on the satellite is given by F_g, and F_c is the satellites centripetal force. If the gravitational force is equal to the centripetal force, then the orbit is stable and the satellite orbits in a circular orbit about the earth."

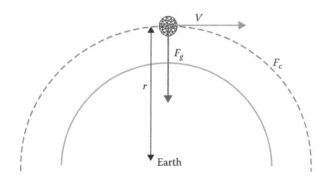

FIGURE 16.20
Standard orbit. (Created by and courtesy of Marty Grove.)

Muz: "If you pull out your high school physics book, the centripetal force is given by

$$F_c = \frac{mV^2}{r},\tag{16.1}$$

where
    m is the mass of the spacecraft in kg
    V is the speed of the spacecraft in a circular orbit around a center of mass (the earth) in meters per second, with a center of mass to spacecraft separation of r meters

The gravitational force exerted on the spacecraft by the earth is given by

$$F_g = \frac{GMm}{r^2},\tag{16.2}$$

where
    G is the gravitational constant given by 6.6738×10^{-11} m³/kg/s²
    M is the mass of the earth (5.9737×10^{24} kg)
    m is the mass of the spacecraft in kg
    r is the distance from the center of mass to the spacecraft

Setting these two equations equal to each other and solving for the speed term V, we get the speed required of a spacecraft to maintain a certain orbit above the earth. Solving for V, we get

$$V = \sqrt{\frac{GM}{r}}.\tag{16.3}$$

This equation is very useful since it tells us the starting velocity of our spacecraft in the LEO. Sometimes, the product of GM is given as the geocentric gravitational constant (GGC) since this number is so commonly used. It is given by 3.986005×10^{14} m³/s². If we wanted to deorbit the satellite, we must slow it down so that its forward momentum is reduced. This can be accomplished by using rocket thrusters that are aimed in the forward direction consequently decreasing

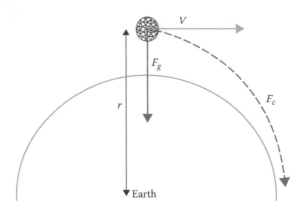

FIGURE 16.21
Deorbit. (Created by and courtesy of Marty Grove.)

　　　　the centripetal force of the satellite (Figure 16.21). The gravitational force now exceeds the centripetal force and the satellite is 'pulled in' to the earth's atmosphere where frictional forces heat up and destroy the satellite."

Tom: "Do we have assurances that the satellite will be completely destroyed? We wouldn't want any critical information or components to make it through to the ground and be retrieved by someone else. Also we wouldn't want any materials staying behind and reentering the LEO orbital zone and contributing to the space debris. Also, we certainly want to mitigate any potential hazards and ensure that there is no collateral damage from parts falling to the ground."

Muz: "Yes, we definitely need to do everything we can to ensure that the satellite deorbit process is safe. For example, there was a climate satellite from NASA that was deorbited in the fall of 2011. This satellite was the Upper Atmosphere Research Satellite (UARS) and it broke into an estimated 26 large pieces during its reentry into the earth's atmosphere on September 23, 2011. The probability of one of these pieces striking a person was estimated by NASA and verified by others, to be 1 in 3200 chance (Malik 2011)."

Wilford: "What happened?"

Muz: "NASA stated that the location of the debris is unknown (Potter et al. 2011). However, no debris-related injuries were ever reported. Even so, these large pieces landed somewhere and so this concern is real. That's why it is important to plan this operation carefully."

Bill: "Excellent example Muz, please continue."

Muz: "Thanks! I now want to consider what it takes to send the satellite to its graveyard orbit. As shown in Figure 16.22, the graveyard orbit is located well above the LEO zone and is referred to as the supersynchronous orbit. Its location is about 300 km (or 186.4 miles) beyond GEO in the HEO zone."

Muz: "In order to send the satellite into its graveyard orbit, there needs to be a change in velocity to accomplish the maneuver. This, of course, necessitates sufficient fuel at the satellites' EOL to change its trajectory for insertion into the supersynchronous orbit."

Karl: "Can you explain the process please?"

Muz: "Sure. We typically use something called the Hohmann transfer, which uses an elliptical orbit to transfer the spacecraft from LEO to GEO. This is the most efficient and cost-effective way to transfer the satellite between these orbits.

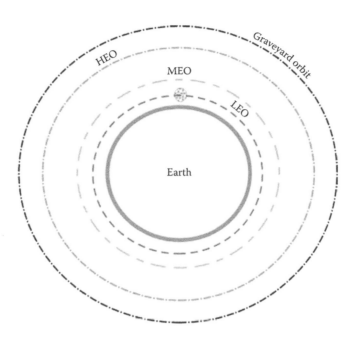

FIGURE 16.22
Graveyard orbit. (Created by and courtesy of Marty Grove.)

The idea is that the spacecraft orbits around the earth at the LEO; it then undergoes an impulsive Delta-v that puts it into an elliptical orbit (the Hohmann transfer orbit). Once it reaches GEO, another impulsive Delta-v event puts it back into a circular orbit at GEO. The total Delta-v to move a spacecraft into a GEO orbit from a LEO depends on the initial and final orbits and orbital mechanics such as whether or not plane changes are required. However, as a ballpark figure, a LEO to GEO transfer that was launched from Kennedy Space Center is on the order of 4330 m/s (Delta-v Budget 2014). The graveyard maneuver is illustrated in Figure 16.23. Regarding security, EOL sanitation methods would be used to erase or eliminate sensitive data and satellite functions would be terminated. The satellite would safely stay in this orbit."

Glen: "Wouldn't this be more expensive than the deorbit method due to the fact that you likely would need more fuel for this maneuver than for the deorbit maneuver?"

Muz: "Not always. Sometimes, some orbital maneuvering and braking is required to deorbit the satellite safely and that requires additional Delta-v. You just have to evaluate each individual case."

Karl: "It looks like there are trade-offs. I can see how it would be beneficial in lowering the risk in harming or damaging something or someone by choosing an alternative to the deorbit procedure. We all know that there is a lot of space junk and debris out there and we don't want to contribute to that problem. If something goes wrong during the deorbit procedure, the satellite can break up and potentially pose a hazard or possibly create more space debris. No one wants more space debris."

Muz: "That's true to some degree. Keep in mind that earlier satellites may not have been optimally designed to completely break up during reentry. With modern design procedures and careful selection of the materials, we can minimize the reentry risks. With regard to the cost savings, if no extra complicated orbital changes are

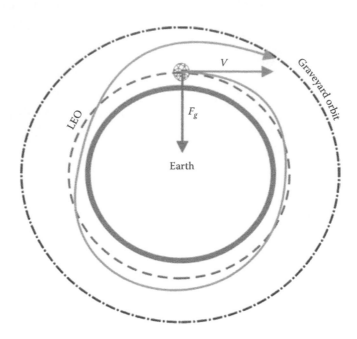

FIGURE 16.23
Graveyard maneuver. (Created by and courtesy of Marty Grove.)

required at the satellite's EOL, the Delta-v required to deorbit is just the amount of orbital maneuvering it takes to lower the perigee into the atmosphere. On the other hand, if a change in orbit is also required before deorbiting the spacecraft, such as changing the inclination angle from a Kennedy launch to an equatorial launch, that can involve almost as much Delta-v as it takes to push a LEO satellite into the GEO orbit (e.g., about 4240 m/s from LEO-Kennedy to LEO Equatorial plus Delta-v for deorbit vs. 4330 m/s from LEO-Kennedy to GEO orbit plus 11 m/s from GEO to graveyard orbit)."

Karl: "So, I understand the graveyard orbit and the deorbit scenarios. What about sending things into deep space or into the sun? That would really get rid of the space junk."

Muz: "Well, for that scenario, we would have to evaluate the escape velocity. This can be found by using conservation of energy. The sum of the initial kinetic energy plus the initial potential energy equals the sum of the final kinetic energy and the final potential energy. Equation 16.4 shows the relationship:

$$\frac{MV_i^2}{2} - \frac{GM}{r_i} = \frac{MV_f^2}{2} - \frac{GM}{r_f}, \tag{16.4}$$

where the first term is the kinetic energy of the satellite, earth system (two-body system) in its initial state; the second term is the potential energy of the two-body system at its initial state; and the terms on the right are the respective kinetic and potential energy of the two-body system in its final state. The parameter V_i is the speed of the satellite in meters per second in its initial orbit at LEO, r_i is the radius from the center of the earth to the satellite in its initial state in its LEO, r_f is the radius from the earth's center to the satellite in its final position, and V_f is the speed of the satellite in its final position. The parameters G and M are, respectively, the gravitational constant and the mass of the earth (Wertz and Larson 2008). For the escape velocity, the final velocity at an infinite radius from the earth is zero, and so both the final kinetic and potential energy terms are zero. Solving for the velocity from the remaining initial kinetic and potential energy terms, we get the following expression for the escape velocity:

$$v_e = \sqrt{\frac{2GM}{r}}, \tag{16.5}$$

where the G and M parameters in Equation 16.5 are the same as before and r is the radius of the satellite in its initial orbit around the earth. If we assume that our satellite is orbiting in a circular orbit 200 km above the equator, then the radius r in Equation 16.5 is given as 6578.1 km (e.g., earth's equatorial radius of 6378.1 km plus 200 km to LEO). If we try to escape the gravitational well from the earth from this position, then the required escape velocity would be about 11.0 km/s. If we are already in LEO at this height, then Equation 16.3 can be used to determine the current orbital speed (about 7784 m/s). This gives a Delta-v of about 3224 m/s required to escape the earth's gravitational well. Figure 16.24 shows the idea of increasing velocity in LEO until escape velocity is reached."

However, the earth is moving through space at a speed of about 29.78 km/s relative to our sun. Once we break the earth's gravitational well, our satellite will be hurling through space at this same speed relative to the sun. If we want to shoot

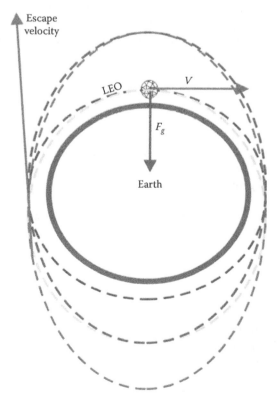

FIGURE 16.24
Escape velocity. (Created by and courtesy of Marty Grove.)

the satellite into the sun, we need to decrease the spacecraft velocity by about 26.9 km/s so that the spacecraft's closest point of approach to the sun is within the sun's capture radius. Otherwise, once our thrusters are turned off, the satellite will orbit around the sun along with the earth and other planets. The total Delta-v required is then 26.9 km/s + 3.224 km/s = 30.124 km/s. This ballpark result is confirmed elsewhere (Delta-v Budget 2014). For a deep space maneuver, it takes approximately 42.1 km/s to escape our solar system, so, escaping in the same direction that the earth is moving, it would take 42.1 km/s – 29.78 km/s (earth's relative velocity through space) = 12.32 km/s. Using the conservation of energy principle with the final spacecraft velocity of 12.32 km/s (after leaving earth's gravitational well) and solving for the initial required velocity of the spacecraft, we get about 16.65 km/s (Adler 2014). We are ignoring the Oberth effect in this discussion (e.g., rockets at higher speeds generate more useful energy than at lower speeds). If the spacecraft is traveling at 11.0 km/s after leaving the earth's influence, then only an additional 16.65 km/s – 11.0 km/s = 5.65 km/s is required to launch the satellite into deep space. From our previous 200 km (equatorial) LEO example, this would be 3.224 km/s + 5.65 km/s = 8.874 km/s, better than shooting into the sun but still higher than the graveyard orbit or the deorbit Delta-v requirements."

Karl: "Hmmm, so the deep space or sunburn options are out?"

Muz: "Yes, pretty much, unless we can do something tricky like use gravitational assist from other planets or some novel propulsion technology. Intercepting the other planets would still require a Delta-v that places the spacecraft on an intercept trajectory with the planet, and so this isn't practical either. Also, to get an idea for how the Delta-v required relates to rocket fuel, we can use the following equation (Wertz and Larson 2008):

$$m_p = m_f \left[e^{(\Delta v / I_{sp} g)} - 1 \right], \tag{16.6}$$

where

m_p is the mass of propellant used for the required Delta-v (Δv) burn
m_f is the final vehicle mass (dry weight plus remaining propellant)
I_{sp} is the specific impulse
g is the standard gravity of 9.806 m/s²

Another way to see the effect of Delta-v on propellant is to use the ideal rocket equation directly (Wertz and Larson 2008):

$$\Delta V = V_e \ln \left(\frac{m_o}{m_f} \right), \tag{16.7}$$

where

m_o is the vehicle initial mass
m_f is the vehicle final mass
V_e is the effective exhaust velocity of the rocket (note that $V_e = I_{sp} g$)

Therefore, for a given specific impulse, a bigger Delta-v means more propellant is required, and more propellant used means more space, weight, and cost to the mission."

Bill: "Ok, so it seems that with careful planning, the deorbit method may be the way to go, unless we need to conduct some orbital maneuvers that gives us a larger Delta-v than what it would take to put our system into the graveyard orbit. We will have to make sure that the optical system memory, data, and critical components are destroyed in the reentry or provide a self-destruct capability to be sure. We also have to ensure that our design doesn't break up too early and scatter pieces back into the LEO. Agreed?"

Rebecca: "That's fine with me."

Wilford: "Me too!"

Bill: "OK. I think that we have everything that we need. Any questions?"

Rebecca: "No. I am glad we went over this."

Bill: "Thanks Muz for the great explanations!"

Muz: "Anytime! Call me if you need anything else."

Bill: "Will do. Oh, by the way, we've also heard about some interesting retrieval methods, in case you want to return some equipment or data back to earth prior to the destruction of the satellite. However, we can save that discussion for another day. Thanks everyone!"

16.4 Conclusion

The disposal and retirement activities are an important phase of any system's life span. It is unlikely that a man-made system will last forever and therefore the system will eventually go through some form of disposal or retirement. Though the systems engineering activities related to a system's EOL require a lot of prediction and contingency plans, a thorough disposal plan can ensure that EOL activities are completed safely, successfully, and within budget.

References

Adler, M. 2014. Space exploration (Beta). http://space.stackexchange.com/questions/3612/calculating-solar-system-escape-and-and-sun-dive-delta-v-from-lower-earth-orbit (accessed July 19, 2014).

Atacama Large Millimeter/Submillimeter Array. 2013. ALMA inauguration heralds new era of discovery. http://www.almaobservatory.org/en/press-room/press-releases/533-alma-inauguration-heralds-new-era-of-discovery (accessed April 2, 2014).

Babecki, A.J. and H.E. Frankel. 2009. Materials problems in satellites. *Eighth Structural Dynamics and Materials Conference*, Greenbelt, MD, p. 430.

Blanchard, B.S. and W.J. Fabrycky. 2011. *Systems Engineering and Analysis*, Upper Saddle River, NJ: Prentice Hall, p. 557.

Chang, K. November 10, 2010. A21 Telescope Is Behind Schedule and Over Budget, Panel Says. *The New York Times*, Section A, p. 21.

Delta-v Budget. 2014. Delta-v budget. http://en.wikipedia.org/wiki/Delta-v_budget (accessed July 18, 2014).

Emanuelli, M. March 28, 2013. Orbit: De-orbit add-on deorbits satellites, minimizes space debris. http://moonandback.com/2013/03/28/d-orbit-add-on-deorbits-satellites-to-minimize-space-debris/ (accessed on April 2, 2014).

Gallahger, D. 1999. Artist's concept of the magnetosphere. Wikimedia Commons. http://commons.wikimedia.org/wiki/File:Magnetosphere_Levels.jpg (accessed December 8, 2014).

Graveyard Orbit. 2014. Graveyard orbit. http://en.wikipedia.org/wiki/Graveyard_orbit (accessed July 18, 2014).

Janovsky, R. et al. 2002. *End-of-Life De-Orbiting Strategies for Satellites.* DGLR-JT2002-028. Stuttgart, Germany: German Aerospace Congress.

Klinkrad, H. 2006. The current space debris environment and its sources. In *Space Debris Models and Risk Analysis.*, H. Kinkrad, Ed. Berlin, Germany: Springer, pp. 5–59.

Kyle, E. September 2014. *Space Launch Report: Worldwide Orbital Launch Summary by Year.* http://www.spacelaunchreport.com/logyear.html (accessed December 8, 2014).

Malik, T. September 16, 2011. Huge defunct satellite falling to earth faster than expected, NASA says. http://www.space.com/12982-dead-nasa-satellite-falling-earth-sept-24.html (accessed April 2, 2014).

Marietta, M. 1993. Geometry of a sun-synchronous orbit. NASA. http://landsat.gsfc.nasa.gov/wp-content/uploads/2013/01/sun-syn_orbit.jpg (accessed April 2, 2014).

Martin, C. 2006. Artist's representation of the Van Allen belt. Wikipedia: http://en.wikipedia.org/wiki/File:Van_Allen_radiation_belt.svg (accessed December 8, 2014).

Military Technology—MILTECH. 2008. *Space Security; Growing Dependence Brings Vulnerability.*

NASA. 1995. NASA Safety Standard 1740.14: Guidelines and Assessment Procedures for Limiting Orbital Debris. Obtained at, http://orbitaldebris.jsc.nasa.gov/mitigate/safetystandard.html. Johnson Space Center, TX: NASA Orbital Debris Program Office (accessed December 8, 2014).

NASA. 2011. Environmental assessment for launch of NASA routine payloads. http://www.nasa.gov/pdf/603832main_FINAL%20NASA%20Routine%20Payload%20EA%20Resized.pdf (accessed July 30, 2014).

NASA. 2014. The James Webb Space Telescope. http://www.jwst.nasa.gov/ (accessed July 16, 2014).

NASA. 2014. Orbital debris. http://orbitaldebris.jsc.nasa.gov/photogallery/beehives.html. Johnson Space Center, TX: NASA Orbital Debris Program Office (accessed December 8, 2014).

OSHA. 2014. Safety and health topics: Cadmium. https://www.osha.gov/SLTC/cadmium/ (accessed July 17, 2014).

Potter, N., G. Sunseri, and K. Dolak. September 24, 2011. NASA UARS Satellite Crashes Into Earth: Location Unknown. http://abcnews.go.com/Technology/nasa-uars-satellite-crashes-earth-location-unknown/story?id=14595092#.UW15gZxmZWg (accessed April 2, 2014).

Scitor Corporation. *Satellite Box Score Count, by Country.* In *Space-Track.* Joint Functional Component Command for Space, United States Strategic Command. Contract No: JFCC SPACE/J35.

U.S. Committee on Extension to the Standard Atmosphere. 1976. *U.S. Standard Atmosphere, 1976.* Washington, D.C.: U.S. Government Printing Office.

U.S. Department of Defense, Defense Logistics Agency. August 1997. *Defense Material Disposition Manual.* http://www.dtic.mil/whs/directives/corres/pdf/416021m.pdf (accessed April 2, 2014).

Wertz, J.R. and W.J. Larson. 2008. *Space Mission Analysis and Design*, 3rd edn. New York, Hawthorne: Springer and Microcosm Press.

Appendix: Mathematical Formulas

A.1 Trigonometric Identities

Pythagorean formulas

$$\sin^2 x + \cos^2 x = 1$$

$$1 + \tan^2 x = \sec^2 x$$

$$1 + \cot^2 x = \operatorname{cosec}^2 x$$

Reciprocal formulas

$$\sin x = \frac{1}{\operatorname{cosec} x} \qquad \cos x = \frac{1}{\sec x} \qquad \tan x = \frac{1}{\cot x}$$

$$\operatorname{cosec} x = \frac{1}{\sin x} \qquad \sec x = \frac{1}{\cos x} \qquad \cot x = \frac{1}{\tan x}$$

Product formulas

$$\sin x = \tan x \cos x \qquad \cos x = \cot x \sin x \qquad \tan x = \sin x \sec x$$

$$\cot x = \cos x \operatorname{cosec} x \qquad \sec x = \csc x \tan x \qquad \operatorname{cosec} x = \sec x \cot x$$

$$\sin \alpha \sin \beta = \frac{1}{2}\cos(\alpha - \beta) - \frac{1}{2}\cos(\alpha + \beta)$$

$$\cos \alpha \cos \beta = \frac{1}{2}\cos(\alpha - \beta) + \frac{1}{2}\cos(\alpha + \beta)$$

$$\sin \alpha \cos \beta = \frac{1}{2}\sin(\alpha + \beta) + \frac{1}{2}\sin(\alpha - \beta)$$

$$\cos \alpha \sin \beta = \frac{1}{2}\sin(\alpha + \beta) - \frac{1}{2}\sin(\alpha - \beta)$$

Angle-sum and angle-difference formulas

$$\sin(\alpha + \beta) = \sin \alpha \cos \beta + \cos \alpha \sin \beta$$

$$\sin(\alpha - \beta) = \sin \alpha \cos \beta - \cos \alpha \sin \beta$$

$$\cos(\alpha + \beta) = \cos \alpha \cos \beta - \sin \alpha \sin \beta$$

$$\cos(\alpha - \beta) = \cos \alpha \cos \beta + \sin \alpha \sin \beta$$

$$\tan(\alpha+\beta) = \frac{\tan\alpha+\tan\beta}{1-\tan\alpha\tan\beta}$$

$$\tan(\alpha-\beta) = \frac{\tan\alpha-\tan\beta}{1+\tan\alpha\tan\beta}$$

$$\sin(\alpha+\beta)\sin(\alpha-\beta) = \sin^2\alpha - \sin^2\beta = \cos^2\beta - \cos^2$$

$$\cos(\alpha+\beta)\cos(\alpha-\beta) = \cos^2\alpha - \sin^2\beta = \cos^2\beta - \cos^2\alpha$$

Double-angle formulas

$$\sin 2\alpha = 2\sin\alpha\cos\alpha = \frac{2\tan\alpha}{1+\tan^2\alpha}$$

$$\cos 2\alpha = \cos^2\alpha - \sin^2\alpha = 2\cos^2\alpha - 1 = 1 - 2\sin^2\alpha = \frac{1-\tan^2\alpha}{1+\tan^2\alpha}$$

$$\tan 2\alpha = \frac{2\tan\alpha}{1-\tan^2\alpha}$$

$$\cot 2\alpha = \frac{\cot^2\alpha - 1}{2\cot\alpha}$$

Power formulas

$$\sin^2\alpha = \frac{1}{2}(1-\cos 2\alpha) \quad \sin^3\alpha = \frac{1}{4}(3\sin\alpha - \sin 3\alpha) \quad \sin^4\alpha = \frac{1}{8}(3 - 4\cos 2\alpha + \cos 4\alpha)$$

$$\cos^2\alpha = \frac{1}{2}(1+\cos 2\alpha) \quad \cos^3\alpha = \frac{1}{4}(3\cos\alpha + \cos 3\alpha) \quad \cos^4\alpha = \frac{1}{8}(3 + 4\cos 2\alpha + \cos 4\alpha)$$

$$\tan^2\alpha = \frac{1-\cos 2\alpha}{1+\cos 2\alpha} \quad \cot^2\alpha = \frac{1+\cos 2\alpha}{1-\cos 2\alpha}$$

Half-angle formulas

$$\sin\frac{\alpha}{2} = \pm\sqrt{\frac{1-\cos\alpha}{2}} \quad \cos\frac{\alpha}{2} = \pm\sqrt{\frac{1+\cos\alpha}{2}}$$

$$\tan\frac{\alpha}{2} = \pm\sqrt{\frac{1-\cos\alpha}{1+\cos\alpha}} = \frac{1-\cos\alpha}{\sin\alpha} = \frac{\sin\alpha}{1+\cos\alpha}$$

$$\cot\frac{\alpha}{2} = \pm\sqrt{\frac{1+\cos\alpha}{1-\cos\alpha}} = \frac{1+\cos\alpha}{\sin\alpha} = \frac{\sin\alpha}{1-\cos\alpha}$$

Euler's formulas

$$e^{\pm i\alpha} = \cos\alpha \pm i\sin\alpha \quad i = \sqrt{-1}$$

$$\sin\alpha = \frac{e^{i\alpha} - e^{-i\alpha}}{2i} \quad \cos\alpha = \frac{e^{i\alpha} + e^{-i\alpha}}{2}$$

$$\tan\alpha = -i\left(\frac{e^{i\alpha} - e^{-i\alpha}}{e^{i\alpha} + e^{-i\alpha}}\right) = -i\left(\frac{e^{i2\alpha} - 1}{e^{i2\alpha} + 1}\right)$$

Function-sum and function-difference formulas

$$\sin\alpha + \sin\beta = 2\sin\frac{1}{2}(\alpha + \beta)\cos\frac{1}{2}(\alpha - \beta)$$

$$\sin\alpha - \sin\beta = 2\cos\frac{1}{2}(\alpha + \beta)\sin(\alpha - \beta)$$

$$\cos\alpha + \cos\beta = 2\cos\frac{1}{2}(\alpha + \beta)\cos\frac{1}{2}(\alpha - \beta)$$

$$\cos\alpha - \cos\beta = -2\sin\frac{1}{2}(\alpha + \beta)\sin\frac{1}{2}(\alpha - \beta)$$

$$\tan\alpha + \tan\beta = \frac{\sin(\alpha + \beta)}{\cos\alpha\cos\beta}$$

$$\tan\alpha - \tan\beta = \frac{\sin(\alpha - \beta)}{\cos\alpha\cos\beta}$$

A.2 Polar Coordinates

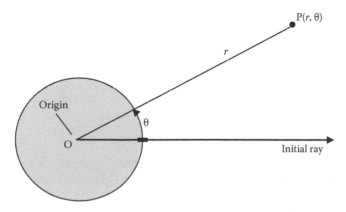

Fix an origin O and an initial ray from O. Then each point P can be located by assigning to it a polar coordinate pair (r, θ), where r is the directed distance from O to P and θ is the angle between the initial ray and OP.

The origin in polar coordinates corresponds to the origin in Cartesian coordinates. We can imagine an x, y plane overlaying the above, with coinciding origins. To change from Cartesian coordinates to polar coordinates, we observe the following identities:

$$x = r\cos\theta,$$

$$y = r\sin\theta,$$

$$\frac{x}{y} = \tan\theta,$$

$$x^2 + y^2 = r^2.$$

A.3 Complex Numbers

Since there is no real number that satisfies the equation $x^2 + 1 = 0$, we develop a set of complex numbers of the form $a + bi$, where a and b are real, and $i = \sqrt{-1}$. We call i the *imaginary unit*, and it has the property that $i^2 = -1$. If $z = a + bi$, we say that a is the *real part* of z and b is called *the imaginary part* of z. We denote them Re{z} and Im{z}, respectively.

Two complex numbers $a + bi$ and $c + di$ are *equal* if and only if $a = c$ and $b = d$. We consider the real numbers as a subset of the complex numbers where all $b = 0$.

The *complex conjugate*, or merely *conjugate*, of a complex number $a + bi$ is $a - bi$. The conjugate of the complex number z is usually denoted $\bar{z}$.

Fundamental operations with complex numbers are

1. Addition: $(a + bi) + (c + di) = a + bi + c + di = (a + c) + (b + d)i$
2. Subtraction: $(a + bi) - (c + di) = a + bi - c - di = (a - c) + (b - d)i$
3. Multiplication: $(a + bi)(c + di) = ac + adi + cbi + bdi^2 = (ac - bd) + (ad + bc)i$
4. Division:

$$\frac{a+bi}{c+di} = \left(\frac{a+bi}{c+di}\right)\left(\frac{c-di}{c-di}\right) = \frac{ac - adi + bci - bdi^2}{c^2 - d^2 i^2}$$

$$= \frac{ac + bd + (bc - ad)i}{c^2 + d^2} = \frac{ac + bd}{c^2 + d^2} + \frac{bc - ad}{c^2 + d^2}i$$

A.3.1 Absolute Value of Complex Numbers

The absolute value of a complex number $a + bi$ is defined as $|a + bi| = \sqrt{a^2 + b^2}$.

If $z_1, z_2, z_3, \ldots, z_n$ are complex numbers, the following properties hold:

1. $|z_1 z_2| = |z_1||z_2|$ and, in fact, $|z_1 z_2 \ldots z_n| = |z_1||z_2|\ldots|z_n|$
2. $\left|\dfrac{z_1}{z_2}\right| = \dfrac{|z_1|}{|z_2|}$ if $z_2 \neq 0$

3. $|z_1 + z_2| \leq |z_1| + |z_2|$ and $|z_1 + z_2 + \cdots + z_n| \leq |z_1| + |z_2| + \cdots + |z_n|$
4. $|z_1 + z_2| \geq |z_1| - |z_2|$ and $|z_1 - z_2| \geq |z_1| - |z_2|$

A.3.2 Ordered Pair Form of Complex Numbers

We can define a complex number as an ordered pair (a, b). This allows us to represent the complex numbers graphically in the way we graph real ordered pairs on a Cartesian coordinate plane, where the x axis becomes the *real* axis and the y axis becomes the *imaginary* axis. We plot our ordered pairs on the (r, i) axes.

We define equality, sum, and product as follows:

$$(a, b) = (c, d) \text{ if and only if } a = c \text{ and } b = d,$$

$$(a, b) + (c, d) = (a + c, b + d),$$

$$(a, b)(c, d) = (ac - bd, ad + bd),$$

$$m(a, b) = (ma, mb).$$

This is merely a different notation than the previous, and all properties of the complex numbers hold. Further, the complex numbers constitute a field.

A.3.3 Polar Form of Complex Numbers

If P is a point in the complex plane with coordinates (x, y) or $x + iy$, then

$$x = r \cos \theta, \qquad y = r \sin \theta,$$

where $r = \sqrt{x^2 + y^2} = |x + iy|$ is called the amplitude, modulus, or absolute value of $z = x + iy$ and θ, called the argument of $z = x + iy$, is the angle that OP makes with the positive real axis. It follows that

$$z = x + iy = r(\cos \theta + i \sin \theta).$$

A.3.3.1 De Moivre's Theorem

If $z_1 = x_1 + iy_1 = r_1(\cos \theta_1 + i \sin \theta_1)$ and if $z_2 = x_2 + iy_2 = r_2(\cos \theta_2 + i \sin \theta_2)$, then

$$z_1 z_2 = r_1 r_2 (\cos(\theta_1 + \theta_2) + i(\sin(\theta_1 + \theta_2))),$$

$$\frac{z_1}{z_2} = \frac{r_1}{r_2}(\cos(\theta_1 - \theta_2) + i \sin(\theta_1 - \theta_2)).$$

We can generalize this to many complex values. Thus, if $z_1 = z_2 = \cdots = z_n = z$, de Moivre's theorem states that $z^n = (r(\cos \theta + i \sin \theta))^n = r^n(\cos n\theta + i \sin n\theta)$.

A.3.4 Dot and Cross Products on Complex Numbers

Let $z_1 = x_1 + iy_1$ and $z_2 = x_2 + iy_2$ be two complex numbers and $\bar{z}$ represents the complex conjugate of z. We define the *dot product* (also called the *scalar product*) of z_1 and z_2 to be

$$z_1 \circ z_2 = |z_1||z_2|\cos\theta = x_1 x_2 + y_1 y_2 = \text{Re}\{\bar{z}_1 z_2\} = \frac{1}{2}\{\bar{z}_1 z_2 + z_1 \bar{z}_2\},$$

where θ is the angle between z_1 and z_2, $0 < \theta < \pi$.

The *cross product* of z_1 and z_2 is defined to be

$$z_1 \times z_2 = |z_1||z_2|\sin\theta = x_1 y_2 - y_1 x_2 = \text{Im}\{\bar{z}_1 z_2\} = \frac{1}{2i}\{\bar{z}_1 z_2 - z_1 \bar{z}_2\}.$$

Putting these together, we find that $\bar{z}_1 z_2 = (z_1 \circ z_2) + i(z_1 \times z_2) = |z_1||z_2|e^{i\theta}$.

We also have the following:

1. A necessary and sufficient condition that z_1 and z_2 be *perpendicular* is that $z_1 \circ z_2 = 0$.
2. A necessary and sufficient condition that z_1 and z_2 be *parallel* is that $z_1 \times z_2 = 0$.
3. The magnitude of the projection of z_1 on z_2 is $|z_1 \circ z_2|/|z_2|$.
4. The area of a parallelogram having size z_1 and z_2 is $|z_1 \times z_2|$.

A.4 Chain Rule

If $q(x) = u(x)v(x)$, then

$$\frac{d}{dx}q(x) = \frac{d}{dx}(uv) = u\frac{dv}{dx} + v\frac{du}{dx}.$$

If $q(x) = \frac{u(x)}{v(x)} v(x) \neq 0$, then

$$\frac{d}{dx}q(x) = \frac{d}{dx}\left(\frac{u}{v}\right) = \frac{v\left(\dfrac{du}{dx}\right) - u\left(\dfrac{dv}{dx}\right)}{v^2}.$$

If $q(x) = \dfrac{c}{u(x)}, c \in \Re$, then

$$\frac{d}{dx}q(x) = \frac{d}{dx}\left(\frac{c}{u}\right) = c\frac{d}{dx}\left(\frac{1}{u}\right) = -\frac{c}{u^2}\frac{d}{dx}(u).$$

If $q(x) = u^n(x)$, then

$$\frac{d}{dx}q(x) = \frac{d}{dx}(u^n) = nu^{n-1}\frac{d}{dx}(u).$$

Let $y = f(u)$ and $u = g(x)$. Then we can write y as a function of a function, $y = f(g(x))$. Now, if y is a differentiable function of u, and if u is a differentiable function of x, then $y = f(g(x))$ is a differentiable function of x, and

$$\frac{dy}{dx} = \frac{dy}{du}\frac{du}{dx}.$$

If $z = f(x, y)$ is a continuous function of the variables x and y with continuous partial derivatives $\partial z/\partial x$ and $\partial z/\partial y$, and if x and y are differentiable functions of the variable t, $x = g(t)$ and $y = h(t)$, then z is also a function of the variable t, and dz/dt (called the *total derivative of z with respect to t*) is given by

$$\frac{dz}{dt} = \frac{\partial z}{\partial x}\frac{dx}{dt} + \frac{\partial z}{\partial y}\frac{dy}{dt}.$$

If $z = f(x, y)$ is a continuous function of variables x and y with continuous partial derivatives $\partial z/\partial x$ and $\partial z/\partial y$, and if x and y are continuous functions $x = g(r, s)$ and $y = h(r, s)$ of the independent variables r and s, then z is a function of r and s with

$$\frac{\partial z}{\partial r} = \frac{\partial z}{\partial x}\frac{\partial x}{\partial r} + \frac{\partial z}{\partial y}\frac{\partial y}{\partial r} \quad \frac{\partial z}{\partial s} = \frac{\partial z}{\partial x}\frac{\partial x}{\partial s} + \frac{\partial z}{\partial y}\frac{\partial y}{\partial s}.$$

Suppose that $f(x) = g(x)h(x)$ where both g and h are continuously differentiable scalar functions of the vector $x \in \mathfrak{R}^{n\times 1}(x = [x_1, x_2, \ldots, x_n]^T)$. Then

$$\nabla_x f(x) = \nabla_x g(x)h(x) + \nabla_x h(x)g(x),$$

where $\nabla_x f(x) = \dfrac{\partial f(x)}{\partial x} = \left[\dfrac{\partial f}{\partial x_1}, \dfrac{\partial f}{\partial x_2}, \ldots, \dfrac{\partial f}{\partial x_n}\right]^T$ is the gradient of f with respect to x.

A.5 Algebra

A.5.1 Vectors and Vector Spaces

A vector (linear) space over a field $\mathcal{F}$ is designated as $(\mathcal{L}, \mathcal{F})$ and contains the set of element $\mathcal{L}$, called vectors (of arbitrary length); a field $\mathcal{F}$; and two operations, which are *scalar multiplication* and *vector addition*.

Two often-used vector spaces are $(\mathfrak{R},\mathfrak{R})$ and $(\mathcal{C},\mathcal{C})$. While $(\mathcal{C},\mathfrak{R})$ is a vector space, $(\mathfrak{R},\mathcal{C})$ is not because scalar multiplication will not, in general, yield vectors whose elements are real numbers.

A.5.2 Matrices and Matrix Operations

A *matrix* is a rectangular array of numbers. The numbers in the array are called the *entries* in the matrix. The rows of entries in a matrix can be viewed as vectors, as can the columns of the matrix:

$$A = \begin{bmatrix} a_{11}\ a_{12}\ a_{13}\ a_{14} \\ a_{21}\ a_{22}\ a_{23}\ a_{24} \\ a_{31}\ a_{32}\ a_{33}\ a_{34} \end{bmatrix}.$$

The subscripts of a indicate the *row* and *column* location of the entry. We can consider matrix A as a set of row vectors:

$$b_1 = (a_{11}, a_{12}, a_{13}, a_{14}),$$

$$b_2 = (a_{21}, a_{22}, a_{23}, a_{24}),$$

$$b_3 = (a_{31}, a_{32}, a_{33}, a_{34}),$$

or as a set of column vectors:

$$c_1 = \begin{vmatrix} a_{11} \\ a_{21} \\ a_{31} \end{vmatrix}, \quad c_2 = \begin{vmatrix} a_{12} \\ a_{22} \\ a_{32} \end{vmatrix}, \quad c_3 = \begin{vmatrix} a_{13} \\ a_{23} \\ a_{33} \end{vmatrix}, \quad c_4 = \begin{vmatrix} a_{14} \\ a_{24} \\ a_{34} \end{vmatrix}.$$

We define the size of a matrix by the number of rows and columns it contains: matrix A is a 3×4 matrix. We denote the size of a matrix using superscripts: $M^{m,n}$ indicates a matrix M with m rows and n columns. We may say a matrix is an element of the real field $\Re^{m,n}$ or of the complex field $C^{m,n}$.

A matrix with n rows and n columns is called a *square* matrix, and the entries $a_{11}, a_{22}, a_{33}, ..., a_{nn}$ are called the *diagonal* of the matrix.

The *sum* of two matrices, $A + B$, is defined *only* if the matrices are the same size and is then obtained by adding together the corresponding entries of the two matrices.

The *product* of a *scalar* c with a matrix A—denoted cA—is the matrix obtained by multiplying each entry of A by c.

The *product* of matrix A with matrix B requires that if A is $m \times r$, then B must be $r \times n$. The entry in row i and column j of matrix AB is calculated as follows:

    a. Single out row i of matrix A and column j of matrix B.

    b. Multiply corresponding entries by each other.

    c. Add all products together. This becomes the ijth entry of AB.

A *null matrix* is one in which all entries are 0s.

A matrix $M^{n,n}$ in which the diagonal entries are all 1s and all other entries are 0s is the *identity matrix*, denoted I_n.

The *transpose* of a matrix M is denoted M^T. To transpose a matrix, the rows and columns of the matrix are interchanged. That is,

$$A^T = [a_{ij}]_{n \times m}^T = [a_{ji}]_{m \times n} \in \Re^{m \times n}.$$

The following are properties of the transpose:

1. $(A^T)^T = A$
2. Given $A \in \Re^{n \times m}$, $B \in \Re^{m \times p}$, $(AB)^T = B^T A^T \in \Re^{p \times n}$
3. $(A + B)^T = A^T + B^T$

The *inverse* of a matrix M is denoted M^{-1}. If a matrix $M \in \Re^{n \times n}$ is *invertible*, it is also called *nonsingular*, and there is a matrix M^{-1} such that $MM^{-1} = M^{-1}M = I_n$.

Let $A \in \Re^{n \times n}$, $B \in \Re^{n \times n}$, $C \in \Re^{m \times m}$, $u \in \Re^{n \times 1}$, and $v \in \Re^{n \times 1}$. Then

1. $(A^{-1})I = A^{-1}$
2. $(AB)^{-1} = B^{-1}A^{-1}$
3. $(A^T)^{-1} = (A^{-1})^T = A^{-T}$

A.6 Eigenvalue Problems

Let A be a square matrix of order n. A number λ is said to be an eigenvalue of A if there exists a nonzero solution vector Λ of the linear system:

$$A\Lambda = \lambda\Lambda.$$

The solution vector Λ is said to be an eigenvector corresponding to the eigenvalue λ. This is typically called the *standard eigenvalue problem*.

If A is real symmetric, the eigenvalue problem has the following desirable features:

a. The eigenvalues λ are all real.
b. The eigenvectors can be chosen to be mutually orthogonal—that is,

$$Az_i = \lambda_i z_i \quad \text{for } i = 1, 2, \ldots, n$$

or, equivalently,

$$AZ = Z\Lambda,$$

where

Λ is a real diagonal matrix whose diagonal elements λ_i are the eigenvalues
Z is a real orthogonal matrix whose columns z_i are the eigenvectors

This implies that

$$z_i^T z_j = 0 \quad \text{if } i \neq j \text{ and } \|z_i\|_2 = 1.$$

Further, we can write $A = Z\Lambda Z^T$.

This is known as the *eigendecomposition* or *spectral factorization* of A.

If A is real nonsymmetric, it may have complex eigenvalues, occurring as complex conjugate pairs. If x is an eigenvector corresponding to a complex eigenvalue λ, then the complex conjugate vector $\bar{x}$ is the eigenvector corresponding to the complex conjugate eigenvalue $\bar{\lambda}$.

A.7 Convolution

If functions f and g are piecewise continuous on $[0, \infty)$, then the convolution of f and g, denoted $f * g$, is given by the integral

$$f * g = \int_0^t f(\tau)g(t - \tau)d\tau.$$

Convolution is commutative. That is, $f * g = g * f$.

A.7.1 Convolution Theorem

Let $f(t)$ and $g(t)$ be piecewise continuous on $[0, \infty)$ and of exponential order. Then

$$\mathfrak{F}\{f * g)\} = \mathfrak{F}\{f(t)\}\mathfrak{F}\{g(t)\},$$

where $\mathfrak{F}$ denotes the Fourier transform described in the following texts. That is, convolution in one domain (i.e., time) equals point-wise multiplication in the other domain (i.e., frequency).

A.7.2 Fourier Series

The Fourier series of a function f defined on the interval $(-p, p)$ is given by

$$f(x) = \frac{a_0}{2} + \sum_{n=1}^{\infty}\left(a_n \cos\frac{n\pi}{p}x + b_n \sin\frac{n\pi}{p}\right),$$

where

$$a_0 = \frac{1}{p} \int_{-p}^{p} f(x)\,dx$$

$$a_n = \frac{1}{p} \int_{-p}^{p} f(x) \cos \frac{n\pi}{p} x \, dx$$

$$b_n = \frac{1}{p} \int_{-p}^{p} f(x) \sin \frac{n\pi}{p} x \, dx$$

A.7.3 Fourier Transforms

Fourier transforms occur in pairs. That is, if

$$F(\alpha) = \int_{a}^{b} F(x) K(\alpha, x)\,dx$$

is an integral that transforms $f(x)$ into $F(\alpha)$, then the function f is recovered by means of another integral transformation

$$f(x) = \int_{c}^{d} F(\alpha) H(\alpha, x)\,d\alpha$$

called the *inversion integral* or *inverse transform*. The functions K and H are called the *kernels* of the transforms.

Some common Fourier transforms are

1. Fourier transform: $\Im\{f(x)\} = \displaystyle\int_{-\infty}^{\infty} f(x) e^{i\alpha x}\,dx = F(\alpha)$

   Inverse Fourier transform: $\Im^{-1}\{F(\alpha)\} = \dfrac{1}{2\pi} \displaystyle\int_{-\infty}^{\infty} F(\alpha) e^{-i\alpha x}\,d\alpha = f(x)$

2. Fourier sine transform: $\Im_s\{f(x)\} = \displaystyle\int_{0}^{\infty} f(x) \sin \alpha x \, dx = F(\alpha)$

   Inverse Fourier sine transformation: $\Im_s^{-1}\{F(\alpha)\} = \dfrac{2}{\pi} \displaystyle\int_{0}^{\infty} F(\alpha) \sin \alpha x \, d\alpha = f(x)$

3. Fourier cosine transform: $\mathfrak{I}_c\{f(x)\} = \int\limits_0^\infty f(x)\cos\alpha x\,dx = F(\alpha)$

   Inverse Fourier cosine transform: $\mathfrak{I}_c^{-1}\{F(\alpha)\} = \dfrac{2}{\pi}\int\limits_0^\infty F(\alpha)\cos\alpha x\,d\alpha = f(x)$

We see that the Fourier transform is a generalization of Fourier series and uses complex numbers. Fourier transforms are particularly useful in analyzing the frequency content of continuous signals, in quantifying optical information, and in studying electrical and magnetic fields and their properties. The Fourier transform takes a signal in the time domain with real variables and transforms it into a function of angular frequency with complex variables. See the following figure:

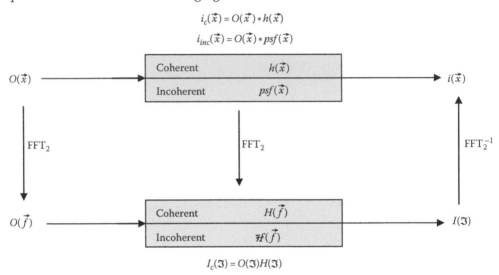

Some common fourier transforms and properties include the following:

Operation	$f(t)$	$F(\omega)$
Addition	$f_1(t) + f_2(t)$	$F_1(\omega) + F_2(\omega)$
Scalar multiplication	$kf(t)$	$kF(\omega)$
Complex Conjugation	$f^*(t)$	$F^*(-i\omega)$
Time Reversal	$f(-t)$	$F(-i\omega)$
Scaling (a is real)	$f(at)$	$\dfrac{1}{\|a\|}F\left(\dfrac{\omega}{a}\right)$
Time shift	$f(t-t_0)$	$F(\omega)e^{-i\omega t_0}$
Frequency shift (ω_0 is real)	$f(t)e^{i\omega_0 t}$	$F(\omega-\omega_0)$
Time convolution	$f_1(t)*f_2(t)$	$F_1(\omega)F_2(\omega)$
Frequency convolution	$f_1(t)f_2(t)$	$\dfrac{1}{2\pi}F_1(\omega)*F_2(\omega)$
Time differentiation	$\dfrac{d^n f}{dt^n}$	$(i\omega)^n F(\omega)$
Time integration	$\displaystyle\int_{-\infty}^{t} f(x)dx$	$\dfrac{F(\omega)}{i\omega} + \pi F(0)\partial(\omega)$

$f(t)$	$F(\omega)$
$u(t)$	$\pi\partial(\omega)+\dfrac{1}{i\omega}$
$\mathrm{Sgn}(t)$	$\dfrac{2}{i\omega}$
$\cos(\omega_0 t)u(t)$	$\dfrac{\pi}{2}\left[\partial(\omega-\omega_0)+\partial(\omega+\omega_0)\right]+\dfrac{i\omega}{\omega_0^2-\omega^2}$
$\sin(\omega_0 t)u(t)$	$\dfrac{\pi}{2i}\left[\partial(\omega-\omega_0)-\partial(\omega+\omega_0)\right]+\dfrac{\omega_0}{\omega_0^2-\omega^2}$
$e^{-at}\sin(\omega_0 t)u(t)$	$\dfrac{\omega_0}{(a+i\omega)^2+\omega_0^2},\quad a>0$
$e^{-at}\cos(\omega_0 t)u(t)$	$\dfrac{a+i\omega}{(a+i\omega)^2+\omega_0^2},\quad a>0$
$\mathrm{Rect}\left(\dfrac{t}{\tau}\right)$	$\tau\,\mathrm{sinc}\left(\dfrac{\omega\tau}{2}\right)$
$\dfrac{W}{\pi}\mathrm{sinc}(Wt)$	$\mathrm{rect}\left(\dfrac{\omega}{2W}\right)$
$\Delta\left(\dfrac{t}{\tau}\right)$	$\dfrac{\tau}{2}\mathrm{sinc}^2\left(\dfrac{\omega\tau}{4}\right)$
$\dfrac{W}{2\pi}\mathrm{sinc}^2\left(\dfrac{Wt}{2}\right)$	$\Delta\left(\dfrac{\omega}{2W}\right)$
$\displaystyle\sum_{n=-\infty}^{\infty}\partial(t-nT)$	$\omega_0\displaystyle\sum_{n=-\infty}^{\infty}\partial(\omega-n\omega_0),\quad \omega_0=\dfrac{2\pi}{T}$
$e^{-t^2/2\sigma^2}$	$\sigma\sqrt{2\pi}e^{-\sigma^2\omega^2/2}$

A.8 Linear Systems Theory

A physical system that changes over time and whose mathematical model is a linear differential equation is said to be a *linear system*. A linear system has the property that the *superposition principle* holds—that is, the response to a superposition of inputs is a superposition of outputs. A linear system may also be *shift invariant*—that is, a phase shift in input leads to an identical phase shift in output.

For a linear differential equation

$$a_n(t)y^n + a_{n-1}(t)y^{n-1} + \cdots + a_1(t)y' + a_0(t)y = g(t),$$

the values of the variables $y(t)$, $y'(t)$, ..., $y^{(n-1)}(t)$ at a specific time t_0 describe the *state* of the system. The function g is called the *input function*, the *forcing function*, or the *excitation function*. A solution $y(t)$ is called the *output* or the *response* of the system.

A.9 Superposition Principle

Simply stated, the superposition principle says that

  a. The sum of two or more solutions of a homogeneous linear differential equation is also a solution
  b. A constant multiple of a solution of a homogeneous linear differential equation is also a solution

References

Anton, H. 1991. *Elementary Linear Algebra*, 6th edn. Hoboken, NJ: John Wiley & Sons, Inc. http://www.amazon.com/Elementary-Linear-Algebra-Howard-Anton/dp/0471509000.

Simmons, G.F. and S.G. Krantz. 2007. *Differential Equations: Theory, Technique, and Practice*. New York: McGraw Hill.

Thomas, G.B. and R.L. Finney. 1990. *Calculus and Analytic Geometry*, 7th edn. Reading, MA: Addison Wesley.

Zill, D.G. and M.R. Cullen. 1986. *Differential Equations with Boundary-Value Problems*, 3rd edn. Boston, MA: PWS-KENT Publishing Co.

Index

A

ABCD matrix method, 184–185
Absorptance, 34
Absorption spectra, 221
Achieved availability (A_a), 376, 708
Achromatic lens, 182–184
Acknowledged SoS, 130
Acousto-optics, 34, 350
Acquisition Categories (ACATs), 14
Acquisition Decision Memorandum (ADM), 14
Active imaging, 34, 80
Active-matrix liquid-crystal display
  (AMLCD), 726
Adaptive optics, 34, 145–146, 683
Administrative delay time (ADT), 374–375
Afterburning process, 245
Agreement category, 541
AHP, *see* Analytical hierarchy process (AHP)
Aircraft
  afterburning, 245
  ramjet, 245
  turbofan, 244–245
  turbojet, 243–244
Algebra
  matrices and matrix operations, 778–779
  vectors and vector spaces, 777
Alternative analysis, 267; *see also* Analytical
  hierarchy process (AHP)
Alternative *versus* alternative (AvA) matrix, 190,
  192–193
Analysis of alternatives (AoA)
  acquisition process, 14
  stakeholder's need, 279
  vs. trade studies, 277
Analytical hierarchy process (AHP)
  detector cooling, 472
  feasibility studies
    consistency ratio (CR) test, 282, 290–291
    cost and risk, 283
    decision making, 282–283
    eigenvector analysis, 285–286
    four-criterion matrix, 288–289
    project stage, 283
    ranking order, 281
    Saaty's test, 291–292
    sensitivity analysis, 289–290
    solution ranking, 289
    stakeholder's preferences, 280–281
    three-criterion pair-wise
      ranking matrix, 284
    transitivity, 282
  MBSE
    alternatives ranking, 193–194
    alternative *versus* criteria matrix,
      193–194
    ATC algorithms, 187
    AvA matrix, 192–193
    CvC matrices, 190–192
    decision consistency, 194–195
    DHS UAV project meeting,
      187–189
    MagicDraw tool, 187
    unique features, 196
  quality loss function, 603
Antireflection coating, 182
AoA, *see* Analysis of alternatives (AoA)
Application architecture, 52, 59
Atmospheric heating, 246
Atmospheric turbulence compensation
  (ATC) system
  area of interest, 145
  definition, 34
  Fried parameter, 35
  GPPP device, 156
  isoplanatic angles, 522
  MATLAB®-based algorithms, 39, 187, 683
  optical remote detection system, 23
  perimeter surveillance OV-1, 144–145, 147
  spatial resolution, 44, 83, 148–149, 152, 355
Atmospheric turbulence-limited spatial
  resolution, 34, 152, 302
Attribute control charts
  C-chart, 615–616
  centerline, 615
  NP-chart, 616
  P-chart
    in control, 649
    out of control, 648
    parameters, 615
    ramp up, 647
    variables, 615–616
  quality characteristics, 614
  U-chart, 616–617
  use, 609

Printed and bound by CPI Group (UK) Ltd, Croydon, CR0 4YY

22/10/2024

01777619-0002